Digital Image Processing and Image Acquisition

Bernd Jähne

Digital Image Processing and Image Acquisition

Seventh Revised and Extended Edition

Springer

Bernd Jähne
Interdisziplinäres Zentrum für
Wissenschaftliches Rechnen (IWR)
Heidelberg University
Heidelberg, Germany

ISBN 978-3-032-04805-9 ISBN 978-3-032-04806-6 (eBook)
https://doi.org/10.1007/978-3-032-04806-6

This Springer imprint is published by the registered company Springer Nature Switzerland AG
The registered company address is: Gewerbestrasse 11, 6330 Cham, Switzerland

Preface

When the first English edition of this book was published in 1991 [154], image processing was only possible for a few specialists in research and technology, as expensive image acquisition and processing systems were required. Today, anyone can take pictures and videos and mobile devices are many times more powerful than the systems back then.

Twenty years after the publication of the sixth edition in 2005 [158], a new edition was overdue. It is based on the eighth German edition, which was published in 2024 [160]. The book has been reworked, printed in full color, many graphics have been revised and more have been added. The chapters have been restructured and the links between the individual topics extended. This has made it clear that image acquisition and image processing are largely based on the same mathematical methods. The image acquisition section of the book has been significantly expanded to take into account the rapid development of image sensor technology, optics and the diverse methods of reconstructing the three-dimensional world from image data. The focus has been placed on working out the basic concepts so that the reader can understand and make optimum use of the variety of image acquisition methods, which at first glance may seem confusing.

Artificial intelligence methods have not been included in this book. Instead, this book contains the necessary knowledge about image acquisition and processing to apply machine learning methods to image analysis effectively. For this reason, the introductory chapter on classification contained in the sixth edition has not been included in the new edition.

The proven structuring of the text into basic and in-depth material has been retained. Further sections are marked with a ♯ sign in the headings. In this way, the reader can first work through the basics and then, depending on their needs and interests, deepen their studies. The most important formulas are framed. Separate text sections with examples and in-depth studies have been systematically expanded. Each chapter is concluded by a section that refers to further topics and literature. For help with reading, please refer to the description of the notation in Appendix A. The reference section at the end of the book (Appendix D) contains a compilation of important data, relationships and formulas. References to these blocks, which are marked with margin labels, are indicated with the number of the block, for example (≻ R13).

I would like to thank all persons, institutions and companies who have provided me with material for this book. The sources are indicated at the relevant places. I would also like to thank my colleagues and collaborators, the listeners of my lectures and the readers of the earlier editions of this book who pointed out errors and suggested additions and improvements. My sincere thanks go to the staff of Springer-Verlag for their unbroken interest in this book over the decades. Special thanks are in order for Dr. Claudia Herrmann. She went to great lengths to thoroughly proofread the eighth German edition. As a biologist, she repeatedly emphasized the importance of explaining the subject intuitively first, which is often easily possible in image processing, before treating it mathematically. Then, even complex subjects will be easier for the reader to understand. This English edition has also benefited from her efforts.

Finally, I would like to address you, the readers of this book. I am grateful for any critical or constructive comments you may have on the book. Please suggest improvements or additions and point out any inconsistencies or printing errors that occurred despite all the care taken during production.

Heidelberg, October 2025 Bernd Jähne

From the preface of the first edition

Digital image processing is a fascinating subject in several aspects. Human beings perceive most of the information about their environment through their visual sense. While for a long time images could only be captured by photography, we are now at the edge of another technological revolution which allows image data to be captured, manipulated, and evaluated electronically with computers. With breathtaking pace, computers are becoming more powerful and at the same time less expensive, so that widespread applications for digital image processing emerge. In this way, image processing is becoming a tremendous tool for analyzing image data in all areas of natural science. For more and more scientists digital image processing will be the key to study complex scientific problems they could not have dreamed of tackling only a few years ago. A door is opening for new interdisciplinary cooperation merging computer science with the corresponding research areas.

Many students, engineers, and researchers in all natural sciences are faced with the problem of needing to know more about digital image processing. This book is written to meet this need. The author — himself educated in physics — describes digital image processing as a new tool for scientific research. The book starts with the essentials of image processing and leads — in selected areas — to the state-of-the art. This approach gives an insight as to how image processing really works. The selection of the material is guided by the needs of a researcher who wants to apply image-processing techniques in his or her field. In this sense, this book tries to offer an integral view of image processing from image acquisition to the extraction of the data of interest. Many concepts and mathematical tools that find widespread application in natural sciences are also applied in digital image processing. Such analogies are pointed out, since they provide an easy access to many complex problems in digital image processing for readers with a general background in natural sciences. The discussion of the general concepts is supplemented with examples from applications on PC-based image processing systems and ready-to-use implementations of important algorithms.

I am deeply indebted to the many individuals who helped me to write this book. I do this by tracing its history. In the early 1980s, when I worked on the physics of small-scale air-sea interaction at the Institute of Environmental Physics at Heidelberg University, it became obvious that these complex phenomena could not be adequately treated with point measuring probes. Consequently, a number of area extended measuring techniques were developed. Then I searched for techniques to extract the physically relevant data from the images and sought for colleagues with experience in digital image processing. The first contacts were established with the Institute for Applied Physics at Heidelberg University and the German Cancer Research Center in Heidelberg. I would like to thank Prof. Dr. J. Bille, Dr. J. Dengler and Dr. M. Schmidt cordially for many eye-opening conversations and their cooperation.

I would also like to thank Prof. Dr. K. O. Münnich, director of the Institute for Environmental Physics. From the beginning, he was open-minded about new ideas on the application of digital image processing techniques in environmental physics. It is due to his farsightedness and substantial support that the research group "Digital Image Processing in Environmental Physics" could develop so fruitfully at his institute. Many of the examples shown in this book are taken from my research at Heidelberg University and the Scripps Institution of Oceanography. I gratefully acknowledge financial support for this research from the German Science Foundation, the European Community, the US National Science Foundation, and the US Office of Naval Research.

La Jolla, California, and Heidelberg, spring 1991 Bernd Jähne

Contents

II Image Processing

1 Introduction

Contents

1.1 A Tool for Science and Technology

Visual observation played an important role from the very beginnings of science. At first, verbal descriptions and manual drawings were the only ways to document experimental results. Objective documentation of results only became possible with the invention of *photography*. Three prominent examples of early scientific applications of photography are *astronomy*, *photogrammetry* and *particle physics*. Astronomers were able to determine the position and size of stars and create photogrammetric topographical maps. The examination of countless images from hydrogen bubble chambers led to the discovery of many elementary particles. However, the manual analysis procedures were very time-consuming. Semi-automatic or even fully automatic optomechanical devices were constructed, but these had the disadvantage that they could only be used for a specific purpose. This is the reason why quantitative image analysis was not widely used. Images were mainly used for documentation, qualitative description and illustration of observed phenomena.

Today, image acquisition and processing technologies are tools that can be used by every scientist and engineer. As a result, image processing has evolved from a few specialized applications to a standard scientific tool. Our everyday life is also characterized by these technologies. Today, every cell phone is equipped with more than one camera and it is possible for anyone

B. Jähne, *Digital Image Processing and Image Acquisition*,
https://doi.org/10.1007/978-3-032-04806-6_1

to capture and intuitively edit images and videos. For some time now, such general applications have been the driving force behind new developments across the entire spectrum of technologies, from optics and image sensors to evaluation algorithms and ease of use.

A simple example shows the importance of visual information. Imagine we had the task of writing an article about a new technical system, e. g. a solar power plant. It would take an enormous amount of effort to describe such a system without the use of pictures and technical drawings. An article consisting only of descriptive text would also be a frustrating experience for the reader. He would have a lot of trouble imagining how such a new power plant works and could only get a vague idea of what the system looks like.

Technical drawings and photographs, on the other hand, are an enormous help. The reader will immediately have an idea of the power plant and can study details in the visual material that catch their attention but are not described in the text. Much more detail can be conveyed by pictures than by text, which is aptly expressed by the saying "A picture is worth a thousand words".

1.2 Application Examples

In this section we discuss examples of scientific and technical applications of digital image processing. The examples show that image processing can be used to investigate complex phenomena that are not accessible using conventional measurement techniques.

1.2.1 Counting and Measuring

A classic task of digital image processing is counting particles and determining their size distribution. Fig. 1.1 shows three examples of very different particles: Gas bubbles under the surface of breaking waves, soap bubbles and pigment particles. The first step in such tasks is to find an appropriate imaging and illumination setup.

The bubble images in Fig. 1.1a were taken with a telecentric illumination and image acquisition system (Section 3.4.5). In this setup, the chief rays are parallel to the optical axis. The size of the imaged bubbles therefore does not depend on the distance to the camera. As the dimension perpendicular to the image plane is lost due to the projection onto the image plane, the measurement volume for concentration measurements must be determined by estimating the degree of blurring of the bubbles.

It is much more difficult to measure the size and shape of the soap bubbles in Fig. 1.1b, as these are transparent and therefore the bubbles behind them are also still visible. In addition, the bubbles show clear deviations from the circular shape, so that suitable parameters must be found to describe the shape.

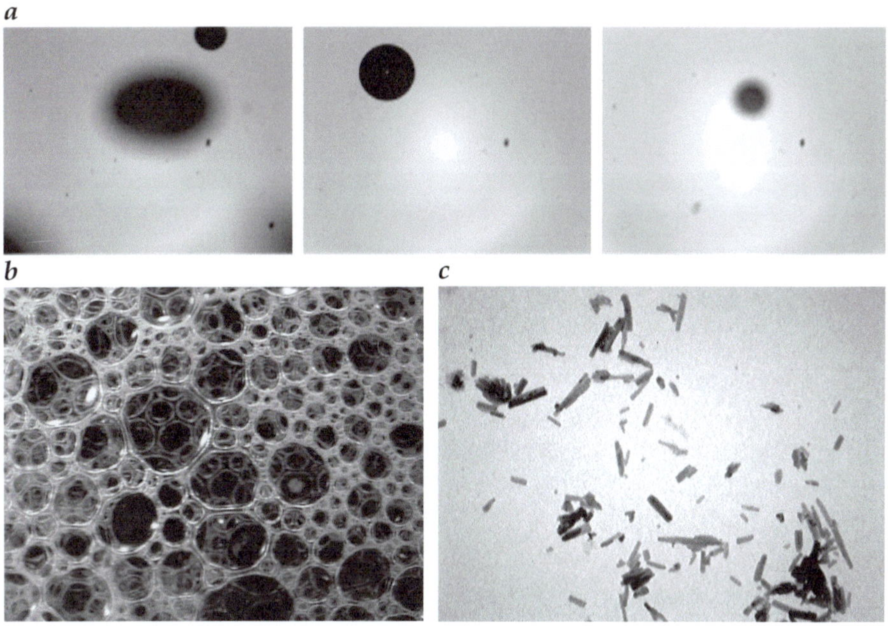

Figure 1.1: *Particle measurement using image processing:* ***a*** *Gas bubbles impacted into the water by breaking waves, visualized by a telecentric illumination and imaging system; from Geißler and Jähne [103].* ***b*** *Soap bubbles.* ***c*** *Electron micrograph of color pigments (provided by Dr. Klee, Hoechst AG, Frankfurt).*

A third application is the measurement of the size distribution of color pigment particles. They influence the properties of a color. Measuring the size distribution is therefore an important quality control task. The image in Fig. 1.1c taken with a transmission electron microscope shows the problematic nature of this task: the particles tend to clump together. These clumps must be identified and — if possible — separated into individual particles so as not to distort the determination of the size distribution.

Most products manufactured today are automatically inspected for defects by a *visual inspection system*. One class of tasks is the optical measurement of parts for correct size and arrangement. Example images are shown in Fig. 1.2. The position, diameter and roundness of the drill holes are checked here. Fig. 1.2c shows that it is not easy to illuminate metallic parts. The edge of the left borehole is partially bright, which makes it difficult to detect and correctly measure the borehole.

1.2.2 3D Measurement

In images, 3D scenes are projected onto a 2D image plane. As a result, depth is lost and special imaging techniques are required to obtain the topography of a surface or volumetric images. In recent years, a variety of techniques

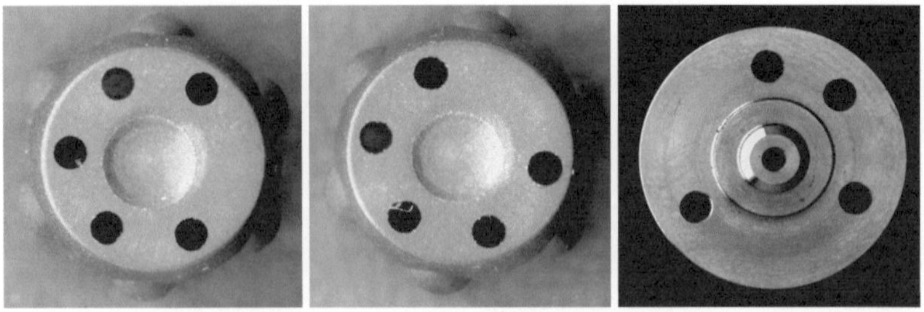

Figure 1.2: Industrial parts where the position and diameter of holes are checked by a visual inspection system (provided by Martin von Brocke, Robert Bosch GmbH).

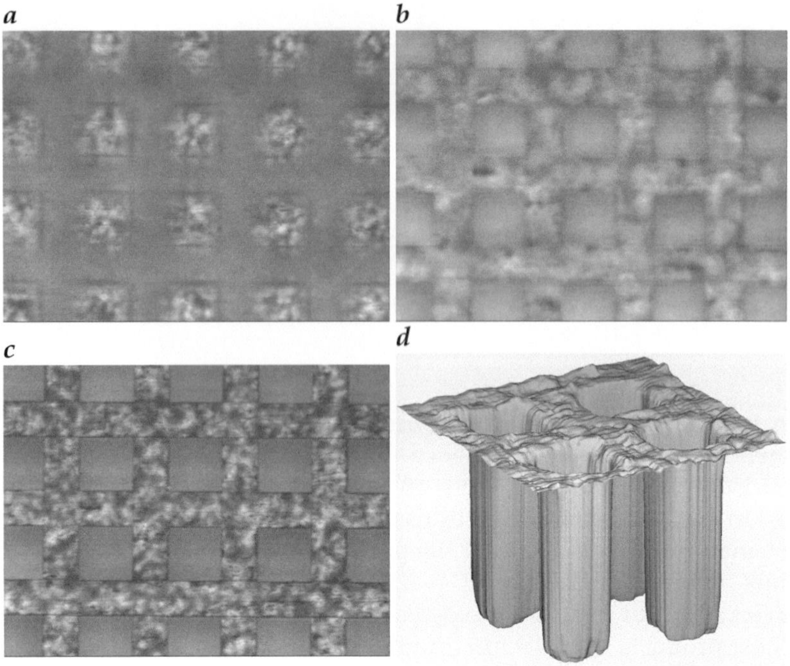

Figure 1.3: Focus series of a mold made of acrylic glass with small rectangular holes, taken with a confocal technique using statistically distributed intensity patterns. The images are focused at the following depths, measured from the bottom of the holes: ***a*** *16 µm,* ***b*** *480 µm and* ***c*** *620 µm (surface of the mold).* ***d*** *3D reconstruction. From Scheuermann et al. [289].*

have been developed to generate *depth maps* and *volumetric images*. Therefore, image processing techniques must also be suitable for such images.

Fig. 1.3 shows the reconstruction of the mold for microstructures, which was taken with a special form of confocal microscopy [289]. The mold is made of acrylic glass, a transparent plastic with a smooth surface that is

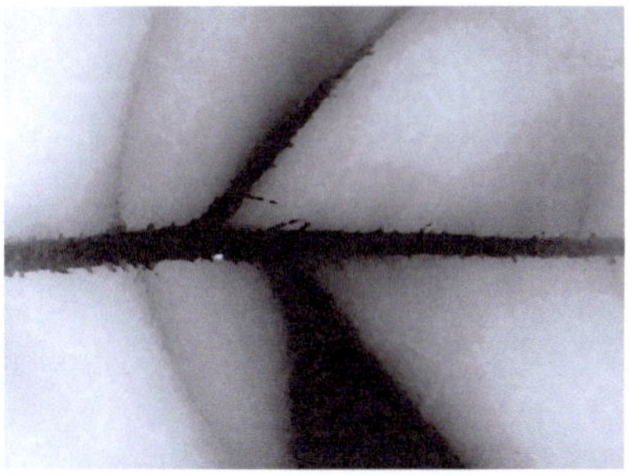

Figure 1.4: *Depth map of a plant leaf recorded using optical coherence tomography (darker areas are closer, lighter areas further away; by Jochen Restle, Robert Bosch GmbH).*

almost invisible in a normal microscope. The mold has narrow rectangular holes about 500 μm deep.

In order to make the transparent material visible, statistically distributed patterns were projected onto the image plane through the microscope optics. This pattern only appears sharp on surfaces that lie in the focused plane. The further away a reflecting surface is from the focused plane, the more blurred the pattern appears. At the beginning of the focus series, the patterns on the bottom of the holes are in focus (Fig. 1.3a), at the end the patterns on the surface of the shape (Fig. 1.3c). The depth of the reflecting surface can be reconstructed by finding the position of maximum contrast in the image stack (Fig. 1.3d).

The depth map shown in Fig. 1.4 shows a section of a plant leaf that was recorded using another modern optical 3D measurement technique, *white light interferometry* or *coherence tomography*. Light with a coherence length of only a few wavelengths is used. Therefore, interference patterns can only be observed with very small path length differences in an interferometer. This effect is used for depth measurement with an accuracy in the order of magnitude of the wavelength of the light used.

The driving force behind the development of modern volumetric imaging techniques, which allow us to look inside 3D objects, was medical diagnostics. Fig. 1.5 shows a sectional image of a human skull with a tumor. While *X-rays* (*computed tomography, CT*) mainly depicts the bony structures, T2-weighted *Magnetic Resonance Imaging* (*MRT*) tends to show the soft tissue, the vitreous body of the eyes and a surgical scar in high contrast. In *positron emission tomography* (*PET*), a high signal is observed at the site of the tumor, as the positron emitter supplied to the body accumulates here.

1.2.3 Analysis of Dynamic Processes

The analysis of *image sequences* enables the investigation of dynamic processes. The enormous potential of this technique is demonstrated by a number of examples in this section.

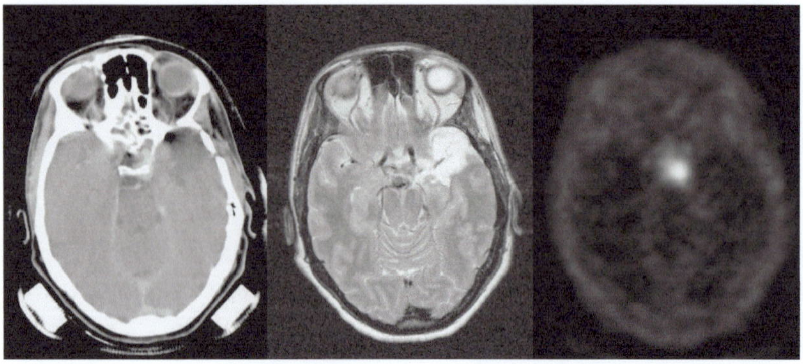

Figure 1.5: Sectional image at eye level through a human skull with a tumor, taken with X-rays (left), T2-weighted magnetic resonance imaging (center), and positron emission tomography (right) (by Michael Bock, German Cancer Research Center (DKFZ) Heidelberg).

Figure 1.6: Detection of pedestrians through motion analysis. People walking to the right are marked in red in the image on the right. (by Christoph Garbe, IWR, Heidelberg University, [318]).

Movement is an important feature for recognizing objects in scenes. One of the classic applications is the recognition of people (Fig. 1.6). In the application shown, the movement field was calculated in image sequences in video real time and only the people walking to the right were marked in red in the right-hand image.

In botany, the study of plant growth and the mechanisms that control it is a central issue. Figure 1.7a shows a castor bean leaf of which the growth rate (percent increase in area per unit time) was determined by time-lapse imaging (Fig. 1.7b), in which an image was taken approximately every minute. This new technique for growth analysis is sensitive enough to be able to determine the diurnal growth rate [292]. In Figure 1.7c growing maize roots can be seen. Time runs from left to right. Growth is greatest near the root tip.

In science, images are often taken at the limits of what is technically possible. Therefore, they are often very noisy. Figure 1.8 shows a *motility assay* with fluorescently labeled motor proteins moving on a support covered with myosin molecules. Such assays are used to elucidate the molecular

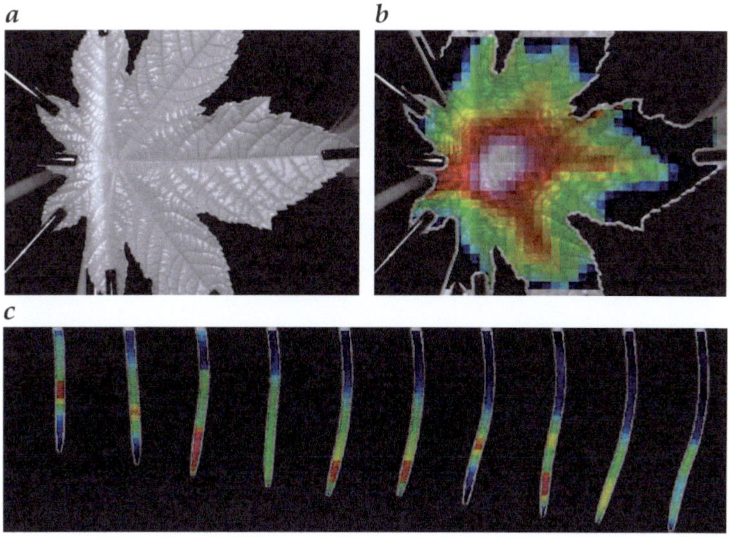

Figure 1.7: *Growth studies in botany:* ***a*** *picture of a castor bean leaf;* ***b*** *related* ***growth map****;* ***c Growth of corn roots****. The superimposed color values indicate the growth rate, red means high growth, green means low growth (by Uli Schurr and Stefan Terjung, Botanical Institute, Heidelberg University).*

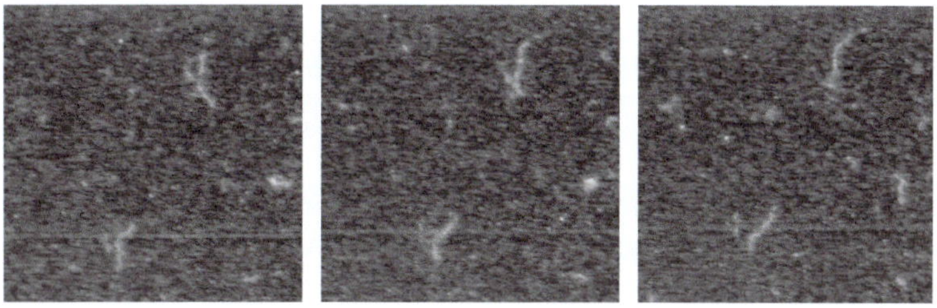

Figure 1.8: Motility Assay *for the movement analysis of motor proteins (by Dietmar Uttenweiler, Institute of Physiology, Heidelberg University).*

mechanisms of muscle cells. Despite the high noise level, the filaments can be seen to move with the naked eye. However, automatic motion analysis of such noisy image sequences is a major challenge.

The next example comes from oceanography. The small-scale processes that take place in the immediate vicinity of the ocean surface are very difficult to measure due to the wave motion. It is also not possible to capture the spatial structure of the waves from measurements of time series at one point. Therefore, image-based measurements are of great importance. Figure 1.9 shows a *space-time image* of wind-generated waves. The vertical coordinate is the spatial coordinate in the wind direction, the horizontal coordinate is the time. A special lighting technique based on the *shape from shading* paradigm

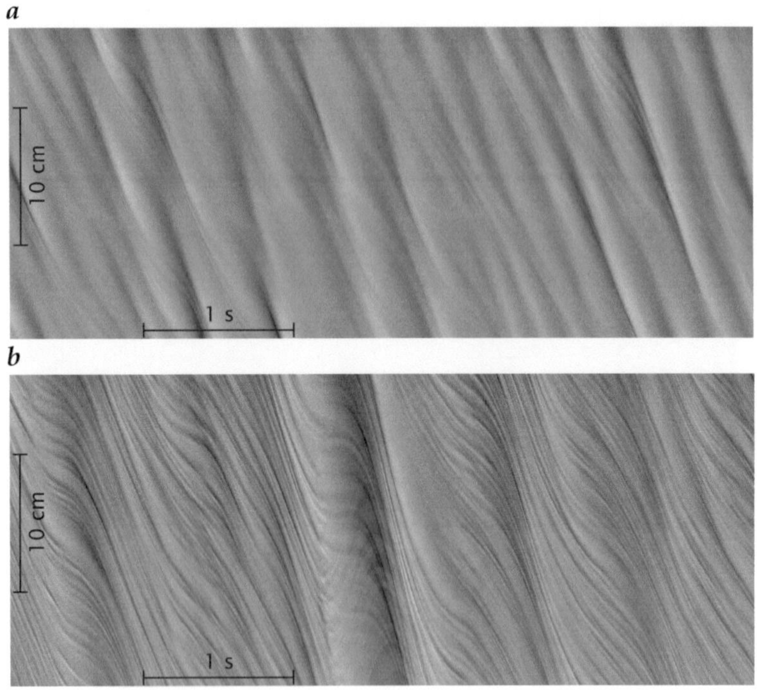

*Figure 1.9: Space-time image of wind-generated waves at wind speeds of **a** 2.5 and **b** 7.5 m/s. The vertical coordinate is the spatial coordinate in wind direction, the horizontal coordinate is the time.*

was used to visualize the inclination of the waves in the wind direction. In such space-time images, movement is shown by the orientation of lines of constant gray values. A horizontal line corresponds to static objects. The greater the angle to the horizontal axis, the faster the object moves. The image sequence provides a direct insight into the complex non-linear dynamics of wind waves. A fast-moving large wave can be seen modulating the movement of the small waves. Sometimes the small waves move at the same speed as the large waves (bound waves), but usually they are much slower and vary greatly in their phase velocity and amplitude.

In the last example, much larger spatial and time scales occur. Figure 1.10 shows the annual variation of the tropospheric column density of NO_2. NO_2 is one of the most important trace gases for ozone chemistry in the atmosphere. The main sources of tropospheric NO_2 are industry and traffic, forest and bush fires (biomass burning), microbiological soil emissions and thunderstorms. Satellite images make it possible to study the regional distribution of NO_2, the different sources and their annual cycles.

The data is calculated from spectroscopic images from the "GOME" instrument on the ERS2 satellite. Each pixel consists of a complete spectrum with 4000 channels covering the ultraviolet and visible range. The total at-

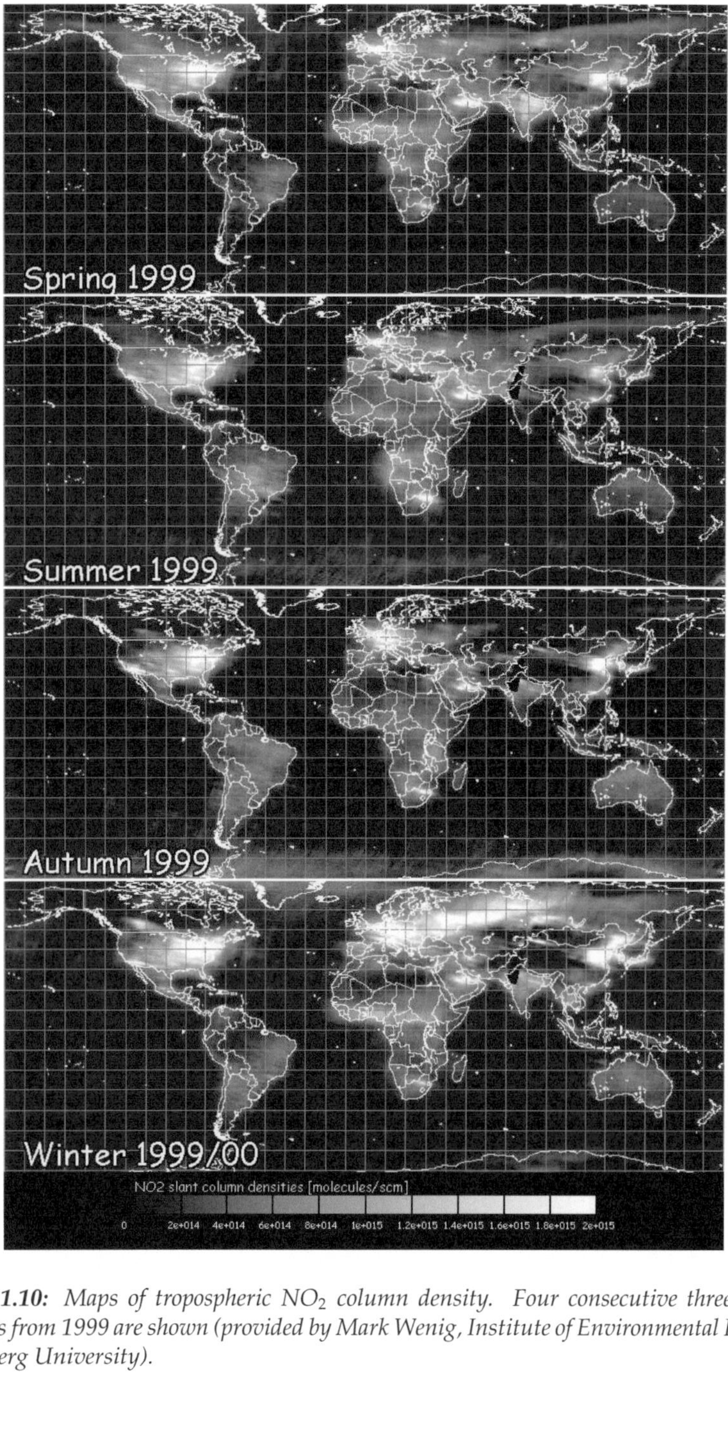

Figure 1.10: *Maps of tropospheric NO_2 column density. Four consecutive three-month averages from 1999 are shown (provided by Mark Wenig, Institute of Environmental Physics, Heidelberg University).*

a *b*

Figure 1.11: *Industrial inspection tasks: **a** Character recognition, **b** Position detection for connectors (provided by Martin von Brocke, Robert Bosch GmbH).*

mospheric column density of the NO_2 concentration can be determined from the characteristic absorption spectrum. However, this is overlaid by absorption spectra of other trace gases and influenced by a number of other effects (e. g. Fraunhofer lines in the solar spectrum and scattering by aerosols). Furthermore, to determine the tropospheric column density, it is necessary to subtract the stratospheric component using suitable image processing algorithms.

The resulting maps of tropospheric NO_2 column density in Fig. 1.10 show a wealth of interesting details. Most emissions come from industrialized countries. In the northern hemisphere, a clear annual trend can be seen with a maximum in winter.

1.2.4 Classification

Another important task is the classification of objects in images. The classic example of a classification task is *optical character recognition* or *OCR* for short. Fig. 1.11a shows a typical industrial OCR application, the recognition of the labeling on an integrated circuit. Object classification also includes the recognition of different positions of objects so that they can be gripped by a robot.

In Fig. 1.11b, connectors lie in random positions on a conveyor belt. In order to grip them correctly, it is necessary to recognize whether they are lying on the belt with their front or back side.

The classification of defects is another important task. Fig. 1.12 shows typical defects in integrated circuits, such as an incorrectly centered component (Fig. 1.12a) and broken and missing wire connections (Fig. 1.12b–f).

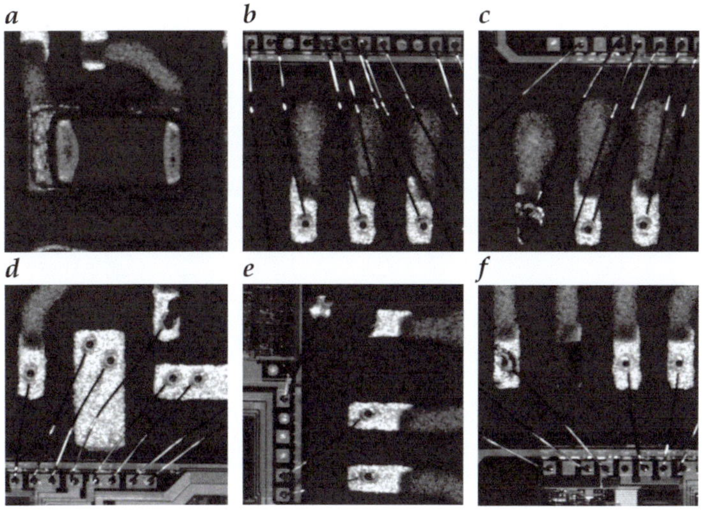

Figure 1.12: *Errors when soldering and bonding integrated circuits (provided by Florian Raisch, Robert Bosch GmbH).*

1.3 Hierarchy of Image Acquisition and Processing

Image acquisition and processing is a multi-stage process. A chain of image processing operations is carried out to extract the data we are interested in from an image. This results in the *hierarchical* scheme in Fig. 1.13. It provides an overview of the different phases of image processing and also summarizes the topics discussed in this book.

All image processing begins with the acquisition of an image with a — not necessarily optical — acquisition system. In order to image the object property of interest in the best way, we can select a suitable image acquisition system, the illumination, the most favorable wavelength range and many other options (Chapter 2). During imaging, the three-dimensional world is projected onto the image sensor plane (Chapter 3). The image sensor converts the radiant energy on the image plane into an electrical signal and a serial data stream that can be transferred to a computer where it can be stored and further processed (Chapter 4). The image signal is also digitized and quantized (Chapter 5). An essential task that combines image acquisition and evaluation processes is the recovery of the third dimension lost through projection onto the image sensor. The term *light field* provides a uniform theoretical approach to this. First, passive methods are discussed (Chapter 8) and then those that actively use illumination for this purpose (Chapter 9).

The first processing stage, *image preprocessing*, can involve a number of different operations. Any non-linear characteristics of a sensor must be corrected. It may also be that the brightness and contrast of the image require

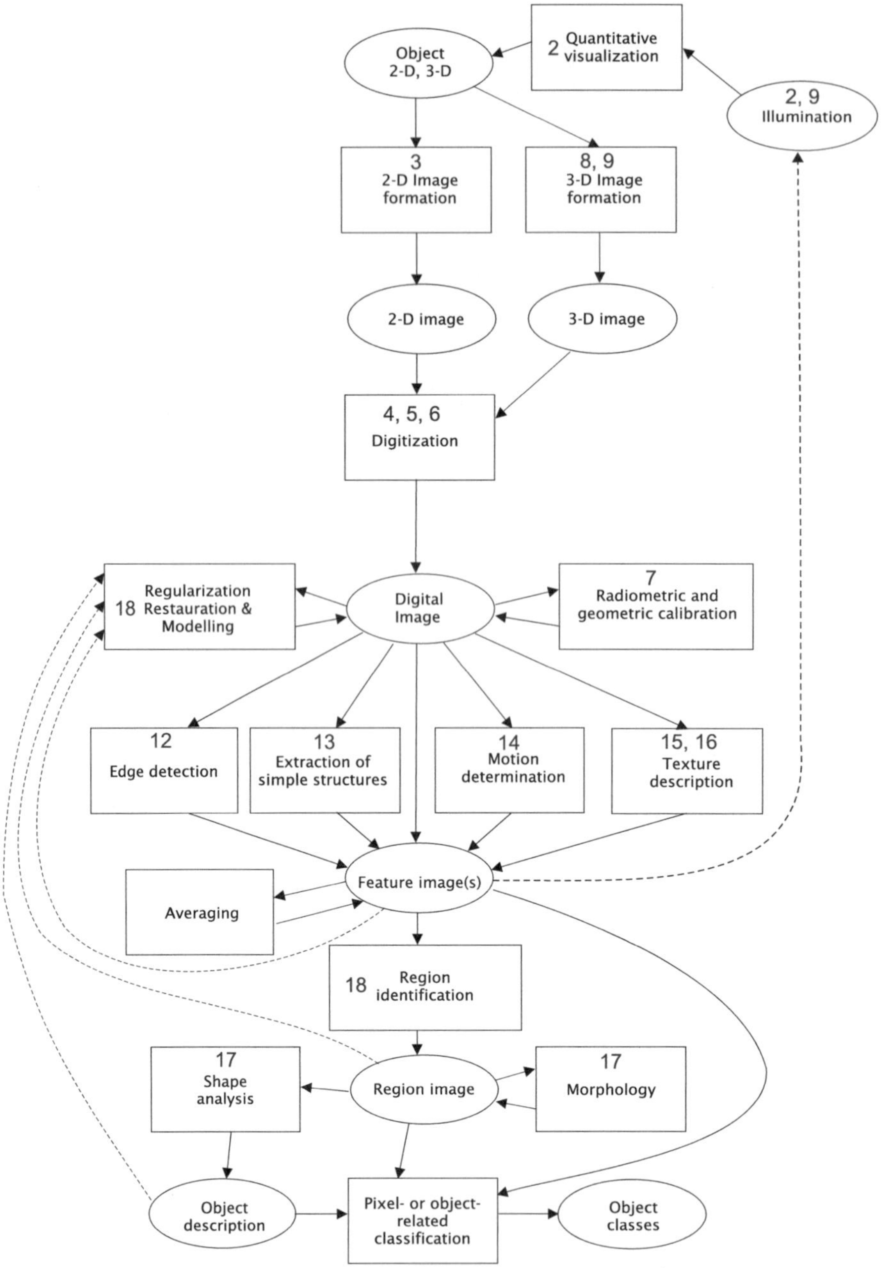

Figure 1.13: *Hierarchy of image processing operations from image acquisition to image comprehension. The blue numbers in the boxes refer to the corresponding chapters of this book. Dashed lines represent possible feedback loops.*

improvement. Coordinate transformations are also a common task for restoring geometric distortions that occur during image acquisition. Radiometric and geometric corrections are elementary operations; they are discussed in Chapter 7.

A whole chain of processing steps is necessary to analyze and identify objects. First, suitable filter operations must be carried out to distinguish the objects of interest from other objects and the background. Essentially, one or more *feature images* are extracted. Basic tools for this task are averaging (Chapter 11), edge detection (Chapter 12), the analysis of simple neighborhoods (Chapter 13) and complex patterns known in image processing as *textures* (Chapter 16). Another important property of an object is its *motion*. Techniques for determining motion are discussed in Chapter 14.

Furthermore, objects must be separated from the background, i. e. regions of constant features and discontinuities must be identified by *segmentation*. This is only a simple task if an object is clearly different from the background. Unfortunately, this is not often the case. Then more complex segmentation techniques are necessary (Chapter 17). They use various optimization strategies to achieve a minimal deviation between the image data and a given model that incorporates global knowledge about the objects in the image.

The same mathematical techniques can be used for other tasks. Image interference, e. g. due to defocused optics, motion blur, sensor errors or errors during image transmission can be corrected (*image restoration*). Likewise, images from indirect image acquisition techniques that do not directly provide an image, such as *tomography*, can be reconstructed (*image reconstruction*).

If objects are segmented, we can use morphological operators to analyze and change their shape or to determine other parameters such as the average gray value, the area, the circumference or other shape descriptions of the object (Chapter 18). These parameters can then be used to classify objects (*classification*).

It seems logical to break down a complex task into a series of simple subtasks. However, it is not clear whether this strategy leads to a solution. Let us consider a simple example. An object that differs only slightly in its gray value from the background is to be found in a noisy image. Here we cannot simply use the gray value to distinguish the object from the background. We could reduce the noise level by averaging neighboring pixels. At the edge of the object, however, we average the pixels of the background and the object and thus obtain incorrect average values. If we knew the position of the edges, we could prevent the averaging from running over the edge. But we can only determine the edge after the averaging, when the object and background can be distinguished by reducing the noise. It may be possible to solve this problem using an iterative approach. We carry out the averaging and make an initial estimate of the object edges. We use this to improve the averaging by not extending it over the object edges, and recalculate the edges, and so on. Although this approach sounds promising, it needs to be examined carefully: does the iteration converge at all? If so, is the limit correct?

Be that as it may, the example makes it clear that more difficult image processing tasks require feedback. Processing steps return parameters to previous processing steps. This means that there is no longer a linear processing chain, but there are also loops that can be run through several times. Figure 1.13 shows some of these possible feedback steps. Non-image processing steps can also be included. If an image processing task cannot be solved with a captured image, we might decide to change the lighting, zoom in closer to the object or look at it from a different angle. In this approach, we speak of *active vision*. In the context of intelligent systems that explore their environment through their sensory organs, we can also speak of an *action-perception cycle*.

1.4 Image Processing and Computer Graphics

For a long time, image processing and computer graphics were regarded as two separate fields of research. In the meantime, knowledge in both areas has grown to such an extent that the close relationship between the two fields is obvious. The aim of computer graphics is to create photorealistic images of a three-dimensional scene using a computer, while image processing attempts to reconstruct the three-dimensional scene from camera images. In this sense, we perform the reverse procedure in *image processing* compared to computer graphics. In computer graphics, we start with knowledge of the shape and properties of an object, i. e. we start at the bottom of the Fig. 1.13 scheme and work our way up until we obtain a two-dimensional image. For image processing and computer graphics, we essentially need the same knowledge. We need to know the interaction between lighting and objects, how a three-dimensional scene is projected onto an image plane, and more.

The calculation methods also have many similarities. It is therefore not surprising that the computing architecture of the *GPU* (*graphical processing unit*) invented for computer graphics is also excellently suited for fast image processing and was an essential prerequisite for the success of learning methods such as *deep learning*.

The term "visual computer" was coined early on to describe the integration of computer graphics, perception and image processing [117].

1.5 Interdisciplinary Nature of Image Processing

Digital image processing is an interdisciplinary science in many respects. Image processing incorporates concepts from various fields. Before we can process an image, we need to know how the digital signal is related to the properties of the recorded object. A number of physical processes are involved, from the interaction of radiation with matter to the geometry and radiometry of the optical image. In one way or another, an image sensor converts the irradiance into an electrical signal. The signal is then converted into digital numbers and processed by a digital computer to extract the relevant parameters. This processing chain (see also Fig. 1.13) involves many fields from *physics*, *computer science* and *mathematics*, including optics, solid state physics, computer architecture, algebra, analysis, statistics, complexity theory, graph theory, systems theory and numerical mathematics. From an engineering perspective, contributions from *technical optics*, *electrical engineering*, *optoelectronics* or *photonics* and *technical* and *applied computer science* are required.

Digital image processing has many points of overlap with other research areas. Image processing tasks can partly be understood as measurement problems and as such are a subfield of the science of *metrology*. Image processing, like *speech recognition*, includes tasks of *pattern recognition*. Other fields with similar relationships are the disciplines of *neuroinformatics*, *artificial intelligence* and *cognitive science*. All of these areas have close links to the life sciences in common.

When we speak of *computer vision*, we are referring to a computer system that performs the same task as a biological system, namely to "discover from images what is in the world and where it is" [215]. In contrast, the term *machine vision* is used for a system that performs a visual industrial inspection task, such as, e. g. checking the size and completeness of parts on a production line. For many years, a vision system has been thought of only as a passive observer. However, as biological systems do, a computer vision system can also actively explore its environment, i. e. move or change its viewing angle. We speak of *active vision*.

There are a number of research areas in the field of image processing that have developed independently of each other for historical reasons. The best-known example is *photogrammetry* (surveying using photographs to create maps and identify objects). Other areas are *remote sensing*, which works with aerial and satellite images, *astronomy*, and *medical image processing*.

The second important aspect of the interdisciplinary nature of image processing has to do with the wide range of application areas. Today, image processing is not only used in the natural sciences and engineering, but also in many areas of the humanities and is therefore part of the research field known as *computational humanities*.

As we have seen from the examples in Section 1.2, image processing has gained crucial importance in many applications. The close links to so many application areas are fertile ground for the further rapid development of image processing, as the application areas result in a constant stream of new techniques and ideas. Interdisciplinarity is not just an interesting side effect. It is a necessity. Lack of knowledge in either machine vision or the application area will inevitably lead to sub-optimal solutions or even complete failure.

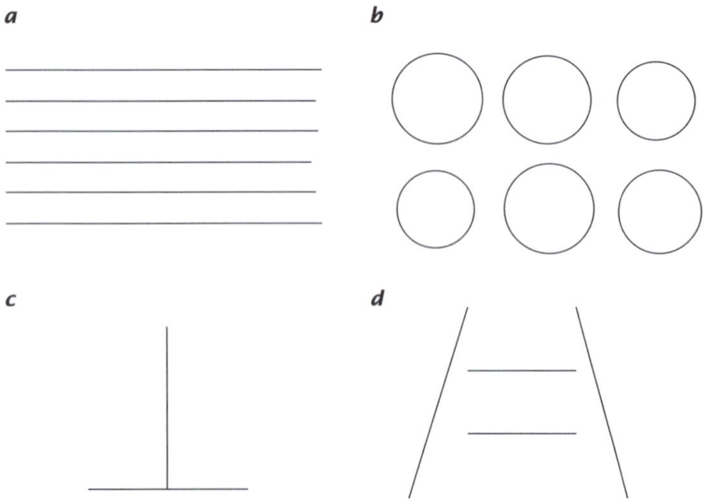

*Figure 1.14: Test images on the question of determining length and area with the human visual system: **a** parallel lines with up to 5 % difference in length; **b** circles with up to 10 % difference in radius; **c** the vertical line appears longer, although it has the same length as the horizontal line **d** perspective illusion: The upper line (in the background) appears longer than the lower line (in the foreground), although both are the same length.*

1.6 Human and Machine Vision

Machine image processing is inconceivable without the human visual system. This seemingly trivial statement has such far-reaching consequences that it should be kept in mind when dealing with digital image processing. Every image, whether captured directly or processed by a computer, can only be evaluated with the help of our visual system. Every image output on a monitor, printer, or other image output device must therefore consider the characteristics of the visual system. The simplest questions we should ask first are:

- What differences in intensity can we recognize?
- What spatial resolution does our eye have?
- How accurately can we estimate the size of lengths and areas and compare them with each other?
- How well can we distinguish colors?
- What properties can we use to recognize and distinguish objects?

It is obvious that a deeper knowledge of the human visual system is an invaluable aid to computer vision. It is not the intention at this point to give an overview of the visual system. Rather, we want to become aware of the elementary relationships between human and computer vision. For this

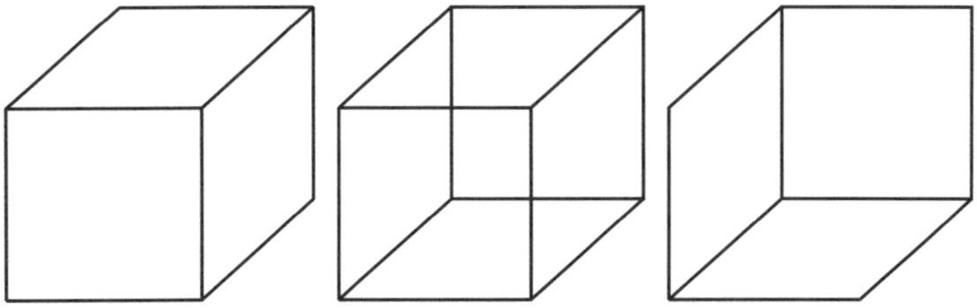

Figure 1.15: *Recognition of three-dimensional objects: three different representations of a cube with identical edges in the image plane.*

reason, we will highlight various characteristics of the human visual system in the corresponding chapters. Here are just a few introductory remarks.

The reader can carry out some experiments on the question of length and area estimation (Fig. 1.14). While small differences in length are easily recognized with parallel lines (Fig. 1.14a), an area comparison with circles seems to be much more difficult (Fig. 1.14b). The other examples show how the estimation can be misguided by surrounding objects. One speaks of *optical illusion*. Shown are examples of length estimation (Fig. 1.14c and d). These examples make it clear how much human vision is based on context. You must therefore be very careful when estimating and comparing lengths and areas in images.

The second topic is the question of recognizing objects in images. Although Figure 1.15 contains only a few lines and although it is a flat image without direct depth information, we immediately recognize a cube and its spatial orientation in the right and left images. We can only draw this conclusion from the hidden lines and our knowledge of the shape of a cube. The middle image, which also shows the hidden lines, is ambivalent. With some practice, you can switch between the two possible orientations in space.

Figure 1.16 shows a remarkable property of the human visual system. We easily recognize sharp boundaries between the different textures in Fig. 1.16a and also the letters I and M, which differ from the background only in the orientation of the texture. In Fig. 1.16b we identify a white isosceles triangle, although parts of the borders do not exist. These few observations show the enormous power of the human visual system in object recognition. In contrast, it is less suitable for the exact determination of gray values, distances and areas.

In the first edition of this book [155], the introductory chapter stated that the "marginal performance of computer vision systems should make us modest". This is still true today — 34 years later — despite the increase in computing speed and storage capacity by around four to five orders of magnitude. In terms of performance, size and energy consumption, computer systems are still worlds away from the performance of biological vision systems in visually capturing and understanding the environment and reacting intelligently and flexibly in real time.

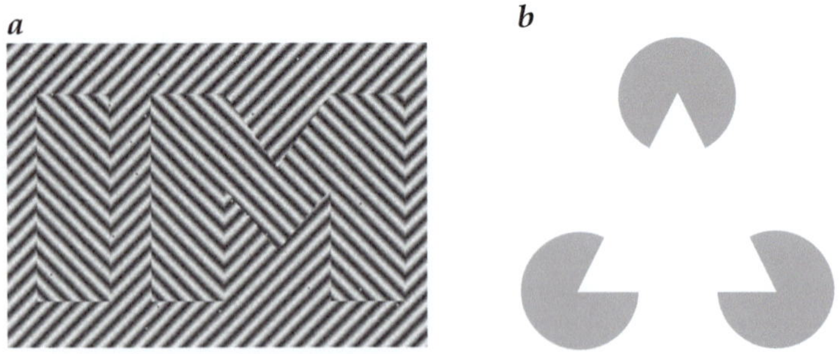

Figure 1.16: a Recognition of boundaries between patterns; b "Interpolation" of object boundaries.

Another relationship between human and machine vision is worth mentioning. Significant developments in machine vision have been made possible by advances in understanding the human visual system. We will encounter some examples in this book: the *pyramids* as an efficient data structure for image processing (Chapter 15), local orientation (Chapter 13) and the determination of motion by filtering techniques (Chapter 14).

1.7 ♯References for Further Reading

This section provides some references to further reading.

Textbooks. Before the first edition of this book was published in 1989, there were only a few German-language monographs and introductory textbooks. The first monograph on the subject from 1980 entitled "Acquisition and machine processing of image data" [176] dealt with both image acquisition and evaluation, as does this textbook today. Otherwise, of the early German-language textbooks, those by Wahl 1984 [346], Haberäcker 1985 [118] and Voss 1988 [345] are still worth mentioning today. Anyone interested in the historical development of image processing will find suitable material in the early English-language monographs and textbooks by Rosenfeld 1969 [270], Rosenfeld and Kak 1976 & 1982 [272, 273], Yaroslavsky 1979, English translation 1985 [358], Ballard "Computer Vision" 1982 [9] and Horn "Robot Vision" 1986 [141].

English-language classics in several editions are Petrou and Petrou [240], Pratt [247], Sonka et al. [311], Umbaugh [335] and Gonzalez and Woods [111]. Well-known textbooks in several editions that have been translated from German into English are Burger and Burge [32] and the book on "Machine Vision" by Beyerer et al. [16]. The textbook by van der Heijden [339] deals with image processing from the point of view of measurement technology. Modern textbooks focus more on computer vision: introductory like Nalwa [230] and Klette [183] or comprehensive like Forsyth and Ponce [88], Szeliski [320] and Prince [249].

Textbooks and monographs on special topics. Due to the interdisciplinary nature of image processing (Section 1.5), there is an abundance of textbooks or monographs that deal in detail with very diverse aspects or are limited to certain areas of application:

Topic	Reference
Image sensors, image acquisition, optics	Burke [33], Janesick [149], Holst and Lomheim [140], Teubner and Brückner [326]
Computational imaging, compressed sensing	Brady [24], Starck et al. [312], Eldar and Kutyniok [69]
Human and machine vision, perception	Korn [189], Levine [198], Mallot [212]
Microscopy and nanoscopy	Diaspro [58], Diaspro [59], Kubitscheck [190]
Imaging flow measurement	Raffel et al. [254], Nitsche and Dobriloff [233], Adrian and Westerweel [4]
Medical image acquisition and processing	Lehmann et al. [195], Dössel [65]
Radar, microwaves and remote sensing	Richards and Jia [263], Schott [294], Robinson [268], Ulaby and G.Long [332]
Mathematical foundation	Trottenberg et al. [331], Paragios et al. [238], Scherzer et al. [288], Bredies and Lorenz [25]
Geometry of computer vision	Faugeras [76], Faugeras et al. [77], Hartley and Zisserman [124]
Computer graphics	Watt [348], Foley et al. [87], Pharr and Humphreys [241]
Pattern recognition, classification, learning	Schürmann [295], Schölkopf and Smola [291], Bishop [20], Goodfellow et al. [112]
Image comprehension	Marr [215], Ullman [333]
Machine vision and industrial image processing	Jain et al. [147], Jähne et al. [164], Steger et al. [313], Demant et al. [54]
Photogrammetry	Bähr and Vögtle [17], Förstner and Wrobel [99], Luhmann et al. [210]

Handbooks. "The Image Processing Handbook" by Russ [275] impresses with a wealth of application examples and thus shows the diverse evaluation possibilities of image processing. However, it is less suitable for systematically learning the methods of image processing. In contrast, the "Practical Handbook on Image Processing for Scientific Applications" [157] takes a strictly methodical approach, but also contains many practical tips. A comprehensive summary of the field of computer vision is provided by the three-volume work "Handbook of Computer Vision and Applications"

[165]. The practice-oriented handbook "Machine Vision" [142] also covers lighting technology, optical and camera systems. Finally, the "Handbook on Machine Vision" published by Fraunhofer Vision [277] clearly presents the system components, processes and image processing standards used in industry for quality assurance in production.

Article collections. The following anthologies provide a good overview of the fundamentals of image processing based on older key publications from journals: "Digital Image Processing" [41], "Readings in Computer Vision: Issues, Problems, Principles, and Paradigms" [82], "Computer Vision: Principles" [174] and "Computer Vision: Advances and Applications" [175].

Part I

Image Acquisition

2 Radiation and Quantitative Imaging

Contents

B. Jähne, *Digital Image Processing and Image Acquisition*,
https://doi.org/10.1007/978-3-032-04806-6_2

2.1 Introduction

An image recording system collects the radiation emitted by objects and makes it visible. Machine or computer vision as a replica of human vision captures scenes with the given lighting and is limited to *light*, i. e. the narrow range of electromagnetic radiation to which the eye is sensitive. However, possible radiation for image-recording systems is not limited to light. It includes not only the entire range of electromagnetic radiation, but also particle radiation, such as electrons or other particles, and acoustic waves.

This is of great importance for vision systems used in industrial and scientific applications. The first task here is to establish a quantitative relationship between the object property of interest and the recorded radiation. Which radiation is best suited for this? Which object properties can actually be measured by the received radiation?

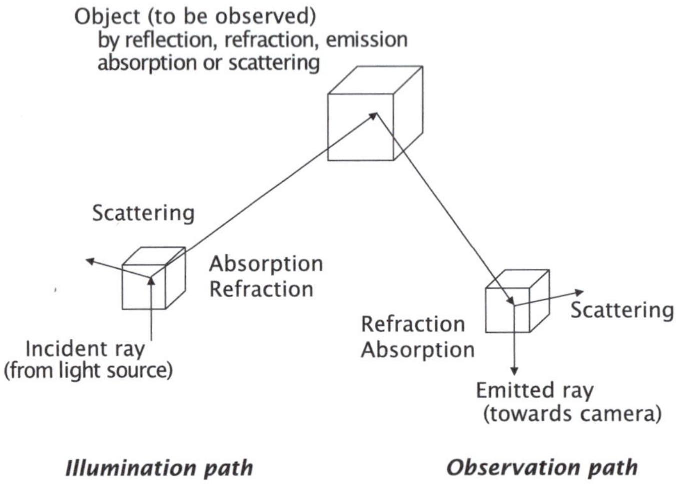

Figure 2.1: *Schematic representation of the interaction between radiation and matter during object visualization. The relationship between the radiation emitted from the light source to the camera and the object can be disturbed by scattering, absorption and refraction of the incident and emitted beam.*

In most cases, objects do not emit radiation on their own, but are illuminated. Figure 2.1 illustrates how the radiation emitted by a light source on

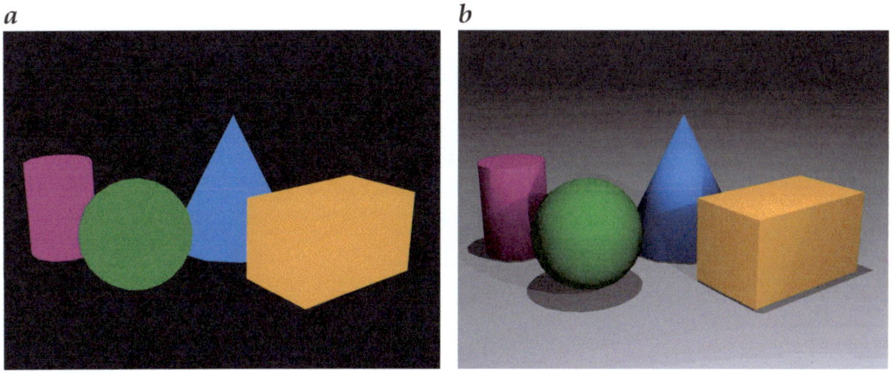

Figure 2.2: *Object detection in an image based on its color:* ***a*** *Ideal but unrealistic case: objects appear in uniform color* ***b*** *Real case: The color value of the objects depends on the ambient brightness and the inclination of the object surfaces relative to the light sources and the camera.*

the surface or in the volume of the object can be directed towards the image acquisition system by reflection, refraction or scattering. The radiation can be influenced by additional processes both on the path from the light source to the object (*illumination path*) and on the path from the object to the camera (*observation path*). Scattering and absorption of the incident and emitted beam not caused by the observed object itself can lead to an attenuation of the radiant flux. The position of the object may appear shifted due to deflection of the emitted beam.

The careful arrangement of all components for image acquisition must ensure that these additional influences are minimized and that the measured radiation correctly reflects the object property under consideration. Even in cases where we have no influence on the illumination, there is still the careful choice of radiation type and wavelength range.

First, we will look at these relationships intuitively using a few examples. A simple example is the detection of an object based on its color, i. e. the dependence of the reflection coefficient on the wavelength. We assume that the object surface is uniformly colored. From an image processing perspective, the objects should also appear in the image with a uniform color, as shown in Figure 2.2a. However, this is not the case in reality (Fig. 2.2b). In fact, the objects appear in different brightness and color depending on how the surface is inclined relative to the direction of irradiation of the light sources and the viewing direction of the camera. How much light hits an object surface in a scene is generally very complicated, as it is not only directly irradiated by the light sources, but also indirectly by the other objects in the scene, which reflect some of the incident radiation back. This is why it is so difficult to create photorealistic images using computer graphics.

For many applications, however, it seems unnecessary to know why something is displayed as it is in an image. Isn't it enough that the different

Figure 2.3: False-color representation of an image from the Landsat satellite (Danube valley with Ingolstadt in the right part of the image), composed of three wavelength ranges.

Figure 2.4: Two plant leaves, acquired with different illumination.

objects look different? On the Landsat satellite image in Figure 2.3, for example, urban areas, forests, rivers, lakes and agricultural regions can be easily recognized.

However, whether this is always possible depends heavily on the variation in lighting conditions. Figure 2.4 shows two plant leaves that look completely different depending on the lighting. Therefore, it is clear that it is not always possible to reliably recognize leaf veins.

To summarize, we can formulate the central question for a successful image acquisition system as follows: Which parameters influence the "brightness" of the object of interest in the image? This question is answered in two stages in this chapter. First, we analyze what types of information can be transmitted from an object to its image. From this point of view, the properties of radiation are considered in general.

Radiation can be regarded as a wave or a stream of particles, depending on which physical phenomenon we are looking at. For image acquisition, we need both approaches. The direction of the wave amplitude can coincide with the direction of propagation of a wave. Then we speak of a *longitudinal wave*. However, the oscillation of the wave can also be perpendicular to the direction of propagation. Such a wave is called a *transverse wave*. With the transverse wave type, there are therefore many possible directions of oscillation. This effect is called *polarization* and is discussed in more detail in Section 2.2.7.

The properties of radiation are considered systematically. We start with the basic description of waves (Section 2.2), before moving on to electromagnetic waves (Section 2.3.1), particle radiation (Section 2.3.2) and acoustic waves (Section 2.3.3). With these considerations, we have gathered all the knowledge about the many ways in which the radiation emitted by an object transmits information that could be recorded by imaging techniques (Section 2.3.4).

Since radiation is a form of energy, we also need to address how to adequately describe radiant energy and power (Section 2.4). This is the field of *radiometry*. In this context, it is also important to understand how we humans perceive the intensity of radiation (*photometry*) and whether this can also be measured. An important aspect of this is how humans see colors and how these are related to the wavelengths of light and thus to the physical description in *spectroscopy* (Section 2.5).

In Section 2.6 we investigate why and how objects emit radiation. Physically speaking, this question deals with the broad field of interaction between radiation and matter. First of all, we ask ourselves under which conditions objects emit radiation of their own accord even without being irradiated, i. e. are themselves a source of radiation (Section 2.6.1). In physical terms, this means how other forms of energy can be converted into radiant energy. It is also important to investigate what happens when the radiation source moves relative to the radiation receiver (Section 2.6.2).

In the second step, we systematically investigate how radiation interacts with matter from the point of view of image acquisition: reflection and refraction of radiation on surfaces (Section 2.6.3 and 2.6.4), absorption (Section 2.6.5), changes in the polarization state (Section 2.6.6), scattering (Section 2.6.7), luminescence (Section 2.6.8) and finally non-linear effects (Section 2.6.9).

2.2 Basic Properties of Waves

Waves are one of the universal and fundamental phenomena in physics [70, 75, 330]. A *wave* is the temporal variation (oscillation) of an arbitrary

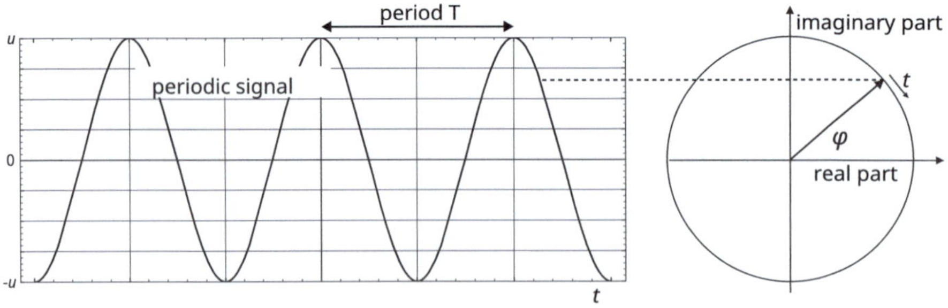

Figure 2.5: Representation of a periodic function in the complex number plane as a rotating pointer with the amplitude u. The periodic variation with a cosine function results as a projection of the pointer onto the real axis.

quantity $u(\boldsymbol{x}, t)$ that propagates in space and is described by the *wave equation*

$$\boxed{\frac{\partial^2 u(\boldsymbol{x}, t)}{\partial t^2} = c^2 \Delta u(\boldsymbol{x}, t) \qquad \text{wave equation}} \tag{2.1}$$

The second time derivatives are equal to the sum of the second spatial derivatives (*Laplace operator* Δ, see Appendix A) except for the factor c^2.

2.2.1 Frequency, Wavenumber and Plane Waves

When we think of waves, we intuitively think of quantities that vary periodically in time and space. How can these be adequately described? A temporally periodic pattern at a location is characterized by the time between two maxima or the *period* T (Fig. 2.5) or equivalently by the number of periods per time unit, the *frequency* ν:

$$\nu = 1/T. \tag{2.2}$$

Similarly, a spatially periodic signal is characterized by the distance between two maxima, the *wavelength* λ (Fig. 2.6). The direction of the pattern is best described by a vector perpendicular to the lines of constant values. Let us give this vector $\boldsymbol{\kappa}$ the length $1/\lambda$:

$$\kappa = |\boldsymbol{\kappa}| = 1/\lambda, \tag{2.3}$$

the wavelength and direction can be expressed by a vector, the *wavenumber vector* $\boldsymbol{\kappa}$. The components of $\boldsymbol{\kappa} = [\kappa_1, \kappa_2]^T$ directly reflect the number of wavelengths per unit length in the relevant direction[1]. The wavenumber vector $\boldsymbol{\kappa}$ can be used to describe arbitrary-dimensional periodic patterns.

[1]The definition of the wavenumber vector with the factor 2π, which is more common in physics, is not used here, i. e. $|\boldsymbol{k}| = 2\pi/\lambda$. Vectors are understood as column vectors. More on the notation in Appendix A.

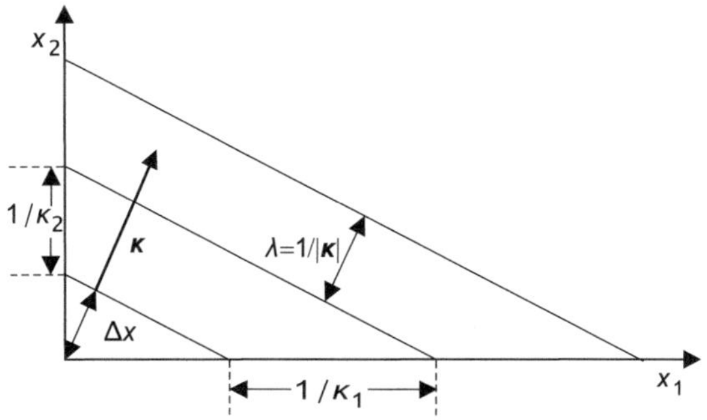

Figure 2.6: *Description of a wave as a spatially periodic pattern that moves with the phase velocity c in the direction of the wave vector* $\boldsymbol{\kappa}$. *Further explanation in the text.*

Two details are still missing for a complete description of a periodic signal: the amplitude u and the relative position of the pattern in the origin of the coordinate system (Fig. 2.6). This is given by the distance Δx of the first maximum from the origin. Since this distance is at most one wavelength, we specify it relative to the period or wavelength by a *phase angle* $\varphi = 2\pi\Delta t/T = 2\pi\nu\Delta t$ or $\varphi = 2\pi\Delta x/\lambda = 2\pi k\Delta x$ (Fig. 2.6). The complete description of a temporally or spatially periodic pattern is thus given by

$$u\cos(2\pi\nu t - \varphi) \quad \text{and} \quad u\cos(2\pi\boldsymbol{\kappa}^T\boldsymbol{x} - \varphi), \tag{2.4}$$

where $\boldsymbol{\kappa}^T\boldsymbol{x}$ is the scalar product of the column vectors $\boldsymbol{\kappa}$ and $\boldsymbol{x}$ (Appendix A).

However, this description is complicated, as you have to specify two parameters with the amplitude u and the phase φ. Ideally, a simple but complete description of a periodic pattern should only require one specification. This is possible using the *complex number* $\mathrm{u} = u\exp(-\mathrm{i}\varphi)$, which combines amplitude and phase, and the complex exponential function $\exp(\mathrm{i}\varphi) = \cos\varphi + \mathrm{i}\sin\varphi$. The real part of $\mathrm{u}\exp(2\pi\mathrm{i}\nu t)$ and $\mathrm{u}\exp(2\pi\mathrm{i}\boldsymbol{\kappa}^T\boldsymbol{x})$ then results in the expressions in Eq. (2.4):

$$\mathrm{Re}\left(\mathrm{u}e^{2\pi\mathrm{i}\nu t}\right) = u\cos(2\pi\nu t - \varphi) \quad \text{and} \quad \mathrm{Re}\left(\mathrm{u}e^{2\pi\mathrm{i}\boldsymbol{\kappa}^T\boldsymbol{x})}\right) = u\cos(2\pi\boldsymbol{\kappa}^T\boldsymbol{x} - \varphi). \tag{2.5}$$

This relationship is shown graphically in Figure 2.5. In the complex number plane, a periodic signal is a circular motion. A vector with the magnitude u makes one revolution in a period T. The phase $\varphi = 2\pi\nu t$ changes linearly with time. The temporal change (derivative) of the phase is the *circular frequency* ω and the spatial gradient of the phase is proportional to the wave vector:

$$\frac{\partial\varphi}{\partial t} = \omega = 2\pi\nu \quad \text{and} \quad \nabla\varphi = 2\pi\boldsymbol{\kappa}. \tag{2.6}$$

The complex exponential function can also be used to describe the spatio-temporal variation of a monochrome wave particularly easily, as the exponents of the two exponential functions to be multiplied add up:

$$\mathrm{u}(\boldsymbol{x}, t) = \mathrm{u}e^{2\pi \mathrm{i}\boldsymbol{\kappa}^T \boldsymbol{x}} e^{-2\pi \mathrm{i}\nu t} = \mathrm{u}e^{2\pi \mathrm{i}(\boldsymbol{\kappa}^T \boldsymbol{x} - \nu t)}. \tag{2.7}$$

Is this wave a solution of the wave equation Eq. (2.1)? You can see this simply by substituting Eq. (2.7) into Eq. (2.1) and thus obtain a relationship between the frequency and the wavenumber, the so-called *dispersion relation*:

$$\boxed{\nu = |\boldsymbol{\kappa}|\, c \quad \text{or} \quad \lambda = \frac{c}{\nu} \qquad \text{dispersion relation.}} \tag{2.8}$$

If the coordinate system is rotated so that the wave vector points in x-direction and the dispersion relation Eq. (2.8) is set to Eq. (2.7)

$$\mathrm{u}(x, t) = \mathrm{u}e^{2\pi \mathrm{i}\kappa(x - ct)}, \tag{2.9}$$

it can be seen that c is the propagation velocity of the wave, the so-called *phase velocity*. A wave that, like Eq. (2.9), has a constant amplitude throughout the entire space and propagates in a straight line in the direction of the wave vector $\boldsymbol{\kappa}$ is called a *plane wave*.

A plane wave with an arbitrary wavenumber and the corresponding frequency determined by the dispersion relation are not the only solutions of the wave equation Eq. (2.1). Since this is linear, the *superposition principle* applies. This means that every superposition of plane waves is also a solution. The most general spatio-temporal structure is any spatial structure that moves with the phase velocity c. This means

$$\mathrm{u}(\boldsymbol{x}, t) = \mathrm{u}\left(\boldsymbol{\kappa}^T \boldsymbol{x} - \nu t\right) = \mathrm{u}\left(\boldsymbol{\kappa}^T \boldsymbol{x} - |\boldsymbol{\kappa}|\, ct\right), \tag{2.10}$$

as can be shown by direct insertion, is also a solution to the wave equation.

This general solution can be thought of as being composed of an infinite number of plane waves, which has a complex-valued amplitude $\mathrm{u}(\boldsymbol{\kappa}, \nu)$ to be determined for each wavenumber. If this decomposition is actually possible and completely describes the spatio-temporal structure, another and at the same time simpler representation of the general solution of the wave equation would be found, namely exactly the complex-valued function $\tilde{u}(\boldsymbol{\kappa}, \nu)$.

The fact that this is possible and that both representations are equivalent is described by the theory of vector spaces and orthogonal transformations such as the *Fourier transform*. The mathematical foundations of vector spaces and the Fourier transform and other orthogonal transformations, insofar as they are necessary for understanding imaging and processing, are described in Appendix B.

This allows a wave to be described as a spatiotemporal structure $\mathrm{u}(\boldsymbol{x}, t)$ on the one hand and as a superposition of plane waves with a wavenumber-dependent complex-valued amplitude distribution $\mathrm{u}(\boldsymbol{\kappa}, \nu)$ in *Fourier space* on the other. Both representations are equivalent. If one is known, the other results from a Fourier transformation.

The representation of waves in Fourier space is particularly simple. A single point in Fourier space represents a plane wave that extends across the entire space-time domain. The temporal and spatial derivatives of a plane wave do not change the wavenumber and frequency since they only change the complex exponential function by a factor. Therefore, the wave equation Eq. (2.1) becomes a simple algebraic equation in wavenumber-frequency space, i. e. the dispersion relation Eq. (2.8).

Although the decomposition of waves into plane waves is essential for understanding optics and image acquisition, it is not sufficient. A plane wave only occurs when radiation from a distant point-like object hits an imaging system. Furthermore, real waves are not infinitely extended in time.

Therefore, other elementary waveforms are important for imaging systems. First we look at other spatial distributions. For a monochrome wave with an arbitrary spatial distribution, i. e. with the approach $\mathrm{u}(\boldsymbol{x}, t) = \mathrm{u}(\boldsymbol{x})\mathrm{e}^{-2\pi\mathrm{i}\nu t}$, the wave equation Eq. (2.1) simplifies to

$$\boxed{\Delta \mathrm{u}(\boldsymbol{x}) + 4\pi\kappa^2 \mathrm{u}(\boldsymbol{x}) = 0 \qquad \text{Helmholtz equation.}} \tag{2.11}$$

This solution can be obtained by applying a Fourier transform in the time domain to the wave equation Eq. (2.1).

2.2.2 Spherical Wave

Now we ask ourselves how to describe the monochrome wave emitted isotropically in all directions by a point source of radiation. This could be done by directly solving the Helmholtz equation (2.11) in *spherical coordinates* $[r, \theta, \phi]$, but unfortunately this is mathematically complex [22, 305]. A solution can be found much more quickly using three basic considerations:

1. Since the spherical wave is an isotropic function, it cannot depend on the angular coordinates $[\theta, \phi]$ but only on r: $\mathrm{u}(r)$.
2. Furthermore, we consider the case of propagation without attenuation, i. e. without energy loss. The energy density of a wave is — as with all oscillations — proportional to the square of the amplitude. The spherical surface $4\pi r^2$, over which the energy is evenly distributed, increases quadratically with the radius r. This means that the energy density must be proportional to r^{-2} (*inverse-square law*) and the amplitude $\mathrm{u}(r) \propto r^{-1}$.
3. We assume a monochromatic spherical wave with wavenumber k_0.

This means that the spherical wave can only have the following solution:

$$\mathrm{u}(r) = \frac{\mathrm{u}}{r}\mathrm{e}^{2\pi\mathrm{i}\kappa_0 r} \quad \text{with} \quad r = \left(x^2 + y^2 + z^2\right)^{1/2}. \tag{2.12}$$

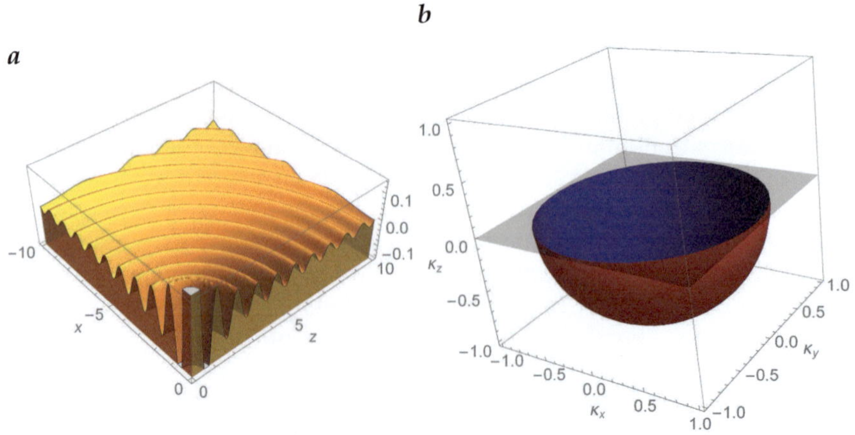

Figure 2.7: A spherical wave propagating in all directions from one point is another elementary waveform. Shown is ***a*** *a spatial 2D section in the xz-plane through a spherical wave with a wavelength $\lambda = 1$ in local space and* **b** *the Fourier transform in wavenumber space, a δ-sphere that has been cut through in the $k_x k_y$-plane.*

Extension 2.1: Derivation of the spherical wave equation in Fourier space

The spherical wave equation Eq. (2.12) can also be derived by simple considerations in Fourier space. Firstly, the Fourier transform of the spherical wave $\mathfrak{u}(\boldsymbol{\kappa})$ can only depend on the magnitude of the wavenumber vector due to isotropy, i. e. $\mathfrak{u}(|\boldsymbol{\kappa}|) = \mathfrak{u}(\kappa)$. Secondly, the spherical wave is monochromatic. Therefore, the Fourier transform must be zero for all wavenumber vectors except those with magnitude κ_0. This fact is expressed mathematically with a *delta distribution*, see Eq. (B.44). The δ function $\delta(x - x_0)$ is zero everywhere except for the value x_0. At this point, the distribution has the value infinity. However, the integral over this infinitely thin and high function is one (Appendix B.1.1). With these two basic facts, the Fourier transform of a spherical wave can only be an infinitely thin spherical shell with the radius κ_0:

$$\tilde{\mathfrak{u}}(\boldsymbol{\kappa}) = \frac{1}{4\pi\kappa_0^2}\delta(\kappa - \kappa_0), \tag{2.13}$$

where the function was normalized with the area of the spherical shell (Fig. 2.7). Now we just have to find the inverse Fourier transform of this function. This can be found in the table of elementary Fourier transform pairs (> R13).

2.2.3 Wave Packets

All elementary waveforms considered so far are both extended over the entire space and unlimited in time and are therefore unrealistic for actually occurring waves. Therefore, waveforms that are limited in time and space are still missing.

In this section, we will first consider time-limited waves, known as *wave packets*. According to the wave equation Eq. (2.10) any wave packet that is arbitrarily shaped and spatially limited, and that moves through space at the phase velocity, is a valid solution. Therefore, if a wave packet is limited to a time interval Δt, it is automatically also limited in space to a distance $\Delta z = c\Delta t$.

So far, we have learned (Section 2.2.1) that any waveform can be composed of plane waves. What is less intuitive, however, is how a wave packet limited in time and space can be composed of infinitely extended plane waves. We therefore proceed step by step.

First, we superimpose two plane waves traveling in the z direction, which differ in wavenumber by $\Delta\kappa$ and in frequency by $\Delta\nu$:

$$\begin{aligned} u(z) &= \left(e^{2\pi i((\kappa+\Delta\kappa/2)z-(\nu+\Delta\nu/2)t)} + e^{2\pi i((\kappa-\Delta\kappa/2)z-(\nu-\Delta\nu/2)t)}\right)\Big/2 \\ &= e^{2\pi i(\kappa z-\nu t)}\left(e^{\pi i(\Delta\kappa z-\Delta\nu)} + e^{-\pi i(\Delta\kappa z-\Delta\nu t)}\right)\Big/2. \end{aligned}$$

The common factor $e^{2\pi i(\kappa z-\nu t)}$ was excluded. The second term is simplified to a cosine function using the definition of the complex exponential function $e^{ix} = \cos x + i\sin x$:

$$\boxed{u(z) = e^{2\pi i(\kappa z-\nu t)}\cos\left(\pi(\Delta\kappa z - \Delta\nu t)\right) \qquad \text{beat.}} \tag{2.14}$$

The superposition of two plane waves results in a wave with the mean wavenumber κ, the amplitude of which is slowly modulated with a wavelength $\Delta\lambda = 1/\Delta\kappa$. This effect is called *beat* and is shown in Fig. 2.8b on the left for a wave with $\kappa = 1$ and $\Delta\kappa = 0.1$. In wavenumber space, the superimposed wave consists of two δ functions with $\kappa = 0.95$ and 1.05 (Fig. 2.8b on the right). By superimposing several plane waves with the same wavenumber spacing $\Delta\kappa$, the period $1/\Delta\kappa$ of the beat does not change, but the wave amplitude disappears over an ever greater distance (Fig. 2.8b and c). The wave amplitudes of the superimposed waves were not the same. For the superposition of P waves, the amplitude of the wave p was set equal to the corresponding *binomial coefficient*, i. e. the following superposition was performed:

$$u(z) = \frac{1}{2^{P-1}}\sum_{p=0}^{P-1}\binom{P-1}{p}\exp\left(2\pi i(\kappa - p\Delta\kappa/2)z\right) = e^{2\pi i\kappa z}\cos^{P-1}(\pi\Delta\kappa z). \tag{2.15}$$

This surprisingly simple result of the summation is due to the fact that the superposition of two waves can be repeated iteratively after Eq. (2.14). This procedure corresponds to the scheme of *Pascal's triangle*. Each averaging over a wave pair results in a further factor $\cos(\pi\Delta\kappa z)$ and the binomial factors are automatically obtained (more on this in Appendix C.4.2).

In the second step, we still have to eliminate the periodic repetition of the wave packets. If the wavenumber difference $\Delta\kappa$ is halved, the period in

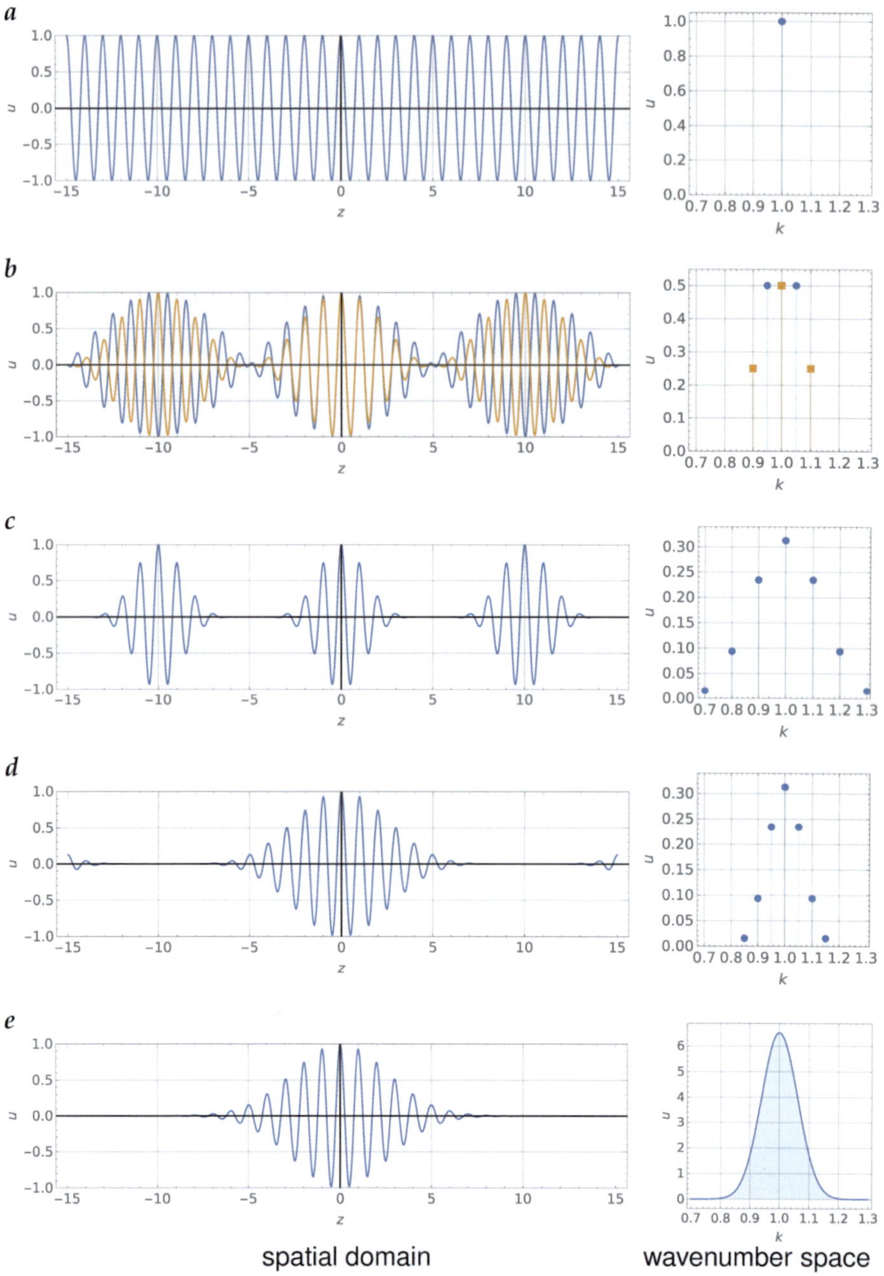

Figure 2.8: Superposition of plane waves traveling in the z-direction, represented in spatial and wavenumber space: ***a*** *a wave with wavenumber 1;* ***b*** *superposition of two waves with wavenumbers 0.95 and 1.05 (blue) and three with wavenumbers 0.9, 1 and 1.1 (orange);* ***c*** *seven superimposed waves around wavenumber 1 with wavenumber differences of 0.1;* ***d*** *also with wavenumber differences of 0.05.* ***e*** *Gaussian wave packet with the same width as* ***d****.*

the spatial domain is doubled. At the same time, however, the wave packet becomes twice as wide (Fig. 2.8d). This means that both the number of superimposed waves must be increased and the wavenumber difference must be further reduced. The limit transition to a single wave packet occurs when an infinite number of plane waves with a wavenumber difference approaching zero are superimposed. Then the discrete superposition changes into a continuous and the binomial distribution into a *normal distribution* (Appendix C.4.2). This brings us to the *Gaussian wave packet* (Fig. 2.8e), which is so important for optics and imaging.

The spatio-temporal limitation of a wave inevitably leads to a new important phenomenon. While the infinitely extended plane wave and the spherical wave are *monochromatic* (consisting of only one wavelength and frequency), this is no longer possible with a wave packet.

The spatial restriction of the expansion of a wave with wavenumber κ_0 by multiplication with a "window" function $w(z)$ leads to a broadening in Fourier space with the Fourier transform $\tilde{w}(\kappa)$ of $w(z)$:

$$\mathrm{u}(z) = \mathrm{u}\,\exp(2\pi \mathrm{i}\kappa_0 z)\,w(z) \quad \circ\!\!-\!\!\bullet \quad \tilde{\mathrm{u}}(\boldsymbol{\kappa}) = \mathrm{u}\,\tilde{w}(\kappa - \kappa_0). \tag{2.16}$$

The symbol $\circ\!\!-\!\!\bullet$ indicates that the right-hand function is the Fourier transform of the left-hand function.

This means that the wave packet is no longer monochromatic. The more it is spatially constricted, i. e. the fewer wave trains it contains, the broader the distribution of wavenumbers.

Extension 2.2: Classical uncertainty principle

Here we take a closer look at wave packets in spatial and Fourier space using the Gaussian wave packet. Remarkably, the Gaussian wave packet is described by the same function in both the spatial and wavenumber domains (Fourier space). The Gaussian bell curve is one of the many functions whose form is preserved during a Fourier transformation (> R14):

$$\exp\left(-\pi z^2\right) \quad \circ\!\!-\!\!\bullet \quad \exp\left(-\pi \kappa^2\right) \quad \text{with} \quad \int_{-\infty}^{\infty} \exp\left(-\pi x^2\right) \mathrm{d}x = 1. \tag{2.17}$$

The factor π was used so that the integrals over both functions add up to one. We have already seen this in the discussion of wave packets (Fig. 2.8): If a wave packet is expanded in spatial space, the Fourier transform is compressed by the same factor in Fourier space. This relationship applies universally and is described by the similarity theorem (scaling property) of the Fourier transform (Appendix B.1.4):

$$\mathrm{u}(sx) \quad \circ\!\!-\!\!\bullet \quad \frac{1}{|s|}\tilde{\mathrm{u}}\left(\frac{\kappa}{s}\right), \quad \text{wobei} \quad s \in \mathbb{R} \neq 0. \tag{2.18}$$

We can now calculate the relationship between the width of the wave packet in spatial and Fourier space. By comparing $\exp(-\pi(sz)^2)$ with $\exp(-z^2/(2\sigma_z^2))$

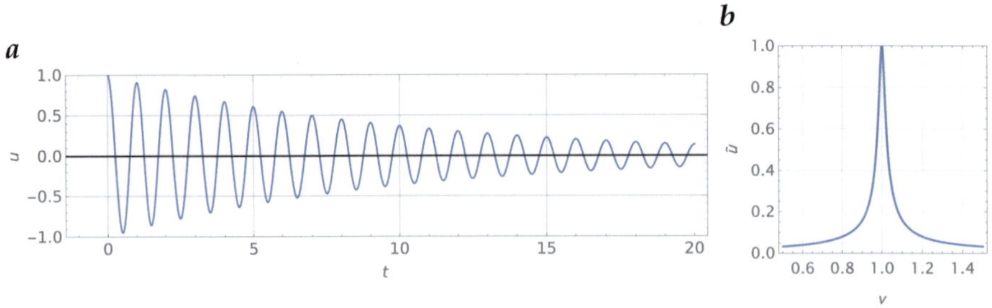

*Figure 2.9: A wave train with a period of one time unit, which decays exponentially with a time constant τ of 10 time units, represented in **a** time and **b** frequency space (magnitude).*

we get $s^2 = 1/(2\pi\sigma_z^2)$. This results in $\exp(-2(\pi\sigma_z\kappa)^2)$ for the Gaussian function scaled in Fourier space according to Eq. (2.18)

$$\sigma_z\sigma_k = \frac{1}{2\pi} \quad \text{and also} \quad \sigma_t\sigma_\nu = \frac{1}{2\pi}. \tag{2.19}$$

In a completely analogous way, the same considerations can be made with the time duration of a wave packet and the frequency broadening and then arrive at the second relation in the above equation. This result is known as the *classical uncertainty relation*, in signal and system theory also as the *time-bandwidth product*, see [106, Section 9.10] or [251, Section 3.8.1]. We have used the standard deviations σ directly here. Normally, with the uncertainty principle, the signals are squared and then the second moments are calculated. Then you get half the value of Eq. (2.19). The Gaussian bell curve is the only function that achieves the specified product, for all other functions it is greater. Further details on the uncertainty principle can be found in Appendix B.1.4.

Example 2.1: Damped oscillation as a wave packet

A damped oscillator is a good model for generating waves. The radiation starts at time zero and then decays exponentially (Fig. 2.9). This signal can be represented as the multiplication of a plane wave of amplitude u by an exponential function that is set to zero for all negative values:

$$\mathrm{u}(t) = u\,\Theta(t)\exp(-t/\tau)\exp(2\pi\mathrm{i}\nu_0) \quad \text{with} \quad \Theta(t) = \begin{cases} 1 & t > 0 \\ 1/2 & t = 0 \\ 0 & t < 0 \end{cases}. \tag{2.20}$$

For the latter, the *unit jump function* or *heaviside function* $\Theta(t)$ is used. The Fourier transform of the truncated exponential function is tabulated in ≻ R13 and the multiplication with the plane wave leads to a shift of the Fourier transform by the frequency ν_0. This results in the following Fourier transform pair:

$$\frac{1}{\tau}\Theta(t)\exp(-t/\tau)\exp(2\pi\mathrm{i}\nu_0) \quad \circ\!\!-\!\!\bullet \quad \frac{\mathrm{i}\tau}{\mathrm{i} - 2\pi\tau(\nu - \nu_0)}. \tag{2.21}$$

The Fourier transform is complex-valued. The absolute value of this

$$\left|\frac{\mathrm{i}\tau}{\mathrm{i} - 2\pi\tau(\nu - \nu_0)}\right| = \frac{1}{1 + 4\pi^2\tau(\nu - \nu_0)} \tag{2.22}$$

is shown in Fig. 2.9. It is noticeable that the decay from the peak frequency is much slower than with the Gaussian wave packet (Fig. 2.8e). Accordingly, the time-bandwidth product (see Appendix B.1.4) is significantly larger.

2.2.4 Group Velocity and Dispersion

In all our considerations so far, we have assumed that all waves travel at the same speed, regardless of their wavelength. But what happens if the phase velocity depends on the wavenumber? The answer to this question is essential for wave packets, as they always have a distribution of wavenumbers.

Since the essential properties of a wave packet, namely the modulation of the amplitude, occur when only two plane waves with different wavenumbers are superimposed, we can already see the basic effect in the simple example discussed at the beginning of Section 2.2.3. In Eq. (2.14), both the spatial and temporal behavior were specified. The equation shows that the temporal behavior of the phase of the waves of the wave packet and for the phase of the amplitude modulation are different:

$$\text{Phase of the wave packet:} \qquad \varphi = \kappa z - \nu t = \kappa\left(z - \frac{\nu}{\kappa}t\right)$$

$$\text{Amplitude modulation phase:} \quad \varphi = \Delta\kappa z - \Delta\nu t = \Delta\kappa\left(z - \frac{\Delta\nu}{\Delta\kappa}t\right)$$

The velocity of the phase of the waves in a packet is $c = \nu/\kappa$, which is already known from the *dispersion relation* Eq. (2.8) known *phase velocity*. However, the amplitude modulation moves at the speed $\Delta\nu/\Delta\kappa$. This is called *group velocity* c_g and is calculated, if the Δ values are allowed to approach zero, from the derivative of the dispersion relation $\nu(\boldsymbol{\kappa})$ to

$$\boxed{c_g = \frac{\mathrm{d}\nu}{\mathrm{d}\kappa}, \quad c = \frac{\nu}{\kappa} \quad \text{group and phase speed.}} \tag{2.23}$$

This means that the group velocity and phase velocity are different if the phase velocity depends on the wavenumber. The difference between the two velocities is shown by the time-space representation of two one-dimensional wave packets in the z direction (Fig. 2.10). Orientation in such a representation means velocity: The greater the angle to the t-axis, the faster the movement (see also Section 14.3.1). In a dispersion-free medium, waves and wave packets have the same orientation (Fig. 2.10a). If the group velocity is different from the phase velocity, the waves travel through the wave packet with the differential velocity (Fig. 2.10b).

Wave packets always contain waves with different wavenumbers. Therefore, one might conclude that the shape of a wave packet in a dispersive

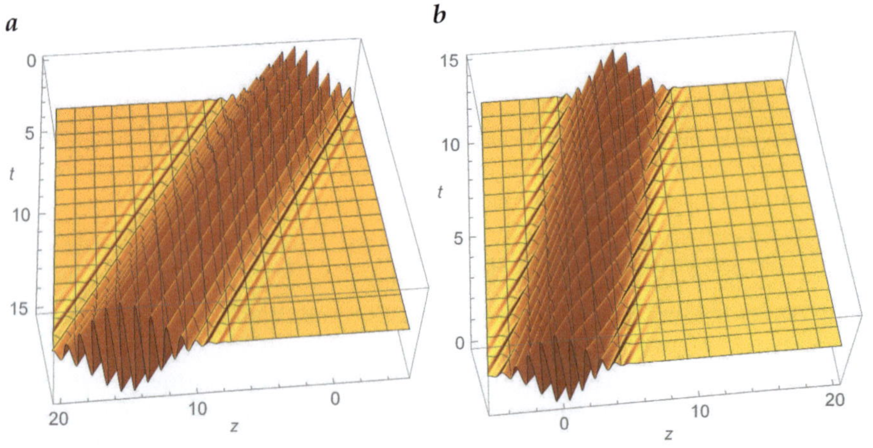

*Figure 2.10: Wave packets and dispersion: one-dimensional wave packet **a** in a dispersion-free medium in which the group velocity is equal to the phase velocity, represented in space-time; **b** in a medium in which the group velocity is half the phase velocity In both cases, wavenumber, frequency and phase velocity are set equal to 1;*

medium must change over time since the wavenumbers propagate at different speeds. However, this is not the case, as we will now analyze by considering the Fourier space (Fig. 2.11a).

In *wavenumber-frequency space* ($\kappa\nu$ spacefor short), a wave packet lies on the dispersion curve $\nu/\nu_0 = \sqrt{\kappa/\kappa_0}$ (blue curve in Fig. 2.11a) and occupies a small area around the central wavenumber κ_0. The ratio ν_0/κ_0 is the phase velocity and the slope of the dispersion curve at wavenumber κ_0 (dashed line) corresponds to the group velocity. Now we move to a coordinate system that moves with the group velocity. To do this, we need to rotate the coordinate system so that the κ axis is parallel to the gradient that corresponds to the group velocity (red coordinate system). With a linear dispersion curve (phase velocity = group velocity), all waves would now lie on the κ axis. The frequencies in the moving system would then be zero for all waves. Phase and amplitude distribution move at the same speed, as shown in Figure 2.10a.

This is no longer the case with a curved dispersion curve. The phase now still has a frequency that corresponds to the distance of the dispersion curve from the κ' axis of the co-moving coordinate system. If the dispersion curve is sufficiently linear in the area of the wave packet, then all components of the wave packet in the co-moving system have the same residual frequency. This means that the phases move without changing the shape of the wave packet, as shown in Figure 2.10b.

If the curvature of the *dispersion relation* is more pronounced, i. e. the wave packet contains significantly different group velocities, the shape of the wave packet changes. This effect occurs with wave packets that are

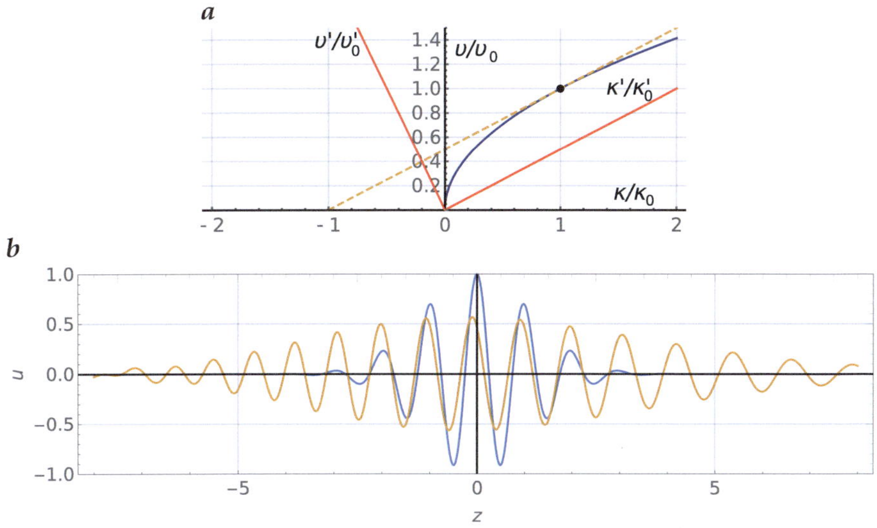

Figure 2.11: *Wave packets and dispersion:* ***a*** *visualization of a wave packet on a curved dispersion curve (blue) around κ_0 in wavenumber-frequency space and in a coordinate system (red) that moves with the group velocity.* ***b*** *Diverging small wave packet with waves around wavenumber one in a medium with a dispersion relation $\nu = a\sqrt{\kappa}$; representation in a coordinate system moving with the group velocity; blue curve at time $t = 0$, orange curve at time $t = 100\,T$.*

spatially confined and contain a wide range of wavenumbers. A wave packet with only seven wave trains in a dispersive medium with $c_g = c/2$ has approximately doubled in width in a time of only 100 periods T. The faster traveling large wavelengths are located at the front of the wave packet, the slower traveling smaller wavelengths at the rear (Fig. 2.11b). This means that analyzing a wave packet on the dispersion curve in the $\kappa\nu$ space allows an intuitive and quick analysis of the dispersion effects.

2.2.5 Beam Bundles and Gaussian Beam

So far, we have only restricted a wave in time and only in the direction of propagation in space. Therefore, the question arises whether and if so, how *beam bundles* can be generated from the elementary waveforms, plane waves or spherical waves considered so far by superposition and what properties these have. To achieve this, spherical waves seem more suitable, as they emanate from a point source. But how can a light source then no longer emit radiation in all directions, but only in a certain angular range?

The basic idea is as follows: We assemble an extended light source from point light sources. Even with two spherical waves that emit waves with the same amplitude and phase position, there is a clear effect if they are placed half a wavelength apart in the x direction. Then the two spherical

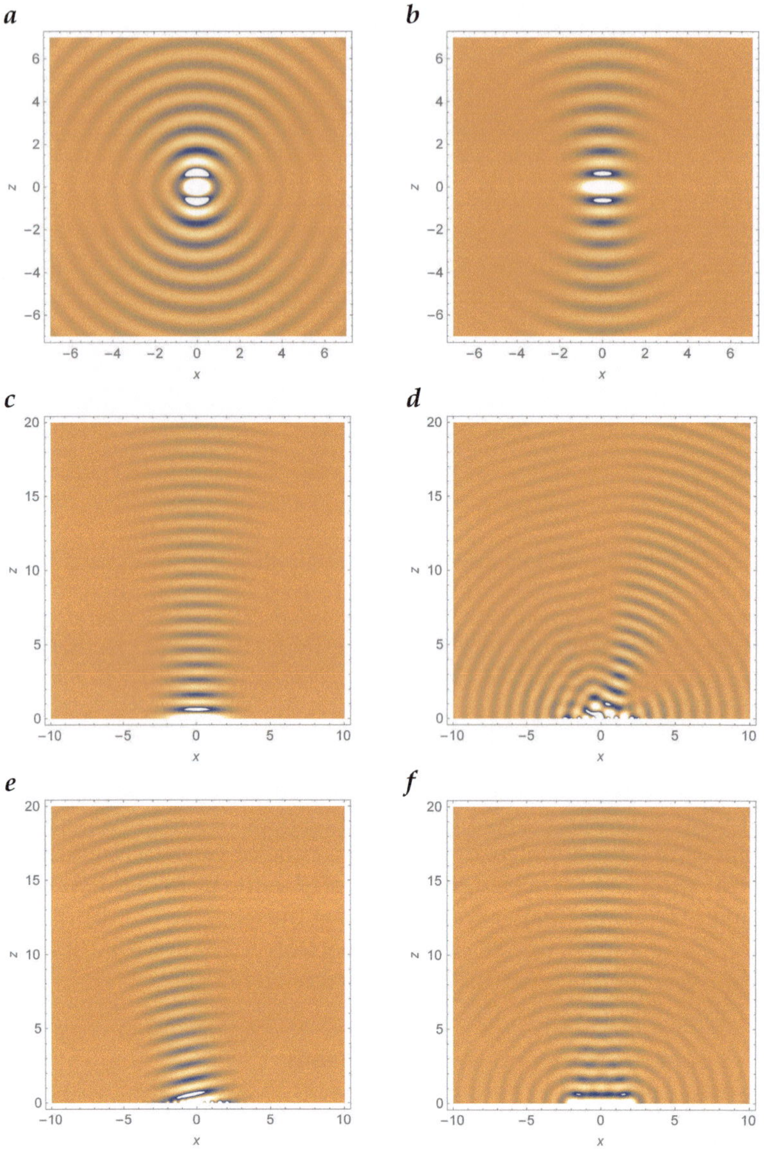

Figure 2.12: Superposition of spherical waves with a wavelength of one, shown in xz sections: ***a*** *Superposition of two spherical waves with the same phase position, shifted by half a wavelength in the x direction (dipole);* **b** *Superposition of a total of 7 spherical waves along the x-axis, each at a distance of half a wavelength and with amplitudes as described in the text;* **c** *Superposition of a total of 25 spherical waves along the x-axis, each at a distance of half a wavelength and with amplitudes calculated from the corresponding binomial coefficients;* ***d*** *like* **c** *only with random phase of the individual spherical waves* **e** *like* **c** *only with a phase shift of $\pi/5$ between the individual spherical waves;* ***f*** *superposition of a total of 9 spherical waves of the same amplitude, each at a distance of half a wavelength along the x-axis.*

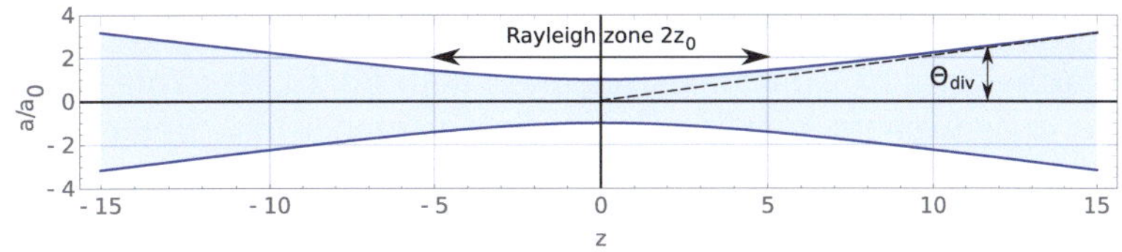

Figure 2.13: *Course of the beam radius of a Gaussian beam.*

waves cancel each other out in the far field in the x direction, while they add up in the z direction (Fig. 2.12a). However, the beam is still very wide. To achieve a narrower beam, we can use the same trick as for wave packets in Section 2.2.3 and superimpose more in-phase spherical waves half a wavelength apart with amplitudes corresponding to a binomial distribution. The more spherical waves we superimpose, i. e. the wider the light source is in the x direction, the narrower the beam becomes (Fig. 2.12b and c).

The significance of the phase position of the individual spherical waves in relation to each other becomes clear if we give each spherical wave a random phase position. This results in a random directional distribution without focusing the beam (Fig. 2.12d). If, on the other hand, the phase between the individual spherical waves is changed by a fixed amount, the direction of the beam can be changed. Figure 2.12e shows the resulting beam if the phase is shifted by $\pi/5$ from spherical wave to spherical wave.

The significance of the amplitude distribution according to a binomial distribution can be seen from a counterexample. If 9 spherical waves are superimposed along the x-axis, each at a distance of half a wavelength and with the same amplitudes, this results in good focusing in the z-direction, but strong secondary maxima in other directions (Fig. 2.12f).

A beam with a binomial distribution of amplitudes has no secondary maxima. This is why it is of particular importance. Since the binomial distribution changes to a Gaussian normal distribution in the limiting case of the superposition of many spherical waves, this beam is known as *Gaussian beam*.

The exact course of a Gaussian beam with a two-dimensional normal distribution of amplitudes can be derived using the Helmholtz equation Eq. (2.11) [280]. As we already know the principle from the simple superposition of spherical waves, only the results are shown here. The amplitude of a Gaussian beam has the following course (Fig. 2.13):

$$\mathrm{u}(r,z) = \mathrm{u}_0 \frac{1}{a(z)} \mathrm{e}^{\left(-\frac{r^2}{a^2(z)}\right)} \quad \text{with} \quad r^2 = x^2 + y^2, \quad a^2(z) = a_0^2\left(1 + \frac{z^2}{z_0^2}\right). \tag{2.24}$$

Within the *Rayleigh zone* of $-z_0 \leq z \leq z_0$ with

$$z_0 = \pm\frac{a_0^2\pi}{\lambda} \tag{2.25}$$

the beam is focused and only changes its beam radius a from a_0 to $\sqrt{2}a_0$, while in the far field ($|z| \gg z_0$) it increases linearly to Eq. (2.24). This behavior is referred to as *beam divergence*. The beam divergence Θ_{div} determines how well the beam can be focused (minimum beam radius a_0) and how wide the focused zone z_0 is:

$$\boxed{\Theta_{\text{div}} = \frac{a}{z} \quad \text{for} \quad z \gg z_0 \quad \rightsquigarrow \quad \frac{a_0}{\lambda} = \frac{1}{\pi\Theta_{\text{div}}} \quad \text{and} \quad \frac{z_0}{\lambda} = \frac{1}{\pi\Theta^2_{\text{div}}}.} \tag{2.26}$$

This immediately makes it clear that radiation cannot be focused arbitrarily well, but at most to a diameter of the same order of magnitude as the wavelength of the radiation.

Example 2.2: Divergence and focusability of a Gaussian beam

A Gaussian beam with a wavelength of 0.5 μm and a divergence $\Theta_{\text{div}} = 0.01$ (10 mm broadening to 1 m) can be focused to a beam radius of 16 μm and the Rayleigh zone has an extension of $2z_0 = 3.2$ mm. With a ten times smaller divergence of 0.001, the minimum beam radius increases tenfold to 160 μm. However, the focused area is a hundred times larger at $2z_0 = 318$ mm.

2.2.6 ♯Coherence, Intensity and Interference

Many radiation sources do not emit a long monochromatic wave train, but many temporally and thus also spatially narrowly limited wave packets. These wave packets have no relation to each other in terms of phase, as they are generated independently of each other.

Therefore, there can only be a defined phase relationship at a point in space for a limited time, which corresponds to the duration of the wave packets. This can be described by the temporal complex *coherence function*:

$$\mathfrak{S}_{uu}(\tau) = \lim_{T\to\infty} \frac{1}{2T} \int_{-T}^{T} \mathfrak{u}(t)^* \mathfrak{u}(t+\tau) \mathrm{d}t = \overline{\mathfrak{u}(t)^* \mathfrak{u}(t+\tau)}. \tag{2.27}$$

This is calculated from the complex wave amplitude $\mathfrak{u}$ and represents an *autocorrelation function* (Appendix C.5.1). With a stationary radiation source, the intensity (energy) is calculated — as with any oscillation — from the mean square of the amplitude:

$$I = \overline{\mathfrak{u}(t)^* \mathfrak{u}(t)}. \tag{2.28}$$

This relationship can be used to normalize the coherence to a *degree of coherence* (autocovariance coefficient):

$$\mathfrak{s}_{uu}(\tau) = \frac{\mathfrak{S}_{uu}(\tau)}{I} = \frac{\overline{\mathfrak{u}(t)^* \mathfrak{u}(t+\tau)}}{\overline{\mathfrak{u}(t)^* \mathfrak{u}(t)}}. \tag{2.29}$$

Two quantities can be calculated from the degree of coherence to characterize the coherence, the *coherence time* τ_c and the *coherence length* d_c:

$$\boxed{\tau_c = \int_{-\infty}^{\infty} |\mathfrak{s}(\tau)|^2 \, \mathrm{d}\tau \quad \text{and} \quad d_c = c\tau_c.} \tag{2.30}$$

The coherence time specifies the time interval over which the radiation is coherent, the coherence length specifies the distance difference over which the radiation propagates at the speed c.

Example 2.3: Degree of coherence of the plane wave and a wave packet

From Eq. (2.29), the coherence of a plane wave is calculated as $\mathfrak{s}_{uu}(\tau) = \exp(2\pi i\nu\tau)$. At time intervals of one period $T = 1/\nu$, the coherence degree $\mathfrak{s}_{uu}(NT) = 1$, and the absolute value of $\mathfrak{s}_{uu}$ is always 1: $|\mathfrak{s}_{uu}(\tau)| = 1$. As expected, the coherence time and length of a plane wave are therefore infinite. A Gaussian wave packet at a point in space can be described by $\exp(-t^2/(2\sigma^2))\exp(2\pi i\nu t)$. The calculation of the degree of coherence according to Eq. (2.29) results in

$$\mathfrak{s}_{uu}(\tau) = \exp\left(-\frac{\tau^2}{2\sigma^2}\right)\exp(2\pi i\nu\tau). \tag{2.31}$$

The degree of coherence oscillates as with the plane wave, but decreases with increasing time difference τ in the form of a Gaussian bell curve. The coherence time is proportional to the width of the bell curve: $\tau_c = \sqrt{\pi}\sigma$.

In many cases, the amplitude of a wave cannot be measured directly, only the energy of the radiation (see Section 2.3.2 and Section 4). However, this means that the phase position of the wave is lost. This can then only be determined from the intensity by superimposing two wave trains with the same frequency. We assume a reference wave train with the amplitude $\mathfrak{u}_r$ and one with the amplitude $\mathfrak{u}$ shifted by a phase shift φ. For the superposition, we can omit the explicit spatial and temporal change of the wave, as it is the same for both waves except for the phase shift φ. The superposition results in the intensity

$$I = \left|\mathfrak{u} + \mathfrak{u}_r \exp(i\varphi)\right|^2 = u^2 + u_r^2 + uu_r \exp(i\varphi) + uu_r \exp(-i\varphi). \tag{2.32}$$

Converted with the intensities of the two wave trains, the so-called *interference equation* follows

$$\boxed{I = I + I_r + 2\sqrt{I I_r}\cos(\varphi) \qquad \text{interference equation}} \tag{2.33}$$

This equation states that the sum intensity, depending on the phase shift, fluctuates with the amplitude $2\sqrt{I I_r}$ around the mean value $I + I_r$. One measurement is not sufficient to determine the phase position, as there are a total of three unknown variables.

A coherent superposition of two wave trains is achieved with an experimental setup in which a wave train is first divided into two partial beams that take different paths and are then superimposed again. We assume that the amplitudes of both partial beams are the same and that a path difference d has occurred between them before the superposition. This results in a phase shift of $2\pi\kappa d = 2\pi d/\lambda$ and the interference equation (2.33) is simplified to

$$I = 2I\,(1 + \cos(2\pi\kappa d))\,. \tag{2.34}$$

Such an arrangement is called an *interferometer*. It allows measurements of distances and all quantities that lead to a change in distance. These can be measured with an accuracy of a small fraction of the wavelength of the radiation wavelength used. The price for these highly accurate measurements is that all components for beam splitting and combining must also be positioned and stable with accuracies corresponding to a fraction of the wavelength. This can make the image-recording system based on interferometry very complex.

The temporal coherence function is the autocorrelation function of the wave amplitude. It follows directly from this that there must be a relationship between the coherence function and the power spectrum of the wave amplitude, as the power spectrum is the Fourier transform of the autocorrelation function (Section C.5.3):

$$\boxed{\tilde{S}(\nu) = \overline{\tilde{u}(\nu)^*\tilde{u}(\nu)} = \int\limits_{-\infty}^{\infty} \mathfrak{S}(\tau)\exp(-2\pi \mathrm{i}\nu\tau)\mathrm{d}\tau \quad \text{power spectrum.}} \tag{2.35}$$

This means that coherence is inextricably linked to the width of the spectral distribution of the radiation. The narrower the bandwidth of the radiation, the higher the coherence. Or from a practical point of view: if a radiation source with a specified coherence length is required, this can be selected by the width of the spectral distribution.

2.2.7 ♯Polarization

With a *transversal wave*, the oscillation can be in any direction perpendicular to the direction of propagation. This directional characteristic of transverse waves is called *polarization*. The direction of polarization is defined by the direction of the oscillation.

Using the *superposition principle*, each individual wave train is therefore composed of a wave with an amplitude $\mathfrak{u}_x$ in the x direction and an amplitude $\mathfrak{u}_y$ in the y direction. It should be noted that these amplitudes of the waves are complex, as they describe both the amplitude itself and the phase position

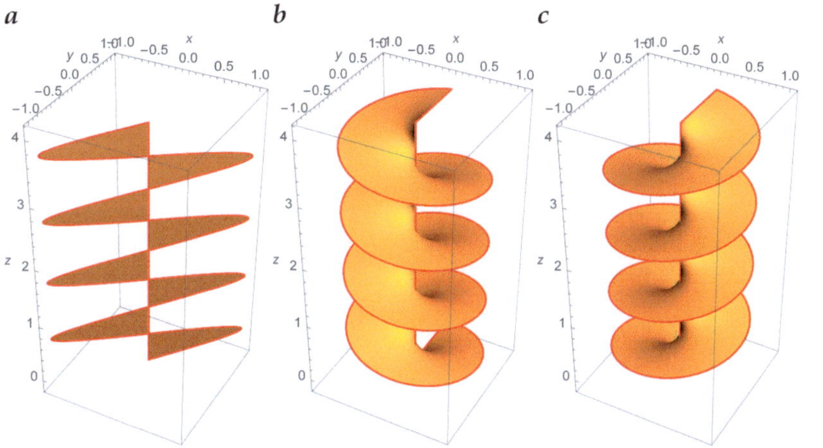

Figure 2.14: *A **a** linear, **b** left circular and **b** right circular polarized plane wave.*

of the wave train (Section 2.2.1). The two complex amplitudes are combined in the *Jones vector* $\mathbf{j}$:

$$\mathbf{j} = \begin{bmatrix} \mathrm{u}_x \\ \mathrm{u}_y \end{bmatrix}. \tag{2.36}$$

The polarization state of a single wave train is therefore described by four variables: the amplitudes in the x and y directions together with the associated phase positions. In contrast, there are only two degrees of freedom for a longitudinal wave: the amplitude and phase position of the longitudinal oscillation in the direction of propagation.

If the absolute phase position of the wave is not important, the description of polarized light is reduced to three degrees of freedom: the amplitudes of the wave in x- and y-direction and their *relative* phase position to each other. In this case, only one degree of freedom remains for a longitudinal wave.

There are two special cases for the polarization state of a single wave train. In the first case, there is no phase shift between the amplitudes in both directions. Then the fields oscillate only in one plane and one speaks of a *linearly polarized* wave (Fig. 2.14a). Its Jones vector is

$$\mathbf{j} = \mathrm{u} \begin{bmatrix} \cos\phi \\ \sin\phi \end{bmatrix}, \tag{2.37}$$

where the angle ϕ indicates the angle of the oscillation plane to the x-axis. In the second special case, the two amplitudes are the same, but shifted in phase by 90°. Then the field vector rotates with the period of one wavelength around the z-axis and the Jones vector is

$$\mathbf{j} = \mathrm{u} \begin{bmatrix} 1 \\ \pm \mathrm{i} \end{bmatrix}. \tag{2.38}$$

Now, depending on the direction of rotation in the direction of propagation, a left or right *circularly polarized* wave is present (Fig. 2.14b and c). In the general case Eq. (2.36), one speaks of an *elliptically polarized* wave.

Example 2.4: Addition of a right and a left circular wave

By adding a right and left circularly polarized wave of the same complex amplitude $u/2$ and with the phase shift 2φ to each other, linearly polarized light is generated. In symmetrical notation, this results in

$$\frac{u}{2}\exp(-i\varphi)\begin{bmatrix} 1 \\ i \end{bmatrix} + \frac{u}{2}\exp(i\varphi)\begin{bmatrix} 1 \\ -i \end{bmatrix} = \frac{u}{2}\begin{bmatrix} \exp(-i\varphi) + \exp(i\varphi) \\ i(\exp(-i\varphi) - \exp(i\varphi)) \end{bmatrix} = u\begin{bmatrix} \cos\varphi \\ \sin\varphi \end{bmatrix}.$$

The linearly polarized radiation has an angle φ to the x-axis.

A single monochromatic wave train is always in one of the described polarization states. Does this also apply in general? No, because a wave, as discussed in Section 2.2.6, can be composed of many individual wave packets. If these are evenly polarized in all directions, then the resulting wave is not polarized. Therefore, the description of the polarization of a wave composed of many wave packets is one level more complicated.

The further consideration of polarization also assumes that not the wave amplitude can be measured directly, but only the intensity of the wave (see Section 2.2.6). The measurement of the absolute phase position of the radiation is then lost. The description of the polarization state must nevertheless contain four parameters, since in addition to the three components for a single wave train, a parameter must be added that takes into account that there can also be unpolarized light and thus any mixture of unpolarized with polarized radiation.

These four real and intensity-based parameters can be summarized in the four-component *Stokes vector*:

$$S = \begin{bmatrix} S_0 \\ S_1 \\ S_2 \\ S_3 \end{bmatrix} = \begin{bmatrix} I_x + I_y \\ I_x - I_y \\ I_{x-y} - I_{x+y} \\ I_r - I_l \end{bmatrix}. \tag{2.39}$$

The indices on the intensities indicate the corresponding polarization state: x: linear, horizontal; y: linear, vertical; $x - y$: linear, 45°; $x + y$: linear, -45°; r: circular, right and l: circular, left.

The intensities of the different polarization states can be measured using suitable filters that are only permeable for one of the polarization states. The Stokes components are constructed in such a way that only the first component is non-zero for unpolarized light and only the first three for linearly polarized light.

Example 2.5: Jones and Stokes vectors for various polarization states

The following table contains the Jones and Stokes vectors for a series of polarization states with a wave amplitude u or an intensity I.

Beschreibung	Jones	Stokes
linear, horizontal, $\varphi = 0°$	$[1,0]^T$ u	$[1,1,0,0]^T I$
linear, vertical, $\varphi = 90°$	$[0,1]^T$ u	$[1,-1,0,0]^T I$
linear, $\varphi = +45°$	$[1,1]^T$ u$/\sqrt{2}$	$[1,0,1,0]^T I$
linear, $\varphi = -45°$	$[1,-1]^T$ u$/\sqrt{2}$	$[1,0,-1,0]^T I$
circular, right	$[1,\mathrm{i}]^T$ u$/\sqrt{2}$	$[1,0,0,1]^T I$
circular, left	$[1,-\mathrm{i}]^T$ u$/\sqrt{2}$	$[1,0,0,-1]^T I$
unpolarized	—	$[1,0,0,0]^T I$

The Stokes vector can be used to calculate two other important parameters for characterizing the radiation, which are important for image acquisition systems, the *degree of polarization* p and the *angle of polarization* φ:

$$p = \frac{\sqrt{S_1^2 + S_2^2 + S_3^2}}{S_0}, \qquad \varphi = \frac{1}{2}\arctan\frac{S_2}{S_1}. \tag{2.40}$$

It should be noted that the polarization angle only covers an angle range of 180°.

2.3 Types of Radiation

2.3.1 Electromagnetic waves

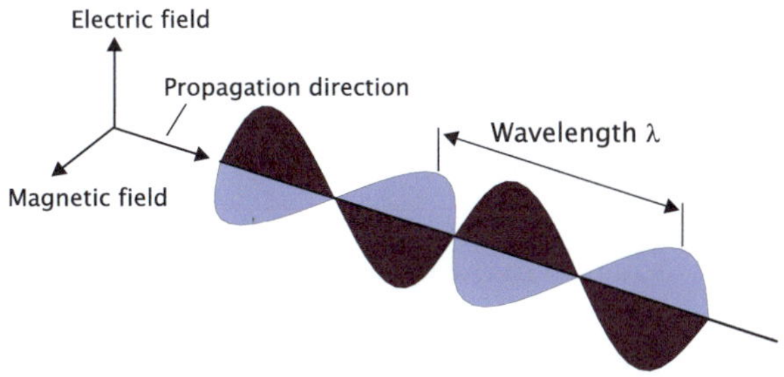

Figure 2.15: *The electric field E and the magnetic field B of an electromagnetic wave are perpendicular to each other and perpendicular to the direction of propagation.*

Electromagnetic radiation consists of alternating *electric* and *magnetic fields*. In an *electromagnetic wave* these fields are perpendicular to each other and perpendicular to the direction of propagation (Fig. 2.15). A wave equation of the form as in Eq. (2.1) for the electric and magnetic fields can be derived directly from Maxwell's equations [22].

Figure 2.16 shows that electromagnetic waves cover the enormous frequency and wavelength range of 24 decades. Only a tiny section of less than an octave with a wavelength of about 400–700 nm is *light*, the part of electromagnetic waves to which the human eye is sensitive. The classification shown in Fig. 2.16 is somewhat artificial. It is largely historically determined by the way in which electromagnetic radiation is generated or detected.

Electromagnetic waves do not require a carrier. In matter-free space (vacuum), they are dispersion-free and propagate at the *velocity of light* $c_0 = 299.792.458$ m/s $\approx 3\times10^8$ m/s. When propagating in matter, electromagnetic waves interact with the electric charges, electric currents and thus the electric and magnetic fields of the medium. Despite these diverse interactions, the fundamental nature of the waves remains the same. Only the speed of propagation is slowed down and the wave is attenuated.

The simplest case is an isotropic medium that reacts linearly to the disturbances of the passing electromagnetic wave. To capture both effects, the phase velocity c is modified in the wave equation Eq. (2.1). We choose the approach

$$c = \frac{c_0/n}{1 + \mathrm{i}\frac{\alpha}{4\pi\kappa}}. \tag{2.41}$$

This means that the influence of the medium is represented by a complex *refractive index* [22]:

$$\mathfrak{n} = n + \mathrm{i}n_i = n\left(1 + \frac{\mathrm{i}\alpha}{4\pi\kappa}\right). \tag{2.42}$$

The real part n, the ordinary refractive index, indicates the ratio of the speed of light c_0 in a vacuum to the phase velocity c in the medium: $n = c_0/c$. Due to the slower propagation, at the same frequency according to the *dispersion relation* Eq. (2.8), the wavelength is also smaller by the same factor n. The imaginary component causes the solutions of the wave equation Eq. (2.1) to have an additional factor of the form

$$\exp(-\alpha z/2). \tag{2.43}$$

This makes the meaning of the constant α clear. It indicates the attenuation of electromagnetic radiation in the medium and is called *absorption coefficient* (Section 2.6.5).

In general, the refractive index depends on the frequency or the wavelength. Therefore, the propagation speed is not constant, but wavelength-dependent. This effect is called *dispersion* and the waves are referred to as dispersive.

Refractive index and absorption are the two elementary optical parameters of a material that is permeable to electromagnetic waves. In the context of image acquisition, they can be used to identify a material or a physical quantity to be measured that influences one of these two parameters.

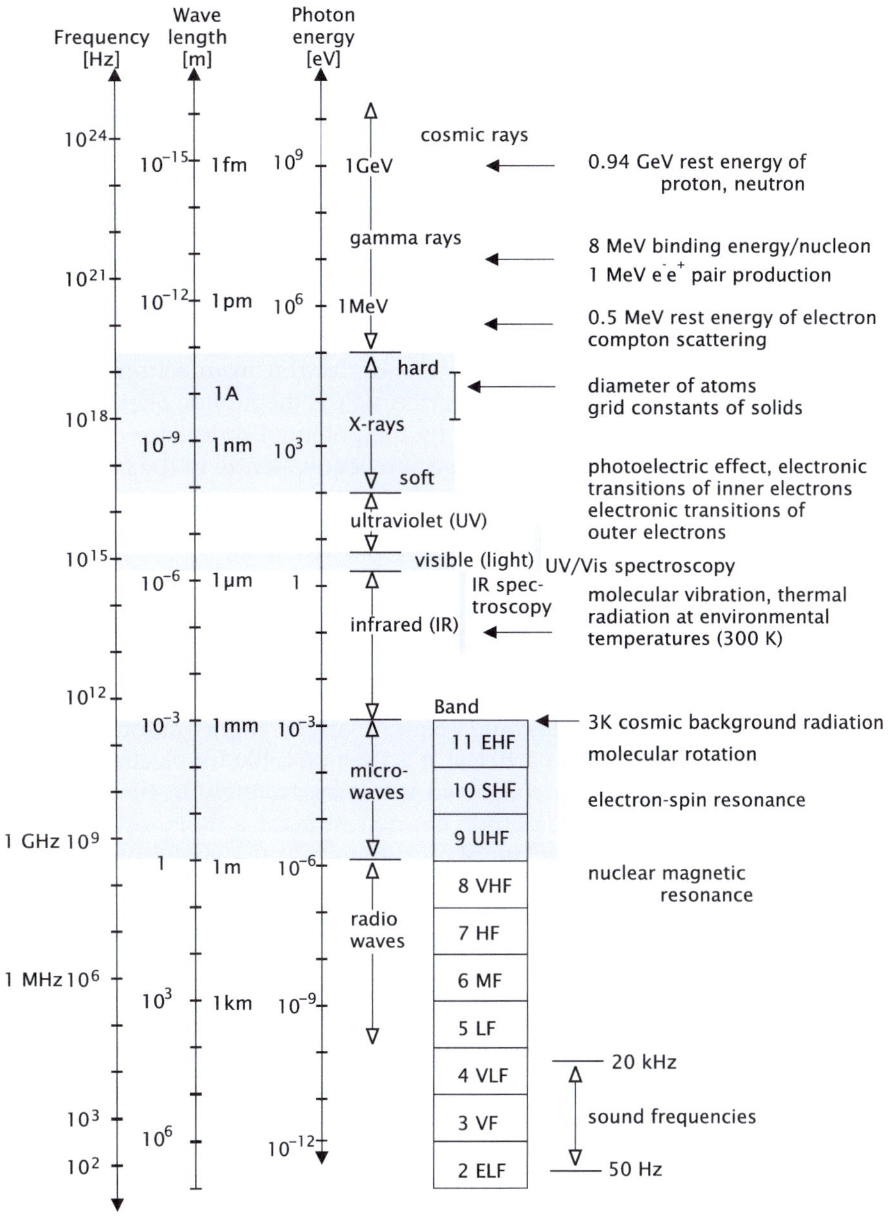

Figure 2.16: *The electromagnetic spectrum with wavelength, frequency and photon energy scales. The frequency range of audible acoustic waves is also shown.*

2.3.2 Photons and Particle Radiation

In addition to wave phenomena, electromagnetic radiation is characterized by a particle property. Electromagnetic energy only occurs in discrete energy levels. For a given frequency, it must be an integer multiple of $h\nu$, where h is the *Planck's constant* or the *quantum of action*:

$$\boxed{Q = h\nu = \frac{hc}{\lambda} = hc\kappa.} \tag{2.44}$$

An electromagnetic quantum is called *photon*.

In any interaction between radiation and matter, be it absorption or emission of radiation, energy can only be exchanged in units of these quanta. The energy of the photon is often expressed in electron volts (eV), as photons can be generated by the transition of a bound electron from a higher-energy state to a lower-energy state. An electron volt is the kinetic energy that an electron gains when it is accelerated by the potential difference of one volt and, according to Eq. (2.44), can be specified equivalently in the energy unit joule (J) or with the wavenumber κ:

$$1\,\text{eV} \equiv 1{,}602 \cdot 10^{-19}\,\text{J} \equiv 0{,}86055\,1/\mu\text{m}. \tag{2.45}$$

A photon for yellow light has, e. g. an energy of approximately 2 eV. Figure 2.16 also shows the photon energy scale in eV.

The higher the frequency of electromagnetic radiation, the clearer the particle nature becomes, as the individual quanta become more energetic. The energy of photons can become greater than the energy associated with the rest mass of elementary particles. It is then possible for electromagnetic energy to be spontaneously converted into mass, namely in the form of a pair of particles. Although a photon has no rest mass, a momentum $p = mv$ is associated with it because the photon travels at the speed of light and thus has energy according to Eq. (2.44). The momentum p then results from Einstein's energy-mass equivalence to

$$E = mc^2 \quad \rightsquigarrow \quad p = E/c \quad \rightsquigarrow \quad p = h\kappa. \tag{2.46}$$

The quantization of the energy of electromagnetic waves is of crucial importance for image acquisition. The frequency of light is in the PHz range (10^{15} Hz, Fig. 2.16). These fast oscillations cannot be measured directly electronically. Today, this is only possible up to frequencies of just under 1 THz. The oscillation of electromagnetic radiation with a wavelength of just under 1 mm and greater (terahertz radiation, microwaves and radio waves, Fig. 2.16) can be converted directly into an alternating current signal using suitable antennas. This makes it possible to measure not only the amplitude (or energy), but also the phase position and — in the case of polarization-dependent antennas — also the polarization state.

None of this is possible with light. Radiation detectors, which "count" photons by absorbing them, can only measure the energy of the radiation; the phase position and polarization state of the radiation are lost. The smallest amount of energy that can be detected is $h\nu$. The number of photons counted per time unit is a *random variable* with a *Poisson distribution* (Section C.4.1). If on average N photons are counted in a time interval, then the standard deviation of the Poisson distribution is $\sigma_N = \sqrt{N}$. You therefore have to count 10,000 photons to be able to measure the radiation flux with a relative standard deviation of 1%.

In contrast to electromagnetic waves, particle radiation moves at a slower speed than light, as the particles have a rest mass. In terms of image acquisition, *electrons* is the most important particle radiation. Electrons are also emitted during the decay of certain radioactive isotopes and are then called *beta radiation*. Also important are the positively charged nuclei of the hydrogen atom, the *protons*, the nuclei of the helium atom (*alpha radiation*), which have a double positive charge, *positrons*, the positively charged antiparticle to the electron, and the charge-free *neutrons*.

Particle radiation also has a wave character. Wavelength λ and frequency ν are directly related to the energy and momentum of the particles:

$$\begin{aligned} \nu &= E/h \quad \text{Bohrsche Frequenzbedingung,} \\ \kappa &= p/h \quad \text{de-Broglie-Beziehung.} \end{aligned} \tag{2.47}$$

These relationships correspond to those of the photon in Eqs. (2.44) and (2.46). The significance for imaging lies in the fact that their particles typically have radiation of a much shorter wavelength than visible light. Electrons with an energy of about 20 keV, for example, have a wavelength of 10^{-11} m or 10 pm. This is about 1/50 000 of the wavelength of light. Since the resolution of an imaging system is of the order of the wavelength of the radiation (Section 3.5.3), imaging systems based on electrons (*electron microscope*) potentially have a much higher resolution than any light microscope.

2.3.3 ♯Acoustic Waves

Unlike electromagnetic waves, *acoustic* or *elastic waves* require a carrier. Acoustic waves propagate elastic deformations. Longitudinal acoustic waves are generated by isotropic pressure. They cause a uniform compression and thus a deformation in the propagation direction of the wave. For the local density, pressure and velocity variation (ρ', p' and $\boldsymbol{u}'$), the wave equation (2.1) applies with the propagation velocity

$$c = \frac{1}{\sqrt{\rho_0 \beta_{\mathrm{ad}}}} \quad \text{with} \quad \beta_{ad} = -\frac{1}{V}\frac{\mathrm{d}V}{\mathrm{d}p}, \tag{2.48}$$

also called *sound velocity*. Where ρ is the mean density and β_{ad} is the *adiabatic compressibility*. The latter is the relative volume change with uniform pressure change under the condition that no heat exchange takes place.

Equation (2.48) universally relates the *sound velocity* to the elastic properties of the medium. The lower the compressibility, the higher the speed of sound. In an ideal gas, the speed of sound is independent of the pressure and only depends on the gas via the molar mass M, the ratio of the specific heat at constant pressure, c_p, or volume, c_v, and the absolute temperature T:

$$c = \sqrt{\frac{c_p}{c_v}\frac{RT}{M}}, \quad R \text{ universal gas constant.} \tag{2.49}$$

Acoustic waves move much slower than electromagnetic waves. Their speed in air, water and iron at 20°C is 344 m/s, 1485 m/s and 5100 m/s respectively. An audible sound wave with 3 kHz has a wavelength of about 10 cm in air. Acoustic waves with a much higher frequency, so-called *ultrasound*, can have small wavelengths down to the micrometer range. With suitable acoustic optics, *ultrasound microscopy* is then possible.

When we use sound or ultrasound for imaging purposes, we must bear in mind that the propagation of sound in solids is much more complicated than in liquids or gases. Firstly, solids are not isotropic and elasticity can no longer be described as scalar compressibility. It is replaced by an elasticity tensor. Secondly, shear forces can also generate transverse acoustic waves in which the deformation is perpendicular to the direction of propagation, as is the case with electromagnetic waves. Sound waves therefore propagate in solids with different modes and different velocities.

Despite these complex relationships, the speed of sound depends only on the elastic properties of the medium. Therefore, acoustic waves do not show *dispersion* in the approximation of continuum mechanics considered here, i. e. for wavelengths that are much larger than the distances between atoms. Waves with different frequencies therefore propagate at the same speed. This is an important fact for *acoustic imaging*.

2.3.4 Radiation as an Information Carrier for Quantitative Imaging

Image recording techniques can only evaluate the information about a scene or an object that is received by the recording system in the radiation emitted by the object. Therefore, we summarize what we have worked out so far from the point of view of what information radiation can transmit from the object to be measured to the image acquisition system:

Intensity
The intensity of the radiation determines the "brightness" of the recorded image (Section 3.6).

Phase
The phase position is lost when measuring the intensity of the radiation. However, it is directly accessible to all recording systems that measure the wave amplitudes. This is the case for radio, micro- and terahertz waves, but also for acoustic waves. For electromagnetic waves with higher frequencies, interferometric systems are required in which the wave to be measured is superimposed with a reference wave (Section 2.2.6).

Polarization State
In addition to the intensity of the radiation, the polarization state of transverse radiation contains three other independent parameters that can be described with the Stokes vector (Section 2.2.7). The full measurement of the polarization state thus contains four times the information compared to a simple measurement of the intensity alone.

Frequency and Wavelength
The measurement of the frequency or wavelength of the radiation is possible with spectroscopic recording techniques. This opens up an additional dimension, as a complete spectrum could be measured at each pixel. However, it is also possible to use all effects for measurement purposes that lead to a shift in wavelength or frequency. Spectroscopic techniques enable the non-contact measurement of a variety of quantities (Sections 2.6.7, 2.6.8 and 2.6.2).

This makes it clear that image acquisition is more than simply taking a picture of an object that we want to detect, identify or measure. Rather, it requires careful consideration of which of the many parameters of the radiation that can be captured from the object with the imaging system are best suited to solve the task at hand.

In this context, a further distinction between imaging systems is important. You can simply record the radiation emitted by the object itself or use the naturally given lighting conditions. Or you can specifically use an illumination method that is best suited to the given task (Chapter 9).

The latter method is the most common for automatic visual inspection in production and for most scientific and technical applications. Even if no active illumination technique is possible, we can still choose wavelength(s) and image acquisition modalities.

2.4 Radiometry and Photometry

2.4.1 Classification of Radiometric Quantities

Radiometry describes and measures radiation and its interaction with matter. Due to the dual nature of radiation, radiometric terms refer to either energy or particles. In the case of electromagnetic radiation, the particles are photons (Section 2.3.2). When it is necessary to distinguish between the two types for radiometric terms, the two indices e and p are used.

Radiometry is actually not a difficult field. However, the very different, imprecise and often even incorrect use of the terms has caused confusion. In addition, radiometry is studied less frequently and in less detail than other topics in optics.

As a result, knowledge of radiometry is less widespread. A good knowledge of radiometry is important for image acquisition. Geometrical optics only tells us where an object's image is located. In contrast, radiometry can calculate how much radiation an imaging system acquires from an object, i. e. how bright its image will be.

Radiometry becomes easier to understand if we keep a clear classification scheme in mind. It is only a matter of adequately describing energy and power (energy/time) for radiation. Since the energy of radiation is not bound to a line, as is the case with electrical energy, but is transmitted freely in space, the following distinction must be made:

- Total radiation or per area?
- Incoming or outgoing radiation?
- From all directions or direction-dependent?
- All wavelengths or per wavelength?

We start with the total energy and power of radiation.

2.4.2 Radiant Energy and Power

Since radiation is a form of energy, it can do work. A body that absorbs radiation heats up. Radiation can release electrical charges in a material suitable for detecting radiation. The *radiation energy* is denoted by Q and has the unit joule (Ws) or is given by the number of photons.

The power of the radiation, i. e. the energy per unit time, is the *radiant flux*. It is denoted by Φ:

$$\Phi = \frac{\mathrm{d}Q}{\mathrm{d}t}. \tag{2.50}$$

This relationship is important for describing the total energy emitted by a light source per unit of time. It is measured in joules/s (J/s), watts (W) or photons per s (s^{-1}).

Example 2.6: Conversions radiant flux

When converting energy-based radiant flux Φ_e into photon-based radiant flux Φ_p, you can proceed as follows. According to Eq. (2.44), a photon has the energy hc/λ. This results in the conversion

$$\Phi_p\,[\mathrm{s}^{-1}] = \frac{\Phi_e \lambda}{hc} \approx 5.034 \cdot 10^{24}\,\lambda\,[\mathrm{m}]\,\Phi_e\,[\mathrm{W}]$$

or in more practical units for light wavelengths

$$\Phi_p\,[\mathrm{s}^{-1}] \approx 5.034 \cdot 10^{18}\,\lambda\,[\mu\mathrm{m}]\,\Phi_e\,[\mathrm{W}] \tag{2.51}$$

An example: A blue high-power LED with a wavelength of around 400 nm and a radiant flux of 500 mW emits around 10^{18} photons per second.

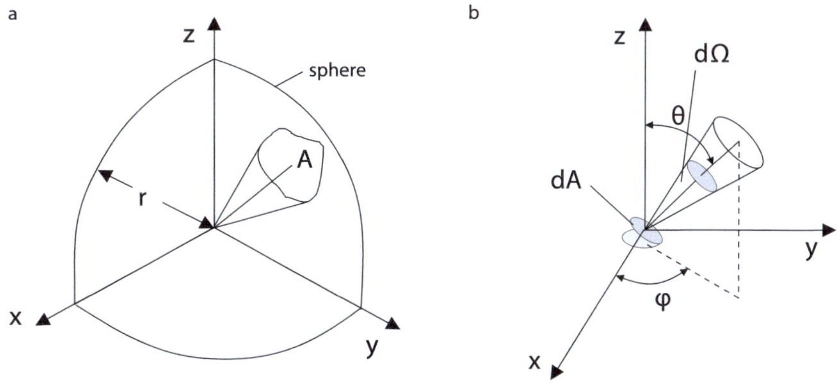

Figure 2.17: ***a*** *Definition of the solid angle.* ***b*** *Definition of the radiance and radiant flux emitted per surface patch* $\mathrm{d}A$*, projected in the direction of propagation, and per solid angle* Ω.

2.4.3 Area-related Radiometric Quantities

A distinction is made between incoming and outgoing radiation in the radiant flux per area,

$$\text{irradiance } E = \frac{\mathrm{d}\Phi}{\mathrm{d}A_0} \quad \text{and} \quad \text{radiant intensity } M = \frac{\mathrm{d}\Phi}{\mathrm{d}A_0}. \tag{2.52}$$

The *irradiance* E is the radiant flux that strikes a surface per unit area, e. g. on a sensor that converts radiant energy into an electrical signal. The unit of irradiance is Wm^{-2}, or photons per m^2 and s ($\mathrm{m}^{-2}\mathrm{s}^{-1}$). The radiant flux emitted per area by a light source is also called *specific excitance* or *emittance* and is denoted by M.

Another important radiometric quantity for image acquisition is the incident radiant energy per area. It is called *exposure* or *irradiation*. We can derive it directly from the radiant energy or define it as the integral of the irradiance over a time interval Δt:

$$\text{irradiation } H = \frac{\mathrm{d}Q}{\mathrm{d}A_0} = \int_t^{t+\Delta t} E(t)\mathrm{d}t. \tag{2.53}$$

The unit of irradiation is Jm^{-2}, or photons per m^2 (m^{-2}).

2.4.4 Solid Angle and Angle-related Quantities

The concept of *spatial angle* is important for understanding the spatial distribution of radiation. Let us imagine a small light source as the center of a sphere of radius r that emits rays in a cone shape (Fig. 2.17a). The boundaries of the cone describe an area A on the sphere. The solid angle (Ω) is measured in *steradian* (sr) and is the area A divided by the square of the radius

($\Omega = A/r^2$). Although steradian is a dimensionless quantity, it is advisable to use it explicitly when a radiometric term referring to a solid angle can be confused with the corresponding non-directional term. The solid angle of a whole sphere is 4π and that of a hemisphere is 2π.

The radiant flux emitted by a light source per solid angle is called *irradiance I*:

$$I = \frac{\mathrm{d}\Phi}{\mathrm{d}\Omega}. \tag{2.54}$$

It is obvious that this expression only makes sense when describing point light sources, e. g. if the distance from the source is much greater than its dimensions. The radiant intensity is also helpful when describing light rays.

For an extended light source, the radiant flux per unit area and per solid angle, the *radiance*, is the appropriate quantity (Fig. 2.17b):

$$L = \frac{\mathrm{d}^2\Phi}{\mathrm{d}A\,\mathrm{d}\Omega} = \frac{\mathrm{d}^2\Phi}{\mathrm{d}A_0\cos\theta\,\mathrm{d}\Omega}. \tag{2.55}$$

Radiation can either be emitted from a surface, pass through it or impinge on it. The radiance L depends on the angle of incidence θ on the surface (Fig. 2.17b) and the azimuth angle ϕ. For a flat surface, $\theta \in [0.\pi/2]$ and $\phi \in [0.2\pi]$. It is important to note that the radiance is related to a unit area in the direction of propagation ($\mathrm{d}A = \mathrm{d}A_0 \cdot \cos\theta$). This means that the effective area from which radiation is emitted increases with the angle of incidence. The units for energy-related and photon-related radiation are $\mathrm{Wm^{-2}sr^{-1}}$ and $\mathrm{s^{-1}m^{-2}sr^{-1}}$.

Incident radiation in particular is often referred to as brightness. However, this term should not be used as it has contributed greatly to the confusion between radiance and irradiance. Although both variables have the same dimension, their meaning is different. Radiance L describes the angular distribution of radiation, while irradiance E integrates the radiation incident on a surface element over a solid angle range in all directions in which the surface element can receive radiation:

$$E = \int_\Omega L(\theta,\phi)\cos\theta\,\mathrm{d}\Omega = \int_0^{\pi/2}\int_0^{2\pi} L(\theta,\phi)\cos\theta\sin\theta\,\mathrm{d}\theta\,\mathrm{d}\phi. \tag{2.56}$$

The factor $\cos\theta$ arises from the fact that the radiation density is defined in relation to a unit area perpendicular to the direction of propagation (Fig. 2.17b), while the irradiance refers to the unit area parallel to the surface.

Example 2.7: Radiant flux and radiant intensity

In this example, we calculate the radiant intensity of two very different light sources: an incandescent lamp with a radiant flux of 50 W and a laser with a radiant flux of 0.5 mW. The incandescent lamp radiates into the full space, i. e. a solid angle $\Omega = 4\pi$ and thus generates a radiant intensity of about

4 W/sr. The laser has a divergence of 2 mm/m. It therefore irradiates a circle with a radius r of 1 mm at a distance d of 1 m. This corresponds to a solid angle $\Omega = \pi r^2/d^2 \approx 3 \cdot 10^{-6}$ sr. The laser therefore has a radiant intensity of $\approx$ 160 W/sr. Although the incandescent lamp has a 100 000-fold higher radiant flux than the laser, the latter exceeds the incandescent lamp in radiant intensity by a factor of 40.

2.4.5 Spectroradiometry

Since any interaction of matter and radiation depends on the wavelength or frequency of the radiation, it is necessary to treat all radiometric quantities as a function of *wavelength*. Therefore, we relate all these quantities to a unit interval of the wavelength λ. Alternatively, unit intervals of frequencies ν or wavenumbers can also be used. can be used. The *wavenumber* κ specifies the number of wavelengths per unit interval (see Eq. (2.3) in Section 2.2.1). To distinguish the different spectral quantities from each other, we explicitly specify the dependencies by corresponding indices: $L(\lambda)$, $L(\nu)$ and $L(k)$.

2.4.6 ♯Photometry

The radiometry discussed in the previous section measures the properties of radiation in units of energy or number of photons. The *photometry* relates these quantities to the response of the human eye to the radiation. Photometry has two meanings for scientific imaging: First, photometry provides quantitative access to radiometric quantities as perceived by a human observer. Secondly, the response of any type of radiation sensor used to convert irradiance into an electrical signal can be described in the same way. The key to understanding photometry is to consider the spectral sensitivity of the human eye. Otherwise, photometry contains nothing new.

The *human visual system* only responds to electromagnetic radiation of wavelengths between 360 and 800 nm. However, even in individuals with normal vision, spectral sensitivity is subject to certain fluctuations. The visible range of the electromagnetic spectrum (light, Fig. 2.16) cannot therefore be precisely defined. Normally, a narrower wavelength range of 400 to 700 nm is used for light.

On the retina of the eye, onto which an image is projected, there are mainly two types of receptors, rods and cones. Photopigments in the outer segments of the receptors absorb radiation. The absorbed energy is then converted into neural electrochemical signals that are transmitted to downstream neurons, the optic nerve and the brain. The cones have three different photopigments. They are therefore sensitive to three different spectral ranges and thus enable color vision (Fig. 2.18a). Cone vision is only active at medium and high illuminance levels and is therefore also called *photopic vision*. At low illuminance levels, only the rods are active. This form of vision is therefore also called *scotopic vision*.

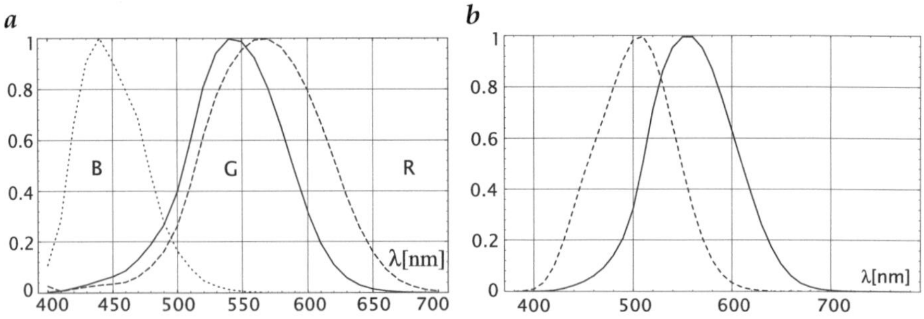

Figure 2.18: ***a*** *Relative spectral sensitivity of the cones of the human retina according to DeMarco et al. [55].* ***b*** *Relative spectral sensitivity of an "average" human eye as defined by the CIE in 1980 at medium to high (photopic vision* $V(\lambda)$*, solid line) and at low illuminance levels (scotopic vision* $V'(\lambda)$*, dashed line); data from [202].*

At first glance, the quantitative measurement of the spectral sensitivity of the eye seems impossible, as we can only rely on a subjective impression of how the human eye "sees" radiation. And yet spectral sensitivity can be measured by taking advantage of the fact that the eye is very sensitive to differences in brightness. In 1924, the International Commission on Illumination (CIE) defined a standard for the spectral sensitivity of the human observer under photopic conditions on the basis of extensive studies on many test subjects. This standard was later slightly revised several times. Fig. 2.18b shows the values defined in 1980. The curve of relative spectral sensitivity for scotopic vision $V'(\lambda)$ is similar in shape, but the maximum is shifted from about 555 nm to 510 nm (Fig. 2.18b).

Physiological measurements can only provide a relative spectral function of the luminous efficacy. It is therefore necessary to define a separate unit for light quantities. This unit, the candela, is one of seven basic units of the international system of units (Système Internationale or SI). The candela is defined as the luminous intensity of a monochromatic light source with a frequency of $5.4 \cdot 10^{14}$ Hz and a radiant intensity of 1/683 W/sr. The crooked factor 1/683 has historical reasons and stems from the fact that the candela used to be defined independently of radiometric quantities.

With this definition of luminous intensity and the eye's ability to detect small changes in brightness, the luminous intensity of any light source can be measured by comparing it with a standard light source. With this approach, however, the measurement of luminous intensity is dependent on the individual observer. It is therefore better to use a standard spectral luminous efficacy function. Then each photometric quantity can be calculated from its

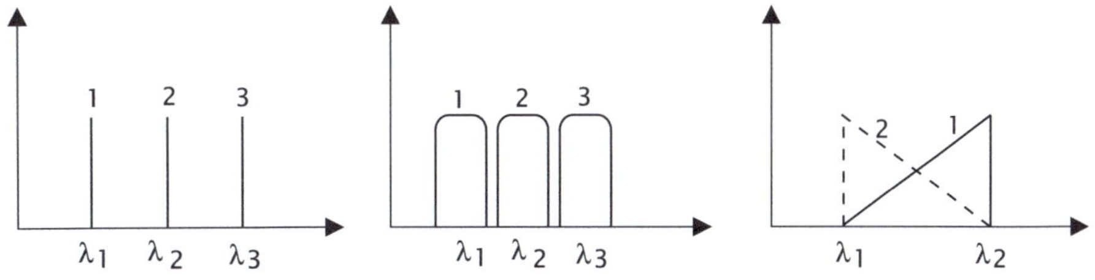

Figure 2.19: *Examples of spectral scanning:* ***a*** *Line-by-line sampling;* ***b*** *Band-by-band sampling;* ***c*** *Sampling according to a specific model of the spectral range, here for a single spectral line of unknown wavelength.*

corresponding radiometric quantity:

$$\begin{aligned} Q_v &= 683\frac{\text{lm}}{\text{W}} \int_{380\,\text{nm}}^{780\,\text{nm}} Q(\lambda)V(\lambda)\,\text{d}\lambda \quad \text{photopisch}, \\ Q_{v'} &= 1754\frac{\text{lm}}{\text{W}} \int_{380\,\text{nm}}^{780\,\text{nm}} Q(\lambda)V'(\lambda)\,\text{d}\lambda \quad \text{skotopisch}, \end{aligned} \tag{2.57}$$

where $V(\lambda)$ is the spectral luminous efficiency for day vision (photopic). A list of all photometric quantities and their radiation equivalents can be found in Appendix D (> R2). The unit of luminous flux, the photometric quantity equivalent to radiant flux (unit *W*), is lumen (lm).

2.5 Spectroscopy and Color

2.5.1 Spectral Sampling Methods

In principle, *spectroscopic image acquisition* is a very powerful tool for identifying objects and their properties, as almost all optical material constants depend on the wavelength of the radiation. The disadvantage of spectroscopic imaging is that an additional coordinate is added and the amount of data increases accordingly. It is therefore important to measure the spectrum with a minimum number of sampling points that is sufficient to fulfill a given task. We describe several measurement strategies here and also discuss human color vision as a form of spectral measurement from this point of view.

Line-by-line sampling is a method in which only a narrow range of the spectrum is recorded per channel (Fig. 2.19a). It is used when processes related to emission or absorption at certain spectral lines are to be recorded. A channel only "sees" a specific wavelength and is insensitive to all other wavelengths, as far as such a narrow bandpass filter can be technically realized. In this way, a very specific effect or a particular chemical substance can be recorded. However, this technique is not suitable for estimating the total radiation of objects, as it is insensitive to most wavelengths.

Band-by-band sampling is the appropriate technique (Fig. 2.19b) if the total radiation is to be measured in a specific wavelength range and a certain wavelength resolution is still required. Ideally, the individual bands have a constant sensitivity and are directly adjacent to each other. Band scanning ensures optimum resolution with few channels, but does not allow the wavelengths within a band to be differentiated. The spectral resolution that can be achieved with this scanning method is limited to the width of the spectral bands of the sensors.

In many cases, it is possible to set up a model of the spectral radiation of a particular object. Then a better technique of spectral sampling can be chosen, which essentially does not sample specific wavelengths, but the parameters of the model. This technique is called *model-based spectral sampling*.

We will illustrate this general approach with a simple example. It shows a method for measuring the mean wavelength and total radiant flux of an arbitrary spectral distribution $\Phi(\lambda)$ within a certain wavenumber range. The quantities sought are defined as follows

$$\Phi = \frac{1}{\lambda_2 - \lambda_1} \int_{\lambda_1}^{\lambda_2} \Phi(\lambda)\, \mathrm{d}\lambda \quad \text{and} \quad \overline{\lambda} = \int_{\lambda_1}^{\lambda_2} \lambda \Phi(\lambda) d\lambda \Bigg/ \int_{\lambda_1}^{\lambda_2} \Phi(\lambda)\, \mathrm{d}\lambda\,. \tag{2.58}$$

In the second equation, the spectral distribution is multiplied by the wavelength λ. We therefore need a sensor with a sensitivity that varies linearly with the wavenumber. We try it with two sensor channels of the following linear spectral sensitivity (Fig. 2.19c):

$$\begin{aligned} R_1(\lambda) &= \frac{\lambda - \lambda_1}{\lambda_2 - \lambda_1} R_0 \\ R_2(\lambda) &= R_0 - R_1(\lambda), \end{aligned} \tag{2.59}$$

where R are the wavelength-dependent sensitivities of the two sensors.

The sum of the sensitivity of the two channels is independent of the wavelength, while the difference is directly proportional to the wavelength and varies between $-R_0$ for $\lambda = \lambda_1$ and R_0 for $\lambda = \lambda_2$. Therefore, the sum signal of the two sensors R_1 and R_2 results in a measurement of the total radiant flux, while the average wavelength can be calculated:

$$\overline{\lambda} = \frac{\lambda_1 R_2 + \lambda_2 R_1}{R_1 + R_2}. \tag{2.60}$$

Apart from these two variables, however, the two sensors do not provide any further details about the spectral distribution of the measured radiation.

2.5.2 ♯Color Vision

Human color vision can be viewed in terms of the spectral scanning techniques discussed above as a mixture of band-pass scanning and model-based scanning. For color perception, the eye has three types of photopigments in the retinal photoreceptors known as cones. These photopigments have different spectral sensitivities (Fig. 2.18b). They comprise spectral bands with maximum sensitivities at 445 nm, 535 nm and 575 nm respectively, which however overlap considerably. In contrast to our model examples, the three sensor channels are of unequal width and cannot simply be brought into a linear relationship with each other. In fact, the color sensitivity of the human eye is uneven, so the associated nonlinearities make the scientific study of color vision quite difficult. Only a few basic facts will be discussed here, insofar as they are necessary for understanding colors.

With three color sensors, color signals span a 3D space. Each point in this space represents a color. It is clear that many spectral distributions, so-called *metameric color stimuli* or *metamers* for short, can be mapped to a point in *color space*. In general, we can write the signal s_i generated by a sensor with a spectral sensitivity $R_i(\lambda)$ as

$$s_i = \int R_i(\lambda)\Phi(\lambda)\,\mathrm{d}\lambda. \tag{2.61}$$

The triples generated by three sensors for primary colors are often called *tristimulus*.

At the heart of *colorimetry* is a system with which colors can be reproduced as linear combinations of basic or *primary colors*. A set of three spectral distributions $\phi_j(\lambda)$ represents a set of primary colors and results in a vector of responses that can be described by the matrix P as follows:

$$p_{ij} = \int R_i(\lambda)\Phi_j(\lambda)\,\mathrm{d}\lambda. \tag{2.62}$$

Each vector $\boldsymbol{p}_j = (p_{1j}, p_{2j}, p_{3j})$ represents the tristimulus of the "primary" colors in 3D color space. It is then obvious that only colors that are a linear combination of the base vectors $\boldsymbol{p}_j$ can be represented:

$$\boldsymbol{s} = R\boldsymbol{p}_1 + G\boldsymbol{p}_2 + B\boldsymbol{p}_3 \qquad \text{with} \qquad 0 \le R, G, B \le 1, \tag{2.63}$$

where the coefficients are denoted by R, G and B for the three primary colors red, green and blue. Only if the three base vectors $\boldsymbol{p}_j$ are orthogonal can all colors be represented as a linear combination of the base vectors. A possible and technically easy to implement primary color system are the monochrome colors red, green and blue with 700 nm, 546,1 nm and 435,8 nm, as defined by the CIE in 1931. In the following, we use the screen primary color system according to the European EBU standard, with red, green and blue phosphor (phosphor) as primary colors, as this is the usual method of displaying color images on RGB color monitors.

From the significant overlap of the spectral responses of the three cone types in the retina, especially in the green region (Fig. 2.18b), it is obvious that there is no set of physical primary colors that can represent all possible colors of the 3D color space. The colors that can be reproduced lie within the parallelepiped formed by the three non-orthogonal basis vectors of the primary colors. The more the primary colors are correlated with each other (i. e. the smaller the angle between two of them), the smaller the color space that can be represented by them. Mathematically, colors that cannot be represented by a mixture of primary colors have at least *a* negative coefficient in Eq. (2.63).

One component in the 3D color space is the intensity. If a color vector is multiplied by a scalar, its intensity changes, but not its color. In this way, all colors can be normalized via the intensity. This operation reduces the 3D color space to a 2D color plane or color map:

$$r = \frac{R}{R+G+B}, \quad g = \frac{G}{R+G+B}, \quad b = \frac{B}{R+G+B} \tag{2.64}$$

with

$$r + g + b = 1. \tag{2.65}$$

It is sufficient to use only the two components r and g. The third component then results from $b = 1 - r - g$ according to Eq. (2.65). This means that all colors that can be represented by the three primary colors R, G and B are restricted to a triangle in rg-space (Fig. 2.20a). As already mentioned, some colors cannot be represented by the primary colors. The limit of possible colors is given by all visible monochromatic colors from deep red to blue. The line of monochromatic colors forms a u-shaped curve in rg-space. Since all colors that lie on a straight line between two colors can be generated by additive mixing of these two, the area bordered by the u-shaped spectral curve and the straight mixing line between its two ends for blue and red color (*purple line*) forms the space of all possible colors.

2.5.3 ♯Color Systems

To avoid negative color coordinate values, a new color coordinate system with virtual primary colors is often selected, i. e. with primary colors that do not correspond to any physical color. This color system is known as *XYZ color system* or *xy color space* and is constructed in such a way that it contains the curve of monochromatic colors with only positive coefficients (Fig. 2.20c). It is described by the following linear coordinate transformation:

$$\begin{bmatrix} X \\ Y \\ Z \end{bmatrix} = \begin{bmatrix} 0{,}490 & 0{,}310 & 0{,}200 \\ 0{,}177 & 0{,}812 & 0{,}011 \\ 0{,}000 & 0{,}010 & 0{,}990 \end{bmatrix} \begin{bmatrix} R \\ G \\ B \end{bmatrix}. \tag{2.66}$$

The reverse transformation from the XYZ color system to the RGB color system results from the inversion of the matrix in Eq. (2.66).

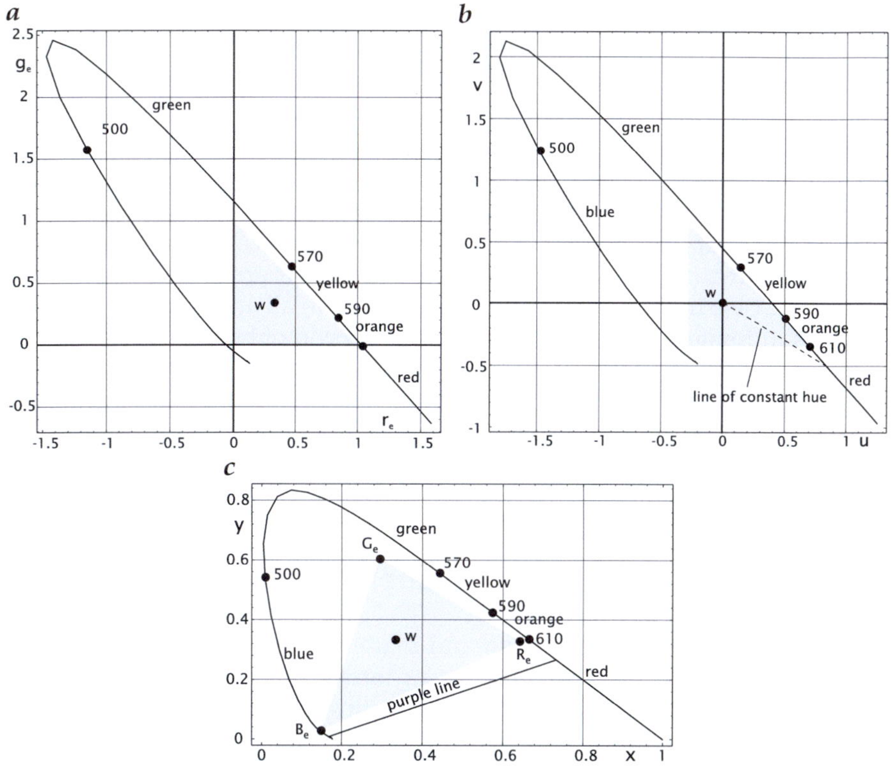

Figure 2.20: *Color map, displayed in the* ***a*** *rg color space;* ***b*** *uv color space;* ***c*** *xy color space; the gray triangles enclose the colors that can be generated by additive color mixing from the primary colors R, G and B.*

The color systems discussed so far are not directly related to human color perception. We cannot directly create colors such as green or blue from the *rg* or *xy* values. A natural description of colors includes not only the *luminosity* or *illuminance* (*intensity*), the color type such as green or blue (*hue*) and the purity of the color (*saturation*). We can create any degree of saturation from a pure color by mixing it with white.

Hue and saturation can be extracted from color maps using simple coordinate transformations. The reference point is the white point in the center of the chart (Fig. 2.20b). If we draw a line from this point to a pure (monochromatic) color, a blend line is created for a pure color with white. It is therefore a line with a constant hue. The saturation increases linearly from this white point to the pure color. The *white point* is given in the *rg* diagram by $w = [1/3, 1/3]^T$.

A color system that has its center at the white point is called *color difference system*. From a color difference system, we can get to a hue, saturation, intensity (HSI) by using a polar coordinate system. Then the radius is proportional to the saturation and the angle to the hue (Fig. 2.20b).

So far, color vision is simple. All difficulties arise from the fact that a color system for screens and printers as well as for transmission by television signals must be adapted as well as possible to human color vision. If there are deviations, we as humans typically see some specific colors distorted by cameras, on screens or in color prints. We only have to consider this problem for an image capture and processing system when it comes to applications in which human color vision must be reproduced as accurately as possible. This is the case in the graphic arts industry and in all applications in which humans ultimately judge the image quality. If you only want to use colors, or in more precise physical terms different wavelengths, to better distinguish objects from others or to identify them more accurately, then it makes sense not to be guided by human color vision, but to think strictly physically and not to limit yourself to the visible wavelength range of electromagnetic waves from the outset.

2.6 Interaction between Radiation and Matter

The interaction between radiation and matter is the basis for every imaging technique. Basically, two classes of interaction can be distinguished (Fig. 2.21). The first class is related to discontinuities in optical properties at the transition between two different materials. The second class is related to the volume and depends on the optical material constants. In this section we give a brief summary of the most important types of interactions. This is intended to provide an overview of the many ways in which material properties can be measured using imaging techniques.

2.6.1 Radiation Emission and Thermography

Emission of electromagnetic radiation occurs at all temperatures. It is the simplest and at the same time ubiquitous form of interaction between matter and electromagnetic radiation. The cause of the spontaneous emission of electromagnetic radiation is thermal molecular motion, which increases with temperature. When radiation is emitted, thermal energy is converted into electromagnetic radiation. According to the law of conservation of energy, matter cools down in the process.

There is an upper limit to thermal emission. For reasons of energy conservation, the proportion of radiation that is absorbed at a certain wavelength must also be emitted [181]. This means that the upper emission limit is an absorptivity of one. A substance with perfect absorptivity — and thus maximum emission — is called *black body*.

The correct theoretical description of the radiation of a black body by *Planck* in 1900 [243] required the assumption that emission and absorption of

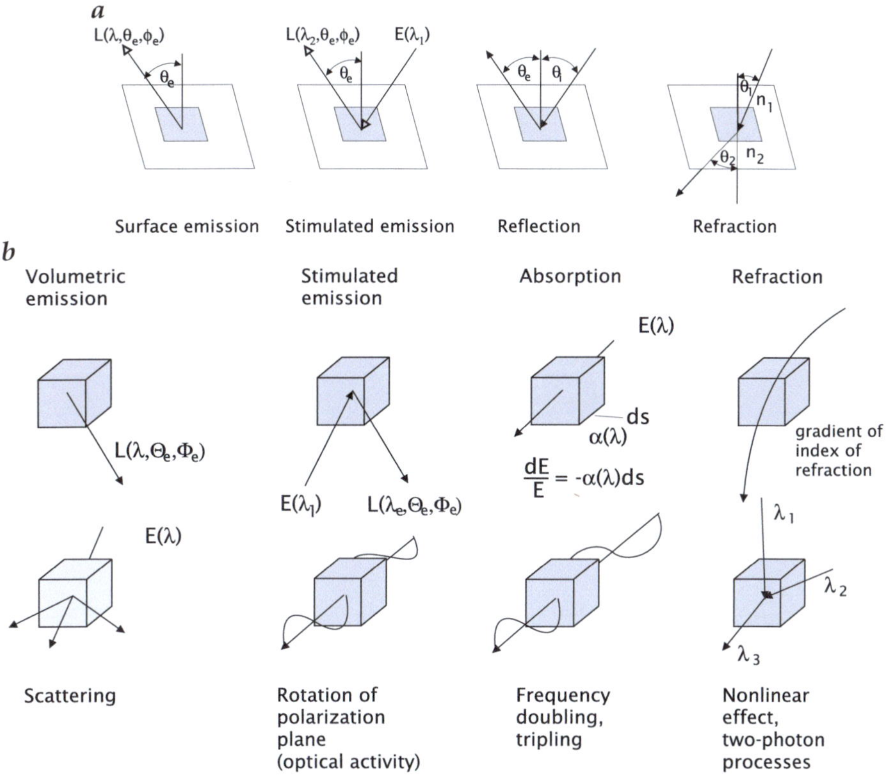

Figure 2.21: *Principle possibilities of interaction between radiation and matter:* ***a*** *at the surface of an object, i. e. due to the discontinuity of optical properties;* ***b*** *volume-related.*

radiation take place in discrete energy quanta $E = h\nu$. The spectral radiation of a blackbody with the absolute temperature T is (Fig. 2.22)

$$L_e(\nu, T) = \frac{2h\nu^3}{c^2}\frac{1}{\exp\left(\frac{h\nu}{k_B T}\right) - 1}, \quad L_e(\lambda, T) = \frac{2hc^2}{\lambda^5}\frac{1}{\exp\left(\frac{hc}{k_B T\lambda}\right) - 1} \tag{2.67}$$

with

$$\begin{array}{ll} h = 6,6262 \cdot 10^{-34}\,\mathrm{Js} & \text{Planck's constant,} \\ k_B = 1,3806 \cdot 10^{-23}\,\mathrm{J/K} & \text{Boltzmann constant and} \\ c = 2,9979 \cdot 10^{8}\,\mathrm{m/s} & \text{speed of light in a vacuum.} \end{array} \tag{2.68}$$

The radiation curves in Fig. 2.22 do not cross each other. At higher temperatures, the radiance for *all* wavelengths is higher. The radiance drops very steeply at short wavelengths and much more slowly at longer wavelengths.

For the long-wave limiting case ($h\nu \ll k_B T$), the following simple approximations result from Eq. (2.67):

$$L_e(\nu, T) = \frac{2h\nu^2 k_B T}{c^2}, \quad L_e(\lambda, T) = \frac{2ck_B T}{\lambda^4} \tag{2.69}$$

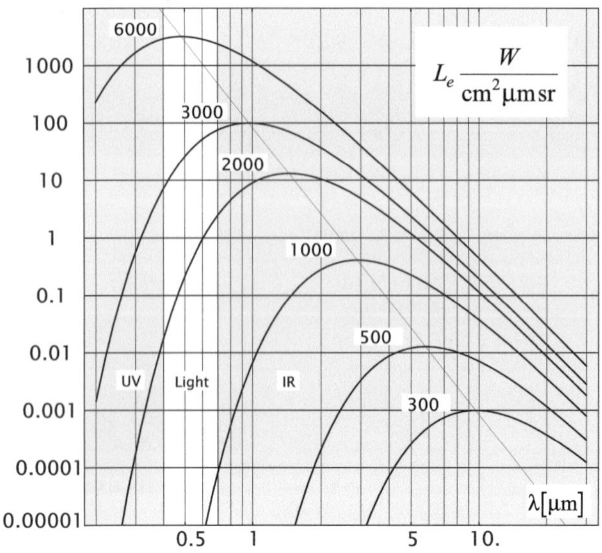

Figure 2.22: *Spectral radiance of a black body at different absolute temperatures T, plotted double logarithmically. The thin line marks the wavelength of maximum emission as a function of absolute temperature.*

The long-wave part of the blackbody radiation is therefore directly proportional to the absolute temperature T and it decreases with λ^4.

The radiation of a black body has the important property that the radiation density is the same in all directions. Such a radiator is called *Lambertian radiator* or isotropic radiator. This means that the spectral radiant flux density (constant radiant density integrated over a hemisphere) is Eq. (2.56) π times higher than the radiance.

The total radiant flux density of a black body, integrated over all wavelengths, is proportional to T^4 according to Stefan and Boltzmann's law:

$$M_e = \int_0^{\pi/2} \int_0^{2\pi} \int_0^{\infty} L_e(\lambda, T) \cos\theta \sin\theta \, d\theta \, d\phi \, d\lambda = \frac{2}{15} \frac{k_B^4 \pi^5}{c^2 h^3} T^4 = \sigma T^4, \qquad (2.70)$$

where $\sigma \approx 5.67 \cdot 10^{-8}\mathrm{Wm^{-2}K^{-4}}$ is the *Stefan-Boltzmann constant*. The wavelength of the maximum emission of a black body is described by *Wien's law*:

$$\lambda_m \approx \frac{2.898 \cdot 10^{-3}[\mathrm{Km}]}{T}. \qquad (2.71)$$

Real objects emit less radiation than a black body. The ratio of the radiation emitted by a real body to the radiation emitted by a black body is called (specific) *emissivity* ϵ and depends on the wavelength.

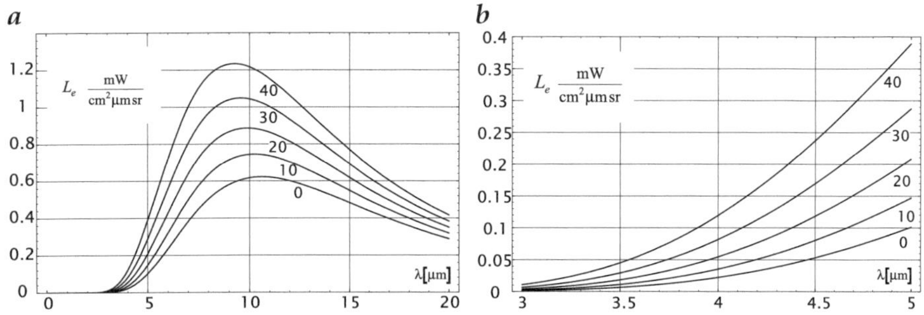

Figure 2.23: *Radiation density of a black body at temperatures between 0 and 40°C in the wavelength range **a** 0–20 μm and **b** 3–5 μm.*

Example 2.8: Radiant intensity at ambient temperature and 3000 K

From Eq. (2.70) you can calculate that a blackbody at ambient temperature (300 K) emits $5.67 \cdot 10^{-8} \cdot 3^4 \cdot 10^8 \approx 460\,\mathrm{W/m^2}$. A high value, even if it is compared with the solar constant (irradiance of the earth by the sun without absorption by the atmosphere) of $1367\,\mathrm{W/m^2}$. The maximum irradiance at room temperature (300 K) in the infrared range is around 10 μm.
An incandescent lamp at 3000 K and an assumed emissivity $\epsilon = 0.39$ emits 3900 times more radiation, i. e. around $1.8\,\mathrm{W/mm^2}$. A high-power LED light source with 3 W electrical power achieves a radiant intensity of 1–$2\,\mathrm{W/mm^2}$. However, since the maximum radiation of the incandescent lamp is in the near infrared at 1 μm (Fig. 2.22), the LED has a significantly higher radiant intensity in the visible spectral range than an incandescent lamp.

Radiation in the *infrared* and *microwave range* can be used to take images of the temperature of objects. This application is called *thermography*. This is one of the few examples of imaging self-luminous objects at ambient temperatures. The wavelength dependence of the radiation density of a black body radiator in the ambient temperature range from 0°C to 40°C is shown in Figure 2.23. The differences are large enough to be able to measure temperatures sensitively. The decisive factor here is that the radiance is largely independent of the inclination of the surfaces due to the Lambertian characteristic of the radiation.

It is also crucial for a correct measurement that no radiation is absorbed on the way from the object to the imaging system. The atmosphere is only partially transparent in the infrared. The two most important atmospheric windows (cf. Fig. 2.34) are at 3-5 μm, with an absorption maximum around 4.2 μm, which is due to CO_2, and one at 8-14 μm, also with an absorption band due to water vapor. Fig. 2.23 shows the radiation of a black body at ambient temperatures between 0 and 40 °C in wavelength ranges of 0-20 μm and 3-5 μm. Although the radiation maximum is at 10 μm and there about 20 times higher than at 4 μm, the relative change in radiation with temperature

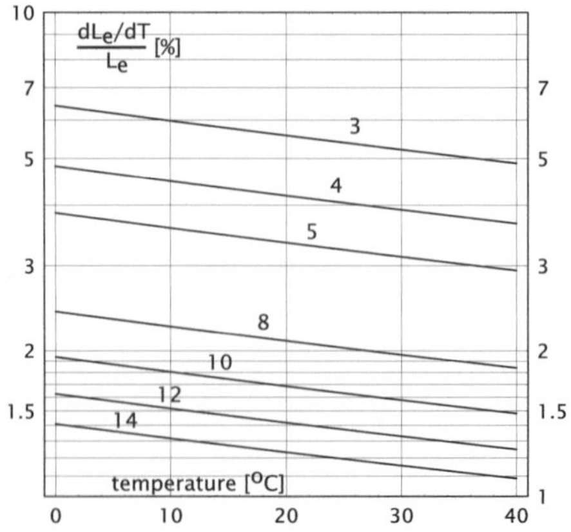

Figure 2.24: Relative change in photon-based radiance in percent per degree in the temperature range between 0 and 40°C.

is much greater at 4 μm than at 10 μm. The relative change in radiance L with absolute temperature T can be calculated directly from Eq. (2.22):

$$\frac{\mathrm{d}L_e(\lambda, T)}{\mathrm{d}T} \Bigg/ L_e = \frac{hc}{k_B T^2 \lambda} \frac{\exp\left(\frac{hc}{k_B T \lambda}\right)}{\exp\left(\frac{hc}{k_B T \lambda}\right) - 1} \tag{2.72}$$

The result shows Figure 2.24 for wavelengths between 3 and 14 μm. While the radiance at 20°C only changes by around 1.7%/K at a wavelength of 10 μm, the change at a wavelength of 4 μm is more than twice as high at around 4%/K. The relative change in radiance is even smaller if one integrates over the entire wavelength range. Then the radiation density increases with T^4 and changes relatively by $4/T$, i. e. 1.36%/K at 20°C. When measuring ambient temperatures with microwaves, i. e. at the long-wave end of the Planck distribution, the radiance is only proportional to T Eq. (2.69) and therefore only changes relatively by $1/T$ or 0.34%/K. If we want to measure small differences in the ambient temperature range of 0 to 40°C as accurately as possible, then the short-wave wavelength range of 3–5 μm (*MWIR* or mid-wave infrared) is the best choice, although the absolute radiation is about five times smaller than the long-wave range of 8–14 μm (*LWIR* or long-wave infrared).

Thermal imaging is complicated by the fact that real objects are not perfect black bodies, but partially reflect radiation from their surroundings. Direct measurements of the temperature therefore require precise knowledge of the emissivity. Objects with high emissivity such as water, snow or skin are less problematic, but metallic surfaces with low emissivity are. If an object has the *emissivity* ϵ, the part $1 - \epsilon$ of the received radiation results from the object environment by reflection of the ambient radiation on the object. This distorts the temperature measurement.

Under the simplified assumption that the environment has a constant temperature T_e, we can estimate the influence of the reflected ambient radiation on the temperature measurement. For small temperature differences ($\Delta T = T - T_e \ll T$), we can approximate the radiation flux density M as follows:

$$M = \epsilon \sigma T^{\gamma} + (1 - \epsilon)\sigma T_e^{\gamma},$$

where the exponent γ depends on the wavelength interval in which we measure. The exponent γ can take values from 1 to approx. 10.

The measured radiation is interpreted as if it came from a black body with the simulated temperature T':

$$T'^{\gamma} = \epsilon T^{\gamma} + (1 - \epsilon)T_e^{\gamma}. \tag{2.73}$$

For the last term, we can introduce the following approximation by Taylor series expansion:

$$T_e^{\gamma} = (T + \Delta T)^{\gamma} \approx T^{\gamma}\left(1 + \gamma \frac{\Delta T}{T}\right).$$

This then leads to the final result that

$$T' - T \approx (1 - \epsilon)\Delta T. \tag{2.74}$$

Surprisingly, this result applies universally, regardless of the exponent γ and thus the wavelength range selected for the measurement. Even a 1 % deviation of ϵ from one causes a temperature measurement error of 0.01 K per 1 K difference between the object temperature and the ambient temperature.

Example 2.9: Temperature measurement error for outdoor water surfaces

Even with an almost perfect *black body* such as the water surface with an average emissivity of 0.97, this leads to considerable errors in the absolute temperature measurement. If water surface temperatures are measured outdoors, such as in lakes or the ocean, the water surface reflects about 3% of the radiation from the sky. A cloudless sky can be up to 80 K colder than a 300 K warm water surface and thus lead to an error of -0.03 · 80 K = -2.4 K in the temperature measurement.

Relative temperature measurements are also subject to error, but to a much lesser extent. If we assume a constant ambient temperature, we can immediately conclude from Eq. (2.73) that

$$\Delta T' \approx \epsilon \Delta T. \tag{2.75}$$

This means that the measured temperature differences are smaller by a factor of ϵ than in reality and can be easily corrected.

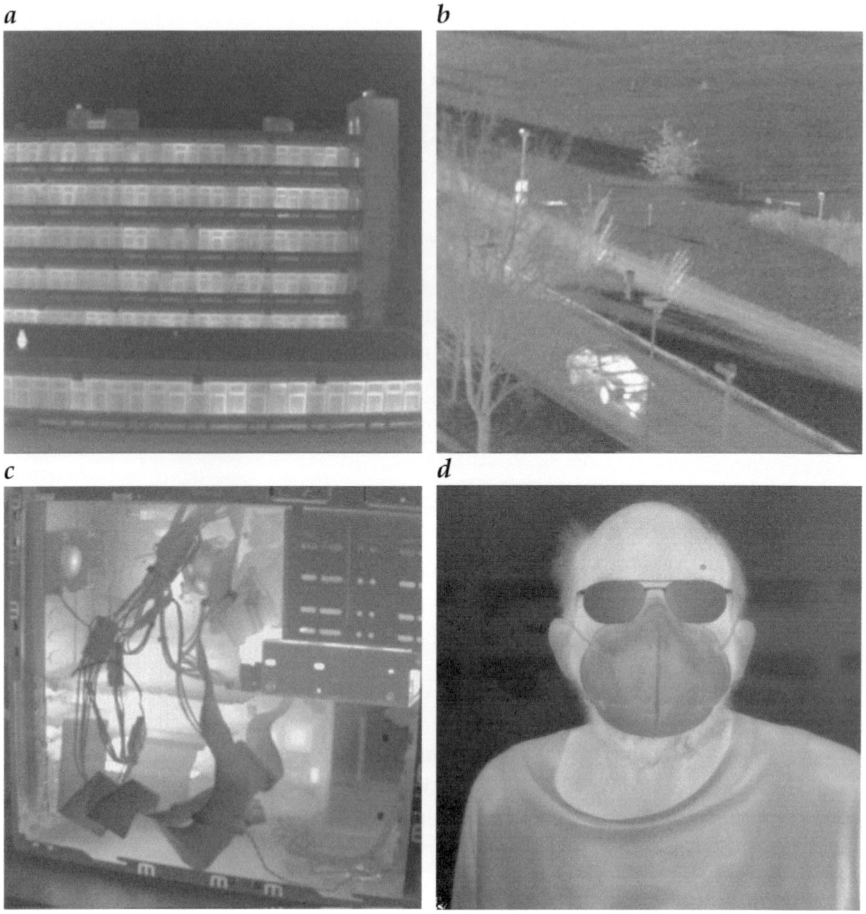

Figure 2.25: Examples of thermal images: ***a*** *Building on a cold winter's day,* ***b*** *Street scene,* ***c*** *View into a PC and* ***d*** *Person with glasses and breathing mask* ***(related video).***

In Figure 2.25 some pictures illustrate the application of thermography. The building (Figure 2.25a) and the street scene (Figure 2.25b) were taken on a cold winter's day. The warm window frames are clearly visible, which means that the heat escapes particularly well from the rooms and these are therefore the weak point in the thermal insulation. The hood is hot, heat escapes from the heated interior through the windows of the car, the left front tire is clearly heated. The heat from the engine compartment is reflected on the road surface in front of the car. A look inside a PC (Figure 2.25c) shows the hot spots on the motherboard, a hot hard disk and a warm fan motor. In the case of the person with glasses (Figure 2.25d), we can see that the glasses are not permeable to IR radiation and in the accompanying **Video**

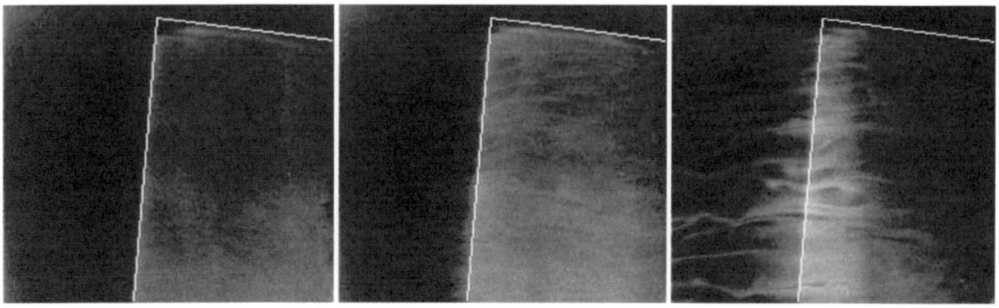

Figure 2.26: *View of a water surface in the Heidelberg Aeolotron at 2 m/s wind speed with an MWIR thermal imaging camera. The area outlined in white was heated with a CO_2 laser (10.6 μm thermal radiation). Thermal images with an image section 25 cm × 25 cm 0.5, 1.0 and 3.0 s after switching on the laser. The wind blows from right to left.*

the warming up and cooling down of the respirator mask when breathing in and out.

Temperature images are also suitable for looking beneath the surface if the surface is heated or cooled in a suitable manner at the same time. How warm the surface becomes depends on how well the heat can flow into the object. If there is a crack or cavity under the surface, heat conduction is hindered and the part of the surface above remains warmer. The heating can be pulsed or modulated. Methods of this type are called *heat flow thermography*, *lock-in thermography* or *active thermography* and have become common methods of non-destructive testing technology [26, 277]. The basic principles of these techniques were developed independently in industrial testing technology [38] and environmental sciences [152], even before thermal imaging cameras existed.

Example 2.10: Active thermography for the study of exchange processes at the sea surface

The Institute of Environmental Physics at Heidelberg University has a ring-shaped wind-wave channel, the Aeolotron, with which exchange processes at the sea surface can be systematically investigated. As it is a closed, thermally well-insulated system, the conditions can be set so that no heat exchange takes place at the water surface. The water surface can be selectively heated with a CO_2 laser over an area of a few square decimeters. The 10.6 μm infrared radiation is absorbed in a thin layer of water only 20 μm thick at the water surface.

When the laser is switched on, the water surface heats up until it reaches equilibrium with the heat transport into the water depth through molecular diffusion and turbulence (Fig. 2.26). After 0.5 s, the heating has only penetrated the water so little that the heating is still uniform because the turbulent transport directly at the water surface is not effective. Only after 3 s do the turbulent structures become clearly visible.

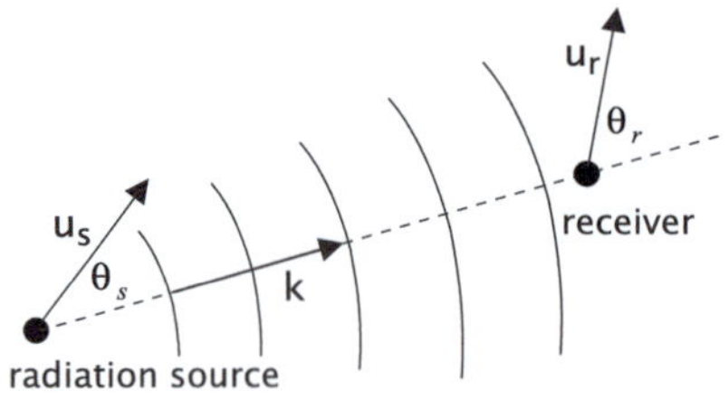

Figure 2.27: The classic Doppler effect: the source and receiver move relative to the medium at the speeds $\boldsymbol{u}_s$ and $\boldsymbol{u}_r$, respectively.

The heated structures are moved out of the heated area by the water flow at the water surface. Image sequence analysis (Chapter 14) can be used to measure the flow field at the water surface through the visible structures. It is also possible to determine how quickly climate-relevant gases, such as carbon dioxide, are transported from the atmosphere into the sea with this technique [191, 226, 227].

2.6.2 ♯Doppler Effect

A difference in speed between the radiation source and the receiver causes the receiver to measure a different frequency than the source emitted. The radiation source can also be an object that is illuminated and reflects or scatters radiation. The geometric arrangement is illustrated in Fig. 2.27. This phenomenon is known as the *Doppler effect*. The frequency shift is directly proportional to the difference in speed:

$$\nu_r = \frac{c - \boldsymbol{u}_r^T \hat{\boldsymbol{k}}}{c - \boldsymbol{u}_s^T \hat{\boldsymbol{k}}} \nu_s \quad \text{or} \quad \Delta\nu = \nu_r - \nu_s = \frac{(\boldsymbol{u}_s - \boldsymbol{u}_r)^T \boldsymbol{\kappa}}{1 - \boldsymbol{u}_s^T \hat{\boldsymbol{k}}/c}, \tag{2.76}$$

where $\hat{\boldsymbol{k}} = \boldsymbol{\kappa}/|\boldsymbol{\kappa}|$, ν_s is the frequency of the source, ν_r is the frequency measured at the receiver, $\boldsymbol{\kappa}$ is the wavenumber of the radiation and c is the propagation velocity of the radiation. $\boldsymbol{u}_s$ and $\boldsymbol{u}_r$ are the velocities of the source and receiver relative to the medium in which the wave propagates. Only the velocity component in the direction of the receiver causes a frequency shift.

If the source moves towards the receiver ($\boldsymbol{u}_s^T \boldsymbol{\kappa} > 0$), the frequency increases as the wave fronts follow each other more quickly. A critical limit is exceeded when the source moves at more than the propagation speed of the radiation. Then the radiation follows behind the source. At low velocities relative to the wave propagation velocity, the frequency shift is directly proportional to the relative velocity between source and receiver:

$$\Delta\nu = (\boldsymbol{u}_s - \boldsymbol{u}_r)\boldsymbol{\kappa}. \tag{2.77}$$

The relative frequency shift $\Delta\nu/\nu$ is given directly by the ratio of the velocity difference between source and receiver $\Delta\boldsymbol{u} = \boldsymbol{u}_s - \boldsymbol{u}_r$ in the direction of the receiver to the wave propagation velocity:

$$\boxed{\frac{\Delta\nu}{\nu} = \frac{\Delta\boldsymbol{u}^T\hat{\boldsymbol{k}}}{c}}. \tag{2.78}$$

For electromagnetic waves, the speed relative to a "medium" is not relevant. From the theory of relativity, the frequency results in:

$$\nu_r = \frac{\nu_s}{\gamma(1-\Delta\boldsymbol{u}^T\hat{\boldsymbol{k}}/c)} \quad \text{with} \quad \gamma = \frac{1}{\sqrt{1-(|\Delta\boldsymbol{u}|/c)^2}}. \tag{2.79}$$

For small velocities, the relationships Eq. (2.77) and Eq. (2.78) also result. In this case, acoustic and electromagnetic waves can be treated equally with regard to the frequency shift resulting from a relative movement between source and receiver.

Example 2.11: Doppler effect in velocity measurements with electromagnetic and acoustic waves

The speed of light at about $3 \cdot 10^8$ m/s is so high that it is doubtful whether we can measure speeds of the order of m/s. The relative frequency shift according to Eq. (2.78) is smaller than 10^{-8} and would be very difficult to measure directly. However, the problem can be solved interferometrically by superimposing a reference radiation and the radiation shifted by the velocity. Then, according to the basic considerations about wave packets in Section 2.2.3, a beat occurs. The beat frequency for light with 10^{15} Hz (Fig. 2.16) is below 10 MHz, so that the resulting intensity fluctuations can be measured directly electronically. In the microwave range at 10 GHz, the beat frequencies are only in the 100 Hz range. Coherent superposition of radiation is therefore of central importance for velocity measurement. In flow measurement technology, such methods are known under the names *laser Doppler* and *phase Doppler* [6, 219]. They also play an important role in the *microwave radar* technique for remote sensing [332].

The situation is much simpler with sound waves. At a sound velocity of 300 m/s, the relative frequency shifts are in the percentage range and can be easily measured.

2.6.3 Reflection, Refraction and Dispersion

Electromagnetic waves are reflected and refracted at the interface between two optical media with different refractive indices. If a plane wave with a *wave vector* $\boldsymbol{\kappa}_1$ hits the interface at an angle θ_1 to the normal of the surface, the wave front reaches the surface on the left sooner than on the right (Fig. 2.28) and the phase is correspondingly smaller there. As the phase velocity of the reflected wave is the same as that of the incident wave, the reflected plane wave travels at the same angle θ_1 in the other direction to the normal. The angle of incidence is therefore equal to the angle of reflection.

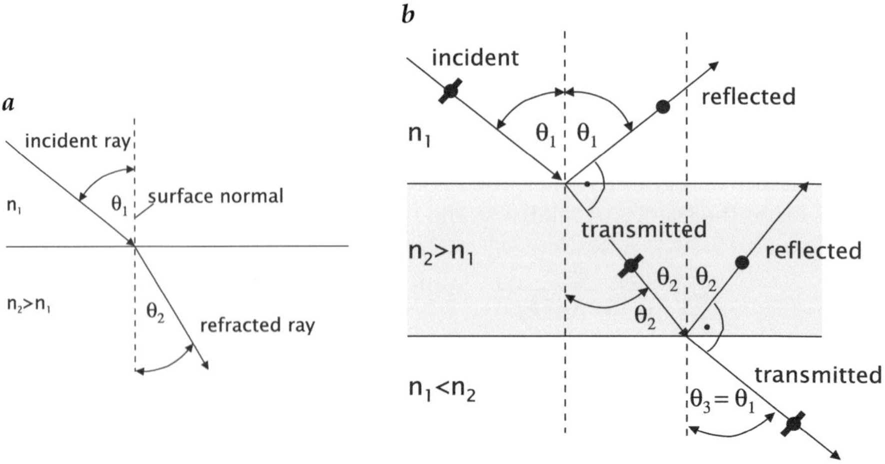

Figure 2.28: ***a*** *Reflection and refraction at the interface between two optical media with a different refractive index.* ***b*** *Parallel polarized light is completely refracted and not reflected if the angle between the reflected and refracted beam would be 90°This condition applies to the transition of both optically thinner and optically thicker media.*

In the second, optically denser medium ($n_2 > n_1$), the wave travels correspondingly slower (Section 2.3.1). As the phase of the wave front is retained, it is refracted towards the normal and the *Snellius refraction law* applies:

$$\boxed{\sin\theta_2 = \frac{n_1}{n_2}\sin\theta_1 \quad \text{Snellius refraction law.}} \qquad (2.80)$$

The surface normal and the wave vectors of the incident, reflected and refracted wave lie in one plane.

Refraction is the basis for transparent optical elements (lenses) that can create an image of an object. This means that all rays that are emitted from a point on the object and travel through the optical element converge at a point on the image plane.

To ensure that the wavefronts are not deformed, the surface must be smooth and even. It is sufficient for the roughness of the surface to be significantly smaller than the wavelength. Depending on the wavelength, a surface therefore appears rough or smooth.

How much radiation is reflected depends not only on the difference between the refractive indices n_1 and n_2 of the two media, but also on the angle of incidence and the polarization of the radiation. *parallel* or *perpendicular* means that the electric field vector is parallel or perpendicular to the plane of incidence in which the surface normal and wave vectors lie.

The *reflection coefficient* ρ indicates the ratio of the reflected to the incident *radiant flux* (Section 2.4.1). These relationships result from continuity considerations of the electric and magnetic field at the interface [22]. The result

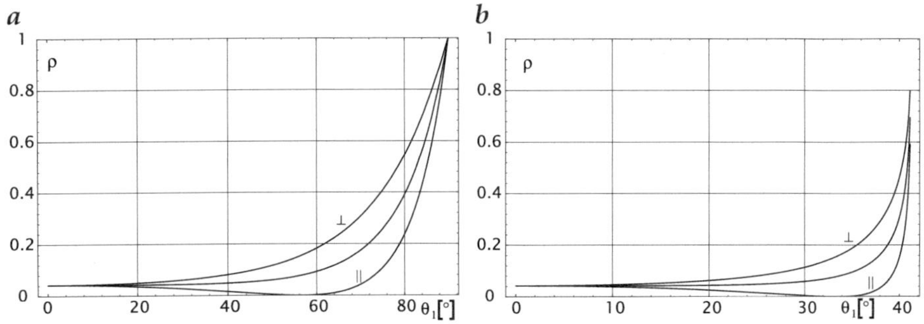

*Figure 2.29: Reflection at the transition **a** from air ($n_1 = 1.00$) to BK7 glass ($n_2 = 1.517$) and **b** from BK7 glass into air for parallel (||), perpendicular ($\perp$) and unpolarized incident light.*

is the *Fresnel equations*:

$$\boxed{\rho_{\parallel} = \frac{\tan^2(\theta_1 - \theta_2)}{\tan^2(\theta_1 + \theta_2)}, \quad \rho_{\perp} = \frac{\sin^2(\theta_1 - \theta_2)}{\sin^2(\theta_1 + \theta_2)}, \quad \rho = \frac{\rho_{\parallel} + \rho_{\perp}}{2}.} \tag{2.81}$$

The angle of incidence θ_1 and the angle of reflection θ_2 are related to each other via Snellius' law Eq. (2.80) (Fig. 2.29).

With perpendicular incidence ($\theta_1 = 0$), the reflection does not depend on the polarization:

$$\rho = \frac{(n_1 - n_2)^2}{(n_1 + n_2)^2} = \frac{(n-1)^2}{(n+1)^2} \quad \text{with} \quad n = n_1/n_2. \tag{2.82}$$

As illustrated in Fig. 2.29, parallel polarized light is not reflected at all at a certain angle (*Brewster angle* θ_b) if refracted and reflected rays are perpendicular to each other (Fig. 2.28b):

$$\theta_b = \arctan\frac{n_2}{n_1}. \tag{2.83}$$

In the case of reflection from an optically denser material ($n_2 > n_1$), the Brewster angle is always greater and, conversely, smaller than 45°.

According to Snell's law (Eq. (2.80)), when light passes from an optically denser medium to a thinner one ($n_1 > n_2$), the sine of the second angle ($\sin\theta_2$) can exceed one. This means that the beam cannot leave the optically denser medium but is completely reflected. At the *critical angle*

$$\theta_c = \arcsin\frac{n_2}{n_1} \quad \text{with} \quad n_1 > n_2. \tag{2.84}$$

is the exit angle 90°, i. e. parallel to the boundary surface. This phenomenon is called *total reflection*.

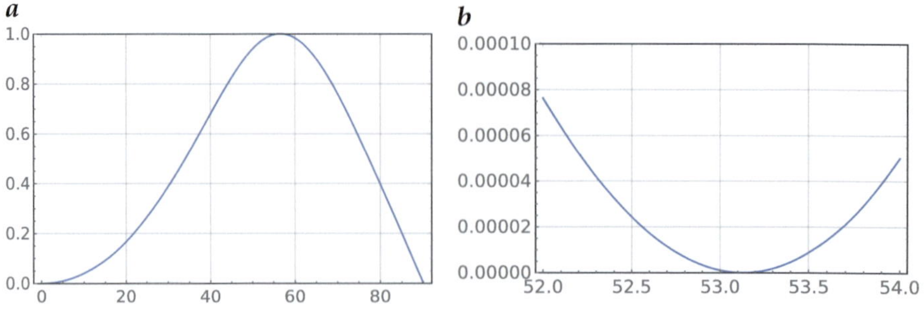

Figure 2.30: ***a*** *Degree of polarization for reflection of unpolarized radiation from a glass surface with a refractive index n* = 1.517. ***b*** *Reflection coefficient around the Brewster angle of 53.1° for water with n* = 4/3

Example 2.12: Total reflection at water and glass interfaces to air

In the case of water and N-BK7 glass with refractive indices of 4/3 and 1.517, all rays with angles of incidence greater than 48.6° and 41.2° are totally reflected at the transition to air ($n = 1.0003$) and are perfect mirrors under these conditions.

Example 2.13: Polarization by reflection

The Fresnel equations and Fig. 2.29 show that unpolarized light is partially polarized by reflection depending on the angle of incidence. At perpendicular incidence, the light remains unpolarized; at the Brewster angle, it is completely perpendicularly polarized (Fig. 2.30a). If the plane of incidence is oriented in the x-direction, the *Stokes vector* (Eq. 2.39) is $[\rho_\parallel + \rho_\perp, \rho_\parallel - \rho_\perp, 0, 0]^T I$ and the *degree of polarization* results according to Eqs. (2.40) and (2.81) to

$$p = \frac{|\rho_\parallel - \rho_\perp|}{\rho_\parallel + \rho_\perp}. \tag{2.85}$$

This allows the two-dimensional orientation of surfaces on which unpolarized light is reflected to be measured using images with polarization degree and polarization angle, although the ambiguity must be taken into account. The same degree of polarization occurs at two angles.

Example 2.14: Brewster angle microscopy

When reflected at the Brewster angle, parallel polarized light is not reflected. The reflection coefficient is therefore zero and increases approximately quadratically around the Brewster angle (Fig. 2.30b). With a deviation of less than 0.5° the reflection coefficient is only around 10^{-5}. However, the Fresnel equations Eq. (2.81) only apply to a sudden change in the refractive index at interfaces. This is not the case with real surfaces. The density of liquids in particular does not change abruptly at the surface. The orientation of the molecules can change and so-called surface-active substances accumulate on the surface. These slight changes in the reflection conditions can be seen with high sensitivity when observed under parallel polarized light at the Brewster angle and thus even the structure of monomolecular films on the water surface can be analysed [133, 135]. Measurement methods that determine the properties of surfaces and thin films on them using illumination with polarized light [48, 100, 359] are called *ellipsometry*. The name results from the fact that polarized light is generally *elliptically polarized* (Section 2.2.7).

2.6.4 Rough Surfaces and BRDF

Most natural and technical objects do not reflect light specularly, but show a more or less complex diffuse reflection. This is due to microscopic irregularities which, depending on the inclination of the individual reflective surfaces, reflect light in different directions. The angular distribution of the backscattered radiation can vary greatly. Some materials show strong forward scattering in the direction of the specular reflection, others scatter almost uniformly in all directions and there are also so-called retroreflective materials (cat's eye effect), which mainly backscatter radiation in the direction from which the radiation hits the surface.

Mixed reflection is also very common. Some of the radiation is reflected directly from the smooth or rough surface. Another part of the radiation penetrates the material, is partially absorbed and scattered by optical inhomogeneities and leaves the object again. While the direct reflection coefficient is determined by the refractive index and therefore depends only slightly on the wavelength, this is generally not the case for the penetrated and then reflected radiation. The absorption in the material can be strongly color-dependent and thus also the backscattering.

The situation becomes even more complex when thin layers, regular surface structures or regular optical inhomogeneities of the same order of magnitude as the wavelength of the radiation occur in the material. Interference effects are then possible. These lead to direction-dependent interference colors, which we can observe in soap bubbles, thin layers of oil on water and many animal species such as dragonflies, butterflies or birds. In technology, these effects are used for security features on banknotes, ID cards and products.

It is obvious that it is not enough to describe the reflection of radiation on surfaces that are not perfectly reflective with a wavelength-dependent reflection coefficient. This has an advantage and a disadvantage. The disadvantage is that we have to find an adequate, more complex description of the reflection properties, while the advantage is that this enables a much better differentiation of materials when taking images.

The reflection coefficient will now depend on both the direction of the incident and the outgoing radiation. The direction is described by two parameters, the polar angle θ and the azimuth angle ϕ. This leads to a four-a

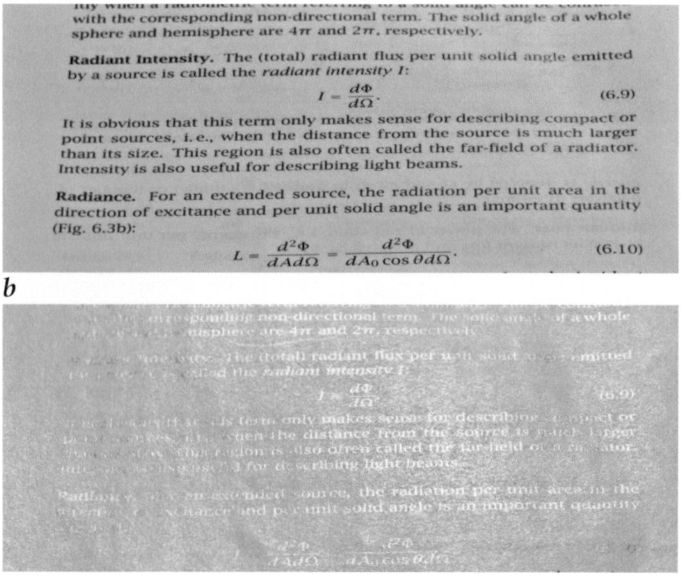

with the corresponding non-directional term. The solid angle of a whole sphere and hemisphere are 4π and 2π, respectively.

Radiant Intensity. The (total) radiant flux per unit solid angle emitted by a source is called the *radiant intensity* I:

$$I = \frac{d\Phi}{d\Omega}. \qquad (6.9)$$

It is obvious that this term only makes sense for describing compact or point sources, i.e., when the distance from the source is much larger than its size. This region is also often called the far-field of a radiator. Intensity is also useful for describing light beams.

Radiance. For an extended source, the radiation per unit area in the direction of excitance and per unit solid angle is an important quantity (Fig. 6.3b):

$$L = \frac{d^2\Phi}{dA\,d\Omega} = \frac{d^2\Phi}{dA_0 \cos\theta\, d\Omega}. \qquad (6.10)$$

b

Figure 2.31: *Example of how the BRDF influences the image acquisition (black does not have to appear darker than white):* ***a*** *Book page illuminated with diffuse incident light* ***b*** *Book page illuminated with directional incident light under the reflection condition (illumination angle equal to exposure angle). As the ink scatters into a narrower angle range, it now appears brighter than the white paper.*

dimensional function, the *bidirectional reflectance distribution function* (BRDF or reflectance distribution for short):

$$f(\theta_i, \phi_i, \theta_e, \phi_e) = \frac{L_e(\theta_e, \phi_e)}{E_i(\theta_i, \phi_i)}. \qquad (2.86)$$

The index i stands for the incident radiation and the index e for the emitted radiation.

For a perfect mirror (specular reflection), the BRDF is zero everywhere, with the exception of $\theta_i = \theta_e$ and $\phi_e = \pi + \phi_i$, i. e.

$$f(\theta_i, \theta_e) = \delta(\theta_i - \theta_e) \cdot \delta(\phi_e - \pi - \phi_i). \qquad (2.87)$$

The other extreme is a perfectly scattering material that reflects incident radiation evenly in all directions regardless of the angle of incidence. Such a surface is called a *Lambertian radiator* or Lambertian reflector. An essential property of this surface is that the reflection is independent of the viewing direction:

$$L_e = \frac{1}{\pi} E_i \quad \text{or} \quad f(\theta_i, \phi_i, \theta_e, \phi_e) = \frac{1}{\pi}. \qquad (2.88)$$

a *b*

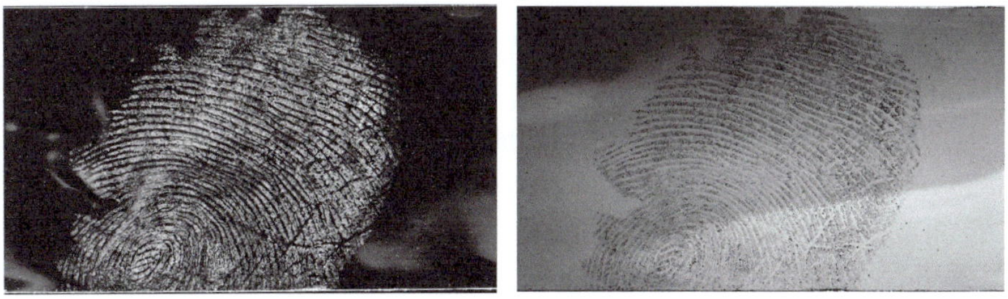

Figure 2.32: *Image of a fingerprint on a strip of adhesive tape:* ***a*** *measurement of the scattered light,* ***b*** *Measurement of the absorbed light.*

a *b*

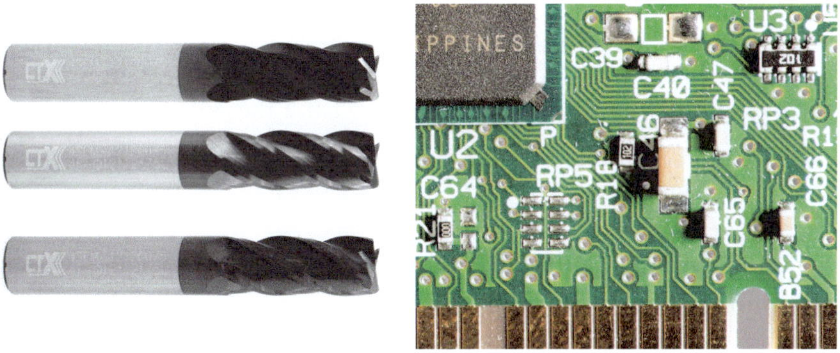

Figure 2.33: *Metallic objects are difficult to inspect due to the complex BRDF of their surfaces:* ***a*** *milling cutter under different lighting conditions,* ***b*** *Section of a circuit board with solder joints and metal contacts.*

The importance of an adequate lighting technique for image capture is illustrated by the following examples. Figure 2.31 shows how a black object with a narrower scattering cone can appear brighter than a white object with a wide scattering cone with appropriate lighting.

Suitable imaging technology makes it possible to capture different scattering behavior. Scattering or absorption makes a fingerprint visible on a scotch tape (Figure 2.32). Due to the complex BRDF of metal surfaces with reflective and scattering components, it is particularly difficult to illuminate them for inspection tasks for geometric measurement or defect detection (Figure 2.33).

2.6.5 Absorption

Radiation that propagates in matter is more or less absorbed and converted into different forms of energy, in particular ultimately into heat. There are two ways of describing absorption. In physics, the molecular one dominates with *cross sections* on individual molecules [244], in chemistry one based on concentrations in units of mol/l [120]. Another difference is that in physics mainly natural (ln) and in chemistry mainly decadic (log) logarithms are used. Both perspectives are considered in parallel here.

The absorption in a thin layer $\mathrm{d}z$ is proportional to the radiation energy. Therefore applies:

$$\frac{\mathrm{d}I(\lambda)}{I} = -\alpha_e(\lambda, z)\,\mathrm{d}z = -\sigma(\lambda)n(z)\,\mathrm{d}z. \tag{2.89}$$

The *absorption coefficient* α is a property of the medium and depends on the wavelength of the radiation. It indicates what fraction of the radiation is absorbed per unit length and therefore has the unit of a reciprocal length (m^{-1}).

If you have individual discrete absorbers such as molecules, you can assign them a cross-sectional area, the *absorption cross-section* σ_a (unit cm^2), on which they absorb the radiation completely. The absorption coefficient is then the product of the cross-sectional area and the *particle density* n

$$n = \frac{N}{V}, \quad [n] = \left[\frac{1}{\mathrm{cm}^3}\right], \tag{2.90}$$

the number of particles N per volume V. Alternatively, it is possible to specify a *molar absorption coefficient* ε and the *substance concentration* c in the units mol/l:

$$\alpha_{10} = \varepsilon \cdot c, \quad [\varepsilon] = \left[\frac{1}{\mathrm{cm\,mol/l}}\right]. \tag{2.91}$$

By integrating Eq. (2.89) we can calculate the attenuation of the radiation on the path from 0 to z:

$$\boxed{\begin{aligned} \ln\frac{I(z)}{I(0)} &= -\int_0^z \alpha_e(\lambda, z')\mathrm{d}z' = -\sigma_a(\lambda)\int_0^z n(z')\mathrm{d}z' \quad \text{or} \\ \log\frac{I(z)}{I(0)} &= -\int_0^z \alpha_{10}(\lambda, z')\mathrm{d}z' = -\varepsilon(\lambda)\int_0^z c(z')\mathrm{d}z' \end{aligned}} \tag{2.92}$$

or if the concentration is homogeneous (n or c independent of the position z):

$$\ln\frac{I(z)}{I(0)} = -\sigma_a(\lambda)n\,z \quad \text{or} \quad \log\frac{I(z)}{I(0)} = -\varepsilon(\lambda)c\,z. \tag{2.93}$$

The exponential attenuation of radiation in a homogeneous medium, as expressed by Eq. (2.93), is often referred to as *Lambert-Beer-Bouguer law*.

By comparing the two equations in (2.93) and using the relationship between particle density and mass concentration, we obtain

$$n = cN_A, \quad N_A \approx 6{,}023 \cdot 10^{23}/\mathrm{mol} \quad \text{Avogadro constant} \tag{2.94}$$

also the relationship between the cross section σ and the molar absorption coefficient ε to

$$\varepsilon\left[\frac{1}{\mathrm{cm\,mol/l}}\right] = \frac{N_A}{10^3 \ln 10}\sigma[\mathrm{cm}^2], \qquad \frac{N_A}{10^3 \ln 10} \approx 2.62 \cdot 10^{20}. \tag{2.95}$$

The *Avogadro constant* indicates the number of molecules per mole.

The path integral over the absorption coefficient

$$A(z_1, z_2)_{e,10} = \int_{z_1}^{z_2} \alpha_{e,10}(z')\mathrm{d}z' \tag{2.96}$$

results in a dimensionless quantity called *absorbance* or *optical density*. The optical density is a logarithmic expression of the radiation attenuation and means that along the path from point z_1 to point z_2 the radiation has been attenuated to $\tau_e = e^{-A_e}$ or $\tau_{10} = 10^{-A_{10}}$, depending on whether natural or decadic logarithms are used. The term *transmissivity* τ indicates the fraction of radiation that remains after the rays have traveled a certain distance in the medium.

When radiation passes through composite media, such as, e. g. the atmosphere, many components contribute to absorption. For this case, Eq. (2.86) must be extended to several absorbers, whereby the logarithmic absorptances Eq. (2.96) simply add up:

$$\boxed{\begin{aligned} \ln\frac{I(z)}{I(0)} &= -\sum_{p=1}^{P}\sigma_p(\lambda)\int_0^z n_p(z')\mathrm{d}z' \quad \text{or} \\ \log\frac{I(z)}{I(0)} &= -\sum_{p=1}^{P}\varepsilon_p(\lambda)\int_0^z c_p(z')\mathrm{d}z' \end{aligned}} \tag{2.97}$$

If the absorption cross-sections σ_p or molar absorption coefficients ε_p of all P absorbers involved are known as a function of the wavelength, this represents a discrete inverse problem (Appendix C.6). This can be solved by measuring with a sufficient number of wavelengths $Q > P$ using the method of least squares (Appendix C.6). However, it is not possible to determine the concentration curve $c_p(z')$ itself, but only the so-called *column density*

$$s = \int_0^z n_p(z')\mathrm{d}z' \quad \text{or} \quad s = \int_0^z c_p(z')\mathrm{d}z'. \tag{2.98}$$

Only the average concentration can be determined from this:

$$\overline{n_p} = \frac{1}{z}\int_0^z n_p(z')\mathrm{d}z' \quad \text{or} \quad \overline{c_p} = \frac{1}{z}\int_0^z c_p(z')\mathrm{d}z', \tag{2.99}$$

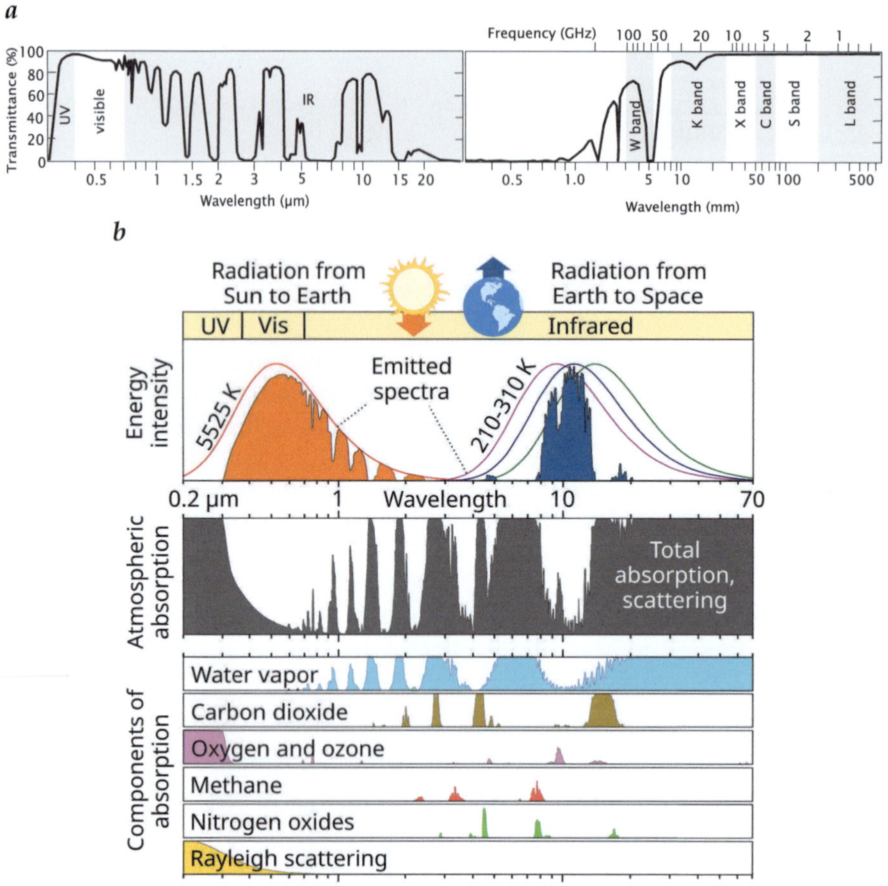

Figure 2.34: *Transmissivity of the atmosphere for electromagnetic radiation **a** from the ultraviolet range to the microwave range and **b** only in the UV, visible and IR range with solar radiation, thermal radiation and main absorbers (modified according to* `https://commons.wikimedia.org/wiki/File:Atmospheric_Transmission.svg`*)*

tomographic methods are required to determine the concentration fields (Section 9.3).

Example 2.15: Transmissivity of the atmosphere

Figure 2.34 shows the transmissivity of the atmosphere when the radiation passes vertically. For radiation from 300–700 nm the atmosphere is largely transparent, in the infrared from 0.7–14 μm interspersed with absorption bands and then opaque up to the terahertz range. In the microwave range there is then the widest window for wavelengths between 7–1000 mm. The most important absorbers can also be found in Figure 2.34b.

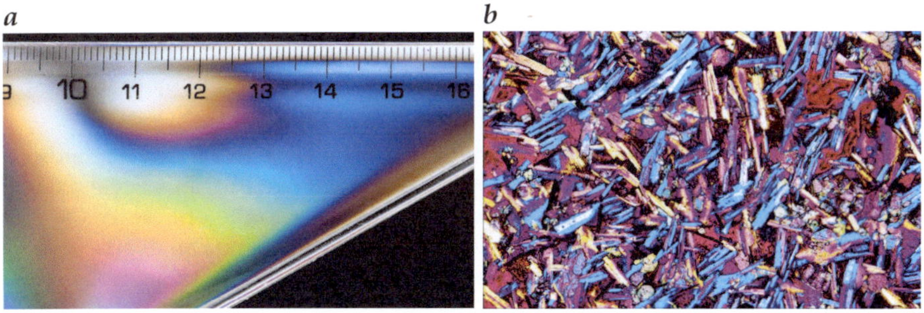

Figure 2.35: *Images of samples between two crossed polarization filters. Without rotation of the polarization plane, the images remain black.* ***a*** *Induced optical activity due to internal stress in a plastic ruler made of Plexiglas.* ***b*** *Thin section of the volcanic rock basalt (source:* `https://commons.wikimedia.org/wiki/File:Basalt_Plagioclase_and_Pyroxene.jpg`*).*

2.6.6 ♯Change in the Polarization State

When radiation passes through matter, the polarization state of the radiation can also be changed. The simplest effect is the rotation of the plane of polarization. Substances that cause this are called *optically active*. This is based on the fact that right-circular and left-circular radiation (Section 2.2.7) have different phase velocities. This effect is therefore also called *circular birefringence*.

The rotation φ is proportional to the concentration of the optically active material c and the distance d:

$$\varphi = \gamma(\lambda) c d. \tag{2.100}$$

The constant γ is the *specific rotation* and has the units [m^2/mol] or [cm^2/g]; it depends on the wavelength of the radiation. In general, the specific rotation is greater at short wavelengths. Two well-known optically active materials are quartz crystals and sugar solutions. Optical activity — including the measurement of wavelength dependence — can be used to identify chemical substances and measure their concentration.

Various external influences such as electric fields (*Kerr effect*), magnetic fields (*Faraday effect*) and internal mechanical stresses in materials influence or induce optical activity. In the production of plastic objects, internal stresses occur during curing even without external mechanical forces, which rotates the plane of polarization. This in Figure 2.35 was taken with white light between crossed linear polarization filters. Without optical activity, the image is black. As the optical activity depends on the wavelength, the result is a colorful image. The color whose polarization direction is rotated by an odd multiple of 90° appears brightest.

Extension 2.3: Double refraction and polarization microscopy in mineralogy

Most crystals are anisotropic. This means that the refractive index depends on the direction. As a result, the crystal is characterized by an optical axis and the refractive indices parallel and perpendicular to the optical axis are different. This effect is called *birefringence*. As crystals are also exposed to mechanical stresses during their formation, this results in additional induced effects. Polarization microscopy in mineralogy [255] is therefore an important tool for identifying minerals and determining their genesis (Fig. 2.35b).

2.6.7 ♯Elastic and Inelastic Scattering

The attenuation of radiation due to scattering can be described in a similar way to radiation loss due to absorption:

$$\beta(\lambda) = -\frac{1}{I}\frac{\mathrm{d}I(\lambda)}{\mathrm{d}x}. \tag{2.101}$$

The *scattering coefficient* β is also a reciprocal length with the unit m^{-1}. If radiation is attenuated by both absorption and scattering, both effects can be summarized in the *extinction coefficient* $\kappa(\lambda)$:

$$\kappa(\lambda) = \alpha(\lambda) + \beta(\lambda). \tag{2.102}$$

Although scattering seems similar to absorption, it is a much more complex phenomenon. Equation (2.102) can only be used if the radiation of the individual scattering centers adds up incoherently at any point far away from the particles. The complexity of scattering is related to the fact that scattered radiation (without additional absorption) is not lost. Scattered light can be scattered more than once. Therefore, part of it can re-enter the original beam more than once. The probability that radiation is scattered more than once on a certain path is directly related to the total attenuation due to scattering along the path of the beam and thus to the transmissivity τ. If τ is less than 0.1, less than 10% of the radiation is scattered.

The total proportion of scattered light and the analysis of the angular distribution is related to the optical properties of the scattering medium. Therefore, the scattering is caused by optical irregularities of the medium. For further discussion, we assume that small spherical particles with radius r and refractive index n are embedded in an optically homogeneous medium.

As with absorption (Section 2.6.5), scattering by a particle is described by an *effective cross section*, the *scattering cross section*. The scattering cross-section also has the unit of an area. It can be regarded as the effective scattering area of the particle that completely scatters the incident *radiant flux*. Therefore, the effectiveness factor Q_s is defined as the effective cross-sectional area related to the geometric cross-sectional area of the particle:

$$Q_s = \sigma_s/(\pi r^2). \tag{2.103}$$

The angular distribution of the scattered radiation is defined by the *differential cross section* $\mathrm{d}\sigma_s/\mathrm{d}\Omega$, the radiation flux density that is scattered per unit solid angle. The total effective cross-section is the integral over all solid angles of the differential effective cross-section:

$$\sigma_s = \int \frac{\mathrm{d}\sigma_s}{\mathrm{d}\Omega}\mathrm{d}\Omega. \tag{2.104}$$

The scattering by small particles is decisively influenced by the ratio of the particle size to the wavelength of the radiation, expressed in the dimensionless particle size $q = 2\pi r/\lambda = 2\pi r\kappa$. If $q \ll 1$ (*Rayleigh scattering*), we have very weak scattering proportional to λ^{-4}:

$$\frac{\sigma_s}{\pi r^2} = \frac{8}{3}q^4 \left|\frac{n^2-1}{n^2+2}\right|. \tag{2.105}$$

For $q \gg 1$ the scattering can be described by geometrical optics. If a particle completely reflects the incident radiation, the effective cross-section is equal to the geometric cross-section ($\sigma_s/\pi r^2 = 1$) and the differential effective cross-section is constant (isotropic scattering, $\mathrm{d}\sigma/\mathrm{d}\Omega = r^2/2$).

In contrast, the scattering of particles of about one wavelength of radiation (*Mie scattering*) is very complex due to diffraction and interference effects of the light scattered by the different parts of the particle surface. The differential cross section varies strongly with the scattering angle and is mainly forward directed, while Rayleigh scattering is rather isotropic.

All scattering effects considered so far, such as Rayleigh and Mie scattering, are elastic, i. e. the scattered radiation has the same wavelength as the incident radiation. However, there is also the case that the scattered radiation has a shifted wavelength if the energy states of the absorber (rotation and/or oscillation) change during scattering. This is referred to as *inelastic scattering* or the *Raman effect*. As with absorption and luminescence (Section 2.6.8), the wavelength shift in Raman scattering is molecule-specific and can therefore also be used to measure concentration. However, the scattering cross-sections of Raman scattering are usually several orders of magnitude smaller than those of elastic scattering [222, 244].

2.6.8 ♯Luminescence

Luminescence is the emission of radiation from materials resulting from a transition from an excited to a lower energy state. The term *fluorescence* is used to describe luminescence that occurs immediately (in the range of nanoseconds) after excitation, while the term *phosphorescence* is used for delayed luminescence (milliseconds to minutes).

Luminescence is a very versatile process as it can be triggered by various processes. In *chemiluminescence*, the energy required to generate the excited state is released during a chemical reaction. Chemiluminescence usually has

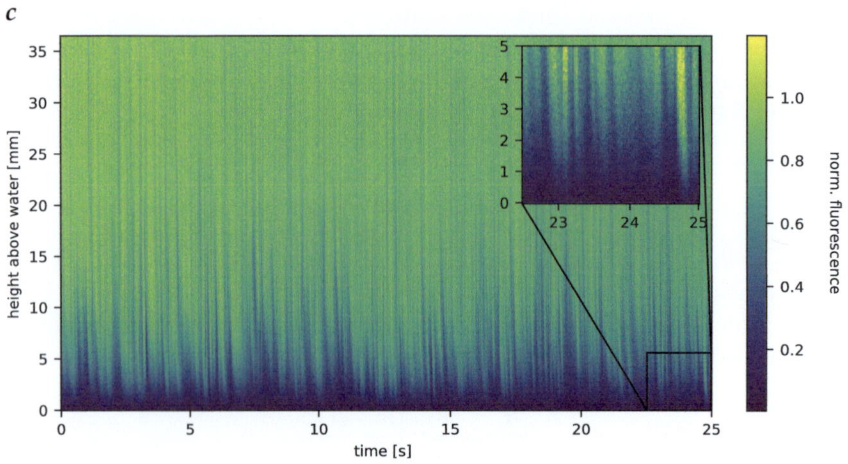

Figure 2.36: Time series of vertical sulphur dioxide profiles in the immediate vicinity of a water surface [96]. The gas is transported to the water surface, where it reacts with water to form hydrogen sulphite. This reaction is so fast that the sulphur dioxide concentration immediately at the water surface is zero.

a low efficiency (i. e. the number of photons emitted per reacting molecule is low) in the range of 1% or less. Flames are the classic example of a low efficiency chemiluminescence process. *Bioluminescence* is chemiluminescence in living organisms. Fireflies and the glow of marine microorganisms are well-known examples. In fireflies, luciferin is oxidized enzymatically. In contrast to most chemiluminescence processes, this reaction converts almost 100% of the chemical energy into radiant energy.

Bioluminescence processes play a role in many biological processes. The imaging of such processes is becoming increasingly important in the investigation of biochemical reactions. Another sophisticated tool in biochemistry is the labeling of biomolecules with fluorescent dyes. Today, even individual chromosomes or gene sequences can be labeled with fluorescent dyes.

In environmental physics, fluorescence imaging can be used to measure concentration fields of many substances in air and water. Fig. 2.36 shows the measurement of vertical profiles of *sulphur dioxide* directly above the water surface. The fluorescence was excited with a laser at 223.7 nm and the UV light around 300 nm was measured with a UV-sensitive camera.

Fluorescence imaging is also suitable for visualizing concentration fields in water with high resolution [237]. This can be used to quantitatively investigate which processes control the exchange of climate-relevant gases such as *carbon dioxide* between the atmosphere and the sea. As these processes take place quickly, the fluorescence in a wafer-thin layer on the water surface must be measured with a short exposure time. To illuminate the water surface over a large area, optical power of over 100 W is therefore required

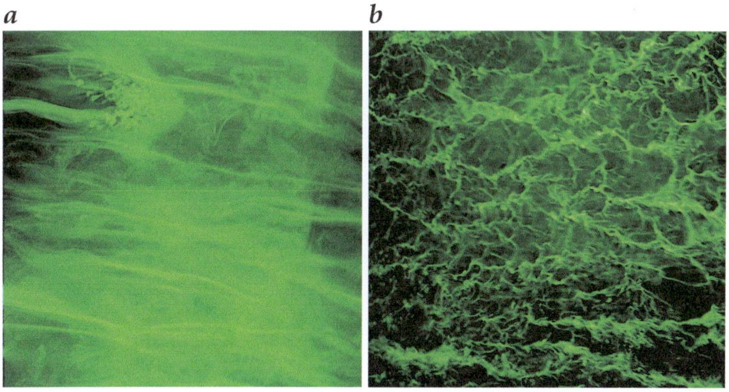

*Figure 2.37: Concentration fields in the wafer-thin water-side boundary layer directly at the water surface, showing the transport of gaseous trace substances from the atmosphere into the water, measured in the large Aeolotron wind/wave channel at Heidelberg University. A section of approx. 20 cm × 20 cm on the water surface at **a** lower (top left shows the effect of a small breaking wave without bubble entry) and **b** at high wind speed. Images from Hofmann [137].*

for fluorescence excitation and high-speed cameras that record 500 images/s. Two example images are shown in Fig. 2.37.

Luminescence processes always compete with other processes that reduce the energetic state of the molecules without emitting radiation. A well-known example of this is the radiationless transfer of energy when molecules collide.

Some molecular forms, especially electronegative molecules such as *oxygen*, inactivate excited states in molecular collisions with high efficiency. This process is known as *quenching*. The presence of quench molecules causes a weakening or quenching of the fluorescence. Therefore, the fluorescence radiation can be used to measure the concentration of quenching molecules. The dependence of the fluorescence intensity on the concentration of these molecules is formulated in the *Stern-Volmer equation*:

$$\frac{L}{L_0} = \frac{1}{1 + kc_q}. \tag{2.106}$$

Where L is the fluorescent radiance, L_0 is the fluorescent radiance without quenching, c_q is the concentration of the quenching molecule and k is the quenching constant, which depends on the lifetime of the fluorescence state. Efficient quenching requires that the excited state has a sufficient lifetime.

Sensitive fluorescent dyes for measuring quenching by dissolved oxygen are for instance ruthenium complexes [40]. The relative fluorescence intensity as a function of the concentration of dissolved oxygen shows Figure 2.38a. The fluorescence is stimulated with a semiconductor laser at 445 nm. Oxygen carried from the atmosphere over the water surface into oxygen-free water by wind-induced turbulence is shown in Figure 2.38b.

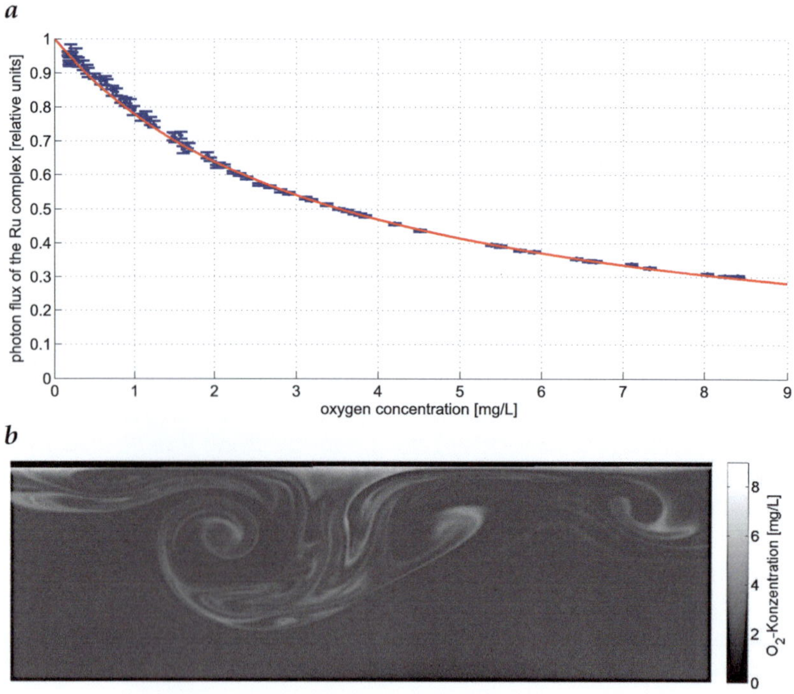

Figure 2.38: ***a*** *Quenching of the fluorescence of a ruthenium complex [94] by dissolved oxygen: Measured values and fit with the Stern-Volmer equation.* ***b*** *Oxygen concentrations immediately below the water surface in an image section of 64 mm × 20 mm in a wind-wave channel with wind-induced turbulence at 2.2 m/s* ***(video about this)*** *and at 4.0 m/s wind speed* ***(video about this).***

Extension 2.4: Pressure measurement through fluorescence quenching

Pressure-sensitive paints [204] are also based on the quenching of fluorescence by oxygen. Suitable fluorescent dyes are used to paint aircraft models in the wind tunnel, for example. Pressure increases on the surface of the models in the flow cause an increased oxygen partial pressure and thus a corresponding decrease in fluorescence. As the lifetime of the fluorescent light decreases with increasing extinction of the fluorescence, it may be more accurate to measure the *fluorescence lifetime* instead of the fluorescence intensity, which depends on the thickness of the paint layer and the homogeneity of the illumination.

2.6.9 ♯Nonlinear Optics

The superposition principle for waves breaks down with very high field strengths in matter. Then the interaction with the medium can no longer be described by linear equations and we enter the field of *nonlinear optics*.

Such phenomena only occur with very intense light sources such as *lasers*, especially when they are pulsed.

A typical non-linear phenomenon is the *frequency doubling* or *frequency tripling* of light. This effect is often used to double the frequency of a laser beam and thus halve its wavelength. For imaging, a similar effect of greater importance is *multi-photon absorption*, in particular *two-photon absorption*. Since this process combines the energy of several photons to induce an electronic transition, it only occurs at high photon fluxes because only then is the probability high that two photons are in the immediate vicinity of an absorbing molecule at the same time. Taking into account the two-photon absorption, Eq. (2.89) in Section 2.6.5 therefore expands to

$$\frac{\mathrm{d}I(\lambda)}{\mathrm{d}z} = -\alpha(\lambda)I - \alpha_2(\lambda)I^2. \tag{2.107}$$

From the point of view of quantitative visualization, nonlinear optical effects offer a wide range of new possibilities for visualizing specific effects and material properties, which can only be mentioned here.

2.7 ♯References for Further Reading

This chapter covers the basics that are essential for understanding image acquisition. To refresh or expand this knowledge, please refer to some of the classic textbooks, e. g. Elmore and Heald [70], Towne [330], Hecht [128], Saleh and Teich [281] or Kenyon [178]. Excellent and theoretically in-depth descriptions are provided by Born and Wolf [22], Goodman [113] (Fourier optics) and Lauterborn and Kurz [194] (coherent optics).

The interaction between radiation and matter with regard to remote sensing is dealt with by Stewart [315], Robinson [268] and Drury [64]. An insight into imaging techniques across the entire electromagnetic spectrum is provided by Richards [262], `www.alienvision.org`. The topic of thermal imaging has now become a field of work in its own right, see e. g. Gaussorgues [102] and Vollmer and Möllmann [343]. Imaging methods using sound waves are indispensable in medical technology [65] and the geosciences. Sound waves, triggered by earthquakes, are one of the few experimental techniques to explore the internal structure of the Earth [74, 92, 321].

A good description of color vision in terms of image processing can be found in Pratt [247] and Wendland [353]. The practical aspects of photometry and radiometry are dealt with in the "Handbook of Applied Photometry" by DeCusaris [53] and McCluney [221]. The oldest application of quantitative visualization is flow visualization. The "Atlas of Visualization" by Nakayama and Tanida [229] provides a fascinating insight with many image examples.

3 2D Image Acquisition

Contents

3.1 Introduction

Image acquisition comprises two important aspects. The first concerns the *geometry*. Where do we find an object in an image? All image acquisition techniques project three-dimensional space onto a two-dimensional image plane in one way or another. Therefore, the image capture can be simplified

B. Jähne, *Digital Image Processing and Image Acquisition*,
https://doi.org/10.1007/978-3-032-04806-6_3

as a projection from the three-dimensional to the two-dimensional space. This involves the loss of a coordinate. This is certainly a significant loss of information.

However, we constantly and unconsciously experience that our visual system gives us a three-dimensional impression of our surroundings, which is sufficient for us to find our way around and interact with it. Technically speaking, the visual system reconstructs the three-dimensional world from two-dimensional images. The ease with which this is done may lead us to believe that it is a simple task. However, as we see in Chapter 8, this is not the case.

We deal with the geometric aspects of image acquisition in three steps after a general introduction to coordinate systems (Section 3.2). First, we consider the ideal optical image in the limiting case of *geometric optics* (Sections 3.3–3.4.5). This means that we use infinitely thin beams based on the basic fact that radiation propagates in a straight line in a homogeneous medium. We neglect the fact that a beam like a Gaussian beam (Section 2.2.5) has a minimum expansion and divergence given by the wave nature of the beams, which is given by Eq. (2.26).

In the second step, we learn that even in the limiting case of geometric optics, the image on the image plane is not perfect (Section 3.4.6), in order to then develop a general concept of how a non-perfect image, whether due to imaging errors or the wave nature of the radiation, can be elegantly described. This leads us to the system theory of imaging in Section 3.5.

The third aspect concerns *radiometry*. How "bright" is an imaged object and how does the brightness in the image depend on the optical properties of the object and the imaging system? The radiometry of an imaging system is the subject of Section 3.6. For the basics of radiometry, please refer to Section 2.4.

3.2 World and Camera Coordinates

3.2.1 Definition

There are two ways to describe the position of objects in 3D space (Fig. 3.1). The so-called *world coordinate system* $\boldsymbol{x}_s = [x_s, y_s, z_s]^T$, is related to the scene under consideration. The coordinates x_s and y_s describe the horizontal positions and z_s the vertical position.

A second, the *camera coordinate system* $\boldsymbol{x}_c = [x_c, y_c, z_c]^T$, is linked to the moving camera. The z_c axis coincides with the optical axis of the camera system (Fig. 3.1). Physicists are familiar with such considerations, as physical phenomena are discussed in different coordinate systems. In mechanics, for example, motion is studied in relation to two observers, one at rest and one moving with the object.

The transition from world coordinates to camera coordinates is described by a *translation* and a *rotation*. First, the origin of the world coordinate system

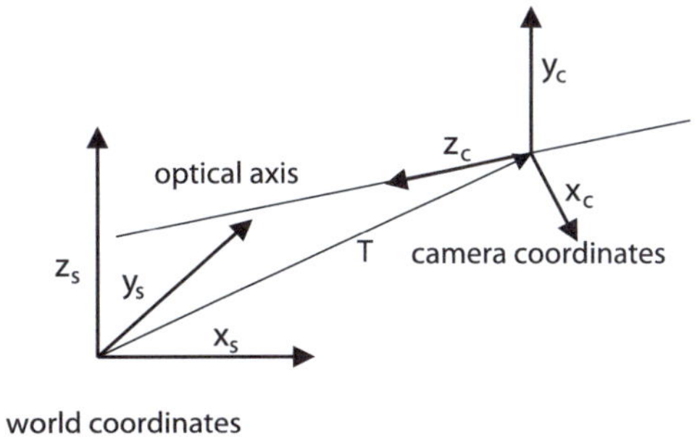

***Figure* 3.1:** *World and camera coordinate systems: The world coordinate system can be converted into the camera coordinate system by rotation around the origin and a translation t.*

is shifted to the origin of the camera coordinate system via the translation vector $\boldsymbol{t}$ (Fig. 3.1). Then the orientation of the shifted world coordinate system is changed by rotation around the corresponding axes so that it coincides with the camera coordinate system. Mathematically, the translation can be described with a vector subtraction and the rotation with a multiplication of the coordinate vector with a matrix:

$$\boldsymbol{x}_c = \boldsymbol{R}(\boldsymbol{x}_s - \boldsymbol{t}). \tag{3.1}$$

3.2.2 Rotation

The *rotation* of a coordinate system has two important properties. It does not change the length or norm of a vector, and it leaves the coordinate system orthogonal. A transformation with these properties is called an *orthonormal transformation* in linear algebra.

The coefficients of a transformation matrix have an intuitive meaning. This becomes clear when we transform unit vectors $\hat{\boldsymbol{e}}_p$ in the direction of the coordinate axis. With $\hat{\boldsymbol{e}}_1$ we obtain, e. g.

$$\hat{\boldsymbol{e}}'_1 = \boldsymbol{A}\hat{\boldsymbol{e}}_1 = \begin{bmatrix} a_{11} & a_{12} & a_{13} \\ a_{21} & a_{22} & a_{23} \\ a_{31} & a_{32} & a_{33} \end{bmatrix} \begin{bmatrix} 1 \\ 0 \\ 0 \end{bmatrix} = \begin{bmatrix} a_{11} \\ a_{21} \\ a_{31} \end{bmatrix}. \tag{3.2}$$

The columns of the transformation matrix indicate the coordinates of the base vectors in the new coordinate system. This makes it easy to formulate the orthonormality condition that the rotation matrix R must fulfill:

$$\boldsymbol{R}^T\boldsymbol{R} = \boldsymbol{I} \quad \text{or} \quad \sum_{p'=1}^{3} r_{p'p}r_{p'q} = \delta_{p-q}, \tag{3.3}$$

where $\boldsymbol{I}$ is the unit matrix whose elements are one on the main diagonal and zero otherwise. With Eq. (3.2), this equation simply states that the transformed basis vectors remain orthogonal:

$$\hat{e}_p'^T \hat{e}_q' = \delta_{p-q}. \tag{3.4}$$

In Equation (3.2), three out of nine matrix elements remain independent. Unfortunately, the relationship between the matrix elements and three parameters to describe the rotation turns out to be quite complex and nonlinear. One way to solve this problem is to use the three *Euler rotation angles* ϕ, θ, ψ. Unfortunately, Euler's angles are not uniformly defined in the literature. We follow the standard mathematical approach, use right-handed coordinate systems and count the counterclockwise rotation angles as positive. The rotation of the displaced world coordinate system into the camera coordinate system is broken down into three steps (see Fig. 3.2, [108].

1. Rotation around the z_s-axis by the angle ϕ, $\boldsymbol{x}_a = \boldsymbol{R}_\phi \boldsymbol{x}_s$:

$$\boldsymbol{R}_\phi = \begin{bmatrix} \cos\phi & \sin\phi & 0 \\ -\sin\phi & \cos\phi & 0 \\ 0 & 0 & 1 \end{bmatrix} \tag{3.5}$$

2. Rotation around the x_a-axis by the angle θ, $\boldsymbol{x}_b = \boldsymbol{R}_\theta \boldsymbol{x}_a$:

$$\boldsymbol{R}_\theta = \begin{bmatrix} 1 & 0 & 0 \\ 0 & \cos\theta & \sin\theta \\ 0 & -\sin\theta & \cos\theta \end{bmatrix} \tag{3.6}$$

3. Rotation around the z_b-axis by the angle ψ, $\boldsymbol{x}_c = \boldsymbol{R}_\psi \boldsymbol{x}_b$:

$$\boldsymbol{R}_\psi = \begin{bmatrix} \cos\psi & \sin\psi & 0 \\ -\sin\psi & \cos\psi & 0 \\ 0 & 0 & 1 \end{bmatrix} \tag{3.7}$$

Cascading the three rotations $\boldsymbol{R}_\psi, \boldsymbol{R}_\theta$ and $\boldsymbol{R}_\phi$ results in the matrix

$$\begin{bmatrix} \cos\psi\cos\phi - \cos\theta\sin\phi\sin\psi & \cos\psi\sin\phi + \cos\theta\cos\phi\sin\psi & \sin\theta\sin\psi \\ -\sin\psi\cos\phi - \cos\theta\sin\phi\cos\psi & -\sin\psi\sin\phi + \cos\theta\cos\phi\cos\psi & \sin\theta\cos\psi \\ \sin\theta\sin\phi & -\sin\theta\cos\phi & \cos\theta \end{bmatrix}.$$

The inverse transformation from camera coordinates to world coordinates is given by the transposition of this matrix. Since the matrix multiplication is not commutative, the rotation is also not commutative. It is therefore important to observe the order in which the rotations are performed.

For infinitesimally small rotation angles ε, however, the rotation is commutative. Then $\cos\varepsilon \approx 1$ and $\sin\varepsilon \approx \varepsilon$ apply. This limiting case has practical significance, as minor rotations occur frequently. A rotation around the z_s-axis can be described as follows:

$$\boldsymbol{x}_c = \boldsymbol{R}_\varepsilon \boldsymbol{x}_s = \begin{bmatrix} 1 & \varepsilon & 0 \\ -\varepsilon & 1 & 0 \\ 0 & 0 & 1 \end{bmatrix} \boldsymbol{x}_s \quad \text{or} \quad \begin{array}{rcl} x_c & = & x_s + \varepsilon y_s \\ y_c & = & y_s - \varepsilon x_s \\ z_c & = & z_s \end{array}.$$

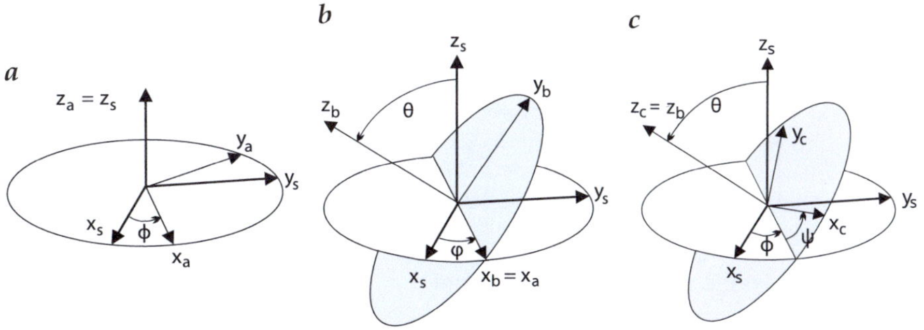

Figure 3.2: Rotation of world coordinates $\mathbf{x}_s$ into camera coordinates $\mathbf{x}_c$ using the three Euler angles ϕ, θ and ψ with successive rotations around the z_s, x_a and z_b axes.

Example 3.1: Accuracy considerations for small rotation angles

To do this, we consider the point $[x_s, 0, 0]^T$. According to the approximation formula, this is converted into the point $[x_s, -\varepsilon x_s, 0]^T$ is rotated. The correct position, however, is $[x_s \cos\varepsilon, -x_s \sin\varepsilon, 0]^T$. If we develop the trigonometric functions into a third-order Taylor series, this results in a position error of $[1/2\varepsilon^2 x_s, -1/6\varepsilon^3 x_s, 0]^T$. Assuming a 1024×1024 pixel image ($x_s < 512$ for a centered rotation) and a maximum acceptable error of 1/10 pixel, ε must be smaller than 0.02 or 1.14°. This is still a significant rotation, as this means a vertical offset of up to $\pm\varepsilon x_s = \pm 10$ pixels at the edge of the image lines.

3.3 Ideal Imaging: the Central Projection

3.3.1 The Pinhole Camera Model

The basic aspects of the imaging geometry of an optical system can be explained using the *pinhole camera* model. The imaging element of this camera is an infinitesimally small pinhole (Fig. 3.3). Only the light beam emanating from the point $[x, y, z]^T$ of the object that passes through the pinhole hits the image plane at the point $[x', y', z']^T$. This creates an image in the image plane. The relationship between the 3D world coordinates and the 2D *image coordinates* $[x', y']^T$ is given by:

$$x' = \frac{z'}{z} x, \quad y' = \frac{z'}{z} y. \tag{3.8}$$

The world coordinates parallel to the image plane are therefore scaled by the factor z'/z. The image coordinates $[x', y']^T$ therefore only contain ratios of the world coordinates, from which neither the distance of an object nor its size can be determined.

A straight line in three-dimensional space that does not run through the projection center is always projected onto a straight line in the image

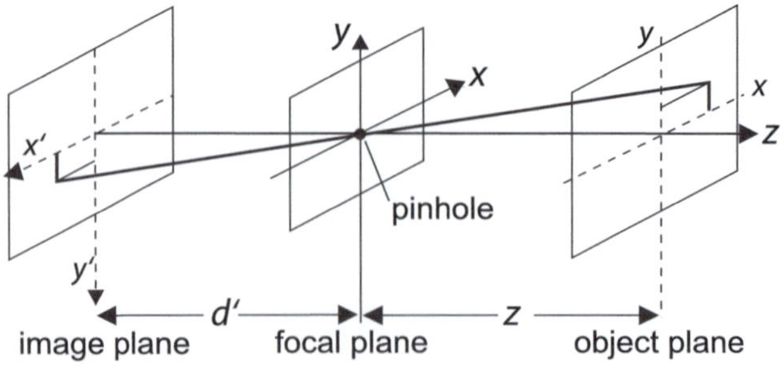

Figure 3.3: *Image generation with the pinhole camera.*

plane. This can be visualized geometrically without any calculations. All rays emanating from a straight line pass through the pinhole, the central point. They therefore lie on a plane that is defined by the straight line in space and the central point. This plane in turn intersects with the image plane in a straight line. This is why triangles are also mapped onto triangles. However, the angles between the sides of the triangles are generally not retained. Therefore, a square is not mapped onto a square but onto a general quadrilateral.

All object points that lie on a beam passing through the pinhole are mapped to a point in the image plane. In a scene with several transparent objects, the objects are therefore projected onto each other and the three-dimensional structure of the scene is lost. Even the shape of individual objects is no longer recognizable. This observation shows how much information is lost by projecting a 3D scene onto a 2D image plane.

In natural image scenes, mainly opaque objects occur. Here the 3D space is reduced to surfaces lying in space which, projected onto the image plane, can be described with two 2D functions $g(x', y')$ and $z(x', y')$ instead of a general 3D gray value function $g(x, y, z)$. The surface is completely projected onto the "image" plane as long as no more than one point of the surface lies on the same central ray. Otherwise, parts of the surface remain invisible. This effect is referred to as *occlusion*. We can show the occluded 3D space if we install a point light source at the position of the pinhole (Fig. 3.4). Then the invisible areas of the scene are in the shadow of the objects that are closer to the camera.

For scenes without occlusions, we only need the depth information $z(x', y')$ to fully reconstruct the 3D structure of a scene. One possibility — which we also find in our visual system — is stereo vision, i. e. viewing a scene with two sensors from different viewpoints (Section 8.5.1).

3.3.2 Perspective Projection

Image capture with a pinhole camera is essentially a *perspective projection*, as all rays must pass through a central point, the pinhole. This makes the

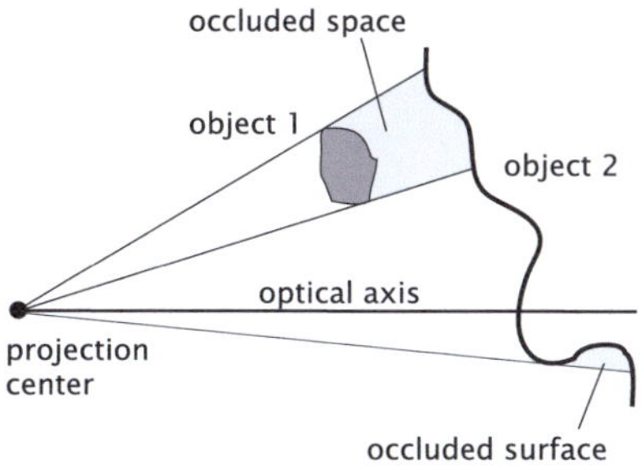

Figure 3.4: Occlusion of deeper objects and surfaces in the central projection.

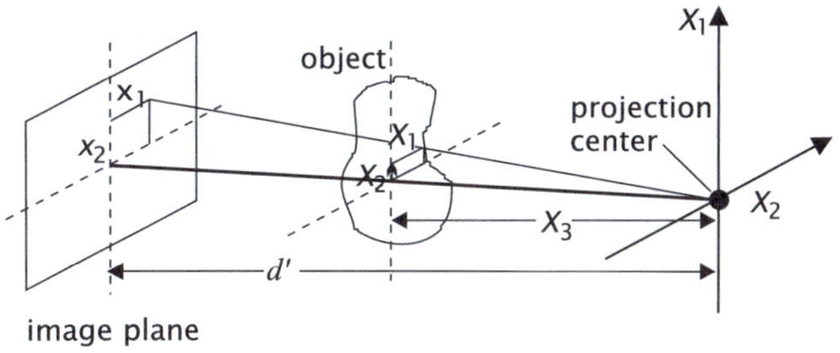

Figure 3.5: Perspective projection with X-rays.

pinhole camera model very similar to an image with penetrating rays such as X-rays emitted by a point source (Fig. 3.5). The only difference to the pinhole camera is that the object now lies between the central point and the image plane.

The mapping equation corresponds to Equation (3.8):

$$\boxed{\begin{bmatrix} x \\ y \\ z \end{bmatrix} \longmapsto \begin{bmatrix} x' \\ y' \end{bmatrix} = \frac{z'}{z}\begin{bmatrix} x \\ y \end{bmatrix}.} \tag{3.9}$$

The image coordinates scaled with the image width are called *generalized image coordinates*:

$$\underset{\sim}{x}' = \frac{x'}{z'}, \quad \underset{\sim}{y}' = \frac{y'}{z'}. \tag{3.10}$$

Generalized image coordinates are dimensionless and are underlined with a tilde symbol. In the system in which an object is observed, they correspond to the tangent of the angle to the optical axis. These coordinates explicitly take into account the limits of the projection onto the image plane. This is because we cannot deduce absolute positions from these coordinates. We only know the angle at which an object is projected onto the image plane. (This is also a method used in astronomy). This reduces the general mapping equation Eq. (3.9) of the perspective projection to

$$\boldsymbol{X} = \begin{bmatrix} x \\ y \\ z \end{bmatrix} \mapsto \underset{\sim}{x} = \begin{bmatrix} \frac{x}{z} \\ \frac{y}{z} \end{bmatrix}. \tag{3.11}$$

3.3.3 ♯Homogeneous Coordinates

Computer graphics uses the elegant formalism of *homogeneous coordinates* to uniformly describe all the transformations discussed so far, i. e. translation, rotation and perspective projection [76, 87, 218]. This means that the entire mapping process can be represented by a single 4×4 matrix.

Homogeneous coordinates are represented by a 4D column vector with an arbitrary factor $f > 0$

$$\boldsymbol{x} = [fx, fy, fz, f]^T \tag{3.12}$$

from which we obtain the original 3D coordinates by dividing the first three components by the fourth component. Any transformation is obtained by multiplying the homogeneous coordinates by a 4×4 matrix $\boldsymbol{M}$. In particular, we obtain the image coordinates

$$\boldsymbol{x}' = [fx', fy', fz', f]^T \tag{3.13}$$

through

$$\boldsymbol{x}' = \boldsymbol{M}\boldsymbol{x}. \tag{3.14}$$

Since matrix multiplication is associative, we can imagine the matrix $\boldsymbol{M}$ composed of many transformation matrices that perform such elementary transformations as *translation* and *rotation* around a coordinate axis, *perspective projection* and *scaling*. The transformation matrices of the elementary transformations result in

$$\boldsymbol{T} = \begin{bmatrix} 1 & 0 & 0 & t_x \\ 0 & 1 & 0 & t_y \\ 0 & 0 & 1 & t_z \\ 0 & 0 & 0 & 1 \end{bmatrix} \quad \text{translation by } \left[t_x, t_y, t_z\right]^T$$

$$\boldsymbol{R}_x = \begin{bmatrix} 1 & 0 & 0 & 0 \\ 0 & \cos\theta & -\sin\theta & 0 \\ 0 & \sin\theta & \cos\theta & 0 \\ 0 & 0 & 0 & 1 \end{bmatrix} \quad \text{rotation around the } x \text{ axis}$$

$$\boldsymbol{R}_y = \begin{bmatrix} \cos\phi & 0 & \sin\phi & 0 \\ 0 & 1 & 0 & 0 \\ -\sin\phi & 0 & \cos\phi & 0 \\ 0 & 0 & 0 & 1 \end{bmatrix} \quad \text{rotation around the } y \text{ axis}$$

$$
\begin{aligned}
\boldsymbol{R}_z &= \begin{bmatrix} \cos\psi & -\sin\psi & 0 & 0 \\ \sin\psi & \cos\psi & 0 & 0 \\ 0 & 0 & 1 & 0 \\ 0 & 0 & 0 & 1 \end{bmatrix} && \text{Rotation around the } z \text{ axis} \\
\boldsymbol{S} &= \begin{bmatrix} s_1 & 0 & 0 & 0 \\ 0 & s_2 & 0 & 0 \\ 0 & 0 & s_3 & 0 \\ 0 & 0 & 0 & 1 \end{bmatrix} && \text{scaling} \\
\boldsymbol{P} &= \begin{bmatrix} 1 & 0 & 0 & 0 \\ 0 & 1 & 0 & 0 \\ 0 & 0 & 1 & 0 \\ 0 & 0 & -1/z' & 1 \end{bmatrix} && \text{perspective projection.}
\end{aligned}
\tag{3.15}
$$

The perspective projection is slightly different than in Eq. (3.11). From the matrix multiplication of the homogeneous vector

$$
\boldsymbol{x} = [fx, fy, fz, f]^T
$$

with $\boldsymbol{P}$ we get

$$
\left[fx, fy, fz, f\frac{z'-z}{z'}\right]^T. \tag{3.16}
$$

The image coordinates result from the division by the fourth coordinate:

$$
\begin{bmatrix} x' \\ y' \end{bmatrix} = \begin{bmatrix} x\dfrac{z'}{z'-z} \\ y\dfrac{z'}{z'-z} \end{bmatrix}. \tag{3.17}
$$

From this equation we can see that the image plane is at the origin, because at $z = 0$ the image and world coordinates are identical. The projection center is located at the point $[0, 0, z']^T$.

Complete transformations from world coordinates to image coordinates can be composed of these elementary matrices. For example, Strat [317] proposed the following decomposition:

$$
\boldsymbol{M} = \boldsymbol{C}\boldsymbol{S}\boldsymbol{P}\boldsymbol{R}_z\boldsymbol{R}_y\boldsymbol{R}_x\boldsymbol{T}. \tag{3.18}
$$

The scaling $\boldsymbol{S}$ and the image shift $\boldsymbol{C}$ are transformations in the two-dimensional image plane. Strat [317] showed how the parameters for the transformation of camera coordinates into world coordinates can be determined completely in a non-iterative way from a set of points whose position in space is known exactly. In this way, an absolute calibration of the external camera parameters position and orientation and the inner camera parameters position of the optical axis, focal length and pixel size can be achieved.

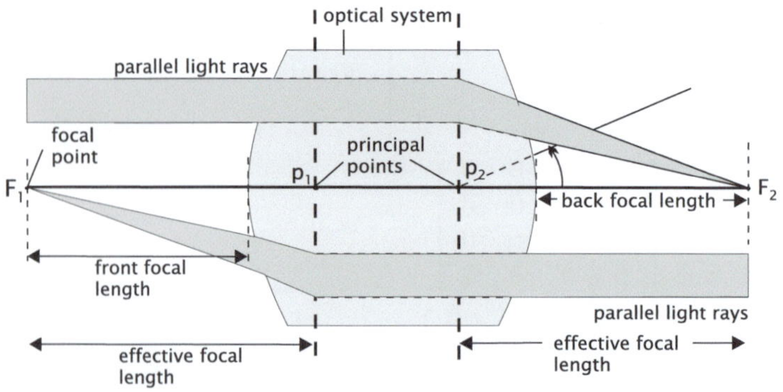

Figure 3.6: An optical system as a black box.

3.4 Real Imaging

3.4.1 Geometry of a Real Optical System

The pinhole camera model oversimplifies the imaging conditions of an optical system. It produces an image of an object at *any* distance, while a real optical system only produces a sharp image of an object at a certain distance range. Fortunately, even for complicated optical systems with a slight change, the geometry of the image can still be modeled by a perspective projection, as in Figs. 3.6 and 3.7 is illustrated. The pinhole is replaced by two *principle planes* with the following considerations.

For imaging through an optical system to work, a ray bundle with parallel light rays, which is also parallel to the optical axis, must converge at the end to a point in the image-side *focal point* on the optical axis. How the ray path runs in detail in optics with many lenses is complicated but unimportant. The decisive factor is that there must be a plane in which we imagine the parallel ray bundle to the convergent ray deflected by the focal point on the image side (Fig. 3.6). This plane is called the image-side principal plane. The same consideration with a parallel ray bundle from the image side leads to the object-side principle plane. The two principal planes intersect the *optical axis* in the *principal points* H' for the image-side principal plane and H for the object-side principal plane.

The *effective focal length* or simply *focal length* f' and f of an optical system result from the distance between the principal point and the corresponding focal point. If the same medium (usually air) is present on both sides of the optical system, the focal length on the image side and the focal length on the object side must be the same for reasons of symmetry. In the following, we therefore set $f' = f$. The object-side and image-side *focal point intersection distances* are the distances from the last lens surface to the respective focal points.

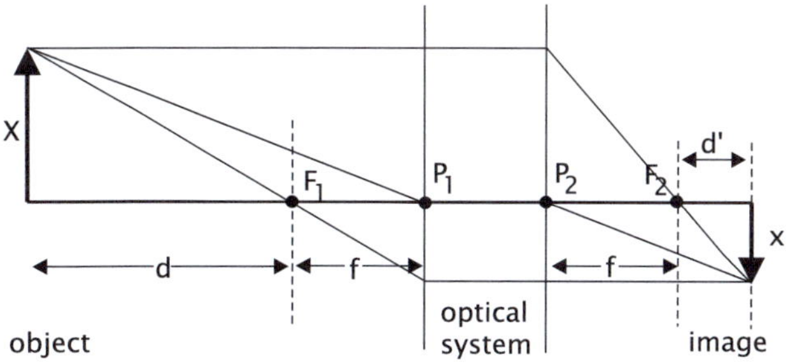

Figure 3.7: Modeling of the optical image by the principal points H and H′ and the focal points F and F′. The system generates an image at a distance d′ behind F′ from an object at a distance d from F.

3.4.2 Image Scale and Image Equation

The *lateral image scale* m_l of an optical system is the ratio of the image size x' to the object size x and results directly by applying the ray theorem to three similar triangular pairs in Fig. 3.7 to

$$\boxed{m_l = \frac{x'}{x} = \frac{f}{d} = \frac{d'}{f} = \frac{f+d'}{f+d}.} \tag{3.19}$$

The lateral image scale m_l is proportional to d': $d' = fm_l$ and inversely proportional to d: $d = f/m_l$.

Example 3.2: Imaging distances for a given magnification

For a given image scale, the distance to the object and the distance of the image plane from the focal plane (object focus d') can be determined very easily. If the object is at infinity ($m_l = 0$), $d = \infty$ and $d' = 0$, if it is reduced by a factor of 10 ($m_l = 1/10$), $d = 10f$ and $d' = f/10$ and for a 1:1 image ($m_l = 1$), $d' = d = f$.

The relationship between the positions of the object and the corresponding image is particularly simple if we specify the distances to the respective focal points (Fig. 3.7). The relationships in Eq. (3.19) immediately result in

$$\boxed{dd' = f^2.} \tag{3.20}$$

This relationship is called Newton's *imaging equation*. The better known Gaussian mapping equation uses the distances to the respective principal points (object distance $g = d + f$ and image distance $b = d' + f$)

$$\frac{1}{b} + \frac{1}{g} = \frac{1}{d'+f} + \frac{1}{d+f} = \frac{1}{f}. \tag{3.21}$$

This equation is of course equivalent to equation Eq. (3.19) and can be converted into this equation. For practical applications, however, Newton's imaging equation is much clearer and easier to handle.

A less well-known quantity is the *axial image scale,* which relates the positions of the image and object planes to each other. The axial image scale therefore reflects the image scale in the direction of the optical axis.It indicates how far we have to move the image plane when we move the object plane by a certain distance. In contrast to the lateral image scale, the axial image scale is not constant along the optical axis. It can therefore only be defined in the limiting case of small displacements. We therefore assume slightly shifted object and image positions $d + \Delta z$ and $d' - \Delta z'$ and use them in Eq. (3.20):

$$(d + \Delta z)(d' - \Delta z') = f^2 \quad \text{and} \quad dd' = f^2.$$

Neglecting the small quadratic term $\Delta z \Delta z'$ results in

$$\frac{\Delta z'}{\Delta z} \approx \frac{d'}{d} \tag{3.22}$$

and the axial image scale m_a becomes

$$\boxed{m_a \approx \frac{d'}{d} = \frac{f^2}{d^2} = \frac{d'^2}{f^2} = m_l^2.} \tag{3.23}$$

For practical purposes, it is important to know the total distance l between the object plane and the image plane. This results from Fig. 3.7 together with Eq. (3.19) to

$$l = \left(2 + \frac{1}{m_l} + m_l\right) f + \Delta h \quad \geq \quad 4f + \Delta h, \tag{3.24}$$

where Δh is the distance between the two main planes. This can also be negative. Good optics manufacturers specify the position of the principal planes in their data sheets. The total distance l is minimal for a symmetrical 1:1 image with a lateral image scale $m_l = 1$.

3.4.3 Scheimpflug Principle

An extended imaging equation can be derived for tilted object and image planes. To do this, we assume that the object plane is inclined by the angle θ in the x direction against the plane perpendicular to the optical axis. Through geometric considerations in Section 3.3.1 we have already shown that straight lines are mapped onto straight lines and thus also planes onto planes. The only thing left to do is to determine the angle of the image plane. At the same time, we have also found an algebraic proof.

A point on the inclined object plane has the distance $d + x \tan\theta$ at the height x and is therefore mapped to the distance $d' = f^2/(d + x\tan\theta)$ at the height x'. It is therefore mapped closer by the distance:

$$\Delta d' = \frac{f^2}{d + x\tan\theta} - \frac{f^2}{d} = -f^2 \frac{x\tan\theta}{d(d + x\tan\theta)} \tag{3.25}$$

Figure 3.8: *Tilt-shift lens for the realization of a Scheimpflug setup.*

From the lateral image scale at this point,

$$m_l = \frac{x'}{x} = \frac{f}{d + x\tan\theta} \tag{3.26}$$

x' can be calculated as

$$x' = f\frac{x}{d + x\tan\theta}. \tag{3.27}$$

From Eqs. (3.25) and (3.27) the inclination of the image plane results in

$$\boxed{\tan\theta' = \frac{\Delta d'}{x'} = -m_l\tan\theta. \qquad \text{Scheimpflug condition}} \tag{3.28}$$

This means that the inclination of the image plane in relation to the inclination of the object plane only depends on the lateral image scale m_l on the optical axis. The minus sign indicates that the image plane is inclined in the opposite direction to the object plane. In a 1:1 image, both planes are inclined in the same direction; in a reduced image, the image plane is inclined less than the object plane by a factor of m_l.

Example 3.3: Illustration of the Scheimpflug condition

The Scheimpflug condition Eq. (3.28) has a clear geometric interpretation. The object and image planes intersect the corresponding principal plane at a distance p or p', for which the following applies

$$\tan\theta = \frac{f+d}{p} \quad \text{and} \quad \tan\theta' = -\frac{f+d'}{p'}.$$

This means that $p = p'$:

$$\frac{p'}{p} = -\frac{\tan\theta}{\tan\theta'}\frac{f+d'}{f+d} = 1.$$

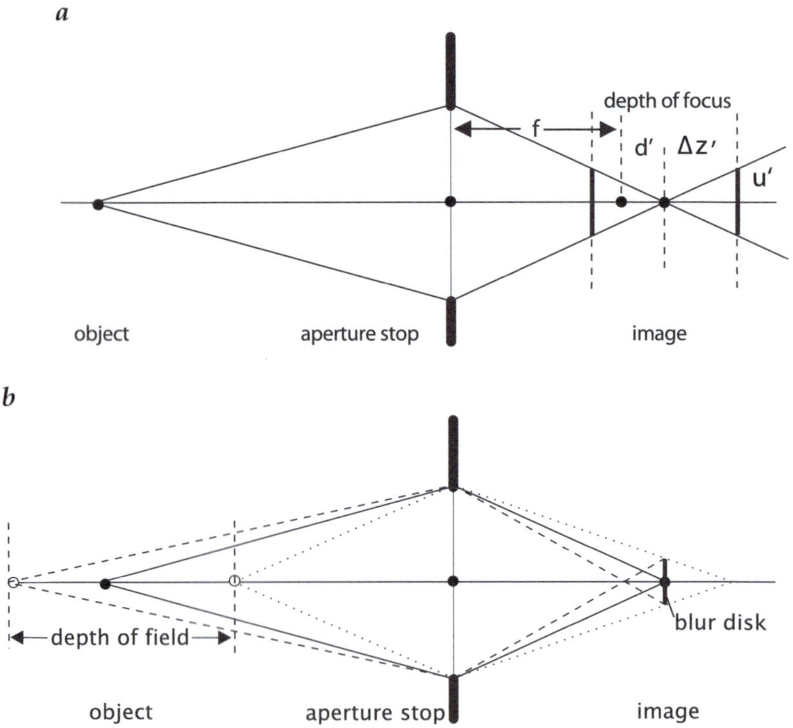

Figure 3.9: *Illustration of the* **a** *depth of focus and* **b** *depth of field with an object lying on the optical axis.*

A Scheimpflug set-up is always useful if you want to look at a plane at an angle and keep to the imaging equation. This is particularly interesting if you want to record a plane from several directions, as with a stereo arrangement (Section 8.5.1). With a tilt-shift lens, a Scheimpflug setup can be realized without additional components (Fig. 3.8).

3.4.4 ♯Depth of Focus and Depth of Field

The imaging equations Eqs. (3.20) and (3.21) relate the positions of object and image to each other. If the image or object planes are shifted slightly, an image is still generated, but it becomes blurred. The degree of blurring depends on the deviations from the positions given by the mapping equations.

The concept of *depth of focus* is based on the fact that a certain amount of blurring does not affect the image quality. For digital images, the size of the sensor elements is a natural measure of acceptable blur, as smaller structures cannot be resolved anyway.

We calculate the blur in the context of geometric optics for a point on the optical axis (Fig. 3.9a). On the image plane, the point is mapped to a point in object space. As the distance increases, it is blurred into an increasingly larger disk with the diameter u'. Using the f-number or *aperture number* k of an optical system as the ratio of the *focal length* f to the aperture a of a lens

$$\boxed{k = \frac{f}{a} \qquad \text{aperture number}} \tag{3.29}$$

we can express the diameter of the *blur circle* using the ray theorem as

$$\frac{u'}{\Delta z'} = \frac{a}{f + d'} \quad \leadsto \quad u' = \frac{1}{k}\frac{f}{f + d'}\Delta z', \tag{3.30}$$

where $\Delta z'$ is the distance from the focused image plane. The range of positions of the image plane $[d' - \Delta z', d' + \Delta z']$ for which the diameter of the blur circle is smaller than u' is called *image-side depth of field*. Equation (3.30) can be resolved according to $\Delta z'$

$$\boxed{\Delta z' = \pm k\left(1 + \frac{d'}{f}\right)u' = k(1 + m_l)u' \qquad \text{depth of focus}} \tag{3.31}$$

where m_l is the *lateral image scale* defined in Eq. (3.19). Equation (3.31) shows the important role of the *f-number* and the magnification for the depth of field. These two parameters determine the depth of focus for a given u'.

Example 3.4: Accuracy requirement for image sensor mounting

The depth of focus determines how precisely an image sensor must be mounted perpendicular to the optical axis. We assume that lenses with a large aperture $k = 1.0$ are used. From Equation (3.31) we can also deduce that the depth of focus is smallest when the lens is set to infinity ($m_l = 0$). The diameter of the blur circle should be at most the size $\Delta x'$ of a pixel. These values directly result in a depth of focus $\Delta z' = \pm\Delta x'$. With pixel sizes between 2-10 μm, this places high demands on the flatness and mounting accuracy of image sensors.

The *depth of field* is of greater importance for practical use in image capture. It is the range of object distances for which the radius of the blur circle on the image plane is smaller than ϵ (Fig. 3.9b). With the help of *axial image scale* in Eqs. (3.22) and (3.23) we obtain from Eq. (3.31)

$$\boxed{\Delta z \approx \pm k \cdot \frac{1 + m_l}{m_l^2}u'. \qquad \text{depth of field}} \tag{3.32}$$

Example 3.5: Object-side depth of field with small image scale

A typical image sensor has sensor elements of the size $6 \times 6\,\mu$m. We therefore also set the diameter of the blur circle to 6 μm in order to avoid any significant additional blurring. For an image taken with an aperture $k = 8$, a focal length $f = 15\,\mathrm{mm}$ and an image scale $m_l = 0.01$ (object distance $d = 1.5\,\mathrm{m}$), Eq. (3.32) results in an object-side depth of field $\Delta z = \pm\, 0.24\,\mathrm{m}$. If we want to achieve a higher resolution under the same shooting conditions (aperture and image scale), we must use an image sensor with the same area but more and smaller pixels. Under this boundary condition, the depth of field decreases with the pixel size. Double the resolution here means halving the depth of field. In general, this example shows that a high depth of field is possible with small image scales.

If a lens is set to infinity, the diameter of the object-side blur circle corresponds to the diameter of the aperture, as the rays run parallel on the object side. Objects are nevertheless imaged without additional blurring as long as the blur circle mapped onto the image plane has a diameter of u': $u' = m_l a$. With $m_l = f/d_{\min}$, the result for the minimum distance is

$$d_{\min} = \frac{fa}{u'} = \frac{f^2}{ku'}. \tag{3.33}$$

At high imaging scales, as in *microscopy*, the depth of field is very shallow. Modern microscopy objectives are set up in such a way that they focus an object to infinity. This means that the image-side and object-side depth of field swap roles. With these lenses, it is common to specify the *numerical aperture NA* instead of the f-number. This is defined by

$$\boxed{NA = n \sin \theta_0 \approx \frac{2n}{k} \qquad \text{numerical aperture.}} \tag{3.34}$$

where n is the object-side refractive index and θ_0 is the maximum angle at which a ray of light passing through the center of the image plane passes through the aperture. In addition, it makes sense to set the permissible blur circle as large as the lateral resolution: $u \approx \Delta x$. This results in the depth of field for the microscopic image

$$\Delta z = \pm k \Delta x \approx \pm \frac{2\Delta x}{NA}. \tag{3.35}$$

The depth of field in microscopic imaging is therefore the product of the lateral resolution Δx and the *aperture number k*. In microscopic imaging, therefore, only a thin plane $2\Delta z$ is in focus, which is not much thicker than the lateral resolution Δx. In general, however, it should be noted that the concept of depth of field, as we have discussed here, only applies in the limiting case of geometric optics. It can therefore only be used as long as the radius of the blur circle is significantly larger than the blur caused by diffraction or aberrations of the optical system (Section 3.5.3). Therefore, Eq. (3.35) is only an approximation.

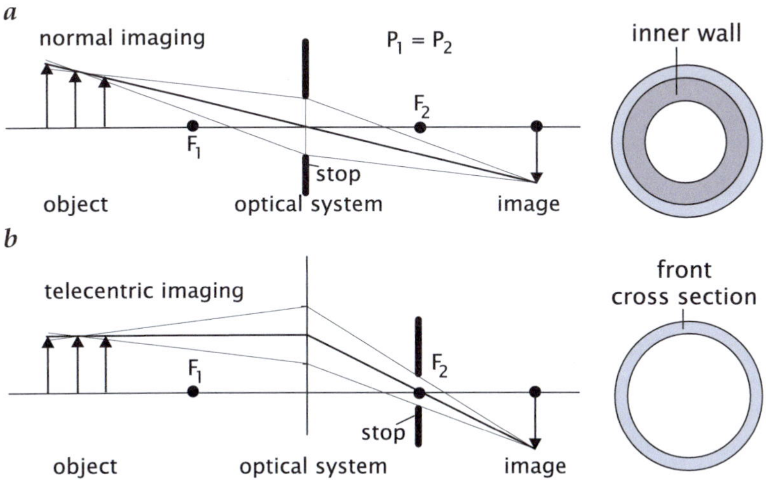

Figure 3.10: ***a*** *Diverging image with an aperture in the principle plane;* ***b*** *Telecentric image with an aperture in the focal point on the image side. The figure on the right illustrates how a short cylindrical tube, whose axis coincides with the optical axis, is imaged by the respective system.*

3.4.5 ♯Telecentric and Entocentric Imaging

A cone-shaped ray bundle is used for imaging from an object point, which is limited by the aperture (Fig. 3.10a). The *chief ray* is the symmetry axis of the cone and passes centrally through the aperture. With a normal or *entocentric lens*, the chief ray runs away from the optical axis if the object point is not on the optical axis. This has negative consequences for optical measurements. The object appears larger when it is less distant from the lens and smaller when it is further away. Since the distance of the object cannot be deduced from its image, either its distance must be known exactly or measurement errors are unavoidable.

Repositioning the *aperture* from the image-side principal point to the image-side *focal point* solves this problem and changes the imaging system to a telecentric one (Fig. 3.10b). By shifting the aperture to the focal point, the main beam in the object space becomes parallel to the optical axis. This means that small changes in the object position do not cause a change in the size of the image. Of course, the further away the object is from the focused object plane, the more blurred the image will be. However, the center of the blur circle does not change its position.

Telecentric lenses have become very important in image processing. Their disadvantage is, of course, that the diameter of the telecentric lens must be at least as large as that of the object to be measured. This makes the precise optical measurement of large parts very expensive.

Fig. 3.10 also shows how a cylinder whose axis coincides with the optical axis of the lens is imaged with an ordinary and a telecentric lens. The ordinary lens sees the end face of the cylinder and the inner wall, the telecentric lens only the end face.

Telecentric lenses illustrate the importance of apertures in the construction of optical systems, a fact that is often not given sufficient consideration.

3.4.6 ♯Optical Aberration

Real optical systems are not perfect, even in the limiting case of *geometrical optics*, i. e. if diffraction-limited effects due to the wave nature of the radiation are neglected. A real optical system does not map a point in object space to a point in image space, but rather to a more or less broadened intensity distribution. This is characterized by the type and width of the distribution and the deviation of the centre of gravity from the image point specified by the projective mapping. The latter does not lead to blurring in the image but to a geometric distortion.

All these *aberrations* have a common cause in the nature of the law of reflection and refraction (Section 2.6.3) and the fact that lenses and mirror surfaces are usually spherical surfaces. For larger inclinations θ and object heights h of the rays, we can expect aberrations. This is because the deflection in refraction depends on $\sin\theta$ Eq. (2.80) and in reflection from mirrors on the angle θ itself, while the distance to the optical axis is proportional to $\tan\theta$.

Spherical lenses and mirrors only provide a good image if the rays have a small angle θ and a small distance to the optical axis, i. e.

$$\tan\theta \approx \sin\theta \approx \theta. \tag{3.36}$$

At higher angles θ, aberrations occur in proportion to θ^3. It is therefore to be expected that the aberrations do not increase linearly relative to the angle but quadratically.

With lenses, there are also so-called *chromatic aberrations*, which are caused by the fact that the *refractive index* — and thus the *focal length* of lenses — is wavelength-dependent. (Section 2.6.3). Five monochromatic aberrations of the third order and two types of chromatic aberrations can be distinguished. They are briefly described below.

Spherical aberration
This aberration already occurs with rays parallel to the optical axis. Each radial zone of a lens has a slightly different focal length. In a simple spherical plano-convex lens, the focal length decreases with the distance from the optical axis. The lack of a common focal point results in a cylindrically symmetrical brightness distribution in the *focal point*. Due to the larger area of the off-axis rays, the distribution with the smallest width is found at a slightly shorter focal length than would be expected from the near-axis rays.

Axial chromatic aberration

This already occurs with parallel beam bundles. Due to the dependence of the refractive index on the wavelength, the focal length of a lens also depends on the wavelength of the radiation. With a simple lens, the smaller the wavelength of the radiation, the shorter the focal length. This means that only *one* color can be focused. Chromatic aberrations can be reduced by using several lenses made of different types of glass. The simplest case is an *achromat* made of two lenses. This allows the focal length for two wavelengths to be brought to coincidence, so that a secondary chromatic aberration still remains. With an *apochromat*, three wavelengths are brought into alignment. Often the chromatic correction is also extended beyond the visible spectral range into either the near UV or IR.

Lateral chromatic aberration

The dependence of the focal length on the wavelength also means that the magnification depends on the wavelength. In addition, obliquely incident rays are refracted to different degrees. These effects become increasingly noticeable towards the edge of the image. As a result, there are increasingly strong shifts between the individual colors at the edge of the image. This leads to color fringing at the edges.

Astigmatism

Astigmatism is an aberration that only occurs when the beam is incident at an angle. For reasons of symmetry, a beam is only oblique in one direction and parallel to the optical axis in the perpendicular direction. A flat beam in one of these two directions is therefore refracted to different degrees. This means that tangential and radial beams have different focal points.

The easiest way to explain this is to use a spoked wheel whose axis is mapped to the center of the image. The spokes (radial structure) and the outer edge of the wheel (tangential structure) are focused at different distances.

Coma

Coma is an aberration that occurs with wide, obliquely incident beams. It is therefore an aberration that occurs to a greater extent with lenses with large apertures (f-numbers smaller than 2). With coma, due to the lack of symmetry at oblique incidence, each radial zone is mapped onto a blur circle whose width increases with increasing distance from the optical axis and whose center of gravity shifts simultaneously with the height of incidence.

This leads to an asymmetrical distribution with a bright center and a triangular, tail-like and very wide smearing in the radial direction on the image plane. The coma is particularly disturbing when imaging point-shaped objects and prevents a faint object from becoming visible in the immediate vicinity of a bright one, as it is superimposed by the coma of the bright one.

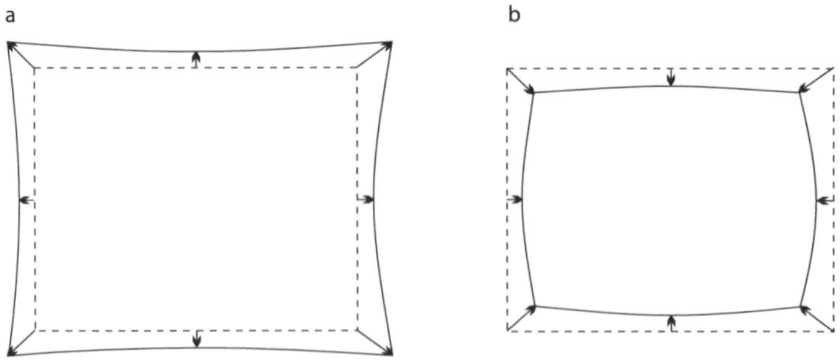

Figure 3.11: Radial symmetric distortion is illustrated by the distortion of a rectangle. Its sides appear curved, as the distortion typically increases with the third power of the distance from the center of the image. ***a*** *Positive or pincushion distortion;* ***b*** *negative or barrel distortion.*

Image field curvature

Optical systems tend to focus an image not on a plane but on a curved surface. To a first approximation, this surface is a spherical surface and the curvature (reciprocal radius) is referred to as field curvature. This means that even if all other aberrations are corrected, an image on a planar image plane becomes increasingly blurred towards the edge.

Geometric distortion

A real lens system deviates from the perfect central projection. The barrel or pincushion distortions of squares (Fig. 3.11) are very obvious. These are particularly pronounced with wide-angle and zoom lenses, but even with well-corrected lens systems, these distortions often still exhibit shifts of several pixels at the edge of the image. Straight lines are then distorted to the same extent.

The fact that the distortion is primarily radial results from the consideration of symmetry. Since lens systems are cylindrically symmetrical, a circle concentric with the optical axis is imaged as a circle, but with a distorted radius (*radial distortion*). This distortion can be approximated by the following equation:

$$x' = \frac{x}{1 + k_3|\boldsymbol{x}|^2}. \tag{3.37}$$

Depending on whether the distortion k_3 is positive or negative, barrel or pincushion distortions are observed in images of squares. With a commercial lens, the deviation in the edge area of the sensor is a few pixels.

If the distortions are corrected with Eq. (3.37), the maximum image error according to Lenz [196] can be reduced to well below the size of one pixel. This strong correction and the stable geometry of modern image sensors enable the use of cameras for sub-pixel accurate measurement tasks without the use of expensive measuring lenses. Lenz and Tsai [197] describe further details that influence the accuracy of geometric measurements with image sensors. If, as in *photogrammetry*, higher accuracies of geometric distortion are required, a simple radially symmetric correction is not sufficient, since centering errors in the lens elements cause asymmetric distortions. For a detailed analysis of the correction of geometric distortion, please refer to standard textbooks on photogrammetry such as Luhmann [206].

Distortions also occur when non-planar surfaces are projected onto an image plane, as is the case with satellite images and aerial photographs. The correction of geometric distortions in images is therefore an important factor in remote sensing and photogrammetry [263]. An exact correction of geometric distortions requires a subpixel-precise shifting of pixels. We will deal with this problem later in Section 5.5 after we have worked out the basics.

3.5 Linear System Theory of Imaging

In this section, we will describe optical mapping from the point of view of linear system theory. This can describe in a very general way the fact that a point in object space is not mapped to a point in image space but to a distribution. This allows us to better understand and quantitatively describe the influence of the optical aberrations discussed in the previous section on the resolution of the optical image.

In Section 3.5.1 we will introduce the concept of the point response and in Section 3.5.2 the optical transfer function. We have thus developed the theoretical concept to be able to investigate the limitation of the resolution due to the wave nature of the radiation in Section 3.5.3.

Linear system theory is of central importance in image acquisition and processing. The application to optical imaging is just one example. We will discuss another one in Section 10.3 in the description of linear shift invariant filters (convolution operators).

3.5.1 Point Spread Function (PSF)

In the discussion of depth of focus and depth of field (Section 3.4.4) and aberrations (Section 3.4.6) we have seen that a point in object space is not mapped onto a point in image space, but onto a more or less extended surface of varying intensities. Obviously, the function that describes the imaging of a point is an essential property of a mapping system. This function is called *point response* or *point spread function* and is abbreviated as *PSF*. We assume that the PSF is not dependent on the position. Under this condition, an optical system can be considered as a linear shift invariant system (*LSI*, Section 10.3).

If we know the PSF, we can calculate how any object is imaged. To do this, we divide the object into individual points. Fig. 3.12 shows this process. Without limiting the generality, we assume that it is a 1:1 mapping, i. e. object and image are at the same positions. A point x in the object plane is

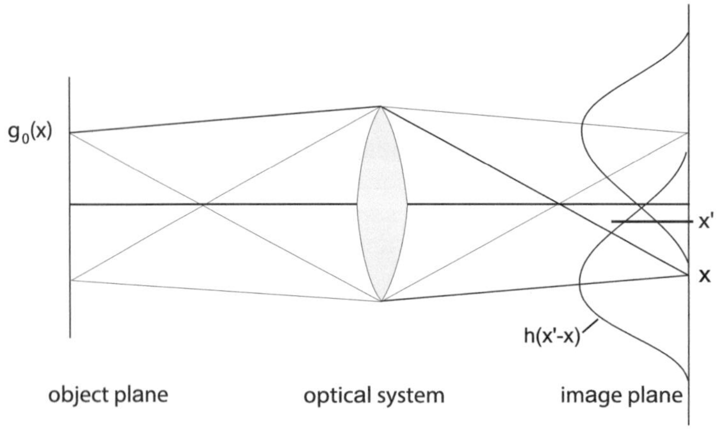

Figure 3.12: *Illustration by convolution with the point response $h(x')$. A point at x in the object plane results in an intensity distribution with a maximum at the position of the corresponding point x' in the image plane. At a point x' in the image plane, the contributions of all points x, i. e. $g_o(x)h(x'-x)$, must be integrated.*

projected onto the image plane with an intensity distribution corresponding to the point response h. We use $g_o(\boldsymbol{x})$ to denote the intensity values from the object plane that are projected onto the image plane without distortion. We can then calculate the intensity of a point $\boldsymbol{x}'$ on the image plane by multiplying the proportions of the point responses integrieren (Fig. 3.12):

$$g_i(\boldsymbol{x}') = \int_{-\infty}^{\infty}\int_{-\infty}^{\infty} g_o(\boldsymbol{x})h(\boldsymbol{x}' - \boldsymbol{x})\mathrm{d}^2x = (g_o * h)(\boldsymbol{x}'). \tag{3.38}$$

The operation in Eq. (3.38) is called *convolution*. Convolutions play an important role in image processing, not only in image generation, but also in many image processing operations. In the case of image generation, a convolution obviously "smears" an image and reduces its resolution.

This effect of convolutions can be most easily demonstrated on image structures with periodic gray value variations. As long as the repetition length, the *wavelength*, of this structure is greater than the width of the PSF, no significant changes occur. However, if the wavelength decreases, the amplitude of the gray values also begins to decrease. Fine structures eventually fade until they are no longer visible. These considerations emphasize the importance of periodic structures and lead naturally to the introduction of the *Fourier transform*, which decomposes an image into its periodic variations (Section 6.3).

3.5.2 Optical Transfer Function and Modulation Transfer Function

Convolution with the PSF in space domain is a fairly complex operation. In Fourier space, however, it is performed as a multiplication of complex numbers. Thus, the convolution of the object $g_o(\boldsymbol{X})$ with the PSF $h(\boldsymbol{X})$ in Fourier space corresponds to a multiplication of the Fourier-transformed object $\tilde{g}_o(\boldsymbol{\kappa})$ with the Fourier-transformed PSF, the *optical transfer function* or *OTF* $\tilde{h}(\boldsymbol{\kappa})$. In this section, we consider the optical transfer function in object space, i. e. we project the imaged object back into object space. Then the image generation can be described as follows:

	Image		Imaging		Object
Space domain	$g_i(\boldsymbol{x})$	$=$	$h(\boldsymbol{x})$	$*$	$g_o(\boldsymbol{x})$
Fourier domain	$\tilde{g}_i(\boldsymbol{\kappa})$	$=$	$\tilde{h}(\boldsymbol{\kappa})$	$\cdot$	$\tilde{g}_o(\boldsymbol{\kappa})$.

(3.39)

Equation (3.39) means that we can describe the optical image using either the point response or the optical transfer function. Both descriptions are complete, and like the PSF, the OTF can also be explained clearly. Since the Fourier transform decomposes an object into repeating structures, the OTF tells us how these periodic structures are modified by the optical image. An OTF of 1 for a particular wavelength means that this periodic structure is not affected at all. In contrast, an OTF of 0 causes the structure to disappear completely. Values between 0 and 1 attenuate accordingly. Since the OTF is complex, not only the amplitude of a periodic structure is generally changed, but also its phase.

The *modulation transfer function* (*MTF*) results from practical measurements of the resolution of optical systems. A sinusoidal gray value pattern with amplitude a and wavenumber κ is imaged and the amplitude a' is measured in the image. The ratio a'/a is then the MTF for this wavenumber. The MTF is therefore the amount of the OTF, as the MTF only indicates the amplitude ratio and not any phase shifts:

$$MTF = |OTF| = \left|\tilde{h}(\boldsymbol{\kappa})\right|. \tag{3.40}$$

3.5.3 ♯Diffraction-limited Optical Systems and Fourier Optics

When a plane wave, emitted by a distant object, enters an optical system, it cannot be focused on a point, even if all the aberrations of the optics have been corrected. The limitation of the wave at the aperture leads to a smearing of the point on the image plane into a diffraction disk that is at least of the same order of magnitude as the wavelength of the light used. An optical system whose aberrations have been corrected to such an extent that they are smaller than the effects of diffraction is referred to as *diffraction-limited*.

An exact treatment of diffraction based on Maxwell's equations is mathematically very complex ([22], [70, Chapter 9 and 10] and [144, Chapter 3]). However, the essence of diffraction of a plane wave at a lens aperture can be

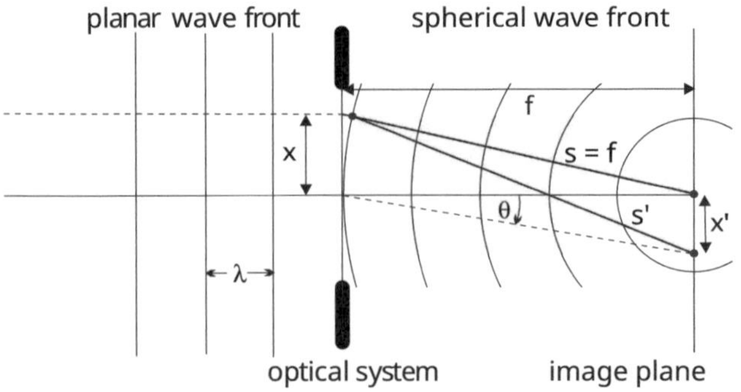

Figure 3.13: Diffraction of a plane wave at the aperture of an optical system. In the aperture, the plane wave can be broken down into spherical waves, which have a path length difference depending on the direction θ and position x'.

captured by an approximation known as *Fraunhofer diffraction*. From this we will derive a fundamental relationship.

We assume that a plane wave from an infinitely distant object impinges on the aperture in the principal plane (Fig. 3.13) and analyze the *PSF* of diffraction-limited optics in this way. A perfect optical system now causes the planar plane wave to bend into a spherical wave that has its center in the focal plane on the optical axis (Fig. 3.13). This occurs because the lens thickness decreases on average with greater distance from the optical axis and the wave therefore passes through less glass towards the edge, where it is slower.

We can understand the effect of the limitation of the wave front by the aperture by applying the *Huygen's principle*. It states that we can take any point on the wavefront as the starting point of an in-phase spherical wave. We then add up all these spherical waves to form a new wavefront in the focal plane. The path lengths from a point x in the aperture plane to the focal point on the image plane all have the same path and therefore the same phase due to the spherical waves with center in the focal point. They therefore overlap constructively to a maximum amplitude. If we move slightly away from the optical axis in the image plane to the point x', the paths are no longer the same and phase shifts occur (Fig. 3.13). As long as these are significantly smaller than one wavelength, there is still a constructive superposition, just with a slightly smaller amplitude. If the phase shift increases to exactly one wavelength at a greater distance from the optical axis, the complex amplitudes u of all waves are evenly distributed in all directions on the complex plane (cf. Fig. 2.5 in Section 2.2.1) and cancel each other out. This allows us to intuitively understand that the wider we open the aperture, the higher the phase shifts on the image plane and the sharper the diffraction slice.

After these introductory considerations, we can calculate the path s from the point $\boldsymbol{x}$ in the aperture plane to the focal point and the path s to the point $\boldsymbol{x}'$ on the optical axis, taking into account that x' is drawn in the negative direction in Fig. 3.13 and we must also take into account the y direction not shown in Figure 3.13:

$$s = f \quad \text{and} \quad s' = \sqrt{(x - x')^2 + (y - y')^2 + f^2 - x^2 - y^2}. \tag{3.41}$$

The difference in the two path lengths, assuming that $x' \ll f$, i. e. the neglect of quadratic expressions in x' and y' leads to

$$s' - s \approx -\frac{xx' + yy'}{f}. \tag{3.42}$$

For an electromagnetic wave with the wavelength λ_e, this path length difference causes a phase shift of

$$\Delta\varphi = \frac{2\pi(s' - s)}{f\lambda_e} = -\frac{2\pi(xx' + yy')}{f\lambda_e} = -\frac{2\pi(\boldsymbol{x}^T\boldsymbol{x}')}{f\lambda_e}. \tag{3.43}$$

Now assume that $\mathrm{u}(x)$ is the complex amplitude of the wavefront in the aperture plane. This is a more general approach than a constant wave $\mathrm{u}(x)$ limited by the aperture, i. e. a box function. We want to treat the problem more generally for an amplitude of the wave front that varies arbitrarily over the aperture of the optics. It should be noted that $\mathrm{u}(x)$ is a complex function (Section 2.2.1). Therefore, phase differences in the aperture plane can also be taken into account.

With this approach, the superposition of all spherical waves $\mathrm{u}(x)$ in the image plane, taking into account the phase shift, results in Eq. (3.43)

$$\mathrm{u}'(\boldsymbol{x}') = \int_{-\infty}^{\infty}\int_{-\infty}^{\infty} \mathrm{u}(\boldsymbol{x}) \exp\left(-2\pi\mathrm{i}\frac{\boldsymbol{x}'^T\boldsymbol{x}}{f\lambda_e}\right) \mathrm{d}^2x. \tag{3.44}$$

This equation states that the wave function $\mathrm{u}'(\boldsymbol{x}')$ on the image plane is given by the 2D Fourier transform (see Appendix B.1.3, equation Eq. (B.20)) of the incident wave function $u(x)$ on the aperture plane. This means nothing other than that a diffraction-limited optical system performs a 2D Fourier transformation.

The further calculations are simplified if we scale the coordinates $\boldsymbol{x}$ with the diameter of the aperture a and $\boldsymbol{x}'$ with the product of the wavelength of the radiation λ_e and the aperture number k in a natural way, i. e.

$$\boxed{\underset{\sim}{\boldsymbol{x}} = \frac{\boldsymbol{x}}{a}, \qquad \underset{\sim}{\boldsymbol{x}}' = \frac{\boldsymbol{x}'}{k\lambda_e}, \qquad \underset{\sim}{\boldsymbol{\kappa}}' = k\lambda_e\boldsymbol{\kappa}'.} \tag{3.45}$$

The wavenumber $\boldsymbol{\kappa}'$ in the image plane scales inversely to the spatial coordinate $\boldsymbol{x}'$ (*similarity theorem* of the Fourier transform, see Appendix B.1.4). The equation for the PSF on the image plane is then simplified to

$$\boxed{\mathrm{u}'(\underset{\sim}{x}') = \int_{-\infty}^{\infty}\int_{-\infty}^{\infty} \mathrm{u}(\underset{\sim}{x}) \exp\left(-2\pi \mathrm{i} \underset{\sim}{x}'^T \underset{\sim}{x}\right) \mathrm{d}^2\underset{\sim}{x}.} \tag{3.46}$$

Example 3.6: PSF and OTF of a lens with square aperture

A square aperture is unusual for an optical system, but it helps us to understand basic principles more easily. In general, the amplitude distribution for an *aperture* is constant, without a location-dependent phase shift and only cut off by the aperture. The amplitude distribution in the aperture plane is therefore the product of two one-dimensional box functions

$$\mathrm{u}(\underset{\sim}{x}', \underset{\sim}{y}') = \mathrm{u}_0 \,\Pi\,(\underset{\sim}{x}')\,\Pi\left(\underset{\sim}{y}'\right) \tag{3.47}$$

and is therefore a separable function. This property is retained in the Fourier transform (Appendix B.1.4), so that we can simply multiply the Fourier transforms of the one-dimensional box functions, the sinc function (> R13), with each other:

$$u'(\underset{\sim}{x}', \underset{\sim}{y}') = \mathrm{u}_0 \frac{\sin(\pi \underset{\sim}{x}')}{\pi \underset{\sim}{x}'} \frac{\sin(\pi \underset{\sim}{y}')}{\pi \underset{\sim}{y}'} = \mathrm{u}_0 \,\mathrm{sinc}(\underset{\sim}{x}')\,\mathrm{sinc}(\underset{\sim}{y}'). \tag{3.48}$$

This function decreases slowly with $1/\underset{\sim}{x}'$ and $1/\underset{\sim}{y}'$ and has zeros at all even values of $\underset{\sim}{x}'$ and $\underset{\sim}{y}'$, except at the zero point. This is easy to understand and always happens when a wave with n wavelengths fits exactly into the box horizontally or vertically.

With incoherent optical imaging, we do not measure the wave amplitude, but the energy, i. e. the magnitude squared of the wave amplitude. When normalized to a maximum of one, this is the point response of the image. This results in the PSF $h(\underset{\sim}{x}')$ nach Eq. (3.48) zu

$$h'(\underset{\sim}{x}') = \mathrm{sinc}^2(\underset{\sim}{x}')\,\mathrm{sinc}^2(\underset{\sim}{y}'). \tag{3.49}$$

The diffraction pattern has a central peak, but pronounced secondary peaks along the axes (Fig. 3.14a).

The OTF results from the 2D Fourier transform of the PSF. Since we already know the Fourier transform of the sinc function, the box function, it is easy to calculate it. If we multiply the sinc function by itself in space, this corresponds to a convolution of the box function with itself in Fourier space (convolution theorem in Appendix B.1.4). Therefore, the OTF is a triangular function

$$\tilde{h}'(\underset{\sim}{\kappa}') = \Lambda(\underset{\sim}{\kappa}'_x)\Lambda(\underset{\sim}{\kappa}'_y). \tag{3.50}$$

The OTF decreases monotonically with the wavenumber to zero (Fig. 3.14b).

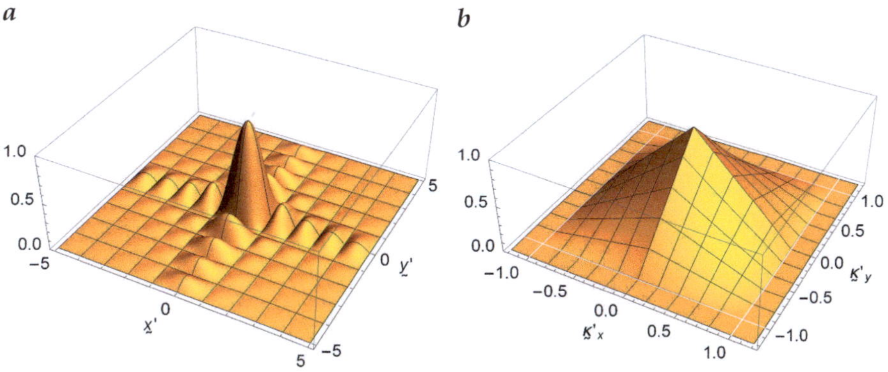

Figure 3.14: *Diffraction-limited imaging through a lens with a uniformly illuminated square aperture:* ***a*** *point response $h(\underset{\sim}{x}')$ in the focal plane according to Eq. (3.49). In order to still see the small amplitudes, $\sqrt{h}$ is shown.* ***b*** *OTF $\tilde{h}(\underset{\sim}{\kappa}')$ according to Eq. (3.50).*

Example 3.7: PSF and OTF of a lens with circular aperture

With the knowledge from the previous example, we can now also tackle the rotationally symmetrical *circular aperture*. The amplitude distribution is now given by

$$u(\underset{\sim}{x}') = u_0 \Pi(\underset{\sim}{x}') \quad \text{with} \quad \underset{\sim}{x}' = |\underset{\sim}{x}'|. \tag{3.51}$$

The Fourier transform of Eq. (3.51) is now a more complex function that often occurs in problems with cylinder geometry, a *Bessel function*, here in first order (> R13):

$$u'(\underset{\sim}{x}') = u_0' \frac{2J_1(\pi \underset{\sim}{x}')}{\pi \underset{\sim}{x}'}. \tag{3.52}$$

The PSF is the absolute value squared of this function

$$h'(\underset{\sim}{x}') = \frac{J_1^2(\pi \underset{\sim}{x}')}{\pi \underset{\sim}{x}'^2} \tag{3.53}$$

and has been scaled so that the integral over it is one. Like the sinc function, the Bessel function J_1 has regular zeros, but these are not at integer values of $\underset{\sim}{x}'$ and only have a distance of one for the higher zeros (Fig. 3.15a):

$$1{,}21967 \quad 2{,}23313 \quad 3{,}23832 \quad 4{,}24106 \quad \ldots \quad 49{,}2492 \quad 50{,}2492 \quad \ldots$$

The central peak of the PSF (Fig. 3.15a) contains 83.9% of the energy. The OTF is like the square aperture, the circular aperture is folded in on itself. As a direct result, the OTF decreases approximately linearly and must be zero for all $\underset{\sim}{\kappa}' > 1$ (Fig. 3.15b and c):

$$\tilde{h}(\underset{\sim}{\kappa}') = \frac{2}{\pi}\left[\arccos \underset{\sim}{\kappa}' - \underset{\sim}{\kappa}'(1 - \underset{\sim}{\kappa}'^2)^{1/2}\right]\Pi(\underset{\sim}{\kappa}'/2) \approx 1 - \frac{4}{\pi}\underset{\sim}{\kappa}' + \frac{2}{3\pi}\underset{\sim}{\kappa}'^3. \tag{3.54}$$

Regardless of the details, the OTF of an optical image is an ideal *low-pass filter* for all apertures, as these have a finite extension and the OTF is the convolution of the aperture function with itself. All wavenumbers greater

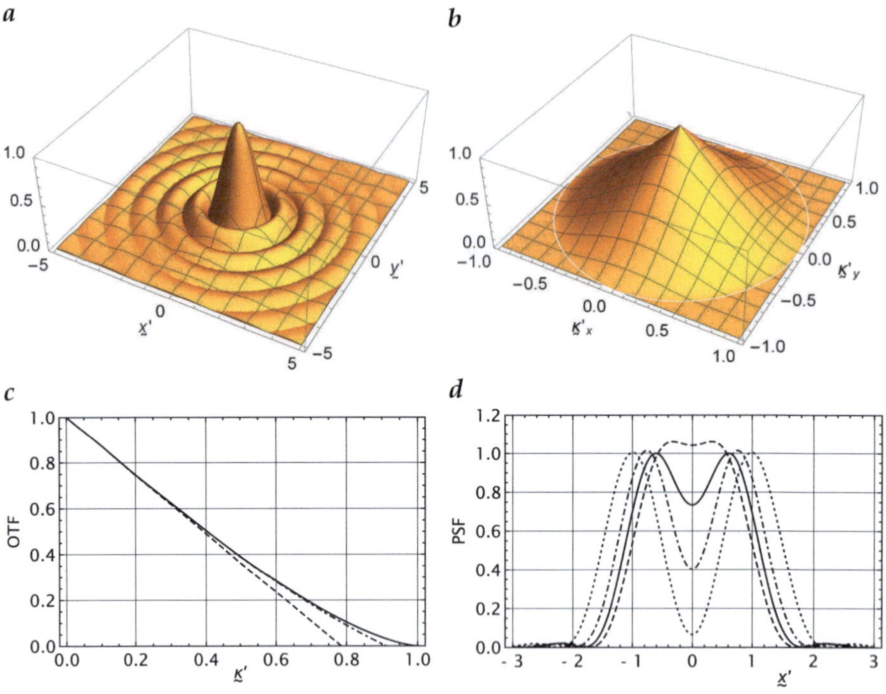

Figure 3.15: *Diffraction-limited imaging through a lens with a uniformly illuminated circular diaphragm:* ***a*** *point response $h'(\mathbf{x}')$ ("airy disk") in the image plane according to Eq. (3.53). To see even small amplitudes, $\sqrt{h}$ is shown.* ***b*** *OTF $\tilde{h}'(\boldsymbol{\kappa}')$ after Eq. (3.54);* ***c*** *radial section of the OTF $\tilde{h}'(\boldsymbol{\kappa}')$ with linear (dashed line) and cubic (dashed/dotted line) approximation according to Eq. (3.54);* ***d*** *Illustration of the image resolution of two points at distances x/λ_{min} 1 (dashed), 1.22 (solid), 1.5 (dashed/dotted), and 2 (dotted).*

than a limit value, and thus all structures with a wavelength smaller than a limit wavelength, are completely canceled out and thus remain invisible. This is a strict limitation of the resolution of optical images.

The great advantage of the rotationally symmetrical circular aperture over all other aperture shapes is that the resolution is the same in all directions. At the same time, the resolution limit is particularly easy to specify. For $\underset{\sim}{x}' > 1$ the OTF is zero. This results in a maximum wavenumber $\kappa'_{\max}$ and minimum wavelength $\lambda'_{\min}$, at which the OTF becomes zero, as follows

$$\boxed{\kappa' = \frac{1}{k\lambda_e} \quad \text{and} \quad \lambda'_{\min} = k\lambda_e \quad \text{resolution limit.}} \tag{3.55}$$

This means that the *f-number k* and the wavelength of the radiation used λ_e determine the resolution limit. The smaller the f-number and the smaller the wavelength of the radiation, the smaller the structures that can be imaged.

Extension 3.1: Coherent imaging of a Gaussian beam

So far, we have only considered the common case of incoherent imaging. In Section 2.2.5, however, we learned about coherent beams and now ask ourselves how such a beam is imaged by an optical system. These considerations are essential when coherent radiation sources are used for imaging, such as laser beams.

The simplest case is a Gaussian beam. We assume that this beam is expanded. It therefore has a small divergence according to Eq. (2.26) and enters a lens almost as a plane wavefront. We assume that the beam cross-section is not cut off by the aperture. Now, for the first time, there is a non-constant distribution of the wave amplitude in the aperture plane.

For the Gaussian beam, the amplitude distribution in the aperture plane corresponds to the Gaussian bell curve Eq. (2.24). In the image plane, the distribution of the wave amplitudes after Eq. (3.44) corresponds to the Fourier transform of the Gaussian bell curve, which is also a Gaussian bell function (> R14). This means that a Gaussian beam does not change its shape during optical imaging.

Of course, this basic property also applies to all other distributions of the wave amplitude of a coherent beam that can be realized with waves and that are invariant under the Fourier transform (> R14).

3.5.4 ♯Resolution Criteria

We conclude the section on linear system theory of optical imaging with a comparative presentation of some resolution criteria.

Minimum Resolvable Wavelength

As already discussed in the previous section, the smallest resolvable wavelength is $\lambda_{\text{min}} = k\lambda_e$, see Eq. (3.55).

Separation of Two Points (Rayleigh Criterion)

At a distance equal to the distance to the first dark ring, two points can be clearly separated from each other (Fig. 3.15d). This is the *Rayleigh criterion* for the resolution of an optical system:

$$\boxed{\Delta x'_{\text{R}} = 1.22k\lambda_e \quad \text{or} \quad \Delta x_{\text{R}} = 0.61\frac{\lambda_e}{NA} \quad \text{Rayleigh criterion.}} \tag{3.56}$$

Instead of the f-number, the *numerical aperture NA* defined in Eq. (3.34) was used in the second equation and applied to the object plane, i. e. for an image to infinity, as is the case in microscopy.

Since a medium other than air is often used between the object and the lens in microscopy, the use of NA is preferable to the f-number k for microscopy and all light-intensive diffraction-limited objects.

Angular Resolution

The resolution of an optical system can be interpreted as the relationship between the angular resolution of the incident plane wave and the spatial resolution on the image plane. With the minimum resolvable wavelength according to Eq. (3.55), the angular resolution is given by $\Delta\theta = \Delta x'/f$

$$\boxed{\Delta\theta = \frac{\lambda_e}{a} \quad \text{angular resolution.}} \tag{3.57}$$

This means that the angular resolution does not depend on the *focal length* f, but only on the aperture a of the optical system in relation to the wavelength λ_e of the radiation used.

Edge Steepness and Blurring

In terms of image processing, it is of interest to know how an intensity jump is mapped. To do this, we approximate the PSF by a Gaussian bell curve (normal distribution) with the standard deviation σ. According to Zhang et al. [362], a regression of the diffraction-limited image results in

$$\sigma = 0{,}42\,\lambda_{\min} = 0{,}42\,k\lambda_e. \tag{3.58}$$

We can easily calculate the edge steepness of a step edge smeared with a normal distribution. The course of the irradiance of a step edge with a difference of $\Delta E'$ results in

$$E'(x') = \frac{\Delta E'}{\sqrt{2\pi}\sigma}\int_{-\infty}^{x'} \exp\left(\frac{-x^2}{2\sigma^2}\right)\mathrm{d}x. \tag{3.59}$$

By derivation, the maximum edge steepness at the inflection point results in

$$\left.\frac{\mathrm{d}E'}{\mathrm{d}x'}\right|_{\max} = \frac{\Delta E'}{\sqrt{2\pi}\sigma}. \tag{3.60}$$

The edge steepness results in the edge width $\Delta x'$, which a straight line with the maximum edge steepness requires to pass through an intensity difference of ΔE

$$\Delta x' = \sqrt{2\pi}\sigma \tag{3.61}$$

and with Eq. (3.55)

$$\boxed{\Delta x'_{\text{edge}} = \sqrt{2\pi}\,0{,}42\,k\lambda_e \approx 1{,}05\,k\lambda_e \quad \text{Edge widening.}} \tag{3.62}$$

This means that the widening of a step edge is approximately as large as the minimum resolvable wavelength of a structure.

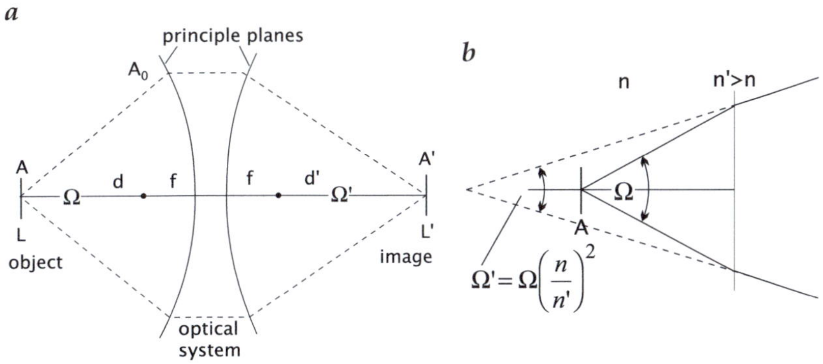

Figure 3.16: *Illustration of the radiation invariance:* ***a*** *The product AΩ is the same in object and image space.* ***b*** *Change in angle when a ray penetrates an optically denser medium.*

3.6 Radiometry of Imaging

It is not enough just to know the imaging geometry. The relationship between the irradiance in the image plane and the radiance of the imaged object and which parameters of an optical system influence this relationship must also be taken into account. The basics of radiometry and in particular the terms used to describe radiation properties can be found in Section 2.4. Here we first consider the invariance of radiance in optical imaging (Section 3.6.1) and then calculate the relationship between object radiance and irradiance on the image plane (Section 3.6.2).

3.6.1 Invariance of the Radiance

Of great practical importance is a fundamental, little-known theorem of radiometry that the radiance of an image is equal to that of an object, only attenuated by the transmissivity of the optical system. For this purpose, we consider the radiation flux Φ through an optical system, which must be maintained for reasons of energy conservation. The *solid angles* at which the optical system appears from the object and from the image (Fig. 3.16a) are

$$\Omega = A_0/(d+f)^2 \quad \text{and} \quad \Omega' = A_0/(d'+f)^2, \tag{3.63}$$

where A_0 is the area of the aperture. The radiant flux of the object emanating from an area A hits the area $A' = A(d'+f)^2/(d+f)$ according to the *image scale* (Section 3.4.2) hits the area $A' = A(d'+f)^2/(d+f)^2$ on the image plane. This results in the radiance

$$L = \frac{\Phi}{\Omega A} = \frac{\Phi}{A_0 A}(d+f)^2, \quad L' = \frac{t\Phi}{\Omega' A'} = \frac{t\Phi}{A_0 A'}(d'+f)^2 = \frac{t\Phi}{A_0 A}(d+f)^2 \tag{3.64}$$

and the following invariance applies:

$$L' = tL \quad \text{and} \quad \Omega A = \Omega' A'. \tag{3.65}$$

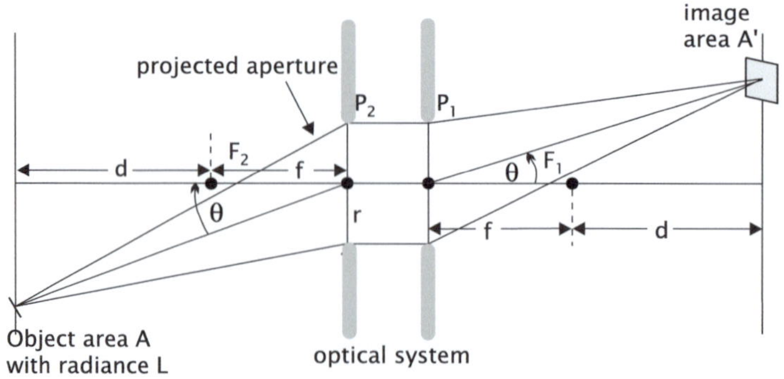

Figure 3.17: *An optical system receives a flux density corresponding to the product of the irradiance of the object and the solid angle at which the projected aperture appears from the object. The radiant flux emitted by the object surface A is mapped onto the image surface A′.*

This invariance only applies if the object and image are in a medium with the same refractive index. If a ray with the radiance L passes into a medium with the refractive index n on the object side into a medium with a higher refractive index n' on the image side, the radiance increases because the rays are refracted towards the optical axis (Fig. 3.16b) and therefore the invariance generally applies

$$\boxed{\frac{L'}{n'^2} = t\frac{L}{n^2} \quad \text{and} \quad n^2\Omega A = n'^2\Omega' A'.} \tag{3.66}$$

3.6.2 Object Radiance and Image Irradiance

From the radiation invariance we can calculate the irradiance in the image plane according to Eq. (3.64):

$$E'\frac{\Omega'}{A'} = tL\Omega' = t\frac{\pi}{4}\left(\frac{a}{f+d'}\right)^2 L = t\frac{\pi}{4}\frac{1}{k^2(1+m_l^2)}L. \tag{3.67}$$

We have used the fact that the area of a circular aperture with diameter a has an area $A_0 = \pi a^2/4$, as well as the definition of the f-number k Eq. (3.29) and the lateral magnification Eq. (3.19).

Now we have to consider rays incident obliquely to the optical axis. An obliquely incident ray sees the circular aperture as an ellipse with a smaller area by $\cos\theta$. Furthermore, the object is not $(d + f)$, but $(d + f)/\cos\theta$ away from the aperture. This leads to a solid angle that is $\cos^2\theta$ smaller. Finally, the radiation falls at an angle θ onto the image plane and thus onto a $1/\cos\theta$ larger area, which leads to an irradiance reduced by $\cos\theta$.

All three effects lead to an irradiance reduced by $\cos^4\theta$:

$$\boxed{E' = t\frac{\pi}{4}\frac{\cos^4\theta}{k^2(1+m_l)^2}L.} \tag{3.68}$$

As with the discussion of the depth of field (Section 3.4.4) and the diffraction-limited resolution (Section 3.5.3), this equation shows the central importance of the *number of apertures* k for the geometry and radiometry of optical systems.

The irradiance E' on the image plane also depends on the *lateral image scale*. As long as $m_l \ll 1$, i. e. the object is much larger than its image (distant image), this does not matter. But even with a 1:1 image (macro image with $m_l = 1$), the irradiance decreases by a factor of 4. With micro images (microscopy with $m_l \gg 1$), the irradiance decreases quadratically with m_l.

For real optical systems, the angle dependence with $\cos^4\theta$ according to Eq. (3.68) does not usually apply, but also depends on the aperture value. When the aperture is open, part of the obliquely incident rays are usually cut off in the lens (*vignetting*), so that the irradiance drops to half or less at the edge of the image. However, it is also possible to construct lenses that have a fall-off smaller than $\cos^4\theta$. One example is *telecentric lenses* (Section 3.4.5). As all the main rays on the lens side are parallel to the optical axis, the irradiance only decreases with $\cos\theta'$, where θ' is the angle at which the chief ray falls on the image plane. If the edge fall-off is important for the application or subsequent image evaluation, we should therefore measure the fall-off of the irradiance towards the edge of the image with a homogeneously illuminated object.

Example 3.8: How much radiation enters a lens?

We assume a flat radiating Lambertian object with the radiance L_{out} and ask ourselves how much of the total radiated power enters the lens. The radiant intensity M_{out} of a flat radiator is, cf. Eq. (2.56)

$$M_{\text{out}} = \int_0^{\pi/2}\int_0^{2\pi} L_{\text{out}}\cos\theta\sin\theta\,d\theta\,d\phi = 2\pi L_{\text{out}}\int_0^{\pi/2}\cos\theta\sin\theta\,d\theta = \pi\,L_{\text{out}}. \tag{3.69}$$

The optics only receive the radiation from the solid angle

$$\Omega_{\text{in}} = \frac{\pi a^2/4}{(d+f)^2} \tag{3.70}$$

through an aperture with a diameter a at a distance $d+f$ from the object. This means that only the following fraction η_{lens} of the radiant intensity is used to capture the image:

$$\eta_{\text{lens}} = \frac{L_{\text{out}}\Omega_{\text{in}}}{M_{\text{out}}} = \frac{1}{4k^2(m_l^{-1}+1)^2}. \tag{3.71}$$

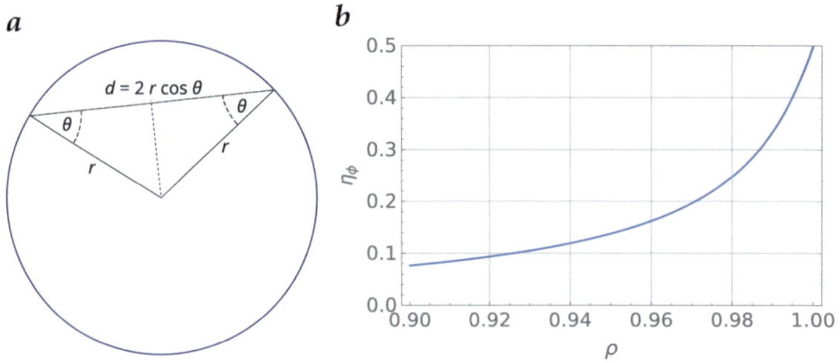

Figure 3.18: ***a*** *Reflection geometry in an integrating sphere;* ***b*** *radiation losses in an integrating sphere with* $f_{out} = 0.01$ *initial aperture,* $f_{in} = 0.01$ *further apertures as a function of the reflection coefficient* ρ.

For long-distance images ($m_l^{-1} \gg 1$), only a vanishing fraction of the radiant intensity is used for imaging, for $m_l = 1/100$ and an aperture value of $k = 4$ it is only about $1.6 \cdot 10^{-6}$. In microscopy ($m_l \gg 1$), on the other hand, η_{lens} depends only on the f-number. At an aperture value of $k = 1$ or $NA = 0.5$, 25% of the light emitted from a surface is recorded by the microscope lens.

Extension 3.2: A homogeneous and isotropic light source: the integrating sphere

Normal light sources are neither homogeneous nor isotropic, especially LEDs. A large-area homogeneous and isotropic light source would be of great practical importance for the radiometric calibration of imaging systems and many other applications. An optimally homogeneous and isotropic light source is the *integrating sphere*. Describing its property is a nice application example of what we have learned about radiometry so far (Section 2.4).

We assume a hollow sphere that is coated on the inside with a diffusely scattering material with a high *reflection coefficient* ρ. A small Lambertian radiator with the radiance $L_{in} = M_{in}/\pi$ is attached to the surface. We now ask ourselves how large the irradiance of this radiator is on all other surfaces of the sphere with radius r (Fig. 3.18). A Lambertian radiator radiates per unit area $L \cos\theta$ in the direction θ. Nevertheless, it looks equally bright from all directions, as the unit area is viewed in the direction of the beam, which, projected onto the radiating surface, is $1/\cos\theta$ in size (Fig. 2.17). The receiving unit area is irradiated at an angle of θ and therefore also receives $\cos\theta$ less. This results in the irradiance E as follows

$$E = M\frac{1}{\pi}\Omega \cos^2\theta. \tag{3.72}$$

The solid angle, related to a unit area on the sphere at the distance d from the radiator on the surface of the sphere to another point on the sphere, is irradiated at the angle θ. In the direction of the beam, the unit area on the sphere therefore only has an area $\cos\theta$ (Fig. 3.18) and the solid angle is

$$\Omega = \frac{\cos^2\theta}{d^2} = \frac{1}{4r^2\cos^2\theta}. \tag{3.73}$$

This results in an irradiance

$$E = L_{in}\frac{1}{4\pi r^2}, \tag{3.74}$$

which is the same everywhere on the sphere, even though it is only irradiated by a point light source at one point on the surface. Each surface element of the sphere radiates with the same radiance. We now take the radiant flux Φ_{in} to be the radiant intensity M_{in} integrated over a small area A_{in} in order to take into account the entire radiant flux introduced into the integrating sphere. The radiance emitted directly from the integrating sphere is then

$$L_1 = \rho\frac{\Phi_{in}}{4\pi^2 r^2}, \tag{3.75}$$

which is reduced by the reflection coefficient ρ. This radiation is now reflected further within the integrating sphere. There are further radiation losses around the fraction f of the inner surface of the integrating sphere that does not reflect, i. e. the inlet openings for the light sources and detectors and the outlet opening. All reflected radiation density must now be added up. This leads to the following sum

$$L = \frac{\Phi_{in}}{4\pi^2 r^2}\frac{1}{1-f}\sum_{n=1}^{\infty}\rho^n(1-f)^n = \Phi_{in}\frac{1}{4\pi^2 r^2}\frac{\rho}{1-\rho(1-f)}, \tag{3.76}$$

where we have used the summation formula for the geometric series.
The repeated reflection leads to a high internal radiation density in the integrating sphere. Nevertheless, for reasons of energy conservation, no more radiation escapes than is radiated into the integrating sphere. The radiant flux emitted through an opening with the area $4\pi r^2 f$ is

$$\Phi_{out} = L\pi\, 4\pi r^2 f_{out}. \tag{3.77}$$

The ratio of radiant flux emitted to radiant flux input results in:

$$\eta_\Phi = \frac{\Phi_{out}}{\Phi_{in}} = \frac{\rho f_{out}}{1-\rho(1-f_{out}-f_{in})}. \tag{3.78}$$

We have divided the fraction f of the non-reflecting surfaces into the exit aperture f_{out} and the entrance apertures for the light source and detectors f_{in}: $f = f_{out} + f_{in}$.
As an example for the losses, we first take an integrating sphere with perfect reflection $\rho = 1$ and 1% area for the exit opening (100 mm diameter at 250 mm sphere radius) and the same area for the other openings. According to Eq. (3.78), this results in an immediately understandable ratio of 0.5. This means that only 50% of the radiation remains. With $\rho = 0.98$ and $\rho = 0.9$, the ratio is reduced to 0.25 and 0.076. With 90% reflection, only just under 8% of the radiation is available. This shows how important a high reflection coefficient of the integrating sphere is.

Example 3.9: Homogeneous irradiation with an integrating sphere

In this example, we investigate the irradiance and how homogeneously a surface can be irradiated with an ideal integrating sphere. The integrating sphere has a circular opening with the diameter a in which the radiation density L is constant and isotropic. A small surface element standing centrally in front of the opening at a distance d sees the opening of the integrating sphere at the solid angle

$$\Omega = \pi \frac{a^2/4}{a^2/4 + d^2} = \pi \sin^2 \theta. \tag{3.79}$$

The distance $s = \sqrt{a^2/4 + d^2}$ is the distance to the edge of the opening seen at the angle θ. Since the radiance L in the aperture is assumed to be constant, the irradiance E is simply given by

$$E = \Omega_{\text{in}} L = \pi \frac{a^2/4}{a^2/4 + d^2} L. \tag{3.80}$$

This means that the irradiance through the integrating sphere corresponds to that of a point light source at a distance s and decreases quadratically with distance s (*inverse-square law*).

Calculating the exact drop in irradiance with increasing distance from the center in front of the opening of the integrating sphere is complex. For small angles, we can make the same considerations as when calculating the irradiance on the image plane in optical imaging (Section 3.6.2) and obtain a $\cos^4 \theta'$ dependence, where θ' is the angle from the edge of the irradiated surface to the center of the aperture:

$$E = \pi \cos^4 \theta' \frac{a^2/4}{a^2/4 + d^2} L \quad \text{with} \quad \cos^4 \theta' = \left(\frac{d^2}{a'^2/4 + d^2} \right)^2 \approx 1 - \frac{a'^2}{2d^2}. \tag{3.81}$$

In order to estimate how large the drop to the edge of the irradiated surface is, we assume that its diameter is one fifth of the distance to the opening of the integrating sphere. Then, according to Eq. (3.81), the irradiance drops by 2% towards the edge of the surface.

3.7 ♯References for Further Reading

This chapter has covered the basic principles of optical imaging. A more detailed description with many details on the construction and use of imaging optics can be found in the book by Teubner and Brückner [326]. The geometry of optical imaging is also important for computer graphics. It is therefore discussed in detail in the standard textbooks on computer graphics, e. g. Watt [348] or Foley et al. [87]. The following two textbooks deal with technical optics: Smith [308] and Iizuka [144] (here the focus is on Fourier optics). Riedl [264] deals with the design of infrared optics. Many practical tips on the use of optical components are provided by Litfin [203], Smith and Brady [307] and Hornberg [142, Chapter 4]. If you want to check the performance of optics yourself or calculate ray paths through optics, we recommend the free optics design software Winlens: `www.winlens.de/`.

In this chapter, the importance of linear system theory for the general description of optical of optical systems was emphasized. Linear system theory has a wide range of applications across all areas of technology and science. technology and science. More details can be found, e. g. in Close and Frederick [43] and Dorf and Bishop [63].

4 Radiation Detection and Image Sensors

Contents

B. Jähne, *Digital Image Processing and Image Acquisition*,
https://doi.org/10.1007/978-3-032-04806-6_4

4.1 Introduction

In the last chapter, we learned how the radiation emitted by an object can be mapped onto an image plane. The question now arises as to how the radiation arriving on the image plane can be detected and converted into a digital signal that can be processed by a computer.

This does not necessarily require a two-dimensional image sensor, an *area sensor*. Even with a single detector, image acquisition is possible if it scans the image field in two directions in a suitable manner or if the receiving optics are swiveled in two dimensions. In this way, images can be obtained, e. g. with large *radio telescopes* in astronomy. Other examples are *radar* or *lidar* systems, which can also be used to capture three-dimensional images (Section 9.4.1).

A combined option is *line sensors*. These are particularly suitable when the object to be recorded moves relative to the image acquisition system, as is the case with satellites, a scanner or objects on a conveyor belt in industrial quality control. Then, a line sensor oriented perpendicular to the direction of movement can be used to capture an image that is virtually infinite in that direction.

In this chapter, we will first focus on the basic possibilities of detecting radiation, i. e. and converting it into an electrical signal (Section 4.2). This is followed by a description of the important class of semiconductor detectors (Section 4.3) and the various architectures of image sensors, including the hardware and software interfaces between a camera and a computer (Section 4.4). In the last Section 4.5 we finally deal with the question of which effects reduce the quality of image signals and how the quality can be described as simply as possible. In all sections, we will only delve as deeply into the subject matter as is necessary to understand radiation detection and image sensors from the perspective of image processing.

4.2 Radiation Detection

4.2.1 Direct Measurement

Measuring the amplitude of a wave is obviously the most direct way of detecting radiation. According to Maxwell's equations, the oscillating electromagnetic field (cf. Section 2.3.1) induces correspondingly oscillating currents and voltages in an appropriately shaped electrical conductor, an *antenna*. This allows not only the frequency of the radiation but also its phases to be measured. With the appropriate geometry of the antenna, the polarization state (Section 2.2.7) is also accessible.

This direct method of radiation detection therefore has the great advantage that it can measure all the information contained in the radiation emitted by an object (Section 2.3.4). The question therefore arises as to why this method is not used for imaging in the visible and adjacent spectral range of electromagnetic radiation.

The first obstacle is the high frequency of light, which is about 10^{15} Hz, i. e. in the petahertz range (Fig. 2.16). There are still no electronic components for such high frequencies. This means that direct measurement is not yet feasible. Initial research in this direction is underway, but still requires the extremely strong electromagnetic fields of short-pulse lasers.[1] [2]

Since the discovery of radio waves in 1886 by Heinrich Hertz at the Technical University in Karlsruhe [130, 131, 132], it has been known that electrical conductors can emit and receive electromagnetic waves. Over time, this method has been extended first into the microwave range and finally into the terahertz range with frequencies of 0.3–3 THz and wavelengths of 0.1–1 mm.

If only the received radiation is measured with an antenna, this is referred to as *radiometry*. The problem with imaging techniques in this wavelength range is the much longer wavelengths in the millimeter to meter range compared to light. From the considerations on resolution in Section 3.5.4 we know that the angular resolution (in radians) results from the ratio of the wavelength λ_e of the radiation to the diameter a (Eq. 3.57). Therefore, correspondingly large antennas are required for microwaves with wavelengths in the millimeter to meter range.

However, as the radiation is measured coherently (the phase position is not lost), several antennas can be interconnected interferometrically and thus the angular resolution can be increased accordingly. This works according to the principle that we learned about in Section 2.2.5 when discussing beams. By appropriately shifting the phase between the individual antennas, the angle from which the radiation is constructively added can be set. The only difference here is that we have radiation receivers rather than radiation sources. However, reversing the direction of radiation does not change the physics of the waves (*Helmholtz reciprocity principle* [344, p. 168ff], [363]).

Extension 4.1: Sending and receiving

The direct measurement of electromagnetic radiation has another essential advantage. Antennas can also emit radiation. This opens up techniques that are summarized under the term *radar* (abbreviation of "radio detection and ranging"). These are active systems that emit and receive radio or micro waves [332]. They can be used to measure a wealth of information.

- The propagation time from the transmitter to the object and back to the receiver indicates how far away an object is. We will discuss *time-of-flight* method in Section 9.4.1.

[1] `https://ufox.cfel.de/research/petahertz_electronics/`
[2] `https://www.mpq.mpg.de/4860291/12_12_05`

- The speed of an object relative to the observer can be determined from the frequency shift between the emitted and received radiation using the *Doppler effect* (Section 2.6.2).
- Due to the coherent emission and detection of radiation, radiation can also be coherently superimposed with radar systems. A particularly important class are systems that can be used to artificially increase the aperture, i. e. the diameter of the antenna and thus increase the resolution accordingly. Such synthetic aperture radar systems, or *SAR* systems for short, have become very important in remote sensing [332]. After all, the micro and radio wave window of the atmosphere with wavelengths between a few millimeters to over 10 m (Fig. 2.34) is by far the widest window. Due to the wavelengths, which are much longer than water droplets and crystals in clouds, these penetrate the clouds with only low absorption and scattering (Section 2.6.7), so that images of the Earth's surface can be taken from satellites even through clouds.

Extension 4.2: Direct measurement of acoustic waves

Like electrical radio waves and microwaves, acoustic waves can be measured directly. The pressure fluctuations of the acoustic wave (Section 2.3.3) transmit forces to sensors that convert them into an electrical signal. Such a sensor is called a *microphone* or, when used in water, a *hydrophone*. As with antennas, acoustic sensors can both receive and emit acoustic waves. This means that everything we have just said about the direct measurement and emission of electromagnetic waves can be transferred to acoustic waves. However, some special features are worth mentioning.

- In contrast to electromagnetic waves, the entire frequency range of acoustic waves from a few hertz up to gigahertz can be measured directly.
- Acoustic waves are particularly suitable for water as they are only slightly attenuated. Acoustic waves can propagate over hundreds of kilometers in the ocean. This is not the case for electromagnetic waves[3] [119]. Blue light penetrates furthest into the water.
- The large difference in the *refractive index* (Section 2.6.3) between air and solids or liquids is problematic for acoustic imaging. This means that most of the acoustic wave is reflected by a liquid or solid. For effective penetration of acoustic radiation, the transmitter must therefore touch the liquid or solid directly without an air cushion in between.

4.2.2 Detection of Photons

There is another reason why visible electromagnetic waves cannot be measured directly, which lies in the way they interact with matter. It is not possible to transfer arbitrary amounts of energy to atoms or electrons. Experimental observations have shown that this is only possible in discrete steps that depend on the wavelength of the light. This was an observation that led to the development of *quantum mechanics* and the term *photon*. We

[3] `https://en.wikipedia.org/wiki/Electromagnetic_absorption_by_water`

have already learned about this fact in Section 2.3. A photon has the energy

$$E = h\nu = \frac{hc}{\lambda}. \tag{4.1}$$

This therefore depends on the frequency ν or the wavelength λ of the electromagnetic radiation.

Which materials are suitable for converting radiation into a current or charge carrier? This is not possible with insulators (dielectrics). Nor is it possible with metals (electrical conductors), as these already contain electrons that move freely and therefore generate a current when a voltage is applied. It would only be possible for an electron to be released from the metal by the incident radiation energy. In this case, we speak of the *external photoelectric effect*. This effect is used in vacuum phototubes and *photomultipliers*. Photomultipliers are sensitive enough to detect single photons as the first electron generated by photoabsorption is accelerated and hits another dynode from which it releases several secondary electrons. This process can be cascaded over several stages to achieve a high amplification factor. The response times of a photomultiplier can be shorter than 10^{-10} s so that single photons can be counted as individual events. Such a device is called a *photon counter*.

While photomultipliers are still used today, phototubes, as they were first used in television cameras and later in scientific cameras, are no longer relevant. This is due to the great progress that has been made with semiconductor image sensors, which are much smaller and easier to handle. Without energy supply, *semiconductors* are non-conductors, as all electrons are firmly bound to atoms. However, electrons can be transferred from the bound state (valence band) to an unbound state (conduction band) by supplying energy This requires a minimum energy E_g to overcome the so-called *band gap energy*. This energy can be provided either by thermal energy (lattice vibrations) or by the absorption of a photon. The transition of an electron from the valence band to the conduction band is called the *inner photoeffect*

If the photon energy is less than this energy threshold, the photon cannot be absorbed and the detector is transparent to the radiation with the corresponding wavelength of the electromagnetic radiation, as it cannot interact with the semiconductor material. Therefore, the maximum detectable wavelength, the *cut-off wavelength* λ_g, is given by

$$\lambda_g = \frac{hc}{E_g} \quad \text{or} \quad \lambda_g(\mu\text{m}) \approx \frac{1{,}238\,\text{eV}\mu\text{m}}{E_g(\text{eV})}. \tag{4.2}$$

For silicon, the most important semiconductor material in electronics, the band gap energy is around 1.1 eV, resulting in a cut-off wavelength of around 1.1 μm. Therefore, silicon is fortunately suitable as a detector for light (0.4–0.7 μm) and near infrared radiation. The penetration depth of the radiation in this wavelength range is around 1–100 μm (Fig. 4.1a).

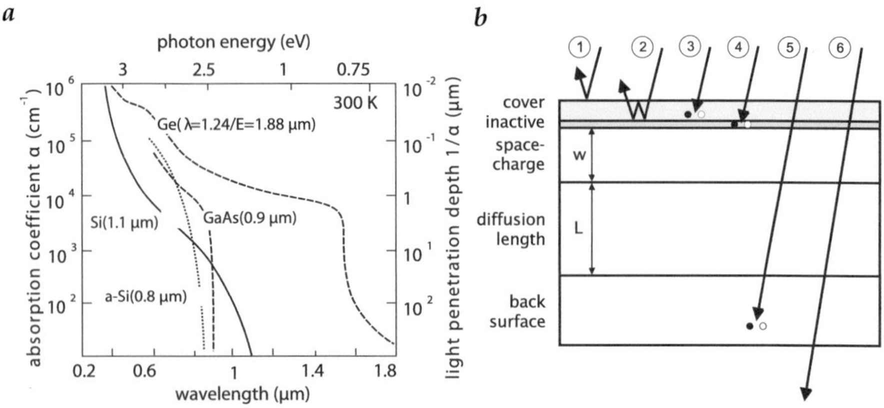

Figure 4.1: ***a*** *Absorption coefficient and penetration depth for some common semiconductor materials as a function of wavelength with data from Sze [319].* ***b*** *Schematic representation of optical losses in semiconductor image sensors: (1) surface reflection, (2) interference at thin layers, (3) absorption in the top layer, (4) loss of photo-induced charges in inactive regions or by recombination, (5) absorption deep in the semiconductor and (6) transmission through the semiconductor.*

The remaining "hole" in the valence band cannot move, while the electron in the conduction band can move freely. As a result, the more electrons there are in the conduction band due to the absorption of radiation, the more conductive the semiconductor material becomes. Detectors that work in this mode are called *photoconductive detectors*.

Photoconductive detectors are unsuitable for image sensors because the electrons move away from the point of origin. The movement is not directional, as the electrons constantly collide with bound electrons and thus change direction and move away from the point of origin in all directions. Such a movement is called *diffusion*. Diffusion processes also play a major role in image processing. They are discussed in detail in this book, starting with Section 15.2.1. From the point of origin, the electron distribution spreads out in a Gaussian distribution whose standard deviation σ becomes wider and wider over time: $\sigma = \sqrt{Dt}$, see (Eq. 15.13). The diffusion constant in silicon at room temperature is about 35 cm^2/s.

If there are electrons in the conduction band, the semiconductor also has positively charged holes that attract electrons. If an electron falls into such a hole, it returns from the conduction band to the valence band. This is known as a recombination process. The lifetime of electrons in the conduction band is therefore limited. It is only about 10 μs. This means that a "charge image" generated by absorbing a flash of radiation has disappeared again in this short time.

In order for an image sensor to function, it is therefore essential that the electrons are trapped in a positively charged area, a so-called potential well, and thus prevented from diffusing freely. The crucial question now is whether the electrons have enough time to move from their point of origin to this area. Fortunately, this is the case. Within 10 µs, the electrons can move a distance of $\sqrt{Dt} \approx 260\,\mu\text{m}$. This is more than enough for the size of image sensor elements in modern silicon image sensors, which are about 1–10 µm in size.

In circuit terms, a charge accumulation corresponds to a capacitor with a capacitance C. According to the definition of capacitance C

$$C = q/U \quad \leadsto \quad U = q/C, \tag{4.3}$$

the voltage U of the capacitor is proportional to the accumulated charge q. Radiation detectors of this type are therefore called *photovoltaic detectors*.

4.2.3 Thermal Detection

Thermal image sensors are not quantum detectors. Instead, they react directly to the absorbed radiation energy, which is converted into heat and thus induces a small rise in temperature. It follows immediately from this principle that these sensors can be used uncooled. If the sensor is in thermal equilibrium with its surroundings, it emits as much radiation as it receives. However, it is necessary for both the sensor and the lens and its mount to be at as constant a temperature as possible.

A common feature of thermal sensors is that they inherently have a broad spectral sensitivity. However, the sensitivity of these detectors is lower than that of semiconductor detectors, as they are based on secondary effects. Three effects can be used for thermal image sensors:

Thermoelectricity. Two compounds of different metals that have been brought to different temperatures show a voltage difference that is proportional to the temperature difference. One of the two compounds must be brought to a reference temperature, the other absorbs the radiation. In order to be as sensitive as possible, the latter must have a low heat capacity and be thermally well insulated. Such a detector is called a *thermopile*.

Pyroelectricity. Pyroelectric materials have a permanent electrical polarization. When the temperature changes, surface charges are generated that can be measured. This phenomenon is known as the *pyroelectric effect*. As the surface charges dissipate slowly, pyroelectric detectors can only measure changes in radiation. This means that a beam interrupter must be used to periodically interrupt the incident radiation. This additional mechanical element is a clear disadvantage. In addition, above a critical temperature, the *Curie temperature*, the pyroelectric effect disappears.

Thermoconductivity. Some materials exhibit very strong changes in electrical conductivity with temperature. This *thermoconductive effect* can be used to measure the absorbed radiation from the change in resistance. Sensors based on this simple principle are called *bolometers*.

Figure 4.2: *Examples of cameras for the thermal radiation range (≻R11):* ***a*** *Uncooled miniature bolometer camera SmartIR640 with a resolution of* 640×512 *pixels and 60 fps. Without lens, the camera has a size of only 30 mm × 30 mm × 23 mm.* ***b*** *High-end InSb cameras ImageIR from InfraTec with a resolution of up to* 1920×1536 *pixels.*

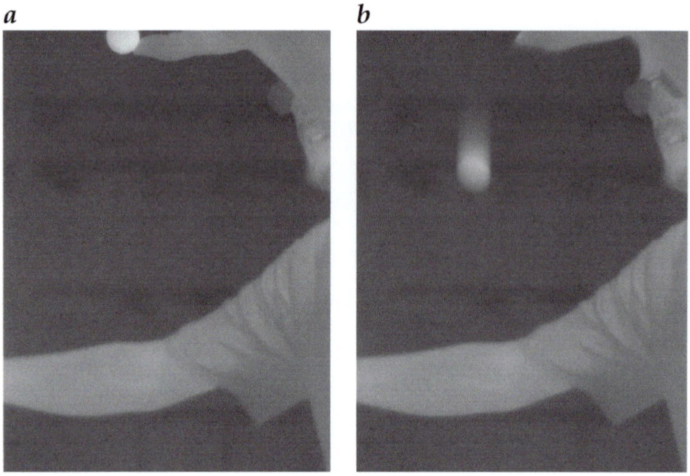

Figure 4.3: *Recording a moving object with an uncooled thermal bolometer camera:* ***a*** *ball at rest,* ***b*** *ball falling.*

Thermal image sensors are dominated by bolometers. Their great advantage is that they are almost as compact as cameras in the visible range, as they do not require complex cryogenic cooling, but only temperature stabilization. As a result, they are also much more robust and have a longer lifetime. Uncooled bolometer cameras are available with a resolution of up to 1280×1024 pixels for the wavelength range of 7.5–14 μm (Fig. 4.2a). The noise level, the *noise-equivalent temperature difference* (*NE*Δ*T*), is 30-100 mK. Cooled semiconductor detectors (Fig. 4.2b) are more sensitive with a NEΔT of 7–30 mK [343].

Although the difference in sensitivity between uncooled bolometer cameras and cooled thermal cameras with semiconductor sensors is not that great, there is another important difference for practical applications. Semiconductor sensors only require exposure times of around one millisecond to measure temperatures in the range up to 30°C. Bolometer cameras, on the other hand, do not have a defined exposure time, but a thermal time constant for how quickly they can react to temperature changes. These are in the range of 10 ms. If moving objects are recorded, they are blurred in the direction of movement due to the time constant and, if they are warmer than their surroundings, they leave an exponentially decaying tail behind them. This effect can be seen in Fig. 4.3.

Another disadvantage of uncooled cameras is that they have lenses with a f-number of 1 in order to achieve the high sensitivity. Cooled cameras have lenses with aperture 2 or larger. As both camera types have comparable pixel sizes between 10 and 20 µm, the depth of field is shallower with uncooled cameras (Section 3.4.4).

4.3 Basics of Semiconductor Radiation Detectors

4.3.1 Quantum Efficiency and Spectral Sensitivity

The first important parameter for characterizing the quality of a radiation detector is the *quantum efficiency* η. If N_p photons are incident on the area A of a sensor element during an exposure time, then the fraction

$$\boxed{\eta(\lambda) = \frac{N_e}{N_p} \quad \text{quantum efficiency}} \tag{4.4}$$

is absorbed and accumulated as N_e charge units. The quantum efficiency defined in this way refers to the total area A of a pixel, not just the part of the area that is light-sensitive. The part of the light-sensitive area in the total area is called *fill factor*.

For low-energy radiation, the quantum efficiency cannot be greater than one, as one absorbed photon generates at most one charge unit. It is less than one because some photons are reflected at the sensor surface or at intermediate layers, are absorbed in regions from which no charge carriers can be accumulated, or penetrate the sensor material without being absorbed (Fig. 4.1b).

The wavelength dependence of the quantum efficiency of some image sensors is shown in Fig. 4.4. The quantum efficiency decreases towards the blue and infrared spectral range. Some image sensors show strong fluctuations with the wavelength. These result from interference from reflections on the top and bottom of thin transparent layers on the image sensor (Section 2.2.6).

When high-energy electromagnetic radiation, such as short-wave UV or X-rays (Fig. 2.16), hits a suitably constructed semiconductor photosensor,

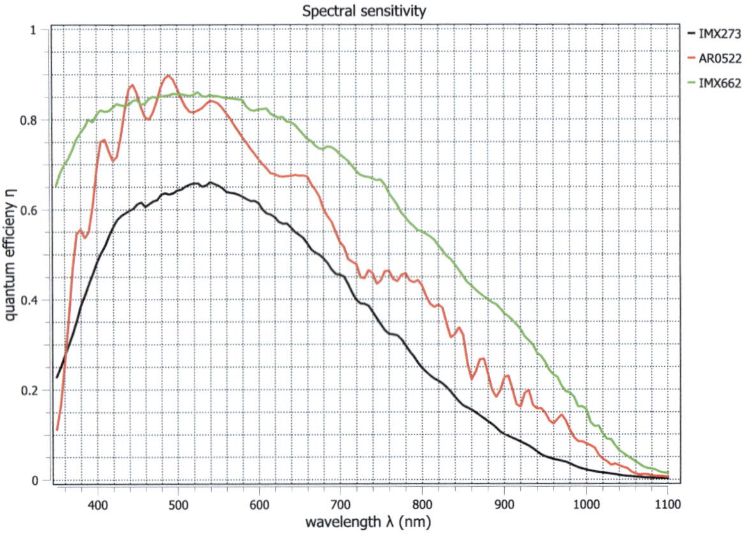

Figure 4.4: *Quantum efficiency of different CMOS image sensors.*

one photon generates a whole cascade of charge units. The number depends on the properties of the sensor material and the energy of the photon. A semiconductor image sensor capable of detecting individual photons can therefore be used to measure their energy or wavelength. Such an image sensor was part of the "high-resolution imager" (HRI) of the X-ray telescope on the ROSAT satellite.

We have already discussed the maximum detectable wavelength in Section 4.2.2. It is given by the band gap of the semiconductor material. Therefore, silicon semiconductor detectors can only detect wavelengths up to a maximum of 1.1 μm. The material is transparent for longer wavelengths. But how does the quantum efficiency behave for shorter wavelengths? Standard silicon image sensors are sensitive from wavelengths of around 350 nm at most (Fig. 4.4). For shorter wavelengths, the photons are absorbed at too short a distance in the semiconductor material, in the passivation layer or in the vapor-deposited transparent electrode material, so that they cannot reach the accumulation range (Fig. 4.1b).

Special measures are required for sufficient sensitivity at shorter wavelengths (Table 4.1), including a cover glass made of quartz glass and a special design of the image sensor. Alternatively, the image sensor can be coated with fluorescent material (Section 2.6.8), in which UV radiation is converted into visible radiation.

The highest quantum efficiencies are achieved with sensors that are back-illuminated ("*back-illuminated*") and have carefully matched anti-reflective coatings. This is necessary because silicon has a high *refractive index* of around 4 in the visible range, so that just under 50% of the light is reflected

Table 4.1: *Spectral sensitivity of different image sensors and "focal plane arrays"* (FPA)*. The last column indicates either the quantum yield or, for sensors in the X-ray range, how many electrons are generated per kV.*

Sensor type	Spectral range	Typical sensitivity
Scintillator, fiber-coupled on Si-CCD	20-100 keV	0,5 e^-/keV
Fluorescent coating fiber-coupled on Si-CCD	3–40 keV	5 e^-/keV
Si-CCD X-ray detector (ROSAT satellite)	800 eV–10 keV	> 0.90
	100 eV–15 keV	> 0.50
Si-CCD, fluorescence coating, MgF_2 window	0.12–1.0 μm	< 0.4
Etched, back-illuminated Si-CCD	0.25–1.0 μm	< 0.80
Standard silicon CCD or CMOS sensor	0.35–1.0 μm	< 0.65
Modern back-illuminated CMOS sensor	0.30–1.0 μm	< 0.90
GaAs FPA	0.9–1.7 μm	
Extrinsic Pt:Si sensor (obsolete)	1.4–5.0 μm	0.001–0.01
InSb FPA	1.0–5.5 μm	< 0.85
HgCdTe (MCT) FPA MWIR	2.5–5.0 μm	
HgCdTe (MCT) FPA LWIR	8.0–12.0 μm	
GaAs/AlGaAs "quantum well" FPA (QWIP)	3.0–5.0 μm	
	8.0–10.0 μm	
InAs/GaSb superlattices FPA (QWIP)	3.0–25.0 μm adjustable	< 0.7

at perpendicular incidence (Eq. (2.82) in Section 2.6.3). This allows high quantum efficiencies of over 0.9 to be achieved. With back-illuminated image sensors, the sensitivity can also be extended far into the ultraviolet range.

In the mid- (3.0–5.0 μm) and long-wave (8.0–14.0 μm) infrared, semiconductor material with a significantly smaller band gap than silicon is required. There is a whole range of binary and ternary semiconductor compounds (Table 4.1). They all have three basic facts in common that make these image sensors expensive.

Firstly, due to the smaller band gap, it is essential to cool these sensors to cryogenic temperatures between 60 and 100 K in order to sufficiently reduce the dark signal through thermal excitation.

Secondly, it requires each pixel to be connected to a silicon readout chip (ROIC). This is usually achieved using indium solder connections (Fig. 4.9). This is why this type of image sensor is also referred to as a *"focal plane array"* or *FPA* for short.

Thirdly, not only the image sensor must be cooled, but also the area surrounding the sensor ("cold shield") so that it only receives radiation from the angular range that corresponds to the aperture of the lens. Without this measure, the sensor would receive radiation from the entire half-space with approximately the same radiance as the lens, so that the useful signal would only be a small fraction of the total signal.

4.3.2 Irradiance and Sensitivity

The number of photons incident on a pixel with the area A during the exposure time t_{exp} at a given *irradiance* E (W/m^2, Section 2.4.1) can be calculated with Eq. (4.1) as follows

$$N_p = \frac{AEt_{\text{exp}}}{h\nu} = \frac{AEt_{\text{exp}}}{hc/\lambda}. \tag{4.5}$$

With the values $c = 2.99792458 \cdot 10^8$ m/s for the speed of light and $h = 6.6260755 \cdot 10^{-34}$ Js for the Planck constant, the photon-based irradiance in units useful for image sensors results in

$$\boxed{N_p[\text{photons}] = 50,34 \cdot A[\mu\text{m}^2] \cdot t_{\text{exp}}[\text{ms}] \cdot \lambda[\mu\text{m}] \cdot E\left[\frac{\mu\text{W}}{\text{cm}^2}\right].} \tag{4.6}$$

This equation can be used to convert the irradiance measured with a calibrated photodiode in W/cm^2 into a photon flux as required for the characterization of an image sensor.

The *sensitivity* R is a quantity equivalent to the quantum efficiency. It is given as the ratio of the electrical signal s to the *radiant power* Φ:

$$R = \frac{s}{\Phi}. \tag{4.7}$$

Depending on whether the signal is a current or a voltage, the sensitivity has either the unit A/W or V/W.

It can be calculated from the quantum efficiency by specifying radiant power as photon flux (see Eq. (2.50) in Section 2.4.1). Then we get

$$\Phi_e = h\nu \frac{\mathrm{d}N_p}{\mathrm{d}t}. \tag{4.8}$$

The resulting current I is the number of elementary charges per time multiplied by the elementary charge e:

$$I = e\frac{\mathrm{d}N_e}{\mathrm{d}t}. \tag{4.9}$$

With Eqs. (4.1) and (4.7) we finally get with the current I as signal

$$\boxed{R(\lambda) = \frac{I}{\Phi_e} = \eta(\lambda)\frac{e}{hc}\lambda \approx 0,8066\, \eta(\lambda)\, \lambda \left[\frac{\text{A}}{\text{W}\mu\text{m}}\right].} \tag{4.10}$$

For green light ($\lambda = 0.5\,\mu$m), a sensor with a quantum efficiency of one has a sensitivity of around 0.4 A/W.

With a constant quantum efficiency, the sensitivity increases linearly with the wavelength. This is because the radiation is quantized in smaller and smaller energy units with increasing wavelength according to Eq. (4.1). Therefore, a detector for infrared radiation in the wavelength range of 3–5 μm is around 10 times more sensitive than a detector for visible radiation.

4.3.3 ♯Dark Current

A photosensor generates a signal even if it does not receive any radiation. This results from the thermal excitation of electrons. The vibrational energy of the atoms in the detector material can also be transferred to an electron, causing it to enter an excited state in the conduction band. The number of thermally excited electrons N_d follows the *Boltzmann distribution*

$$N_d \propto \exp\left(-\frac{E_g}{k_B T}\right), \tag{4.11}$$

where $k_B \approx 0.86173 \cdot 10^{-4}$ eV/K is the *Boltzmann constant*, T is the *absolute temperature* and E_g is the band gap. From Eq. (4.11) it follows that the number of thermally excited electrons increases strongly with temperature. The temperature difference T_d, at which the number of thermally excited electrons in the conduction band doubles, results in

$$T_d \approx \ln 2 \frac{k_B T}{E_g} T. \tag{4.12}$$

A band gap of $E_g = 1.1$ eV for a silicon photodetector results in a doubling temperature of around 7 K at room temperature (300 K). This means that the dark signal is the only property of a photodetector that is strongly dependent on temperature. The dark current can be greatly reduced by cooling the image sensor. Cooling from 50 to -20°C already reduces the dark signal by three orders of magnitude.

4.4 Architecture of Image Sensors

4.4.1 From Photons to Digital Signal; Ideal Image Sensor

By combining all the steps shown in Fig. 4.5, the photons that hit a sensor element during the exposure time are ultimately converted into a digital number. The different architectures of image sensors for arranging the sensor elements and for reading out the image data generated in parallel in serial data streams are discussed in Section 4.4, including CCD image sensors (Section 4.4.2) and CMOS image sensors (Section 4.4.3). It also covers the design of image sensors with sensor elements that are sensitive to different

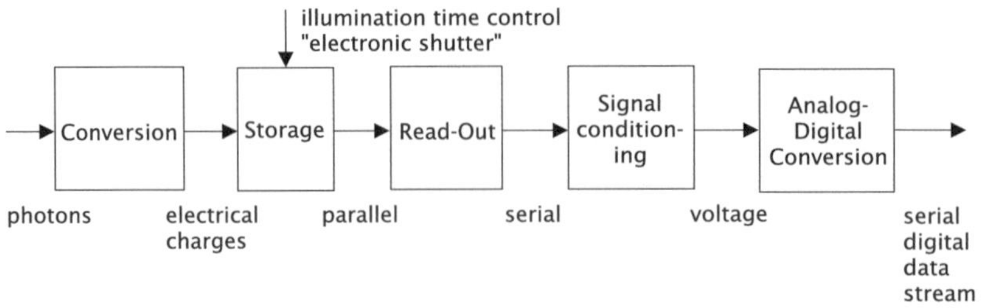

Figure 4.5: *Chain of processes that take place in an image sensor to convert the incident photon stream into a digital data stream.*

wavelengths, i. e. *color image sensors* (Section 4.4.5), and standard interfaces for connecting digital cameras to computers (Section 4.4.6).

An image sensor converts an "image", i. e. a spatially continuously varying *irradiation* that falls through an optical system onto the image plane, into a digital data stream that can be stored and processed in a digital computer. This involves a whole chain of individual processes (Fig. 4.5):

Conversion
The incident radiant energy is converted into an electrical signal. This can be an electrical charge, an electrical current or a voltage. Each individual element of the image sensor integrates the irradiation via the light-sensitive part of its surface.

Accumulation
The generated electrical charges or the generated electrical signal is integrated over a certain time interval, the *exposure time* t_{exp}. With many image sensors, an external electrical signal controls the exposure time. This property is called *electronic shutter*.

Read out
After exposure, the electrical charges, or the averaged signals, must be read out. This step converts the signal recorded simultaneously (in parallel) at all elements of the image sensor into a serial data stream with one or more channels. These data streams are then further processed by one or more processing units.

Signal processing
Finally, the signals are further processed to correct all types of interference and amplify the signal for analog-to-digital conversion.

Analog-to-digital conversion
In the final step, the analog voltages are then quantized by an *analog-to-digital converter* (*ADC*) and the digital values are transferred to the computer where they are displayed and/or stored.

Since each image sensor consists of an array of sensor elements, *digitization* is already performed there, i. e. the conversion of a spatially and temporally continuous signal into a discrete one, as already discussed in Section 5.2. We discuss digitization in Section 5.2 and the influence of the last step, *quantization* by the analog-to-digital converter, in Section 5.6.

An image sensor contains many sensor elements. Therefore, one of the biggest challenges in the design of image sensors is to tap the charges or voltages accumulated in parallel at all these sensor elements, read them out and transform them into one or more serial data streams.

From the processing chain described above, we can easily determine the characteristics of an ideal or best-possible image sensor:

- Each photon is converted into a charge unit. The quantum efficiency is one.
- No charge unit is lost during accumulation. No charge units from other processes are added either. The dark current is therefore zero.
- The image sensor is not insensitive to light outside the set exposure time. This means that the *parasitic light sensitivity* is zero. This property is particularly important for short exposure times
- The readout process and signal processing is noise-free
- The analog-to-digital conversion is ideal with perfectly even intervals, strictly monotonic and without missing digital values ("missing code").
- These properties imply that all image sensor elements deliver the same signal with the same irradiation. The ideal image sensor is therefore perfectly homogeneous.

4.4.2 ♯CCD Image Sensors

The invention of the *Charge-Coupled Device* or *CCD* for short) in the mid-1970s was the breakthrough for semiconductor image sensors. Before this time, image sensors were only available on the basis of photocathodes in vacuum tubes that were scanned by an electron beam. These image sensors no longer have any significance today. In addition to charge accumulation, a CCD contains analog shift registers with which the accumulated charge packets are shifted to the edge of the chip where they can be amplified and read out.

Using the example of a four-phase CCD, Fig. 4.6 illustrates how this works. Negatively charged areas separate the charge packets accumulated at each pixel. In the first step, the potential in the left part of the three-part memory area is raised negatively so that the charges are narrowed down from three to two areas. Then the potential in the area directly to the right of the storage area can be lowered so that the charge carriers can again occupy three areas and have simultaneously moved a third of a pixel further. This procedure can be repeated until the charges reach the edge of the sensor. There, the charge packet must then be converted into a voltage by

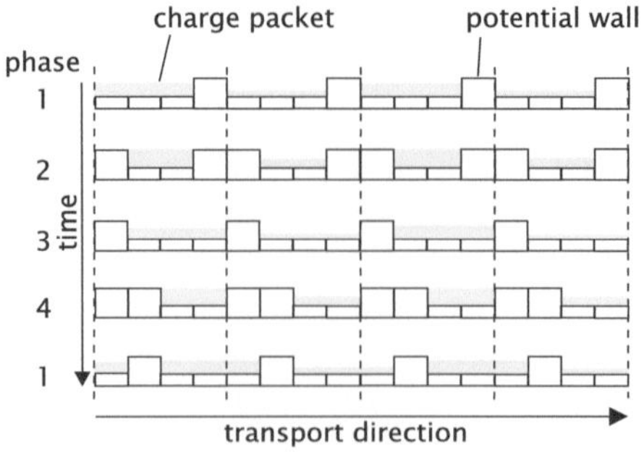

Figure 4.6: Principle of charge transport in a four-phase CCD.

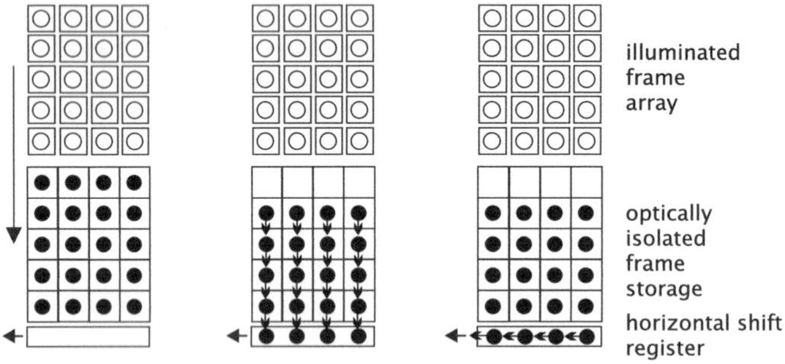

Figure 4.7: Illustration of frame transfer with a frame transfer CCD image sensor.

an external charge amplifier. Due to the manufacturing technology, it is not possible to directly integrate amplifiers or any other evaluation electronics on a CCD sensor.

As the maximum number of charge transport processes that can take place is the number of pixels the image sensor has in one direction, the transport must be so effective that no significant losses occur. If the small portion ϵ remains after a transport step, the following portion remains after N steps:

$$(1-\epsilon)^N \approx 1 - N\epsilon, \quad \text{if} \quad \epsilon << 1/N,$$

For an image sensor with $N = 1000$ pixels in one direction, the loss during charge transport must therefore be less than 0.001% if you want to have lost a maximum of 1% of the charge after a thousand transport steps. This results in a clear limitation of the frequency at which the charge carriers can be transported and therefore the frame rate that a CCD image sensor can achieve.

The transport mechanism discussed in the last section is the basis for all one- and two-dimensional CCD image sensors. It can be realized in various ways.

Frame Transfer
Frame transfer CCD sensors (Fig. 4.7) use the accumulation memory on each sensor element simultaneously as a transport register. At the end of the exposure time, the entire image is shifted over the exposed area into a second, darkened area of the same size. During this transfer phase, the image sensor must be darkened by a shutter if it is continuously illuminated, or it may only be illuminated by a flash during the exposure phase. The dark period between the exposure times is relatively short, as all columns are read out in parallel into the darkened area. With a megapixel sensor with 1000×1000 pixels and a readout rate of 20 MHz, the transfer to the dark area takes only 50 μs.

A repeated two-stage transfer process then takes place in the darkened area (Fig. 4.7), while the next exposure can take place in the illuminated part at the same time. In the first stage, all lines are shifted down one line, whereby the bottom line is transferred to an additional horizontal shift register. In the second stage, as many transfer cycles follow in this register line only until all N charge packets have been read from it. This process is repeated M times until all M lines have been transferred to the horizontal shift register and read out there.

For large CCD image sensors, it is no longer possible to afford a second darkened area. The image area is then read out directly. This variant is called *full-frame transfer*. However, the phase during which the image sensor must be darkened now takes significantly longer, as all pixels must be read out and exposure and readout can no longer overlap. For the numerical example above, this is slightly more than 50 ms. Many CCD image sensors for scientific use and sensors in digital cameras work according to this principle. However, they are now almost irrelevant, as all new developments are based on CMOS sensors.

Interline Transfer
As frame transfer is not suitable for continuous illumination, almost all CCD sensors are based on a more complex readout scheme, the *interline transfer*. With this sensor, each pixel has a second memory area for image data transfer in addition to the accumulation memory. At the end of the exposure time, the accumulated charges are pushed into the memory registers with a transfer cycle. This time is therefore very short. At a readout rate of 20 MHz it is only 50 ns. The next exposure can then start immediately and the last image can be read out in parallel. This is a two-stage process, just like frame transfer. The only difference is that it now does not take place in the darkened area (Fig. 4.7), but from darkened transport registers located between the lines. Hence the name interline transfer.

A CCD sensor only requires a single output amplifier. This is both an advantage and a disadvantage. The advantage is that all pixels pass through the same amplifier and the signal therefore shows hardly any differences in zero point and amplification from pixel to pixel. The disadvantage is the speed limitation and the sequential readout principle. Due to the limited mobility of the electrons in the conduction band imposed by the detector material, the charge transport via the transfer registers is also limited in frequency. More than a few 10 MHz are not possible.

Parallelization is limited. Without additional transfer register lines in the image sensor, which would lead to black stripes in the image, the CCD sensor can be read out at all four corners simultaneously at most. This allows the image refresh rate to be quadrupled. The price for this is that the four amplifiers must be very well aligned in terms of zero point and gain, otherwise you will see steps between the quadrants in the image.

Finally, we will discuss some properties of CCD image sensors that are important for practical handling.

Exposure Time and Electronic Shutter

The *exposure time* does not necessarily have to correspond to the readout time. The exposure time can take longer due to corresponding signals to the sensor. The parallel image transfer of the previous image is then completed earlier and the transfer process stops until the exposure is complete. But how can shorter exposure times be achieved? This is possible by applying an additional positive voltage to all pixels so that the charge carriers are not collected in the potential wells, but are constantly flowing away. This voltage is only removed at the start of the desired exposure time. This enables precise control of the exposure time. Short exposure times of up to a few μs are possible with standard sensors.

Partial Image Read Out

It is often desirable to be able to read out only part of the image, but faster. This is only possible to a limited extent with CCD sensors, as all pixels have to be read out. If a pixel is not read out, it constantly accumulates charge carriers. If the memory area can no longer hold them, the surplus charge carriers migrate to the neighboring pixels and ultimately across the entire image sensor. This effect is called *blooming*. Therefore, the only trick is to read out the image areas that are not of interest faster than the part of interest. Of course, this leads to some transfer losses, but these are not significant. However, the increase in speed is limited, as the transfer rate in the non-interesting image areas cannot be more than quadrupled.

Synchronization and External Triggering

In many applications, more than one camera is used to capture images at the same time, or the image capture must be synchronized with other events. Almost all cameras today emit signals ("integrate enable") that indicate when the exposure begins and when it ends. This can be used to control a flashlight. Conversely, most cameras accept external trigger signals that only trigger the image capture. This means that the image capture of several cameras can be started simultaneously with one signal. It should be noted that there may be delays until the image is captured, which may also be subject to fluctuations (*jitter*). For time-critical tasks, you must therefore carefully check whether the delays and their fluctuations are sufficiently small.

4.4.3 ♯CMOS Image Sensors

Active Pixels

Despite the great success of CCD image sensors, they have now been overtaken by CMOS image sensors in almost all properties. With the rapidly decreasing structure size in integrated circuits, it has now become possible to accommodate a separate preamplifier and further evaluation electronics on each pixel. Such a pixel is called *active pixel sensor* or *APS* for short [299]. This means that it is no longer necessary to transport the accumulated charges across the chip. The voltages generated by the preamplifiers at each pixel can, in principle, be read out in any order. In particular, this makes it possible to read out partial areas of the sensor without any loss of speed, i. e. with a frame rate that is essentially inversely proportional to the number of pixels read out.

APS technology is also very attractive because image sensors can be manufactured using CMOS technology. This makes it possible to integrate additional analog and digital circuits on the image sensor, right up to a complete camera on a chip, including analog-to-digital conversion and transmission of the image signals via a standard interface. Compared to CCD sensors, an incredible variety of CMOS image sensors (*CIS* for short) has therefore developed in a short space of time.

One important feature of CMOS image sensors compared to CCD image sensors should be noted. With CCD image sensors, all pixels are exposed at the same time. This is not the case with the simplest CMOS image sensor. As the pixels are read out one after the other, the exposure begins after the read-out and therefore at a different time for each pixel.

Due to the two-dimensional geometry, this leads to an effect as if a *slit shutter* (as is the case with SLR cameras) were moving across the image sensor. This is therefore referred to as a *rolling shutter*. This leads to distortions with moving objects. Such a sensor can only be illuminated with a flash if the exposure time is long enough so that there is a time range in which all lines are exposed.

Some cameras output a corresponding trigger signal for this time range, which is called *flash window*. In this way, the entire image can be exposed simultaneously using flash lighting, even with rolling shutter image sensors.

By buffering the charge carriers accumulated during the exposure, it is also possible with CMOS image sensors to expose all pixels simultaneously, although they are read out one after the other (*global shutter*). This requires additional transistors and capacitors at each pixel.

High-speed CMOS Image Sensors

CMOS image sensors have another decisive advantage. The image signal can be read out using many parallel readout channels. This is why CMOS sensors dominate in high-speed cameras (> R9,> R10). Due to the parallelization of the readout, however, it is no longer possible to read out partial images faster. If there is a separate readout unit with corresponding amplifiers and analog-to-digital converters at each column of a CMOS image sensor, then the frame rate cannot be increased if the length of the line is reduced, but only if the number of lines is reduced.

The frame rates that can be achieved today are not limited by the CMOS technology, but by the interfaces with which the image data can be transferred to the computer (Section 4.4.6).

Extension 4.3: Image sensors with high dynamic range.

The contrast range of an image sensor is limited by the fact that only a limited number of charge carriers can be stored on a pixel. This means that the contrast range is significantly lower than that of the human eye (see Section 5.6.4). CMOS image sensors offer a wealth of possibilities to increase the contrast range by combining images that were taken under different conditions at the same time or immediately one after the other:

- Images taken in immediate succession with different exposure times.
- Readout of the accumulated charge with different gains. This leads to an increase in the contrast range, as the dark noise is lower with high amplification. This technology was first implemented in expensive image sensors for scientific applications. The term "*Scientific CMOS*" (*sCMOS*) was coined.
- Parallel acquisition with photodiodes of different sizes on the same pixel.

Another possibility are image sensors with a logarithmic characteristic curve. Despite a long development history [136], these sensors still lead a niche existence.

4.4.4 ♯Micro Lenses

With many image sensors, not the entire area of a pixel is light-sensitive. With interline transfer CCD sensors, the *fill factor* (ratio of the light-sensitive area to the total area of the pixel) is relatively low at only 20–30%, as the rest of the area is required for the interline memory and wiring. Initially, the fill factor of CMOS image sensors was even smaller, but due to the increasing miniaturization of the components and the stacking of several wafers with contacts up to individual pixels on top of each other (*stacked image sensor*), it has now become much better than that of interline CCD sensors. A micro lens on each pixel can overcome this disadvantage and effectively improve sensitivity by a factor of two to three (Fig. 4.8).

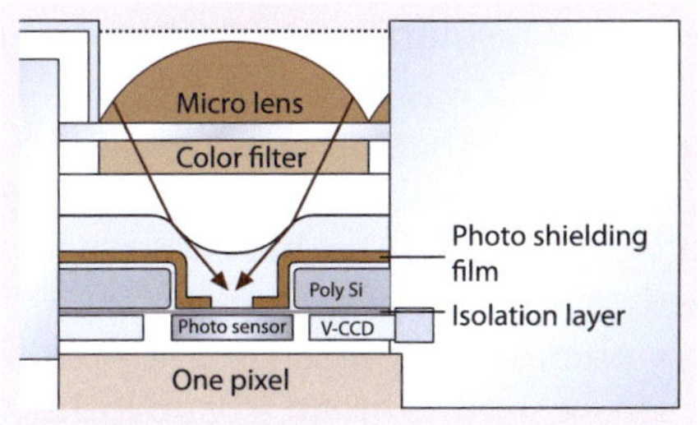

Figure 4.8: *Schematic representation of a pixel of a modern image sensor with micro lens and color filter.*

However, this improvement comes at the price of restricting the receiving cone for incident radiation. Therefore, an image sensor with a micro lens array for a lens with a large aperture (f-number smaller than 2) can have a significantly lower sensitivity than expected. There may also be a sharp drop in the effective irradiation of the sensor elements towards the edge of the image sensor if the chief ray are incident at an angle. To counter this effect, image-side telecentric lenses (Section 3.4.5) are of interest. This problem can be avoided with *image-side telecentric lenses* (Section 3.4.5). Alternatively, it is also possible to adapt the image sensors to the chief ray angle, which becomes more oblique towards the edge, by placing the micro lenses on the pixels with a corresponding shift towards the edge. In addition, micro lenses make the image sensors largely insensitive to UV radiation, as they are not transparent in this wavelength range.

4.4.5 Color, Spectral and Polarization Image Sensors

For an image sensor that should see colors like humans (Section 2.5.2), three color channels are required that should have the same spectral sensitivity as the human eye. This can be achieved by taking three image sensors — one each for the colors red, green and blue — and splitting the light from a lens into these three color components using two dichroic beam splitters. This procedure has two advantages. The full local resolution is retained and hardly any light is lost. The disadvantages are also evident. With three image sensors, the camera becomes larger and special lenses must be used for optimum image quality, which are corrected taking into account the beam splitters between the lens and the image sensors.

It is much easier to apply different color filters directly to the image sensor elements. By far the most commonly used color filter system is the *Bayer pattern*. In a block of 2×2 pixels, the upper left and lower right are provided with a green filter, the upper right with a red filter and the lower left with a blue filter. The simplicity of this setup is impressive, but it also has disadvantages. The larger distances when sampling the individual color channels reduce the spatial resolution and lead to additional *aliasing* effects (Section 5.1).

For technical applications where objects need to be differentiated based on their optical properties, it is neither necessary nor useful to imitate human color vision. Here, it is best to select the spectral ranges with which this task can best be performed.

In principle, two different approaches are possible. You can use a monochrome image sensor with as broad a bandwidth as possible and select the spectral ranges using bandpass filters on the optics or using appropriate lighting. LED lighting is best suited for the latter, as efficient LEDs with bandwidths of 10–50 nm are available for wavelengths from UV with 265 nm to 1050 nm in the near IR. Alternatively, multispectral mosaic image sensors can also be used, which not only have three color channels, but much more spectral channels, e. g. 16 in a block of 4×4 pixels [62].

In remote sensing from satellites or flying objects, so-called hyperspectral push-broom camera systems have been in use for a long time. These contain a dispersive element so that one coordinate of the image sensor is used for the spectrum and the second for the spatial resolution perpendicular to the direction of flight. The second local coordinate in the direction of flight results from the recording of image sequences.

Another important characteristic of electromagnetic radiation is the polarization state discussed in Section 2.2.7. This cannot be detected by normal image sensors. However, it is possible to provide pixels with polarization filters instead of color filters. Sony has presented image sensors for the first time in which polarization filters have been integrated on a 2×2 grid, which can simultaneously record the proportion of linearly polarized light in four directions rotated by 45° each [217].

According to the considerations in Section 2.2.7, the first three components of the *Stokes vector* can be determined. However, circularly polarized light remains unknown. Polarization image sensors are, e. g. very useful for determining the orientation of smooth surfaces made of dielectric materials. To do this, it is sufficient to irradiate them with non-polarized light [12, 104, 361].

Extension 4.4: Better color image sensors

Current solutions for color image sensors are far from ideal. It would be ideal if we knew the wavelength of every absorbed photon. However, this is not possible when measuring with the *inner photoeffect*. You only know after (Eq. 4.2) that the wavelength was smaller than the cut-off wavelength λ_g. There were attempts to exploit the fact that blue light is absorbed much more strongly than green light, and the latter more strongly than red light (cf. Fig. 4.1), and so to arrange three photodiodes on top of each other. The *Foveon sensor* is the best-known example of this. The problem with this is that the three color channels are spectrally too wide, making the color space obtained too small and deviating greatly from human color vision (Fig. 2.18).

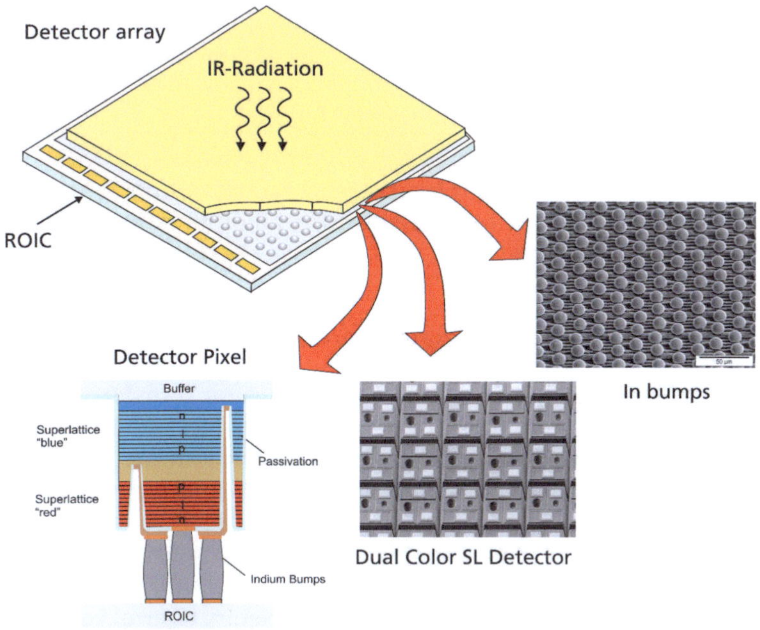

Figure 4.9: *Design details for a two-color "focal plane array" (from the Fraunhofer Institute for Applied Solid State Physics, Freiburg).*

Figure 4.10: *IR two-color image, the radiation in the 3–4-µm range is shown in cyan, that in the 4–5-µm range in red (from the Fraunhofer Institute for Applied Solid State Physics, Freiburg [258]).*

There is a better solution in thermal infrared. By alternating many thin layers of InAs and GaSb, each only a few monolayers thick, it has been possible to create superlattices that can only absorb photons with a high quantum efficiency in a narrow wavelength range and are otherwise transparent. This allows two pixels with different wavelength ranges to be arranged on top of each other (Fig. 4.9). The Fraunhofer Institute for Applied Solid State Physics has used this technique to develop a two-color FPA of size 384×288 for the spectral ranges 3–4 μm and 4–5 μm [259]. A panoramic image taken with it, composed of several partial images, shows Fig. 4.10.

In summary, we can conclude at the end of this section that an image sensor with microlenses, color filters and anti-reflective coatings is a complex opto-electronic component today. Therefore, an optimal image acquisition system requires careful matching between lenses and image sensors.

4.4.6 ♯Digital Standard Hardware Interfaces

Unfortunately, there is no standardized interface for transferring the serial image data stream of a digital camera to the computer and simultaneously configuring and controlling the camera from the computer. This is due to the fact that there are too many different cameras and the requirements of the various image processing applications are too diverse to be covered by a single standard. With some interfaces, only the physical level is standardized, i. e. the connectors and the protocols for transmitting the image data, while each manufacturer still has its own proprietary control commands for controlling the camera.

Camera Link (CL)
One of the oldest standardized interfaces is Camera Link[4]. It is a high-end standard for high-speed transmission of image data with precise trigger options. The standard is based on the serial transmission protocol called *Channel Link* and uses multiple serial channels to transmit the image data and control signals. Depending on the number of serial channels used (base, medium or full configuration), 24, 48 or 64 bits can be transmitted in parallel at clock rates of up to 80 MHz. At a clock rate of 80 MHz, this corresponds to data rates of 240, 480 and 640 MB/s.

In 2009, Camera Link was the fastest available standard for image transmission. Standard PC bus systems such as the PCI bus can be used in base configuration, while the medium and full configuration requires the faster PCI Express bus. This fact underlines a serious difficulty of high-speed image transmission. There are camera systems today that deliver more than 4000 Mpixel/s (> R10), while standard interfaces for peripheral devices cannot handle this data stream. This problem makes high-speed image capture expensive and impractical. The camera must have a fast storage system of several GB, in which the high data stream is temporarily stored and then transferred to the computer much more slowly.

The Camera Link standard only regulates the physical level of image data transfer, but not the way in which the camera is controlled. Although there is an additional serial channel with an adjustable baud rate, each camera

---

[4] `https://www.automate.org/vision/vision-standards/vision-standards-camera-link`

manufacturer uses its own commands to set the camera parameters, such as the exposure time.

Camera Link HS (CLHS)
The high demands on image data transmission have triggered two initiatives with the aim of further increasing the transmission bandwidth. One of these is an extension of the Camera Link standard to the *Camera Link HS*[5] standard. The transfer rates are between 300 and 6000 MBytes/s. In addition, it is planned to transmit the data via fiber optics, which will enable significantly longer cable connections than with Camera Link, where no 10 m are possible.

CoaXPress (CXP)
The standard *CoaXPress*[6] was presented at the Vision trade fair in Stuttgart in 2009 and was awarded the Vision Prize. It is an asymmetrical high-speed connection between camera and computer that can be distributed via one to four coaxial cables. Image data can be transmitted from the camera at 1.25–6.25 Gbit/s per cable, while control commands and configuration data are transmitted from the computer to the camera at 20 Mbit/s. This roughly corresponds to the speed of Camera Link, but a simple coaxial cable is sufficient for the connection. The power supply for the camera can also be transmitted via the coaxial cable ("Power-over-Coax") and cable lengths of over 100 m are possible.

Firewire (IEEE 1394)
This standard connects digital cameras to the computer via a standard interface for peripheral devices, which is also used by other devices such as external hard disks. The big advantage is the low cost. The camera can also be powered via the Firewire bus. The disadvantages are the low transfer rate of around 30–40 MB/s and the short connection cables with a maximum length of 5 m without bus repeater.

The *DCAM* standard was also defined for the first time with *Firewire* or *IEEE1394* in order to set the basic functions of cameras, such as exposure time, offset, and gain, in a standardized way. With the extension to IEEE1394b (Firewire 800), the transfer rate was doubled and fiber optic cables can be used to connect the camera to the computer. Due to the unfavorable licensing conditions for use, the standard has now been replaced by the *Universal Serial Bus* and has become obsolete.

USB3 Vision
Another widely used standard for connecting peripheral devices to computers is the "Universal Serial Bus Interface" or *USB* for short. In addition to Firewire, this is a second cost-effective solution for connecting cameras to a computer. The transmission speed of USB 2.0 is comparable to Firewire, but

[5] `https://www.automate.org/vision/vision-standards/vision-standards-camera-link-hs`
[6] `https://www.coaxpress.com/`

there is no standard for controlling the cameras. Each camera manufacturer has its own proprietary drivers.

Fortunately, this changed with the introduction of the USB 3.0 standard. With the initiative *USB3 Vision*[7], the USB 3.0 interface is integrated into the generic programming interface GenICam (see below) for cameras.

USB3 Vision brings with it a significant increase in the data transfer rate to more than 350 Mbytes/s. This makes it by far the fastest standard interface for connecting cameras to computers.

As anyone who uses USB devices knows, the USB standard is in a rather chaotic state today. This applies above all to the variety of possible further developments in the direction of USB4 or Thunderbolt, including the naming. Accordingly, the further development of the standard from USB3 Vision to higher image data transfer speeds has also stalled.

GigE Vision
The *GigE Vision*[8] standard utilizes the high data rates that are possible via Gigabit Ethernet network connections. The biggest advantage of this standard is the cheap and up to 100 m long cables. However, at just under 100 Mbytes/s, the GigE interface is more than three times slower than USB3 Vision. These restrictions will hopefully soon be lifted. With the development of version 3 of the GigE Vision standard, which had not yet been completed at the time this book was finalized, transmission rates of up to 400 Gbps, i. e. up to 400 times faster than before, should be possible. Cameras with a 100 GigE interface are already available, although they are still proprietary due to the pending standardization.

4.4.7 ♯Standard Software Interfaces

Due to the very different requirements for cameras, there will always be different standardized interfaces that are used to transfer image data from the camera to a computer or an embedded system. This makes it all the more important that standardized software controls the image data transfer and allows the camera parameters to be set.

The standard *GenICam*[9] fulfills precisely this task. This provides a generic interface for connecting cameras to computers. The following are supported Camera Link, GigE Vision, HSLink and USB3 Vision. The rapid integration of new physical interfaces such as USB3.0 is proof of the success of this approach.

However, the GenICam standard is not yet ideal from a user's perspective. One reason for this is that cameras can have very different properties, which have to be set using corresponding optional or proprietary parameters. For this reason alone, you often have to adapt your software individually to specific cameras. Secondly, even elementary parameters such

[7] `https://www.automate.org/vision/vision-standards/usb3-vision-standard`
[8] `https://www.automate.org/vision/vision-standards/vision-standards-gige-vision`
[9] `https://www.GenICam.org`

as exposure time, signal offset (called black level) or gain are not handled uniformly and therefore make life difficult for the user. Furthermore, many parameters that are actually necessary do not have to be implemented by a manufacturer but are only optional or recommended.

Almost all camera manufacturers offer GenICam-based viewers and software development tools, most of which are supplied free of charge with the cameras or can be downloaded from the Internet. The ambivalent relationship of manufacturers to standards is expressed in the fact that only their own cameras can be operated with these tools. Cameras from other manufacturers, which could also be operated without any problems due to the standard, are simply hidden in almost all tools, with a few laudable exceptions. This makes it difficult for the user to operate cameras from several manufacturers at the same time, although this is often necessary as one manufacturer does not offer all the cameras required. As a user, one would actually wish for an openly available code for the GenICam standard. This would also make it easier for the entire community to work on it.

4.5 Characterization of the Image Signal Quality

4.5.1 What Limits the Signal Quality?

In this section, we first discuss which effects limit the quality of an image sensor signal on the basis of what we have worked out so far. Based on the considerations on the architecture of image sensors in connection with the question of an ideal image sensor in Section 4.4.1, we can easily summarize which effects limit the quality of an image sensor signal

Temporal Noise
As with all measurements, the accuracy of measurements with image sensors is limited by the fact that repeated measurements always produce slightly different values. This effect is caused by the temporal *noise*. We will therefore have to investigate where temporal noise can occur during a measurement with an image sensor.

Spatial Inhomogeneity
With an image sensor, we do not just take one measurement, but simultaneously on a large number of pixels. Typically, these are millions of measurements. With the same irradiation, each pixel will deliver slightly different results. This effect causes *spatial inhomogeneity*.

Dark Signal
Even without light, an image sensor delivers a signal with temporal noise. This will limit the minimum measurable irradiation. Furthermore, this signal will depend on the exposure time and thus also limit the maximum exposure time that is possible with an image sensor.

Figure 4.11: *General system-theoretical model for a digital camera. The camera is regarded as a "black box" from which only the input and output signals are measured (according to the EMVA 1288 standard, release 4.0 General of June 21, 2021 [72],* `www.standard1288.org`*).*

Maximum Signal
Due to the limited number of charge carriers that can be stored on a pixel, there is also a maximum possible irradiation that can be measured.

This summarizes the four basic limitations of an image signal. In addition, there is a wealth of other artifacts, i. e. deviations from ideal behavior. This can be, for example, a non-ideal electronic shutter that still lets light through, or the mutual interference of neighboring pixels.

4.5.2 Systems Theory Approach

We now ask ourselves how the four basic limitations of an image signal can be measured most easily. Simple in this context also means that it should not be necessary to open the camera and carry out internal measurements. In other words, anyone should be able to carry out these measurements.

This brings us directly to a system-theoretical approach that we have already successfully used to describe optical imaging (see Fig. 3.6 in Section 3.4.1). We consider the camera as a "black box" about which we have no or only minimal internal information (Fig. 4.11). The input signal is the number of photons that fall on each pixel during the exposure time, the output signal is the digital gray value for all pixels of the camera. The input signal should be selected for the measurements in such a way that all pixels are irradiated equally. In order to calculate the input signal correctly, an absolute calibration of the irradiance is necessary after (Eq. 4.6) in Section 4.3.2 and we need to know the pixel size and the exposure time. The output signal, a matrix $\boldsymbol{G}$ of values for all pixels, is slightly different for each pixel due to temporal noise and the inhomogeneity of the image sensor.

For a statistical analysis, it is therefore necessary to measure at least the *mean* $\mathrm{E}(\boldsymbol{G}) = g$, the *variance* of the temporal noise σ_g^2 and the variance of the inhomogeneity, which we denote here as s_g^2 to clearly distinguish it from the standard deviation of the temporal noise. We must calculate these values for a series of irradiations from the minimum signal in darkness to the maximum, where the sensor saturates.

For the practical implementation of these measurements, a difficulty arises when calculating the mean values and standard deviations. Since each pixel has different values due to the inhomogeneities, the following complex approach seems unavoidable, in which we have to take many images for each illumination level in order to be able to measure the variance accurately enough. According to Eq. (C.63) in Appendix C.4, the determination of the variance still has a relative error of 14% even when averaging over 100 values.

We can find a better solution if we consider that an image typically has millions of pixels. Therefore, it could be an effective solution to calculate the variance of the temporal noise from the deviations of the values of all pixels from the mean value of only one image. Unfortunately, it is not that simple, as the gray values vary due to both temporal noise and spatial inhomogeneity:

$$\operatorname{var}(\boldsymbol{G}) = \frac{1}{MN-1}\sum_{m=0}^{M-1}\sum_{n=0}^{N-1}(g_{mn} - g)^2 = \sigma_g^2 + s_g^2. \tag{4.13}$$

Therefore, by averaging over only one image, we obtain the sum of the variances of the temporal noise and the spatial inhomogeneity. Nevertheless, this approach leads to a solution, as the inhomogeneities are stationary by definition. Should they nevertheless vary over time, this would be part of the temporal noise. This means that the inhomogeneities in the difference between two images taken with the same irradiation are eliminated. The variance of the temporal noise of the difference image is twice as high as that of a single image (see Eq. (C.31)):

$$\sigma_g^2 = \operatorname{var}\left({}^1\boldsymbol{G} - {}^2\boldsymbol{G}\right)/2 = \frac{1}{2MN-2}\sum_{m=0}^{M-1}\sum_{n=0}^{N-1}\left[\left({}^1g_{mn} - {}^2g_{mn}\right) - \left({}^1g - {}^2g\right)\right]^2. \tag{4.14}$$

To be on the safe side, we have assumed that the mean values of the two images are not the same. This can occur if either the irradiation is slightly different or the mean value of the camera fluctuates slightly over time. Both changes then have no effect on the calculation of the temporal variance. Otherwise this would be distorted towards higher values.

This results in the surprisingly simple relationship for the variance of the spatial inhomogeneity after multiplying out all terms in the sums

$$s_g^2 = \left[\operatorname{var}\left({}^1\boldsymbol{G} + {}^2\boldsymbol{G}\right) - \operatorname{var}\left({}^1\boldsymbol{G} - {}^2\boldsymbol{G}\right)\right]/2 = \mathrm{E}\left[\left({}^1\boldsymbol{G} - {}^1g\right)\left({}^2\boldsymbol{G} - {}^2g\right)\right]. \tag{4.15}$$

This result is easy to understand and represents the *covariance* (see Eq. C.20 in Appendix C.3.2) of the two mean-free images. Since the temporal noise is independent in both images, it is not included in the covariance and what remains is the spatial inhomogeneity, which is the same in both images.

We have thus found an efficient method for calculating both the variance of temporal noise and that of spatial inhomogeneity from just two images taken under the same conditions. With a megapixel sensor, the relative standard deviation of the variance determination is 0.14% according to Eq. (C.63). The accuracy of s_g^2 only becomes critical if $s_g^2 \ll \sigma_g^2$. This is due to the difference formation in Eq. (4.15). If s_g^2 is 0.07% of σ_g^2, then the difference between the variances is as large as the statistical error and therefore the relative error is 100%. This means the following for the *standard deviation* as an important comparison variable to the signal: If s_g is 2.6% of σ_g, then the relative error of s_g is 100%. This means that with a megapixel sensor, it will not be possible to measure spatial inhomogeneities that are smaller than a few percent of the standard deviation of the temporal noise using this method. This is hardly the case, as both disturbances are of approximately the same order of magnitude.

4.5.3 Signal-to-Noise and Signal-to-Nonuniformity Ratio (SNR)

Now we come to the important question of which quantity best describes the signal quality. This is not the size of the signal itself, but how high the signal is in relation to the disturbance variables.From the basic considerations in Section 4.5.1 we know that there are only two effects, namely temporal noise and spatial inhomogeneity.

This consideration leads to the definition of *signal-to-noise ratio* or, abbreviated *SNR*, if we first consider only the effect of temporal noise:

$$\mathrm{SNR} = \frac{g - g_0}{\sigma_g}. \tag{4.16}$$

We must not use the squared value (variance) for the comparison. Instead, we must use the square root of the variance, the *standard deviation*. In addition, the dark value of the signal, g_0, must be subtracted so that only the light-induced signal is taken into account in the ratio.

The consideration of only the SNR is justified by the fact that the spatial inhomogeneity is stationary and should therefore actually be completely correctable. Unfortunately, this is not the case in practice for two reasons:

- The inhomogeneity change slightly with temperature, exposure time and ageing processes.
- The inhomogeneity may depend on the wavelength of the irradiation. If this is the case, the inhomogeneities of an image sensor can only be corrected for a specific spectral composition.

For these reasons, it makes sense to also consider the influence of the inhomogeneity, i. e. the *signal-to-inhomogeneity ratio*, which could also be called SNR after signal-to-nonuniformity ratio. If you combine both effects,

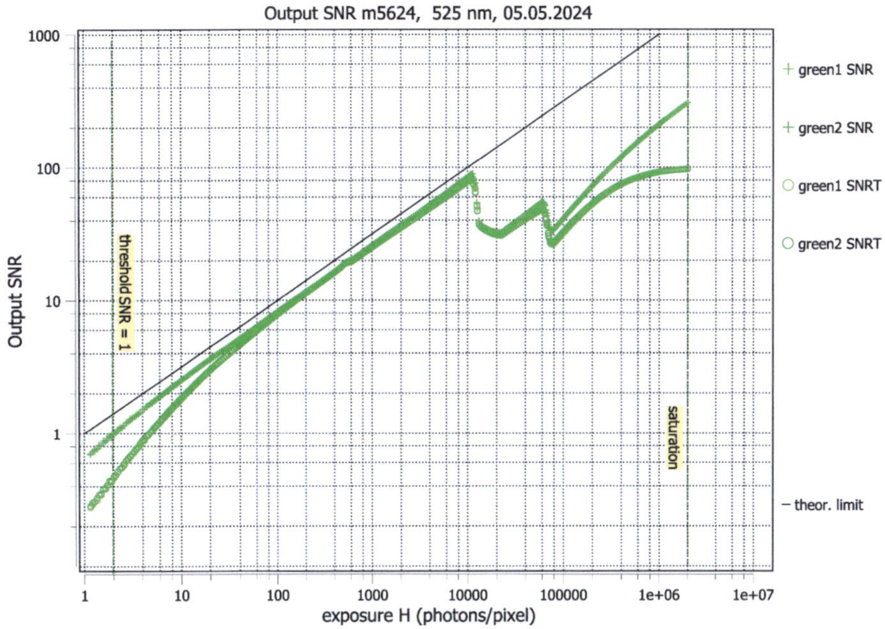

Figure 4.12: *Illustration of the SNR and the total SNR for two green channels of the Sony IMX490 HDR color image sensor with linear 24-bit output signal in double-logarithmic representation. The SNR of an ideal sensor is shown as a black line.*

you get the total SNR or *SNRT*:

$$\mathrm{SNRT} = \frac{g - g_0}{\sqrt{\sigma_g^2 + s_g^2}}. \tag{4.17}$$

Note that the variances (Appendix C.3.3) rather than the standard deviation must be added for statistical errors.

Even with an ideal image sensor that has no internal noise sources, the SNR is not infinite. This is because the input signal is already noisy (Fig. 4.11). The noise of the input signal is Poisson distributed (Appendix C.4.1). If on average N_p photons are incident on a pixel during the exposure time, then the variance of the temporal noise is equal to the mean value Eq. (C.42) and the standard deviation is therefore the square root of the mean value. Therefore, the SNR of an ideal image sensor, which also has no inhomogeneity,

$$\mathrm{SNR}_{\mathrm{ideal}} = \sqrt{N_p}. \tag{4.18}$$

This means that the SNR of an ideal image sensor increases with the square root of the number of photons per pixel. At 10,000 photons/pixel, a typical number of a sufficiently irradiated image sensor, the SNR is therefore a

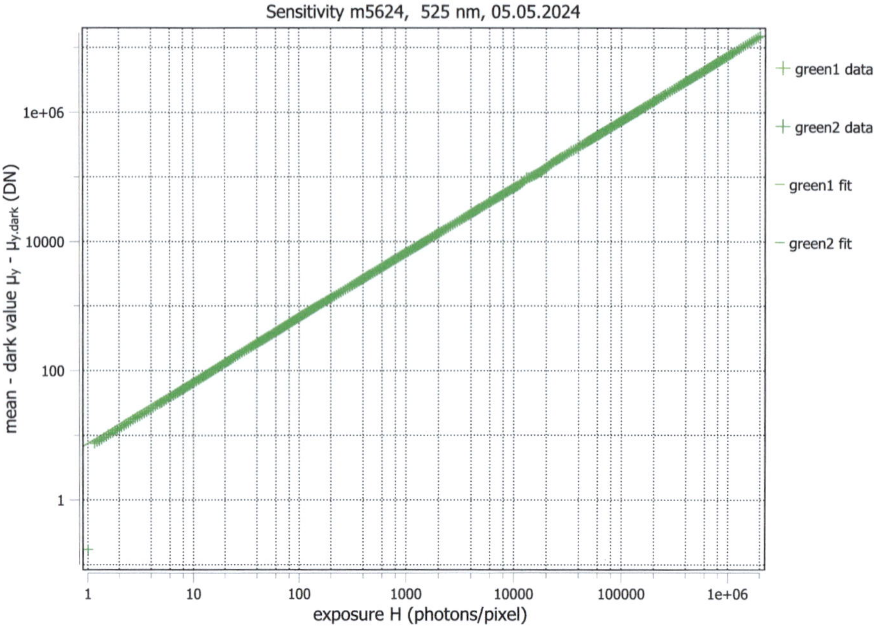

Figure 4.13: *Characteristic curve of the green pixels of the HDR color image sensor IMX490, whose SNR is shown in Fig. 4.12, in double-logarithmic representation over six decades. In this mode, the sensor has an output signal of 24 bits.*

maximum of 100 (1% standard deviation of the signal fluctuation), at 100 photons/pixel only 10 (10% signal fluctuation).

If the SNR is plotted double-logarithmically against the irradiation in photons/pixel (Fig. 4.12), there are two advantages. Firstly, the SNR of an ideal sensor in this representation is a straight line with a gradient of 0.5, which makes it easy to compare with a real sensor. Its SNR must always be below that of an ideal sensor. The distance to the ideal SNR directly indicates the factor by which the sensor is worse than the ideal one. On the other hand, the entire signal range from one photon/pixel to saturation can be represented over several orders of magnitude in the logarithmic representation.

Figure 4.12 also shows the SNR and SNRT of a highly dynamic camera over an irradiation range of more than six decades. The signal of this camera is composed of four signals. Each pixel has a large photodiode, which is used for low irradiation, and a small one, which is used for high irradiation. Furthermore, the signal of each photodiode is read out with high and low amplification. As long as the large photodiode is used, the SNR is very close to the SNR of an ideal image sensor. This is the case for irradiations up to about 10,000 photons/pixel. Then the system first switches to the small photodiode with high amplification. The SNR first drops abruptly and then rises again parallel to the ideal SNR. At around 60,000 photons/pixel, the system switches to the low gain. The SNR drops sharply again and then continues to rise until the sensor reaches saturation at 2,000,000 photons/pixel and has a high SNR of 300.

At this high saturation, the SNR is of course much worse than with an ideal sensor. However, if so much light is available, further improvement is not necessary. The SNR is high enough. It would be ideal for image processing if the SNR were independent of the irradiation. The sensor achieves an SNR of greater than 30 over more than three decades. The SNR of a good industrial image sensor with only one photodiode per pixel and a fixed minimum gain only achieves an SNR of greater than 30 over around 1.5 decades (Fig. 4.15).

If you look at the SNRT in Fig. 4.12, you will notice that the inhomogeneities of the small photodiode dominate the signal quality at high saturations. The SNRT is then a factor of three worse than the SNR and only reaches a value of 100. With the large photodiode, there is no significant effect from the inhomogeneities at high saturations. These only worsen the SNR at very low irradiations.

The signals of the four channels of the HDR sensor are interconnected in such a way that a linear characteristic curve results over the entire irradiation range of six decades, from which the measured values show only slight deviations (Fig. 4.13).

4.5.4 Input and Output SNR

So far, we have only looked at the SNR of the output signal. This is sufficient if the characteristic curve of a sensor is linear, as in the example of an HDR camera considered so far. Then the signal and noise are amplified by the same factor and the SNR of the output signal is identical to the SNR of the input signal.

However, this is no longer the case if the characteristic curve is non-linear. Then the input and output SNR are different and must be distinguished. What we are actually interested in is not the measured output SNR, but the input SNR, because this indicates the accuracy with which we can measure the input signal, the irradiance.

We therefore need to find out how we can calculate the input SNR from the measured output SNR. To do this, we first look at the definitions:

$$\mathrm{SNR}_{\mathrm{out}}(N_p) = \frac{g - g_0}{\sigma_g} \qquad \mathrm{SNR}_{\mathrm{in}}(N_p) = \frac{N_p}{\sigma_p}. \tag{4.19}$$

In this equation, the measurable variables are marked in blue and the unknown variables in red. The quantity σ_p is the standard deviation for measuring the number of photons, which includes not only the Poisson-distributed noise of the input signal, but also all other noise sources in the camera, projected back onto the input signal.

Using error propagation, we can calculate how the standard deviation of the output signal depends on that of the input signal. According to Eq. (C.16), this is related to the slope of the characteristic curve $(g - g_0)(N_p)$ (light-induced output signal $g - g_0$ as a function of the input signal N_p):

$$\sigma_g = \left| \frac{\mathrm{d}(g - g_0)}{\mathrm{d}N_p} \right| \sigma_p. \tag{4.20}$$

This results in the input SNR of

$$\mathrm{SNR}_{\mathrm{in}} = \frac{N_p}{\sigma_p} = \frac{N_p}{\sigma_g} \frac{\mathrm{d}(g - g_0)}{\mathrm{d}N_p} = \frac{N_p}{g - g_0} \frac{\mathrm{d}(g - g_0)}{\mathrm{d}N_p} \mathrm{SNR}_{\mathrm{out}} = \frac{\mathrm{d}\ln(g - g_0)}{\mathrm{d}\ln N_p} \mathrm{SNR}_{\mathrm{out}}. \tag{4.21}$$

Example 4.1: Relationship between input and output SNR for different characteristic curves

First, we consider the linear characteristic $g = \eta K N_p$ or $\ln g = \ln(\eta K) + \ln N_p$ and verify that the input and output SNR are identical:

$$\frac{\mathrm{d}\ln g(\ln N_p)}{\mathrm{d}\ln N_p} = 1 \quad \rightsquigarrow \quad \mathrm{SNR}_{\mathrm{in}} = \mathrm{SNR}_{\mathrm{out}}.$$

The second example is the gamma characteristic $g = a_1 (N_p)^\gamma$ or $\ln g = \ln a_1 + \gamma \ln N_p$:

$$\frac{\mathrm{d}\ln g(\ln N_p)}{\mathrm{d}\ln N_p} = \gamma \quad \rightsquigarrow \quad \mathrm{SNR}_{\mathrm{in}} = \gamma\, \mathrm{SNR}_{\mathrm{out}}.$$

With the gamma characteristic curve, the input and output SNR differ by the factor γ.

We have thus found a method for calculating the input SNR from the measured output SNR for any characteristic curve. It is important to note that the input SNR must be considered and displayed for a non-linear characteristic curve. In the world of non-linear systems, it is essential to strictly differentiate between the two variables in order to avoid incorrect conclusions. It is of course possible to compare the input SNR of non-linear and linear cameras without any problems. The non-linearity does not play a role here.

4.5.5 SNR for Linear Cameras

We were unable to calculate the quantum efficiency of the sensor from the previous signal analysis. This is because we did not use a model. From the point of view of an image sensor expert, this is of course unsatisfactory, as they want to know which parameters determine the signal quality of the sensor. From the user's point of view, this is irrelevant. Ultimately, the user is only interested in what quality the sensor delivers at what irradiation and whether this quality is limited by temporal noise or spatial inhomogeneity.

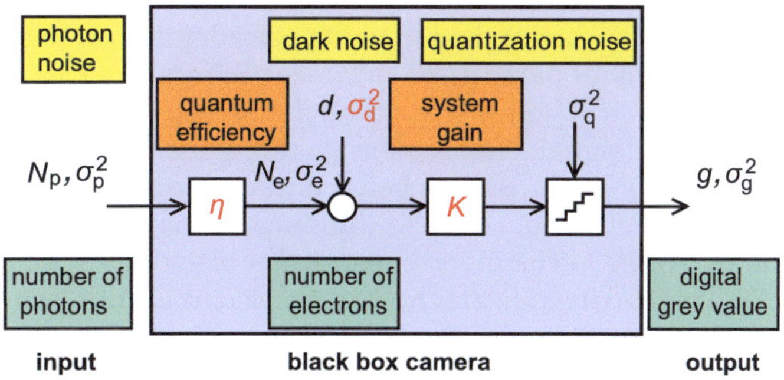

Figure 4.14: Model for a digital camera with a linear characteristic curve. The camera is shown as a black box containing only three process steps: the conversion of photons into charge carriers, their linear amplification and quantization. The unknown camera parameters are marked in red (according to the EMVA 1288 standard, release 4.0 Linear from June 21, 2021 [71], `www.standard1288.org`*).*

For most industrially used cameras, a model can be used to determine the quantum efficiency and other model parameters. The following prerequisites are necessary for this:

1. The camera must have a linear characteristic curve. This is the only way to summarize the many different noise sources in the camera in two simple parameters, the dark noise with the variance σ_d^2 and the quantization noise with the variance σ_q^2 (Fig. 4.14). This is possible because with a linear characteristic curve, noise sources that occur at different points in the processing chain are amplified in the same way as the signal itself. We can therefore think of all these noise sources at the input of the electronic processing chain and summarize them in the dark noise. Only the quantization noise must be excluded, as quantization is a non-linear process (Section 5.6). Therefore, the quantization noise must remain at the end of the process chain where it occurs.
2. A photon may only generate one charge carrier. This is the case in the near UV, visible range and near IR, but not in the deeper UV. In the latter, the photon energy is much greater than the band gap. This means that two charge carriers are generated with a certain probability, which implies a further source of noise.
3. The raw data must be available. Correction of defective pixels is not a problem, but any pre-processing that changes the temporal noise is. This includes operations such as debayering, i. e. interpolation of missing color values, edge sharpening and noise suppression.

The simple linear model is based on the signal processing chain described in this chapter (Section 4.4.1). The internal photoelectric effect converts some of the incident photons into charge carriers, which are accumulated over the exposure time and linearly amplified together with the dark signal and then digitized (Fig. 4.14). The aim of the following consideration is to determine the three unknown model parameters *quantum efficiency* η, variance of *dark noise* σ_d^2 and *system gain* K solely from the relationship between the input signal and the output signal. The system gain is actually a dimensionless number, but the unit DN/e$^-$ is used for better understanding. An ideal image sensor that would directly count the photo-induced charge units would have a system gain of 1 DN/e$^-$. The inverse gain K^{-1} is also often specified with the unit e$^-$/DN. This clearly indicates how many electrons must be generated to achieve one digit.

To carry out the measurement, it is necessary to homogeneously illuminate the image sensor without a lens with a series of irradiances. These must be absolutely calibrated in order to be able to calculate the average number of photons according to Eq. (4.6). In addition, the pixel area A must be known. If this is not specified by the manufacturer, it can be determined by a geometric calibration using a lens with a known focal length and a test target (cf. Section 3.4.2).

Our considerations are based on the standard 1288 of the European Machine Vision Association (*EMVA*), only the nomenclature has been adapted to that of this book. We first look at the mean values. From Eqs. (4.4) and Fig. 4.14 we know the dependence of the output signal g on the irradiation of the sensor element, i. e. the number of photons N_p that fall on the image sensor during the exposure time. The output signal g is obtained by first converting N_p photons with the quantum efficiency η into N_e electrons and then amplifying these linearly by the factor K to the output signal. This results in the following relationship:

$$g = g_0 + K N_e = g_0 + K\eta N_p. \tag{4.22}$$

After subtracting the dark value g_0, the result is

$$\boxed{g - g_0 = \underbrace{K\eta}_{\text{slope } a} N_p. \quad \text{characteristic curve}} \tag{4.23}$$

The slope of the linear *camera characteristic curve* is the product of the system gain K and the quantum efficiency η.

For further analysis, it is necessary to examine the variance of the temporal noise of the output signal For this purpose, it is necessary to perform an error propagation calculation to determine how the individual noise sources contribute to the noise of the output signal. The theoretical principles are described in the appendices C.2.3 and C.3.3. The two rules we need from this are that a) when two statistically independent signals are added, their variances add up — regardless of the probability distribution —, see Eq. (C.30), and b) when a signal is amplified by a factor of K, the variance must be multiplied by K^2, see Eq. (C.17).

We start with the noise of the photo-induced electrons at the beginning of the electronic chain. According to the laws of quantum mechanics, the number of measured events in a given time interval is subject to a *Poisson distribution* (Appendix C.4.1). Therefore, the variance of the fluctuations is equal to the mean number of accumulated charge carriers:

$$\sigma_e^2 = N_e. \tag{4.24}$$

This noise is unavoidable and is the same for all image sensors.

All other noise sources, such as readout noise and amplifier noise, depend on the specific design of the image sensor and the camera electronics. As already discussed in the last section, these can be summarized in an image signal-independent normally distributed noise source with the variance σ_d^2 due to the linear camera characteristic curve and considered as additional noise at the input (Fig. 4.14). The only noise source that cannot be shifted and must therefore be considered separately is the pseudo noise due to the final *quantization*, which we will discuss in detail in the next chapter (Section 5.6). After Eq. (5.54), additional uniformly distributed noise with the variance $\sigma_q^2 = 1/12\,\text{DN}^2$ comes from this noise source.

With the aforementioned laws of error propagation and using Eq. (4.24) we obtain

$$\sigma_g^2 = K^2(\sigma_d^2 + \sigma_e^2) + \sigma_q^2 = K^2(\sigma_d^2 + N_e) + \sigma_q^2. \tag{4.25}$$

Since the mean number of electrons N_e cannot be measured directly, we replace this quantity after Eq. (4.22) with $(g - g_0)/K$ and obtain

$$\boxed{\sigma_g^2 = \underbrace{K^2\sigma_d^2 + \sigma_q^2}_{\text{Offset}} \quad + \underbrace{K}_{\text{slope } b}(g - g_0) \quad \text{photon transfer curve.}} \tag{4.26}$$

This equation is of central importance for the characterization of image sensors. From the linear relationship between the variance of the noise of the photoinduced output signal σ_g^2 and the mean photoinduced signal $g - g_0$, the system gain can be determined directly from the slope and the variance of the dark noise σ_d^2 from the offset. This method is known as *photon transfer method* [148, 150].

The variance of the dark signal consists of the two terms

$$\sigma_0^2 = K^2\sigma_d^2 + \sigma_q^2 \tag{4.27}$$

and has the unit DN^2. From this quantity and the known system gain K, σ_d can be determined as follows

$$\sigma_d = \frac{\sqrt{\sigma_0^2 - \sigma_q^2}}{K}. \tag{4.28}$$

This standard deviation has the unit number of electrons, i. e. e^-, and reflects the equivalent noise at the output of the image sensor before any amplification and without being influenced by the quantization noise.

In addition, we can determine the quantum efficiency η, since the slope a of the sensitivity curve according to Eq. (4.22) is equal to the product ηK. Therefore, η results in

$$\eta = \frac{a}{b}. \tag{4.29}$$

The linear model also enables the calculation of the SNR Eq. (4.16). It results as a function of the irradiation in number of photons (N_p) using Equations (4.22) and (4.26):

$$\boxed{\mathrm{SNR}(N_p) = \frac{\eta N_p}{\sqrt{\sigma_d^2 + \sigma_q^2/K^2 + \eta N_p}}.} \tag{4.30}$$

Apart from the small effect caused by the quantization noise, the system gain K is reduced as expected, so that the SNR depends only on the quantum efficiency $\eta(\lambda)$ and the dark noise σ_d (unit e^-).

Two limiting cases are of interest in order to better understand the SNR: firstly the limiting case of high irradiation with $\eta N_p \gg \sigma_d^2$ and secondly the limiting case of low irradiation with $\eta N_p \ll \sigma_d^2$. These approximations result in

$$\mathrm{SNR}(N_p) \approx \begin{cases} \sqrt{\eta N_p}, & \eta N_p \gg \sigma_d^2 + \sigma_q^2/K^2, \\[2ex] \dfrac{\eta N_p}{\sqrt{\sigma_d^2 + \sigma_q^2/K^2}}, & \eta N_p \ll \sigma_d^2 + \sigma_q^2/K^2. \end{cases} \tag{4.31}$$

This means that the SNR curve initially rises linearly and at high irradiation levels rises more slowly only with the square root of the irradiation.

An ideal image sensor would have no dark and quantization noise and the quantum efficiency $\eta = 1$. Therefore, for an ideal sensor, Eq. (4.30) results in an SNR that we have already derived in general, see Eq. (4.18).

The spatial variance of the inhomogeneity of an image sensor with a simple linear model can also be easily described analytically. With a linear characteristic curve, each pixel can have a different dark value and a different slope. The standard deviation of the spatial inhomogeneity of the dark value is called *dark signal nonuniformity*, or abbreviated *DSNU*, that of the slope *photo response nonuniformity*, or *PRNU* for short. According to the EMVA 1288 standard, DSNU and PRNU are defined as follows

$$\begin{aligned} \mathrm{DSNU}_{1288} &= s_{g0}/K \qquad (e^-), \\ \mathrm{PRNU}_{1288} &= \frac{\sqrt{s_g^2 - s_{g0}^2}}{g - g_0} \qquad (\%). \end{aligned} \tag{4.32}$$

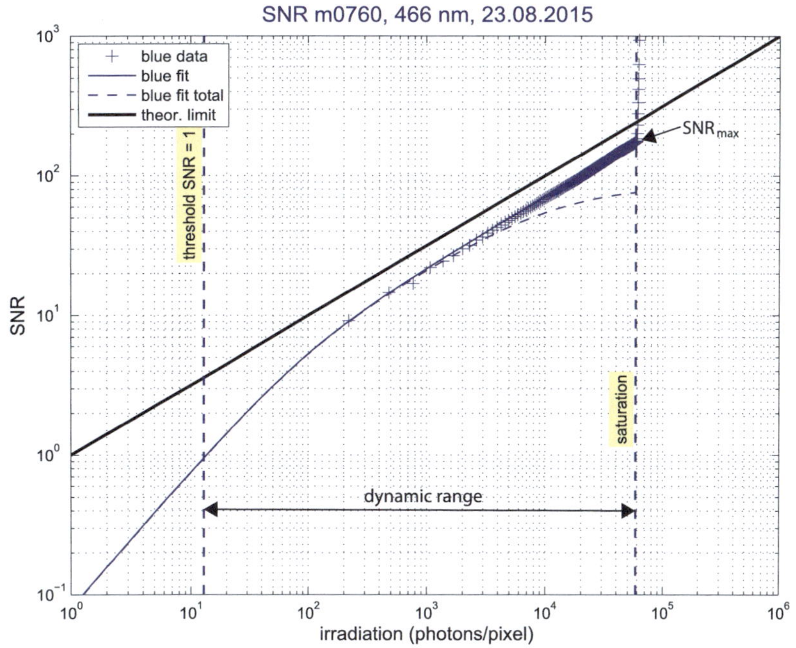

Figure 4.15: *Representation of the SNR in double-logarithmic representation. Measured values for a modern CMOS sensor are plotted together with the theoretical curve determined from the measured quantum efficiency and the dark noise according to Eq. (4.30). The SNR of an ideal sensor is shown as a thick line, the SNRT as a dashed line.*

The index "0" refers to the dark image, the index 1288 was used to distinguish these definitions according to the standard from other definitions. The DSNU_{1288} has the unit e^-. Multiplication by the system gain K gives the corresponding value in the unit DN. The PRNU_{1288} is defined relative to the mean gray value and thus indicates the percentage variation of the sensitivity variation. The inverse SNR indicates the temporal variation of the signal relative to the mean value. In this way, the two variables temporal and spatial variance can be compared directly.

The variances resulting from the DSNU and PRNU can be added to the variances of the temporal noise as in Eq. (4.17). This also results in the following for the SNRT

$$\boxed{\mathrm{SNRT}(N_p) = \frac{\eta N_p}{\sqrt{\sigma_{g0}^2/K^2 + \mathrm{DSNU}_{1288}^2 + \eta N_p + \mathrm{PRNU}_{1288}^2\,(\eta N_p)^2}}. \tag{4.33}$$

In the equation, we have abbreviated $\sigma_d^2 + \sigma_q^2/K^2$ by the variance σ_{g0}^2 of the dark noise measured directly in DN.

While the DSNU behaves like the temporal dark noise, the PRNU results in a new asymptotic behavior for the SNRT. For high saturation, the PRNU becomes the largest term in the denominator. This means that the SNRT no longer increases with the square root of the irradiation, but asymptotically reaches the value $1/\mathrm{PRNU}_{1288}$. If the PRNU is large enough, it limits the quality of the image signal at high saturations.

Figure 4.15 shows the SNR of a linear camera. As long as the constant dark noise has no influence, the SNR runs parallel to the SNR of the ideal sensor (theoretical limit) with a slope of one half. It is lower by the factor $\sqrt{\eta}$, which means a constant distance of $\log(\eta)/2$ in a logarithmic representation. At low saturation, the dark noise becomes noticeable and the SNR decreases steeper with a slope of one and more and more relative to the SNR of an ideal sensor without dark noise.

When comparing the SNR with the SNRT in Fig. 4.15, it is noticeable that the DSNU has no significant influence on the signal quality in the example shown. At low irradiation, SNR and SNRT are identical. At high saturations, the SNRT is significantly smaller than the SNR. Here the PRNU dominates the signal quality.

Extension 4.5: Photon-counting image sensors

In recent years, *photon-counting image sensors* have been developed and first operational sensors have come onto the market. These are based on two different approaches. On the one hand, there are arrays of single-photon avalanche photodiodes, or *SPAD image sensors* for short, [66, 211] which generate a cascade of charge carriers from an electron generated by a photon, similar to a *photomultiplier* (Section 4.2.2). The associated electrical signal is then large enough that it can be detected directly and photons can be counted. Despite their high sensitivity, SPAD image sensors can be used at high irradiance levels since exposure times as short as about 10 ns are possible.

On the other hand, a new paradigm for image acquisition was developed with the *"quanta" image sensors* [90]. The basic idea is to make the pixels so small and sensitive that they can detect a single photon.

With both types of image sensors, much higher frame rates are possible than conventional image sensors, as only binary images need to be output. The binary images can then be flexibly added together to form arbitrarily adjustable bit depths.

In the following, we will therefore examine the characteristic curve, photon transfer curve and SNR of a sensor that can record and accumulate such binary images [89, 91]. This is a good example for practicing Poisson statistics. We assume an average number of photons N_p that hit a pixel during the set exposure time. Of these, an average of ηN_p is then detected with a quantum efficiency η. With a binary image, either none or just one photon can be detected, but not more.

We therefore only need two probabilities, which we can calculate using the *Poisson distribution* (Appendix C.4.1). One is the probability of detecting no photons and the other is the probability of detecting one or more photons.

These result in

$$\begin{aligned} g = 0: &\quad \text{no photon} && f_0 = \exp(-\eta N_p), \\ g = 1: &\quad \text{one or more photons} && f_{>0} = 1 - \exp(-\eta N_p). \end{aligned} \tag{4.34}$$

From this, according to Eq. (C.3), the mean value is calculated as

$$g(N_p) = 1 - \exp(-\eta N_p) \approx \eta N_p \quad \text{for} \quad \eta N_p \ll 1. \tag{4.35}$$

This equation represents the non-linear *characteristic curve* of a photon-counting image sensor that outputs binary images. The characteristic curve initially rises linearly with the slope ηN_p, from which the quantum efficiency can be determined directly, then quickly flattens out and reaches a maximum value of one. This leads to an expansion of the signal range, with an average value of ≈ 0.63 for one detected photon and ≈ 0.95 for three.
With the help of Eq. (C.6), the variance is given by

$$\sigma_g^2(N_p) = \exp(-\eta N_p)\left[1 - \exp(-\eta N_p)\right]. \tag{4.36}$$

If Eq. (4.35) is inserted into Eq. (4.36), the *photon transfer curve* results in

$$\sigma_g^2(g) = g\,[1 - g] \approx g \quad \text{for} \quad g \ll 1. \tag{4.37}$$

In the second step, we can now add the K binary images and thus flexibly set the bit depth. For K additions, Appendix C.3.3 for the mean value, variance and photon transfer curve of a gray value image with values from 0 to K results in

$$g(N_p) = K\left[1 - \exp(-\eta N_p/K)\right] \approx \eta N_p \quad \text{for} \quad \eta N_p/K \ll 1, \tag{4.38}$$

$$\sigma_g^2(N_p) = K \exp(-\eta N_p/K)\left[1 - \exp(-\eta N_p/K)\right] \tag{4.39}$$

and

$$\sigma_g^2(g) = g\,[1 - g/K] \approx g \quad \text{for} \quad g \ll K. \tag{4.40}$$

The photon count now refers to K binary images. This means that on average only $\eta N_p/K$ photons are detected per binary image. The photon transfer curve is a parabola that has maximum variance $\sigma_g^2 = K/4$ at $g = K/2$. Measurements of the characteristic curve and the photon transfer curve at an exposure time of 2.09 μs and addition over 127 images (7-bit resolution) correspond remarkably well with the derived model equations (Fig. 4.16).[10] This proves that under the measured conditions, the thermal generation of charge carriers (dark signal, Section 4.3.3) is negligible.
The only free parameter in the characteristic curve is the quantum efficiency. Therefore, this can be determined by regression with Eq. (4.38). The determined quantum efficiencies are smaller than 6%. This is partly due to the fact that a color sensor was irradiated with white light. With monochromatic illumination the quantum efficiency would be about 12 – 24%. It also shows that the SPAD image sensor technology, which is still in its early stages, has significant potential for development in terms of quantum efficiency compared

[10]The author would like to thank S. Bauer from Ubicept for lending the SPAD camera free of charge and for his assistance with the measurements, and A. Manakov for carrying out and evaluating the measurements.

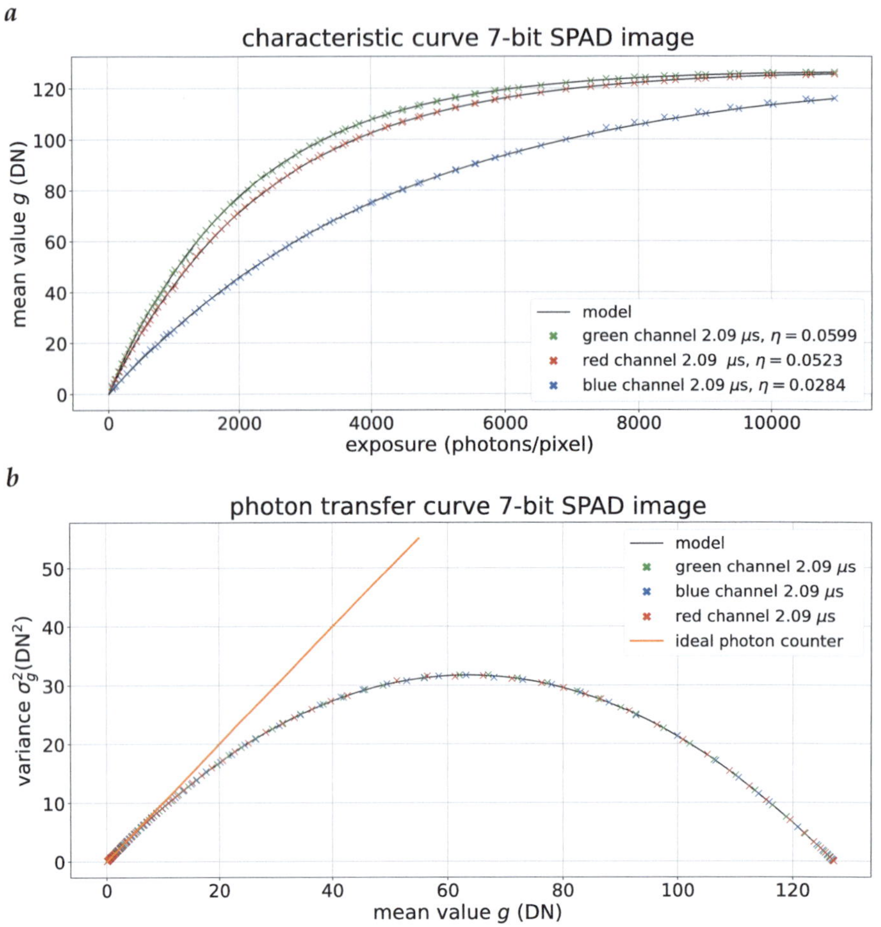

Figure 4.16: ***a*** *characteristic curve and* ***b*** *photon transfer curve of a SPAD image sensor with 127 binary images added.*

to conventional CMOS image sensors, which have peak quantum efficiencies of more than 80% (Fig. 4.4).

According to Eq. (4.40), there are no free parameters in the photon transfer curve, so no regression is necessary. Nevertheless, the measured values for all three color channels correspond exceptionally well with the theoretical curve. As expected for a photon-counting sensor, the photon transfer curve has a slope of one for low mean values g according to the Poisson statistics (Appendix C.4.1): $\sigma_g^2 = g$.

In the third step, we now calculate the SNR. Since there is a nonlinear characteristic curve, the SNR for the input and output signals are different. $\mathrm{SNR}_{\mathrm{out}}$ is directly derived from the mean and variance according to Eqs. (4.38) and

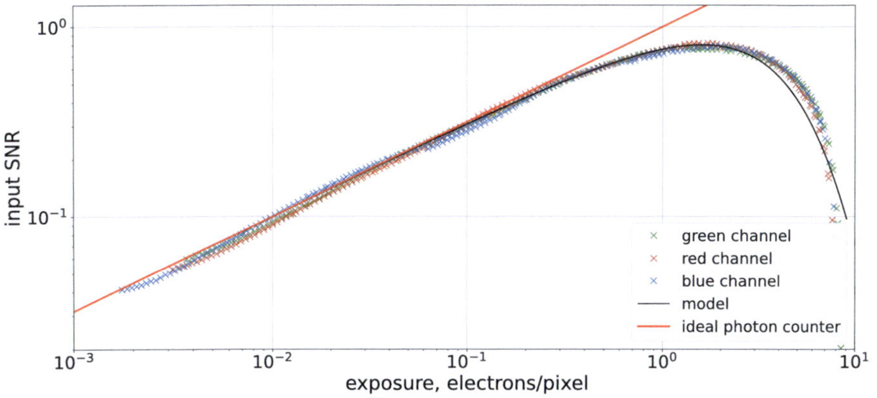

Figure 4.17: *SNR$_{in}$ of a SPAD image sensor for a single binary image shown as a function of the mean number of detections (electrons), $\eta N_p/K$.*

(4.39)

$$\mathrm{SNR}_{\mathrm{out}}(N_p) = \sqrt{K}\sqrt{\frac{1-\exp(-\eta N_p/K)}{\exp(-\eta N_p/K)}}. \tag{4.41}$$

With the help of the derivative of the characteristic curve, Eq. (4.38), we can calculate the more important SNR$_{\mathrm{out}}$ according to Eq. (4.21) and obtain

$$\mathrm{SNR}_{\mathrm{in}}(N_p) = \frac{\eta N_p}{\sqrt{K}}\sqrt{\frac{\exp(-\eta N_p/K)}{1-\exp(-\eta N_p/K)}} \approx \sqrt{\eta N_p} \quad \text{for} \quad \eta N_p/K \ll 1. \tag{4.42}$$

For small saturations, this results in an increase of SNR$_{\mathrm{in}}$ with $\sqrt{\eta N_p}$ as with conventional linear CMOS image sensors (Fig. 4.17). SNR$_{\mathrm{in}}$ reaches a maximum value of approximately 0.8 for a binary image with an average of two detection events and then drops sharply. When K binary images are added, the SNR$_{\mathrm{in}}$ increases by a factor of $\sqrt{K}$. This allows the desired or required SNR$_{\mathrm{in}}$ to be flexibly adjusted.

4.5.6 Quality Parameters Derived from the SNR

Four quality parameters that are important for the application can be derived from the SNR. These indicate the smallest and largest irradiation that can be measured, their ratio and the maximum SNR.

To do this, we must first check at which output signals we can carry out measurements. For a digital camera whose signal has been digitized with Q bits, the values range from 0 to $2^Q - 1$. However, the practically usable range of values is smaller. On the one hand, the average dark value g_0 must be greater than zero so that the full fluctuation range of the signal due to the dark noise can still be captured by the value range. If values were truncated (*underflow*), the mean dark value would be distorted. Similarly, the maximum usable value is less than $2^Q - 1$ so that temporal fluctuations are fully captured by the noise and no *overflow* occurs.

Therefore, the saturation irradiation $N_{p.\mathrm{sat}}$ cannot be specified for the maximum value $2^Q - 1$. The saturation irradiation is determined by the fact that the distribution of values around the mean value is not significantly truncated by *overflow*. This question was examined in more detail for the EMVA 1288 standard. The overflow should be between 0.1 and 0.2% [71].

From the saturation irradiation $N_{p.\mathrm{sat}}$ the *saturation capacity* $N_{e.\mathrm{sat}}$ (unit e^-) can be calculated:

$$\boxed{N_{e.\mathrm{sat}} = \eta N_{p.\mathrm{sat}}.} \tag{4.43}$$

The saturation capacity must not be confused with the so-called *full-well capacity*. The latter represents the maximum number of electrons that a sensor element can hold and thus saturate the signal. The full-well capacitance is usually slightly higher than the saturation capacity, as the maximum possible value $2^Q - 1$ is reached sooner than the physical saturation of the sensor. This setting is made by the manufacturers so that there are no excessive deviations from the linearity of the signal.

The *minimum detectable irradiation*, also called *sensitivity threshold!absolute* $N_{p.\mathrm{min}}$, can also be defined with the help of the SNR. It is the number of photons required to achieve an SNR of one.

For this purpose, we need to calculate the inverse function of Eq. (4.30), i. e. the number of photons as a function of the SNR. This function results in

$$\boxed{N_p(\mathrm{SNR}) = \frac{\mathrm{SNR}}{2\eta}\left(\mathrm{SNR} + \sqrt{\mathrm{SNR}^2 + 4(\sigma_d^2 + \sigma_q^2/K^2)}\right).} \tag{4.44}$$

The absolute sensitivity threshold for SNR = 1 results in

$$N_p(\mathrm{SNR} = 1) = N_{p.\mathrm{min}} = \frac{1}{2\eta}\left(1 + \sqrt{1 + 4(\sigma_d^2 + \sigma_q^2/K^2)}\right). \tag{4.45}$$

The saturation irradiation and the absolute sensitivity threshold can also be specified in the associated energy-related irradiations instead of the number of photons per pixel. The *noise-equivalent irradiation* or *noise-equivalent exposure*, in short *NEE*, results with Eq. (4.5) to

$$H_0 = \frac{hc}{\lambda A} N_{p.\mathrm{min}} \quad (\mathrm{NEE}). \tag{4.46}$$

In the same way, we obtain the *saturation-equivalent irradiation* or *saturation-equivalent exposure*, in short *SEE*, to

$$H_s = \frac{hc}{\lambda A} N_{p.\mathrm{sat}} \quad (\mathrm{SEE}). \tag{4.47}$$

Example 4.2: Saturation capacity and SEE

To become more familiar with the quantities involved, here is a small example. We take the frequency $\nu = 5.4 \cdot 10^{14}$ Hz and wavelength $\lambda = 555$ nm, for which the human eye has the highest sensitivity, a saturation capacity of $N_{e.\mathrm{sat}} = 100\,000$ and a quantum efficiency of 80 %. The SEE is then calculated to be approximately $4 \cdot 10^{-14}$ J. For a pixel with a size of $10\,\mu\mathrm{m} \times 10\,\mu\mathrm{m}$, this results in an irradiance of 10^{-6} W/cm^2 with an exposure time of 40 ms.

A third quality criterion for an image sensor is the *dynamic range*, or *DR* for short. It is a measure of the signal range that an image sensor can record and is calculated as the ratio of the saturation capacity to the absolute sensitivity threshold at

$$\mathrm{DR} = \frac{N_{p.\mathrm{sat}}}{N_{p.\mathrm{min}}}. \tag{4.48}$$

The signal range is often not specified as a direct ratio, but either in dB ($20 \log_{10} \mathrm{DR}$) or in bits ($\mathrm{ld}\,\mathrm{DR} = \log_2 \mathrm{DR}$).

In this context, it is important to emphasize that image sensors usually have a small signal range compared to other measuring instruments, which is in the range of 1000. Other measuring devices, such as, e. g. a photodiode, easily have a thousand times higher signal range of over 10^6. Only special image sensors, so-called *high-dynamic range* or *HDR* image sensors have a signal range of up to 10^6 [51, 136].

Finally, the maximum achievable SNR is an important criterion for the quality of an image sensor. For a linear sensor, this results directly from Eq. (4.30) in a good approximation to

$$\mathrm{SNR}_{\max} = \sqrt{\eta\, N_{p.\mathrm{sat}}} = \sqrt{N_{e.\mathrm{sat}}}. \tag{4.49}$$

For good image quality, it is therefore crucial to accumulate as many charge units as possible per pixel. For a non-linear image sensor, $\mathrm{SNR}_{\max}$ must be determined numerically from the input SNR curve.

4.5.7 ♯Spatial Inhomogeneity and Defective Pixels

The description of the spatial *inhomogeneity* of image sensors would be simple if it were purely random. Then it would be sufficient to specify the variance of the spatial variations. However, this is not the case. With spatial inhomogeneity, a number of other effects occur, which we will analyze below in spatial and Fourier domains.

Random Variation

If the spatial variations are purely random, i. e. do not show any spatial correlation, then the *power spectrum* of the variations is flat (Appendix C.5.3), the variations are evenly distributed over all wavelengths. This is also referred to as *white noise*.

Gradual Variation

Imperfections during production can lead to image sensor parameters slowly changing spatially over the area of the image sensor. In the case of the PRNU, these effects are not easy to measure as they require very homogeneous illumination of the image sensor, which is difficult to achieve. Fortunately, these variations are not of decisive importance as they do not significantly impair the image quality. A human viewer will not notice them at all, as they escape the eye, which is not built to see absolute irradiance levels, but to orient itself to the context (Section 1.6).

In addition, uneven illumination of a scene and the edge fall-off of the irradiance in an optical system (Section 3.6.2) sometimes lead to significantly higher gradual spatial changes in irradiance on the image plane. Therefore, in practice, all these effects must be corrected together. The gradual variations in the image sensor parameters are then also corrected.

Periodic Variation and Patterns

This type of interference is caused by electronic crosstalk within the camera electronics or the structure of the sensor. Such interference is a significant impairment of image quality, as we can see it very sensitively. Many image processing operators are also sensitive to such interference. It is therefore important to analyze periodic variation in a targeted manner. This is easiest to do in Fourier space. There, periodic variations appear as sharp peaks with specific wavenumbers.

Outliers

Outliers are individual pixels, rows, columns or clusters of pixels that deviate significantly from the expected probability distribution. These can be dead pixels that show no sensitivity to light, pixels with significantly reduced sensitivity to light or pixels with a high dark current (so-called hot pixels). Outliers correspond to singularities in the spatial domain which, like delta functions in Fourier space, have a constant power spectrum. Therefore, outliers cannot be distinguished from purely random fluctuations in the power spectrum. However, a good method for analyzing outliers is the analysis of histograms.

As the requirements of the various image processing applications are very different from one another, it is not possible to generally define what an outlier or a defective pixel is. It therefore makes much more sense to provide statistical information about the pixel properties in the form of a *histogram*. These can be used to derive application-specific criteria for defining defective pixels.

In order to also see individual outliers, the histogram must have a logarithmic frequency axis including the occurrence of a single pixel in the histogram (Fig. 4.18). The asymmetrical deviation towards higher dark values is typical for dark images. These are pixels with a higher dark current. The extent to which these deviations affect the image quality ultimately depends on the application.

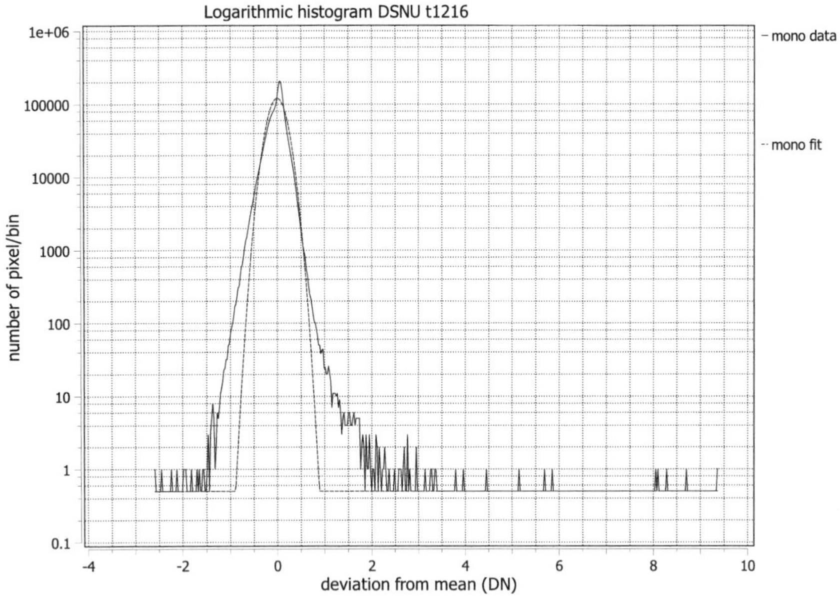

Figure 4.18: *Histogram with logarithmic scaling of the frequencies of the dark values for a modern 12-bit CMOS camera measured with a short exposure time of 0.984 ms as a deviation from the average dark value of 5.35 DN. The dark image was averaged over 2000 individual images.*

Analysis of Nonuniformity

Suitably calculated profiles clearly show all the types of inhomogeneities discussed (Figs. 4.19 and 4.20). Four types of profiles are shown. The green line shows the profile of the middle line. All other profiles show profiles calculated from all lines.

The red profile shows the maximum value found in the corresponding column at each position. In this profile, individual upward outliers can be clearly seen in the DSNU image (so-called *hot pixels*). The blue profile shows the minimum value and the black profile shows the average value across all rows.

The difference between the red and blue curves illustrates the absolute fluctuation range in the corresponding column at each position. The black curve's vertical averaging eliminates almost all small-scale variations, showing primarily the large-scale fluctuations and vertical stripes sometimes present in CMOS image sensors. Overall, all four profiles provide a good impression of the relative importance of different types of inhomogeneities.

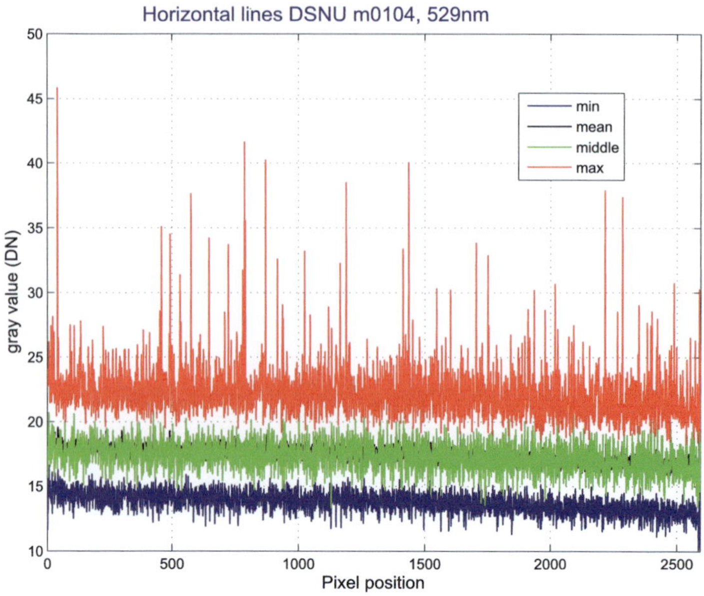

Figure 4.19: *Horizontal profiles of the dark image (absolute units DN) to illustrate the inhomogeneities of a modern CMOS camera.*

Extension 4.6: Inhomogeneity versus "fixed-pattern noise"

The inhomogeneity is often referred to as *fixed-pattern noise* or *FPN* for short. However, this term is misleading, as spatial inhomogeneities do not represent noise that changes over time but are stationary by definition. Any temporal changes in the inhomogeneity would be assigned to temporal noise.

The main difference between temporal noise and spatial inhomogeneity is as follows: Temporal noise is unpredictable. We do not know which value will be measured within the possible probability distribution.

Therefore, the only way to suppress temporal noise is to average over as many images as possible. To analyze the inhomogeneity patterns, an average image must be used that was calculated over many individual images under the same recording conditions. Even then, the standard deviation of the temporal noise only decreases with the square root of the number of averaged images, see Eq. (C.29) in Appendix C.3.3. Therefore, temporal noise can only be reduced in practice, but not eliminated.

The situation is completely different for inhomogeneities. Since they are stationary, it is sufficient to take two images under the same recording conditions. The inhomogeneities are then eliminated in the difference image (see Section 4.5.2).

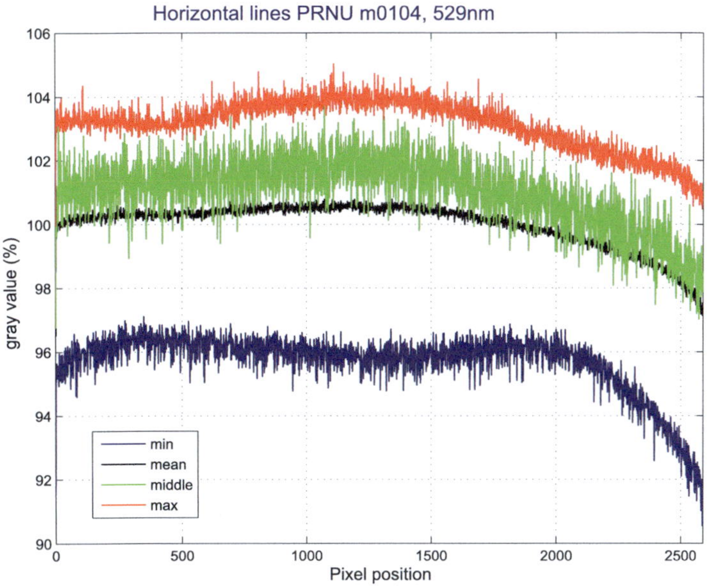

Figure 4.20: *Horizontal profiles of the sensitivity image (relative units in % relative to the mean value) to illustrate the inhomogeneities of a modern CMOS camera.*

Finally, it should be noted that the PRNU may strongly dependent on the wavelength for some image sensors. In Fig. 4.21 the deviation of the PRNU for individual wavelengths is shown when the PRNU has been fully corrected on average over all wavelengths shown. The sensor in Fig. 4.21a shows strong large-scale variations of up to ±4% and the small-scale variations from pixel to pixel are also at least ±1%. The second sensor (Fig. 4.21b), on the other hand, shows only minimal variations both on a large scale and from pixel to pixel, which are no greater than ±0.2%.

4.5.8 ♯Dark Current and Time for Self-saturation

The dark signal Eq. (4.22) introduced in Section 4.5.5 is not constant. The reason for this is that it depends on the exposure time how many charge carriers are thermally excited. Therefore, the dark signal increases linearly with the exposure time t_{exp}:

$$g_d = g_{d.0} + g_{\mathrm{therm}} = g_{d.0} + g_I t_{\mathrm{exp}}. \tag{4.50}$$

In this equation, the dark signal is given in units of electrons per pixel (e^-/pixel). This is obtained by dividing the directly measured values in the unit DN by the system gain K, see Eq. (4.22).

The quantity g_I is called *dark current* and has the unit e^-/(pixel s). Since the thermally excited electrons are also Poisson distributed, $\sigma^2_{\mathrm{therm}} = \mu_{\mathrm{therm}}$

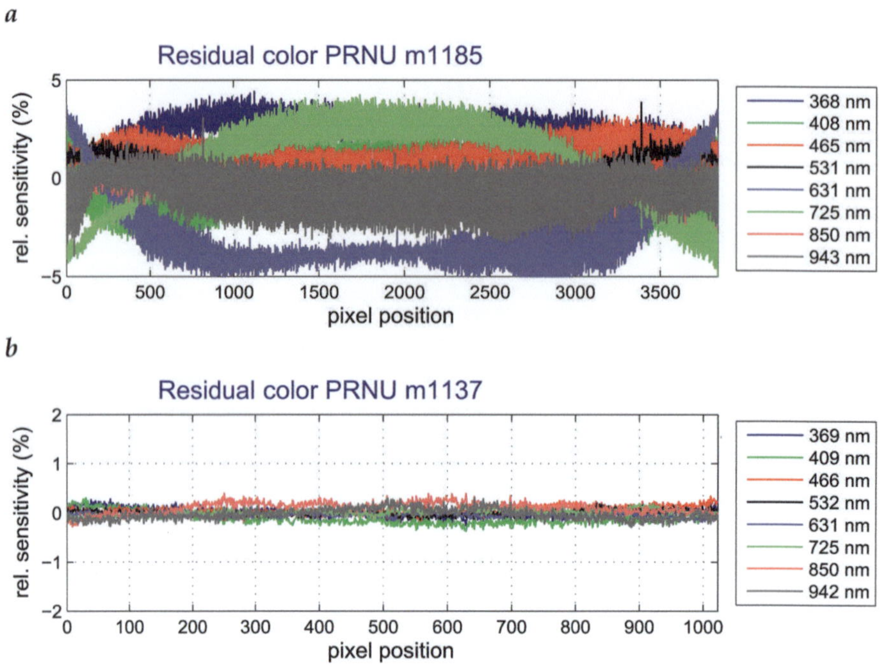

Figure 4.21: *Dependence of the PRNU on the wavelength for two different CMOS image sensors.*

applies to the variance of the dark signal

$$\sigma_d^2 = \sigma_{d.0}^2 + \sigma_{\text{therm}}^2 = \sigma_{d.0}^2 + g_I t_{\text{exp}}. \tag{4.51}$$

This relationship is important for sensors that have dark current compensation. In this case, the dark current can still be measured by measuring the variance according to Eq. (4.51).

The *self-saturation time* t_{sat} can be derived from the dark current. This is the exposure time required to saturate the image sensor with the dark current alone:

$$t_{\text{sat}} = \frac{g_{\text{max}}}{g_I}. \tag{4.52}$$

This quantity is very useful to determine the influence of the dark current as a function of the exposure time. The fraction of saturation caused by the dark current is given by the simple relationship

$$\frac{g_{dark}}{g_{\text{max}}} = \frac{t_{\text{exp}}}{t_{\text{sat}}}. \tag{4.53}$$

Depending on the fraction that can be tolerated, this also results in the maximum possible exposure time with the image sensor.

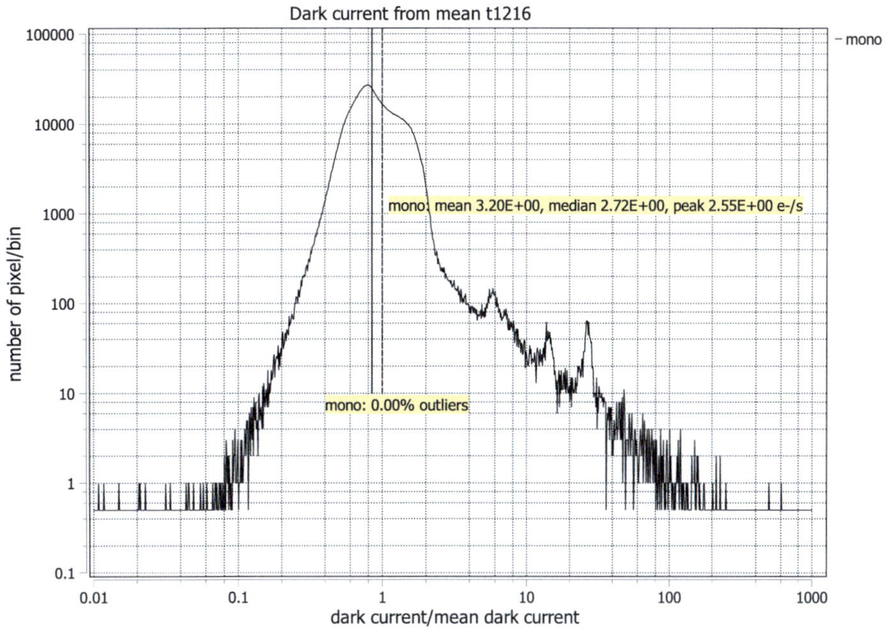

Figure 4.22: *Histogram with logarithmic scaling of the frequencies of the dark current and logarithmic scale for the dark current in relation to the mean dark current of 3.2 e⁻/s for the same 12-bit CMOS camera as in Fig. 4.18. For the measurement, 11 exposure times between the minimum exposure time of 0.02 ms to 600 ms were used. The dark current was calculated for each pixel from over 2000 images averaged at each exposure time.*

The strong increase in the dark signal with temperature has already been discussed in Section 4.3.3. At room temperature, the dark current doubles approximately every 7°C.

The dark current also exhibits inhomogeneity. This effect is called *dark current nonuniformity* or *DCNU* for short. An example is shown in Fig. 4.22. The measurements were carried out at an ambient temperature of 16.8°C. The temperature of the camera housing was 26°C, that of the image sensor 40°C. The distribution is strongly asymmetrical even with logarithmic scaling of the dark current.

There are three further peaks in the distribution at approximately 6, 14 and 25 times the mean dark current of $3.2\,e^-/s$. As a result, the *median* and the peak value of the distribution are significantly lower at $2.72\,e^-/s$ and $2.55\,e^-/s$, respectively. At the highest secondary peak value, the dark current increases to $80\,e^-/s$, which leads to an increase in the dark value by 31 DN at a system gain of $0.387\,DN/e^-$. Although this is still a moderate increase of less than 1% of the 4095 DN signal range, these pixels are clearly visible as white dots in the dark image. One speaks of *hot pixels*.

However, the average dark noise also increases significantly. Using Eq. (4.51), you can calculate that the standard deviation of the dark noise roughly triples from $2.1\,e^-$ at 0.894 ms to $6.0\,e^-$ at 10 s exposure time. The *absolute sensitivity threshold* is correspondingly higher at this maximum exposure time of the camera.

4.6 ♯References for Further Reading

A good introduction to radiation detection is provided by Rieke [265], Saleh and Teich [279]. Detailed descriptions of the functionality and construction of CCD image sensors can be found in the monographs by Theuwissen [327] and Janesick [149]. The monograph by Holst and Lomheim [140] is written from the perspective of camera system users and covers CMOS and CCD image sensors and the characterization of their performance. The monograph published by Nakamura [228] deals extensively with digital cameras. The handbook by Teubner and Brückner [326] covers the entire range from analog cameras with photographic films to camera systems in mobile devices.

The following books are recommended for various special areas: CCD image sensors in the deep UV [199], use of CCD image sensors in astronomy (low irradiance) [143], highly dynamic image sensors [51, 136, 220] and "Smart Cameras", i.e. sensors and cameras that already have integrated pre-processing [14].

A detailed description of the measurement procedures for characterizing image sensors and cameras according to the European Machine Vision Association (EMVA) standard 1288, which is currently being converted into ISO standard 24942, is publicly available at `www.standard1288.org`. Further information on the standard, including data sheets for cameras that have been measured in accordance with it, can be found at `https://zenodo.org/communities/emva1288`.

5 Sampling and Interpolation

Contents

5.1 Introduction and Overview

In Section 3.5 we described the irradiation on the image plane as a continuous function. However, a digital computer can only store and process a matrix of

B. Jähne, *Digital Image Processing and Image Acquisition*,
https://doi.org/10.1007/978-3-032-04806-6_5

pixels. Therefore, the irradiation values must be sampled in a suitable way in the horizontal and vertical direction. This process is called *digitization*. The gray values are scanned at discrete points, which can be represented by a matrix. In the case of an area image sensor that has a matrix of discrete sensors, the scanning already takes place there.

At first glance, digitizing a continuous image appears to be an enormous loss of data, as we reduce the continuous gray value information to a function on a grid of dots. However, this consideration is not entirely correct, as optical imaging reduces the image resolution. Even with a diffraction-limited object, all structures whose wavelength is below a cut-off wavelength are completely deleted (compare (Eq. 3.55) in Section 3.5.3).

This raises two important questions:

1. How does the image signal change when it is sampled and is there a condition under which no change in the image signal occurs?
2. What is the best ratio of the pixel size of an image sensor to the resolution of the object used?

We will answer both questions in Section 5.2. In the following Section 5.3 we will then examine how the continuous image signal can be reconstructed from the sampled signal. With this knowledge, it is then also possible to display images on arbitrary grids with other sampling points. We will discuss this important practical aspect in Section 5.5 about interpolation. Before that, we will deal with the question of whether and how it is possible to scan on non-square and multidimensional grids (Section 5.4).

Intuitively, we understand that scanning leads to a reduction in resolution, i. e. that structures of the size of the scanning step size or smaller are lost. However, the fact that considerable interference occurs when scanning an image with fine details may initially come as a surprise. Fig. 5.1 shows a simple example in which digitization is simulated by superimposing a 2D grid on an object with two linear grids with different grid constants. After scanning, the two grids appear with altered periodicity and direction. Such image distortions are referred to as the *Moiré effect*.

The same phenomenon is known as *aliasing* for one-dimensional signals, especially for time series. A signal with sinusoidal oscillation can be seen in Fig. 5.2. It is sampled in steps that are slightly shorter than one wavelength. The result of the sampling is a signal with a much longer wavelength. These problems always occur when digitizing analogue data; they are a general problem of signal processing. In this context, image processing is merely a special case of signal theory.

Since the aliasing effect occurs with periodic signals, the key to understanding and avoiding it lies in an analysis of the digitization process in Fourier space. The aim of this analysis is to determine the conditions under which the sampling points reproduce a continuous image correctly and completely. These conditions can be formulated in the so-called *sampling theorem*.

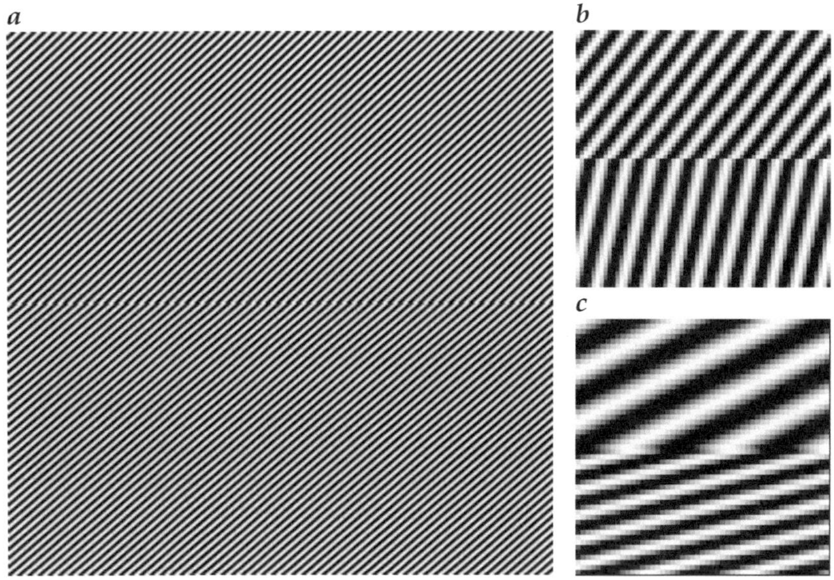

Figure 5.1: The Moiré effect: ***a*** *Original image with two periodic patterns (top* $\hat{\mathbf{k}} = [0.21\ 0.22]^T$*, bottom* $\hat{\mathbf{k}} = [0.21\ 0.24]^T$*).* ***b*** *Every fourth and* ***c*** *every fifth point scanned in each direction.*

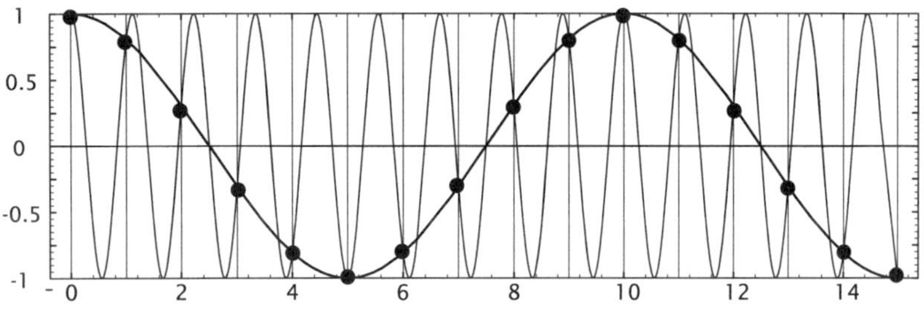

Figure 5.2: Illustration of the aliasing effect: An oscillating signal is sampled with a step width of Δx *equal to 9/10 of the wavelength. The result is a signal whose wavelength is ten times the sampling step size.*

The following considerations are more of an illustrative explanation than a strict mathematical proof of the sampling theorem.

A computer cannot process numbers with arbitrary precision. These are always quantized. This *quantization* is also already carried out in modern image sensors, as they contain an *analog-to-digital converter*, and a digital camera sends a stream of digital numbers to the computer (Section 4.4.1).

At first glance, quantization also looks like a drastic degradation of the signal. This is not the case either, as the image sensor signal is limited in its quality by temporal noise and inhomogeneities (Section 4.5.1). Therefore,

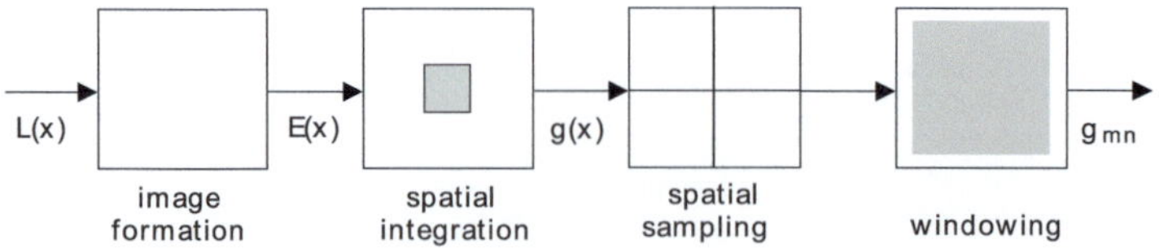

Figure 5.3: The chain of processing steps that lead from the object to a digital image, an image matrix G.

the crucial question is what is the best ratio between standard deviation of temporal noise and quantization. This is directly linked to the question of how well we can measure the mean and standard deviation of the signal for a given quantization and standard deviation of the temporal noise. We will learn in Section 5.6 that a noise-free signal is not good, but that the right amount of noise is optimal. It will also turn out that quantization can also be regarded as a sampling, namely the sampling of a probability density function. Since we are looking at the quality of an image with our human eyesight, some observations on our eye's perception of brightness are also important in this section.

5.2 Steps of the Sampling Process

We assume an infinitely extended, continuous image $g(\boldsymbol{x})$ that we want to map onto a matrix $\boldsymbol{G}$. Here we take into account what was discussed in Section 3.5 regarding image generation. This allows us to distinguish three steps of digitization: Image generation, sampling and limitation to a finite image matrix (Fig. 5.3).

5.2.1 Image Generation

We cannot consider digitization without image generation. The optical system, including the sensor, influences the image signal, so we must include image generation.

Digitization means scanning the image at defined points of a discrete grid $\boldsymbol{r}_{m,n}$ (we will examine discrete grids further in Section 6.2.3). On a rectangular grid, we can write these points according to Eq. (6.2) as:

$$\boldsymbol{r}_{m,n} = [m\,\Delta x_1, n\,\Delta x_2]^T \quad \text{with} \quad m, n \in \mathbb{Z}. \tag{5.1}$$

In general, we do not record the illumination intensity exactly at these points, but from a certain area around them. Let's take an ideal CCD camera as an example, which consists of a matrix of photodiodes without light-insensitive gaps. We also assume that the photodiodes are equally sensitive over the entire surface. Then the signal at the grid points is the integral over the area of the individual photodiodes:

$$g(\boldsymbol{r}_{m,n}) = \int\limits_{(m-1/2)\Delta x_1}^{(m+1/2)\Delta x_1} \int\limits_{(n-1/2)\Delta x_2}^{(n+1/2)\Delta x_2} g'(\boldsymbol{x})\,\mathrm{d}x_1\,\mathrm{d}x_2. \tag{5.2}$$

This represents a convolution with a *rectangular function* and a sampling at the grid points. The two steps can be mathematically separated. First we can perform the continuous *convolution* and then the sampling. In this way, the image generation process is generalized and separated from the sampling process. Since convolution is an associative process, we can combine the averaging of the CCD sensor with the PSF of the optical system (Section 3.5.1) in a single convolution process. Therefore, the image formation process can be described by the following operation:

$$g(\boldsymbol{x}) = \int\limits_{-\infty}^{\infty} g'(\boldsymbol{x}')h(\boldsymbol{x}-\boldsymbol{x}')\mathrm{d}^2x' \quad \circ\!\!-\!\!\bullet \quad \tilde{g}(\boldsymbol{\kappa}) = \tilde{g}'(\boldsymbol{\kappa})\tilde{h}(\boldsymbol{\kappa}), \tag{5.3}$$

where $h(\boldsymbol{x})$ and $\tilde{h}(\boldsymbol{\kappa})$ are the resulting *PSF* $g'(\boldsymbol{x})$ can be regarded as the gray value image that would result from a perfect sensor, i. e. an optical system (including sensor) whose OTF is one and whose PSF is a *δ-distribution* (> R13).

In general, the image is blurred by the image generation; fine details are lost. In Fourier space, this leads to an attenuation of high wavenumbers, and the resulting image is referred to as *bandlimited*.

5.2.2 Sampling

Next, we look at the *sampling*. Sampling means that only the information at the grid points is retained. Mathematically, this is a multiplication by a function that is only non-zero at the grid points. This operation can be performed by multiplying the image function $g(\boldsymbol{x})$ by a function that represents the sum of the δ functions located at the grid points $\boldsymbol{r}_{m,n}$. This function is called a two-dimensional δ comb or *"nailboard function"*. Thus, we can express the sampling process as follows:

$$g_s(\boldsymbol{x}) = g(\boldsymbol{x})\sum_{m,n}\delta(\boldsymbol{x}-\boldsymbol{r}_{mn}) \quad \circ\!\!-\!\!\bullet \quad \tilde{g}_s(\boldsymbol{\kappa}) = \sum_{\tilde{m},\tilde{n}}\tilde{g}(\boldsymbol{\kappa}-\tilde{\boldsymbol{r}}_{\tilde{m}\tilde{n}}), \tag{5.4}$$

where

$$\tilde{\boldsymbol{r}}_{\tilde{m}\tilde{n}} = \begin{bmatrix} \tilde{m}\Box\kappa_1 \\ \tilde{n}\Box\kappa_2 \end{bmatrix} \quad \text{with} \quad \tilde{m},\tilde{n} \in \mathbb{Z} \quad \text{and} \quad \Box\kappa_d = \frac{1}{\Delta x_d} \tag{5.5}$$

are the points of the so-called *reciprocal lattice*, which play an essential role in solid state physics and crystallography.

According to the convolution theorem (Theorem B.4, p. 630), the multiplication of the image with the 2D δ comb corresponds to a convolution of the Fourier transform of the image, the image spectrum, with another 2D δ comb whose lattice constants are reciprocal to the lattice constants in x space (see Eqs. (5.1) and (5.5)). A dense sampling in x-space leads to a wide grid in κ-space and vice versa. Thus, the sampling leads to a repetition of the image spectrum at each grid vector $\tilde{\boldsymbol{r}}_{\tilde{m}\tilde{n}}$ in Fourier space.

The effect of periodic replication can be explained clearly. It simply says that there is an ambiguity in the interpretation of the sampled signal. It is consistent with wavenumbers from each of the replicated copies and we cannot distinguish which wavenumber the signal actually has.

5.2.3 The Sampling Theorem

Now we can formulate the condition we need to avoid distortion of the signal during sampling. This condition is called *sampling theorem*. If the image spectrum is extended, the periodically repeating copies overlap. We cannot distinguish whether the spectral amplitudes originate from the original spectrum in the center or from one of the copies. To avoid distortions, we must exclude overlaps.

A reliable method of preventing overlaps is as follows: We must restrict the spectrum to the area around the central point of the reciprocal lattice up to the lines separating the central lattice point from all other lattice points. In solid-state physics, this zone is referred to as the first *Brillouin zone* [49, 182].

On a rectangular lattice, this results in the simple condition that the maximum wavenumber at which the image spectrum is not zero must be limited to less than half the lattice constant of the reciprocal lattice. Quantitatively, this can be expressed as follows: If the spectrum $\tilde{g}(\boldsymbol{\kappa})$ of a continuous function $g(\boldsymbol{x})$ is band-limited, i. e.

$$\tilde{g}(\boldsymbol{\kappa}) = 0 \;\; \forall \, |\kappa_d| \geq \square\kappa_d/2, \tag{5.6}$$

then it can be reconstructed exactly from sampled points at a distance of

$$\Delta x_d = 1/\square\kappa_d. \tag{5.7}$$

In other words, we only obtain a correct periodic structure if we set at least two sampling points per wavelength. The maximum wavenumber that can be sampled without error is called *Nyquist wavenumber* or *limit wavenumber*. In the following, we will often use dimensionless wavenumbers that are normalized to the limit wavenumber. This scaling is marked with a tilde:

$$\boxed{\underset{\sim}{\kappa}_d = \frac{\kappa_d}{\square\kappa_d/2} = 2\kappa_d\Delta x_d.} \tag{5.8}$$

All components $\underset{\sim}{\kappa}_d$ of the wavenumber fall into the interval $]-1, 1[$.

Now we can explain the *Moiré*- and aliasing effects. We assume a periodic structure that violates the sampling theorem. The original spectrum contains a single peak marked with the long vector $\boldsymbol{\kappa}$ in Fig. 5.4. Due to the periodic

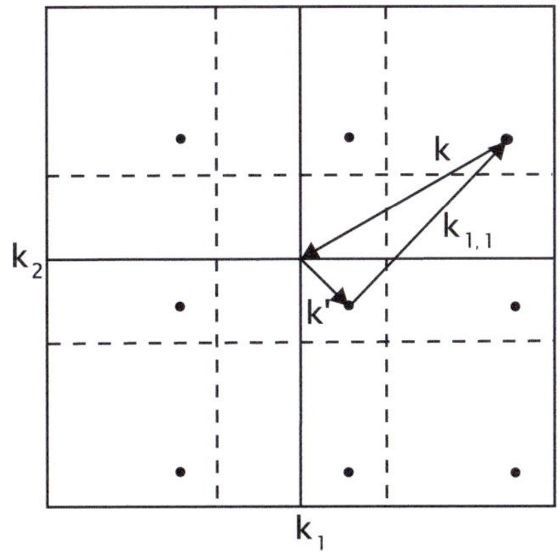

Figure 5.4: Explanation of the Moiré effect with a periodic structure that does not correspond to the sampling theorem.

repetition of the sampled spectrum, there is exactly one peak at $\boldsymbol{\kappa}'$, which is located in the central cell. This peak not only has a different wavelength (see Fig. 5.4), but generally also a different direction, as shown in Fig. 5.1.

The observed wavenumber $\boldsymbol{\kappa}'$ differs from the actual wavenumber $\boldsymbol{\kappa}$ due to a lattice translation vector $\tilde{\boldsymbol{r}}_{\tilde{m},\tilde{n}}$ on the reciprocal lattice. The indices $\tilde{m}$ and $\tilde{n}$ must be selected so that they meet the following condition:

$$\begin{aligned} |\kappa_1 + \tilde{m}\,\square\kappa_1| &< \square\kappa_1/2, \\ |\kappa_2 + \tilde{n}\,\square\kappa_2| &< \square\kappa_2/2. \end{aligned} \tag{5.9}$$

According to this condition, we obtain, as just shown, a distorted wavenumber

$$\kappa_1' = \kappa_1 - \square\kappa_1 = 9/10\,\square\kappa_1 - \square\kappa_1 = -1/10\,\square\kappa_1 \tag{5.10}$$

for the one-dimensional example in Fig. 5.2.

In fact, the sampling theorem, as we have now described it, is too strict. The condition that the periodically repeating image spectra must not overlap is both necessary and sufficient.

5.2.4 Limitation to a Finite Window

Up to this point, the sampled image still has an infinite size. In practice, however, we can only work with finite image matrices. So the last step is to limit the image to the size of a finite window. In the simplest case, we multiply the sampled image g_s by a rectangular function. More generally, we can use any *window function* $w(\boldsymbol{x})$ that is zero for sufficiently large values of $\boldsymbol{x}$ and then obtain the bounded image

$$g_l(\boldsymbol{x}) = g_s(\boldsymbol{x}) \cdot w(\boldsymbol{x}) \quad \circ\!\!-\!\!\bullet \quad \tilde{g}_l(\boldsymbol{\kappa}) = \tilde{g}_s(\boldsymbol{\kappa}) * \tilde{w}(\boldsymbol{\kappa}). \tag{5.11}$$

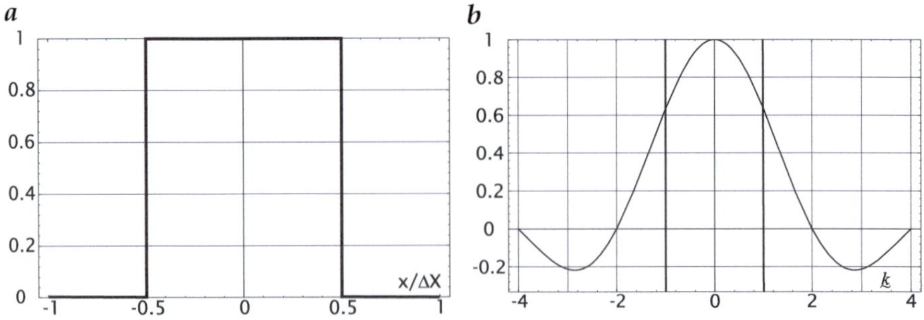

*Figure 5.5: **a** PSF and **b** transfer function of standard sampling.*

The spectrum of the sampled image is convolved in Fourier space with the Fourier transform of the window function. Let's take a closer look at the example of a rectangular window function. If the window in x-space consists of $M \times N$ sampling points, its size is $M\Delta x_1 \times N\Delta x_2$. The Fourier transform of the 2D rectangular function is a 2D sinc function (> R13). The main peak of the sinc function has a half-width of $1/(M\Delta x_1) \times 1/(N\Delta x_2)$. A needle-shaped peak in the spectrum of the image thus becomes a 2D sinc function. In general, the resolution in the spectrum is reduced in order of magnitude to the half-width of the sinc function.

In summary, the sampling leads to a limitation of the wavenumber, while the limitation of the image size determines the wavenumber resolution. The scales in spatial and wavenumber space are reciprocal to each other. The resolution in spatial space determines the size in wavenumber space and vice versa.

5.2.5 Standard Sampling

The sampling type discussed in Section 5.2.1 using the example of an ideal CCD camera is called *standard sampling*. The mean value of an elementary cell is assigned to an associated sampling point. This is a type of regular sampling, as each point in continuous space is weighted equally. We may be tempted to assume that standard sampling satisfies the sampling theorem. Unfortunately, this is not the case (Fig. 5.5). At *Nyquist wavenumber*, the Fourier transform of the rectangular function is still $2/\pi$. The first zero crossing occurs at twice the Nyquist wavenumber. Accordingly, we observe Moiré effects with CCD cameras. These effects become even more pronounced as only a small part — usually 20 % of the chip area in interline transfer cameras — is light sensitive [197].

Smoothing over larger areas with a rectangular function does not help much, since the Fourier transform of the rectangular window only decreases with k^{-1} (Fig. 5.5). The ideal window function for sampling is identical to the ideal interpolation formula Eq. (5.15) discussed in Section 5.3, since its Fourier transform is a rectangular function with the width of the unit cell of the reciprocal grid. However, this window function is not practically applicable. A detailed discussion of interpolation follows in Section 5.5.

Example 5.1: Avoidance of aliasing with diffraction-limited optics

By using optics with diffraction-limited resolution (Section 3.5.3), it is possible to ensure that no aliasing occurs. With a given pixel size $\Delta x'$, the minimum resolvable wavelength can be set by selecting the aperture so that it is scanned just twice per wavelength. Since all structures with a wavelength shorter than $\lambda_{\min}$ are completely erased with diffraction-limited resolution, aliasing effects are reliably excluded. The following applies: $2\Delta x' < \lambda_{\min} = k\lambda_e$. This means that the *number of apertures* must be $k > 2\Delta x'/\lambda_e$. Example: With a pixel size of 2.0 μm and a wavelength of light of 0.5 μm, the f-number must be greater than 8.

5.3 Reconstruction from Sampling Points

5.3.1 Perfect Reconstruction

The sampling theorem provides the conditions under which we can reconstruct a continuous function from sampling points. However, we still do not know how this reconstruction, i. e. the reversal of sampling, works.

The reconstruction is an *interpolation* of the sampling points. In general, the interpolation points $g_r(\boldsymbol{x})$ are calculated from the sampled values $g(\boldsymbol{r}_{m,n})$ weighted with corresponding factors (depending on the distance from the interpolation point):

$$g_r(\boldsymbol{x}) = \sum_{m,n} h(\boldsymbol{x} - \boldsymbol{r}_{m,n}) g_s(\boldsymbol{r}_{m,n}). \tag{5.12}$$

Using the integral properties of the δ function, we can replace the sample points on the right-hand side with the continuous values:

$$\begin{aligned} g_r(\boldsymbol{x}) &= \sum_{m,n} \int_{-\infty}^{\infty} h(\boldsymbol{x} - \boldsymbol{x}') g(\boldsymbol{x}') \delta(\boldsymbol{r}_{m,n} - \boldsymbol{x}') \mathrm{d}^2 x' \\ &= \int_{-\infty}^{\infty} h(\boldsymbol{x} - \boldsymbol{x}') \sum_{m,n} \delta(\boldsymbol{r}_{m,n} - \boldsymbol{x}') g(\boldsymbol{x}') \mathrm{d}^2 x'. \end{aligned}$$

The last integral is a convolution of the weighting function h with the product of the image function g and the 2D δ comb. In Fourier space, the convolution is replaced by a complex multiplication:

$$\tilde{g}_r(\boldsymbol{\kappa}) = \tilde{h}(\boldsymbol{\kappa}) \sum_{\tilde{m},\tilde{n}} \tilde{g}(\boldsymbol{\kappa} - \tilde{\boldsymbol{r}}_{\tilde{m}\tilde{n}}). \tag{5.13}$$

The interpolated function cannot be the same as the original image if the periodically repeated image spectra overlap. This is not new, but corresponds exactly to the statement of the sampling theorem. The interpolated image function only corresponds to the original image function if the weighting function is a rectangular function of the width of the elementary cell of the reciprocal grid. Then the sampling effects — all replicated and shifted spectra — are eliminated. Only the band-limited original spectrum remains, and Eq. (5.13) becomes

$$\tilde{g}_r(\boldsymbol{\kappa}) = \Pi(k_1\Delta x_1, k_2\Delta x_2)\tilde{g}(\boldsymbol{\kappa}). \tag{5.14}$$

Then the interpolation function is the inverse Fourier transform of the rectangular function, a sinc function (> R13):

$$h(\boldsymbol{x}) = \text{sinc}(x_1/\Delta x_1)\,\text{sinc}(x_2/\Delta x_2). \tag{5.15}$$

5.3.2 Oversampling

Unfortunately, this function only goes to zero with $1/x$. Therefore, we need a large image area for correct interpolation; mathematically, it must be infinitely large. We can restrict this condition if we "overfulfill" the sampling theorem, i. e. ensure that $\tilde{g}(\boldsymbol{\kappa})$ is already zero before the Nyquist wavenumber is reached. After Eq. (5.13) we can then choose $\tilde{h}(\boldsymbol{\kappa})$ in the range in which $\tilde{g}$ disappears. We can use this freedom to create an interpolation function that decreases faster to zero in the spatial domain, resulting in a minimum-length interpolation mask.

We can also start from a given interpolation formula. Then the deviation of its Fourier transform from the rectangular function tells us to what extent structures will be disturbed as a function of wavenumber. Possible interpolation functions are discussed in Section 5.5.

The principle of *oversampling* is not only essential for the construction of effective interpolation functions. It also enables the construction of precise filters of all kinds with small filter masks (see Chapters 11 and 12). In general, a balance must be found between oversampling, which leads to more data points, and the requirements for the filters. Practical experience shows that sampling with around 3 to 6 samples per wavelength, i. e. 1.5–3 times oversampling, is a good compromise.

5.4 ♯Multidimensional and Non-orthogonal Grids

So far, we have only investigated sampling on rectangular 2D grids. In this section, we extend our considerations to higher dimensions and non-orthogonal grids.

5.4.1 ♯Non-orthogonal Bases

This requires two extensions. First, we need to define D-dimensional lattice vectors with a set of not necessarily orthogonal basis vectors $\boldsymbol{b}_w$ that span the D-dimensional space. A vector on the lattice can then be written as

$$\boldsymbol{r}_n = [n_1\boldsymbol{b}_1, n_2\boldsymbol{b}_2, \ldots, n_D\boldsymbol{b}_D]^T \quad \text{with} \quad \boldsymbol{n} = [n_1, n_2, \ldots, n_D],\ n_D \in \mathbb{Z}. \tag{5.16}$$

In image sequences, one of the coordinates is time. Furthermore, we must note that for some grids, e. g. the triangular grid, more than one point per cell is required. Therefore, for a general grid, we must allow P grid points per unit cell. Each of these grid points is identified by an offset vector $\boldsymbol{s}_p$. This results in an additional sum over all points of the unit cell in the sampling integral and Eq. (5.4) must be expanded to

$$g_s(\boldsymbol{x}) = g(\boldsymbol{x}) \sum_p \sum_n \delta(\boldsymbol{x} - \boldsymbol{r}_n - \boldsymbol{s}_p). \tag{5.17}$$

In this equation, the limits of the sums have been omitted for the sake of simplicity.

5.4.2 ♯Extended Sampling Theorem

The extended sampling theorem results directly from the Fourier transform of Eq. (5.17). In this equation, the continuous signal $g(\boldsymbol{x})$ is multiplied by a sum of delta combs. According to the convolution theorem (Theorem B.4, p. 630), this becomes a convolution of the Fourier transform of the signal with the sum of the delta combs. The Fourier transform of a delta comb is in turn a delta comb (≻ R13). Since the convolution of a function with a *delta distribution* Eq. (B.44) replicates it at the zero point of the delta distribution, the Fourier transform of the sampled signal is a sum of shifted copies of the Fourier transform:

$$\tilde{g}_s(\boldsymbol{\kappa}) = \sum_{\tilde{n}} \tilde{g}(\boldsymbol{\kappa} - \tilde{\boldsymbol{r}}_{\tilde{n}}) \sum_{p=1}^{P} \exp\left(-2\pi \mathrm{i} \boldsymbol{\kappa}^T \boldsymbol{s}_p\right). \tag{5.18}$$

The phase factors $\exp(-2\pi \mathrm{i}\boldsymbol{\kappa}^T\boldsymbol{s}_p)$ result from the shift of the points in the unit cell by the vectors $\boldsymbol{s}_p$ according to the shift theorem (Theorem B.3, p. 629). The vectors $\tilde{\boldsymbol{r}}_v$

$$\tilde{\boldsymbol{r}}_{\tilde{n}} = \tilde{n}_1\tilde{\boldsymbol{b}}_1 + \tilde{n}_2\tilde{\boldsymbol{b}}_2 + \ldots + \tilde{n}_D\tilde{\boldsymbol{b}}_D \quad \text{with} \quad \tilde{n}_d \in \mathbb{Z} \tag{5.19}$$

are the points of the *reciprocal lattice*. The fundamental translation vectors of the grids in spatial and Fourier domain are linked via the relationship

$$\boldsymbol{b}_d^T \tilde{\boldsymbol{b}}_{d'} = \delta_{d-d'}. \tag{5.20}$$

This equation states that a translation vector in Fourier space is perpendicular to all vectors of the spatial domain with the exception of the corresponding vector. Furthermore, the magnitudes of the corresponding vectors are reciprocal to each other so that their scalar product is one. In three dimensions, the basis vectors of the reciprocal lattice can therefore be calculated using

$$\tilde{\boldsymbol{b}}_d = \frac{\boldsymbol{b}_{d+1} \times \boldsymbol{b}_{d+2}}{\boldsymbol{b}_1^T(\boldsymbol{b}_2 \times \boldsymbol{b}_3)}. \tag{5.21}$$

The indices in this equation are to be taken modulo 3, and $\boldsymbol{b}_1^T(\boldsymbol{b}_2 \times \boldsymbol{b}_3)$ gives the volume of the primitive unit cell in space. All these equations are familiar to solid state physicists and crystallographers [49, 182]. Mathematicians know the lattice in Fourier space as the *dual basis* or *reciprocal basis* of a vector space spanned by a non-orthogonal basis. For an orthogonal basis, all vectors of the dual basis point in the same direction as the corresponding vectors and their magnitudes are calculated from $\left|\tilde{\boldsymbol{b}}_d\right| = 1/\left|\boldsymbol{b}_d\right|$. The length of the basis vector $\tilde{\boldsymbol{b}}_d$ is Δx_d and that of the corresponding dual vector □$k_d = 1/\Delta x_d$. Therefore, an orthonormal basis is dual to itself.

5.4.3 ♯Reconstruction of the Continuous Signal

The continuous signal is reconstructed by suitable *interpolation* of the values at the sampled points. The interpolated values $g_r(\boldsymbol{x})$ are calculated from the sampled values at the points $\boldsymbol{r}_n + \boldsymbol{s}_p$ with weighting factors that depend on the distance to the point to be interpolated:

$$g_r(\boldsymbol{x}) = \sum_p \sum_n g_s(\boldsymbol{r}_n + \boldsymbol{s}_p) h(\boldsymbol{x} - \boldsymbol{r}_n - \boldsymbol{s}_p). \tag{5.22}$$

Using the integral property of the δ*-distribution* Eq. (B.45), we can replace the sampled points on the right-hand side of the equation with the values of the continuous signal and then swap the summation and integration:

$$\begin{aligned} g_r(\boldsymbol{x}) &= \sum_p \sum_n \int_{-\infty}^{\infty} g(\boldsymbol{x}')h(\boldsymbol{x} - \boldsymbol{x}')\delta(\boldsymbol{r}_n + \boldsymbol{s}_p - \boldsymbol{x}')\mathrm{d}^D x' \\ &= \int_{-\infty}^{\infty} h(\boldsymbol{x} - \boldsymbol{x}') \sum_p \sum_n \delta(\boldsymbol{r}_n + \boldsymbol{s}_p - \boldsymbol{x}') g(\boldsymbol{x}')\mathrm{d}^D x'. \end{aligned}$$

The last integral contains a convolution of the weighting function h with a function that represents the sum of products of the continuous signal g with shifted δ combs. In Fourier space, the convolution is replaced by a multiplication and vice versa. Taking into account the shift theorem and the fact that the Fourier transform of a δ comb is again a δ comb, we finally obtain

$$\tilde{g}_r(\boldsymbol{\kappa}) = \tilde{h}(\boldsymbol{\kappa}) \sum_p \sum_{\tilde{n}} \tilde{g}(\boldsymbol{\kappa} - \tilde{\boldsymbol{r}}_{\tilde{n}}) \exp\left(-\mathrm{i}2\pi\boldsymbol{\kappa}^T \boldsymbol{s}_p\right). \tag{5.23}$$

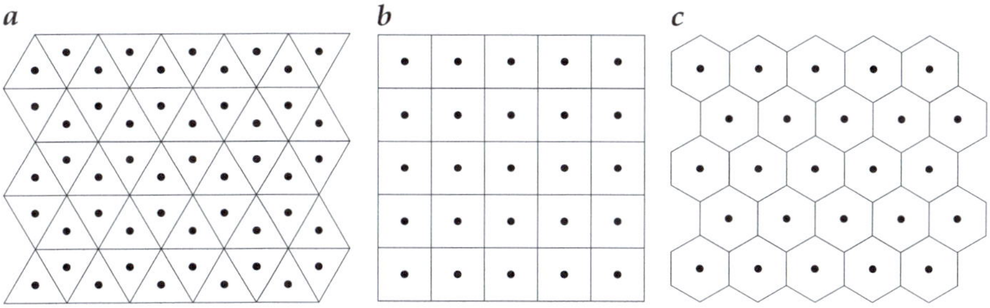

Figure 5.6: Representation of 2D signals by area-filling regular polygons: ***a*** *triangles;* ***b*** *squares;* ***c*** *hexagons. Also shown are the centers of the polygons, which correspond to the grid points.*

The interpolated signal $\tilde{g}_r$ can only be equal to the original signal $\tilde{g}$ if its periodic repetitions do not overlap. This is precisely the statement of the sampling theorem. The Fourier transform of the ideal interpolation function is a box function that is 1 inside the first *Brillouin zone* and zero outside. This eliminates all replications and leaves the band-limited original signal $\tilde{g}$ unchanged.

5.4.4 ♯Examples

Finally, we will look at all 2D grids that can be created using area-filling regular polygons as an example. There are only three possibilities: Triangles, squares and hexagons (Fig. 5.6). With all other regular polygons, a plane cannot be filled without gaps. With these grids, the grid points result from the midpoints of the polygons, which are also their centers of symmetry. With a square grid, the grid points in turn form a square grid, with a hexagonal grid a triangular grid. The situation is more complicated with a triangular grid. Here, the grid points consist of two triangular grids that are offset from each other. The second grid is offset from the first by a third of the edge lengths of the triangular grid.

The basis vectors described in Section 3.5 span elementary cells. These are shown dashed in Fig. 5.7. The basis vectors are not orthogonal in the triangular and hexagonal lattice and the primitive unit cell is a parallelogram in each case.

Only in the rectangular grid are the basis vectors orthogonal and the primitive unit cell is a rectangle. The unit cells of the rectangular and hexagonal grid contain only one grid point. This is not possible with the triangular grid, as the grid points are not arranged at regular intervals in only two directions. Therefore, the elementary cell of the triangular grid spanned by the basis vectors contains two grid points.

As described in Section 3.5, the basis vectors in Fourier space are perpendicular in pairs to the corresponding basis vectors in space (Fig. 5.7).

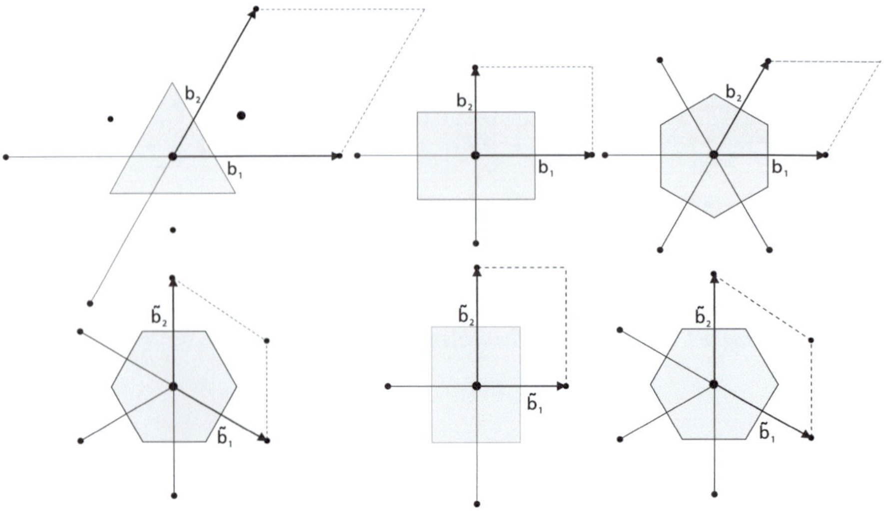

*Figure 5.7: Grid with the fundamental basis vectors **b** and* $\tilde{\mathbf{b}}$*, the elementary cells spanned by the basis vectors (dashed), the grid points and the primitive elementary cell containing only one grid point (gray; called Wigner-Seitz cell in local space and first Brillouin zone in Fourier space). Local space (top) and Fourier space (bottom) for two-dimensional grids: Triangular lattice (left), rectangular lattice (center) and hexagonal lattice (right).*

The primitive elementary cells in Fourier space (first Brillouin zone) of the triangular and hexagonal lattice are each hexaogonal.

The hexagonal lattice has a certain optimality, as its primitive unit cell is closest to the circle in both spaces. We will use an example for filters in Section 11.4.3 to show the advantages of the hexagonal grid.

5.5 Interpolation

5.5.1 Basics

There are two reasons why interpolation is necessary for signal and image processing. Firstly, the original, continuous signal needs to be reconstructed from sampled signals. Secondly, it often happens that one discrete image has to be converted to another, whereby the grids of the two images do not coincide. This is, e. g. the case with the correction of geometric distortions discussed in Section 3.4.6. The two cases differ in the type of interpolation. In the first case, interpolation to a continuous function is necessary, in the second only to a few discrete values. Both types of interpolation are dealt with in this section.

The basis of interpolation is the sampling theorem (Section 5.2.3). It states that the discrete image fully reflects the continuous image, provided that the sampling conditions are met. In other words, this means that each periodic structure occurring in the image must be sampled at least twice per wavelength. From this basic fact, a general procedure for interpolation can easily be derived, at least in principle: One first reconstructs the continuous image and then performs a resampling on new grid points. This procedure only works as long as the new grid is not coarser than the old one. If it is coarser, aliasing will occur. In this case, the image must be pre-filtered before it can be resampled.

Although these procedures sound simple and straightforward, they are actually very difficult. The problem is that the reconstruction of the continuous image from the sampled image is practically very complex and can only be done approximately. Therefore, we have to try to optimize the interpolation by introducing some constraints. In this section, we will first explain why ideal interpolation is generally impossible and then discuss practical methods in Sections 5.5.2–5.5.5.

In Section 5.3.1 we established that the reconstruction of a continuous function from sampling points can be regarded as a convolution operation

$$\boxed{g_r(\boldsymbol{x}) = \sum_{m,n} g(\boldsymbol{x}_{mn}) h(\boldsymbol{x} - \boldsymbol{x}_{mn}),} \tag{5.24}$$

where the continuous interpolation mask h is the sinc function:

$$h(\boldsymbol{x}) = \frac{\sin \pi x_1/\Delta x_1}{\pi x_1/\Delta x_1} \frac{\sin \pi x_2/\Delta x_2}{\pi x_2/\Delta x_2}. \tag{5.25}$$

The transfer function of the point response in Eq. (5.25) is, according to Eqs. (5.8) and (5.14), a rectangular function with the width $2\underset{\sim}{k}_w = 1/\Delta x_w$:

$$\hat{h}(\boldsymbol{\kappa}) = \Pi(\underset{\sim}{k}_1/2, \underset{\sim}{k}_2/2) \quad \text{with} \quad \underset{\sim}{k}_d = 2k_d \Delta x_d. \tag{5.26}$$

The value interpolated in Eq. (5.24) at the grid points $\boldsymbol{x}_{mn}$ should reproduce the grid points and not depend on other grid points. From this we can derive the *interpolation condition*:

$$h(\boldsymbol{x}_{m,n}) = \begin{cases} 1 & m = 0, n = 0 \\ 0 & \text{otherwise.} \end{cases} \tag{5.27}$$

The interpolation mask in Eq. (5.25) fulfills these conditions. Each interpolation mask must therefore have zero crossings at all grid points with the exception of the zero point, where it has the value 1.

Because interpolation is a convolution operation and can therefore be described by a transfer function in Fourier space Eq. (5.26), we have a tool for classifying the errors that occur in the interpolation technique. The rectangular shape of the transfer function for the ideal interpolation function simply means that all wavenumbers within the range $|k_w| \leq 1/(2\Delta x_w)$ of possible wavenumbers experience neither a phase shift nor an amplitude attenuation. Outside the permitted interval, the transfer function is zero. Therefore, no distorted wavenumbers are generated during interpolation.

The ideal interpolation function in Eq. (5.24) is separable. Therefore, the interpolation can easily be formulated for higher dimensional images. It is to be expected that all solutions of the interpolation problem are also separable. Therefore, we only need to discuss the one-dimensional interpolation problem. Once it is solved, we also have a solution for the n-dimensional interpolation.

An important special case is the interpolation only for points that lie exactly between the existing grid points. This interpolation scheme doubles resolution and image size in all directions in which it is applied. The continuous interpolation kernel is then reduced to a discrete convolution mask. Since the interpolation kernel Eq. (5.25) is separable, we can first interpolate the intermediate points of a row in the horizontal direction and then the intermediate rows vertically. In higher-dimensional images, further 1D interpolations are added in the z and/or t direction. The interpolation kernels are the same in all directions. We only need the continuous kernel $h(x)$ for half-integer values of $x/\Delta x$. From Eq. (5.25) we therefore obtain the interpolation convolution mask

$$h = \left[\cdots \frac{(-1)^{m-1}\,2}{(2m-1)\pi} \cdots -\frac{2}{3\pi}\;\frac{2}{\pi}\;\frac{2}{\pi}\;-\frac{2}{3\pi} \cdots \frac{(-1)^{m-1}\,2}{(2m-1)\pi} \cdots \right] \tag{5.28}$$

with coefficients of alternating sign.

5.5.2 Interpolation in Fourier Space

In Fourier space, interpolation is reduced to a simple operation. As Eq. (5.26) shows, the transfer function of an ideal interpolation kernel is a rectangular function that is zero outside the wavenumbers that can be represented. This basic fact leads to the following interpolation procedure in Fourier space:

1. Enlargement of the matrix of the Fourier-transformed image. If a $M \times M$ matrix is enlarged to a $M' \times M'$ matrix, the image in space is also enlarged to a $M' \times M'$ image. However, due to the reciprocity of the Fourier transformation, the image size itself remains unchanged. Only the distance between the pixels is reduced. This results in a higher spatial resolution:

$$M\Delta k \rightarrow M'\Delta k \quad \circ\!\!-\!\!\bullet \quad \Delta x = \frac{1}{M\Delta k} \rightarrow \Delta x' = \frac{1}{M'\Delta k}. \tag{5.29}$$

2. Filling the supplemented area in Fourier space with zeros and calculating an inverse Fourier transform.

Theoretically, this procedure results in a perfectly interpolated image. Unfortunately, however, it has three disadvantages:

1. The Fourier transform of a finite image implies a cyclic repetition of the image in both spatial and Fourier space. Therefore, the convolution performed by the Fourier transform is also cyclic.
   This means that at the right or left edge of the image, the convolution continues with the opposite side of the image. Since reality is not periodic and interpolation masks are large, this can lead to sensitive interpolation disturbances even at a relatively large distance from the edges of the image.
2. The Fourier transform can only be calculated efficiently for certain values of M'. The best known are the fast base-2 algorithms, which can only be applied to images of size $M' = 2^{N'}$. Therefore, interpolation based on the Fourier transform is slow for numbers M' that cannot be represented as a product of many small factors (we will learn this in Section 6.4.2).
3. As the Fourier transformation is a global transformation, it can only be used for scaling. In principle, rotations and affine transformations are also possible according to the extended similarity theorem (Theorem B.1, p. 628). However, this only shifts the interpolation problem from spatial to wavenumber domain.

5.5.3 Linear Interpolation

The *linear interpolation* is the classic interpolation approach. The interpolated points lie on straight lines that connect neighboring grid points. To simplify the calculations, we use normalized spatial coordinates $\tilde{x} = x/\Delta x$ in the following. For reasons of symmetry, we set the two existing grid points to $-1/2$ and $1/2$. This results in the interpolation equation

$$g(\tilde{x}) = \frac{g_{1/2} + g_{-1/2}}{2} + (g_{1/2} - g_{-1/2})\tilde{x} \quad \text{für} \quad |\tilde{x}| \leq 1/2. \tag{5.30}$$

If we compare Eq. (5.30) with Eq. (5.24), we can infer the continuous interpolation mask for linear interpolation:

$$h_1(\tilde{x}) = \begin{cases} 1 - |\tilde{x}| & |\tilde{x}| \leq 1 \\ 0 & \text{otherwise.} \end{cases} \tag{5.31}$$

Their interpolating nature is illustrated in Fig. 5.8. The interpolation mask Eq. (5.31) for linear interpolation is a triangular function, $h_1(x)$, whose transfer function is the squared sinc function (> R13)

$$\hat{h}_1(\tilde{k}) = \frac{\sin^2 \pi\tilde{k}/2}{(\pi\tilde{k}/2)^2}. \tag{5.32}$$

A comparison with the ideal transfer function for the interpolation Eq. (5.26) shows that two disturbances are introduced by the linear interpolation:

1. While small wavenumbers, in particular the mean value $\tilde{k} = 0$, are interpolated correctly, the amplitude of high wavenumbers is reduced slightly,

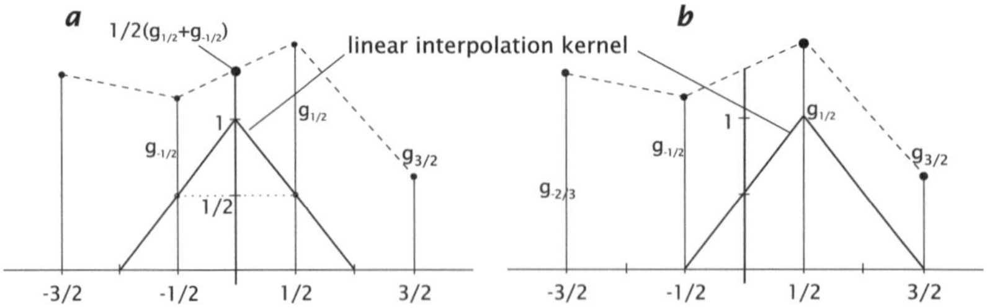

Figure 5.8: *Illustration of linear interpolation:* ***a*** *At $x = 0$, the mean value of $g_{1/2}$ and $g_{-1/2}$ is taken,* ***b*** *at $x = 1/2$, $g_{1/2}$ is replicated.*

which leads to a slight smoothing. At $\underset{\sim}{k} = 1$, the transfer function is reduced to around 40% with $\tilde{h}_1(1) = (2/\pi)^2 \approx 0.4$.

2. As $\tilde{h}_1(\underset{\sim}{k})$ is not zero for wavenumbers $\underset{\sim}{k} > 1$, some false high wavenumbers are generated. If the continuously interpolated image is resampled, this results in slight aliasing effects. The first secondary maximum has an amplitude of $(2/3\pi)^2 \approx 0.045$.

Obviously, both types of error also express that there are discontinuities in the first derivative at the grid points, since the linear interpolation leads to a piecewise linear interpolation (Fig. 5.8).

If we only interpolate the intermediate grid points at $\underset{\sim}{x} = 0$, the continuous interpolation function Eq. (5.31) becomes a discrete convolution mask with values at $\underset{\sim}{x} = [\ldots\ -3/2\ \ -1/2\ 1/2\ 3/2\ \ldots]$. Since Eq. (5.31) is zero for $|\underset{\sim}{x}| \geq 1$, the discrete interpolation mask $H = 1/2[11]$ results with the transfer function

$$\tilde{h}_1(\underset{\sim}{k}) = \cos \pi\underset{\sim}{k}/2. \tag{5.33}$$

The transfer function is real, so that no phase shift occurs. However, the significant amplitude attenuation towards high wavenumbers shows that these are not interpolated correctly.

Phase shifts also occur at all points other than the symmetrical interstitial point at $\underset{\sim}{x} = 0$. We now examine the phase shift and the amplitude attenuation of the linear interpolation for any points ϵ between the grid points, i. e. $\epsilon \in [-1/2, 1/2]$. Then for a point ϵ the interpolation mask is $[1/2 - \epsilon, 1/2 + \epsilon]$. The mask contains a symmetrical part $[1/2, 1/2]$ and an antisymmetrical part $[-\epsilon, \epsilon]$. The transfer function is therefore complex and results in

$$\tilde{h}_1(\epsilon, \underset{\sim}{k}) = \cos \pi\underset{\sim}{k}/2 + 2\mathrm{i}\epsilon \sin \pi\underset{\sim}{k}/2. \tag{5.34}$$

To determine the error in the phase position, we refer the phase position to the point at $\epsilon = 0$, since no phase shift should occur at this point. According to the shift theorem (Theorem B.3, p. 629, > R12), we must multiply Eq. (5.34) by $\exp(-\mathrm{i}\epsilon\pi\underset{\sim}{k})$:

$$\tilde{h}_1(\epsilon, \underset{\sim}{k}) = (\cos \pi\underset{\sim}{k}/2 + 2\mathrm{i}\epsilon \sin \pi\underset{\sim}{k}/2) \exp(-\mathrm{i}\epsilon\pi\underset{\sim}{k}). \tag{5.35}$$

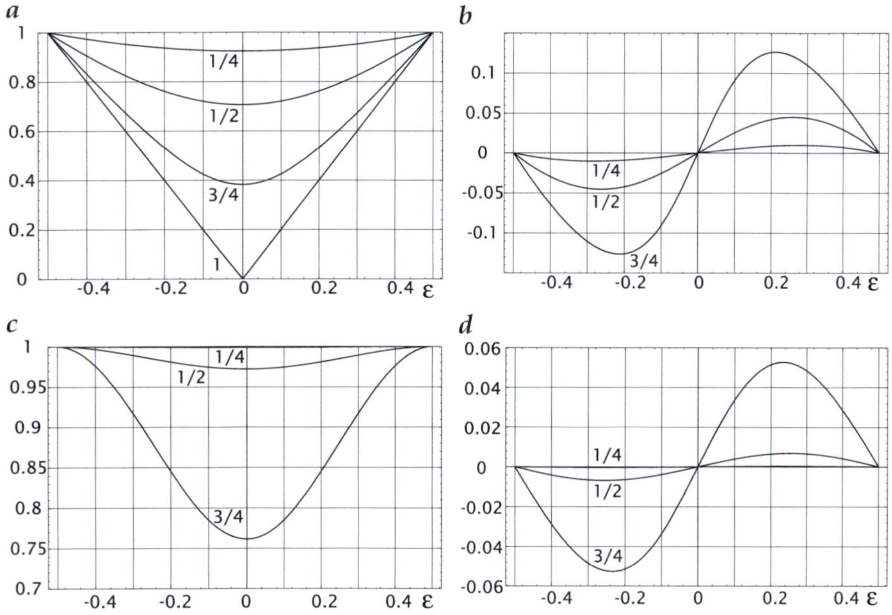

Figure 5.9: *Amplitude attenuation (left column) and phase shift, expressed as position shift $\Delta x = \Delta\varphi\lambda/2\pi$ in radians (right column), for wavenumbers $\tilde{k} = 1/4, 1/2, 3/4$, represented as a function of the position ϵ of the point to be interpolated between $-1/2$ to $1/2$ for linear interpolation (**a** and **b**) and cubic B-spline interpolation (**c** and **d**).*

The transfer function is only real for $\epsilon = 0$ ($\tilde{h}_1(0, \tilde{k}) = \cos \pi\tilde{k}/2$) and $\epsilon = 1/2$ ($h_1(1/2, \tilde{k}) = 1$). For all other points, there is a non-negligible phase shift (Fig. 5.9). The phase shift $\Delta\varphi$ is expressed as the shift $\Delta x = \Delta\varphi\lambda/2\pi = \Delta\varphi/(\pi\tilde{k})$ of the associated periodic structure.

5.5.4 ♯Interpolation with Polynomials

Given the considerable limitations of linear interpolation, as discussed in Section 5.5.3, we ask ourselves whether higher order interpolations are advantageous. Linear interpolation connects two neighboring points by a straight line. Similarly, we can place a P-degree polynomial with $P + 1$ unknown coefficients a_p through $P + 1$ points:

$$g_r(\tilde{x}) = \sum_{p=0}^{P} a_p \tilde{x}^p. \tag{5.36}$$

For reasons of symmetry, we set the positions of an even number of grid points — P is therefore odd — to half-integer values:

$$\tilde{x}_p = \frac{2p - P}{2}. \tag{5.37}$$

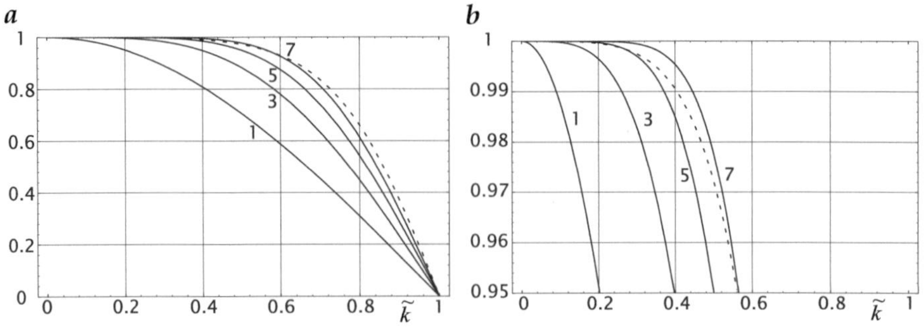

Figure 5.10: *Transfer functions of discrete polynomial interpolation filters for interpolating a value between two grid points. The degree of the polynomial (1 = linear, 3 = cubic etc.) is marked in the graphic. The dashed line represents the transfer function for cubic B-spline interpolation (Section 5.5.5).* ***a*** *shows the value range from 0 to 1 and* ***b*** *a 5 % range below the ideal transfer function* $\tilde{h}(\underset{\sim}{k}) = 1$.

With the interpolation condition for the grid points, $g_r(\underset{\sim}{x}_p) = g_p$, the following linear system of equations with $P+1$ equations and $P+1$ unknowns a_P results if P is odd:

$$\begin{bmatrix} g_0 \\ \vdots \\ g_{(P-1)/2} \\ g_{(P+1)/2} \\ \vdots \\ g_P \end{bmatrix} = \begin{bmatrix} 1 & -P/2 & P^2/4 & -P^3/8 & \cdots \\ \vdots & & & & \\ 1 & -1/2 & 1/4 & -1/8 & \cdots \\ 1 & 1/2 & 1/4 & 1/8 & \cdots \\ \vdots & & & & \\ 1 & P/2 & P^2/4 & P^3/8 & \cdots \end{bmatrix} \begin{bmatrix} a_0 \\ \vdots \\ \vdots \\ a_P \end{bmatrix}. \tag{5.38}$$

From this we can determine the coefficients of the polynomial. For a cubic polynomial ($P = 3$), for example, this results in the system of equations

$$\begin{bmatrix} g_0 \\ g_1 \\ g_2 \\ g_3 \end{bmatrix} = \begin{bmatrix} 1 & -3/2 & 9/4 & -27/8 \\ 1 & -1/2 & 1/4 & -1/8 \\ 1 & 1/2 & 1/4 & 1/8 \\ 1 & 3/2 & 9/4 & 27/8 \end{bmatrix} \begin{bmatrix} a_0 \\ a_1 \\ a_2 \\ a_3 \end{bmatrix} \tag{5.39}$$

with the solution

$$\begin{bmatrix} a_0 \\ a_1 \\ a_2 \\ a_3 \end{bmatrix} = \frac{1}{48} \begin{bmatrix} -3 & 27 & 27 & -3 \\ 2 & -54 & 54 & -2 \\ 12 & -12 & -12 & 12 \\ -8 & 24 & -24 & 8 \end{bmatrix} \begin{bmatrix} g_0 \\ g_1 \\ g_2 \\ g_3 \end{bmatrix}. \tag{5.40}$$

From this solution, we can deduce that the gray value at the point $\underset{\sim}{x} = 0$ is interpolated by $g_r(0) = a_0 = -1/16 g_0 + 9/16 g_1 + 9/16 g_2 - 1/16 g_3$, which corresponds to the interpolation mask 1/16[-1, 9, 9, -1].

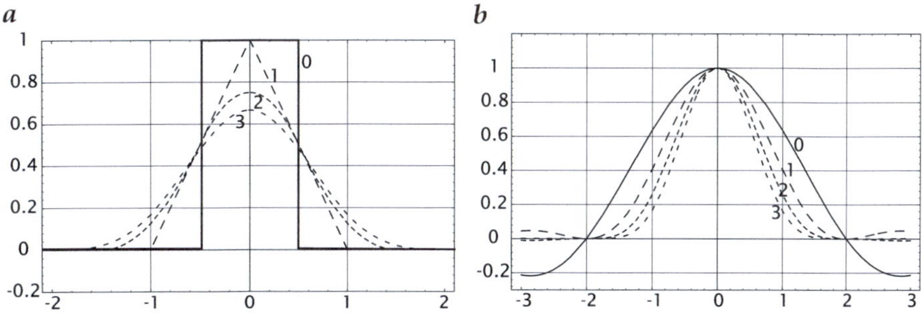

Figure 5.11: *a B-spline interpolation kernels, generated by a cascaded convolution of the rectangular kernel: zeroth order (nearest neighbor), first (linear interpolation), second (quadratic B-spline) and third order (cubic B-spline);* ***b*** *associated transfer functions.*

Fig. 5.10 shows the transfer functions for interpolations with polynomials of different degrees. As the degree P of the interpolating polynomial increases, the transfer function increasingly approaches the rectangular function. However, it converges slowly. For an exact interpolation, we must therefore use a large interpolation mask.

5.5.5 ♯Interpolation with Splines

In addition to the limited accuracy, interpolation with polynomials has other significant disadvantages. The interpolated curve is not continuous even in the first derivative at the grid points. This is due to the fact that a different polynomial is used for each interval between grid points. In general, therefore, only the interpolated function is continuous at the grid points, but its derivatives are not.

Splines avoid these disadvantages by imposing additional continuity conditions on the derivatives at the grid points. From the many classes of spline functions, we will only discuss one class here, the so-called *B-splines*. Since B-splines are separable, it is again sufficient to consider the properties of the one-dimensional functions. Against the background of image processing, the easiest way to access B-splines is their convolution property. The core of a B-spline curve of order P is generated by convolving the rectangular function $(P+1)$ times itself (Fig. 5.11a):

$$\beta_P(\underset{\sim}{x}) = \underbrace{\Pi(\underset{\sim}{x}) * \ldots * \Pi(\underset{\sim}{x})}_{(P+1)\text{-mal}} \quad \circ\!\!-\!\!\bullet \quad \tilde{\beta}_P(\underset{\sim}{k}) = \left(\frac{\sin \pi \underset{\sim}{k}/2}{(\pi \underset{\sim}{k}/2)}\right)^{P+1}. \tag{5.41}$$

The B-spline function is not a suitable interpolating function, as the transfer function (Fig. 5.11b) decreases too quickly. This means that it smoothes too much. In addition, the B-spline kernel for $P > 1$ does not fulfill the interpolation condition Eq. (5.27).

B-splines can only be used for interpolation if the discrete grid points are first transformed in such a way that a subsequent convolution with a B-spline kernel restores the original image at the grid points. This transformation is called B-spline transformation and is constructed from the following condition:

$$g_p(x) = \sum_n c_n \beta_P(x - x_n) \qquad \text{with} \qquad g_p(x_n) = g(x_n). \tag{5.42}$$

A cubic B-spline interpolation kernel centered around a grid point is only non-zero for three grid points. The coefficients $\beta_3(-1) = \beta_{-1}, \beta_3(0) = \beta_0$ and $\beta_3(1) = \beta_1$ are 1/6, 2/3, 1/6. The convolution of this kernel with the unknown B-spline transformation values c_n should result in the original values g_n at the grid points:

$$\boldsymbol{g} = \boldsymbol{c} * \boldsymbol{\beta}_3 \quad \text{or} \quad g_n = \sum_{n'=-1}^{1} c_{n+n'} \beta_{n'}. \tag{5.43}$$

Equation (5.43) corresponds to the sparse linear system of equations

$$\begin{bmatrix} g_0 \\ g_1 \\ \vdots \\ g_{N-1} \end{bmatrix} = \frac{1}{6} \begin{bmatrix} 4 & 1 & 0 & \ddots & 0 & 1 \\ 1 & 4 & 1 & 0 & \ddots & 0 \\ 0 & 1 & 4 & 1 & 0 & \ddots \\ \ddots & & & & \ddots & \ddots \\ \ddots & \ddots & 1 & 4 & 1 & 0 \\ 0 & \ddots & 0 & 1 & 4 & 1 \\ 1 & 0 & \ddots & 0 & 1 & 4 \end{bmatrix} \begin{bmatrix} c_0 \\ c_1 \\ \vdots \\ c_{N-1} \end{bmatrix}, \tag{5.44}$$

whereby cyclical boundary conditions were assumed. A linear system of equations must therefore be solved. However, the special form of the system of equations as a convolution results in a more effective solution. In Fourier space, Eq. (5.43) reduces to

$$\tilde{\boldsymbol{g}} = \tilde{\boldsymbol{\beta}}_3 \tilde{\boldsymbol{c}}. \tag{5.45}$$

The transfer function of $\boldsymbol{\beta}_3$ is $\tilde{\beta}_3(\tilde{k}) = 2/3 + 1/3\cos(\pi\tilde{k})$. This function has no zero. Therefore, $\boldsymbol{c}$ is obtained by inverse filtering (Section 10.3.5), i.e. by convolution of $\boldsymbol{g}$ with a convolution mask that has the transfer function

$$\tilde{\beta}_3^{-1}(\tilde{k}) = \tilde{\beta}_T(\tilde{k}) = \frac{1}{2/3 + 1/3\cos\pi\tilde{k}}. \tag{5.46}$$

Such a transfer function belongs to a recursive filter (Section 10.4.4), which is applied first forwards and then backwards with the following recursion [338]:

$$\begin{aligned} g'_n &= g_n - (2 - \sqrt{3})(g'_{n-1} - g_n) \\ c'_n &= g'_n - (2 - \sqrt{3})(c_{n+1} - g'_n). \end{aligned} \tag{5.47}$$

The entire operation only requires two multiplications and four additions. The B-spline transformation is followed by a B-spline interpolation. In the continuous case, using Eqs. (5.41) and (5.46), we obtain the effective transfer function for the B-spline interpolation

$$\tilde{\beta}_I(\underset{\sim}{k}) = \frac{\sin^4(\pi \underset{\sim}{k}/2)/(\pi \underset{\sim}{k}/2)^4}{(2/3 + 1/3 \cos \pi \underset{\sim}{k})}. \qquad (5.48)$$

Essentially, the B-spline transformation Eq. (5.46) amplifies high wavenumbers (at $\underset{\sim}{k} = 1$ by a factor of approximately 3), which largely compensates for the smoothing effect of the B-spline interpolation.

We investigate this compensation at the grid points and at the intergrid points. From the equation of the cubic B-spline interpolation kernel Eq. (5.41) (see also Fig. 5.11a), the interpolation coefficients for grid and intergrid points are obtained as

$$1/6\,[1\;4\;1] \qquad \text{and} \qquad 1/48\,[1\;23\;23\;1]\,, \qquad (5.49)$$

with the transfer functions

$$2/3 + 1/3 \cos \pi \underset{\sim}{k} \qquad \text{and} \qquad 23/24 \cos(\pi \underset{\sim}{k}/2) + 1/24 \cos(3\pi \underset{\sim}{k}/2). \qquad (5.50)$$

As expected, the transfer function compensates for the application of the B-spline transformation Eq. (5.46) at the grid points. The interpolation curve therefore passes through the grid points. At the intermediate grid points, the effective transfer function for the cubic B-spline interpolation is then

$$\tilde{\beta}_I(1/2, \underset{\sim}{k}) = \frac{23/24 \cos(\pi \underset{\sim}{k}/2) + 1/24 \cos(3\pi \underset{\sim}{k}/2)}{2/3 + 1/3 \cos \pi \underset{\sim}{k}}. \qquad (5.51)$$

The amplitude attenuation and the phase shift, expressed as a position shift in pixel distances, are shown in Fig. 5.9c and d. It should be noted that the shifts are related to the interstitial points. Shift and amplitude attenuation are zero at the grid points $[-1/2, 1/2]^T$. While the amplitude attenuation is maximum at the interstitial point, the position shift there is also zero due to the symmetry. The phase shift at the wavenumbers $\underset{\sim}{k} = 3/4$ is unfortunately only about 50 % smaller than with linear interpolation (Fig. 5.9b). It is still significant with a maximum value of around 0.13. This value is far too high for algorithms that should be accurate in the range of 1/100 pixels. If no better interpolation techniques can be used, this means that the maximum wavenumber should be less than 0.5. Then the maximum shift is less than 0.01 and the amplitude attenuation is less than 3%.

This discussion of the phase shift only applies to arbitrary shifts. For intermediate grid points, no position shift occurs at all.

In this special case — which often occurs in image processing, e. g. in the calculation of pyramids (Chapter 15) — the optimization of interpolation filters is quite simple, since only the amplitude attenuation over the wavenumber range of interest must be minimized.

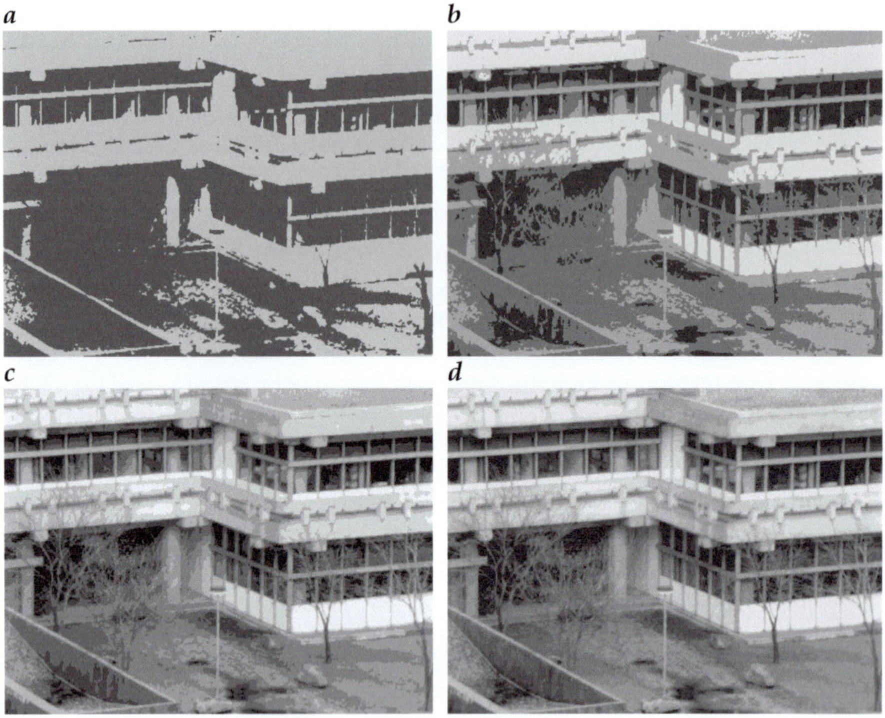

Figure 5.12: Representation of an image with different numbers of quantization levels: ***a*** *2,* ***b*** *4,* ***c*** *8,* ***d*** *16. Too little quantization produces false edges and objects with low contrast may disappear partially or completely.*

5.6 Quantization

For representation with digital numbers, the measured irradiance must be mapped to a limited number Q of discrete gray values. This process is referred to as *quantization*. The number of quantization levels required can be discussed under two criteria.

On the one hand, we can argue that the human eye should not be able to recognize gray value levels, just as individual pixels in digital images should not be perceptible. The images in Fig. 5.12 were quantized with 2 to 16 gray value levels. It is clear from them that incorrect edges are generated if there are too few gray values and that it is difficult to recognize objects that differ only slightly from the background in terms of their gray values. For printed images, 16 grey values are usually sufficient, whereas on a monitor with the same resolution, the grey value levels can still be recognized. By default, image data is quantized with 8 bits in 256 grey values. This fits in well with the architecture of standard computers, which address the main memory byte by byte. In addition, this resolution is good enough to simulate

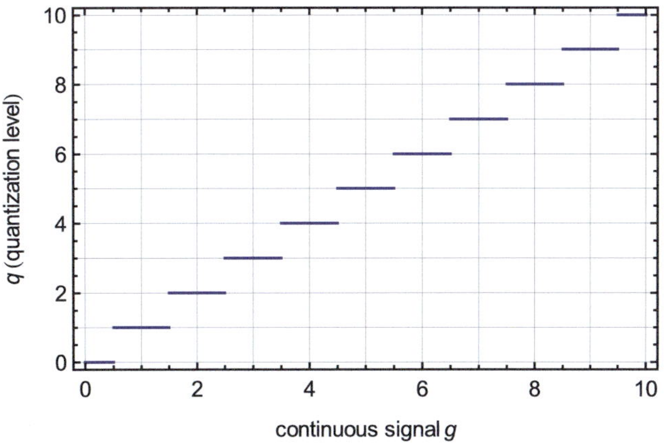

Figure 5.13: Quantization is a non-linear step function.

a continuous transition of gray values, as the resolution of the human visual system is only around 2%.

The other criterion is based on the task at hand. For a simple application, two quantization levels may suffice. Examples include machine vision systems in which objects are to be recognized and measured with uniform illumination and against a homogeneous background. Other applications, such as spectroscopy or X-ray imaging for medical diagnosis, require the resolution of much smaller differences in intensity. In this case, the standard resolution (8 bit) would be insufficient.

5.6.1 Equidistant Quantization

After digitization (Section 5.2), the pixels still show continuous values. To be able to process them with a computer, we need to map them to a limited number Q of discrete values:

$$[0, \infty[\stackrel{Q}{\longrightarrow} \{g_0, g_1, \ldots, g_{Q-1}\} = G.$$

If the quantization levels are evenly spaced Δg, the quantized values result in

$$q = \text{floor}\left(\frac{g + 0.5}{\Delta g}\right). \qquad (5.52)$$

As simple as this operation looks — it is not easy to describe its influence. This is because quantization is a non-linear function (Fig. 5.13) and — in contrast to sampling — the original value cannot be reconstructed. It is therefore not possible on a deterministic basis to find a theorem for quantization that corresponds to the sampling theorem (Section 5.2.3).

However, this is not necessary as every signal contains noise. The crucial question is therefore, what is the optimal ratio between the standard deviation σ_g and the quantization interval Δg. This can be expressed in the factor

$$\gamma = \frac{\sigma_g}{\Delta g}. \tag{5.53}$$

What is the optimal value of γ? We first discuss the extremes. If the noise is much smaller than the quantization interval ($\gamma \ll 1$), then any information about the noise in the signal is lost. This means that the standard deviation of the noise cannot be estimated and the mean of the value can be wrong up to $\pm\Delta g/2$. Further statements are not possible. If the noise is much larger than the quantization interval ($\gamma \gg 1$), then the noise can be captured well, but the question remains as to how finely the noise must be sampled.

In other words, a theory of quantization must answer the question of whether it is possible to reconstruct the probability density function (PDF, see Appendix C) of the continuous signal from the PDF of the quantized signal. If this is possible, both the mean value and the standard deviation of the noisy signal can be calculated from it without additional errors due to quantization.

This issue was studied in detail by Widrow and Kollar [354] and resulted in quantization theorems that are very similar to the sampling theorem when the signal g is replaced by its PDF. The more detailed analysis shows that there are several *quantization theorems*. The first one specifies the conditions how the continuous PDF *exact* can be reconstructed from the discrete PDF of the quantized signal, the second one specifies the conditions how only the moments (this includes mean and standard deviation) of the PDF can be reconstructed exactly. The conditions for the second theorem are less strict. In both cases, however, it is necessary that the PDF of the continuous signal is band-limited. For details, please refer to Widrow and Kollar [354, Sect. 4.3].

If this condition is met, the mean value can be calculated exactly from the PDF and a pseudo-noise is added to the noise of the signal by the quantization, which is uniformly distributed in the quantization interval (Appendix C.4.3). The variance of this additional noise results in

$$\sigma_q^2 = \frac{1}{\Delta g} \int\limits_{g_q-\Delta g/2}^{g_q+\Delta g/2} (g - g_q)^2 \mathrm{d}g = \frac{1}{12}(\Delta g)^2. \tag{5.54}$$

The standard deviation σ_q corresponds to approximately 0.3 times the distance between the quantization levels Δg. The variance of the continuous signal g then results from the measured to

$$\sigma_g^2 = \sigma_{\text{measured}}^2 - \sigma_q^2. \tag{5.55}$$

The problem is that normally distributed smoking is not band-limited. Therefore, the quantization theorems can never be fulfilled exactly. Estimates by Widrow and Kollar [354] show, however, that with a standard deviation

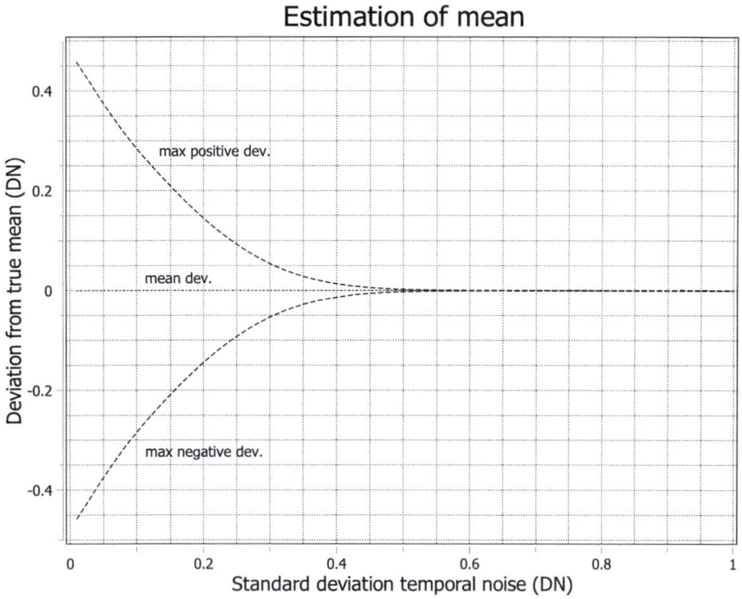

Figure 5.14: *Results of Monte Carlo simulations for the accuracy of the computation of mean values from noisy quantized signals. The values are calculated for the specified standard deviations of normally distributed noise and are given in units of quantization levels (DN = digital number).*

of half the width of the quantization interval ($\gamma = 0.5$) the moments of the PDF can still be estimated with sufficient accuracy.

5.6.2 Accuracy of Quantized Values

The theoretical studies by Widrow and Kollar [354] were verified by Monte Carlo simulations [159]. For this purpose, signals with standard deviations of the noise between 0.0 and 1.0 DN were simulated, in which the mean value took all possible values between the quantization values and in each case the mean errors and also the maximum deviations in the determination of the *mean value* and the *standard deviation* were calculated.

The systematic errors in the determination of the mean value are completely uncritical (Fig. 5.14). With a standard deviation of 0.4 DN, the maximum error remains below 0.015 DN. If, on the other hand, the signal is noise-free, there is no deviation on average over all possible positions of the mean value relative to the quantization interval, whereas the maximum deviation from the true value can be up to ±0.5 DN. This happens when the mean value lies in the middle of a quantization interval, i. e. has a half-integer value.

This is an astonishing result. It means that there are significantly higher errors with a noise-free signal than with a standard deviation of the noise

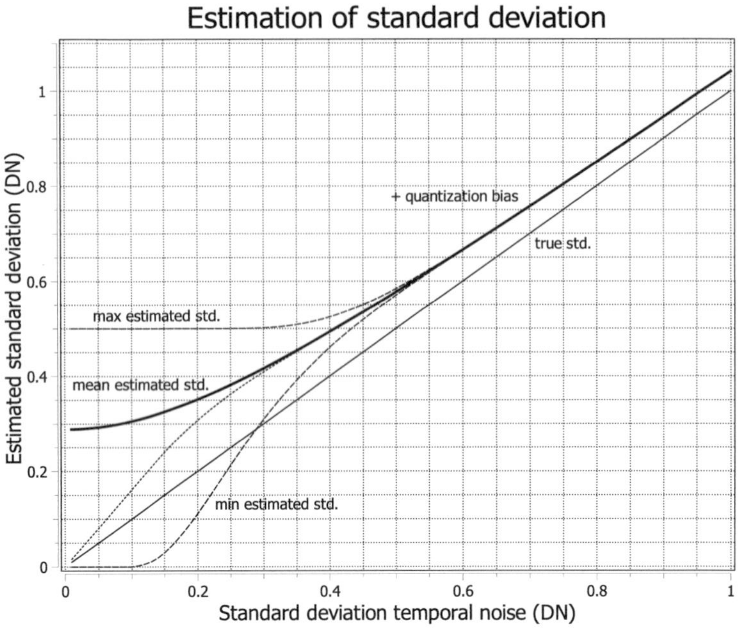

Figure 5.15: *Results of Monte Carlo simulations for the accuracy of the computation of standard deviations from noisy quantized signals. The values are calculated for the specified standard deviations of normally distributed noise and are given in units of quantization levels (DN = digital number).*

of 0.5 DN, i. e. about half the width of the quantization interval. This does not make a significant difference for a single measured value, but only when averaged.

The estimate of the standard deviation shows the expected increase by the quantization noise (Fig. 5.15) according to (Eq. 5.55). However, as this is known, this deviation can be easily corrected. The estimates only diverge for standard deviations below 0.5 DN. For very small noise with a standard deviation below 0.1 DN, any value between 0.0 and 0.5 DN can be estimated. This standard deviation value, which is at least five times higher, is obtained when the mean value lies exactly on the boundary between two quantization intervals. In this case, a value either 0.5 DN higher or lower than the correct value is determined with the same probability. In contrast, with a standard deviation of 0.4 DN, the deviations in the estimate of the standard deviation are limited to ± 0.04 DN. The error is therefore a maximum of 10%.

This confirms the theoretical estimates of Widrow and Kollar [354] and a quantization with $\Delta g = 2\sigma_g$ ($\gamma = 0.5$) is optimal. These statements are particularly relevant for images and higher-dimensional signals, as it is easy to reduce the statistical errors by averaging over many values. The error of the mean value is reduced for P statistically independent measurements to (see Eq. (C.29) in Section C.3.3)

$$\sigma_{\mathrm{mean}} = \frac{1}{\sqrt{P}} \sigma_g. \tag{5.56}$$

This equation states that the standard deviation of 100 measurements should be about 1/10 of that of the individual measurements and thus the mean value can be determined much more accurately than the width of the quantization interval. The statistical uncertainty of the mean value is then just $0.05\,\Delta g$.

Finally, however, it should be noted that the accuracy of the determination of the mean value and the standard deviation is also limited by other systematic errors. The most important source of error is the non-uniformity of the quantization levels. In a real quantization e. g.by an analogue-to-digital converter, the quantization steps are not uniform, but show systematic deviations that can be up to half of the nominal quantization interval. Therefore, if high demands are placed on the accuracy of signal values, a careful examination of the analog-to-digital converter is necessary in order to estimate what actually limits the accuracy of the measurement.

5.6.3 Signed Image Representation

Normally, brightness (irradiance) is understood as a positive quantity. It therefore seems natural to represent it in an unsigned 8-bit representation as a number in the range from 0 to 255. However, this causes problems as soon as we perform arithmetic operations with the images. The subtraction of two images is a simple example that produces negative numbers. As negative values cannot be displayed, they appear as high positive values (in the two's complement representation, the highest bit is set to one). For example, the number −1 becomes the gray value 255 with −1 modulo 256 = 255.

This confronts us with the problem of two different representations of values as signed or unsigned 8-bit numbers. Accordingly, each algorithm must be available in several versions, one for unsigned values, one for signed values and others for mixed cases in which, for example, unsigned images are filtered with signed filter coefficients.

One solution is to treat values *stets* as signed numbers. In an 8-bit representation, unsigned numbers can be converted by subtracting 128 from them:

$$q' = (q - 128) \mod 256, \quad 0 \le q < 256. \tag{5.57}$$

In this representation, the average gray value intensity of 128 corresponds to the value zero and lower values represent negative values. Thus, values in this representation are interpreted as a deviation from the mean value 128.

The signed values resulting from this operation can be represented as such. Only for the image display do the values have to be converted back into unsigned gray values using the reverse point operation:

$$q = (q' + 128) \mod 256, \quad -128 \le q' < 128. \tag{5.58}$$

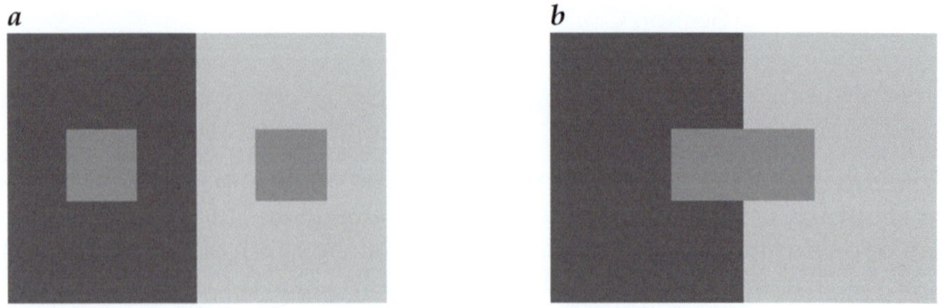

Figure 5.16: *The context determines the brightness with which our visual system perceives an object. Both squares are equally bright.* ***a*** *However, the square on the dark background appears brighter than the one on the light background.* ***b*** *The two objects only appear equally bright when they touch each other.*

This addition is identical to the subtraction performed in Eq. (5.57), as all calculations are performed modulo 256. This representation is used throughout this book when images with negative values are shown, e. g. when first order derivatives are performed. For an example, see Fig. 12.4.

5.6.4 Human Visual System Brightness Perception

With regard to quantization, it is important to know how the human eye perceives luminosities and what differences it can detect. The small squares in Fig. 5.16 with the same luminosity appear brighter on a dark background than on a light background, although their absolute brightness is identical. This optical illusion only disappears when the two squares touch.

The resolution of the human visual system has a logarithmic rather than a linear characteristic. This means that we can perceive relative differences in brightness. In a wide range of luminance values, we can distinguish relative differences of about 2%. This limit depends on a number of factors, in particular the spatial frequency (wavelength) of the pattern used for the experiment. At a certain wavelength, the resolution is optimal.

The properties of the human visual system described above differ significantly from those of machine systems. These typically only have a resolution of 256 gray values. This means that a digitized image has a much lower dynamic range than images perceived by humans. This is the reason why the quality of digitized images, especially of scenes with high contrast, is significantly worse for us than the images we see directly. In a digital image of a scene captured with a linear image sensor, either the bright areas are overflowing or the dark areas are underflowing. This problem is illustrated in Fig. 5.17 using a scene with high contrast (looking at a desk towards a window).

Although the *relative* resolution in the bright image areas is far better than 2%, it is significantly lower in the dark areas. At a value of 10, the resolution is only 10%.

a *b*

Figure 5.17: *A scene with high contrast, captured by a CCD camera with linear contrast and* ***a*** *a small or* ***b*** *a large aperture.*

A solution for dealing with high dynamics in scenes can be found in video sensors that convert the irradiance E into the value g using an exponential function rather than linearly:

$$g = E^{\gamma}. \tag{5.59}$$

The exponent γ is referred to as the *gamma value*. Typically, γ has a value of 0.4. With this exponential conversion, cameras can approximate the logarithmic character of the human visual system and achieve a significant increase in the contrast range. Assuming a minimum relative resolution of 10% and 256 gray values, $\gamma = 1$ results in a contrast range of 25. With $\gamma = 0.4$ it is more than twelve times higher with a value of 316.

However, many scientific applications require a linear relationship between the luminance of the recorded object and the resulting gray value in the camera. For such applications, the gamma value must therefore be set to one.

5.7 ♯References for Further Reading

The sampling theorem is presented in detail in Poularikas [245, Sect. 1.6]. Sampling of stochastic processes, also with random intervals, is discussed in Papoulis [236, Sect. 11.5], while Fliege [86] discusses the sampling theorem with respect to multirate signal processing. Modern extensions of the sampling theorem, including methods for irregular or random sampling points, are discussed in Unser [337], Benedetto and Ferreira [15] and Eldar [68].

A detailed discussion of interpolation filters with tables of filter coefficients can be found in Jähne [157, Chapter 8]. Readers interested in the mathematical background of interpolation are referred to Davis [52], Lancaster and Salkauskas [192] and Thévenaz et al. [328].

6 Discrete Image Representation

Contents

6.1 Introduction

We imagine images as continuous functions $g(x)$ in 2D space. However, computers cannot store continuous images. In the last chapter we have already dealt in detail with the sampling of continuous functions and under which conditions and how they can be reconstructed from the sampled values. In this chapter we will now examine how two-dimensional images or — more generally — one-dimensional time series and multidimensional signals can be adequately discretely represented and which mathematical methods are required for this. Several aspects are essential in this context.

B. Jähne, *Digital Image Processing and Image Acquisition*,
https://doi.org/10.1007/978-3-032-04806-6_6

First, we have to deal with the elementary cells of discrete representations and their neighborhood relations (Section 6.2). In two dimensions these are pixels, in three dimensions voxels. Discrete or digital signals can be represented in completely different ways, the most important being the temporal and spatial and the frequency and wavenumber representation (Section 6.3). These differ only in the way they view spatial data. As the different representations each fully represent the data and are therefore equivalent, they can be converted into each other. The conversion from the spatial to the wavenumber representation is known as *Fourier transform*. Its properties for discrete signals are also illustrated in this chapter. It is an example of a more general class of operations, the *unitary transformations* (Appendix B).

Of particular interest is the question of how the transformation from one representation to another can be calculated quickly. Are there fast algorithms that can accomplish this? A key example of this is the fast Fourier transform, which is discussed in Section 6.4 and where some important basic principles for the design of fast algorithms are discussed, which are of general importance for fast signal processing.

6.2 Digital Images in the Space Domain

Images represent a distribution of *irradiance* in a plane. Mathematically, this can be described as a continuous function of two spatial variables:

$$E(x_1, x_2) = E(\boldsymbol{x}). \tag{6.1}$$

6.2.1 Pixel and Voxel

Computers cannot process continuous images, only digital number fields. For this reason, images must be stored as two-dimensional arrays of values. A point on the 2D grid is referred to as *pixel* or *pel*. Both terms are short forms of the English term *picture element*. A pixel or pixel represents the irradiance at the corresponding grid position. Matrix notation is commonly used to specify the position of the pixel. The first index, m, indicates the position of the row (*row index*), the second, n, the position of the column (*column index*) (Fig. 6.1a). If a digital image consists of $M \times N$ pixels, it is represented by a $M \times N$ matrix. The column index, n, runs from 0 to $N - 1$ and the row index, m, from 0 to $M - 1$. M indicates the number of rows, N the number of columns. In accordance with the matrix notation, the vertical axis (y-axis) runs from top to bottom and not the other way round, as is often the case. The horizontal axis (x-axis) runs from left to right as usual.

Each pixel actually represents not just a point, but a rectangular region, the *elementary cell* of the grid. The value associated with the pixel must appropriately represent the average irradiance of the associated grid cell.

Fig. 6.2 shows an image, displayed with different numbers of pixels. If only a few pixels are used (Fig. 6.2a, b), not only is the spatial resolution

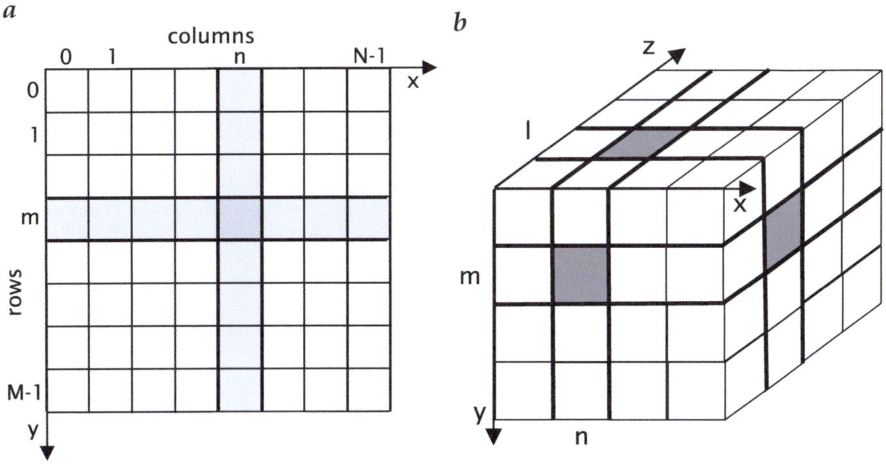

Figure 6.1: *Representation of digital images by fields of discrete points on a square grid:* ***a*** *2D image,* ***b*** *3D image.*

poor, but the gray value differences at the pixel edges also create artifacts that distract from the actual image content. If the pixels become smaller, these effects are also less noticeable, until we have the impression of a spatially continuous image. This is the case when the pixels are smaller than the spatial resolution of our visual system. You can see this relationship for yourself by looking at Fig. 6.2 from different distances.

How many pixels are necessary? There is only one general answer to this question: no more than necessary, as the effort required for subsequent processing increases with the number of pixels. For viewing a digital image, the pixel size should be smaller than the spatial resolution of the observer's visual system at a given distance. For a given task, the pixel size should be smaller than the smallest objects to be examined. In terms of metrology, it should be noted that even a high-resolution sensor with 1000×1000 pixels still has a relative spatial resolution of only 10^{-3}. This is a poor resolution compared to other methods such as length measurement, electrical voltage measurement or frequency determination. These quantities can be determined with relative resolutions of far better than 10^{-6}. However, they only provide a measurement at a single point, whereas a 1000×1000 pixel image contains a million pixels. This gives us insight into the spatial variations of a measured variable. Recording image sequences also reveals the temporal changes and thus the kinematics and dynamics of the processes under investigation. In this way, image processing opens up a completely new quality of measurement.

A square grid is the simplest geometry of a digital image. Other geometric arrangements of the pixels and shapes of the elementary cells are possible. This question corresponds to the investigation of crystal structures in 3D

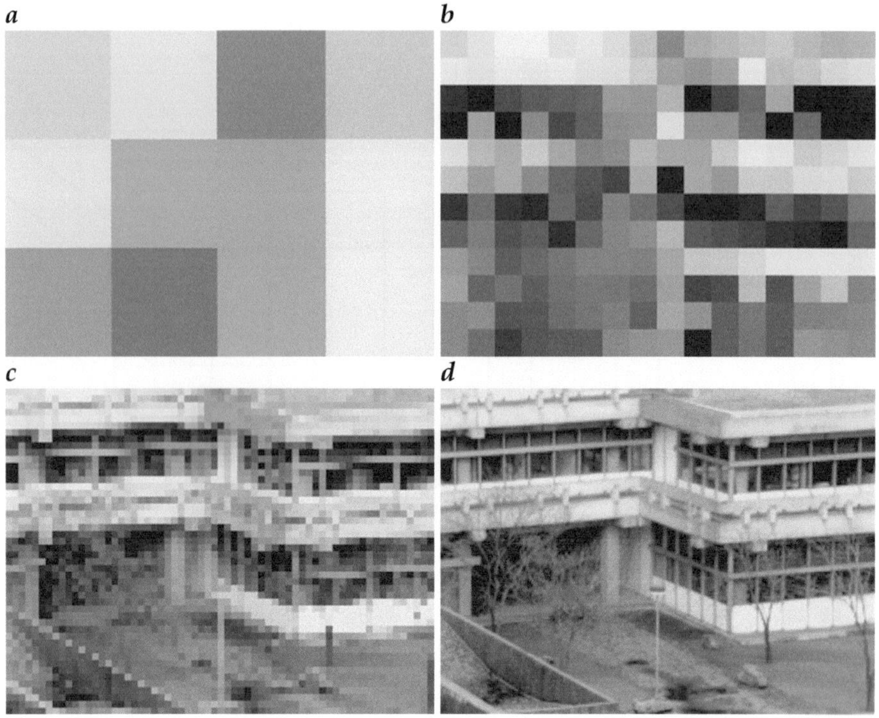

Figure 6.2: *On a square grid, each pixel represents a square region of the image. The same image is shown with* ***a*** 3×4, ***b*** 12×16, ***c*** 48×64 *and* ***d*** 192×256 *pixels. If the image contains a sufficient number of pixels, it appears as a continuous image to the human visual system.*

space, which solid-state physicists, mineralogists and chemists deal with. Crystals show periodic 3D patterns of the arrangement of their atoms, ions or molecules, which can be classified on the basis of their symmetries and the geometry of a unit cell. In two dimensions, the classification of digital lattices is much easier than in three-dimensional space. If we restrict ourselves to regular polyhedra, there are only three possible arrangements: Triangles, squares and hexagons (Fig. 6.3).

In image processing, three-dimensional and even higher-dimensional signals are of interest in addition to two-dimensional signals. In a 3D image, a pixel becomes a *voxel* (short form of the term *volume element*). On a grid that is square in every direction, each voxel represents the average gray value of a cube. The position of a voxel is given by three indices. Here l stands for the depth, m is again the row vector and n the column vector (Fig. 6.1b). A Cartesian grid is the most general solution for digital data of dimension n because it is the only geometric shape that can be easily extended for any dimension (hyperpixel).

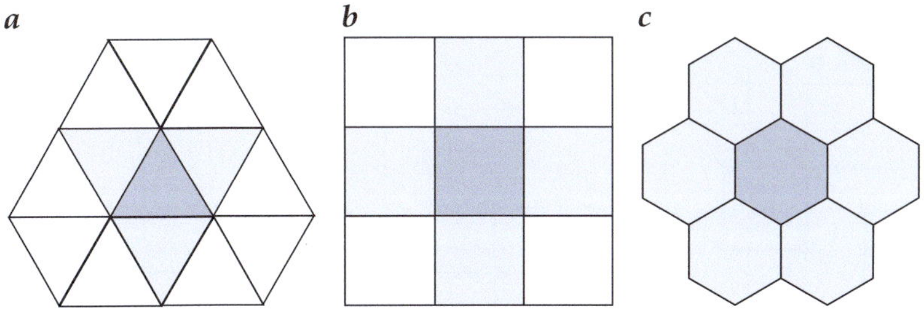

Figure 6.3: *The three possible regular 2D grids:* ***a*** *triangular grid,* ***b*** *square grid,* ***c*** *hexagonal grid.*

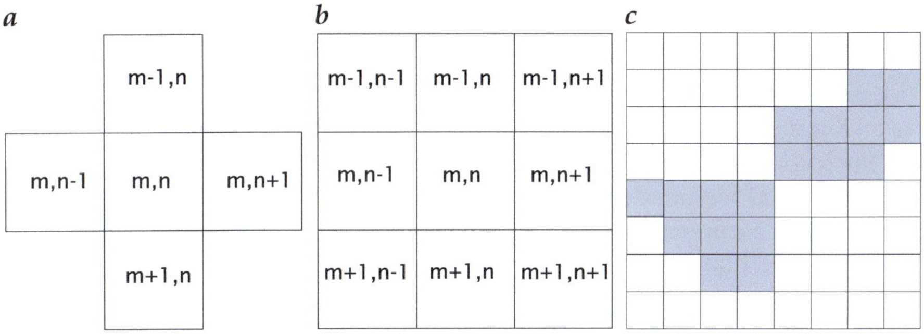

Figure 6.4: *Neighborhoods on a square grid:* ***a*** *Neighborhood of 4 and* ***b*** *Neighborhood of 8.* ***c*** *The gray area counts as one object (contiguous region) in a neighborhood of 8, but as two objects in a neighborhood of 4.*

6.2.2 Neighborhood Relations

An important property of *discrete images* are neighborhood relations, as they define what we regard as *contiguous region* and thus as *digital object*. A two-dimensional square grid has the unfortunate property that there are two ways to define neighboring pixels (Fig. 6.4a, b). We can consider pixels as neighboring if they have a common edge or at least one common corner. So there are either 4 or 8 neighbors. We then speak of a *4-neighborhood* or a *8-neighborhood*.

Both neighborhood types are necessary for the definition of objects as contiguous regions. A region or object is said to be contiguous if we can reach every pixel in the region from every other pixel via a path that leads from one neighboring pixel to the next. The gray object in Fig. 6.4c is considered as one object in an 8-neighborhood, but in a 4-neighborhood it splits into two objects as the two areas do not have a common edge. The white background shows the same properties in both neighborhoods. In the 8-neighborhood, therefore, two connected regions intersect, whereas in the 4-neighborhood, two separate regions result. These inconsistencies can be

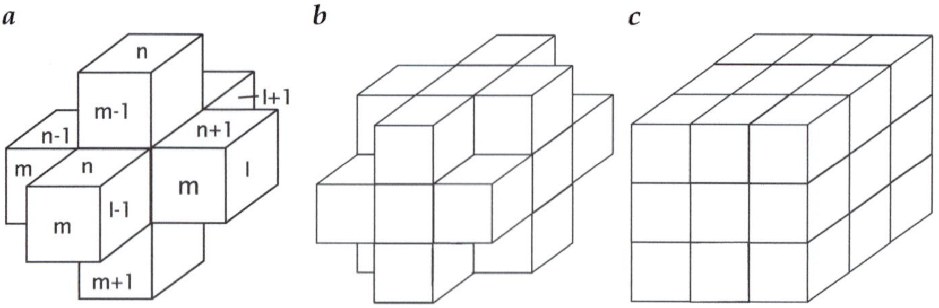

Figure 6.5: *The three neighborhood types on a three-dimensional cubic grid.* ***a*** *6-neighborhood: voxels with connected faces;* **b** *18-neighborhood: voxels with connected edges;* **c** *26-neighborhood: voxels with connected corners.*

overcome if the objects are defined as a 4-neighborhood and the background as an 8-neighborhood or vice versa.

Such complications do not only occur with *square grids*. With a *triangular grid* (Fig. 6.3a) we can define a 3- and 12-neighborhood where the neighbors have either a common edge or corner. With a *hexagonal grid* (Fig. 6.3c), on the other hand, we can only define a *6-neighborhood*, as there are no pixels that have a common corner but no common edge. Neighboring pixels always have one common edge and two common corners. Despite this advantage, hexagonal grids are rarely used in image processing, as almost all commercially available image sensors provide pixels on a square grid. The photo sensors of the retina of the human eye, on the other hand, tend to have a hexagonal shape [347].

In three-dimensional space, the neighborhood relations are even more complex. There are now three ways to define neighborhoods. There are voxels with connected faces, edges and Ecken. These definitions lead to neighborhoods of 6, 18 and 26 (Fig. 6.5). Here, too, we have to define different neighborhoods for objects and background in order to obtain a consistent definition of connected regions. Objects must be defined as a 6-neighborhood and the background must be defined as a 26-neighborhood or vice versa.

6.2.3 Discrete Geometry

The discrete nature of digital images makes it necessary to redefine elementary geometric properties such as distance, slope of a line and coordinate transformations such as translation, rotation and scaling. These quantities are necessary for the definition and measurement of geometric parameters of objects in digital images.

In order to discuss discrete geometry, we introduce the *grid vector*, which describes the position of a pixel. The following discussion is limited to rectangular grids. The lattice vectors in 2D, 3D and 4D images or image sequences is defined as

$$r_{m,n} = \begin{bmatrix} n\Delta x \\ m\Delta y \end{bmatrix}, \quad r_{l,m,n} = \begin{bmatrix} n\Delta x \\ m\Delta y \\ l\Delta z \end{bmatrix}, \quad r_{k,l,m,n} = \begin{bmatrix} n\Delta x \\ m\Delta y \\ l\Delta z \\ k\Delta t \end{bmatrix}. \tag{6.2}$$

To measure distances, the *euclidean distance* can also be transferred from continuous space to a discrete grid with the following definition:

$$d_e(\boldsymbol{r}, \boldsymbol{r}') = \|\boldsymbol{r} - \boldsymbol{r}'\| = \left[(n - n')^2 \Delta x^2 + (m - m')^2 \Delta y^2\right]^{1/2}. \tag{6.3}$$

Corresponding formulas apply for distance measurement in higher dimensions. Two further measurement systems are often used in digital images. The *block distance*

$$d_b(\boldsymbol{r}, \boldsymbol{r}') = |n - n'| + |m - m'| \tag{6.4}$$

specifies the length of the path between two points on the discrete grid if you can only walk in a horizontal and vertical direction. This corresponds to a path in a 4-way neighborhood. The *chessboard distance*, on the other hand, is defined as the maximum of the distance in horizontal and vertical direction:

$$d_c(\boldsymbol{r}, \boldsymbol{r}') = \max(|n - n'|, |m - m'|). \tag{6.5}$$

However, only the Euclidean metric is relevant for practical applications (surveying tasks, etc.). It is the only metric for digital images that preserves the isotropy of continuous space. With the block distance metric, e. g. distances in the direction of the diagonals are longer than with the Euclidean distance. The curve of all points with the same distance to a given point is not a circle, but a rhombus corresponding to a square rotated by 45°.

The *translation* on a discrete grid is only defined in multiples of the pixel or voxel distance, which can be expressed by adding a grid vector $\boldsymbol{t}_{m',n'}$:

$$\boldsymbol{r}'_{m,n} = \boldsymbol{r}_{m,n} + \boldsymbol{t}_{m',n'}. \tag{6.6}$$

Similarly, resizing is only possible for integer multiples of the grid constants, by using only every qth pixel on every pth row. However, as this discrete scaling operation only uses a portion of the grid points, it is questionable whether the image is still displayed correctly. *Rotation* by any angle is not possible on a discrete grid with the exception of a few angles, as all points of the rotated grid must coincide with those of the non-rotated grid. On a rectangular, square and hexagonal grid, only rotations by a multiple of 180°, 90° and 60° respectively are possible. The correct representation of even simple geometric objects such as lines and circles is not easy either. Lines are only well-defined for angles that are multiples of 45°. Otherwise, this results in a staircase-like arrangement of pixels with unequal steps (Fig. 6.6).

These limitations of the discrete geometry cause errors in the position, size and orientation of objects. The effects of these errors on subsequent processing must be carefully examined (Chapter 5).

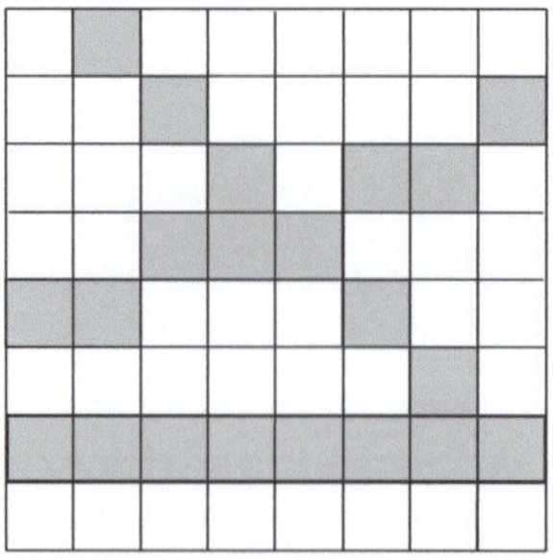

Figure 6.6: *A digital line is only well-defined in the direction of the axes and diagonals. In all other directions, it appears as a staircase-like sequence of pixels with uneven steps.*

6.3 Discrete Multidimensional Signals in Fourier Space

6.3.1 Vector Spaces

Section 6.2 dealt with the spatial representation of digital images. This meant, without explicitly mentioning it, the composition of an image from individual pixels. Each image can be thought of as being composed of basic images in which only one pixel has the value one, while all other pixels have the value zero (Fig. 6.7). We denote such a *basic image* with a value of one in row m, column n with

$$ {}^{m,n}\boldsymbol{P}: \quad {}^{m,n}p_{m',n'} = \begin{cases} 1 & m = m' \wedge n = n' \\ 0 & \text{otherwise.} \end{cases} \tag{6.7}$$

This means that any image can be assembled from the basic images in Eq. (6.7):

$$\boldsymbol{G} = \sum_{m=0}^{M-1}\sum_{n=0}^{N-1} g_{m,n}\, {}^{m,n}\boldsymbol{P}, \tag{6.8}$$

where $g_{m,n}$ denotes the value at position m, n.

We can easily convince ourselves that the basic images ${}^{m,n}\boldsymbol{P}$ form an *orthonormal basis*. To do this, we need an *inner product*, which corresponds to the *scalar product* for vectors. The inner product of two images $\boldsymbol{G}$ and $\boldsymbol{H}$ is defined as

$$\langle \boldsymbol{G} \,|\, \boldsymbol{H} \rangle = \sum_{m=0}^{M-1}\sum_{n=0}^{N-1} g_{m,n} h_{m,n}. \tag{6.9}$$

The notation for the inner product is borrowed from quantum mechanics to distinguish it from matrix multiplication, which is denoted by $\boldsymbol{GH}$. From Eq. (6.9) we can directly derive the orthonormality relation for the basis images ${}^{m,n}\boldsymbol{P}$:

$$\sum_{m=0}^{M-1}\sum_{n=0}^{N-1} {}^{m',n'}p_{m,n}\, {}^{m'',n''}p_{m,n} = \delta_{m'-m''}\delta_{n'-n''}. \tag{6.10}$$

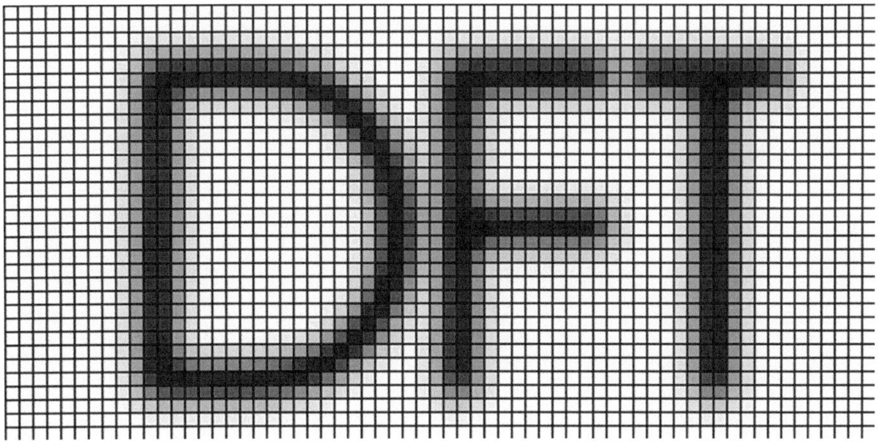

Figure 6.7: *An image can be thought of as being composed of basic images in which only one pixel is not equal to zero.*

The equation says that the inner product of two basic images is zero if two different basic images are used. In contrast, the scalar product of a basic image with itself is one. The *MN* basis images thus span a $M \times N$-dimensional vector space over the body of real numbers.

The analogy to the known two- and three-dimensional vector spaces $\mathbb{R}^2$ and $\mathbb{R}^3$ helps us to understand the transfer to other image representations. A $M \times N$ image represents a point in the $M \times N$ vector space. If we change the coordinate system, the image does not change, only its coordinates. This means that we see the same information from a different angle. This fundamental fact leads to two important conclusions. Firstly, all image representations are equivalent to each other and represent an image completely, and secondly, we can use suitable coordinate transformations to get from one image representation to another and back again using the corresponding inverse transformation.

From the variety of possible other image representations, only one other has gained outstanding importance for image processing in addition to the spatial representation. Their basic images are periodic patterns. The "coordinate transformation" from which they result is the *Fourier transformation*. Fig. 6.8 shows how the same image that consists of individual pixels in Fig. 6.7 is composed of periodic patterns. How such periodic patterns in space and time can be described using complex numbers and the complex exponential function has already been described in detail in Section 2.2.1. In this way, the simple decomposition of a signal into periodic patterns requires the expansion from real to complex numbers. We understand a real image as a complex image with a vanishing imaginary part.

The rest of Section 6.3 provides a clear introduction to the discrete Fourier transform and its most important properties for image and signal processing.

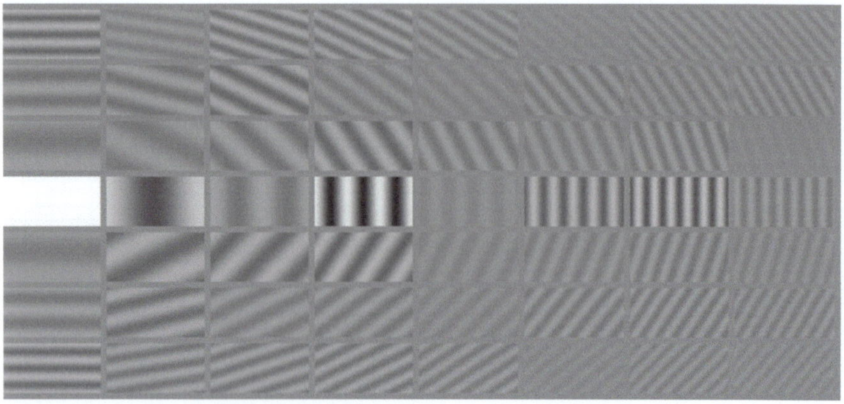

Figure 6.8: The first 56 periodic patterns, the basic images of the Fourier transform, from which the image in Fig. 6.7 is composed.

A more formal presentation including the continuous Fourier transform can be found in Appendix B.1 and a compilation of the most important formulas in Appendix D > R12, > R13, > R14, > R15.

6.3.2 One-dimensional Discrete Fourier Transform

In the *discrete Fourier transform* (*DFT*), the wavenumber is an integer and has a very descriptive meaning, as it indicates how many wavelengths fit into an interval of length N.

The DFT maps a *complex-valued column vector*

$$g = [g_0, g_1, \ldots, g_{N-1}]^T \tag{6.11}$$

to another vector $\tilde{g}$ of a vector space with the same dimension N:

$$\boxed{\tilde{g}_{\tilde{n}} = \frac{1}{N}\sum_{n=0}^{N-1} g_n \exp\left(-\frac{2\pi i n\tilde{n}}{N}\right), \quad 0 \le \tilde{n} < N.} \tag{6.12}$$

The reverse transformation is given by

$$\boxed{g_n = \sum_{\tilde{n}=0}^{N-1} \tilde{g}_{\tilde{n}} \exp\left(\frac{2\pi i n\tilde{n}}{N}\right), \quad 0 \le n < N.} \tag{6.13}$$

Why we use an asymmetric definition of the DFT here is explained in Section 6.3.6.

The equations of the DFT are very similar to those of the continuous Fourier transform:

$$\boxed{\tilde{g}(k) = \int_{-\infty}^{\infty} g(x) \exp(-2\pi i k x) \, \mathrm{d}x.} \tag{6.14}$$

Example 6.1: Transition from continuous to discrete Fourier transform

The transition from the continuous to the discrete Fourier transform can be explained by the following simple considerations. The Fourier integral in Eq. (6.14) is approximated by a sum and the spatial and Fourier space is sampled with $x = n\Delta x$ and $k = \tilde{n}\Delta k$:

$$\tilde{g}(\tilde{n}\Delta k) \approx \sum_{n=0}^{N-1} g_n \exp(-2\pi i n \tilde{n} \Delta x \Delta k) \, \Delta x.$$

If we now set $\Delta x \Delta k = 1/N$ (we learned why this is the case in Section 5.2), then the result is

$$\tilde{g}_{\tilde{n}} = \frac{\tilde{g}(\tilde{n}\Delta k)}{\Delta x} \approx \sum_{n=0}^{N-1} g_n \exp\left(-\frac{2\pi i n \tilde{n}}{N}\right).$$

It is useful to use an abbreviation for the core of the DFT:

$$\mathrm{w}_N = \mathrm{w}^{1/N} = \exp\left(\frac{2\pi i}{N}\right). \tag{6.15}$$

The DFT can also be understood as the scalar product of the vector $\boldsymbol{g}$ with a set of N orthonormal basis vectors:

$$\boldsymbol{b}_{\tilde{n}} = \frac{1}{\sqrt{N}} \left[\mathrm{w}_N^0, \mathrm{w}_N^{\tilde{n}}, \mathrm{w}_N^{2\tilde{n}}, \ldots, \mathrm{w}_N^{(N-1)\tilde{n}}\right]^T. \tag{6.16}$$

Then

$$\tilde{g}_{\tilde{n}} = \frac{1}{N} \sum_{n=0}^{N-1} \mathrm{w}_N^{-nv} g_n = \frac{1}{\sqrt{N}} \left\langle \boldsymbol{b}_{\tilde{n}} \,\middle|\, \boldsymbol{g} \right\rangle = \frac{1}{\sqrt{N}} \boldsymbol{b}_{\tilde{n}}^{*T} \boldsymbol{g}. \tag{6.17}$$

Note the second compact notation of the *scalar product* with the superscript $*$ on the right-hand side of the equation, which means that the first vector is conjugated and transposed. Equation (6.17) means that the coefficient $\tilde{g}_{\tilde{n}}$ is obtained in Fourier space by projecting the vector $\boldsymbol{g}$ onto the basis vector $\boldsymbol{b}_{\tilde{n}}$. The N basis vectors $\boldsymbol{b}_{\tilde{n}}$ are orthogonal to each other and have a magnitude of one:

$$\boldsymbol{b}_{\tilde{n}}^T \boldsymbol{b}_{\tilde{n}'} = \delta_{\tilde{n}-\tilde{n}'} = \begin{cases} 1 & \tilde{n} = \tilde{n}' \\ 0 & \text{otherwise.} \end{cases} \tag{6.18}$$

Therefore, the $\boldsymbol{b}_{\tilde{n}}$ form an orthonormal basis of the vector space, i. e. each vector of the vector space can be represented as a linear combination of these

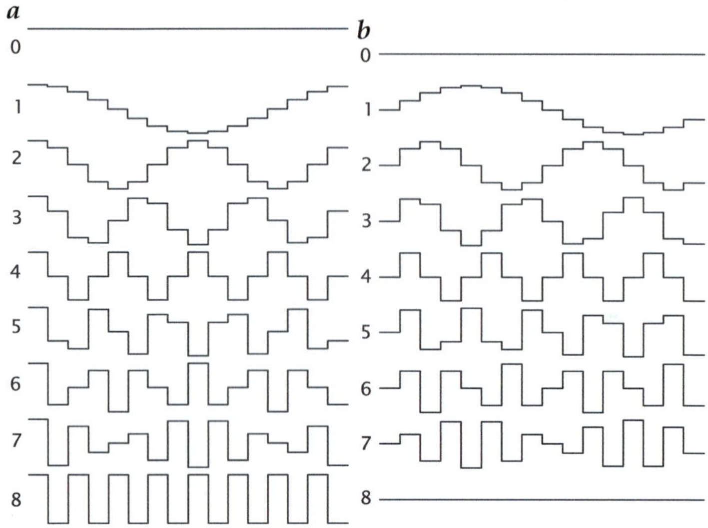

Figure 6.9: *The first 9 basis functions of the DFT for N = 16; **a** real part (cosine function), **b** imaginary part (sine function).*

basis vectors. The DFT calculates the projections of the vector $\boldsymbol{g}$ onto all basis vectors of the Fourier space, i. e. the components of $\boldsymbol{g}$ in the direction of the basis vectors.

In this sense, the DFT can be understood as a special case of a coordinate transformation in an N-dimensional vector space. Mathematically, the DFT differs from more familiar coordinate transformations, such as, e. g. the rotation in a three-dimensional vector space, only in that it is a vector space over the body of complex instead of real numbers.

The real and imaginary parts of the basis vectors are sampled sine and cosine functions of different wavelengths (Fig. 6.9). The index $\tilde{n}$ denotes how often the wavelength of the function fits into the interval $[0, N]$. The basis vector for wavenumber zero, $\boldsymbol{b}_0$, is a real, constant vector. The projection onto this vector provides the mean value of the elements of the vector $\boldsymbol{g}$ multiplied by $\sqrt{N}$.

6.3.3 Multidimensional Discrete Fourier Transform

The discrete multidimensional Fourier transform is discussed using the 2D DFT as an example; the extension to higher dimensions is simple. The 2D DFT maps complex-valued $M \times N$ matrices to complex-valued $M \times N$ matrices:

$$\hat{g}_{\tilde{n}\tilde{m}} = \frac{1}{MN} \sum_{m=0}^{M-1} \sum_{n=0}^{N-1} g_{mn} \exp\left(-\frac{2\pi i m \tilde{m}}{M}\right) \exp\left(-\frac{2\pi i n \tilde{n}}{N}\right) \quad (6.19)$$

or abbreviated

$$\tilde{g}_{\tilde{m}\tilde{n}} = \frac{1}{MN}\sum_{m=0}^{M-1}\left(\sum_{n=0}^{N-1} g_{mn} w_N^{-n\tilde{n}}\right) w_M^{-m\tilde{m}}. \tag{6.20}$$

In the second line, the abbreviation introduced in Eq. (6.15) is used. Analogous to 1D DFT, a matrix is expanded into a set of NM basis matrices that span the $M \times N$-dimensional vector space over the body of complex numbers. The basis matrices have the form

$$\underbrace{\boldsymbol{B}_{\tilde{m}\tilde{n}}}_{M\times N} = \frac{1}{\sqrt{MN}} \begin{bmatrix} w^0 \\ w_M^{\tilde{m}} \\ w_M^{2\tilde{m}} \\ \vdots \\ w_M^{(M-1)\tilde{m}} \end{bmatrix} \left[w^0, w_N^{\tilde{n}}, w_N^{2\tilde{n}}, \ldots, w_N^{(N-1)\tilde{n}}\right]. \tag{6.21}$$

In this equation, the basis matrices are represented as an *outer product* of a column and row vector, which represent the basis vectors of the 1D DFT Eq. (6.16). This shows the separability of the 2D DFT.

We can also express the 2D DFT as a scalar product

$$\tilde{g}_{\tilde{m}\tilde{n}} = \frac{1}{\sqrt{MN}} \langle \boldsymbol{B}_{\tilde{m}\tilde{n}} \,|\, \boldsymbol{G} \rangle, \tag{6.22}$$

where the scalar product of two complex-valued matrices is defined as

$$\langle \boldsymbol{G} \,|\, \boldsymbol{H} \rangle = \sum_{m=0}^{M-1}\sum_{n=0}^{N-1} g_{mn}^{*} h_{mn}. \tag{6.23}$$

The inverse 2D DFT is given by

$$g_{mn} = \sum_{\tilde{n}=0}^{M-1}\sum_{\tilde{m}=0}^{N-1} \tilde{g}_{\tilde{m}\tilde{n}} w_M^{m\tilde{m}} w_N^{n\tilde{n}} = \sqrt{MN}\left\langle \boldsymbol{B}_{-\tilde{m},-\tilde{n}} \,\middle|\, \tilde{\boldsymbol{G}} \right\rangle. \tag{6.24}$$

6.3.4 Properties of the Fourier Transform

In this section, we discuss the most important properties of the continuous and discrete Fourier transform. We focus on those properties of the FT that are most important for image processing. Together with some elementary Fourier transform pairs (> R13), these properties (> R12, > R15) form a powerful tool with which further Fourier transform properties and the transformation of many functions can be easily derived.

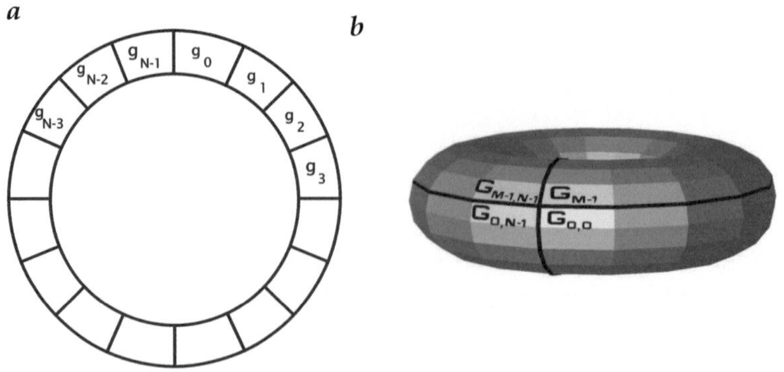

Figure 6.10: *Geometric interpretation of the periodicity of the 1D and 2D DFT with* ***a*** *the Fourier ring and* ***b*** *the Fourier torus.*

Periodicity

The core of the DFT shows a characteristic *periodicity*:

$$\exp\left(-\frac{2\pi \mathrm{i}(n+lN)}{N}\right) = \exp\left(-\frac{2\pi \mathrm{i}n}{N}\right), \mathrm{w}_N^{(n+lN)} = \mathrm{w}_N^n \quad \forall\, l \in \mathbb{Z}. \tag{6.25}$$

The definitions of the DFT limit the space and the Fourier space to a finite number of values. If we ignore this limitation and calculate the forward and backward transformations for all integers, we find the same periodicities in the functions in the spatial and Fourier space directly from the definitions in Eqs. (6.22) and (B.28):

$$\begin{array}{lll} \text{Wavenumber domain} & \tilde{g}_{\tilde{m}+kM,\tilde{n}+lN} = \tilde{g}_{\tilde{m}\tilde{n}}, & \forall\, k,l \in \mathbb{Z} \\ \text{Spatial domain} & g_{m+kM,n+lN} = g_{mn}, & \forall\, k,l \in \mathbb{Z}. \end{array} \tag{6.26}$$

These equations describe a periodic replication in all directions in both spaces beyond the original domain of definition. The periodicity of the DFT leads to an interesting geometric interpretation. In the one-dimensional case, the boundary points g_{N-1} and $g_N = g_0$ are neighbors. We can represent this property geometrically if we think of the points of the vector as being arranged not on a finite line but on a circle, the *Fourier ring* (Fig. 6.10a). This representation has a deeper meaning if we consider the Fourier transform as a special case of the *z-transform* [234, 235]. In two-dimensional space, we have to periodically close a two-dimensional grid in the x and y directions. This leads to a mapping of the grid to the *Fourier torus* (Fig. 6.10b).

Symmetry
Four types of *symmetries* are important for the Fourier transform:

$$
\begin{array}{llcl}
\text{even} & g(-\boldsymbol{x}) & = & g(\boldsymbol{x}), \\
\text{odd} & g(-\boldsymbol{x}) & = & -g(\boldsymbol{x}), \\
\text{hermitian} & g(-\boldsymbol{x}) & = & g^*(\boldsymbol{x}), \\
\text{antihermitian} & g(-\boldsymbol{x}) & = & -g^*(\boldsymbol{x})
\end{array}
\qquad (6.27)
$$

The symbol $*$ denotes the complex conjugate. The *hermitian symmetry* is important because the kernels of FT Eq. (6.14) and DFT Eq. (6.12) are hermitian.

Each function $g(\boldsymbol{x})$ can be decomposed into its even and odd parts by

$$
{}^{e}g(\boldsymbol{x}) = \frac{g(\boldsymbol{x}) + g(-\boldsymbol{x})}{2} \quad \text{and} \quad {}^{o}g(\boldsymbol{x}) = \frac{g(\boldsymbol{x}) - g(-\boldsymbol{x})}{2}. \qquad (6.28)
$$

In this way, the Fourier transform can be broken down into a cosine and a sine transform:

$$
\tilde{g}(\boldsymbol{\kappa}) = 2\int_0^\infty {}^{e}g(\boldsymbol{x}) \cos(2\pi \boldsymbol{k}^T\boldsymbol{x}) \mathrm{d}^D x + 2\mathrm{i}\int_0^\infty {}^{o}g(\boldsymbol{x}) \sin(2\pi \boldsymbol{k}^T\boldsymbol{x}) \mathrm{d}^D x. \qquad (6.29)
$$

It follows directly from this that the Fourier transform of an even or odd function is also even or odd. The complete symmetry relationships for the Fourier transform are as follows:

$$
\begin{array}{lcl}
\text{real} & \circ\!\!-\!\!\bullet & \text{hermitian} \\
\text{imaginary} & \circ\!\!-\!\!\bullet & \text{antihermitian} \\
\text{hermitian} & \circ\!\!-\!\!\bullet & \text{real} \\
\text{antihermitian} & \circ\!\!-\!\!\bullet & \text{imaginary} \\
\text{even} & \circ\!\!-\!\!\bullet & \text{even} \\
\text{odd} & \circ\!\!-\!\!\bullet & \text{odd} \\
\text{real and even} & \circ\!\!-\!\!\bullet & \text{real and even} \\
\text{real and odd} & \circ\!\!-\!\!\bullet & \text{imaginary and odd} \\
\text{imaginary and even} & \circ\!\!-\!\!\bullet & \text{imaginary and even} \\
\text{imaginary and odd} & \circ\!\!-\!\!\bullet & \text{real and odd}
\end{array}
\qquad (6.30)
$$

The DFT shows the same symmetry properties as the FT, see Eqs. (6.27) and (6.30). In the definition for even and odd functions $g(-\boldsymbol{x}) = \pm g(\boldsymbol{x})$, only the continuous function must be replaced by the corresponding vector ($g_{-n} = \pm g_n$) or matrix elements ($g_{-m,-n} = \pm g_{m,n}$). Due to the periodicity of the DFT, these symmetry relations can also be written as

$$
g_{-m,-n} = \pm g_{mn} \equiv g_{M-m,N-n} = \pm g_{mn} \qquad (6.31)
$$

with the sign $+$ for even functions and the sign $-$ for odd functions. This rewriting shifts the center of symmetry from the origin to the point $[M/2, N/2]^T$.

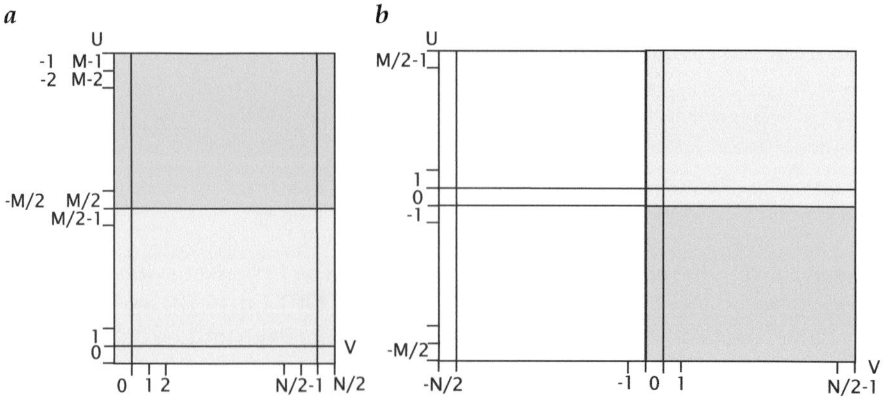

Figure 6.11: ***a*** *Half space as calculated with the overwriting Fourier transform of a real image; the wavenumber zero is in the lower left corner;* ***b*** *FT with the added missing half and displayed so that the wavenumber zero is in the center.*

The symmetry properties are of great importance for the practical application of the Fourier transform. Paying careful attention to them makes it possible to save memory and computing time. Real images are an important application.

Real-valued vectors require only half as much memory as complex-valued vectors. From the symmetry relations in Eq. (6.30) we conclude that the DFT of a real-valued signal is hermitian:

$$\begin{array}{lcl} g_n = g_n^* & \circ\!\!-\!\!\bullet & \tilde{g}_{N-\tilde{n}} = \tilde{g}_{\tilde{n}}^*, \\ g_{mn} = g_{mn}^* & \circ\!\!-\!\!\bullet & \tilde{g}_{M-\tilde{m},N-\tilde{n}} = \tilde{g}_{\tilde{m}\tilde{n}}^*. \end{array} \tag{6.32}$$

Therefore, only half of the values of the complex DFT of a real vector need to be stored. The other half is obtained by mirroring at the center of symmetry $N/2$. Therefore, we need the same amount of memory for the complex Fourier transform as for the corresponding real vector.

In two or more dimensions, the symmetry properties are somewhat more complicated to handle. Again, the Fourier transform is completely determined by a half-space. But now there are several ways to select the half-space. This means that only *one* component of the wavenumber vector is restricted to the positive semi-axis.

We can represent the Fourier transform of a real $M \times N$ image with M rows and $N/2 + 1$ columns (Fig. 6.11) under the assumption that N is even. Unfortunately, $N/2 + 1$ columns are required because the first ($m = 0$) and last column ($m = M/2$) are symmetrical to themselves according to Eq. (6.32).

At first glance, it does not seem possible to overwrite a real image with its complex transformation, as one more column is required. However, this is deceptive, as the first and last column must be real for reasons of symmetry

($\tilde{g}_{0,N-v} = \tilde{g}^*_{0,v}$ and $\tilde{g}_{M/2,N-v} = \tilde{g}^*_{M/2,v}$). Therefore, the real part of column $M/2$ can be stored in the imaginary part of column 0.

For real-valued image sequences, only one half-space is required for the Fourier transform. From a physical point of view, it makes the most sense to choose the half-space that contains the positive frequencies. In contrast to single images, we now obtain the full wavenumber space, as we can distinguish the spatially identical wavenumber vectors $\boldsymbol{\kappa}$ and $-\boldsymbol{\kappa}$ as structures propagating in opposite directions.

Separability

The core of the Fourier transform is *separable*. Therefore, the Fourier transform of separable functions is also separable:

$$\prod_{d=1}^{D} g_d(x_d) \circ\!\!-\!\!\bullet \prod_{d=1}^{D} \tilde{g}_d(k_d). \tag{6.33}$$

Since many significant multidimensional functions are separable, this property is important for effectively calculating the transformation of multidimensional functions from 1D transformations.

Similarity

The similarity indicates how a Fourier transform behaves when the coordinate system is scaled. In one dimension, a function can only be scaled ($x' = ax$). In higher dimensions, a more general transformation of the coordinate system is possible using an affine transformation ($\boldsymbol{x}' = \boldsymbol{A}\boldsymbol{x}$), i. e. the new basis vectors are a linear combination of the old ones. The *rotation* of the coordinate system is a special case.

If a function is compressed in local space, it is stretched accordingly in Fourier space, and vice versa:

$$\boxed{g(a\boldsymbol{x}) \circ\!\!-\!\!\bullet \frac{1}{|a|^D}\tilde{g}(\boldsymbol{\kappa}/a).} \tag{6.34}$$

A rotation of the coordinate system in local space causes the same rotation of the coordinate system in Fourier space:

$$\boxed{g(\boldsymbol{R}\boldsymbol{x}) \circ\!\!-\!\!\bullet \tilde{g}(\boldsymbol{R}\boldsymbol{\kappa}).} \tag{6.35}$$

The above similarity theorem cannot be used for the discrete Fourier transform, as arbitrary scaling and rotations are not possible here. A discrete function can be stretched by an integer factor K (*upsampling*) by filling the newly created points with zeros:

$$(g_{\uparrow K})_n = \begin{cases} g_{n/K} & n = 0, K, 2K, \ldots (N-1)K) \\ 0 & \text{otherwise.} \end{cases} \tag{6.36}$$

Upsampling results in a vector $g_{\uparrow K}$ with KN elements and the following applies

$$g_{\uparrow K} \quad \circ\!\!-\!\!\bullet \quad \frac{1}{K}\tilde{g} \quad \text{with} \quad \tilde{g}_{kN+v} = \tilde{g}_v. \tag{6.37}$$

An upward sampling by the factor K therefore only causes a K-fold replication of the Fourier transform. In Eq. (6.37) it should be noted that due to the periodicity of the discrete Fourier transforms discussed at the beginning of this section: $\tilde{g}_{kN+v} = \tilde{g}_v$.

Shift

In Section 6.3.1 some properties of the basic images of the Fourier space, the complex exponential functions $\exp\left(2\pi i \boldsymbol{\kappa}^T \boldsymbol{x}\right)$, were discussed. A shift of these functions causes multiplication by a phase factor:

$$\exp\left(2\pi i(\boldsymbol{x} - \boldsymbol{x}_0)^T \boldsymbol{\kappa}\right) = \exp\left(-2\pi i \boldsymbol{x}_0^T \boldsymbol{\kappa}\right) \exp\left(2\pi i \boldsymbol{\kappa}^T \boldsymbol{x}\right). \tag{6.38}$$

As a direct consequence of the linearity of the Fourier transform, a shift in the spatial domain therefore does not change the amplitude of the Fourier transform, but only the phase, whereby the occurring phase change $-2\pi i \boldsymbol{x}_0^T \boldsymbol{\kappa}$ is dependent on the wavenumber vector.

Example 6.2: Shift by an integer multiple of a wavelength

If a periodic function with the wavenumber $\boldsymbol{\kappa}$ is shifted by an integer multiple of the wavelength in the direction of the wave vector, i. e. $\boldsymbol{x}_0 = K\lambda \boldsymbol{\kappa}/|\boldsymbol{\kappa}| = K\boldsymbol{\kappa}/|\boldsymbol{\kappa}|^2$, then the phase position does not change, as was to be expected: $-2\pi K(1/|\boldsymbol{\kappa}|^2)\boldsymbol{\kappa}^T\boldsymbol{\kappa} = -2\pi K$.

A shift can also be applied in Fourier space. A shift in Fourier space, $\tilde{g}(\boldsymbol{\kappa} - \boldsymbol{\kappa}_0)$, modulates a signal in spatial space with the wavenumber $\boldsymbol{\kappa}_0$: $\exp(2\pi i \boldsymbol{\kappa}_0^T \boldsymbol{x}) g(\boldsymbol{x})$.

Convolution and Derivation

The *convolution* is one of the most important operations in signal processing. For continuous D-dimensional signals, it is defined by

$$(g * h)(\boldsymbol{x}) = \int\limits_{-\infty}^{\infty} h(\boldsymbol{x}') g(\boldsymbol{x} - \boldsymbol{x}') \mathrm{d}^D x'. \tag{6.39}$$

In signal processing, the function $h(\boldsymbol{x})$ is usually zero except in a small range around the zero point. Such a function is then referred to as *convolution mask*. Convolution with $h(\boldsymbol{x})$ results in a new function $g'(\boldsymbol{x})$ whose values are a kind of weighted averaging of $g(\boldsymbol{x})$ in a small area around $\boldsymbol{x}$. The signal is changed by the convolution with $h(\boldsymbol{x})$ in a defined way. For example, it can become smoother. This is why it is also referred to as *filtering*.

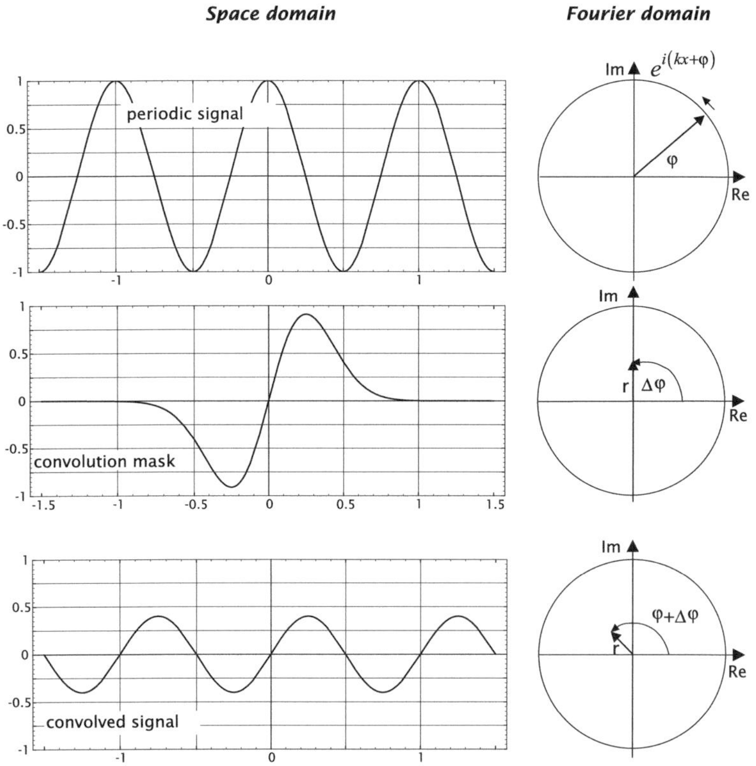

***Figure 6.12:** Convolution of a periodic signal with a function again results in a periodic signal with the same wavelength but changed amplitude and phase position The phase shift $\Delta\varphi$ and amplitude attenuation r can be described by multiplication with the complex number $r\exp(\mathrm{i}\Delta\varphi)$.*

One- and multi-dimensional discrete convolutions are defined analogously to the continuous convolution in Eq. (6.39) by

$$g'_n = \sum_{n'=0}^{N-1} h_{n'} g_{n-n'}, \quad g'_{m,n} = \sum_{m'=0}^{M-1}\sum_{n'=0}^{N-1} h_{m'n'} g_{m-m',n-n'}. \tag{6.40}$$

A convolution of two functions in local space is equivalent to a complex multiplication in Fourier space:

$$\begin{array}{llll} \text{FT:} & h(\boldsymbol{x}) * g(\boldsymbol{x}) & \circ\!\!-\!\!\bullet & \tilde{h}(\boldsymbol{\kappa})\tilde{g}(\boldsymbol{\kappa}), \\ \text{1D DFT:} & \boldsymbol{h} * \boldsymbol{g} & \circ\!\!-\!\!\bullet & N\,\tilde{\boldsymbol{h}}\tilde{\boldsymbol{g}}, \\ \text{2D DFT:} & \boldsymbol{H} * \boldsymbol{G} & \circ\!\!-\!\!\bullet & MN\,\tilde{\boldsymbol{H}}\tilde{\boldsymbol{G}}. \end{array} \tag{6.41}$$

The factors N and MN in the discrete convolution result from the asymmetric definition of the DFT in Eq. (6.12).

Conversely, a convolution in Fourier space also corresponds to a multiplication in local space. Mathematically, the simplicity of convolution in Fourier space results from the fact that the basis functions of Fourier space, the complex exponential functions $\exp\left(2\pi i \boldsymbol{\kappa}^T \boldsymbol{x}\right)$, the common *eigenfunctions all* convolution operations. A convolution with an arbitrary function leaves the exponential function unchanged except for one factor, the *eigenvalue*.

The reduction of convolution in Fourier space to a complex multiplication can also be clearly illustrated by applying convolution to a periodic function (Fig. 6.12). If a periodic signal is convolved, the result must be the same if the function used for convolving is shifted by an integer multiple of the wavelength, as the same values are multiplied together. A shift by half a wavelength must produce the negated result, as all values in the periodic function are replaced by their negative values. It is therefore clear that the result of the convolution must be a periodic function with the same wavelength. Only the amplitude and the phase position can change.

Example 6.3: Phase shift and amplitude change of a periodic signal

Let $g(x)$ be a periodic signal with an amplitude of one and a phase of φ. It can then be described as the real part of the complex E-function $\exp i(kx + \varphi)$. Now we multiply this signal by a complex number

$$a + ib = r\exp(i\Delta\varphi) \quad \text{with} \quad r = \sqrt{a^2 + b^2} \quad \text{and} \quad \varphi = \arctan(b/a)$$

and obtain

$$r\exp(i\Delta\varphi)\exp[i(kx + \varphi)] = r\exp[i(kx + \varphi + \Delta\varphi).$$

This shifts the phase of the periodic signal by $\Delta\varphi$ and multiplies the amplitude by r.

The following properties result directly from the convolution theorem. The convolution is

$$\begin{array}{ll} \text{commutative} & h * g = g * h, \\ \text{associative} & h_1 * (h_2 * g) = (h_1 * h_2) * g, \\ \text{distributive with regard to addition} & (h_1 + h_2) * g = h_1 * g + h_2 * g. \end{array} \tag{6.42}$$

The importance of convolution is further emphasized by the fact that two classes of operations that at first glance do not look like convolution operations, namely the shift and all derivative operations, can also be regarded as convolution operations. In both cases, the Fourier transform is only multiplied by a complex factor. For the shift, this results directly from Eq. (6.38).

The following applies to the partial derivative of a function in local space

$$\boxed{\frac{\partial g(x)}{\partial x_d} \quad \circ\!\!-\!\!\bullet \quad 2\pi i k_d \tilde{g}(\boldsymbol{\kappa}).} \tag{6.43}$$

A first-order derivative in the direction d thus corresponds to a multiplication in Fourier space by the factor $2\pi i k_d$. This property can also be explained clearly. The derivative of a periodic function is in turn a periodic function. Zero points become extreme values and extreme values become zero points. This corresponds to a phase shift by 90° and thus a multiplication by the factor i. The smaller the wavelength (or the larger the wavenumber) of a periodic structure, the higher the derivative:

$$\frac{\mathrm{d}}{\mathrm{d}x}\exp(2\pi \mathrm{i} x k) = 2\pi \mathrm{i} k \exp(2\pi \mathrm{i} x k).$$

With this knowledge and the properties summarized in Eq. (6.42), we can draw some further important conclusions. Since any convolution operation can be interchanged with the shift operation, any convolution is a shift invariant operation. Furthermore, we get the same result if we first differentiate and then convolve a signal or perform these operations in reverse order. These properties are helpful for a deeper understanding of image processing operations and for a faster calculation of convolution operations.

Central Limit Theorem

The *central limit theorem* is known for its importance in probability theory and statistics [236]. However, it also plays an important role in signal processing, as it states under very general conditions that cascaded convolution operations lead to a convolution with the *Gaussian bell curve* ($\propto \exp(-ax^2)$). Since the Fourier transform of the Gaussian function is also a Gaussian function (≻ R14), this means that with continued convolution both the convolution mask and the *transfer function* take the form of a Gaussian bell curve. Therefore, the central limit theorem is an expression of the central role of the Gaussian function for image processing.

It is of great practical importance because — especially if the functions are smooth — the Gaussian curve is approximated with sufficient accuracy even with only a few convolutions.

Smoothness, Compactness, Uncertainty Principle

The smoother a function is, the more compact its Fourier transform is. This general rule can be quantified by expressing the smoothness by the number of derivatives that are continuous and the compactness by the asymptotic behavior for large values of the wavenumber k: If a function $g(x)$ and its first $n-1$ derivatives are continuous, then its Fourier transform for large wavenumbers k decays at least with $|k|^{-(n+1)}$, that is $\lim_{|k|\to\infty} |k|^n g(k) = 0$.

As simple examples, we consider the rectangular and triangular functions (≻ R13). The rectangular function is discontinuous ($n = 0$). Therefore, its Fourier transform, the sinc function, decreases asymptotically with $|k|^{-1}$. The triangular function is continuous, but its first derivative is not continuous. Therefore, its Fourier transform, the sinc^2 function, decreases more steeply with $|k|^{-2}$.

We can understand the relationship between smoothness and compactness as an extension of the reciprocity between local and Fourier space. What is strongly localized in one space has a large extension in the other space and vice versa. The general reciprocity between local and Fourier space can also be found in another theorem, the classical *uncertainty principle*. The theorem relates the mean square width of a function and its Fourier transform to each other. The mean square width $(\Delta x)^2$ is defined as

$$(\Delta x)^2 = \frac{\int\limits_{-\infty}^{\infty} x^2 \left|g(x)\right|^2 \mathrm{d}x}{\int\limits_{-\infty}^{\infty} \left|g(x)\right|^2 \mathrm{d}x} - \left(\frac{\int\limits_{-\infty}^{\infty} x \left|g(x)\right|^2 \mathrm{d}x}{\int\limits_{-\infty}^{\infty} \left|g(x)\right|^2 \mathrm{d}x} \right)^2 . \tag{6.44}$$

This quantity is therefore the variance of $\left|g(x)\right|^2$ and thus a measure of the width of the distribution of the signal's energy. The uncertainty principle states that

$$\Delta x \Delta \kappa \geq \frac{1}{4\pi}. \tag{6.45}$$

Some criteria for the construction of linear (convolution) filter operators can be derived from the relationships between compactness and smoothness and from the uncertainty principle (Chapters 11 and 12).

6.3.5 Phase and Amplitude

According to the explanations in the previous sections, the DFT can be regarded as a coordinate transformation in a finite-dimensional vector space. Therefore, the image information is completely preserved, and the inverse transformation results in the original image again. In Fourier space, we merely view the image from a different angle. Each point of the Fourier space contains two pieces of information, the *amplitude* and the *phase*, i. e. the relative position of a periodic structure. This raises the question of whether the phase or the amplitude contains the more important information about the structures of the image or whether both components are equally important.

To answer this question, we conduct a simple experiment. Fig. 6.13a, b shows two images, one with a building of the University of Heidelberg, the other with a few lines of text. If both images are Fourier transformed and then phase and amplitude swapped and back-transformed, the result is Fig. 6.13c, d. The surprising result of this swap is that it is the phase that dominates the image content of both images. Both images contain massive perturbations, but essential information is preserved.

This experiment shows that the phase of the Fourier transform contains essential information of the image structure. The amplitude only reveals *that* the image contains a periodic structure, but not *where*. This important

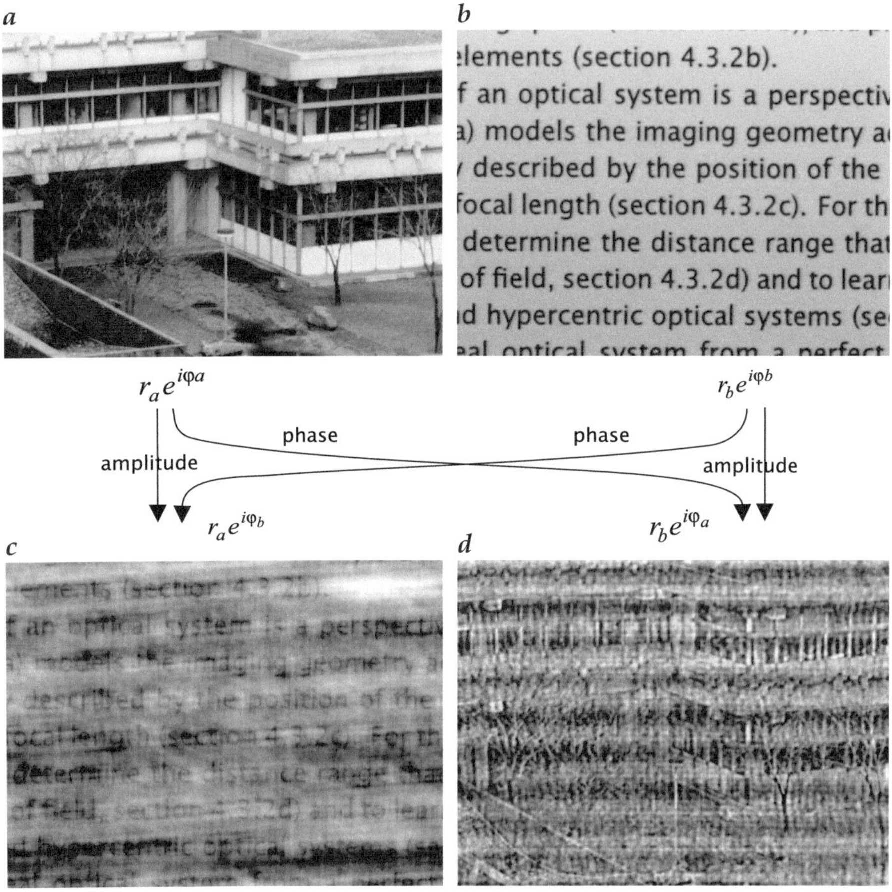

Figure 6.13: *Illustration of the significance of phase and amplitude in Fourier space for the image content:* **a**, ***b*** *Original images;* ***c*** *image composed of the phase of image* **b** *and the amplitude of image* ***a*** *;* ***d*** *image composed of the phase of image* **a** *and the amplitude of image* ***b*** *.*

fact can also be shown with the shift theorem (Theorem B.3). The shift of an object in spatial space only leads to a shift of the phase in wavenumber space; the amplitude does not change. If we do not know the phase of the Fourier component of an object, we do not know what it looks like or where it is located.

It also becomes clear that the *power spectrum*, i. e. the squared amplitude of the Fourier components (Sect. C.5.3), contains very little information because the phase information is missing. If the gray values are associated with the amplitude of a physical process, say a harmonic oscillation, the power spectrum merely reflects the energy distribution with the wavenumber.

6.3.6 Practical Application of the DFT

Units

With the DFT, the problem of the different factors in the definition and their descriptive interpretation should be pointed out again. In addition to the definition in Eq. (6.17), two others are common:

$$
\begin{array}{llcl}
(a) & \tilde{g}_{\tilde{n}} = \dfrac{1}{\sqrt{N}}\displaystyle\sum_{n=0}^{N-1} w_N^{-n\tilde{n}}\, g_n & \bullet\!\!-\!\!\circ & g_n = \dfrac{1}{\sqrt{N}}\displaystyle\sum_{n=0}^{N-1} w_N^{n\tilde{n}}\, \tilde{g}_v, \\
(b) & \tilde{g}_{\tilde{n}} = \dfrac{1}{N}\displaystyle\sum_{n=0}^{N-1} w_N^{-n\tilde{n}}\, g_n & \bullet\!\!-\!\!\circ & g_n = \displaystyle\sum_{n=0}^{N-1} w_N^{n\tilde{n}}\, \tilde{g}_{\tilde{n}}, \\
(c) & \tilde{g}_{\tilde{n}} = \displaystyle\sum_{n=0}^{N-1} w_N^{-n\tilde{n}}\, g_n & \bullet\!\!-\!\!\circ & g_n = \dfrac{1}{N}\displaystyle\sum_{n=0}^{N-1} w_N^{n\tilde{n}}\, \tilde{g}_{\tilde{n}}.
\end{array}
\qquad (6.46)
$$

The symmetric definition (a) is mathematically the most elegant, since it corresponds to the scalar product with the orthonormal basis vectors in Eqs. (6.16) and (6.17) for the back and forth transformation. In practice, however, definition (b) is usually used, since $\tilde{g}_0$ then directly represents the mean value of the vector in the space domain, regardless of its length:

$$
\tilde{g}_0 = \frac{1}{N}\sum_{n=0}^{N-1} w_N^{-n0}\, g_n = \frac{1}{N}\sum_{n=0}^{N-1} g_n. \qquad (6.47)
$$

For this reason, definition (b) will be used almost everywhere in this book.

For the practical application of the DFT, it is important to know at which spatial or temporal distance the signals were sampled in order to be able to correctly compare Fourier transforms calculated with different sampling rates. The quickest way to see the relationship is to approximate the Fourier integral in Eq. (6.14) by a sum and sample the values in spatial and Fourier space with $x = n\Delta x$, $k = v\Delta k$ and the condition $\Delta x \Delta k = 1/N$ (Section 6.3.2). This then results in

$$
\begin{aligned}
\tilde{g}(\tilde{n}\Delta k) &= \int_{-\infty}^{\infty} g(x) \exp\left(-2\pi i v \Delta k x\right) \mathrm{d}x \\
&\approx N\Delta x \frac{1}{N}\sum_{n=0}^{N-1} g_n \exp\left(-\frac{2\pi i n v}{N}\right) = N\Delta x \tilde{g}_{\tilde{n}}.
\end{aligned}
\qquad (6.48)
$$

Therefore, the Fourier transforms calculated with the DFT must be multiplied by the factor $N\Delta x = 1/\Delta k$ in order to relate them to a unit interval of the wavenumber. Without this normalization, the Fourier transforms are related to the interval $\Delta k = 1/(N\Delta x)$ and therefore differ at different sampling rates.

Corresponding relationships apply for two-dimensional and higher-dimensional signals:

$$
\tilde{g}(v\Delta k_1, u\Delta k_2) \approx N\Delta x M \Delta y \tilde{g}_{uv} = \frac{1}{\Delta k_1 \Delta k_2}\tilde{g}_{uv}. \qquad (6.49)
$$

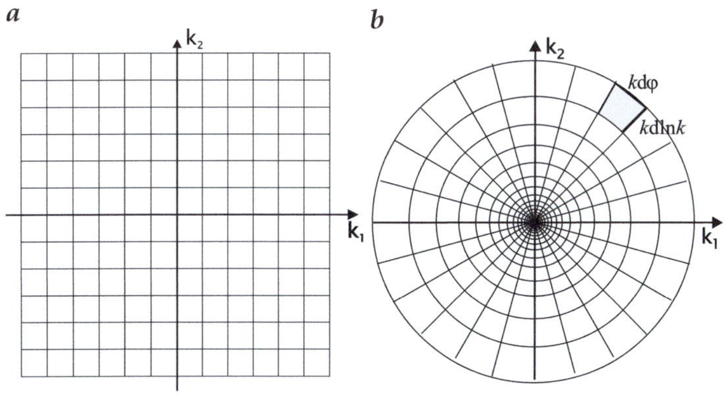

Figure 6.14: *Division of the Fourier space into* **a** *Cartesian and* **b** *logarithmic-polar intervals.*

The same normalization must be applied for the magnitude squared (*energy*) of the signals and not for the squared factors from Eq. (6.48). This results from the *Rayleigh theorem* (conservation of energy) for continuous and discrete signals (> R12, > R15):

$$\begin{array}{ll} \text{continuous:} & \displaystyle\int_{-\infty}^{\infty} |g(x)|^2 \,\mathrm{d}x = \int_{-\infty}^{\infty} |\tilde{g}(k)|^2 \,\mathrm{d}k, \approx \sum_{v=0}^{N-1} |\tilde{g}(v\Delta k)|^2 \,\Delta k \\ \text{discrete:} & \displaystyle\frac{1}{N}\sum_{n=0}^{N-1} |g_n|^2 = \sum_{v=0}^{N-1} |\tilde{g}_v|^2 . \end{array} \tag{6.50}$$

The Rayleigh theorem states that the energy can be integrated either in local or Fourier space. For discrete signals, this means that the mean energy can be calculated either by averaging in local space or by summing in Fourier space (if we use the definition (b) of the DFT in Eq. (6.46)). From the approximation of the integral over the continuous FT by a sum over the DFT in Eq. (6.50) it immediately follows that $|\tilde{g}(v\Delta k)|^2 \approx |\tilde{g}_v|^2 / \Delta k$. The unit of the magnitude squares determined in this way in Fourier space is $\cdot/\mathrm{m}^{-1}$ or $\cdot/\mathrm{Hz}$ for *time series*, where $\cdot$ stands for the unit of the squared signal.

Dynamic Range

While in most cases it is sufficient to render an image with 256 gray values, i. e. one byte per pixel, the Fourier transform of an image requires a much larger dynamic range. Typically, we have a sharp drop in the Fourier components with the size of the wavenumber (Fig. 6.11). Accordingly, at least 16-bit integer or 32-bit floating point numbers are required to represent an image in Fourier space without significant rounding errors.

The reason for this behavior is not the insignificance of high wavenumbers in an image. If we simply omit them, the image would become blurred.

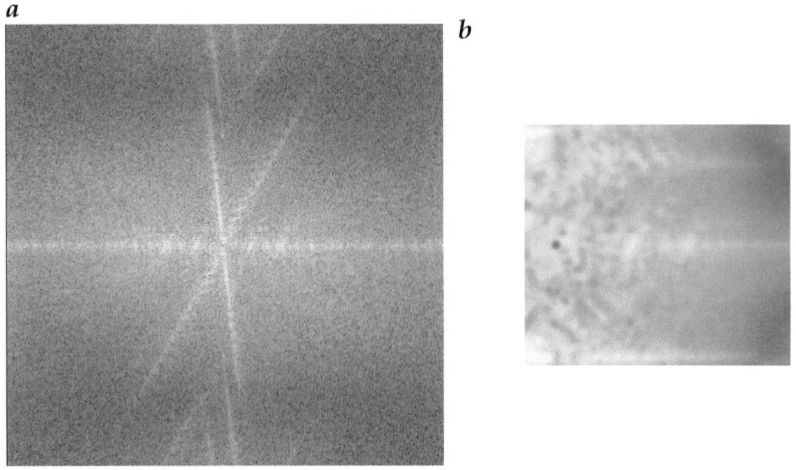

Figure 6.15: *Representation of the Fourier transform (power spectrum $|\tilde{g}_{uv}|^2$) of the image in Fig. 5.12a in **a** Cartesian coordinates and **b** logarithmic polar coordinates (Fig. 6.14); here the k-axis is horizontal and the φ-axis is vertical with an angular range of -90° to 90° to the x-axis. Both spectra were multiplied by k^2 and displayed logarithmically. The gray value scale covers 6 decades.*

This drop in amplitude is caused by the fact that the *relative* resolution increases. We usually think of relative resolutions because we can perceive relative differences in distance better than absolute ones. For example, we can easily see the difference of 10 cm at a distance of one meter, but not at one kilometer. If we apply this principle to Fourier space, it is better to represent the Fourier-transformed images (spectra) in so-called *log polar coordinates* (Fig. 6.14). In such a coordinate system, we divide the space into angular intervals and logarithmic k intervals. This means that the interval areas grow proportionally to k^2. In order to take this increase in area into account, we must multiply the Fourier components in this representation by k^2:

$$\int_{-\infty}^{\infty} |\tilde{g}(\boldsymbol{\kappa})|^2 \mathrm{d}k_1 \mathrm{d}k_2 = \int_{-\infty}^{\infty} k^2 |\tilde{g}(\boldsymbol{\kappa})|^2 \mathrm{d}\ln k \mathrm{d}\varphi. \tag{6.51}$$

If we assume that the power spectrum $|\tilde{g}(\boldsymbol{\kappa})|^2$ is evenly distributed over all wavenumbers in logarithmic polar coordinates, it decreases in Cartesian coordinates with k^{-2}.

For the representation of power spectra, the logarithms of the gray values are usually used to compress the high dynamic range. The discussion above shows that multiplication by k^2 significantly reduces the dynamic range of the spectrum. The representation in logarithmic polar coordinates also allows a much better evaluation of the directions of the spatial structures. In addition, the smaller wavenumbers are spread over a larger range (Fig. 6.15).

6.4 Fast Computation of Fourier Transforms

6.4.1 The Importance of Fast Algorithms

Without a *fast algorithm* to calculate the discrete Fourier transform, its use in image processing would not be possible. If Eq. (6.22) is used directly, this results in a prohibitively high computational effort. Each point of the transformed image requires N^2 complex multiplications and $N^2 - 1$ complex additions (without taking into account the calculation of the cosine and sine functions in the kernel). In total, N^4 complex multiplications and $N^2(N^2 - 1)$ complex additions are required. This results in about $8N^4$ floating point operations. For a 512×512 image, that's 5×10^{11} operations. A 2 GHz Pentium Pro processor can perform about 500 *MFLOPS* (million floating point operations per second) when programmed in a high-level language with an optimizing compiler. A single Fourier transformation of a 512×512 image with 5×10^{11} operations would therefore take about 1,000 s or 0.3 h. This is too slow to be practically usable. These figures show the need to find a suitable algorithm that drastically reduces the computing time of the DFT. Efficient algorithms are a central topic in computer science. To develop suitable algorithms, we need to examine the inner structure of a given task, its *computational complexity*, to find out how it can be solved with a minimum number of operations.

The following simple *search task* as an instructive example: A friend lives in a high-rise building with N floors. We need to find out which floor his apartment is on. Each of our questions is only answered with yes or no. How many questions do we need to find out where the friend lives? The simplest and most direct approach is the question: "Does he live on floor n? In the best case, our first attempt will be correct. However, it is more likely that we will guess wrong, so we will have to ask the same question again and again. In the worst case, we have to ask exactly $N - 1$ questions, on average $N/2$. With each question we can only rule out one of N possibilities, a rather ineffective approach.

In contrast, with the question: "Does he live in the upper half of the house?" we can already rule out half of the possibilities with just one question. Now we know whether he lives in the upper or lower half of the floors. Further questions can be asked in the same way by dividing the remaining possibilities into two halves. This strategy reduces the number of questions considerably. If the number of floors is a multiple of 2, say 2^l, exactly l questions are necessary. So for N floors we need $\operatorname{ld} N$ questions, where ld is the logarithm to the base 2. This strategy, which was used recursively to solve the search task more efficiently, is called the *divide-and-conquer* strategy.

A measure for the complexity of a problem with the problem size N, e. g. N components, results from how the dominant term, which arises when counting the necessary operations, depends on N. This approximation is helpful because for large N the number of operations is dominated by this term with the highest power of N.

We speak of a zero-order problem, $O(N^0)$, if the number of operations does not depend on this size, or of a linear-order problem, $O(N^1)$, if the number of computations increases linearly with the size. The complexity of the direct solution to the search problem in the above example is $O(N)$, while that of the divide-and-conquer strategy is $O(\mathrm{ld}\, N)$.

6.4.2 The 1D Radix-2 FFT Algorithm

Fast algorithms for the Fourier transform are generally abbreviated as *FFT* algorithms, an abbreviation for *fast Fourier transform*). First, we consider a fast algorithm for the one-dimensional DFT. Let us assume that the dimension N of the vector is a multiple of two, i. e. $N = 2^l$.

Since the direct solution according to Eq. (6.17) has the complexity $O(N^2)$, it seems reasonable to apply the divide-and-conquer strategy. If the transformation could be divided into two steps with two vectors of length $N/2$, the number of operations would be halved from N^2 to $2(N/2)^2 = N^2/2$.

If this decomposition of the DFT is possible in principle, it can be continued successively $\mathrm{ld}\, N$ times until finally a vector of length 1 remains, whose DFT is trivial, since no more transformation has to be performed. Of course, this approach only works if such a decomposition is possible and the number of additional operations for merging the partial results is not of a higher order than $O(N)$.

The result of the recursive decomposition is revealing because we no longer need to perform a Fourier transform at all. The entire algorithm for calculating the DFT has been transformed into recursive compositing of partial results.

If this composition is an operation of order $O(N)$, then the order of the entire DFT calculation is $O(N \,\mathrm{ld}\, N)$, since $\mathrm{ld}\, N$ times the partial transforms must be combined. Compared to the direct calculation ($O(N^2)$), this means an enormous saving in arithmetic operations. For $N = 2^{10} = 1024$ they are reduced to about one hundredth. We divide the vector into two halves by transforming the even and odd elements separately (Fig. 6.16):

$$
\begin{aligned}
\tilde{g}_{\tilde{n}} &= \sum_{n=0}^{N-1} g_n \exp\left(-\frac{2\pi i n\tilde{n}}{N}\right) \\
&= \sum_{n=0}^{N/2-1} g_{2n} \exp\left(-\frac{2\pi i 2n\tilde{n}}{N}\right) + \sum_{n=0}^{N/2-1} g_{2n+1} \exp\left(-\frac{2\pi i(2n+1)\tilde{n}}{N}\right) \qquad (6.52) \\
&= \sum_{n=0}^{N/2-1} g_{2n} \exp\left(-\frac{2\pi i n\tilde{n}}{N/2}\right) + \exp\left(-\frac{2\pi i\tilde{n}}{N}\right) \sum_{n=0}^{N/2-1} g_{2n+1} \exp\left(-\frac{2\pi i n\tilde{n}}{N/2}\right).
\end{aligned}
$$

The two partial sums again represent a DFT with $N' = N/2$. The second sum is multiplied by a phase factor that depends only on the wavenumber

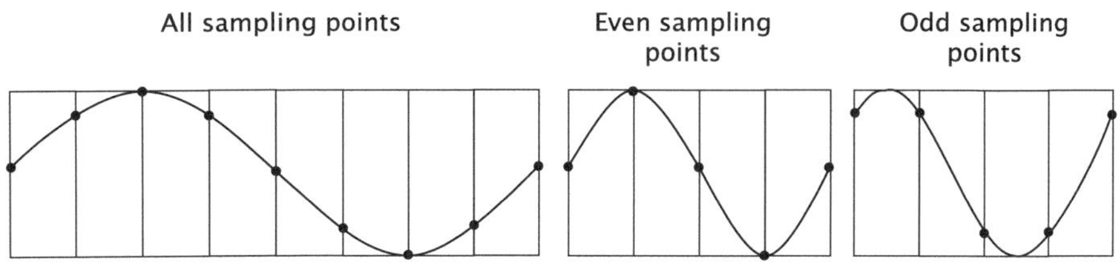

Figure 6.16: Decomposition of a vector into two vectors, consisting of the even and odd sampling points respectively.

v. This phase factor results from the shift theorem, as the odd elements have been shifted one place to the left.

This is illustrated using the base function $v = 1$ for $N = 8$ (Fig. 6.16). At the odd sampling points, the function is phase-shifted by $\pi/4$. Exactly this phase shift is compensated by the phase factor in Eq. (6.52):

$$\exp(-2\pi \mathrm{i}\tilde{n}/N) = \exp(-\pi \mathrm{i}/4)$$

The operations for combining the partial transformation results are each a complex multiplication and an addition, i. e. of constant effort $O(N^1)$. However, the DFT only provides $N/2$ values for the halved vector, while twice the number of values is required for the entire vector.

So we have to think again carefully whether the decomposition really works. To see how the composition of the N values works, we split the DFT for the whole vector into two parts and examine the values for $\tilde{n}$ from 0 to $N/2-1$ and from $N/2$ to $N-1$ separately. At the same time, we simplify the notation for the DFT by abbreviating the Fourier transform over the even and odd sample points with ${}^e\tilde{g}_{\tilde{n}}$ and ${}^o\tilde{g}_{\tilde{n}}$ (*even* and *odd*) respectively.

We now rewrite Eq. (6.52), separately for the two halves. For the first half, we can immediately use the decomposition from Eq. (6.52). For the second half, $\tilde{n}' = \tilde{n} + N/2$, only the phase factor changes. The addition of $N/2$ results in a change of sign:

$$\exp\left(-\frac{2\pi \mathrm{i}(\tilde{n}+N/2)}{N}\right) = -\exp\left(-\frac{2\pi \mathrm{i}\tilde{n}}{N}\right) \quad \text{or} \quad w_N^{-(\tilde{n}+N/2)} = -w_N^{-\tilde{n}}.$$

Using this symmetry, we write:

$$\left.\begin{array}{rcl} \tilde{g}_{\tilde{n}} & = & {}^e\tilde{g}_{\tilde{n}} + w_N^{-\tilde{n}}\,{}^o\tilde{g}_{\tilde{n}} \\ \tilde{g}_{\tilde{n}+N/2} & = & {}^e\tilde{g}_{\tilde{n}} - w_N^{-\tilde{n}}\,{}^o\tilde{g}_{\tilde{n}}. \end{array}\right\} \quad 0 \le \tilde{n} < N/2. \tag{6.53}$$

The Fourier transforms for the indices $\tilde{n}$ and $\tilde{n} + N/2$ differ only in the sign of the second term. Therefore, we only need *one* complex multiplication for the composition of *two* terms. The FFT algorithm now consists of the successive application of the halving of the vector length: The two

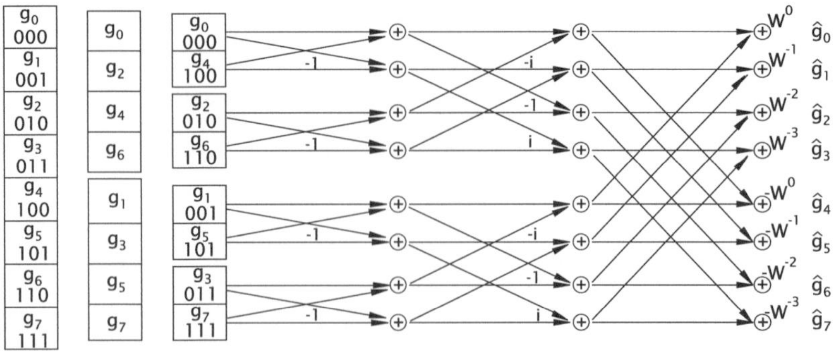

Figure 6.17: *Signal flow diagram of the base-2 FFT algorithm with decimation in time (radix-2 decimation-in-time FFT) for N = 8. Explanations in the text.*

transformations of the vectors of length $N/2$ are again divided into two transformations each over half the vector length. This results in similar expressions as in Eq. (6.52) with the only difference that the phase factor has doubled to $\exp[-(2\pi i\tilde{n})/(N/2)]$. The even and odd parts of the even vector contain the points $\{0, 4, 8, \cdots, N/2-4\}$ and $\{2, 6, 10, \cdots, N/2-2\}$ respectively.

In the last step, we decompose a vector with two elements into two vectors with one element each. Since the DFT of a vector of length 1 is equal to the identity function Eq. (6.17), no further calculations are necessary.

Once the decomposition is complete, we can gradually apply Eq. (6.53) with appropriate phase factors to rebuild the original image step by step in the opposite direction. In the first step, we create vectors with only two elements. We therefore only need one phase factor for $v = 0$, which is equal to 1. Accordingly, the first reconstruction step has a very simple form:

$$\begin{array}{rcl} \tilde{g}_0 & = & g_0 + g_1 \\ \tilde{g}_{0+N/2} = \tilde{g}_1 & = & g_0 - g_1. \end{array} \tag{6.54}$$

The algorithm discussed is a *radix-2 FFT algorithm* with decomposition in space. The individual steps of the FFT algorithm are summarized in Fig. 6.17 in a signal flow diagram for $N = 8$. The left half of the diagram shows the step-by-step decomposition into partial transformations while halving the vector length, the right half shows the calculation scheme for the multiplication with the phase factors and the addition of the partial vectors to a vector of length N. The first column shows the vector to be transformed. The second column is the result of the first decomposition step into two vectors. The vectors with the even and odd elements are separated into the upper and lower halves respectively. This step is repeated until vectors with only one element are obtained.

The vector elements are reordered during this decomposition. That is all that happens in this step; no further calculations are required. We can easily understand the new ordering scheme if we represent the indices of the vectors as dual numbers. In the first decomposition step, we order the elements according to the least significant bit, first the even elements (the bit is zero), then the odd elements (the bit is one). With each further decomposition step, the bit that determines the sorting is shifted one place to the left. At the end, the result is a sorting in which the new ordinal number has been created from the old one by bitwise inversion. For example, the element with the index $1 = 001_2$ will be at position $4 = 100_2$ and vice versa. This means that the chain of decomposition steps can be carried out in a single operation, a bit inversion operation of the indices. This reordering is called *bit reversal*.

In further steps on the right-hand side of the signal flow diagram, the successive composition into vectors follows, the length of which doubles again in each step. The composition into a two-dimensional vector is formulated in Eq. (6.54) and represented in Fig. 6.17 with arrows and dots. The symbols have the following meaning: Points represent a number, an element of a vector. These points are referred to as *nodes* of the signal flow diagram. The arrows take the value of the node at which they start; we multiply the value by the factor near the arrow and transfer the product to the node at which they end. If the factor associated with the arrow is missing, no multiplication takes place. All values arriving at a node are added together. The value of a node is therefore the sum of the values transferred from the previous level.

Only two nodes are involved in the basic operation of the FFT algorithm. The lower node value is multiplied by a phase factor. The sum and the difference of the two node values are then transferred to the upper and lower nodes respectively. Due to the crossing of the signal paths, this operation is referred to as *butterfly operation*.

We can gain a deeper insight into the FFT algorithm by tracing the calculation of a single element. Fig. 6.18 shows the signal paths for $\tilde{g}_0$ and $\tilde{g}_4$. In each level, we trace the number of nodes that contribute to the calculation of a node. In the last level, all elements are involved. The signal paths for $\tilde{g}_0$ and $\tilde{g}_4$ are identical except for the last step. This demonstrates the efficiency of the FFT algorithm very well. All phase factors of the signal paths for $\tilde{g}_0$ are one. As can be expected from Eq. (6.17), $\tilde{g}_0$ contains the sum of all elements of the vector $\boldsymbol{g}$,

$$\tilde{g}_0 = [(g_0 + g_4) + (g_2 + g_6)] + [(g_1 + g_5) + (g_3 + g_7)],$$

while in the last step for $\tilde{g}_4$ the addition is replaced by a subtraction:

$$\tilde{g}_4 = [(g_0 + g_4) + (g_2 + g_6)] - [(g_1 + g_5) + (g_3 + g_7)].$$

The DFT is an example of a unitary transformation (Appendix B), which is usually calculated by multiplication with a unitary matrix. What is the significance of the FFT algorithm in this context? The signal flow diagram in Fig. 6.17 shows that the vector is transformed in several steps. The unitary transformation matrix is therefore divided into several partial transformation matrices, which are applied one after the other. If we consider the

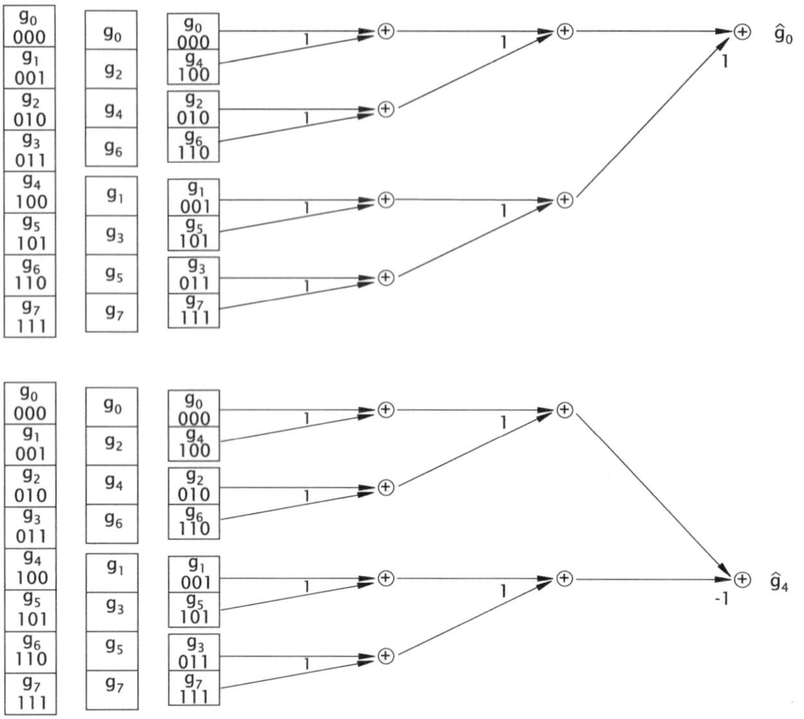

Figure 6.18: Signal flow diagram of the calculation of $\tilde{g}_0$ and $\tilde{g}_4$ with the radix-2 FFT algorithm for an eight-dimensional vector.

algorithm from Fig. 6.17 for $N = 8$, unitary transformations are saved by splitting the unitary matrix into three simpler transformations:

$$
\begin{bmatrix} \tilde{g}_0 \\ \tilde{g}_1 \\ \tilde{g}_2 \\ \tilde{g}_3 \\ \tilde{g}_4 \\ \tilde{g}_5 \\ \tilde{g}_6 \\ \tilde{g}_7 \end{bmatrix}
=
\begin{bmatrix}
1 & 0 & 0 & 0 & 1 & 0 & 0 & 0 \\
0 & 1 & 0 & 0 & 0 & w^{-1} & 0 & 0 \\
0 & 0 & 1 & 0 & 0 & 0 & w^{-2} & 0 \\
0 & 0 & 0 & 1 & 0 & 0 & 0 & w^{-3} \\
1 & 0 & 0 & 0 & -1 & 0 & 0 & 0 \\
0 & 1 & 0 & 0 & 0 & -w^{-1} & 0 & 0 \\
0 & 0 & 1 & 0 & 0 & 0 & -w^{-2} & 0 \\
0 & 0 & 0 & 1 & 0 & 0 & 0 & -w^{-3}
\end{bmatrix}
$$

$$
\begin{bmatrix}
1 & 0 & 1 & 0 & 0 & 0 & 0 & 0 \\
0 & 1 & 0 & \mathrm{i} & 0 & 0 & 0 & 0 \\
1 & 0 & -1 & 0 & 0 & 0 & 0 & 0 \\
0 & 1 & 0 & -\mathrm{i} & 0 & 0 & 0 & 0 \\
0 & 0 & 0 & 0 & 1 & 0 & 1 & 0 \\
0 & 0 & 0 & 0 & 0 & 1 & 0 & \mathrm{i} \\
0 & 0 & 0 & 0 & 1 & 0 & -1 & 0 \\
0 & 0 & 0 & 0 & 0 & 1 & 0 & -\mathrm{i}
\end{bmatrix}
\begin{bmatrix}
1 & 0 & 0 & 0 & 1 & 0 & 0 & 0 \\
1 & 0 & 0 & 0 & -1 & 0 & 0 & 0 \\
0 & 0 & 1 & 0 & 0 & 0 & 1 & 0 \\
0 & 0 & 1 & 0 & 0 & 0 & -1 & 0 \\
0 & 1 & 0 & 0 & 0 & 1 & 0 & 0 \\
0 & 1 & 0 & 0 & 0 & -1 & 0 & 0 \\
0 & 0 & 0 & 1 & 0 & 0 & 0 & 1 \\
0 & 0 & 0 & 1 & 0 & 0 & 0 & -1
\end{bmatrix}
\begin{bmatrix} g_0 \\ g_1 \\ g_2 \\ g_3 \\ g_4 \\ g_5 \\ g_6 \\ g_7 \end{bmatrix}
$$

These transformation matrices each represent a level of the FFT algorithm. This matrix decomposition shows that the FFT algorithm can also be understood as a method of decomposing unitary transformation matrices into partial transformation matrices that are as sparse as possible.

6.4.3 Criteria for Efficient Algorithms

If our algorithm could avoid all trivial multiplications, the number of multiplications would be reduced to $(N/2)(\mathrm{ld}\, N - 3)$.

The FFT algorithm is a classic example of a *fast algorithm*. The reduction in calculation time is enormous. For a vector with 512 elements, only 1536 instead of 262 144 complex multiplications are required compared to the direct method according to Eq. (6.17). This reduces the number of multiplications to 1/170.

Using the FFT algorithm, the discrete Fourier transform can no longer be described as a computationally intensive operation, as it only requires a few operations per vector element. A vector with 512 elements requires only 3 complex multiplications and 8 complex additions per element, which corresponds to 12 real multiplications and 24 real additions.In terms of the number of arithmetic operations required, there are many other faster and more effective FFT algorithms. Most of them are based on polynomial algebra and number theory. A detailed discussion of these algorithms can be found at Blahut [21]. However, the mere number of arithmetic operations is not the only measure of an efficient algorithm. Other factors must also be taken into account.

Accessing data requires additional operations. Consider the simple example of adding two vectors. In addition to the addition, the following operations are performed: Calculating the addresses of the corresponding elements, reading the two elements into the registers and writing the result of these additions back into the memory. Depending on the architecture of the hardware used, these additional operations can consume more time than the actual addition. Thus, an algorithm with a complicated access to the vector elements can lead to a considerable amount of additional work in addition to the arithmetic operations. Therefore, a simple algorithm that requires more arithmetic operations but fewer memory accesses may be more effective.

Another consideration when evaluating algorithms is the amount of memory required. This applies not only to the code, but also to memory space for intermediate results or constant tables. For example, a so-called in-place FFT algorithm, which can perform the Fourier transformation of an image without having to buffer the image, is very effective. Speed and storage space are often related. For example, many integer FFT algorithms first calculate the complex phase factors $w_N^{\tilde{n}}$ and store them in static tables.

The efficiency of algorithms depends to a large extent on the architecture of the computer on which they are implemented. If a multiplication is executed either by software or by a microcoded instruction, it is much slower than an addition or a memory access implemented directly as hardware functions. In this case, the criterion of an efficient algorithm is to minimize the number of multiplications, even at the cost of a larger number of additions or a more complex memory access. However, such a strategy makes no sense on modern high-speed architectures, as additions and multiplications only require one clock cycle each. However, the faster the processors work, the more critical memory access becomes. Fast algorithms must realize effective memory accesses. It is crucial that as many calculation steps as possible are carried out with the same data. As these are relatively small amounts of data, they can be stored in a fast cache, the so-called *cache*, and do not have to be constantly fetched from the much slower main memory.

After this detailed discussion of the algorithm, we can now estimate the number of operations required. In each step of the composition, $N/2$ complex multiplications and N complex additions are performed. In total, we need $N/2\,\mathrm{ld}N$ complex multiplications and $N\,\mathrm{ld}N$ complex additions.

A more detailed analysis shows that even more multiplications can be saved. In the first two steps of the composition, only trivial multiplications with 1 or i are performed (cf. Fig. 6.17). In the subsequent steps, the number of trivial multiplications decreases by half.

6.4.4 ♯Radix-4 FFT Algorithm

Now that we have worked through a fast algorithm in detail, we still do not know whether this algorithm is optimal or whether there are even more efficient algorithms. In fact, we have only learned about a special case of the *divide-and-conquer* strategy. Instead of dividing the initial vector into two parts, we could have used a different division, for example P Q-dimensional vectors at $N = PQ$. Such algorithms are called *Cooley-Tukey algorithms* [21]. A frequently used decomposition is the *base-4 FFT algorithm*. This decomposes a vector into four components:

$$
\begin{aligned}
\tilde{g}_{\tilde{n}} &= \sum_{n=0}^{N/4-1} g_{4n} w_N^{-4n\tilde{n}} + w_N^{-\tilde{n}} \sum_{n=0}^{N/4-1} g_{4n+1} w_N^{-4n\tilde{n}} \\
&+ w_N^{-2\tilde{n}} \sum_{n=0}^{N/4-1} g_{4n+2} w_N^{-4n\tilde{n}} + w_N^{-3\tilde{n}} \sum_{n=0}^{N/4-1} g_{4n+3} w_N^{-4n\tilde{n}}.
\end{aligned}
$$

We simplify the equations with similar abbreviations as for the base-2 algorithm and denote the partial transformations with ${}^0\tilde{g}, \cdots, {}^3\tilde{g}$. Taking into account the symmetry of $w_N^{\tilde{n}}$, the transformations over a quarter of the vector each result from

$$
\begin{aligned}
\tilde{g}_{\tilde{n}} &= {}^0\tilde{g}_{\tilde{n}} + w_N^{-\tilde{n}}\,{}^1\tilde{g}_{\tilde{n}} + w_N^{-2\tilde{n}}\,{}^2\tilde{g}_{\tilde{n}} + w_N^{-3\tilde{n}}\,{}^3\tilde{g}_{\tilde{n}} \\
\tilde{g}_{\tilde{n}+N/4} &= {}^0\tilde{g}_{\tilde{n}} - \mathrm{i}w_N^{-\tilde{n}}\,{}^1\tilde{g}_{\tilde{n}} - w_N^{-2\tilde{n}}\,{}^2\tilde{g}_{\tilde{n}} + \mathrm{i}w_N^{-3\tilde{n}}\,{}^3\tilde{g}_{\tilde{n}} \\
\tilde{g}_{\tilde{n}+N/2} &= {}^0\tilde{g}_{\tilde{n}} - w_N^{-\tilde{n}}\,{}^1\tilde{g}_{\tilde{n}} + w_N^{-2\tilde{n}}\,{}^2\tilde{g}_{\tilde{n}} - w_N^{-3\tilde{n}}\,{}^3\tilde{g}_{\tilde{n}} \\
\tilde{g}_{\tilde{n}+3N/4} &= {}^0\tilde{g}_{\tilde{n}} + \mathrm{i}w_N^{-\tilde{n}}\,{}^1\tilde{g}_{\tilde{n}} - w_N^{-2\tilde{n}}\,{}^2\tilde{g}_{\tilde{n}} - \mathrm{i}w_N^{-3\tilde{n}}\,{}^3\tilde{g}_{\tilde{n}}
\end{aligned}
$$

or, in matrix notation, from

$$\begin{bmatrix} \tilde{g}_{\tilde{n}} \\ \tilde{g}_{\tilde{n}+N/4} \\ \tilde{g}_{\tilde{n}+N/2} \\ \tilde{g}_{\tilde{n}+3N/4} \end{bmatrix} = \begin{bmatrix} 1 & 1 & 1 & 1 \\ 1 & -\mathrm{i} & -1 & \mathrm{i} \\ 1 & -1 & 1 & -1 \\ 1 & \mathrm{i} & -1 & -\mathrm{i} \end{bmatrix} \begin{bmatrix} {}^{0}\tilde{g}_{\tilde{n}} \\ w_N^{-\tilde{n}}\, {}^{1}\tilde{g}_{\tilde{n}} \\ w_N^{-2\tilde{n}}\, {}^{2}\tilde{g}_{\tilde{n}} \\ w_N^{-3\tilde{n}}\, {}^{3}\tilde{g}_{\tilde{n}} \end{bmatrix}.$$

This decomposition requires 12 complex additions and 3 complex multiplications. The number of additions can be further reduced by splitting the matrix into two simpler submatrices:

$$\begin{bmatrix} \tilde{g}_{\tilde{n}} \\ \tilde{g}_{\tilde{n}+N/4} \\ \tilde{g}_{\tilde{n}+N/2} \\ \tilde{g}_{\tilde{n}+3N/4} \end{bmatrix} = \begin{bmatrix} 1 & 0 & 1 & 0 \\ 0 & 1 & 0 & -\mathrm{i} \\ 1 & 0 & -1 & 0 \\ 0 & 1 & 0 & \mathrm{i} \end{bmatrix} \begin{bmatrix} 1 & 0 & 1 & 0 \\ 1 & 0 & -1 & 0 \\ 0 & 1 & 0 & 1 \\ 0 & 1 & 0 & -1 \end{bmatrix} \begin{bmatrix} {}^{0}\tilde{g}_{\tilde{n}} \\ w_N^{-\tilde{n}}\, {}^{1}\tilde{g}_{\tilde{n}} \\ w_N^{-2\tilde{n}}\, {}^{2}\tilde{g}_{\tilde{n}} \\ w_N^{-3\tilde{n}}\, {}^{3}\tilde{g}_{\tilde{n}} \end{bmatrix}.$$

The first matrix multiplication leads to intermediate results that can be used by several arithmetic operations in the second step. In this way, we save four additions. In a successive decomposition, there are now $\log_4 N$ steps. Since only trivial multiplications occur in the first step, as in the base-2 algorithm, and multiplication is only required for 3/4 of the points in the subsequent steps, this results in a total of $3/4N(\log_4 N - 1) = 3/8N(\mathrm{ld}N - 2)$ complex multiplications and $2N \log_4 N = N\mathrm{ld}N$ complex additions. While the number of additions remains the same, the multiplications are reduced by 25 % compared to the base-2 algorithm.

6.4.5 ♯Decomposition in Fourier Space

The FFT algorithm with *decomposition in Fourier space* is another example of a *Cooley-Tukey algorithm*. Here we split the N-dimensional input vector into two consecutive blocks. This splits the vector in Fourier space into even and odd components:

$$\begin{aligned} \tilde{g}_{2\tilde{n}} &= \sum_{n=0}^{N/2-1} (g_n + g_{n+N/2}) w_{N/2}^{-n\tilde{n}} \\ \tilde{g}_{2\tilde{n}+1} &= \sum_{n=0}^{N/2-1} W_N^{-n} (g_n - g_{n+N/2}) w_{N/2}^{-n\tilde{n}}. \end{aligned} \tag{6.55}$$

The recursive application of this decomposition causes a bit inversion of the elements in the output vector, but not in the input vector. Fig. 6.19 shows the signal flow diagram for $N = 8$ as an example. A comparison with the flow diagram for decomposition in time (Fig. 6.17) makes it clear that all steps are carried out in reverse order here. Even the elementary butterfly operation Eq. (6.53) has been reversed.

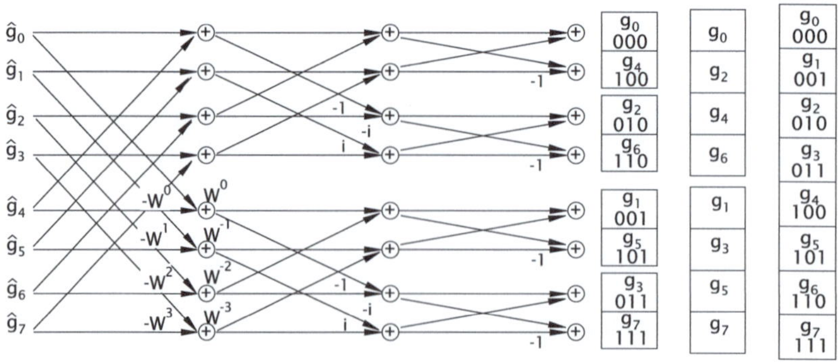

Figure 6.19: *Signal flow diagram of the base-2 FFT algorithm for N = 8 with decomposition in Fourier space.*

6.4.6 ♯Multidimensional FFT Algorithms

In general, there are two ways of developing fast algorithms for *multidimensional discrete Fourier transforms*. On the one hand, we can decompose the multidimensional DFT into 1D DFTs and use fast algorithms for these. On the other hand, we generalize the approach for the decomposition of 1D-FFT to higher dimensions. We show examples for both possibilities.

Decomposition into One-dimensional Transformations

Due to the separability of the DFT kernel, a two-dimensional DFT can be decomposed into two one-dimensional DFTs. From Eq. (6.22) one obtains

$$\tilde{g}_{\tilde{m}\tilde{n}} = \frac{1}{\sqrt{MN}} \sum_{m=0}^{M-1} \left[\sum_{n=0}^{N-1} g_{mn} \exp\left(-\frac{2\pi i n\tilde{n}}{N}\right) \right] \exp\left(-\frac{2\pi i m\tilde{m}}{M}\right). \tag{6.56}$$

The inner sum represents M 1D DFT of the rows, the outer sum represents the subsequent transformation of the N columns. The 2D DFT can thus be executed as a sequence of two 1D DFTs.

Row transformation $$\breve{g}_{m\tilde{n}} = \frac{1}{N} \sum_{n=0}^{N-1} g_{m,n} \exp\left(-\frac{2\pi i n\tilde{n}}{N}\right)$$

Column transformation $$\tilde{g}_{\tilde{m}\tilde{n}} = \frac{1}{M} \sum_{m=0}^{M-1} \breve{g}_{m\tilde{n}} \exp\left(-\frac{2\pi i m\tilde{m}}{M}\right).$$

Similarly, a D-dimensional DFT can be composed of D one-dimensional DFTs.

Multidimensional Decomposition

A decomposition can also be performed directly in multidimensional spaces. We will show such an algorithm in the simple case of a two-dimensional base-2 algorithm with decomposition in spatial domain. We decompose a $M \times N$

0	1	0	1	0	1	0	1
2	3	2	3	2	3	2	3
0	1	0	1	0	1	0	1
2	3	2	3	2	3	2	3
0	1	0	1	0	1	0	1
2	3	2	3	2	3	2	3
0	1	0	1	0	1	0	1
2	3	2	3	2	3	2	3

Figure 6.20: *Decomposition of an image matrix into four partitions for the 2D Basis 2 FFT algorithm.*

matrix into four submatrices by using only every second pixel from every second row (Fig. 6.20). This results in

$$\begin{bmatrix} \tilde{g}_{\tilde{m}\tilde{n}} \\ \tilde{g}_{\tilde{m},\tilde{n}+N/2} \\ \tilde{g}_{\tilde{m}+M/2,\tilde{n}} \\ \tilde{g}_{\tilde{m}+M/2,\tilde{n}+N/2} \end{bmatrix} = \begin{bmatrix} 1 & 1 & 1 & 1 \\ 1 & -1 & 1 & -1 \\ 1 & 1 & -1 & -1 \\ 1 & -1 & -1 & 1 \end{bmatrix} \begin{bmatrix} {}^{0,0}\tilde{g}_{\tilde{m}\tilde{n}} \\ w_N^{-\tilde{n}}\ {}^{0,1}\tilde{g}_{\tilde{m}\tilde{n}} \\ w_M^{-\tilde{m}}\ {}^{1,0}\tilde{g}_{\tilde{m}\tilde{n}} \\ w_M^{-\tilde{m}} w_N^{-\tilde{n}}\ {}^{1,1}\tilde{g}_{\tilde{m}\tilde{n}} \end{bmatrix}.$$

The exponents preceded by $\tilde{G}$ denote the associated partial transformation. The 2D basis 2 algorithm differs only slightly from the one-dimensional variant. As with the 1D radix-4 algorithm (Section 6.4.4), we can reduce the number of additions from 12 to 8 by factorizing the matrix:

$$\begin{bmatrix} 1 & 1 & 1 & 1 \\ 1 & -1 & 1 & -1 \\ 1 & 1 & -1 & -1 \\ 1 & -1 & -1 & 1 \end{bmatrix} = \begin{bmatrix} 1 & 0 & 1 & 0 \\ 0 & 1 & 0 & 1 \\ 1 & 0 & -1 & 0 \\ 0 & 1 & 0 & -1 \end{bmatrix} \begin{bmatrix} 1 & 1 & 0 & 0 \\ 1 & -1 & 0 & 0 \\ 0 & 0 & 1 & 1 \\ 0 & 0 & 1 & -1 \end{bmatrix}.$$

The 2D base-2 algorithm of a $N \times N$ matrix requires $(3/4N^2)\,\mathrm{ld}\,N$ complex multiplications, i. e. 25 % less than the separation into two 1D base-2 FFTs. However, the multidimensional decomposition has the disadvantage that access to the memory is more complex than with the one-dimensional Fourier transform. With the separation into 1D transforms, memory access is local. This results in a higher cache hit rate compared to the split access of the multidimensional decomposition.

6.4.7 ♯Fourier Transform of Real Images

So far, we have only discussed the Fourier transformation of complex-valued signals. The same algorithms can also be used for real-valued signals. However, they lose efficiency because the Fourier transform of a real image is hermitian (Section 6.3.4) and thus only half of the Fourier coefficients are independent. This corresponds to the fact that half of the signal, namely the imaginary part, is zero.

It is obvious that the computing speed can be doubled for the DFT of real data. The simplest way is to calculate two real 1D sequences simultaneously. This concept is easy to implement for DFT of images since many 1D DFTs need to be calculated. Therefore, we can assign the first line x to the real part and the second line y to the imaginary part. This results in the complex vector $z = x + iy$.

The symmetry properties of the DFT discussed in Section 6.3.4 suggest that the real and imaginary parts are mapped into the Hermitian and anti-Hermitian parts of the Fourier space during the transformation. Thus, the Fourier transforms of the two real M-dimensional vectors result in

$$\tilde{x}_{\tilde{n}} = 1/2(\tilde{z}_{\tilde{n}} + \tilde{z}^*_{N-\tilde{n}}), \quad i\tilde{y}_{\tilde{n}} = 1/2(\tilde{z}_{\tilde{n}} - \tilde{z}^*_{N-\tilde{n}}). \tag{6.57}$$

6.5 ♯References for Further Reading

The classic textbook on the Fourier transform — and still one of the best — is Bracewell [23]. An exhaustive description of various transforms can be found in the "Transforms and Applications Handbook" by Poularikas [245]. For the basics of linear algebra, especially the unitary transformations, please refer to modern textbooks on linear algebra, for example Ryan [276] and Strang [316].

Even today, it is still worth reading the historical journal article by Cooley and Tukey [44], which describes the discovery of the fast Fourier transform. The monographs by Brigham [27] and Blahut [21] deal with various algorithms for the fast Fourier transform.

If you are generally interested in the design and analysis of algorithms, the classic textbook by Aho et al. [5] is highly recommended; it is still one of the clearest and most comprehensible presentations. Of the more modern textbooks, the very comprehensive textbook by Cormen et al. [45] is highly recommended. Both textbooks also deal with the FFT. A fast, freely usable and the most widely used implementation of FFT algorithms is the FFTW library[1], see also Frigo and Johnson [95].

[1] `www.fftw.org`

7 Radiometric and Geometric Image Corrections

Contents

7.1 Introduction

For the first processing steps after capturing a digital image, we need two classes of operations, point operations and geometric operations. These two types essentially modify the "what" and the "where" of an image point. They neither extract specific information from the image nor do they perform operations to recognize objects. However, they are essential pre-processing

B. Jähne, *Digital Image Processing and Image Acquisition*,
https://doi.org/10.1007/978-3-032-04806-6_7

steps for this, as they correct radiometric and geometric errors caused by the image acquisition. *Point operations* modify the values of individual image points only depending on the value itself and possibly on the position of the image point. Such an operation is generally written as

$$g'_{mn} = P_{mn}(g_{mn}). \tag{7.1}$$

The indices of the function P denote the possible dependency of the point operation on the position of the image point. In contrast, *geometric operations* only modifies the position of a pixel. A pixel is moved from the position $\boldsymbol{x}$ to a new position $\boldsymbol{x}'$. The relationship between the two coordinates is determined by the geometric mapping function:

$$\boldsymbol{x}' = M(\boldsymbol{x}). \tag{7.2}$$

Point and geometric operations are complementary. They are useful for correcting elementary disturbances in the imaging process, such as non-linear and inhomogeneous radiometric sensitivity of the image sensor or geometric distortions of the optical system. We therefore use point operations to correct and optimize the illumination, to detect underflow and overflow, to enhance and stretch contrast, to average the image, to correct inhomogeneous illumination or for radiometric calibration (Sections 7.2.3–7.3.3). Geometric operations essentially contain two steps. In most applications, the mapping function Eq. (7.2) is not explicitly specified, but must be derived from the comparison between the original and its image (Section 7.5.4). If an image is distorted by a geometric transformation, the transformed pixels usually no longer lie on grid points. Therefore, the gray values at these points must be interpolated from neighboring pixels. We have already discussed this important task in detail in Section 5.5, as correct interpolation is not trivial.

Point and geometric operations are not only important for the elementary pre-processing steps. They are also integrated into much more complex image operations, especially during feature extraction (Chapters 11–16). However, it is important to note that point and geometric operations are not suitable for correcting the effects of an optical system described by its point response. This requires complex reconstruction techniques, which are discussed in Chapter 17. The application of point and geometric operations is limited to simple radiometric and geometric corrections.

7.2 Homogeneous Point Operations

7.2.1 Definitions and Basic Properties

We call a point operation that is independent of the position of the pixel *homogeneous point operation* and write

$$g'_{mn} = P(g_{mn}). \tag{7.3}$$

A point operation maps gray values onto themselves. Since two or more different values can also be mapped onto a single value, point operations are generally irreversible and part of the image content is irretrievably lost. The dot operation

$$P(q) = \begin{cases} 0 & q < t \\ Q-1 & q \geq t \end{cases} \tag{7.4}$$

for example, performs a simple global threshold operation. All gray values below the threshold t are set to 0 (black), those above and on the threshold to the highest value $Q-1$ (white). It is obvious that this point operation is not reversible. An example of a reversible point operation is negative formation. Here, an image with an inverted gray value scale is calculated according to

$$P_N(q) = Q - 1 - q, \tag{7.5}$$

and you get a negative as in photography. The reversal of this operation is again a negative formation:

$$P_N\left(P_N(q)\right) = Q - 1 - (Q - 1 - q) = q. \tag{7.6}$$

The conversion between the signed and unsigned representation of gray values (Section 5.6.3) is another example of a reversible point operation.

7.2.2 Lookup Tables

The direct calculation of homogeneous point operations according to Eq. (7.3) can be very time-consuming. This is shown in the following example. The 14-bit gray values of the 1024×1024 image of a high-resolution CCD camera are to be converted into a logarithmic 8-bit gray value scale that spans 4.3 decades from 0 to 16,383. This conversion is achieved by the point operation

$$P(q) = 59.30 \log q. \tag{7.7}$$

A direct implementation would require the following operations per pixel: A data type conversion from integer to floating point, the calculation of the logarithm, a multiplication by 59.30 and another data type conversion from floating point to 8-bit integer. All these operations must be performed more than a million times for a 1024×1024 image.

The key to a more efficient implementation of this operation lies in the observation that the domain of definition of each point operation consists of only a limited number of Q quantization levels. With a logarithmic conversion of 14 bits into 8 bits, there are at most 16,384 different input values. This means that most of the one million calculations are merely repetitions, on average 64 times. Such unnecessary repetitions can be avoided if we first calculate $P(q)$ for all 16,384 possible gray values and store the results in a table with 16,384 elements. Then the calculation of the point operation is reduced to a replacement of the gray values by the corresponding table element, which is indexed by the respective gray value.

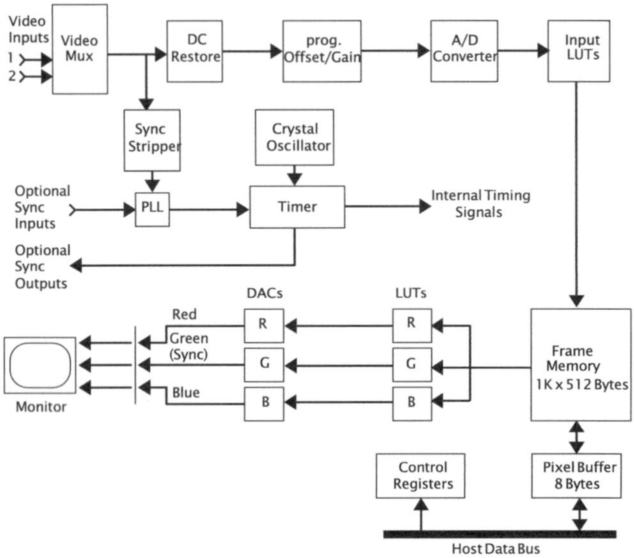

Figure 7.1: *Block diagram of the early frame grabber PCVISIONplus from Imaging Technology, Inc. for PCs from the 1990s with monochrome analog video input. Lookup tables are located between the A/D converter and the frame buffer (input LUT) and between the frame buffer and the video monitor (output LUT).*

Such a table is called *lookup table* or *LUT*. This means that homogeneous point operations are equivalent to *lookup table operations*. Lookup tables are all the more efficient the fewer quantization levels they have. With 8-bit standard images, the tables only contain 256 values. In most cases, however, lookup tables with 65,536 entries, as required for 16-bit images, are still efficient.

Lookup tables were already integrated into the hardware of the first add-on cards for PCs for recording and digitizing analogue video image data (so-called *frame grabber*). As Fig. 7.1 shows, there are two options for the placement of lookup tables. The *input LUT* is located between the *analog-to-digital converter* and the frame buffer and the *output LUT* is located between the frame buffer and the digital-to-analog converter, which converts the signal for the image output, e. g. on a monitor into an analog video signal. The input LUT enables point operations to be performed *before* the image is stored in the image memory. In the first PCs, the 16-bit *ISA bus* (Industry Standard Architecture[1]) was not fast enough to transfer digitized image data to the computer in real time. For this reason, the frame grabber included a three-channel analog RGB output for a separate video monitor. With the output LUT, point operations can be performed before output and observed

[1] https://de.wikipedia.org/wiki/Industry_Standard_Architecture

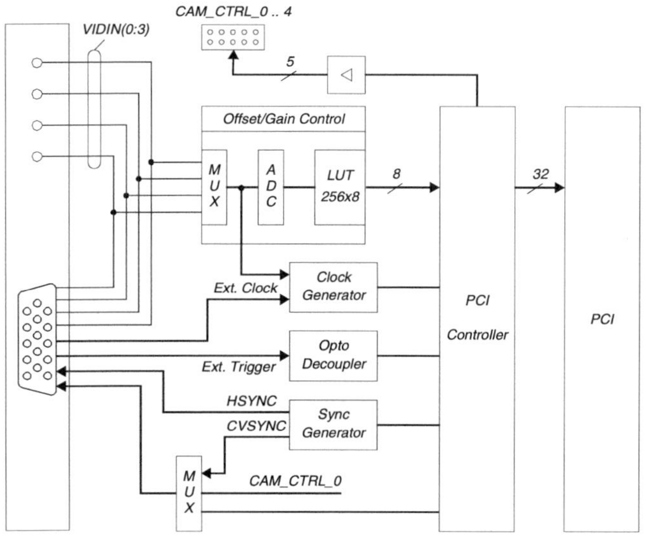

Figure 7.2: *Block diagram of the PCEYE_1 frame grabber from ELTEC Elektronik GmbH with monochrome analog video input from 1995 as an example of one of the first PCI bus frame grabbers without image memory. The image data is transferred in real time via direct memory access (DMA) to the PC main memory for display and further processing.*

on this monitor. In this way, interactive point operations *without* changing the stored image are possible.

Only with the introduction of the 32-bit *PCI bus* (Peripheral Component Interconnect[2]), proposed by Intel in 1991 and then standardized, was it possible to transfer digitized image data in real time. Frame grabbers then no longer required image memory and the digitized image data could be transferred to the main memory at up to 132 MB/s (Fig. 7.2) and the image could be displayed via the computer's graphics card. Accordingly, the frame grabber now only requires an input LUT.

Today, CMOS image sensors already provide a digitized data stream and LUT tables are already integrated into the camera along with many other processing steps. This is illustrated in Fig. 7.3 using the example of an industrial camera with a color image sensor.

The use of input lookup tables is limited, as non-linear LUT functions lead to missing values or map consecutive values to a single value (Fig. 7.4). This results in artifacts that lead to further errors in subsequent processing steps such as the calculation of mean values and edge detection. This particularly affects the steepness of edges and the accuracy of the determination of gray value changes. However, an input LUT is useful for non-linear point operations if the 8-bit input values are mapped to higher-resolution output values, for example 16-bit integer numbers or 32-bit floating point numbers,

[2]https://de.wikipedia.org/wiki/Peripheral_Component_Interconnect

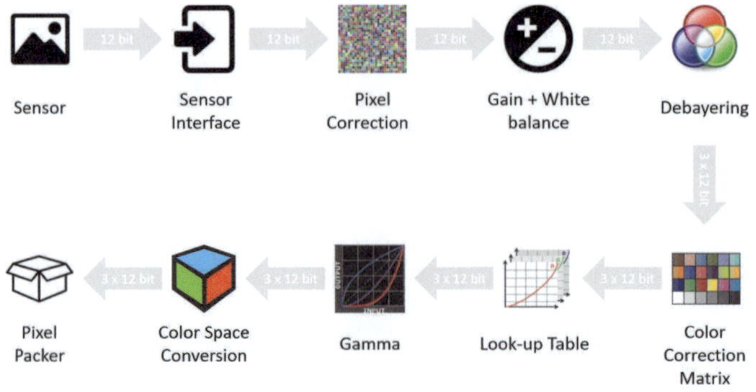

Figure 7.3: Signal flow diagram of the Triton camera series from Lucid Vision Labs for a color image sensor with RGB Bayer pattern.

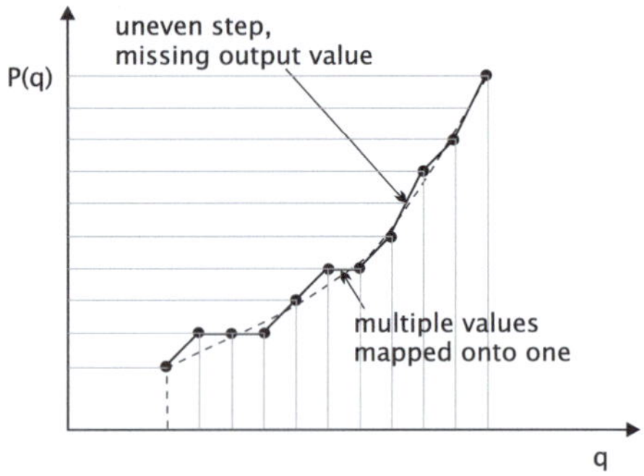

Figure 7.4: Illustration of a non-linear lookup table. Several values are mapped to one, and missing initial values lead to irregular steps.

or if the camera signal is digitized with a higher resolution, for example with 12 bits, and then only 8-bit values are output. This reduces the rounding errors.

At the same time, the gray values can be converted into a calibrated signal, for example in the case of an infrared camera into a temperature. Unfortunately, such generalized LUTs have hardly been implemented in hardware to date. However, they are easy to implement in software.

In contrast to the input LUT, the output LUT is much more commonly used as a tool because it does not change the stored image. LUT operations can also convert gray-scale images to *pseudocolor images*, a technique that is common even with the simplest digitizer cards (Fig. 7.1). It does not require much additional hardware. A digital-to-analog converter is used for each of the primary colors red, green and blue. In addition, each channel has its own LUT with 256 entries for an 8-bit representation. This means that each individual gray value q can be mapped to any color by assigning the corresponding color values to the LUT addresses $r(q)$, $g(q)$ and $b(q)$. Formally, this is a *point vector operation*.

$$\boldsymbol{P}(q) = [r(q), g(q), b(q)]^T. \tag{7.8}$$

If all three point functions $r(q)$, $g(q)$ and $b(q)$ are identical, a shade of gray is displayed. If two of the point functions are zero, the image has the color of the remaining point function.

7.2.3 Interactive Evaluation

Homogeneous point operators implemented via lookup tables are very useful tools for image inspection and image quality assessment. Since lookup table operations can be performed in real time, interactive image modification is possible. If only the output LUT is changed, the original image content remains unchanged.

These operations are important because we should not blindly trust our visual system for quantitative analysis of image brightness. In Section 1.6 we learned that our eye has a limited contrast resolution and is not suitable for estimating absolute brightness.

In Section 4.5.7 we learned about profiles as a helpful tool for assessing the inhomogeneities of an image. Here we will learn to appreciate another simple but very useful tool, the frequency distribution of the values of an image, called *histogram* for short.

We will now discuss LUT operations for

- control and optimization of homogeneous illumination,
- detection of underflow and overflow,
- contrast enhancement,
- contrast spreading and
- compression of the range.

Control and Optimization of Homogeneous Illumination

With the naked eye, we can hardly estimate the homogeneity of an illuminated surface like the one in Fig. 7.5a. A histogram (Fig. 7.5b) shows the frequency of the individual gray values, but not their spatial distribution. It therefore does not help much with the interactive optimization of the lighting.

We have to modify the display so that absolute values become visible to the human eye. Equidensities are helpful for a continuous brightness distribution. This technique uses a stepped lookup table in which certain ranges of values are mapped to one value.

The simplest way to do this is to set the lowest bits to zero using a logical AND operation:

$$q' = P(q) = q \wedge \overline{(2^p - 1)}, \tag{7.9}$$

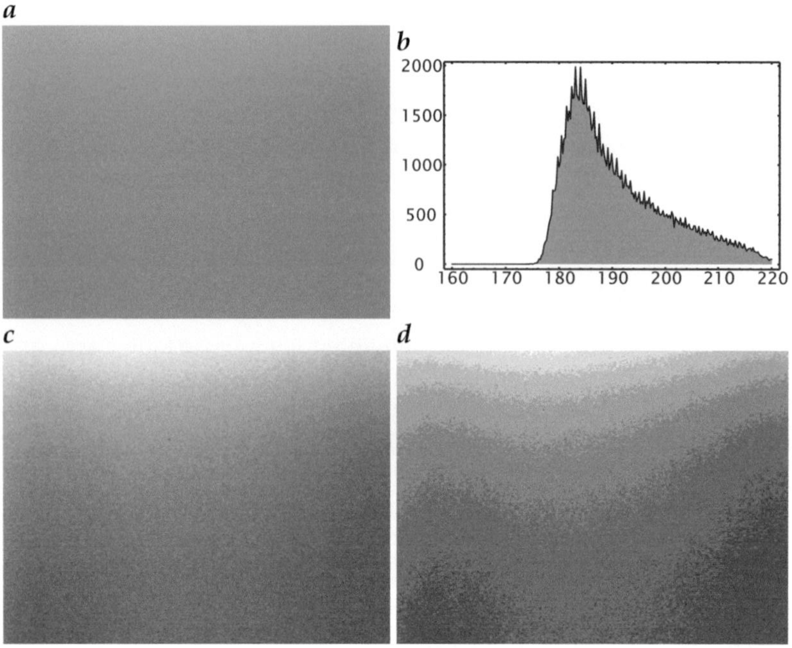

Figure 7.5: ***a*** *A gradual decrease in brightness from top to bottom that is barely perceived by the eye. Values of 160–220, contrast enhanced by a factor of four. Floating point image averaged over 100 images.* ***b*** *Histogram of* ***a****;* ***c*** *and* ***d*** *(contrast enhanced, gray value range 184–200): Artificially created edges by step LUTs with a step height of 1.0 and 2.0 make contour lines of constant irradiation visible.*

where $\wedge$ stands for the logical (bitwise) AND operation and the overline for the inversion. This point operation limits the resolution to $Q - p$ bits and thus to 2^{Q-p} quantization levels.

Now the jumps between the remaining few levels are large enough to be perceived by the eye. We see contour lines of equal absolute values in the image (Fig. 7.5d). We can try to make the lighting more homogeneous by increasing the distances between the contour lines as much as possible.

Another way to mark absolute values is the *pseudo color representation* already mentioned in Section 7.2.2. With this technique, an area q is mapped to an RGB value triple for display. As color differences are much better recognized by the eye, this method helps to mark absolute values.

Detection of Underflow and Overflow

Underflow occurs when the offset of the digital image output signal is set too close to zero, so that values below zero are cut off by the temporal noise or the *DSNU* (Section 4.5.7). *Overflow* occurs when too much light hits the image sensor. Both effects cause errors during further processing, for example with regard to the average gray value or the center of gravity of objects.

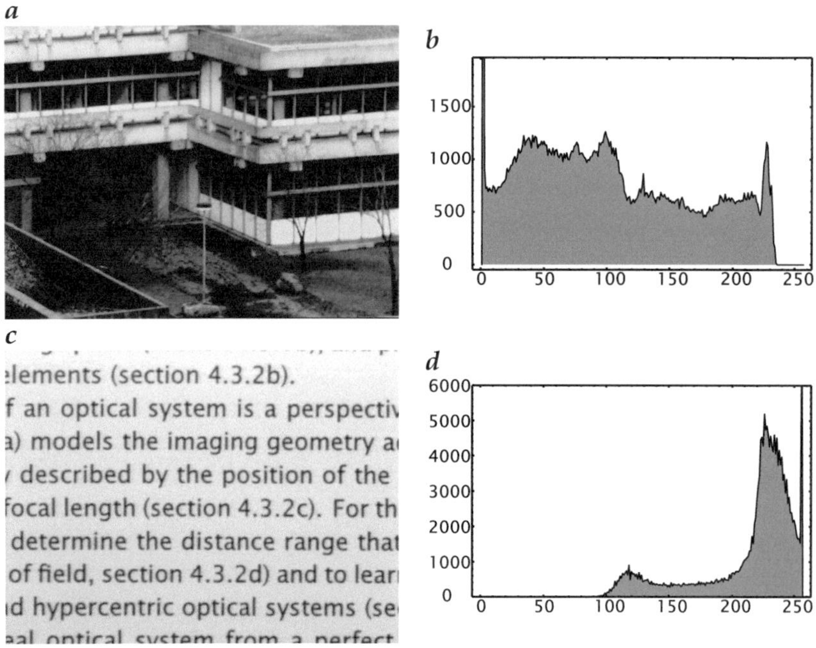

Figure 7.6: *Detection of underflow and overflow in digitized images using histograms:* ***a*** *image with underflow and* ***b*** *its histogram.* ***c*** *image with overflow and* ***d*** *the corresponding histogram.*

Our eyes overlook these incorrect settings in almost all cases. They are only noticed when the corresponding areas no longer show a pattern, but have constant gray values. However, overflow and underflow are recognizable in histograms by pronounced peaks at the minimum and/or maximum value (Fig. 7.6). With pseudo color mapping, the lowest and highest ranges can be displayed in, e. g. blue and red. Then you can immediately recognize values that are "dangerously" close to the limits

They can be avoided by changing the aperture setting of the optics or adjusting the lighting accordingly. It makes little sense to increase the gain of the camera. In Section 4.5.3 we learned that increasing the gain usually only makes the image brighter, but does not increase its quality, which is described by the *signal-to-noise ratio* (*SNR*).

In order to obtain optimal image quality, it is important to adjust the optics and/or the lighting conditions so that we obtain the highest possible values, but without overflow. Close to the *saturation capacity* the SNR is maximum (Section 4.5.6).

Contrast Enhancement

Images are often underexposed due to poor lighting conditions. They are

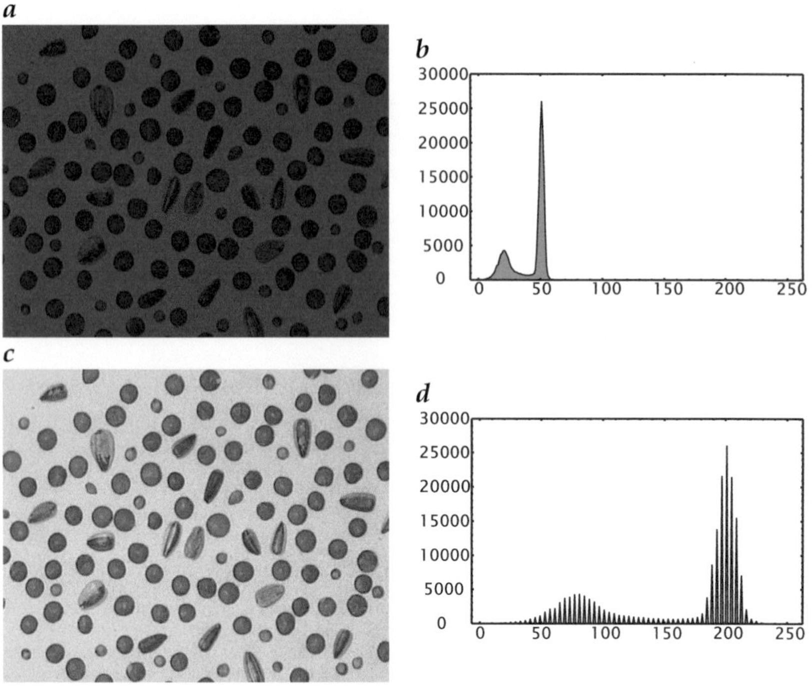

Figure 7.7: Contrast enhancement: ***a*** *underexposed image and* ***b*** *its histogram;* ***c*** *contrast-enhanced image and* ***d*** *its histogram.*

too dark and have a low contrast (Fig. 7.7a). The histogram (Fig. 7.7b) shows that the image only has a small gray value range in the lower gray values.

The appearance of the image improves considerably with a point operation that maps the small gray value range to the full contrast range (for example with the operation: $q' = 4q$ for $q < 64$ and $q' = 255$ for $q \geq 64$) (Fig. 7.7c). However, this only improves our visual impression of the image, but not the *image quality* itself. The histogram in Fig. 7.7d shows that the gray value resolution is still the same.

The best way to improve the *image quality* is to optimize the object lighting with a stronger light source or a more favorable lighting setup. If this is not possible, we can always increase the analog video gain. All modern image processing cards contain an amplifier whose gain and offset can be controlled via software (see Figs. 7.1 and 7.2). Increasing the gain improves the brightness and resolution of the image, but without any gain in information, as the signal-to-noise ratio remains unchanged (Section 4.5.3).

Contrast Spread

It is often necessary to analyze slight differences in illuminance that fall below the resolution of the human visual system. In order to be able to recognize subtle differences, we expand the relevant gray value range.

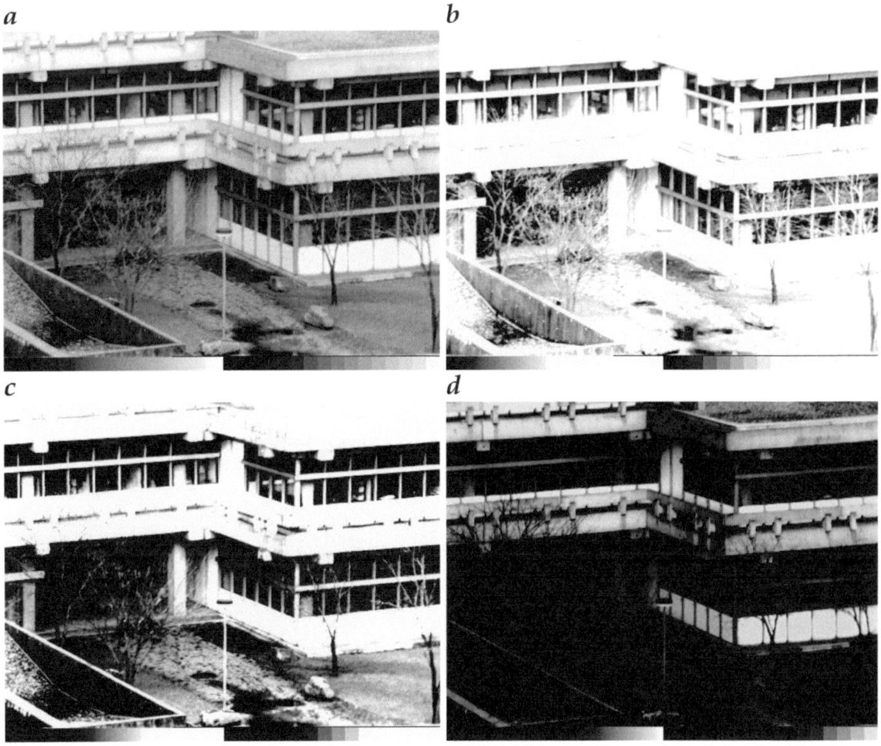

Figure 7.8: ***b** - **d** Contrast spread of the image in **a**. The spread area can be read from the transformation of the gray value wedge at the bottom of the image.*

Then, of course, all gray values outside this range are set to the minimum or maximum gray value. Therefore, the gray values of the object to be analyzed must fall within the range selected for the contrast spread. Examples of contrast spreading are shown in Fig. 7.8. The wedge at the bottom of the image, which ranges from 0 to 255, shows directly which part of the gray value scale has been contrast-enhanced. A continuous wedge can be seen on the left and a wedge with 16 levels on the right.

Compression of the Range

Compared to the human visual system, a digital image has a considerably smaller dynamic range. For a minimum resolution of 10 %, the values must not be less than 10. Therefore, for an 8-bit image, the dynamic range with this minimum resolution is only $255/10 \approx 25$. Due to the low contrast range, digital images have poor quality when scenes with high contrast are recorded. Either the bright parts are bleached out or dark areas are so dark that no details can be recognized.

The dynamic range can be increased using a transformation, which was introduced in Section 5.6.4 as *gamma transformation*. This non-linear homo-

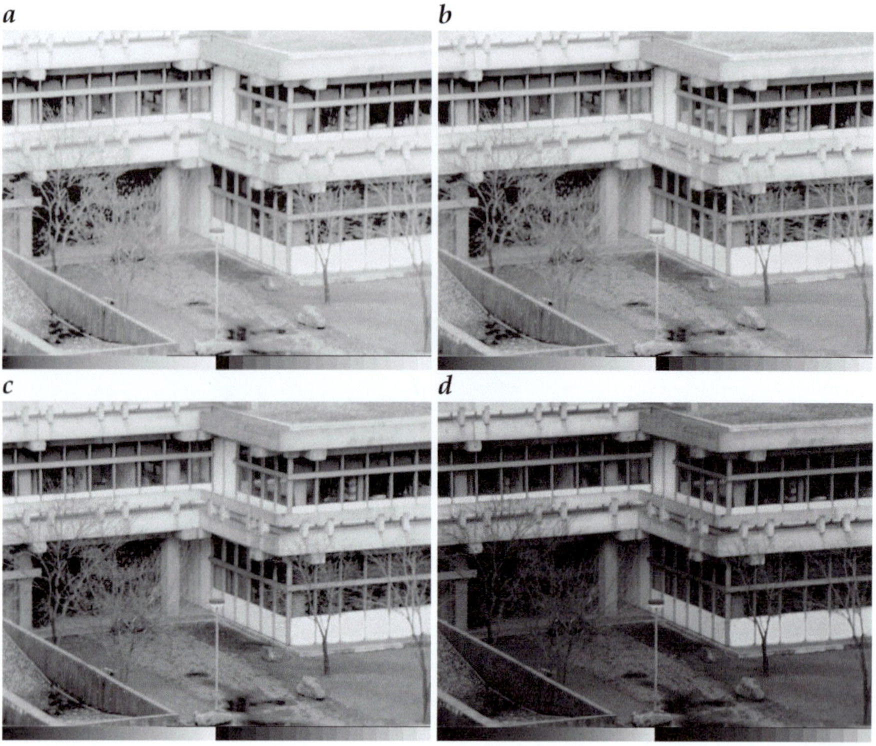

Figure 7.9: *Display of an image with different gamma values:* ***a*** *0.5,* ***b*** *0.7,* ***c*** *1.0 and* ***d*** *2.0.*

geneous point operation has the form

$$q' = \frac{255}{255^{\gamma}} q^{\gamma}. \tag{7.10}$$

The factors in Eq. (7.10) were chosen so that the range $[0, 255]$ is mapped onto itself. This transformation enables the detection of a larger dynamic range at the expense of the resolution in the bright image areas. The dark areas become brighter so that more details can be recognized.

This contrast transformation is more similar to the human visual system's logarithmic perception of brightness. An image displayed with different gamma factors shows Fig. 7.9 together with a linear scale below the image.

7.2.4 Equivalization of the Variance of the Noise

From Section 4.5.5 we know that the variance of the noise in a camera with a linear characteristic curve depends on the image intensity and increases with the photo-induced signal Eq. (4.26), which can be rewritten as follows:

$$\sigma_g^2(g) = \sigma_0^2 + Kg = \sigma_0^2 + (\sigma_{\max}^2 - \sigma_0^2)\frac{g}{g_{\max}}, \tag{7.11}$$

with gray values in the interval $[0, g_{\max}]$ and with $\sigma_{\max}^2 = \sigma_g^2(g\max)$. However, a statistical analysis of the image data and operations is much easier if the noise is independent of the gray value. Only then are the error propagation techniques discussed in Section C.3.3 valid.

A non-linear gray value transformation $h(g)$ can be used to make the variance of the noise independent of the gray value [97]. In a first-order approximation, according to Eq. (C.38), the variance of $h(g)$ results in

$$\sigma_h^2 \approx \left(\frac{\mathrm{d}h}{\mathrm{d}g}\right)^2 \sigma_g^2(g). \tag{7.12}$$

If we set σ_h^2 constant, we get

$$\mathrm{d}h = \frac{\sigma_h}{\sqrt{\sigma^2(g)}}\mathrm{d}g.$$

The integration of the equation results in

$$h(g) = \sigma_h \int_0^g \frac{\mathrm{d}g'}{\sqrt{\sigma^2(g')}} + C. \tag{7.13}$$

With the linear variance model from Eq. (7.11), the integration of Eq. (7.13) results in

$$h(g) = \frac{2\sigma_h}{K}\sqrt{\sigma_0^2 + Kg} + C. \tag{7.14}$$

We use the two free parameters σ_h and C to map the values of h to the interval $[0, h_{\max}]$. This results in the conditions $h(0) = 0$ and $h(g_{\max}) = h_{\max}$ and we obtain

$$\frac{h(g)}{h_{\max}} = \frac{\sqrt{\sigma_0^2 + (\sigma_{\max}^2 - \sigma_0^2)g/g_{\max}} - \sigma_0}{\sigma_{\max} - \sigma_0}, \quad \sigma_h = \frac{h_{\max}}{2g_{\max}}(\sigma_{\max} + \sigma_0). \tag{7.15}$$

The discussion of *quantization* in Section 5.6 showed that a standard deviation of the noise of about half the width of the quantization interval is optimal. It therefore makes sense to specify σ_h. This then results from Eq. (7.15) for $h_{\max}$

$$h_{\max} = 2\sigma_h \frac{g_{\max}}{\sigma_{\max} + \sigma_0} = \frac{2\sigma_h \mathrm{SNR}_{\max}}{1 + \sigma_0/\sigma_{\max}} \approx 2\sigma_h \mathrm{SNR}_{\max}. \tag{7.16}$$

This is an interesting relationship. It states that the necessary range of values in a noise-equivalent signal is given by the maximum *signal-to-noise ratio*. The small correction due to the dark noise is usually negligible.

Example 7.1: Noise equivalization of an sCMOS camera

sCMOS cameras (Section 4.4.3) have image sensors with a higher signal range than standard linear CMOS cameras. How many bits are necessary for such a camera after noise equivalization? We assume the following characteristics:

$$\sigma_d = 3.0\,\text{DN}, \quad \sigma_{\max} = 377\,\text{DN}, \quad \text{SNR}_{\max} = 172, \quad g_{\max} = 65\,000\,\text{DN}.$$

With these values and $\sigma_h = 0.5$, the maximum value for the transformed signal with constant noise is $h_{\max} = 171$. This leads to the astonishing result that, after noise equivalization, less than 8 bit are required to display the signal of a camera that requires 16 bit with a linear signal output without loss [162]. Since the noise fluctuates only a few bits independent of the value and is therefore optimally quantized (Section 5.6), noise equivalent is also a good starting point for almost lossless compression algorithms without affecting the subsequent image processing [159].

Example 7.2: Noise equivalization of an ideal camera

The non-linear gray value transformation is particularly simple for an ideal image sensor with $\sigma_0 = 0$. Then Eqs. (7.15) and (7.16) simplify to:

$$\frac{h(g)}{h_{\max}} = \left(\frac{g}{g_{\max}}\right)^{1/2} \quad \text{and} \quad h_{\max} = 2\sigma_h \frac{g_{\max}}{\sigma_{\max}} = 2\sigma_h \text{SNR}_{\max}. \tag{7.17}$$

If $\sigma_h = 0.5$ is selected, cameras with a maximum SNR of up to 255 can be reduced to 8 bits. If you do not want to determine the standard deviation of the temporal noise from the transformed images, but are only interested in a sufficiently accurate determination of mean values with an error of less than 0.1 DN, you can select the much smaller value $\sigma_h = 0.25$ according to Fig. 5.14. Then even cameras with a maximum SNR of 510 can be reduced to 8 bits.

The *noise equivalization* to a fixed value has another decisive advantage for subsequent image processing. All cameras have the same noise and fixed threshold values can be used for noise-critical evaluations.

The inverse relation $g(h)$ is required if you want to reverse the transformation and display the gray values linearly again. The inversion of Eq. (7.15) results in

$$\frac{g}{g_{\max}} = \frac{h}{h_{\max}} \cdot \frac{(\sigma_{\max} - \sigma_0)h/h_{\max} + 2\sigma_0}{\sigma_{\max} + \sigma_0}. \tag{7.18}$$

It is also interesting to know which interval width Δg of the original signal falls within an interval of the transformed signal. This is obtained by differentiating Eq. (7.18) and with some conversions to

$$\Delta g = \frac{\sqrt{(\sigma_{\max}^2 - \sigma_0^2)g/g_{\max} + \sigma_0^2}}{\sigma_h}, \quad \Delta g \in \left[\frac{\sigma_0}{\sigma_h}, \frac{\sigma_{\max}}{\sigma_h}\right]. \tag{7.19}$$

The equations show that the intervals are wider than one ($\Delta g > 1$) if $\sigma_0 > \sigma_h$.

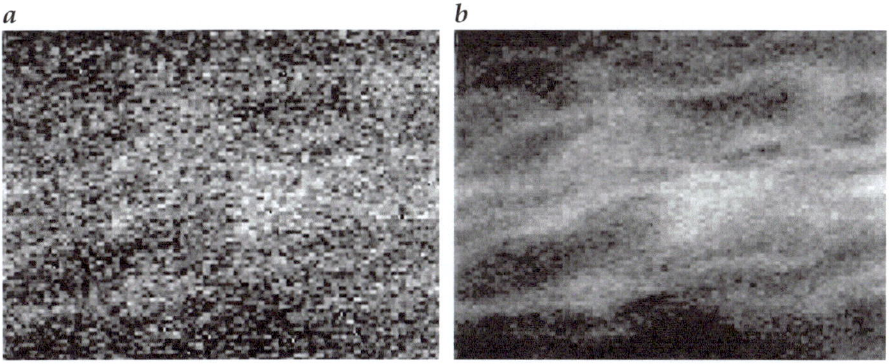

Figure 7.10: *Reduction of noise through image averaging:* ***a*** *Thermal image of small temperature differences that occur on the water surface when water cools down due to evaporation;* ***b*** *the same image, averaged over 16 images; the full gray value range corresponds to a temperature range of 1.1 K.*

7.3 Inhomogeneous Point Operations

Homogeneous point operations, with their independence from the pixel position, are only a subclass of point operations. In general, a point operation also depends on the position of the pixel in the image, and we speak of an *inhomogeneous point operation*. Examples of such functions are calibrations, which are usually point operations.

In general, the calculation of an inhomogeneous point operation is much more time-consuming than that of a homogeneous one. Lookup tables cannot be used because of the dependence on the pixel position. We are forced to calculate the function for each pixel.

The subtraction of a background image without objects or lighting is a simple example of an inhomogeneous point operation, which is written as follows:

$$g'_{mn} = P_{mn}(g_{mn}) = g_{mn} - b_{mn}, \tag{7.20}$$

where b_{mn} is a pixel of the background image.

7.3.1 Image Averaging

One of the simplest inhomogeneous point operations is *image averaging*. A number of image acquisition techniques result in high noise levels. Well-known examples are *thermal images* (Section 2.6.1) and all applications in which only a small number of photons are detected (see Fig. C.2 and Appendix C.4.1).

Fig. 7.10a shows the temperature differences of the water surface caused by evaporation in a wind/wave channel at a wind speed of 1.8 m/s. Due to the considerable noise level, the small temperature differences can hardly be detected. If the average value is calculated from several images, the noise level is significantly reduced (Fig. 7.10b). The error of the mean value (Section C.3.3) from K images is

$$\sigma^2_{\overline{G}} \approx \frac{1}{(K-1)}\sigma^2_G = \frac{1}{K(K-1)}\sum_{k=0}^{K-1}(G_k - \overline{G})^2. \tag{7.21}$$

If we calculate the average of K images, the noise level is reduced to $1/\sqrt{K}$ compared to a single image. When averaging over 16 images, the noise level is therefore reduced to 1/4. From our considerations on quantization in Section 5.6.2, we know that average values in digital signals can be calculated under optimal conditions with an accuracy far below the width of the quantization interval.

7.3.2 Correction of Inhomogeneous Lighting

Every application is affected by *uneven lighting* of the observed scene. Even if we go to great lengths to optimize the lighting setup, it is difficult to achieve perfectly uniform object lighting. Another annoying problem is small dust particles in the optical path, especially on the glass window in front of the CCD sensor. As the window is a certain distance from the sensor, these particles, if they are not too large, are projected onto the sensor plane so blurred that they are not directly visible. However, they absorb light and thus reduce the illumination in a small circular area. These effects are not easy to see in a scene with high contrast and many details. However, they are clearly visible on a uniform background (Fig. 7.12b). Some image sensors, especially cheap CMOS sensors, also show considerable uneven sensitivity of the individual photoreceptors, which also contributes to the unevenness of the image.

These disturbances can significantly impair the image quality. They make it difficult to separate an object from the background and lead to systematic errors in the subsequent image evaluation. Nevertheless, it is possible to correct these interference effects if we know the nature of the interference and are able to take suitable reference images. In the following, we consider a simple example. We assume that the gray value in the image is the product of the inhomogeneous illuminance and the object reflectivity or transmissivity. We also assume that we can take a reference image without an absorbing object or with an object of constant reflectivity. Such a reference image can also be obtained if small objects are randomly distributed in the images by calculating an average image from many images. The non-uniform illumination can now be corrected by dividing by the reference image $\boldsymbol{R}$:

$$\boldsymbol{G}' = c \cdot \boldsymbol{G}/\boldsymbol{R}. \tag{7.22}$$

Multiplication by the constant c is necessary to display the normalized image in integer numbers again. If the objects absorb light, c is normally selected close to the maximum integer number. Fig. 7.11e shows that an effective suppression of uneven lighting is possible with this simple method.

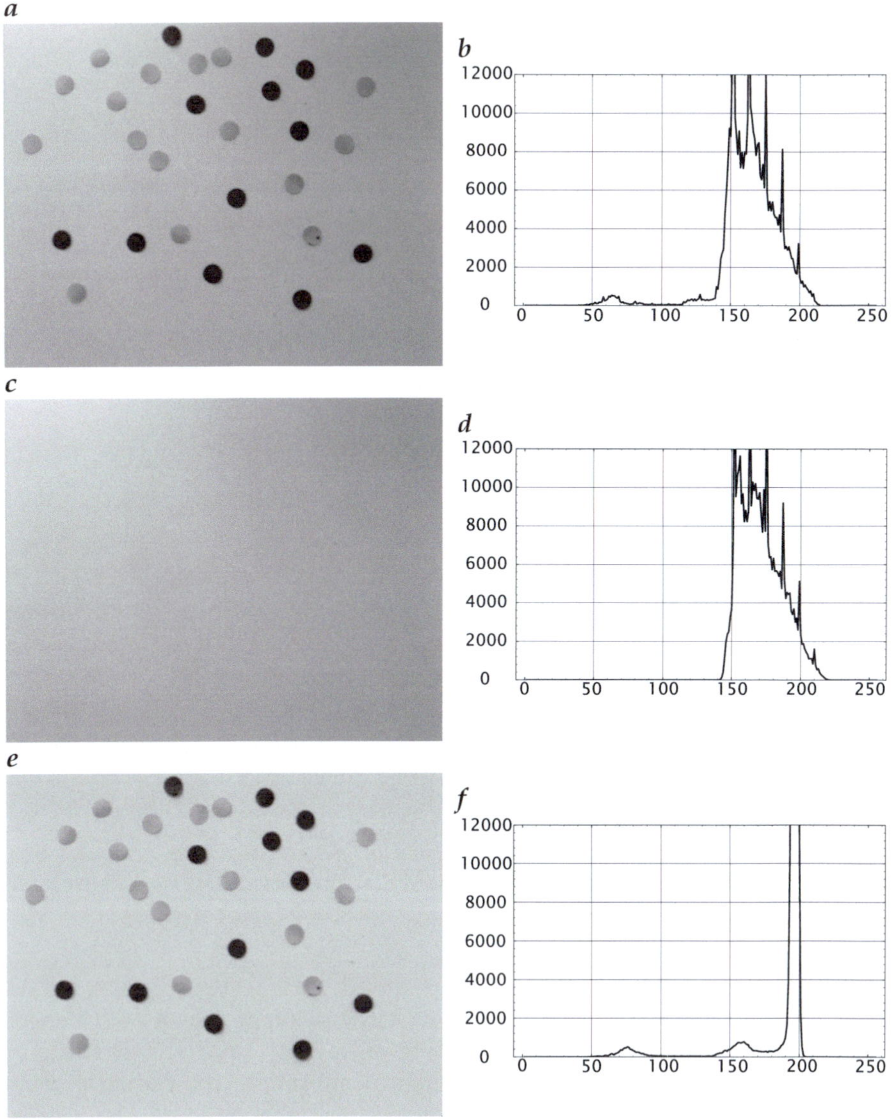

Figure 7.11: *Correction of uneven lighting via an inhomogeneous point operation:* ***a*** *original image and* ***b*** *its histogram;* ***c*** *background image (reference image) and* ***d*** *the corresponding histogram;* ***e*** *image divided by the background image and* ***f*** *the histogram of this result image.*

7.3.3 Radiometric Two-point Calibration

The simple ratio formation can no longer be used if the image is not zero even without illumination, but shows a residual pattern ("fixed-pattern noise"), which may be caused by a zero current that differs from pixel to pixel. In

a b

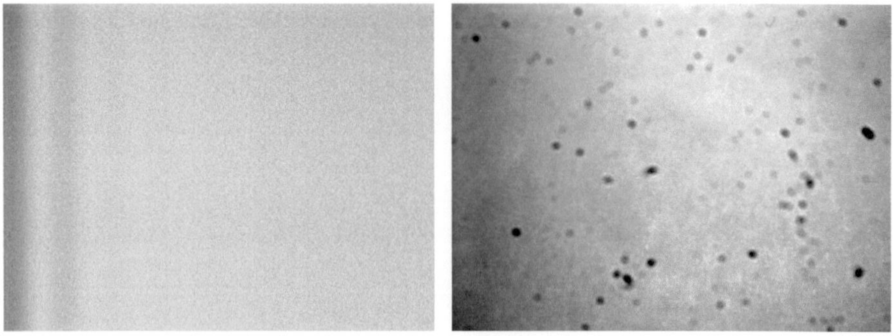

Figure 7.12: *Contrast-enhanced **a** dark image and **b** reference image of a CCD camera with analog output signal for a radiometric two-point calibration.*

a b

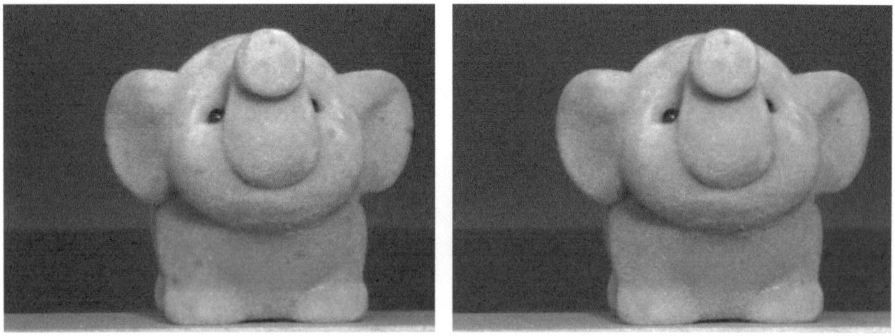

Figure 7.13: *Radiometric two-point calibration with the dark and reference image from Fig. 7.12: **a** original image and **b** calibrated image; the dark spots are no longer visible in the latter.*

such a case, two reference images are required. This technique is also used for a simple radiometric *two-point calibration* for an image sensor with a linear characteristic. Some image processing applications require an absolute or relative *radiometric calibration.* With such a calibration, the radiance of an object can be calculated from the measured gray value.

First we take a dark image $\boldsymbol{B}$ without illumination and then a reference image $\boldsymbol{R}$ with an object that has a constant radiant intensity, for example an *integrating sphere.* Using these two images, we can then obtain a radiometrically calibrated image using the following operation

$$\boldsymbol{G}' = c\frac{\boldsymbol{G} - \boldsymbol{B}}{\boldsymbol{R} - \boldsymbol{B}}. \tag{7.23}$$

Fig. 7.12 shows a contrast-enhanced dark and reference image of a CCD camera with analog output. Typical signal interference can be seen. The signal oscillation on the left edge of the dark image is the result of electronic signal interference. The dark spots in the reference image are caused by dust on the glass window in front of the CCD sensor. The improvement due to the radiometric calibration can be clearly seen in Fig. 7.13.

7.3.4 ♯Non-linear Radiometric Calibration

Sometimes the size that is determined with an image sensor has a non-linear relationship to the measured gray value. A good example of this is thermography. Here, the temperature of the emitting object is determined from the radiation density using Planck's equations (Section 2.6.1).

Here we show a practical calibration procedure for ambient temperatures. Due to the non-linear relationship between radiation and temperature, a simple two-point calibration with linear interpolation is not sufficient for such an application. Haußecker [125] shows that a quadratic relationship is sufficiently accurate for a small temperature range of 0 to 40° C. Therefore, three calibration temperatures are necessary, which are generated by a special *blackbody calibration unit*.

From the recorded calibration images $\boldsymbol{G}_1$, $\boldsymbol{G}_2$ and $\boldsymbol{G}_3$ with known temperatures T_1, T_2 and T_3, respectively, the temperature image $\boldsymbol{T}$ of any image $\boldsymbol{G}$ can be calculated using quadratic interpolation:

$$\boldsymbol{T} = \frac{\Delta\boldsymbol{G}_2 \cdot \Delta\boldsymbol{G}_3}{\Delta\boldsymbol{G}_{21} \cdot \Delta\boldsymbol{G}_{31}} T_1 - \frac{\Delta\boldsymbol{G}_1 \cdot \Delta\boldsymbol{G}_3}{\Delta\boldsymbol{G}_{21} \cdot \Delta\boldsymbol{G}_{32}} T_2 + \frac{\Delta\boldsymbol{G}_1 \cdot \Delta\boldsymbol{G}_2}{\Delta\boldsymbol{G}_{31} \cdot \Delta\boldsymbol{G}_{32}} T_3 \qquad (7.24)$$

with

$$\Delta\boldsymbol{G}_k = \boldsymbol{G} - \boldsymbol{G}_k \quad \text{and} \quad \Delta\boldsymbol{G}_{kl} = \boldsymbol{G}_k - \boldsymbol{G}_l. \qquad (7.25)$$

The symbol $\cdot$ stands for the point-by-point multiplication of the image to distinguish it from the matrix multiplication. Fig. 7.14a, b and c show three calibration images. The infrared camera looks at the calibration body via a mirror, which slightly cuts off the field of view at the edges of the image. This is the reason for the sharp temperature changes that can be seen at the image boundaries in Fig. 7.14a, c. The calibration procedure removes the remaining inhomogeneities (Fig. 7.14d, f) that are visible in the original image.

7.3.5 Window Functions

The so-called *window function* is another important application of inhomogeneous point operations. Before we can calculate the DFT of an image, the image must be multiplied by a window function. If we omit this step, the spectrum is perturbed by the convolution of the image spectrum with the Fourier transform of the rectangular function, the sinc function (see Section 6.3, ≻ R13). As a result, peaks in the spectrum become star-like patterns along the coordinate axes in Fourier space (Fig. 7.15b). We can also explain these perturbations with the virtual periodic repetition of finitely extended

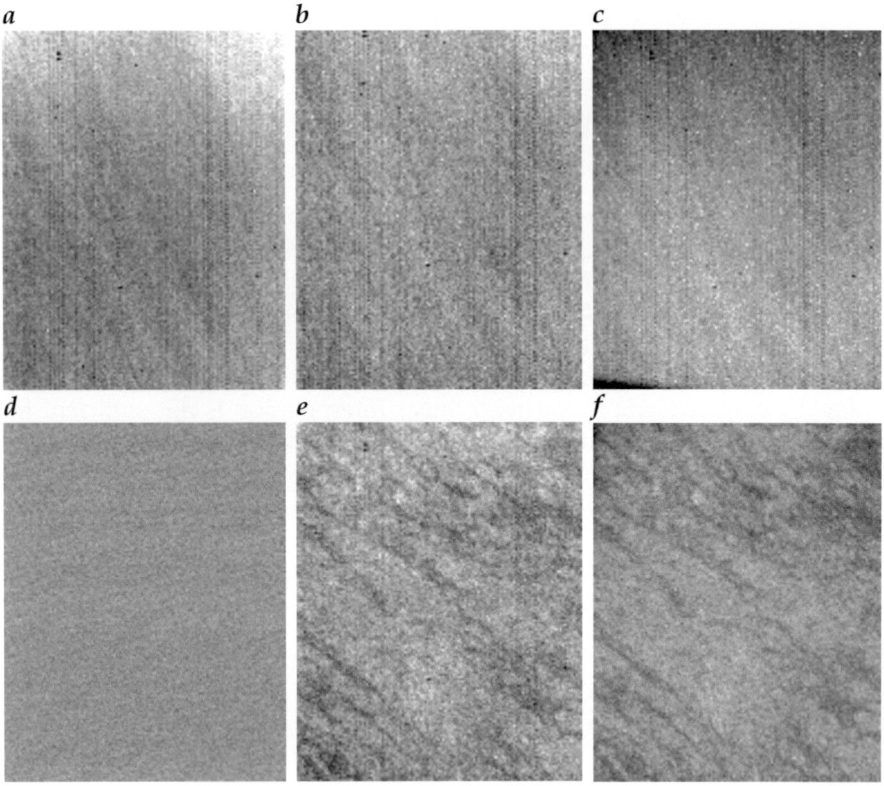

Figure 7.14: *Three-point calibration of infrared temperature images:* ***a*** *to* ***c*** *show images of calibration bodies in the form of aluminum blocks at temperatures of 13.06, 17.62 and 22.28 °C. The gray values of the images are spread over a small range of the infrared camera's 12-bit digital output range of* ***a*** *1715–1740,* ***b*** *1925–1950,* ***c*** *2200–2230 and still show residual inhomogeneities, especially vertical stripes.* ***d*** *Image calibrated by means of quadratic interpolation from the three images* ***a*** *to* ***c*** *.* ***e*** *Original and* ***f*** *calibrated image of the small-scale temperature differences at the ocean surface on an area of about* $0.8 \times 1.0\,m^2$.

images, as explained in more detail in connection with the sampling theorem in Section 5.2.3. The periodic repetition leads to discontinuities at the horizontal and vertical edges of the image, which generate correspondingly high spectral densities along the $x-$ and $y-$ axes in wavenumber space.

To avoid these discontinuities, we need to multiply the image by a window function that gradually decreases to zero towards the edges of the image. An optimal window function should preserve high spectral resolution while minimizing the perturbations of the spectrum, i. e. its DFT should decrease as steeply as possible. However, these are contradictory requirements, as good spectral resolution requires a wide window function. However, such a window falls off steeply at the edges and thus causes a slow drop in the side maxima of its spectrum.

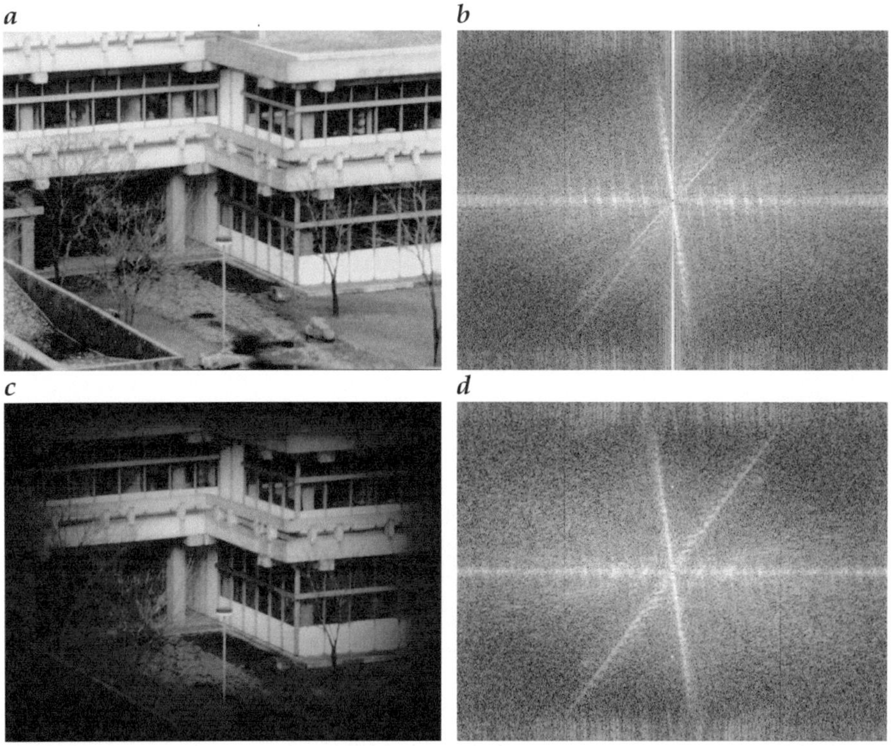

Figure 7.15: *Effect of the window function on the discrete Fourier transform:* ***a*** *original image;* ***b*** *DFT of* ***a*** *without using a window function;* ***c*** *image multiplied by a cosine window;* ***d*** *DFT of* ***c*** *using a cosine window.*

A carefully chosen window is critical for the spectral analysis of time series [214, 234]. However, it is less critical in digital image processing because of the much smaller dynamic range of the gray values. The simple cosine window

$$W_{mn} = \sin\left(\frac{\pi m}{M}\right)\sin\left(\frac{\pi n}{N}\right), \quad 0 \le m < M, \ 0 \le n < N \tag{7.26}$$

is well suited as a window function (Fig. 7.15c and d).

A direct implementation of the window operation is very time-consuming, as the trigonometric function has to be calculated $2MN$ times. It is much more efficient to calculate the window function once, save the window image and then use it to calculate many DFTs. The memory requirements are reduced if we take into account that the window function Eq. (7.26) is separable, i. e. a product of two functions: $W_{m,n} = {}^c w_m \cdot {}^r w_n$. Then we only need to calculate and store the $M+N$ values for the column function ${}^c w_m$ and the row function ${}^r w_n$. The price for the reduced memory space is an additional multiplication per pixel for the window operation.

7.4 ♯Multi-channel Point Operations

7.4.1 ♯Definitions

Point operations can be generalized to *multichannel point operations*. The operation still depends only on the value of the single pixel, but the input value can be a vector instead of a scalar, and likewise the output image can be a multichannel image. For homogeneous point operations that do not depend on the position of the pixel in the image, we can write

$$\boldsymbol{G}' = \boldsymbol{P}(\boldsymbol{G}) \quad \text{with} \quad \begin{array}{l} \boldsymbol{G}' = \left[G'_0,\ G'_1,\ \ldots,\ G'_l,\ \ldots,\ G'_{L-1}\right], \\ \boldsymbol{G} = \left[G_0,\ G_1,\ \ldots,\ G_k,\ \ldots,\ G_{K-1}\right], \end{array} \tag{7.27}$$

where G'_l and G_k are the components l and k of the multichannel images $\boldsymbol{G}'$ and $\boldsymbol{G}$ with L and K channels, respectively.

An important subclass of multi-component point operations are linear operations. Each component of the output image $\boldsymbol{G}'$ in Eq. (7.27) is a linear combination of the components of the input image $\boldsymbol{G}$:

$$G'_l = \sum_{k=0}^{K-1} P_{lk} G_k. \tag{7.28}$$

Here P_{lk} are constant coefficients. Therefore, a general linear multi-component point operation is given by a matrix of coefficients P_{lk}. Then we can write Eq. (7.28) abbreviated with a coefficient matrix $\boldsymbol{P}$:

$$\boldsymbol{G}' = \boldsymbol{P}\boldsymbol{G}. \tag{7.29}$$

If the components of the multi-channel images are not dependent on each other in a point operation, all coefficients in $\boldsymbol{P}$ except those on the main diagonal become zero. For K-channel input and output images, only K different point operations remain, one for each channel. A vector point operation eventually reduces to a standard scalar point operation when the same point operation is applied to each channel of a multi-component image.

With the same number of output and input images, linear point operations can be interpreted as coordinate transformations. If the matrix of coefficients in Eq. (7.29) has a rank $R < K$, the multichannel point operation projects the K-dimensional space onto an R-dimensional subspace.

In general, linear multichannel point operations are relatively easy to handle, as they can be described using the concepts of linear algebra. For quadratic matrices, for example, we can easily specify the condition when an inverse operation exists for a multichannel point operation and calculate it.

For nonlinear multicomponent point operations, the linear coefficients in Eqs. (7.28) and (7.29) must be replaced by nonlinear functions:

$$G'_l = P_l(G_0, G_1, \ldots, G_{K-1}). \tag{7.30}$$

Non-linear multi-component point operations cannot be treated like linear operations in general, but must be treated individually. The complexity can be reduced considerably if a multi-channel point operation can be separated into its linear and non-linear parts.

7.4.2 ♯Dyadic Point Operations

Operations involving two images, $\boldsymbol{G}$ and $\boldsymbol{H}$, are called *dyadic point operations*. Each dyadic image operation can be written as

$$G'_{mn} = P(G_{mn}, H_{mn}). \tag{7.31}$$

Dyadic homogeneous point operations can be implemented as LUT operations. If the gray values of the two input images Q assume different values, there are Q^2 combinations of input parameters and thus just as many different output values. This means that 64k values must be calculated for 8-bit images. For a 512×512 image, this is still a quarter less than if each pixel were calculated directly. All possible results of the dyadic operation can be stored in a large LUT L with $Q^2 = 64\,\text{k}$ entries as follows:

$$L(2^8\, p + q) = P(p, q), \quad 0 \leq p, q < Q. \tag{7.32}$$

The upper and lower byte of the LUT address are given by the gray values of the images G and H respectively.

Some image processing systems contain a 16-bit LUT as a modular processing element. The calculation of a dyadic point operation either via a hardware integrated or software LUT is often significantly faster than a direct implementation, especially if the operation is complex. In addition, it is then easier to monitor exceptions such as division by zero or underflow and overflow.

A dyadic point operation can be used to perform two point operations simultaneously. For example, the magnitude and phase of a complex-valued image (r, i) can be calculated simultaneously with a single dyadic LUT operation if we limit the output to 8 bits:

$$L(2^8\, r + i) = 2^8\, \sqrt{r^2 + i^2} + \frac{128}{\pi} \arctan\left(\frac{i}{r}\right), \quad 0 \leq r, i < Q. \tag{7.33}$$

The magnitude is stored in the upper byte and the phase — scaled to the interval $[-128, 127]$ — in the lower byte.

7.5 Geometric Transformations

In the remaining part of this chapter, we deal with geometric operations as operations complementary to point operations. After basic considerations in Section 7.5.1, we discuss elementary geometric transformations such as affine mapping (Section 7.5.2) and perspective projection (Section 7.5.3) and investigate how to obtain the transformation parameters by point mapping methods (Section 7.5.4). Finally, we briefly discuss fast algorithms for geometric transformations in Section 7.5.5.

Geometric operations have many applications. These include, among others, the

- Correction of geometric distortions through optical imaging (Section 3.4.6),
- Scaling and rotation of an object in the image to a standard position for further analysis of its shape
- Alignment of the epipolar lines during stereo image processing (Section 8.5.1),
- image or partial image shifting, rotation and scaling for better image sequence analysis (Chapter 14) and
- comparable operations for various other 3D image acquisition methods and
- projection of textures onto 3D objects in *computer graphics*.

7.5.1 Forward and Backward Mapping

Geometric transformations define the relationship between the points of two images. This relationship can be expressed in two ways. Either the coordinates of the output image $\boldsymbol{x}'$ are specified as a function of the input coordinates $\boldsymbol{x}$ or vice versa:

$$\boldsymbol{x}' = M(\boldsymbol{x}) \quad \text{or} \quad \boldsymbol{x} = M^{-1}(\boldsymbol{x}'), \tag{7.34}$$

where M stands for the mapping function and M^{-1} for its inverse function. The two equations in Eq. (7.34) show two principal ways of geometric transformation, the *forward* and the *backward mapping*.

With *forward mapping*, a point of the input image is mapped onto the output image (Fig. 7.16a). In general, the image point lies between the points of the output image. With this method, it is not sufficient to simply assign the value of the input image point to the nearest image point in the output image (point-to-point or nearest neighbor mapping). It can then happen that the transformed image contains holes, as no single value is assigned to a point in the output image, or that a point in the output image is assigned a value several times. It is therefore necessary to distribute the value of the input image point to several output image points. The easiest way to do this is to consider the pixels as squares and use the proportion of the input pixel's area that falls on the output pixel as a weighting factor. For each output pixel, the individual proportions of the input pixels that fall on the output pixel are then added up. If the image is continuous, the output pixels are completely covered.

With *inverse mapping*, the coordinates of a point in the output image are mapped back to the input image (Fig. 7.16b). The advantage is that this method avoids holes and overlaps in the output image, as all pixels are

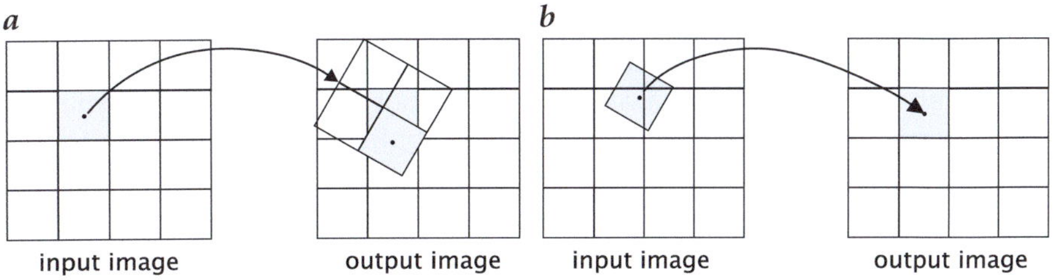

Figure 7.16: *Illustration* ***a*** *of the forward and* ***b*** *of the backward mapping in a spatial image transformation.*

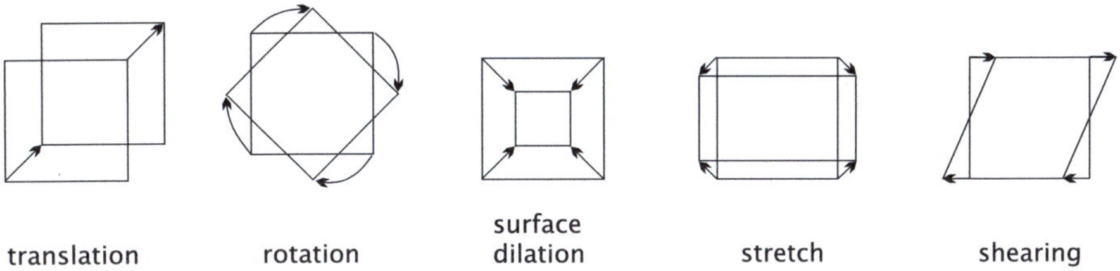

Figure 7.17: *Elementary geometric transformations of a planar surface element.*

scanned one after the other. However, the interpolation problem now occurs in the input image. The coordinates of the output image generally do not hit a point in the input image, but lie in between. The correct values must therefore be interpolated from the surrounding pixels. As a rule, backward mapping is the more flexible technique, as it is easier to implement different interpolation techniques.

7.5.2 Affine Mapping

An *affine mapping* is a linear coordinate transformation that comprises the elementary transformations *translation, rotation, dilation, compression* and *shear* (Fig. 7.17). It can be expressed by vector addition and matrix multiplication:

$$\begin{bmatrix} x' \\ y' \end{bmatrix} = \begin{bmatrix} a_{11} & a_{12} \\ a_{21} & a_{22} \end{bmatrix} \begin{bmatrix} x \\ y \end{bmatrix} + \begin{bmatrix} t_x \\ t_y \end{bmatrix}. \tag{7.35}$$

With *homogeneous coordinates* (Section 3.3.3) the affine mapping can be described with a single matrix multiplication:

$$\begin{bmatrix} x' \\ y' \\ 1 \end{bmatrix} = \begin{bmatrix} a_{11} & a_{12} & t_x \\ a_{21} & a_{22} & t_y \\ 0 & 0 & 1 \end{bmatrix} \begin{bmatrix} x \\ y \\ 1 \end{bmatrix}. \tag{7.36}$$

An affine mapping has six degrees of freedom: two for the translation (t_x, t_y) and four (a_{11}, a_{12}, a_{21} and a_{22}), which include *rotation*, dilation, compression and shear. The affine mapping maps a triangle onto a triangle and a rectangle onto a parallelogram. It is therefore also referred to as *tripoint mapping*. However, these properties also limit the use of the affine transformation. More general deformations, such as the mapping of a rectangle onto any quadrilateral, are not affine mappings.

7.5.3 Perspective Projection

The *perspective projection* is, as discussed in Section 3.3, the basis of all optical imaging systems. The affine image corresponds to a parallel projection and can only be used as a model of optical imaging for the limiting case of a small field of view. The more general perspective projection is most simply formulated with homogeneous coordinates:

$$\begin{bmatrix} w'x' \\ w'y' \\ w' \end{bmatrix} = \begin{bmatrix} a_{11} & a_{12} & a_{13} \\ a_{21} & a_{22} & a_{23} \\ a_{31} & a_{32} & 1 \end{bmatrix} \begin{bmatrix} wx \\ wy \\ w \end{bmatrix} \quad \text{or} \quad X' = PX. \tag{7.37}$$

The two coefficients a_{31} and a_{32} not contained in the affine mapping Eq. (7.36) describe the perspective projection Eq. (3.15) in Section 3.3.3).

Written in standard coordinates, it can be seen that the perspective projection — in contrast to the affine mapping — is a non-linear transformation according to Eq. (7.37):

$$\begin{aligned} x' &= \frac{a_{11}x + a_{12}y + a_{13}}{a_{31}x + a_{32}y + 1} \\ y' &= \frac{a_{21}x + a_{22}y + a_{23}}{a_{31}x + a_{32}y + 1}. \end{aligned} \tag{7.38}$$

However, the perspective projection can be reduced to a linear transformation when using homogeneous coordinates. A perspective projection maps lines onto lines; however, only lines that are parallel to the projection plane remain parallel. A rectangle is mapped onto a general quadrilateral. Therefore, the perspective projection is also referred to as *quadrilateral mapping*.

7.5.4 ♯Determination of the Transformation Coefficients

The coefficients of a transformation as described in Sections 7.5.2 and 7.5.3 are generally not known. Instead, we usually have a set of corresponding points from the object space and the image space. In this section, we will learn how to calculate the coefficients of a transformation using the corresponding points. For an affine mapping, we need three non-collinear points to be able to map a triangle onto a triangle. With these three points, Eq. (7.36) results in the following system of linear equations:

$$\begin{bmatrix} x'_1 & x'_2 & x'_3 \\ y'_1 & y'_2 & y'_3 \\ 1 & 1 & 1 \end{bmatrix} = \begin{bmatrix} a_{11} & a_{12} & t_x \\ a_{21} & a_{22} & t_y \\ 0 & 0 & 1 \end{bmatrix} \begin{bmatrix} x_1 & x_2 & x_3 \\ y_1 & y_2 & y_3 \\ 1 & 1 & 1 \end{bmatrix} \tag{7.39}$$

or

$$\boldsymbol{P}' = \boldsymbol{A}\boldsymbol{P}. \tag{7.40}$$

From this, $\boldsymbol{A}$ is calculated as

$$\boldsymbol{A} = \boldsymbol{P}'\boldsymbol{P}^{-1}. \tag{7.41}$$

The inverse of the matrix $\boldsymbol{P}$ exists if the three points $\boldsymbol{X}_1$, $\boldsymbol{X}_2$ and $\boldsymbol{X}_3$ are linearly independent. In the geometric interpretation, this means that they must not lie on a line, i. e. they are not collinear.

With more than three corresponding points, the parameters of the affine mapping can be solved by the following system of equations using the *least squares* method (Section C.6):

$$\boldsymbol{A} = \boldsymbol{P}'\boldsymbol{P}^T(\boldsymbol{P}\boldsymbol{P}^T)^{-1} \tag{7.42}$$

with

$$\begin{aligned} \boldsymbol{P}'\boldsymbol{P}^T &= \begin{bmatrix} \sum x'_n x_n & \sum x'_n y_n & \sum x'_n \\ \sum y'_n x_n & \sum y'_n y_n & \sum y'_n \\ \sum x_n & \sum y_n & N \end{bmatrix} \\ \boldsymbol{P}\boldsymbol{P}^T &= \begin{bmatrix} \sum x_n^2 & \sum x_n y_n & \sum x_n \\ \sum x_n y_n & \sum y_n^2 & \sum y_n \\ \sum x_n & \sum y_n & N \end{bmatrix}. \end{aligned}$$

The inverse of an affine mapping is itself an affine mapping. The transformation matrix of the inverse transformation is given by the inverse matrix A^{-1}.

Determining the coefficients for the perspective projection is somewhat more complex. If there are four or more corresponding points, the coefficients of the perspective transformation can be calculated. We then reshape Eq. (7.38) and obtain

$$\begin{aligned} x' &= a_{11}x + a_{12}y + a_{13} - a_{31}xx' - a_{32}yx' \\ y' &= a_{21}x + a_{22}y + a_{23} - a_{31}xy' - a_{32}yy'. \end{aligned} \tag{7.43}$$

For N points, this leads to a linear system of equations with $2N$ equations and 8 unknowns of the form

$$
\begin{bmatrix} x'_1 \\ y'_1 \\ x'_2 \\ y'_2 \\ \vdots \\ x'_N \\ y'_N \end{bmatrix}
=
\begin{bmatrix}
x_1 & y_1 & 1 & 0 & 0 & 0 & -x_1x'_1 & -y_1x'_1 \\
0 & 0 & 0 & x_1 & y_1 & 1 & -x_1y'_1 & -y_1y'_1 \\
x_2 & y_2 & 1 & 0 & 0 & 0 & -x_2x'_2 & -y_2x'_2 \\
0 & 0 & 0 & x_2 & y_2 & 1 & -x_2y'_2 & -y_2y'_2 \\
 & & & & \vdots & & & \\
x_N & x_N & 1 & 0 & 0 & 0 & -x_Nx'_N & -y_Nx'_N \\
0 & 0 & 0 & x_N & y_N & 1 & -x_Ny'_N & -y_Ny'_N
\end{bmatrix}
\begin{bmatrix} a_{11} \\ a_{12} \\ a_{13} \\ a_{21} \\ a_{22} \\ a_{23} \\ a_{31} \\ a_{32} \end{bmatrix}
$$

or

$$
\boldsymbol{d} = \boldsymbol{M}\boldsymbol{a}. \tag{7.44}
$$

The equation can then be solved as follows using the method of *least squares*:

$$
\boldsymbol{a} = (\boldsymbol{M}^T\boldsymbol{M})^{-1}\boldsymbol{M}^T\boldsymbol{d}. \tag{7.45}
$$

7.5.5 ♯Fast Algorithms for Geometric Transforms

After the detailed treatment of interpolation, we have the basics to develop fast algorithms for the various geometric transformations. Basically, all fast interpolation algorithms are based on the following two principles: Efficient computation and utilization of interpolation coefficients and partitioning into one-dimensional geometric transformations.

First, many calculations are required to compute the interpolation coefficients for partial displacements, since each displacement requires different interpolation coefficients. So we have to construct the transformations in such a way that we only need constant displacements for a particular transformation process. If this is not possible, it may still be efficient to first calculate the interpolation coefficients for different partial displacements and save them for later use.

Then we have seen in Section 5.5.1 that interpolation is a separable procedure. If this basic fact is used, the number of operations can be reduced considerably. In most cases, it is possible to decompose the two- and higher-dimensional geometric transformations into a series of 1D transformations.

7.6 ♯References for Further Reading

Holst [138, 139] and Biberman [18] deal with the radiometric calibration of image sensors and cameras in the visible and infrared wavelength range. A detailed treatment of geometric transformations can be found in the monograph by Wolberg [355].

8 3D Image Acquisition and Light Fields

Contents

8.1 Introduction: Depth Maps versus Volume Images

There is a well-known saying that "a picture is worth a thousand words". But how much does a two-dimensional picture of a three-dimensional scene say about it? In this chapter we will learn: surprisingly little.

Here we will look at various recording techniques that can be used to reconstruct the depth information that is lost when a 3D object is projected onto an image plane. There are basically two types of 3D imaging techniques. They can either only determine the depth of a surface in space or completely

B. Jähne, *Digital Image Processing and Image Acquisition*,
https://doi.org/10.1007/978-3-032-04806-6_8

reconstruct a 3D object. Often, both the acquisition of a *depth image* and that of a *volume image* are called *3D image acquisition*.

This has led to a lot of confusion. Even more serious is the confusion caused by the multitude of methods for obtaining depth and volume images. Therefore, in this chapter we will not deal with the details of different techniques, but focus on the basic principles. It may sound surprising: Despite the great variety of 3D imaging techniques, they are ultimately based on only a few fundamental principles. Once you know them, it is easy to understand how a particular technique works, how accurate it can be and what its limitations are.

We start with a discussion of the basic limits of projective mapping for three-dimensional vision (Section 8.2). Then we deepen these relationships theoretically by introducing the light field. This allows us to understand that all techniques to either recover the third dimension can be understood as different sparse sampling of the light field (Section 8.3). This then allows us to give a brief overview of the basic principles of depth imaging and volume imaging (Section 8.4).

The further presentation of 3D imaging techniques is divided into two parts. In this chapter, we will only deal with passive methods, i. e. methods in which the lighting conditions of a scene are taken as they are. These include methods of passive triangulation measurement (Section 8.5), which include stereo methods (Section 8.5.1), as well as the passive measurement of the inclination of surfaces (Section 8.6). Methods for direct passive depth measurement are not possible. The majority of 3D imaging techniques actively manipulate the illumination. These methods are discussed in Chapter 9.

8.2 Limits of Projective Mapping

8.2.1 3D PSF and 3D OTF

From Section 3.5 we know that the generation of a 2D image in the image plane is completely described by its point response (PSF) or optical transfer function (OTF). In the following, we will extend this concept into space and explicitly calculate the 3D PSF and 3D OTF within the limits of geometric optics, i. e. with a perfect lens system and without distortion. This approach is based on the need to understand three-dimensional imaging, especially in microscopy. The key question is how a point in 3D object space is mapped not only onto a 2D image plane, but into a 3D image space.

First, we consider how a point in object space is projected into image space. From Fig. 3.9 and Eq. (3.30), the diameter of the blur circle results in

$$u' = \frac{a}{f + d'} \Delta z'. \tag{8.1}$$

Then we replace the diameter a of the aperture by the maximum angle at which the lens captures light from the point under consideration and obtain

$$u' = \tan\theta' \, \delta z'. \tag{8.2}$$

This equation provides the edge of the PSF in image space. This boundary has the shape of a double cone with the z-axis in the center. The tips of the two cones meet at the origin. Outside the cones the PSF is zero, inside we can derive the intensity from the conservation of radiant energy.

As the radius of the cone increases linearly with the distance to the focal plane, the intensity inside the cone decreases quadratically. This results in the PSF $h_i(\boldsymbol{x})$ in image space as follows

$$\begin{aligned} h_i(\boldsymbol{x}') &= \frac{I_0}{\pi(\Delta z' \tan\theta')^2} \Pi\left(\frac{(x'^2 + y'^2)^{1/2}}{2\Delta z' \tan\theta'}\right) \\ &= \frac{I_0}{\pi(\Delta z' \tan\theta')^2} \Pi\left(\frac{r'}{2\Delta z' \tan\theta'}\right). \end{aligned} \tag{8.3}$$

I_0 is the light intensity that the lens collects from the point object; Π is the *rectangular function*, which is defined as follows:

$$\Pi(x) = \begin{cases} 1 & |x| \leq 1/2 \\ 0 & \text{sonst} \end{cases}. \tag{8.4}$$

The last expression in Eq. (8.3) is written in cylindrical coordinates (r', θ, z') so that the rotational symmetry of the PSF around the z' axis is taken into account.

In the second step, we discuss how the mapping effects are projected back into the object space. The lateral and axial magnification must be taken into account (Section 3.4.2). The simplest way to solve the equations is to use the angle θ at which the lens receives radiation from the object:

$$\tan\theta = \frac{a}{f + d}. \tag{8.5}$$

Then the diameter of the blur circle on the object side results in

$$u = \tan\theta \, \Delta z \tag{8.6}$$

and the PSF projected back into the object space to

$$h_o(\boldsymbol{X}) = \frac{I_0}{\pi(\Delta z \tan\theta)^2} \Pi\left(\frac{(x^2 + y^2)^{1/2}}{2\Delta z \tan\theta}\right) = \frac{I_0}{\pi(\Delta z \tan\theta)^2} \Pi\left(\frac{r}{2\Delta z \tan\theta}\right). \tag{8.7}$$

The double cone of the PSF, which is projected back into the object space, has the same aperture angle as the lens (Fig. 8.1a).

We will not calculate the 3D OTF directly, but use the linearity and separability of the Fourier transform to decompose the 3D OTF into suitable

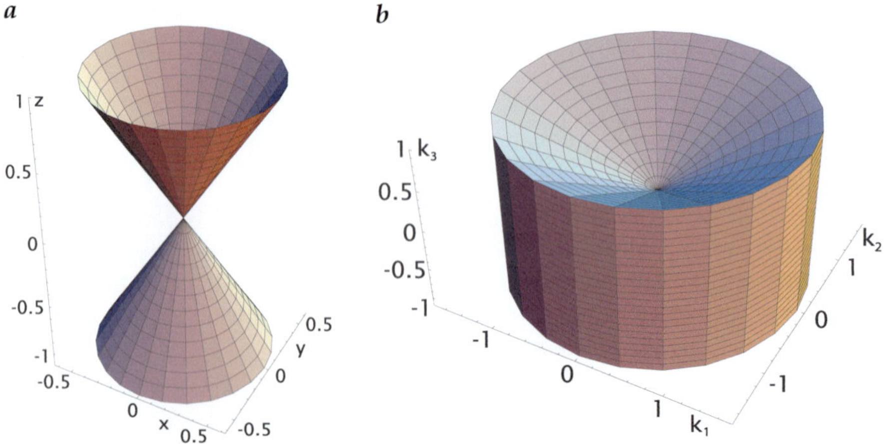

Figure 8.1: ***a*** *Three-dimensional point response (PSF) and* ***b*** *optical transfer function (OTF) in imaging neglecting diffraction effects.*

subfunctions whose Fourier transforms are easy to determine. Two possibilities are presented, which also illustrate some important properties of the Fourier transform.

The first method for calculating the OTF takes advantage of the linearity of the Fourier transform and decomposes the PSF into a bundle of δ lines that pass through the origin of the coordinate system and are evenly distributed in the cross-section of the double cone. Let's think of each of the δ lines as a ray of light. Without further calculations, we know that this decomposition gives the correct quadratic decrease of the PSF, since the same number of δ lines intersect a quadratically increasing area. The Fourier transform of a δ-line is a δ-plane that is perpendicular to the line (> R13). The OTF therefore consists of a bundle of δ planes. These pass through the origin of the k-space and have a maximum angle of θ to the k_1k_2-plane, half the opening angle of the OTF double cone. As the rotational symmetry is retained during the Fourier transformation, the OTF is also rotationally symmetrical to the k_3 axis.

The OTF thus fills the entire Fourier space except for a double cone with the opening angle $\pi/2 - \theta$ (Fig. 8.1b). In this area, the OTF is zero. The exact course of the OTF within the body of rotation is difficult to understand with this decomposition.

We will therefore pursue a second approach based on the separability of the Fourier transform. We think of the double cone as being divided into slices with different diameters that increase with $|x_3|$. In the first step, we perform the Fourier transform only in the x_1x_2 plane. This results in a function with two coordinates in the k-space and one in the x-space (k_1, k_2, x_3 or in cylindrical coordinates q, φ, z). Since the PSF Eq. (8.7) only depends on r (rotational symmetry around the z-axis), the two-dimensional Fourier transform corresponds to a (one-dimensional) *Hankel transform of zero order* [23]:

$$\begin{aligned} h(r,z) &= \frac{I_0}{\pi(z\tan\theta)^2}\Pi(\frac{r}{2z\tan\theta}) \\ \check{h}(q,z) &= I_0\frac{\mathrm{J}_1(2\pi zq\tan\theta)}{\pi zq\tan\theta}. \end{aligned} \tag{8.8}$$

The Fourier transform of the rotationally symmetric disk is also a rotationally symmetric function that contains the *Bessel function* J_1. In the second step, we carry out the missing one-dimensional Fourier transform in the z-direction. Equation Eq. (8.8) shows that $\check{h}(q,z)$ is also a Bessel function in z.

However, since the Fourier transform is one-dimensional this time, we do not obtain a disk function, but a circular function (> R13):

$$\frac{\mathrm{J}_1(2\pi x)}{x} \quad \circ\!\!-\!\!\bullet \quad 2\left(1-k^2\right)^{1/2}\Pi\left(\frac{k}{2}\right). \tag{8.9}$$

Let us now apply the *similarity theorem* of the Fourier transform (Theorem B.1, p. 628, > R12):

$$\begin{aligned} &\text{if} \quad f(x) \quad \circ\!\!-\!\!\bullet \quad \tilde{f}(k), \\ &\text{then} \quad f(ax) \quad \circ\!\!-\!\!\bullet \quad \frac{1}{|a|}\tilde{f}\left(\frac{k}{a}\right), \end{aligned} \tag{8.10}$$

so we get:

$$\tilde{h}(q,k_3) = \frac{2I_0}{\pi|q\tan\theta|}\left(1-\frac{k_3^2}{q^2\tan^2\theta}\right)^{1/2}\Pi\left(\frac{k_3}{2q\tan\theta}\right). \tag{8.11}$$

Due to the box function in this equation, the OTF is zero in a large range. This means that spatial structures with the associated directions and wavelengths disappear completely. This is particularly the case for all structures in the z direction, i. e. perpendicular to the image plane. Such structures cannot be reconstructed without additional information.

3D structures are only visible if they also contain structures that are parallel to the image plane. It is then possible, for example, to distinguish points or lines that lie on top of each other. This can be visualized both in spatial space and in Fourier space. In the spatial domain, the superposition of the corresponding PSFs still results in a depth-dependent focus series in which the two points or lines can be distinguished, albeit blurred. In Fourier space, lines or points are extended objects that are distorted by the OTF, i. e. they partially coincide with the parts of the OTF that are not equal to zero, and are therefore not completely erased.

The limit for structures that disappear during projection is at an angle θ to the k_1k_2 plane that just corresponds to the maximum aperture angle of the lens. Intuitively, we can say that we can recognize all 3D structures that we can actually "look into". At least one ray is required that is perpendicular to the wave vector of the periodic structure and thus runs in the direction of constant values.

8.2.2 Limits of Projective Volume Imaging

As we discussed in detail in Sections 3.5.1 and 3.5.2, a projective optical system is, to a good approximation, a linear displacement invariant system that can be described by a point response and the optical transfer function (OTF).

The 3D OTF for geometrical optics best shows the limits of projective imaging (see Section 8.2.1):

$$\tilde{h}(q, k_3) = \frac{2I_0}{\pi|q \tan \theta|}\left(1 - \frac{k_3^2}{q^2 \tan^2 \theta}\right)^{1/2} \Pi\left(\frac{k_3}{2q \tan \theta}\right). \tag{8.12}$$

The symbols q and k_3 denote the radial and axial components of the wavenumber vector respectively. Two fundamental limitations for 3D image acquisition can be determined from the shape of the 3D ODF:

Complete Loss of Wide Wavenumber Ranges. As shown in Fig. 8.1b, the 3D OTF is rotationally symmetric about the k_3-axis (z-direction) and non-zero only within an angular range of $\pm\theta$ about the xy-plane. Image structures with a wide range of wavenumbers, especially in the direction of the optical axis, are completely lost. To illustrate, we can say that it is only possible to "see" structures in those directions from which the optical system also collects rays.

Loss of Contrast with Large Wavenumbers. According to Eq. (8.12), the OTF is inversely proportional to the radial wavenumber q. Therefore, the amplitude of a periodic structure is damped proportionally to its wavenumber.

8.2.3 Limits of Projective Depth Imaging

The limitations of volume imaging discussed in the last section apply to all optical systems, including the human eye. This raises the question of why we can recognize fine structures at all.

The answer arises from a closer examination of the geometric structure of the observed objects. Most objects in the environment are opaque. Therefore, we only see their surface, i. e. we do not really see 3D objects, but only 2D surface structures. If we map a 2D surface onto a 2D image plane, the PSF is also reduced to a 2D function.

Mathematically, this means multiplying the PSF by a δ plane that is parallel to the observed surface. As a result, the PSF just corresponds to the blur circle at the corresponding distance of the surface from the camera. The reduction to 2D surfaces thus preserves the intensity of all structures that are larger than this blur circle. They are visible to us with the same contrast.

By considering the problem in Fourier space, we arrive at the same result: Multiplying the PSF by a δ plane in spatial space corresponds to the convolution of the OTF with a δ line perpendicular to it in Fourier space, i. e. an integration over the coordinate in the corresponding direction. If we integrate the OTF over the k_3 coordinate, we actually obtain a constant that is independent of the wavenumber q:

$$\frac{2I_0}{\pi} \int\limits_{-q\tan\theta}^{q\tan\theta} \frac{1}{|q\tan\theta|} \left[1 - \left(\frac{z'}{q\tan\theta}\right)^2\right]^{1/2} \mathrm{d}z' = I_0. \tag{8.13}$$

To solve the integral, the following substitution was used: $z'' = z'/(q\tan\theta)$.

To summarize, we can state that there is a fundamental difference between imaging surfaces and volumes. The OTF for surface structures is independent of the wavenumber.

However, for three-dimensional structures, the problem of the OTF decreasing with the radial wavenumber is significant. It will not be possible to discern fine structures with the naked eye or a camera. Projective imaging systems are not designed to recognize three-dimensional objects. This is why other techniques are required to capture volume images.

8.3 Light Fields

It has already become clear from the previous section that projective imaging is not suitable for 3D vision. In this section, we will first see that the limitations of 2D image acquisition are not only limited to geometric aspects, but are much more profound (Section 8.3.1). In the following section (Section 8.3.2) we will introduce a new term, the light field, and learn that these deficits are due to the fact that the light field was not sampled well enough.

8.3.1 General Limits of 2D Image Acquisition

First, we will summarize the geometric limitations of projective image acquisition by starting from the simple *pinhole camera* model (Section 3.3.1).

Loss of Depth

This elementary fact is based on the equations of the projective mapping Eq. (3.9) in Section 3.3.2. The position in the image only gives us the ratio of object size to distance. An object point could be located at any point along a projection ray (Fig. 3.3). We do not know where it is actually located.

Loss of Object Size

The loss of depth information is accompanied by the loss of object size.

Loss of Surface Slope
With the lack of position information, it is also unknown how the surfaces of the object are inclined. It may be surprising that this point is mentioned separately, but it is of particular importance because the orientation of a surface relative to the direction of illumination determines the brightness with which the object is seen by a camera.

Loss of 3D Structure
So far we have only considered the projective mapping of surfaces in space. If a semi-transparent world — as with X-rays — is imaged projectively, all objects in the projection beam are superimposed and the 3D structure is completely inaccessible in one projection. It is not even possible to separate 2D silhouettes of objects from each other.

The second class of limitations is not geometric but radiometric in nature. The irradiance on the image plane says nothing about the "brightness" of the object, i. e. the optical properties of the object. Here, too, we can state a multiple lack of knowledge.

Image Irradiance is not an Object Feature
This elementary fact expresses one of the basic problems of image processing, which was already mentioned at the beginning of the book in Section 2.1 and demonstrated using Fig. 2.2: The brightness of an object in an image is a complex function that depends on its reflective properties, the tilt of its surfaces and the illumination that hits its surfaces. This makes it clear why it is also so difficult to find robust features for objects.

Missing Optical Properties of Object Surfaces
It is clear from the previous considerations that a picture tells us nothing about the optical properties of an object. Even the color is not unambiguous. If an object looks red in a picture, it could actually be an object of red color or a white object illuminated by a red light source. If we were able to extract the optical properties, in the complex case the *BRDF* (Section 2.6.4), from image data, we would have obtained robust features for further image processing.

Missing Irradiation of Object Surfaces
Even less than the optical properties, we can deduce the illumination of the scene from an image. However, if we know the image irradiance and the BRDF of the objects, we can also deduce the irradiance at the object surfaces.

8.3.2 Sparse Light Field Sampling

Where do all the limitations discussed in the last section come from? From the fact that *one* image says far too little about the scene. But what if we could take *all* possible images of a scene, i. e. take them from all directions? This leads to the size called *light field*. This function can be defined in different ways. On the one hand, you could specify the *radiance* (compare Section 2.4.4) emanating from each point in space. The radiance will generally be different in each direction, so we get a five-dimensional function:

$$L(x, y, z, \theta, \phi). \tag{8.14}$$

This definition is necessary if there are semi-transparent objects in the room.

However, if we are dealing with opaque surfaces in space, we know that light can only come from one point in space on a ray, and we can reduce the light field to a four-dimensional function. There are two ways to do this. Either we place a reference sphere around the scene to be observed and observe the radiation at each point of the sphere from all directions or we take a reference plane and then obtain

$$L(x_r, y_r, \theta, \phi), \tag{8.15}$$

where x_r and y_r are the coordinates of the reference plane.

What is the relationship between a classic projective image taken with a pinhole camera and the light field? Of the many rays that emanate from a point on an object, only the ray that passes from the object through the hole of the pinhole camera and hits the image plane is taken. This means that only *one* 2D subspace of the 4D light field is used. Since we are considering the limiting case of geometric optics here, all objects are imaged sharply regardless of their distance.

Example 8.1: Image acquisition by a camera with a lens

A real camera with an optical system not only captures one ray of the light field, but a cone-shaped ray bundle, whereby the aperture of the cone is determined by the aperture of the lens. However, like the pinhole camera, this integration over an entire beam uses only a 2D subspace of the 4D light field and has the further disadvantage that the angular resolution of the light field is limited by the aperture angle of the cone. As we have already seen in Section 3.4.4, this integration also means that objects are only sufficiently sharp in a certain distance range, the *depth of field*.

It is therefore clear that 2D imaging is far too sparse a sampling of the light field and the exciting question arises as to how best and from how many directions under given boundary conditions the light field can be scanned, i. e. images must be captured in order to be able to reconstruct the depth of the objects. This is the basic question of 3D imaging. If you take images from different directions, you also obtain more information about the optical properties of the captured objects, as you can see an object point from several directions.

8.4 Basic Principles of 3D Image Acquisition

In order to obtain the depth of an opaque surface in space, only one additional piece of information per pixel is required to generate a *depth image* in addition to a brightness image. Four basic principles can be distinguished, which we refer to as *depth from paradigms*. In addition, depth can be determined from the slope of surfaces using a paradigm known as *shape from shading*.

Any depth-from-technique that is capable of capturing multiple depths simultaneously is also suitable for volume image acquisition. Thus, the ability to acquire multiple depths simultaneously is another important feature of a depth measurement technique. Beyond these techniques, however, there are two other principles for volumetric imaging.

8.4.1 Direction: Triangulation and Tomography

Depth from Triangulation

If we view an object from two or more positions that are separated by a baseline b, it appears from different angles. This technique is called *triangulation* and is one of the essential techniques in *geodesy* and *cartography*. For opaque objects, two positions are obviously sufficient to determine their distance from the recording systems based on the angular differences.

3D Structure from Tomography

Triangulation cannot work with transparent three-dimensional objects, as all light emitted by objects along a projection beam in its direction is integrated on the image plane. This means that each pixel contains a certain amount of partial information about the object radiances scattered in the three-dimensional object space. This raises the question of whether, and if so how, it is possible to combine the partial information into a complete 3D image using projections from many directions. Such a technique of *depth from multiple projections* is known as *tomography*.

8.4.2 Depth: Time of Flight and Interferometry

Depth from Time of Flight

The direct measurement of the transit time, i. e. the time it takes for the radiation to travel from the light source to the object and back to the recording system, is a second immediately obvious principle for distance measurement. The delay time between sending and receiving the signal is proportional to the sum of the distances between transmitter and object and object and camera.

With this method, it should be possible to measure the distance of several objects simultaneously. One way to do this is to emit only a short flash from the light source and then measure the reflected light as a function of the delay time and thus the distance. This immediately raises the question of whether short flashes are the only way to measure back reflections in three dimensions and not just the distance of a single reflecting object.

Depth from Phase: Interferometry

Interferometry can be seen as a special case of time-of-flight measurement. This technique measures distances in fractions of the wavelength of the radiation by measuring not only the amplitude (energy) of the radiation, but also its phase position.

Phase measurements are possible by superimposing coherent radiation (Section 2.2.6), as this leads to intensity variations. High intensities result when the superimposed wavefronts are in phase (constructive interference) and low intensities with a phase shift of 180° (π, destructive interference). Light has wavelengths between 400 and 700 nm (Section 2.3.1 and Fig. 2.16). Therefore, light can be used to measure distances with nanometer accuracy (10^{-9} m) — a small fraction of the wavelengths. However, the measuring range is limited to distances within one wavelength.

Measuring the phase shift has another serious disadvantage. If two objects reflect radiation from different depths, two periodic signals of the same frequency with different phase shifts are superimposed (neglecting the *Doppler effect*). However, this in turn results in only one periodic signal with a different phase shift. This means that interference methods are not able to determine whether one or more objects are reflecting.

Depth from Coherence

An important property of radiation is the *coherence length* (Section 2.2.6), i. e.the maximum path difference at which coherent superposition is still possible. The coherence length can be measured by the ability to generate interference patterns. If the displacement of the overlapping beams is greater than the coherence length, there are no more interference fringes.

As we learned in Section 2.2.6, the coherence length is directly related to the spectral bandwidth of the radiation. For correspondingly broadband radiation, the coherence length can only be a few wavelengths short, for narrowband radiation it can be many wavelengths short. This means that depth from coherence techniques close the gap in depth resolution between interferometric and time-of-flight methods.

8.4.3 Surface Slope

The shape of a surface can also be determined from the local orientation of the surface elements. Mathematically, the orientation is expressed by vectors that are perpendicular to the surface. With this method, of course, we lose the absolute depth of the surface, but a depth profile can be obtained by integrating the surface slope. We can determine the surface normal from the radiance of the surface, as this depends on the angle between the surface normal and the direction of irradiation.

8.4.4 Space Coding through Lighting

With projective imaging, we do not know from which depth the energy flux density incident on the image plane comes. It can be from any position along the projection beam (see Section 3.3.1 and Fig. 3.3). However, this can be changed using depth-dependent lighting. The simplest case occurs when only one plane of the room is illuminated. In this case, it is obvious that only objects lying in this plane can be seen. A volume image can be obtained by scanning the light plane in depth.

This method also raises the question of whether such *illumination slice images* are the only way to encode space using illumination. This alone will not be the case, since in addition to the radiation intensity, other properties of the radiation can be varied with depth. Another immediately obvious possibility is an irradiation in which the wavelength of the radiation depends on the depth.

8.4.5 Uniform Characterization

Given the diversity of the methods discussed so far, the question arises as to whether it is possible to characterize them using a uniform scheme. To do this, it is first necessary to make a strict distinction between depth image acquisition (depth maps) and volume image acquisition. At the same time, the first characteristic of a 3D image acquisition method is whether it is capable of generating depth maps or volume images or even both.

In depth image acquisition, the first basic parameter is which *depth range* $\square z$ is captured with it, which can also be characterized more precisely by a minimum distance $z_{\min}$ and a maximum distance $z_{\max}$ with $\square z = z_{\max} - z_{\min}$. The depth range can be a limit specified by the method itself or be determined by the measurement uncertainty. For example, the maximum distance may be given by the fact that it can no longer be distinguished from an infinite distance due to the measurement error.

The second basic parameter is therefore the statistical uncertainty or *resolution* δ_z of the measured depth. This does not have to be constant over the entire depth range but can also increase to a greater or lesser extent with distance z. However, the measurement error can also depend on the *reflection coefficient* (Section 2.6.3) or other optical properties of an object.

With many 3D imaging techniques, it is not possible to measure a depth at every pixel. Therefore, a third basic parameter for characterization is how dense the depth map is and on which object properties this depends.

Unfortunately, the uniform characterization of 3D image acquisition outlined here has not yet been implemented in industrial practice. This makes it all the more difficult to objectively compare the different techniques with each other. A first step in this direction was taken with the Fair Datasheet[1] for optical surface measuring devices. Unfortunately, this initiative has not been continued since 2015.

[1] `http://optassyst.de/fairesdatenblatt/`

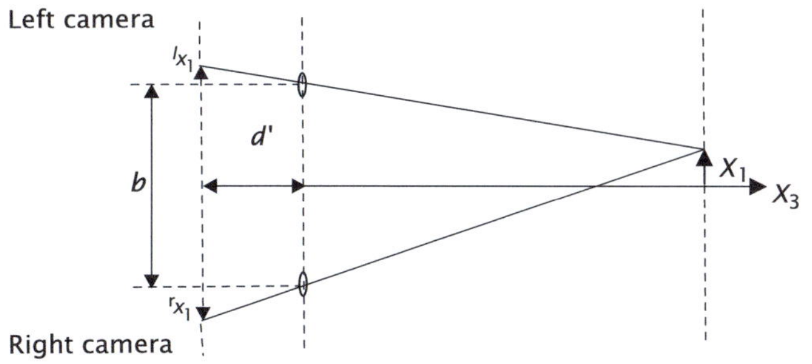

Figure 8.2: *Schematic representation of a stereo camera arrangement.*

The characterization of volume images is first given by discretization, which we have already discussed in detail in Chapter 6. This determines both the size of the volume image and the number of voxels in all directions. A special feature may be that the voxel depth Δz in the depth (axial direction) differs from that in the two lateral directions ($\Delta x, \Delta y$).

8.5 Passive Direction Measurement

If an object is viewed from different positions that are separated by a basis vector $\boldsymbol{b}$, this results in different viewing angles. In one way or another, the differences in viewing angle result in a shift on the image plane known as *disparity* or *parallax* and from which the distance of the object can be determined.

Triangulation-based depth measurement techniques include a wide range of different methods that — at first glance — have nothing in common with each other and yet are based on the same principle. In this section we look at stereoscopy (Section 8.5.1) and depth from focus (Section 8.5.2). Active triangulation, where a camera is replaced by a light source (Section 9.2.2), and confocal microscopy (Section 9.2.3) are part of the next chapter (Chapter 9). With stereoscopy, we also discuss the basic geometry of triangulation.

8.5.1 Stereoscopy

Viewing a scene from two different angles makes it possible to determine the distance between objects and the camera plane. A setup with two image sensors is called a *stereo system*. In many biological systems, depth vision is achieved in this way. Fig. 8.2 illustrates the depth determination with a stereo camera arrangement. Two cameras are placed next to each other with parallel optical axes. The distance vector $\boldsymbol{b}$ of the two axes is referred to as *stereoscopic base*.

An object is imaged at different positions in the image plane because it is seen at slightly different angles. The difference in distance between the positions on the image plane is called *parallax* or disparity. With the help of Fig. 8.2, the parallax p' can be easily calculated as

$$p' = {}^r x' - {}^l x' = (f + d')\frac{x + b/2}{z} - (f + d')\frac{x - b/2}{z} = (f + d')\frac{b}{z}. \tag{8.16}$$

The parallax is inversely proportional to the distance z of the object (zero for an infinitely distant object) and directly proportional to the stereoscopic base and the image distance of the lenses used ($f + d' \approx f$ for distant objects).

We can also express the ratio of the stereo base b to the distance z as the difference of the tangent function of the angles at which the object is seen by the two cameras:

$$\boxed{\frac{p'}{f + d'} = \frac{b}{z} = \tan\theta_r - \tan\theta_l, \qquad \text{triangulation.}} \tag{8.17}$$

This explicitly expresses what stereo has to do with triangulation. This is the basic equation of triangulation.

The further away an object is, the more difficult it is to estimate the distance. This becomes clearer when we use the law of error propagation to calculate the error of distance δz from the error of parallax $\delta p'$:

$$z = \frac{b(f + d')}{p'} \quad \text{with} \quad \delta z = \left|\frac{\mathrm{d}z(p)}{\mathrm{d}p}\right| \delta p' \quad \rightsquigarrow \quad \delta z = \frac{z^2}{b(f + d')}\delta p'. \tag{8.18}$$

Therefore, the error in the depth determination increases with the square of the distance. Even the relative error of the depth determination

$$\frac{\delta z}{z} = \frac{z\,\delta p'}{b(f + d')} \tag{8.19}$$

still grows linearly with distance.

If an object is infinitely far away, the parallax is zero. The greatest distance that we can just detect results from the smallest parallax p', which just corresponds to the measurement error δ_p. Since in this case $d' \ll f$, this results in the following relationship according to (Eq. 8.17)

$$z_{\mathrm{max}} = \frac{bf}{\delta p'} = \frac{b}{\delta\theta'}, \tag{8.20}$$

where for the right-hand side of the equation it was possible to simplify the angle $\tan\theta_r - \tan\theta_l$ to δ_θ due to its smallness. The smallest measurable distance z_{min} is determined by the fact that the parallax is not too large so that the object can still be seen in both cameras (cf. Fig. 8.2).

With regard to the depth resolution, it is now interesting to see how this relates to the lateral image resolution. We assume that the parallax is measured with an accuracy that is proportional to the pixel size: $\delta_p = \epsilon_a \Delta x'$.

With a pixel-accurate stereo algorithm, $\epsilon_a = 1$. In the same way, we set the lateral resolution on the image plane with $\delta x' = \epsilon_l \Delta x'$. Using Eq. (8.18) and Eq. (3.19) for the *lateral image scale* m_l to convert the resolution from the image plane to the object plane then results in the surprisingly simple relation

$$\boxed{\frac{\delta z}{\delta x} = \frac{\epsilon_a}{\epsilon_l}\frac{z}{b}.} \tag{8.21}$$

This means that a depth resolution comparable to the lateral resolution can only be achieved if the stereo base is approximately as large as the distance or the accuracy of the parallax determination is much more accurate than the pixel size, i. e. $\delta_p \ll \Delta x'$ or $\epsilon_a \ll 1$.

Example 8.2: Depth resolution of a stereo camera

As an example, we assume a stereo system with a stereoscopic base of 1 m. The focal length of the lenses is 100 mm. At a distance of 10 m, the parallax is then 1 mm, at a distance of 100 m 100 μm/m and at 1000 m only 10 μm/m. Assuming an accuracy of the parallax determination of 10 μm/m, the maximum distance that can be determined is therefore just 1000 m. Furthermore, (Eq. 8.21) deteriorates assuming the same resolution accuracy in the axial and lateral direction ($\epsilon_a = \epsilon_l$), so that the depth accuracy is a hundred times worse than the lateral resolution at a distance of 10 m and a thousand times worse at a distance of 100 m.

It now remains to be clarified whether dense depth maps can be calculated with stereoscopy. The parallax is a vector quantity and, according to Fig. 8.2, is always parallel to the stereoscopic base $\boldsymbol{b}$ if the lenses have no geometric *distortion* (Section 3.4.6). On the one hand, this simplifies stereo algorithms considerably, as an object can only be shifted in the direction of the stereo base from the right to the left image. On the other hand, it is necessary for the image area to have gray value changes in the direction of the stereo base b. If an image area has no such structures, it is not possible to determine a parallax. This problem is a special case of the *aperture problem*, which we will discuss in detail in the motion analysis (Section 14.1.2). This means that stereo methods can only work if the observed objects in the image have visible, suitable gray value structures. This is not directly possible in all other image areas.

However, there are further limitations. First of all, it should be noted that stereo methods cannot reconstruct 3D volume images. Only the depth of an opaque surface can be calculated. This requires that every part of the surface is seen by both cameras. If a part of the surface is obscured by an object in front of it as seen by one of the two cameras (*occlusion*, cf. Fig. 3.4 in Section 3.3.1), it is impossible to determine the depth.

Furthermore, the surface must look the same from the direction of both cameras. Strictly speaking, this is only the case with a *Lambertian reflector* (see Section 2.6.4). Stereo methods cannot work with smooth, specularly reflecting objects [163]. Even if objects, like many natural and technical surfaces, are partially scattering and specularly reflecting, determining the parallax is problematic.

8.5.2 ♯Depth from Focus

The limited *depth of field* of an optical system (Section 3.4.4) is the basis for another technique for obtaining depth images. An object is only imaged without blurring if it lies within the object's depth of field range. At first glance, this technique does not look like a triangulation method. However, it has exactly the same basic geometry as a triangulation method. The difference is that there are not just two rays of different directions, but a cone-shaped bundle of rays. The triangulation angle now corresponds to the opening angle of the cone and the stereo base obviously corresponds to the diameter of the optics. The depth-of-focus technique therefore has all the characteristics of a triangulation technique.

This can also be shown directly by converting Equation (3.32) for the depth of field into (Eq. 3.19) using the definition for the lateral image scale:

$$\Delta z \approx \pm k \cdot \frac{1+m_l}{m_l^2} u' \quad \rightsquigarrow \quad \delta z = \frac{z^2}{a(f+d')} 2u'. \tag{8.22}$$

As in (Eq. 8.18), the depth resolution decreases with the square of the distance. Twice the value of the minimum determinable diameter of the blur circle $2u'$ corresponds to the accuracy $\delta p'$ with which the parallax can be determined. The stereo base b now corresponds to the diameter a of the aperture.

The two images of the left and right stereo camera now correspond to the recording of many images in a so-called *focus series* with only one camera, whereby the distance between the object and the camera is changed step by step. This also determines the depth range at the same time. The step size must be smaller than the height accuracy δz according to (Eq. 8.22) so that the depth of maximum focus can be determined accurately enough. The advantage over the stereo method is that structures in any direction make it possible to find the depth of maximum sharpness. With the stereo method, parallax determination with structures parallel to the stereo base is not possible, as we found out in the last section.

The disadvantage of the method of depth determination from focusing is that the stereo base is limited to the diameter of the lens aperture. This means that resolutions comparable to the lateral resolution can only be achieved for small distances that are not much larger than the aperture diameter. The technique is therefore limited to macro photography.

From the discussion of the limitations of projective imaging at the beginning of this chapter in Section 8.2.2, it can also be concluded that the depth-of-focus technique is not suitable for capturing volume images, as

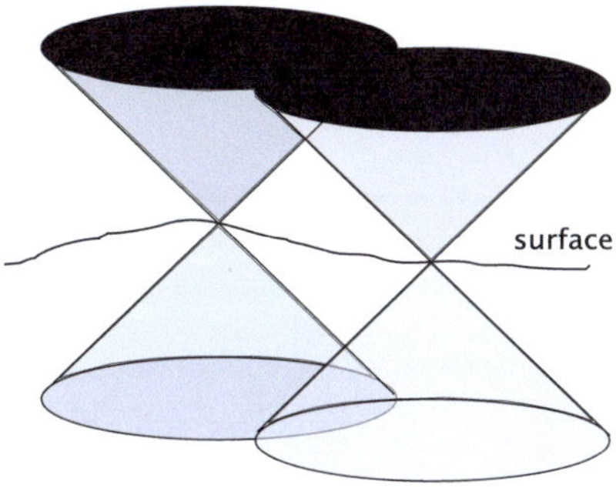

Figure 8.3: *Superposition of the double-cone-shaped point response of two neighboring points on a surface.*

most structures, especially those in the depth direction, disappear in projective imaging. However, it is a useful and simple technique for determining the depth of opaque surfaces.

Steurer et al. [314] developed a simple method for reconstructing a *depth map* from a light microscopic focus series. A depth map is a two-dimensional function that specifies the depth of an object point d — relative to a reference plane — as a function of the image coordinates $[x, y]^T$.

With the given restrictions, only one depth value needs to be found for each pixel. We can take advantage of the fact that the three-dimensional point response of optical images, which we have considered in detail in Section 3.5.1, has a pronounced maximum in the focal plane, since the intensity decreases with the square of the distance from the focal plane. The contrast — i. e. the gray value differences between adjacent pixels — of distinctive gray value structures such as corners, lines or local extrema will therefore be at a maximum in the focal plane. Fig. 8.3 shows that the point responses of neighboring pixels influence each other by superimposition far away from the focused plane, but not in its vicinity.

Steurer's method is based on the fact that the 3D PSF has a pronounced maximum in the focal plane. His algorithm consists of the following four steps:

1. Assumes a focal series with constant depth steps.
2. The local contrast is determined using a suitable operator, e. g. the *variance operator* (Section 16.2.2). The filtered images are segmented to create a mask for the regions with significant gray value changes.
3. In the masked regions, the maximum contrast is searched for at one image point in all depths. The image in which the maximum occurs provides a depth value for the depth map. By interpolating the values, the depth position of the maximum can also be determined more precisely than the depth resolution of the image stack [289].

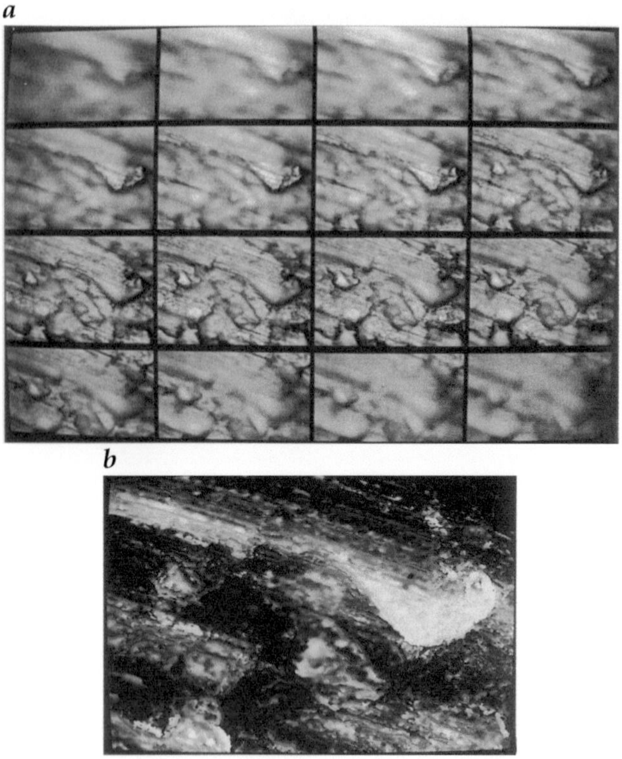

Figure 8.4: ***a*** *focus series with 16 images of a metallic surface taken in steps of 2 µm: The focal plane is shifted from left to right and from top to bottom towards deeper layers.* ***b*** *Depth map calculated from a focus series. Depth is coded by intensity, with objects closer to the observer appearing brighter. From Steurer et al. [314].*

4. As the depth map determined in this way is not dense, the missing depth values must be interpolated. Steurer uses a region growing method and then an adaptive low-pass filtering, which he only applies to the interpolated regions in order not to change the directly calculated depth values. Any other regularization technique can be used here, such as e. g. the *normalized convolution* (Section 11.6.2) or the methods described in Section 17.2.

This method has been successfully used to determine the surface structure of machined metal parts. Fig. 8.4 shows that it leads to good results. A metal chip can be seen protruding from the surface. In addition, the surface shows clear traces of the grinding process.

8.5.3 Epipolar Geometry and Light Field Imaging

From the discussion in the last two sections, one could prematurely conclude that depth-from-direction is a rather limited method and, in particular, is not suitable in principle for generating volume images. However, this overlooks the fact that the methods discussed so far only use a minimal part of the light field discussed in Section 8.3 and therefore only scan the light field very incompletely. The stereo method uses only two directions and the depth-of-focus method uses only one beam cone, which is varied in distance to the object.

In this section, we will now expand the scanning by moving from two cameras to a light field image with many cameras. The easiest way to do this is to imagine a stationary scene. Then one camera is sufficient, which can be moved accordingly. To do this, we move a camera horizontally along the stereo base b from the position of the left to the right camera (cf. Fig. 8.2). This allows us to view the stereo base as a linear function of a coordinate s and we now obtain an image sequence in which the objects move slowly from image to image. An object at infinity always remains at the same position. However, the smaller the distance to the camera, the faster the position changes.

In the same way, we can also move the camera in a vertical direction (coordinate t). With both movements, the objects in the image move more strongly the closer they are to the camera. If we move the camera along the entire st plane, we obtain a four-dimensional light field

$$L(x, y, s, t). \tag{8.23}$$

This replaces the angular coordinates θ and ϕ in Eq. (8.14) with the coordinates s and t and the coordinate z disappears, as we do not move the camera in the direction of the optical axis, i. e. away from the scene.

Such a four-dimensional light field can best be represented as in Fig. 8.5. This shows an image from the center position and below it how the image content changes along the horizontal line marked dark in the image when the camera is moved horizontally in the s direction. In the same way, the vertical strip on the right-hand edge shows how the image content changes along the vertical line marked in black when the camera is moved in the t direction.

The display in Fig. 8.5 has another special feature. Nearby objects, such as the bicycle, move to the right in the s direction, while the patterns on the back wall of the room move to the left. This is due to a trick that is often used in lightfield imaging. The image is shifted from line to line in the s and t directions (shear transformation, see Section 7.5) so that objects no longer move at infinity, but in the center of the image. This is the case in Fig. 8.5 for the telescope tripod, which has vertical lines in the s direction.

Regardless of this shear transformation, in light fields the distance to the st recording plane corresponds to an *orientation* in the s and t directions in

Figure 8.5: Illustration of the image of a four-dimensional light field. Further explanation in the text. From Schilling [290].

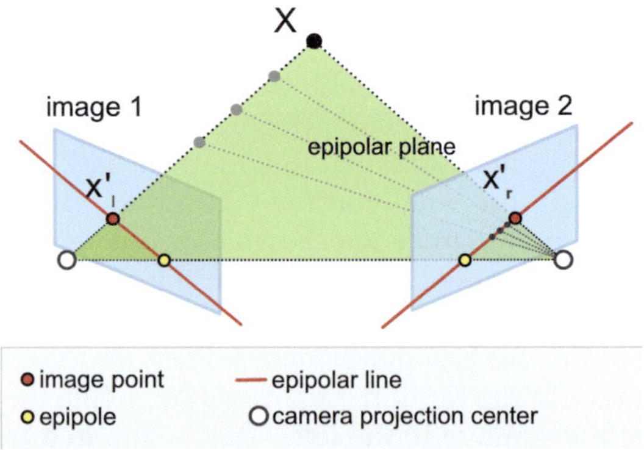

Figure 8.6: Stereo camera arrangement with inclined axes for a general explanation of epipolar geometry (image after Arne Nordmann from `https://de.wikipedia.org/wiki/Epipolargeometrie`*).*

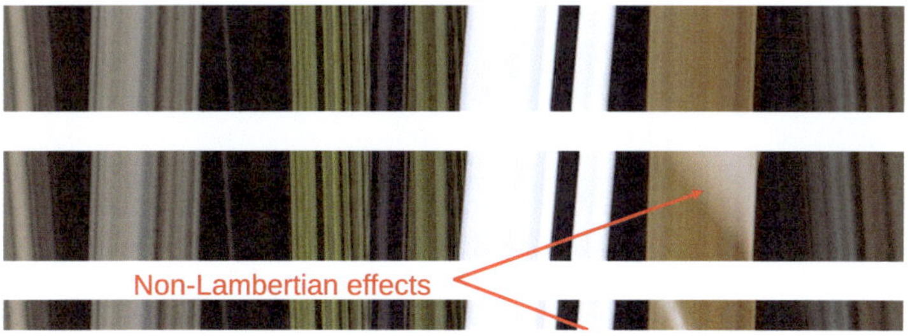

Figure 8.7: *Specular reflections in the light field.*

the four-dimensional light field. This is a completely different situation from the analysis of stereo images.

Let's take another look at a generalized stereo camera arrangement in Fig. 8.6 compared to Fig. 8.2 with parallel axes. In the figure, the image plane is mirrored at the projection center in the direction of the object, which corresponds to the same projection (compare Fig. 3.3 with Fig. 3.5). The object point x and the camera projection centers form a plane that intersects the image planes in the *epipolar lines*. Depending on how far away the object point x is, it shifts in the right camera on the epipolar line. We must therefore search for the structure in the pixel x_l' along the epipolar line in the right-hand image. This works if the gray values change spatially along the epipolar line sufficiently and look the same from the direction of the right camera as in the left camera. However, this means a search algorithm for corresponding structures.

It is obvious that the orientation analysis in light fields provides much more accurate depth in comparison, as this is determined from many images. Furthermore, the determination will be more robust, as the orientations are still present even if specular reflections are superimposed on them, as shown in Fig. 8.7. If you know the position of the light source in the scene, you can even calculate the orientation of the surface from the orientation of the reflections.

If transparent objects are present, several orientations are superimposed. In general, the great importance of analyzing oriented structures in light fields is one of the reasons why we will deal with this in detail in Chapter 13.

8.6 Depth from Surface Slope

The method of *shape from shading* does not determine the depth directly, but the surface normal of surfaces and thus represents a separate class of methods for the 3D reconstruction of surfaces. By integrating the slope, the height profile of a surface can also be calculated. However, the absolute distance is lost in the process.

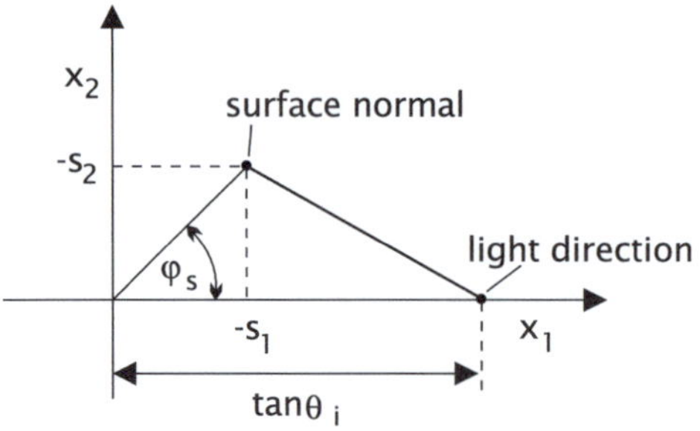

Figure 8.8: Calculation of the radiance, illustrated in gradient space for a Lambertian surface that is irradiated with a distant light source at the angle of incidence θ_i. The azimuth angle of the light source ϕ_i is zero.

8.6.1 Shape from Shading

We first apply this technique to the diffuse reflection of opaque objects. For the sake of simplicity, we assume that the surface of a Lambertian object (Section 2.6.4) is illuminated by parallel light. This gives the radiance L of the surface:

$$L = \frac{\rho(\lambda)}{\pi} E \cos\theta, \tag{8.24}$$

where E is the irradiance and γ is the angle between the surface normal and the direction of light. The easiest way to understand the relationship between the surface normal and the incident and emitted radiation is in *gradient space*. This space is spanned by the gradient of the surface height $a(x, y)$:

$$\boldsymbol{s} = \boldsymbol{\nabla} a = \left[\frac{\partial a}{\partial x}, \frac{\partial a}{\partial y}\right]^T = [s_1, s_2]^T. \tag{8.25}$$

The gradient is directly related to the surface normal if its z component is set to one:

$$\boldsymbol{n} = \left[-\frac{\partial a}{\partial x}, -\frac{\partial a}{\partial y}, 1\right]^T = [-s_1, -s_2, 1]^T. \tag{8.26}$$

The gradient space can therefore be understood as a plane parallel to the xy-plane at the height $z = 1$ if the x, y-axis directions are reversed. The x, y-coordinates at which the surface normals and other direction vectors penetrate this plane are then the coordinates of the gradient space.

The geometry in gradient space is illustrated in Fig. 8.8. Without limiting the generality, we can rotate the direction of the light source in the x-direction. The direction of light is then given by the vector $\boldsymbol{l} = (\tan\theta_i, 0, 1)^T$, and the radiance L of the surface can be expressed as follows:

$$L = \frac{\rho(\lambda)}{\pi} E \frac{\boldsymbol{n}^T \boldsymbol{l}}{|\boldsymbol{n}||\boldsymbol{l}|} = \frac{\rho(\lambda)}{\pi} E \frac{-s_1 \tan\theta_i + 1}{\sqrt{1 + \tan^2\theta_i}\sqrt{1 + s_1^2 + s_2^2}}. \tag{8.27}$$

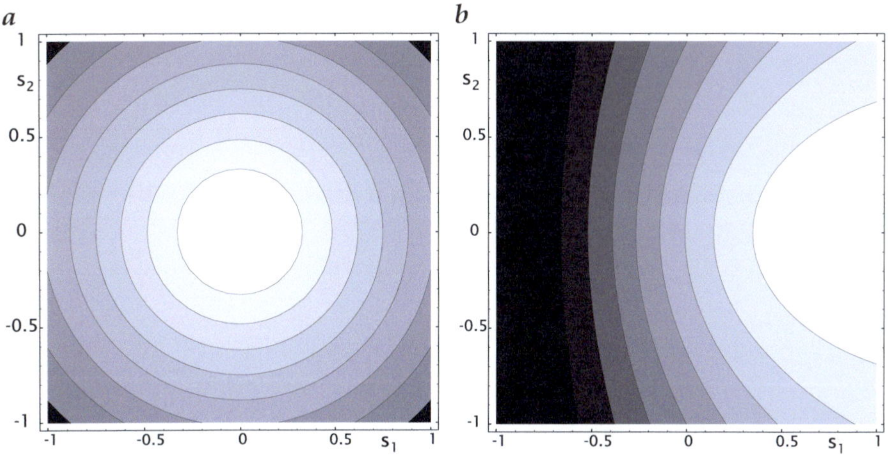

Figure 8.9: *Contour plot of the radiance of a Lambertian surface with a homogeneous reflection coefficient illuminated by parallel light, shown in gradient space for surface slopes between -1 and 1. The radiance is normalized to that of a flat surface.* ***a*** *Angle of incidence $\theta_i = 0°$; the distance of the contour lines is 0.05.* ***b*** *Oblique illumination with an angle of incidence of 45° and an azimuth angle of 0°; the distance of the contour lines is 0.1.*

Illustrations of contour plots of the radiance distribution in gradient space show Fig. 8.9a for a light source with an angle of incidence $\theta_i = 0°$. In this case, the contour lines of equal radiance are concentric circles and lines with constant absolute gradient $s = (s_1^2 + s_2^2)^{1/2}$.

However, the radiance hardly changes at small surface slopes. Oblique illumination causes a higher contrast in the radiance (Fig. 8.9b). With this type of illumination, however, the maximum surface inclination in the opposite direction to the light source is limited to $\pi/2 - \theta$ if the surface normal is perpendicular to the direction of light. With a single illumination source, the surface normal is only partially determined, even if the surface reflectivity is known. We then only know the component of the surface normal in the direction of the illumination change. Surface reconstruction with only one illumination source is a difficult mathematical problem that will not be pursued further here. Because of their limited possibilities, shape-from-shading methods have so far found virtually no practical application.

8.7 ♯References for Further Reading

A large part of the "Handbook of Computer Vision and Applications" deals with 3D image acquisition [165, Vol. I, Part IV]. [184] focuses on the acquisition of depth images using stereo analysis, shape from shading, photometric stereo analysis and structured illumination, while Jiang and Bunke [169] includes the further processing of depth images and the recognition of 3D objects. 3D imaging techniques are also important in close-range photogrammetry, see Luhmann et al. [210].

9 3D Image Acquisition with Active Lighting

Contents

9.1 Introduction

This chapter continues the considerations from the last chapter on 3D image acquisition. After the last chapter only presented passive methods, this one deals with active 3D image acquisition methods. This refers to methods that actively illuminate the scene to be recorded or manipulate or change its lighting.

The *light field* discussed in Section 8.3 still does not provide complete information about the scene. So far, we have simply taken the lighting as it

B. Jähne, *Digital Image Processing and Image Acquisition*,
https://doi.org/10.1007/978-3-032-04806-6_9

is and not illuminated the scene from all possible directions. This approach extends the light field in Eq. (8.23) by two dimensions to

$$L(x_r, y_r, s, t, \theta_i, \phi_i), \tag{9.1}$$

where θ_i and ϕ_i define the direction from which the light is illuminated. It is obvious that this extended light field includes the complete information about the bidirectional reflectance distribution function (*BRDF*) (Section 2.6.4), since now a surface point is illuminated and viewed from all directions. This observation shows how much more powerful 3D imaging techniques with active illumination can be compared to passive methods.

We start with directional measurement techniques (Section 9.2), which include very different methods. Methods with structured illumination (Section 9.2.1), active triangulation (Section 9.2.2) and confocal microscopy (Section 9.2.3) are presented. Time-of-flight measurement is only possible actively (Section 9.4), since the light that is recorded with an imaging system must in one way or another carry information about the distance it has traveled. We discuss time-of-flight methods (Section 9.4.1), interferometric methods (Section 9.4.2) and coherence methods (Section 9.4.3).

Active measurements of the inclination of surfaces are the subject of Section 9.5. This includes two classes of methods. On the one hand, those that are suitable for diffusely reflecting surfaces, such as the photometric stereo technique (Section 9.5.1), a shape-from-shade method (Section 8.6.1) with active illumination, and those for specular surfaces, in which either light reflection (deflectometry) or light refraction (refractometry, Section 9.5.2) is used.

9.2 Active Direction Measurement

9.2.1 Structured Illumination

Passive depth measurement from focusing (Section 8.5.2) can only be used if a surface has fine details. For featureless, smooth surfaces, the confocal illumination technique of Scheuermann et al. [289] can be used, which projects statistical patterns onto the focal plane and then uses the same evaluation method (see Section 1.2.2 and Fig. 1.3).

9.2.2 Depth from Active Triangulation

Instead of a stereo camera arrangement, a camera can be replaced by a light source. For a depth reconstruction, it is then necessary to identify the direction from which the illumination for each pixel comes. This information is equivalent to determining a disparity. Thus, a depth measurement method based on *active triangulation* has the same properties as stereoscopy, which we discussed in the last section.

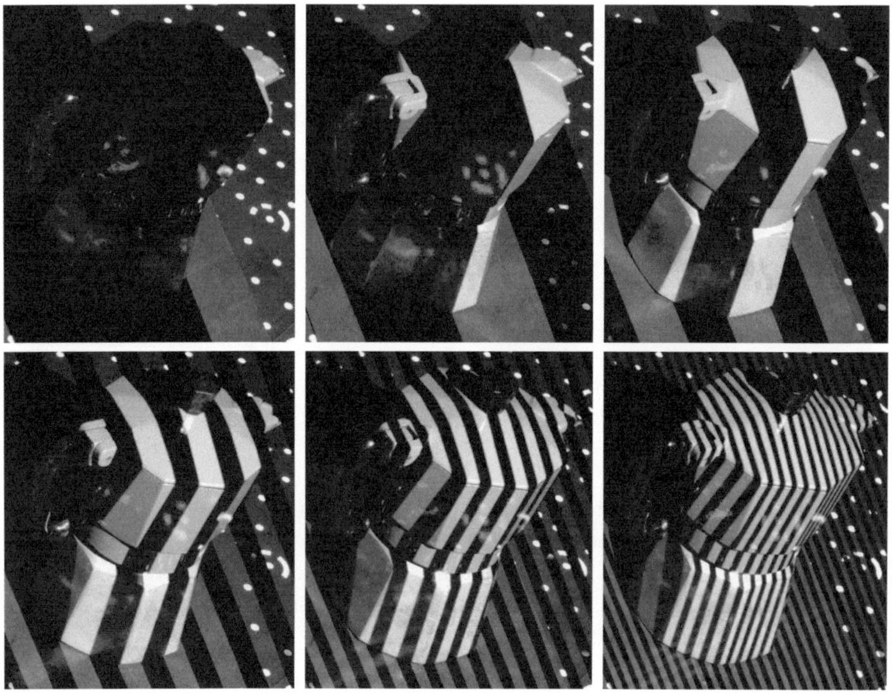

Figure 9.1: Active triangulation by projecting a series of fringe images of different wavelengths for binary coding of the horizontal position.

In recent years, unique encoding techniques for illumination have been developed. Most commonly, a light projector is used to illuminate the scene with stripe patterns in which the stripes are perpendicular to the triangulation base. A single stripe pattern is not sufficient to clearly determine the position of the pattern on the image plane. However, with a suitable sequence of stripe patterns of different wavelengths, the position in the image plane of the projector can be clearly coded by a sequence of light and dark values. Part of such an illumination sequence shows Fig. 9.1.

Such an illumination sequence also has the advantage that — within the limits of the dynamic exposure range of the camera — the detection of the fringe patterns becomes independent of the reflection coefficient of the object and the distance-dependent illuminance of the projector. The *occlusion problem*, which can be seen in the shadow behind the espresso machine in Fig. 9.1, remains.

Binary coding with a sequence of stripe patterns no longer works for fine stripe patterns. These will have to be at least one, more likely several pixels wide. Therefore, the accuracy of the disparity using binary coding is not even pixel-precise and therefore worse than even with a stereo method that is only pixel-precise.

Figure 9.2: Active triangulation using phase-shifted fringe images of the same wavelength. Three of the four patterns with a phase shift of 0, 90 and 180° are shown.

However, phase-shifted periodic patterns with the same wavelength, as shown in Fig. 9.2, can be used for a more precise position determination. With a hybrid method, first the rough position can be determined using binary coding and then the exact positioning using phase shifting. This results in a periodic pattern in the image

$$g = g_0 + g_m(\cos(2\pi\kappa x' + \Delta\varphi) \tag{9.2}$$

with three unknowns: the background brightness g_0, the amplitude of the modulation g_m and the unknown phase shift $\Delta\varphi$, which corresponds to a disparity or parallax $p' = \Delta\varphi/(2\pi\kappa)$, from which the exact depth can be determined.

To determine the three unknowns of the pattern, it is necessary to project at least three of the same patterns onto the objects with equidistant phase shifts. Here, we consider the general case of equally sampled phase shifts of P within one period, ranging from 0 to 2π:

$$\varphi_n = \frac{2\pi p}{P} \quad \text{with} \quad p = \{0, 1, \ldots, P-1\}. \tag{9.3}$$

This results in a sequence with P brightness values g_p at every point in the image. This means that we sample the periodic function equidistantly over a wavelength of P values. The three unknowns result directly from the fact that the signal in Eq. (9.2) is composed of a constant component g_0 and a periodic one with the discrete wavenumber one. This allows us to directly apply the discrete Fourier transform (Section 6.3.2). For the constant part, this results in

$$g_0 = \frac{1}{P}\sum_{p=0}^{P-1} g_p \tag{9.4}$$

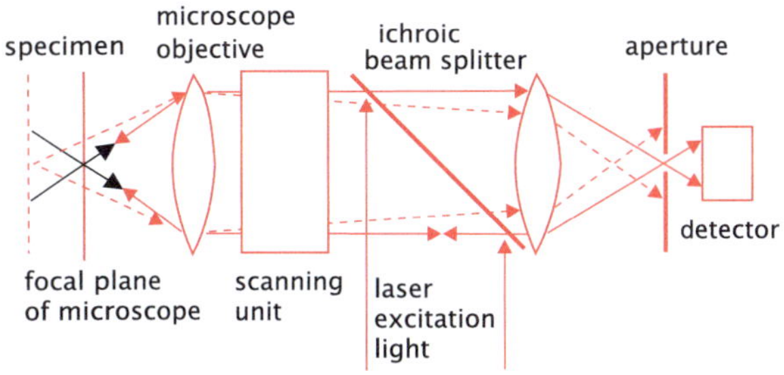

Figure 9.3: *Principle of confocal laser scanning microscopy.*

and for wavenumber one an amplitude and phase shift of

$$g_m = \frac{2}{N}\left|\sum_{p=0}^{P-1} g_p \exp\left(\frac{2\pi i p}{P}\right)\right| \quad \text{and} \quad \Delta\varphi = \arg\left[\sum_{p=0}^{P-1} g_p \exp\left(\frac{2\pi i p}{P}\right)\right]. \tag{9.5}$$

As the wavelength of the periodic pattern is only a few pixels wide, this method can be used to determine the disparity with sub-pixel accuracy.

9.2.3 Confocal Microscopy

At the beginning of the previous chapter, we analyzed in detail that no volume images can be recorded by projective optical imaging (Section 8.2). This is due to the fact that the three-dimensional optical transfer function is zero by a wide range in the direction of the optical axis (Fig. 8.1b). This problem also applies to microscopy. Even if one tries to reconstruct three-dimensional objects from microscopic focus series, it is impossible to see structures lying in the image plane (Fig. 17.12).

Three-dimensional microscopy is of great importance in the life sciences and medicine. Intensive research has therefore been carried out into whether it is possible to change the imaging process — and thus the point response — so that volume images can be recorded.

Finally, special methods of confocal fluorescence microscopy have led to overcoming the diffraction-limited resolution limit (Sections 3.5.3 and 3.5.4). The best-known technique is called stimulated emission depletion or *STED* for short and led to the 2014 Nobel Prize in Chemistry. Today, there are a number of such techniques that are summarized under the term *nanoscopy* [60, 329].The solution is called *confocal laser scanning microscopy, CLSM*) and is based on a special illumination technique in which points in the focal plane are irradiated much more strongly than outside of it. This is achieved by scanning a laser beam across the image plane, which is focused on the focal plane via the microscope optics (Fig. 9.3). Since the same optics are used for image acquisition and illumination, the intensity distribution in the

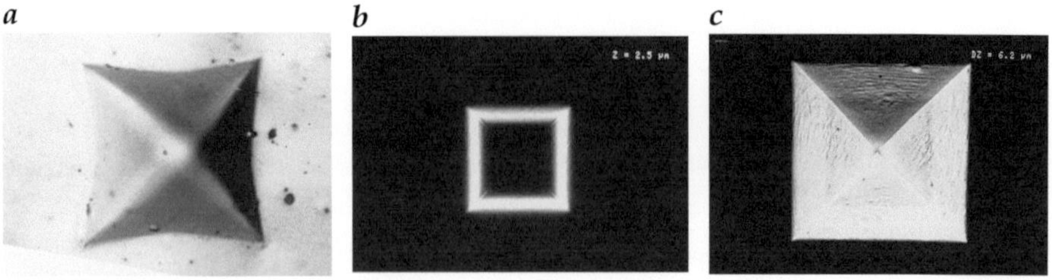

Figure 9.4: *Illustration of confocal laser scanning microscopy:* ***a*** *A square, pyramid-shaped crystal, imaged with a standard microscope and focus on the base of the pyramid.* ***b*** *Similar object, photographed with the CLSM. Only a narrow height contour area 2.5 μm above the base of the square pyramid is visible.* ***c*** *Image reconstructed from a 6.2 μm wide depth scan. The images were kindly provided by Carl Zeiss Jena GmbH.*

object space is approximated by the microscope's point response. (Small differences occur due to the coherence of the laser light).

Only a small circular area in the focal plane is brightly illuminated. Outside the focal point, the light intensity decreases with the square of the distance from the focal plane. In this way, interference from out-of-focus objects outside the focal plane is greatly suppressed. But is it possible to achieve a completely interference-free reconstruction? We will follow two different lines of thought to answer this question.

First, let us imagine a repeating structure in the z-direction. In conventional microscopy, this structure is lost because all depth ranges are irradiated with the same intensity. With confocal microscopy, however, we still see a periodic variation in the z direction due to the sharp drop in light intensity, provided that the wavelength in the z direction is not too small.

We can show the same using the PSF. The PSF of confocal microscopy results from the product of the spatial intensity distribution and the PSF of the optical image. Since both functions decrease with z^{-2}, the PSF of the confocal microscope decreases with z^{-4}. The squaring of the PSF in the spatial domain corresponds to a convolution of the OTF with itself. This convolution also makes the OTF non-zero in the z direction up to the resolution limit.

The 3D resolution of confocal laser scanning microscopy is illustrated by Fig. 9.4. An image taken with a standard microscope shows a crystal in the shape of a square pyramid in sharp focus only at the base of the pyramid (Fig. 9.4a). The edges become increasingly blurred towards the tip of the pyramid. In contrast, in a single image taken with confocal laser scanning microscopy, only a narrow depth range is imaged at all (Fig. 9.4b). However, an image created by superimposing all partial images obtained in a wide range of 6.2 μm shows a sharp object for the entire depth range (Fig. 9.4c). This allows many fine details to be observed that cannot be seen in the image taken with conventional microscopy.

Laser scanning microscopy has now become an indispensable tool in the life sciences [58, 59, 190]. Confocal illumination is used to analyze *fluorescence*

(Section 2.6.8) and thus to specifically image certain substances or parts of a cell. Using different excitation wavelengths and fluorescence spectra, several components can be quantitatively imaged simultaneously.

The strong focusing of the laser beam in combination with the high peak power of short laser pulses allows non-linear optical effects to be used (Section 2.6.9). Such strong photon currents are achieved that the combined energy of two photons with a longer wavelength acts like a photon with twice the energy or half the wavelength. This makes it possible to excite fluorescence in the visible spectrum with radiation in the near infrared, which is absorbed to a lesser extent by biological samples.

Finally, special methods of confocal fluorescence microscopy have led to overcoming the diffraction-limited resolution limit (Sections 3.5.3 and 3.5.4). The best-known technique is called stimulated emission depletion or *STED* for short and led to the 2014 Nobel Prize in Chemistry. Today, there are a number of such techniques that are summarized under the term *nanoscopy* [60, 329].

9.3 Tomography

In the last chapter, when discussing the light field recording technique, we saw that recording a scene from different directions not only has the potential to measure the depth of surfaces, as is the case with stereo methods, but also to distinguish objects at different depths due to their different orientation in the light field. *Tomographic* Methods that we have already mentioned in Section 8.4.1) now cleverly combine image acquisition from different directions with irradiation. They do not directly generate a 3D image of an object, but allow the three-dimensional shape to be reconstructed using appropriate methods.

9.3.1 Basic Principle

In tomographic procedures, the object is irradiated layer by layer from different directions. If we use a point light source (e. g.a X-ray tube, Fig. 9.5b), we see a perspective or fan-shaped projection (*fan beam projection*) on the screen (detector) behind the object, just as in optical vision (Section 3.3). If the point source and the projection screen are rotated around the object, a profile perpendicular to the respective projection axis is obtained for each projection direction.

We will first consider only the parallel projection (Fig. 9.5a), as this leads more easily to clear solutions. If the object absorbs radiation, the intensity loss measured in the projection on the screen is proportional to the path length of the rays in the object. The 3D shape of the object cannot be reconstructed from a single projection, but must be measured by rotating the radiation source and the projection screen around the object from all directions.

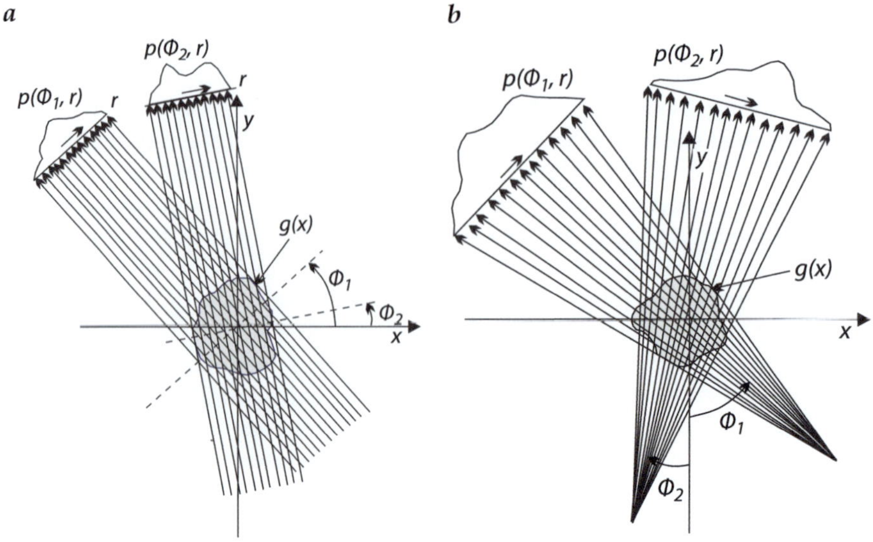

Figure 9.5: *a parallel projection and* ***b*** *fan-shaped projection in tomography.*

Like other methods, tomography can make use of the various interactions between matter and radiation. The most widely used is *transmission tomography*, which is based on the absorption of radiation, e. g. X-rays. Other methods include emission tomography, reflection tomography, time-of-flight tomography (especially with ultrasound) and magnetic resonance imaging (*MRI*).

From the point of view of the light field discussed in Section 8.3, tomography can also be regarded as a complete image of the 4D light field.

9.3.2 Radon Transform and Fourier Disk Theorem

With regard to the reconstruction, it is important that the projection can be viewed as a different 2D representation of the image at all angles ϕ. One coordinate is the position in the projection profile r, the other is the angle ϕ (Fig. 9.6). Accordingly, we can consider the parallel projection as a transformation of the image into another 2D representation. Reconstruction then merely means applying an inverse transformation. The crucial task is therefore to describe the tomographic transformation mathematically and to investigate whether an inverse transformation exists.

A projection beam is described by the projection angle ϕ and the offset r (Fig. 9.6). ϕ is the angle between the projection plane and the x-axis. We also assume that we intersect the 3D object parallel to the xy plane. Then the scalar product of a vector $\boldsymbol{x}$ on the projection ray and a unit vector

$$\hat{\boldsymbol{n}} = \left[\cos\phi, \sin\phi\right]^T \tag{9.6}$$

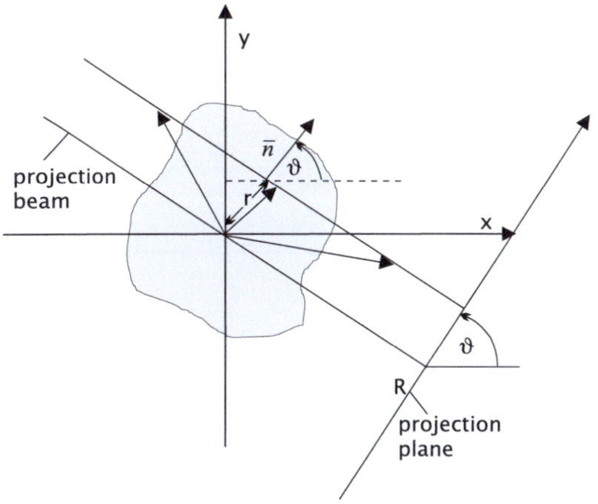

Figure 9.6: *Geometry of a projection beam.*

is constant perpendicular to the projection beam and equal to the offset r of the beam:

$$\boldsymbol{x}\hat{\boldsymbol{n}} - r = x\cos\phi + y\sin\phi - r = 0 \tag{9.7}$$

The projected intensity $p(r, \phi)$ is given by the integration along the projection ray:

$$p(r, \phi) = \int\limits_{\text{path}} g(\boldsymbol{x})\mathrm{d}s = \int\limits_{-\infty}^{\infty}\int\limits_{-\infty}^{\infty} g(\boldsymbol{x})\delta(x_1\cos\phi + x_2\sin\phi - r)\mathrm{d}^2x. \tag{9.8}$$

The δ*-distribution* Eq. (B.44) in this equation reduces the double integral to a projection ray in the direction ϕ at a distance r from the origin of the coordinate system. This projective transformation of a 2D function $g(\boldsymbol{x})$ to $p(r, \phi)$ is called *Radon transform* after the mathematician Radon.

To better understand the properties of the Radon transform, we analyze it in Fourier space. We can understand the Radon transform as a special linear shift invariant filter operation. Since it adds all values along the projection ray, the point response of the projection operator is a δ line (2D) or δ plane (3D) in the direction of the projection ray. In Fourier space, this convolution operation corresponds to a multiplication with the transfer function, a δ-line normal to the δ-line in spatial space (see > R13). In this way, the projection operator cuts a slice out of the spectrum in a direction perpendicular to the projection beam.

The easiest way to calculate this elementary relationship without limiting its generality is to rotate the coordinate system so that the projection direction coincides with the y' axis. Then the r-coordinate in $p(r, \phi)$ becomes identical to the x'-coordinate, and ϕ becomes zero. In this particular case, the Radon transformation is reduced to an integration in the y' direction:

$$p(x',0) = \int_{-\infty}^{\infty} g(x',y')\mathrm{d}y'. \tag{9.9}$$

We can write the Fourier transform of the projection function as

$$\tilde{p}(k_{x'},0) = \int_{-\infty}^{\infty} P(x',0)\exp(-2\pi \mathrm{i} k_{x'}x')\mathrm{d}x'. \tag{9.10}$$

If we replace $P(x',0)$ with the simplified Radon transformation in Eq. (9.9), we get

$$\tilde{p}(k_{x'},0) = \int_{-\infty}^{\infty}\left[\int_{-\infty}^{\infty} g(x',y')\mathrm{d}y'\right]\exp(-2\pi \mathrm{i} k_{x'}x')\mathrm{d}x'. \tag{9.11}$$

If we insert the factor $\exp(-2\pi \mathrm{i} 0y') = 1$ into this double integral, we see that it is the Fourier transform of $g(x',y')$ for $k_{y'} = 0$:

$$\begin{aligned}\tilde{p}(k_{x'},0) &= \int_{-\infty}^{\infty}\int_{-\infty}^{\infty} g(x',y')\exp(-2\pi \mathrm{i} k_{x'}x')\exp(-2\pi \mathrm{i} 0y')\mathrm{d}x'\mathrm{d}y' \\ &= \tilde{g}(k_{x'},0).\end{aligned} \tag{9.12}$$

By back transformation into the original coordinate system, we obtain

$$\tilde{p}(q,\phi) = \tilde{g}(\boldsymbol{\kappa})\delta(\boldsymbol{\kappa} - (\boldsymbol{\kappa}\bar{\boldsymbol{n}})\bar{\boldsymbol{n}}), \tag{9.13}$$

where q is the wavenumber in the direction of ϕ and $\bar{\boldsymbol{n}}$ is the normal vector introduced in Eq. (9.6). The spectrum of the projection is identical to the spectrum of the original object on a ray normal to the direction of the projection ray. This important result is known as the *Fourier disk* or *projection theorem*.

9.3.3 ♯Filtered Back Projection

If projections are available from all directions, a complete reconstruction is possible, as the resulting slices of the spectrum cover the entire spectrum of the object. The inverse Fourier transformation then provides the original image.

The filtered back projection method follows exactly this path with a slight modification, which is necessary for the following reason: If we were to simply add up the spectra of the individual projection beams to obtain the complete object spectrum, the spectral density for small wavenumbers would be far too high, as the beams are denser at a small radius than at a larger radius. We must therefore apply a suitable weighting factor to the spectrum of the projection beams before adding it up to the total spectrum.

In the continuous case, the geometry is very simple. The density of the projection rays is proportional to $|\boldsymbol{\kappa}|^{-1}$. Accordingly, the spectra of the projection beam must be multiplied by $|\boldsymbol{\kappa}|$. The filtered back projection method therefore consists of two steps. First, the individual projections must be filtered. Then the reconstruction is carried out by adding the back-projected projections.

In the first step, the spectrum of each projection direction is multiplied by a suitable weighting function $\tilde{w}(|\boldsymbol{\kappa}|)$. This operation can of course also be performed directly in the spatial domain by convolution with the inverse Fourier transform $w(r)$. Because of this filter operation, the method is called *filtered back projection*.

The second step is the back projection. Each projection direction produces a slice of the spectrum. The total spectrum is obtained by summing all parts. Since the Fourier transform is a linear operation, we can sum the filtered projections in the spatial domain. Each filtered projection contains the part of the object that is constant in the direction of the projection beam. Therefore, we can project the corresponding gray value of the filtered projection back along the direction of the projection ray and add it to the contributions of the other projection rays.

After this illustrative description of the principle of the filtered back projection algorithm, we derive the method for the continuous case, starting with the Fourier transform of the object and writing the inverse Fourier transform in polar coordinates (q, ϕ) in order to use the Fourier slice theorem:

$$g(\boldsymbol{x}) = \int_0^{2\pi}\int_0^{\infty} q\tilde{g}(q,\phi)\exp[2\pi \mathrm{i} q(x_1\cos\phi + x_2\sin\phi)]\mathrm{d}q\mathrm{d}\theta. \tag{9.14}$$

In this formula, the spectrum is already multiplied by the wavenumber q due to the transition to polar coordinates. However, this equation cannot be applied directly to the Fourier disk theorem Eq. (9.13) due to the integration limits. The coordinate q should run from $-\infty$ to ∞ and ϕ only from 0 to π. In Eq. (9.14) we only integrate over half a ray from the origin to infinity. We can compose a full ray from two half rays with the angles ϕ and $\phi + \pi$. We therefore split the integral in Eq. (9.14) into the angular ranges $[0, \pi[$ and $[\pi, 2\pi[$ and obtain

$$\begin{aligned} g(\boldsymbol{x}) &= \int_0^{\pi}\int_0^{\infty} q\tilde{g}(q,\phi)\exp[2\pi \mathrm{i} q(x_1\cos\phi + x_2\sin\phi)]\mathrm{d}q\mathrm{d}\phi \\ &+ \int_0^{\pi}\int_0^{\infty} q\tilde{g}(-q,\phi')\exp[-2\pi \mathrm{i} q(x_1\cos\phi' + x_2\sin\phi')]\mathrm{d}q\mathrm{d}\phi'. \end{aligned}$$

We have used the identities $\phi' = \phi + \pi$, $\tilde{g}(-q, \phi) = \tilde{g}(q, \phi')$, $\cos(\phi') = -\cos(\phi)$ and $\sin(\phi') = -\sin(\phi)$. Now we can summarize the two partial integrals if we replace q by $-q$ and $\tilde{g}(q, \phi)$ by $\tilde{p}(q, \phi)$ in the second integral because of the Fourier slice theorem Eq. (9.13):

$$g(\boldsymbol{x}) = \int_0^{\pi}\int_{-\infty}^{\infty} |q|\tilde{p}(q, \phi) \exp[2\pi i q(x_1 \cos\phi + x_2 \sin\phi)] \mathrm{d}q \mathrm{d}\phi. \tag{9.15}$$

Equation Eq. (9.15) represents the inverse Radon transformation and is the basis for the *filtered back projection algorithm*. The inner integral over q is the back projection of a single projection:

$$P' = \mathcal{F}^{-1}(|q|\mathcal{F}P). \tag{9.16}$$

Where $\mathcal{F}$ represents the 1D Fourier transform operator. P' is the projection function P multiplied by $|\boldsymbol{q}|$ in Fourier space. If we perform this operation as a convolution in space, we can formally write it as

$$P' = [\mathcal{F}^{-1}(|q|)] * P. \tag{9.17}$$

With the outer integral in Eq. (9.15) over the angle ϕ

$$g(\boldsymbol{x}) = \int_0^{\pi} p'(r, \phi)\,\mathrm{d}\phi, \tag{9.18}$$

the object is built up by the individual back-projected filtered projections from all directions. It is important to note that the filtered projection profile $p'(r, \phi)$ in Eq. (9.18) must be understood as a 2D function in order to build the 2D object $g(\boldsymbol{x})$. This means that the projection profile must be projected back in the projection direction.

9.3.4 ♯Discrete Filtered Back Projection

There are some details that have not been discussed so far, but which are responsible for serious problems in reconstruction in the continuous case. First, we note that it is impossible to reconstruct the mean value of an object because $\tilde{g}(\boldsymbol{0})$ is eliminated due to the multiplication by $|\boldsymbol{\kappa}|$ in the Fourier space Eq. (9.15). Secondly, it is impossible to reconstruct an object of infinite size, since the projection rays then always provide infinite values.

Fortunately, all these problems disappear when we switch from the infinite continuous case to the finite discrete one. Then the objects only have a finite size. In practice, the size is limited by the distance between the radiation source and the detector. The resolution of the projection profile is limited by the combined effects of the expansion of the radiation source and the resolution of the detector in the projection plane. Finally, we can only

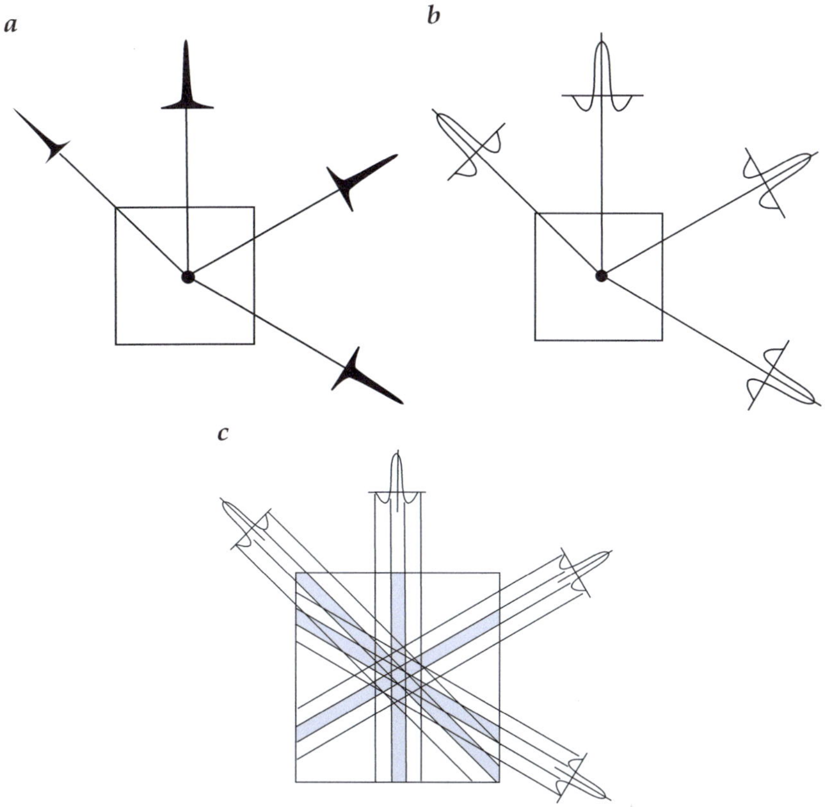

Figure 9.7: *Illustration of the filtered back projection with a point object:* ***a*** *Projections from different directions;* ***b*** *Filtering of the projections;* ***c*** *Summation of the back-projected filtered projections.*

use a limited number of projections. This corresponds to sampling the angle ϕ.

We illustrate these issues in this section with an illustrative example. We can learn the essentials about projection and reconstruction by looking at the reconstruction of the simplest object, a point. This is because the radon transformation Eq. (9.8) and its inverse Eq. (9.15) are linear transformations. The projections of a point look the same from all directions (Fig. 9.7a) and show a sharp peak in the projection function $p(r, \phi_i)$. In the first step of the filtered back projection algorithm, this projection function p is convolved with the $|k|$ filter. This results in a modified projection function p' that is identical to the point response (PSF) of the filter $|k|$ (Fig. 9.7b).

In the second step, the back projections from different directions in the image are added together. From Fig. 9.7c we can see that the peaks of all projection functions only add up at the position of the point in the image. At all other positions in the image, positive and negative values are added

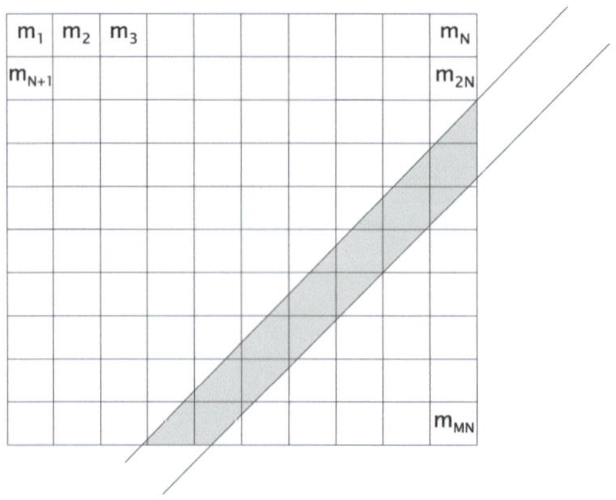

Figure 9.8: *Illustration of algebraic reconstruction from projections: A projection ray d_k crosses the image matrix. All pixels that are hit by the ray contribute to the projection.*

depending on the position of the projection direction. If the projection directions are sufficiently close to each other, they cancel each other out with the exception of the point in the center of the image. Fig. 9.7c also shows that an insufficient number of projections leads to a star-shaped interference pattern surrounding the point.

Using the simple example of reconstructing a point from its projections, we can also show the importance of filtering the projections. Let us imagine what happens if we omit this step. Then we add δ lines as back projections, which rotate around the position of the point. Accordingly, we do not get a point, but a rotationally symmetric function that decreases with $|\boldsymbol{x}|^{-1}$. The result is an inferior reconstruction with heavily smeared structures.

9.3.5 ♯Algebraic Tomographic Reconstruction

The reconstruction of volume images from projections is also possible using algebraic methods. In this section we discuss that the reconstruction from projections can also be considered as a linear *inverse problem* with image data. At the same time, we learn a lot about the solvability of such problems in an illustrative way. The mathematical basics of discrete inverse problems are presented in Appendix C.6.

In order to apply the inverse methods, image data must be mapped to vectors, the *image vectors*. This mapping is easy to perform by renumbering the pixels of the image matrix row by row (Fig. 9.8). In this way, a $M \times N$ image matrix is transformed into a column vector with the dimension $P = M \times N$:

$$\boldsymbol{p} = \left[m_1, m_2, \ldots, m_p, \ldots, m_P\right]^T. \tag{9.19}$$

Now we think of a projection ray that passes through the image matrix (Fig. 9.8). We can then assign a weighting factor to each pixel of the image vector, which represents the contribution of the pixel to the projection ray. We can combine these factors in another P-dimensional vector g_q:

$$g_q = \left[g_{q,1}, g_{q,2}, \ldots, g_{q,p}, \ldots, g_{Q,P}\right]^T. \tag{9.20}$$

The total emission or absorption along the q-th projection beam d_q then results as the scalar product of the two vectors g_q and p:

$$d_q = \sum_{p=1}^{P} g_{q,p} m_p = g_q p. \tag{9.21}$$

If Q projection rays cross the image matrix, we obtain a linear system of equations with Q equations and P unknowns:

$$\underbrace{d}_{Q} = \underbrace{M}_{Q\times P} \underbrace{p}_{P}. \tag{9.22}$$

The *data vector* d contains all measured projections and the *parameter vector* p all pixels of the image matrix to be reconstructed. The *model matrix* M describes the relationship between these two vectors by specifying how the projection rays traverse the image matrix for a given arrangement. With corresponding weighting factors, the limited resolution of the detector and the size of the radiation source can be directly taken into account.

Algebraic tomographic reconstruction is a very general and flexible method. In contrast to the filtered back projection (Section 9.3.3), it is not limited to a parallel projection. The rays can cross the image matrix in any way and in principle even be curved. As an additional advantage, we obtain an estimate of the reconstruction error.

However, it should be borne in mind that algebraic reconstruction requires the solution of very large systems of equations. The model vector contains pixels of an image. Even at moderate resolution with 256×256 pixels, the inverse of a 65536×65536 matrix must be calculated. This matrix contains $4 \cdot 10^9$ points. Therefore, special techniques must be developed to solve such large systems of equations.

9.4 Active Depth Measurement

9.4.1 Time-of-Flight Techniques

Time-of-flight methods measure the time it takes for radiation to travel a certain distance. If the radiation is emitted from the position of the camera, it must cover the distance z between the camera and the object reflecting the radiation twice. Therefore, the delay time τ results in

$$\tau = \frac{2z}{c}, \tag{9.23}$$

where c is the propagation speed of the radiation.

From Eq. (9.23) we see that the statistical error of the depth measurement is independent of the distance to the object. It only depends on the accuracy with which the delay time can be measured:

$$z = \frac{c\tau}{2} \quad \rightsquigarrow \quad \sigma_z = \frac{c}{2}\sigma_\tau. \tag{9.24}$$

At first glance, this looks like a clear advantage over triangulation methods (see Eq. (8.18) in Section 8.5.1), where the measurement uncertainty increases quadratically with the distance. However, this only applies if the measurement error of the transit time is independent of the distance.

The greatest difficulty of the time-of-flight method is the distance squared law. If the radiation emitted by the light source is in the far field, the irradiance of the object decreases quadratically with distance. A far field is present when the emitted area is significantly larger than the diameter of the light source.

With transit times, the first thing that comes to mind is *pulse modulation*, i. e. the direct measurement of the delay time between the transmission and reception of a short pulse. The maximum measurable distance depends on the pulse repetition frequency. With electromagnetic waves, measuring the delay time is a challenge. Since the speed of light c is so high at $3 \cdot 10^8$ m/s, the delay time is only 6.7 ns per meter. A direct transit time measurement has the advantage that the measurement uncertainty could actually be largely independent of the distance. In the best case, one photon is sufficient to measure the transit time. The maximum measurement distance will therefore be relatively sharply limited.

Pulse modulation is just one of many ways to modulate a signal for transit time measurements. The signal amplitude can also be modulated periodically. In this method, the transit time is measured as a phase shift between the outgoing and incoming signal:

$$z = \frac{c}{4\pi\nu}\phi \quad \rightsquigarrow \quad \sigma_z = \frac{c}{4\pi\nu}\sigma_\phi, \tag{9.25}$$

where ν is the frequency of the modulation. The depth range of this technique results from the fact that the phase can clearly only be measured in a range of $\pm\pi$:

$$\Delta z = \frac{c}{2\nu} = \frac{cT}{2}. \tag{9.26}$$

One disadvantage of periodic modulation is therefore the limited depth range. Another is that signals of several depths cannot be distinguished from one another, as they overlap again to form a periodic signal.

It is still unclear how a time-of-flight image sensor with periodic modulation could function. The basic principle could be described as a charge swing. Within a period of radiation modulation, the charge carriers at each pixel are collected in different memories depending on the phase position of the outgoing radiation [297, 357]. In this way, the received signal is sampled as a function of the phase, as we have already seen with active triangulation methods with periodic patterns (Section 9.2.2). The easiest to handle is a pixel with 4 memories [296], in which a period is scanned with four points at 90° intervals. Then the equations for amplitude and phase position Eq. (9.5) simplify to

$$g_m = \frac{1}{2}\sqrt{(g_0 - g_2)^2 + (g_1 - g_3)^2} \quad \text{and} \quad \Delta\varphi = \arctan\frac{g_3 - g_1}{g_2 - g_0}. \tag{9.27}$$

A third disadvantage becomes clear from the equations. In the far field, the measured modulated amplitude g_m decreases quadratically with distance. If there is no background light and the sensor with a SNR $\propto \sqrt{g_m}$ (compare Section 4.5.3) is ideal, then the measurement uncertainty only decreases linearly with distance. If the background light is high, the sensor noise will be dominated by this constant signal. Then the measurement uncertainty even decreases quadratically with distance, as with triangulation methods.

9.4.2 ♯Interferometry

Interferometry can be regarded as a special case of periodic modulation. The modulation results directly from the frequency of the electromagnetic radiation. Nevertheless, it makes sense to consider interferometry as a separate class of depth measurement technology because, unlike the other methods, it relies on *coherent light* (Section 2.2.6). Due to the high frequency of the light, the phase of the outgoing and incoming light cannot be measured directly, but only by the amplitude variation resulting from the coherent optical superposition of the incoming and outgoing light. The error and range of the depth determination from interferometric measurements result from Eqs. (9.25) and (9.26) and the relationship $c = \nu\lambda$ (Section 2.3.1) to

$$z = \frac{\lambda}{4\pi}\phi, \quad \sigma_z = \frac{\lambda}{4\pi}\sigma_\phi, \quad \Delta z = \frac{\lambda}{2}. \tag{9.28}$$

Due to the small wavelength of light (0.4–0.7 μm), interferometric depth measurements are extremely accurate. The very limited depth range of only half a wavelength can be overcome by *multi-wavelength interferometry*.

9.4.3 ♯Coherence Method

A second type of interferometric depth measurement technique is possible if radiation with a low coherence length of only a few wavelengths is used. In this case, interference patterns are also only possible in a depth range of a few wavelengths. This effect can be used for depth measurement by a recording system that passes through a depth range. This type of interferometry is called *white light interferometry* or *coherence radar*.

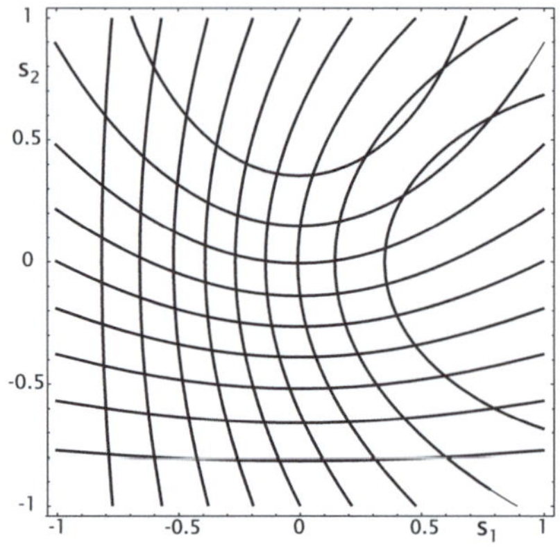

Figure 9.9: Superimposed contour graphics of the radiance of a Lambertian surface with a homogeneous reflection coefficient. The surface is illuminated at an angle of incidence of 45° and an azimuth angle of 0° and 90°.

9.5 Active Measurement of Surface Slope

9.5.1 Photometric Stereo

This section ties in with the considerations on shape-from-shading behaviors in Section 8.6. There we saw that an unambiguous reconstruction of the surface slope is not possible with illumination because there is only one measurement for the two unknown surface directions. With active illumination methods, more than one illumination can be realized.

But even with two illuminations, there are still problems due to the nonlinearity of the method. The curved contour lines in Fig. 9.9 mean that the relationship between the surface slope and the radiance is not linear. As a result, the surface slope cannot be determined unambiguously, even if we illuminate a surface one after the other with two differently oriented light sources (Fig. 9.9). This is the case if the curved contour lines intersect at more than one point. You therefore need three different types of lighting to arrive at a clear solution.

The use of three light sources has a decisive advantage: the reflection coefficient of the surface can be eliminated by calculating ratios. To look at an example, let's illuminate a Lambertian surface with parallel light rays from three different directions:

$$\boldsymbol{l}_1 = [0,\ 0,\ 1]^T, \quad \boldsymbol{l}_2 = [\tan\theta_i,\ 0,\ 1]^T, \quad \boldsymbol{l}_3 = [0,\ \tan\theta_i,\ 1]^T. \qquad (9.29)$$

This results in the ratio of the radiances L (cf. Section 8.6):

$$L_2/L_1 = \frac{-s_1\tan\theta_i + 1}{\sqrt{1+\tan^2\theta_i}}, \quad L_3/L_1 = \frac{-s_2\tan\theta_i + 1}{\sqrt{1+\tan^2\theta_i}}. \qquad (9.30)$$

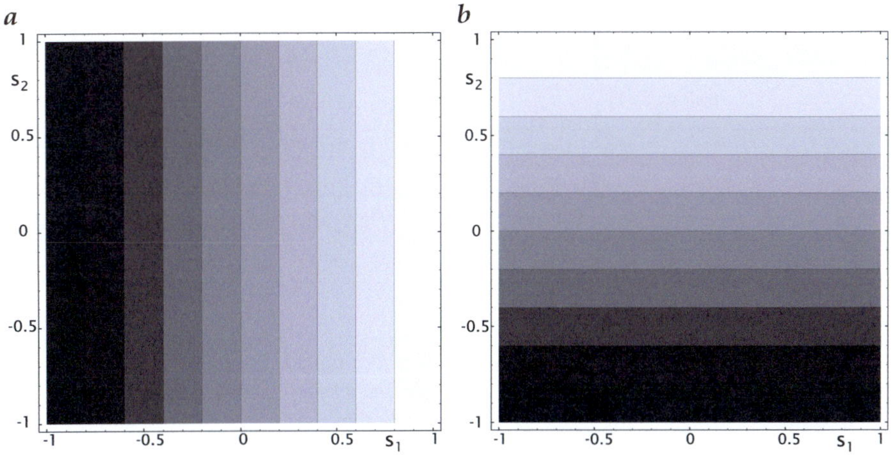

Figure 9.10: *Contour plot of the radiance of a Lambertian surface illuminated with parallel light with an angle of incidence* $\theta_i v$ *on 45° and an azimuth angle* ϕ_i *of* ***a*** *0° or* ***b*** *90°. The radiance was normalized to Eq. (9.30) at an angle of incidence of 0°. The distance of the contour lines is 0.1. Note the perfect linear relationship between the normalized radiance and the components of the surface slope in the x and y directions.*

The equations are now linear in the slopes s_1 and s_2 and — another decisive advantage — decoupled: s_1 and s_2 are now only dependent on L_2/L_1 and L_3/L_1 respectively (Fig. 9.10). In addition, the ratios of the radiation densities in Eq. (9.30) also do not depend on the reflection coefficient of the surface. This is contained in Eq. (8.24) as a factor and disappears when the ratio of two radiance distributions of the same surface is calculated.

9.5.2 ♯Defractometry

Photometric stereo (Section 9.5.1) does not work with shiny surfaces, as the light is only reflected in the direction of the camera if the angle of incidence from the light source is equal to the angle of reflection towards the camera. For this reason, extended light sources are required. It turns out that for transparent shiny surfaces, the technique of *shape from refraction* offers advantages over the shape-from-reflection techniques. The radiance is higher, steeper surfaces can be determined and the non-linearity of the relationship between slope and radiance is considerably lower.

The shape-from-refraction method, or *defractometry*, requires a special illumination technique because — except for a small portion of the light that is reflected at the surface — no significant variations in radiance occur. The basis of the shape-from-refraction technique is *telecentric illumination*, which converts a spatial distribution of radiance into an angularly dependent radiance distribution. Then we need to calculate the relationship between the surface slope and the angle of the refracted beam and use an extended

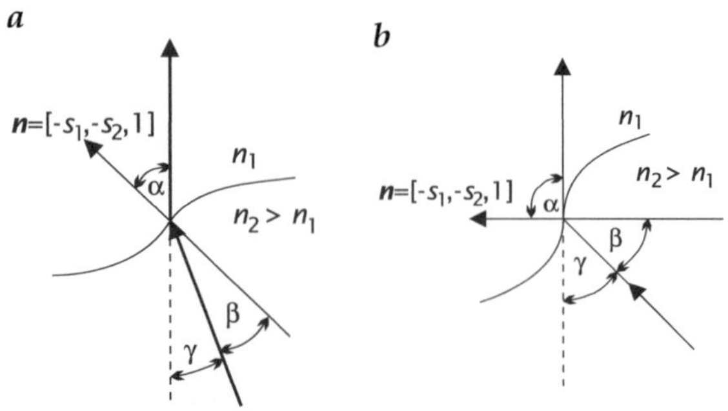

Figure 9.11: The basis of the shape-from-refraction technique is the refraction of light on an inclined surface. The camera is located far above the surface. ***a*** *rays emanating from the light source at the angle γ are refracted in the direction of the camera.* ***b*** *Even for an infinite gradient (vertical surface, $\alpha = 90\,°$), rays from the light source hit the camera.*

light source with a suitable beam distribution. We need to investigate this first.

Fig. 9.11 illustrates the optical geometry for the simple case that the camera is located far above and a light source below a transparent medium with a higher refractive index. The relationship between the surface slope s and the angle γ results from Jähne et al. [163] to

$$s = \tan\alpha = \frac{n\tan\gamma}{n - \sqrt{1+\tan^2\gamma}} \approx \frac{n}{n-1}\tan\gamma\left(1 + \frac{1}{2(n-1)}\tan^2\gamma\right) \qquad (9.31)$$

with $n = n_2/n_1$. The inverse relation is

$$\tan\gamma = s\frac{\sqrt{n^2+(n^2-1)s^2}-1}{\sqrt{n^2+(n^2-1)s^2}+s^2} \approx \frac{n-1}{n}s\left(1 - \frac{n-1}{2n^2}s^2\right). \qquad (9.32)$$

In principle, the shape from refraction technique even works for infinite slopes (vertical surfaces). In this limiting case, the beam going to the camera grazes the surface (Fig. 9.11b) and

$$\tan\gamma = \sqrt{n^2-1}. \qquad (9.33)$$

Through refraction, we achieve that light rays come from a certain direction for a certain inclination. This gives us differences in brightness if we make the intensity of the light rays dependent on the direction. How this is done is described in detail in Jähne et al. [163]. Here we simply assume that the light rays change proportionally to $\tan\gamma$ in the x_1 direction. Then the following applies:

$$L \propto s_1\frac{\sqrt{n^2+(n^2-1)s^2}-1}{\sqrt{n^2+(n^2-1)s^2}+s^2}. \qquad (9.34)$$

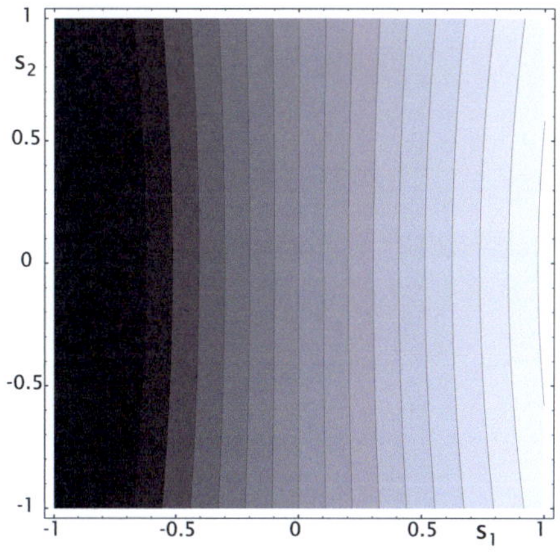

Figure 9.12: *Radiance map for the shape-from-refraction technique. The radiance of a telecentric light source varies linearly in the x_1 direction.*

Of course, the problem also arises here that no vector such as the surface inclination can be derived from a scalar quantity such as the radiation density. However, the shape-from-refraction technique comes very close to an ideal measurement setup. If, as assumed, the radiance varies linearly only in the x_1 direction, then the radiance map in gradient space is also almost linear (Fig. 9.12). Only with very steep slopes does a certain influence of the slope in the x_2 direction become visible, which results from the terms with s^2 in Eq. (9.34).

Ratio images can also be used with the shape-from-refraction technique. This can be implemented with a color image, as it contains three independent channels, red, green and blue (Section 2.5.2). With a total of three channels, we can clearly determine the surface inclination and still have one degree of freedom left for corrections. Color images have the advantage that all three channels are recorded simultaneously, so that the technique can also be applied to moving objects.

A clear color coding of surface inclinations can be achieved with the following color wedge illuminations, where the green wedge runs in the x direction and the red and blue in the two diagonal directions. Then

$$\begin{aligned}
G(\boldsymbol{s}) &= (1/2 + cs_1)E_0(\boldsymbol{s}) \\
R(\boldsymbol{s}) &= [1/2 - c/2(s_1 + s_2)]E_0(\boldsymbol{s}) \\
B(\boldsymbol{s}) &= [1/2 - c/2(s_1 - s_2)]E_0(\boldsymbol{s}),
\end{aligned} \quad (9.35)$$

where we have assumed a relationship of the form $s_1 E_0(\boldsymbol{s})$ as in Eq. (9.34) and c is a calibration factor between the brightness and the surface slope.

Now we have three illuminations available to determine two slope components. So we can use one to compensate for unwanted variations in irradiance E_0. This is possible by normalizing the three color channels with

Figure 9.13: *Application of defractometry to the measurement of wind-generated water waves in a wind/wave channel. The wave height was reconstructed from the slope field and the measured slope was visualized in color using appropriate colour coding [269].*

the sum of all channels $G + R + B$:

$$\frac{G}{G+R+B} = \frac{2}{3}\left(\frac{1}{2} + cs_1\right), \quad \frac{B-R}{G+R+B} = \frac{2}{3}cs_2. \tag{9.36}$$

The position on the color wedge from which the light originates then results in

$$s_1 = \frac{1}{2c}\frac{2G-R-B}{G+R+B}, \quad s_2 = \frac{3}{2c}\frac{B-R}{G+R+B}. \tag{9.37}$$

From these position values, the x- and y-components of the slope can be calculated according to Eq. (9.34). The height can then also be reconstructed from the two components of the surface slope up to a constant. As an example of defractometry, Fig. 9.13 shows the measurement of the slope and height of wind-generated water waves.

9.6 ♯References for Further Reading

The "Manual on industrial image processing" [277] contains several chapters on optical 3D measurement technology, which also show the variety of devices available. The "Handbook of Optical Metrology" (Schuth, 2017) provides an even more detailed description. It describes optical methods for 3D shape detection and surface analysis in great detail, including coherent irradiation. The focus is on practical applications, so the descriptions are not very systematic or in-depth. However, after reading this chapter and the previous one, the reader should be able to categorize the different techniques and rate their performance and limitations.

The monograph by Saleh [278] has an interesting common denominator. It describes imaging techniques that can see below the surface in very different ways.

3D imaging techniques are also important in close-range photogrammetry, see Luhmann et al. [210]. The conference proceedings of the Oldenburg 3D Days are also a good source of information on 3D imaging techniques. The 3D Days have been held annually since 2002 with interruptions in 2021 and 2023 [205, 207, 208] and are continued from 2024 together with the Oldenburg BIM Day [209].

Part II

Image Processing

10 Neighborhood Operators and Local Structure

Contents

B. Jähne, *Digital Image Processing and Image Acquisition*,
https://doi.org/10.1007/978-3-032-04806-6_10

10.1 Basic Properties and Purpose

10.1.1 Features for Object Recognition

An analysis of the spatial relationships of the gray values of a small environment provides us with initial information for object recognition in images. As a simple example, let's take a scene containing objects with uniform radiance. If the values within a small neighborhood do not change, it lies within an object. However, if the values change significantly, the edge of an object lies within the neighborhood. In this way, areas of constant values can be distinguished from edges.

Processing individual pixels of an image using *point operations* alone does not provide this information. Such operations are only useful as a first step in image processing to compensate for inhomogeneous and nonlinear responses of the image sensor, to interactively inspect and manipulate images or to improve the visual impression (Chapter 7).

New classes of operations, *neighborhood operations*, are necessary to combine pixels of a small neighborhood in the right way and provide a new image as a result. Neighborhood operations are a central tool in the first stage of image processing. This is the reason why we discuss possible classes and their properties in this chapter.

The result of each neighborhood operation is again an image, but with modified content. For example, a neighborhood operation to detect edges should result in high values for pixels that belong to an object edge. In contrast, low values should be generated for all pixels that lie within an object, regardless of their gray value. This simple example also shows that information is generally lost when using a neighborhood operator. We can no longer deduce the original gray values. For this reason, neighborhood operations are also called *filter*. Filters extract certain properties of interest from images. The resulting image is therefore called *property image*.

Operators that combine neighboring pixels to create a new image can perform a variety of operations and extract properties:

- Averaging and suppression of noise (Chapters 11)
- Detection and differentiation of simple local structures such as constant gray value areas, surfaces, edges, lines or corners (Chapters 12 and 13).
- Determination of motion in image sequences (Chapter 14)
- Texture analysis and complex features (Chapter 16)
- Correction of disturbances caused by errors during image acquisition or transfer. Such errors result in distorted gray values of a few individual pixels (Chapter 17).
- *reconstruction* of images generated with indirect image acquisition techniques such as *tomography* (Chapter 17).
- *restoration* of images that are disturbed by incorrect focusing, motion blur or similar errors (Chapter 17).

Therefore, this chapter contains important basics for the entire image processing. First, in the rest of this section, we will discuss general properties of neighborhood operators that are valid for all classes of neighborhood operators. Then we will discuss by far the most important class of neighborhood operators, linear shift invariant filters (Section 10.3). An additional section deals with a special subclass of these operators, recursive filters (Section 10.4). Finally, the last section introduces different classes of non-linear neighborhood operators (Section 10.5).

10.1.2 General Definitions and Properties

A neighborhood operator N links the values in a neighborhood around a point using a suitable operation and writes the result back to the point. This operation is performed for all points of the signal.

Definition 10.1 (*Continuous neighborhood operator*) *A continuous neighborhood operator maps a multidimensional signal $g(\boldsymbol{x})$ onto itself using the operation*

$$g'(\boldsymbol{x}) = N(\{g(\boldsymbol{x}')\}, \forall(\boldsymbol{x} - \boldsymbol{x}') \in \mathbb{M}), \tag{10.1}$$

where $\mathbb{M}$ is a compact area.

The area $\mathbb{M}$ is referred to as *mask*, *window* or *structur element*. The size and shape of $\mathbb{M}$ influence the neighborhood operation to the extent that they determine the input values from g for the calculation of $g'(\boldsymbol{x})$. These lie in the area $\mathbb{M}$ whose reference point has been moved to the point $\boldsymbol{x}$. The neighborhood operation N itself is not yet defined. It can contain any calculation rule. For reasons of symmetry, the shape of the mask is often symmetrical and the reference point is at the center of symmetry.

Definition 10.2 (Discrete neighborhood operator) *A discrete neighborhood operator maps a $M \times N$ matrix onto itself using the operation*

$$g'_{mn} = N(g_{m'-m,n'-n}, \forall\, [m', n']^T \in \mathbb{M}), \tag{10.2}$$

where $\mathbb{M}$ is now a set of points.

Expressions that are equivalent to Def. 10.2 can easily be set up for other dimensions. Even if the equations Eqs. (10.1) and (10.2) do not specify the calculation rule for the neighborhood operation, they describe the common structure of all classes of neighborhood operations. A decisive general property of all classes of neighborhood operators is the *shift invariance*. This is given if the calculation rule N in Eq. (10.1) or Eq. (10.2) is not explicitly dependent on the position but only on the position difference.

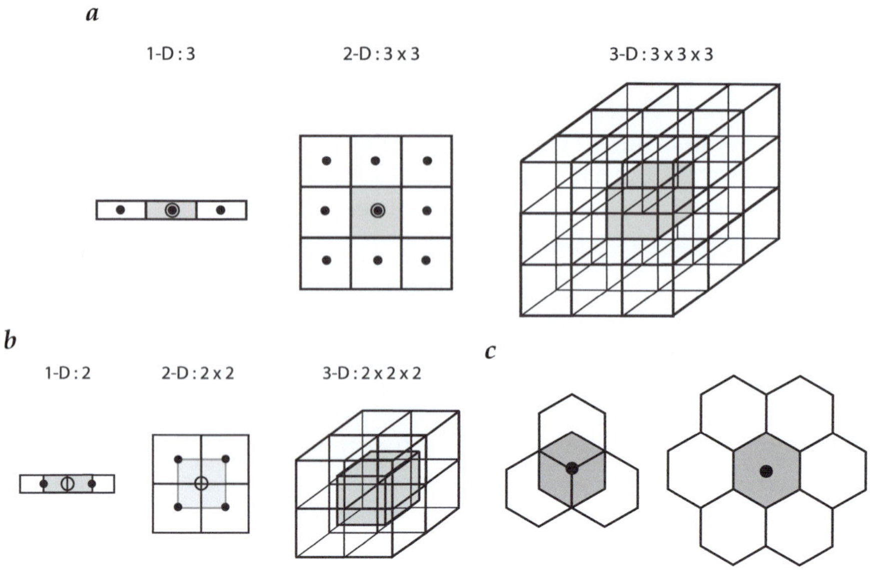

Figure 10.1: Smallest mask with **a** *even and* **b** *odd size for neighborhood operators. The result of the operation is written to the gray-shaded pixel. The center of this point is marked with an open circle. For masks of even size, the result is often written to an intermediate grid that is shifted half a point distance in all directions, as shown here.* **c** *Masks of even and odd size on a hexagonal grid.*

10.1.3 Mask Size and Symmetry

The first determinant of a neighborhood operation is the size of the neighborhood. We also need to determine the reference point relative to the neighborhood where the result of the neighborhood operation is stored. For symmetry reasons, the most natural reference point is the center of a $(2R + 1) \times (2R + 1)$ large mask with an odd number of points in all directions, see Fig. 10.1a.

Masks with an even number of points (Fig. 10.1b) appear less suitable, as there is no point that lies in the center of the mask. However, if the result of the neighborhood operation is returned to a point that lies between the original points in the center of the mask (Fig. 10.1b), it may still make sense to use masks with an even number of points. In this case, however, the resulting image is shifted by half a distance between the points in all directions. Because of this shift, we must never combine property images calculated with a mask with an even number of points with the original image. This would lead to considerable errors. When applying several neighborhood operations in parallel and then combining the results, all masks must have an even or odd number of points in the same direction. Otherwise the grids of the resulting feature images will not coincide.

Example 10.1: Poorly chosen masks of even size

In order to work effectively with the smallest possible masks, you might be tempted to combine the results of a horizontal 1×2- mask with those of a vertical 2×1- mask. However, this does not work well, as the results lie on grids that are shifted in relation to each other. The result of the horizontal neighborhood operation is shifted horizontally by half a point distance, the result of the vertical operation is shifted vertically by half a point distance. Therefore, only results from masks that have an even or odd size in all directions can be combined.

Extension 10.1: Masks on a hexagonal grid

There are also masks with even and odd sizes on a hexagonal grid (Fig. 10.1c). Only here there are three equivalent directions, each rotated by 60°, while on the two-dimensional Cartesian grid there are two or four directions, depending on whether only the axis directions are taken into account (neighbors with a common edge) or also the diagonals (neighbors with a common corner).

10.1.4 Possible Neighborhood Operators in Principle

After the generic considerations on neighborhood operations, we can ask which arithmetic operations are suitable for this. In general, it is conceivable to make the arithmetic operation dependent on the position within the mask or the corresponding value of the data.

The simplest position-dependent operation gives each element in the mask a weighting factor by which the underlying value in the signal is multiplied. All products are summed up. This operation could be described somewhat casually as "weighting and summing up". This operation is linear. If the same weighting factors are used at all positions, this is called a *shift-invariant filter*, for time series it is called an *time-invariant filter*. Linear shift-invariant operators, called *LSI* or *LTI* for short, are the most important classes of filter operations and are therefore discussed in detail in Section 10.3.

In the simplest case, all pixels within the mask are given the same weighting factor. This calculates an average of all values within the mask. This operation is therefore also called *moving average*. If the weighting factor only depends on the distance to the central pixel to which the result is saved, the neighborhood operation is called *isotropic*. It is obvious that an isotropic neighborhood operation cannot be realized easily or only approximately on a discrete grid.

Alternatively, the weighting factor can be dependent on the gray value difference to the central pixel, for example to limit the neighborhood operation to similar values. Such an operation is usually combined with a position-dependent weighting because more distant data points should have a smaller influence. A neighborhood operator with this double weighting is called *bilateral filter*. Each filter with a gray value-dependent weighting factor is no longer shift-invariant.

Filters that select a value from a neighborhood based on certain criteria are completely independent of the position in the mask. For example, you could select the minimum or maximum gray value. Such an operation is a *rank-order filter*. As a value is selected, these filters are also suitable for robust averaging that is not influenced by outliers. The gray values within the neighborhood are sorted and the value in the middle position is selected. The mean value determined in this way is called the median. This operator is therefore a *median filter*. The mode of operation of all ranking filters can generally be described as "sorting and selecting".

What all non-shift-invariant filter operations have in common is that the weighting coefficients have to be recalculated at each position. Such filters are much more computationally expensive. At the same time, however, it is to be expected that filters that depend on the local gray value progression are much more powerful tools than simple shift-invariant filters. It is therefore worth thinking about an effective calculation.

One possibility is to apply a set of suitable shift invariant filters and then select one of the filters or a suitable linear combination of the filters based on the local gray value gradient. The combination of filters with point operations can also be dependent on the local gray value. Of course, this does not apply to a simple addition or linear combination of linear shift-invariant filters. But even a simple multiplication of the results results in a non-linear operator.

Finally, one more consideration that is of crucial importance for the entire consideration of neighborhood operators in this book. The larger the neighborhood, the more complex the operation and the more difficult it will be to describe it analytically. It therefore makes sense to try to compose complex neighborhood operations as combinations or successive executions of the smallest possible operators. If this is successful, they are much easier and more systematic to describe.

10.1.5 Operator Notation

From the final considerations in the last section, it follows that it is helpful to introduce an *operator notation* for neighborhood operators. In this way, even complex compound neighborhood operators can be written in an understandable way. Operators are denoted with calligraphic letters such as $\mathcal{B}$, $\mathcal{D}$, $\mathcal{H}$, $\mathcal{S}$. The operator $\mathcal{H}$ transforms the image $\boldsymbol{G}$ into the image $\boldsymbol{G}'$: $\boldsymbol{G}' = \mathcal{H}\boldsymbol{G}$. This notation can be used for continuous and discrete signals of any dimension and represents a *representation-independent notation* of signal processing operators.

If several operators are executed in succession, they are simply written one after the other, with the operator furthest to the right being executed first. Repeated use of the same operator is abbreviated with an exponent:

$$\underbrace{\mathcal{H}\mathcal{H}\ldots\mathcal{H}}_{\text{p-mal}} = \mathcal{H}^p. \tag{10.3}$$

If a sequence of operators is applied to a single image, this can be omitted. In this way, we can write *operator equations* without a target object. We also use brackets to control the order in which operators are executed in the usual way. In this way, operator properties can also be written in a compact and easy-to-understand way, for example

$$\begin{array}{ll} \text{Commutativity} & \mathcal{H}_1\mathcal{H}_2 = \mathcal{H}_2\mathcal{H}_1 \\ \text{Associativity} & \mathcal{H}_1(\mathcal{H}_2\mathcal{H}_3) = (\mathcal{H}_1\mathcal{H}_2)\mathcal{H}_3 \\ \text{Distributivity with addition} & (\mathcal{H}_1 + \mathcal{H}_2)\mathcal{H}_3 = \mathcal{H}_1\mathcal{H}_3 + \mathcal{H}_2\mathcal{H}_3 \end{array} \tag{10.4}$$

Other operators, such as addition, can also be inserted in this notation, but we must be careful with non-linear operators, as these cannot be interchanged with linear operators. For this reason, the order in which the operators are executed must be strictly adhered to and, if necessary, changed using brackets.

A simple example of a non-linear operator is the pointwise multiplication of two images, a *dyadic point operator* with two input images. We denote it with a special symbol, a centered point (·). This symbol is necessary to distinguish multiplication from the successive execution of operators. For example, the operator expression $\mathcal{B}(\mathcal{D}_p \cdot \mathcal{D}_q)$ means: the operators $\mathcal{D}_p$ and $\mathcal{D}_q$ are applied to the same image, the resulting images are multiplied point by point and the operator $\mathcal{B}$ is applied to the product image. Without brackets, the expression $\mathcal{B}\mathcal{D}_p \cdot \mathcal{D}_q$ would mean: The operator $\mathcal{D}_q$ and the operator combination $\mathcal{D}_p$ and $\mathcal{B}$ are applied to the same image and the results of both operations are multiplied together. In the operator notation, dyadic operators are always treated as subordinate to *monadic operators* with only one input image: All monadic operators are executed first and then the dyadic operators.

10.2 Efficient Neighborhood Operators

The computational effort for neighborhood operators in higher-dimensional signals can be very high. When implementing a neighborhood operator directly, the number of arithmetic operations is directly proportional to the size of the mask and the size of the image. This leads to a number of operations of order $O((NR)^D)$ in a D-dimensional image, where R is the number of elements in the mask and N is the number of pixels in each direction. With respect to a single point, the number of arithmetic operations of a neighborhood operator is of order $O(R^D)$.

This means that the number of arithmetic operations increases with the dimension D of the signal. A neighborhood of size $R = 5$ contains only 5 points in a 1D signal, 25 in a 2D signal, 125 in a 3D signal and even 625 points in a 4D signal. For efficient neighborhood operators in higher-dimensional signals, it is therefore of utmost importance to reduce the order as a function of the mask size. Neighborhood operations of linear order $O(R)$ would be ideal. Then the computational effort per point would not depend on the dimension of the signal, but only on the size of the mask. The ultimate speed would be an algorithm with $O(R^0)$ that does not depend on the size of the mask. Then the calculation speed would only depend on the number of points in the discrete signal. It cannot be faster if you want to generate a feature at every point.

In the following, we discuss general strategies for efficiently calculating neighborhood operators. Three classes of optimization strategies can be distinguished:

- general strategies that are possible for all classes of neighborhood operators,
- strategies that are only possible for a specific class of operators, and
- strategies for optimizing for a very specific neighborhood operator.

The preferred strategy is of course the generic one, as it can be used to accelerate all classes of neighborhood operations. Accelerating a very specific operation is the worst strategy, as it involves a relatively high amount of effort with possibly little benefit. This is only justified if it is a very frequently used operator with a high computational cost.

On modern computer architectures with a hierarchy of memories that can be accessed with different speeds and waiting times, the processing speed depends not only on the number of arithmetic operations, but also on effective memory accesses. This is particularly true for the processing of large blocks of data, such as images, volume images or image sequences. In addition, the problem is becoming increasingly acute, as the increase in speed for memory accesses is growing more slowly than the computing power. The main reason for this is the increasing parallelization, both fine-grained with SIMD instructions (single instruction and multiple data) and coarse-grained with multi-core processors.

10.2.1 Minimization of the Computing Effort

There is no doubt that a neighborhood operator should have as few elements as possible for a fast calculation. Thus, minimizing the computational effort is first of all a design problem for the neighborhood operation. The question is how we can perform a neighborhood operator, which is supposed to fulfill a fixed purpose, with as few elements as possible. It may be that this is simply not possible or that the accuracy or other quality parameters are also reduced by reducing the number of elements. However, it is also clear

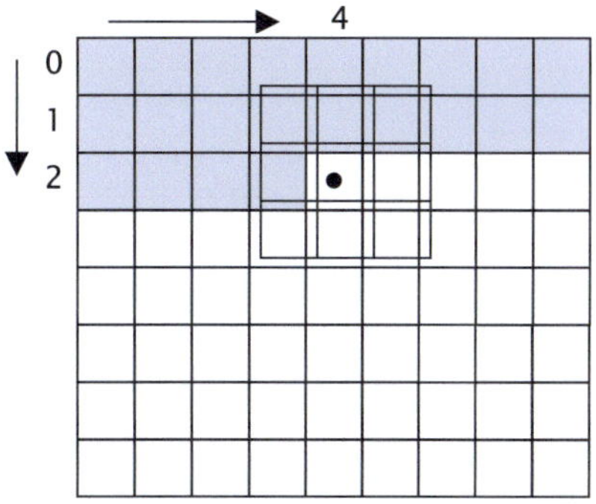

Figure 10.2: Execution of a neighborhood operator by moving the mask line by line. The gray values of the dark pixels have already been replaced by the calculated value. This means that the gray values of the dark pixels that fall within the area of the filter mask must be saved in a separate memory.

that reducing the number of elements by a constant factor does not help very much, as the neighborhood operation still has the order $O(R^D)$.

A better approach is derived from the *divide-and-conquer* paradigm for fast algorithms, which we have already seen in Section 6.4 for the fast Fourier transform. The most effective division of a multidimensional neighborhood operator of size R in all directions is to split it into a sequence of D one-dimensional neighborhood operators of size R. This also reaches all R^D points of the neighborhood, but the number of arithmetic operations is reduced from $O(R^D)$ to a linear order with $O(DR)$ operations independent of the dimension.

The reduction in the number of arithmetic operations is significant and becomes higher the higher the dimension of the signal. While for a 5×5 mask the number of operations can only be reduced from 25 to 10, for a $5 \times 5 \times m$ask there is already a reduction from 125 to 15 operations, almost 10 times less. This approach also has the advantage that neighborhood operations can easily be designed for arbitrary-dimensional signals. If this decomposition is possible, the neighborhood operator is called *separabel*.

Another technique that is often not considered is multigrid data structures, which are discussed in Chapter 15. The idea is simple: If a neighborhood operation with a large mask still works when it can be performed on data with half the resolution, then the number of data points is reduced by a factor of 2^D. A separable neighborhood operation also results in only half the number $R/2$ of mask elements, so that 2^{D+1} fewer arithmetic operations are required in total. This is an 8-fold increase in calculation speed for a 2D image and a 16-fold increase for a volume image.

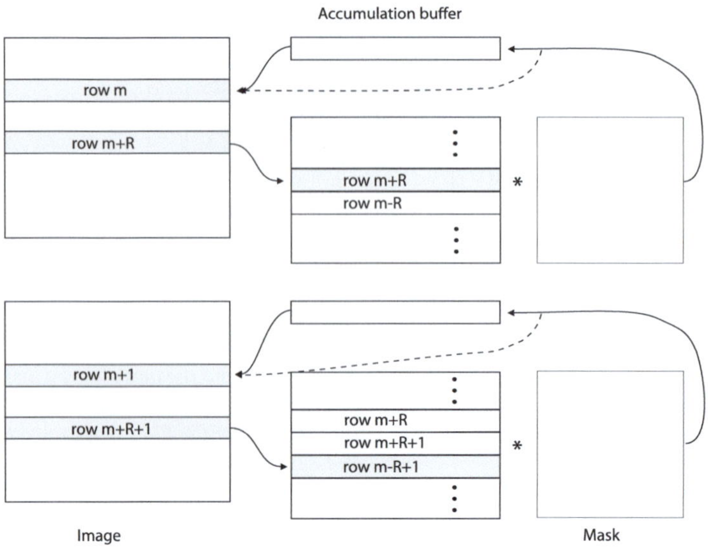

Figure 10.3: *An "in-place" calculation of a neighborhood operator with a* $(2R + 1) \times (2R + 1)$ *mask using* $2R + 1$ *row memories and an accumulation memory. The calculation of the rows* m *(top) and* $m + 1$ *(bottom) is shown.*

10.2.2 ♯Efficient Storage Schemes

An "in-place" calculation requires intermediate storage, as the operation still requires values that have already been overwritten by the result of the neighbor operation (Fig. 10.2). It would of course be easiest to save the overall result of the neighborhood operation in a new data set. Sometimes this is also useful if, for example, several neighborhood operators are calculated simultaneously from the same data set or if it is needed later. Often, however, there is simply not enough memory available to enable such a simple approach. Therefore, we will discuss two effective storage schemes for arbitrary neighborhood operators in the following two subsections.

Cyclic Buffers

Each neighborhood operation with a 2D mask on a 2D signal requires a total of four loops: two to scan the two-dimensional mask and accumulate the results for a single point, and two additional ones to move the mask through all image rows and columns. If the mask runs through the image line by line, it is clear that a buffer of as many image lines as the mask is high is required. If the mask is moved down one line, not all lines need to be read into the buffer. The top line of the buffer is no longer required and can be overwritten by the new bottom line.

Figure 10.3 illustrates the details of how the caches are arranged and used for a neighborhood operator with a mask of height $2R + 1$. To perform

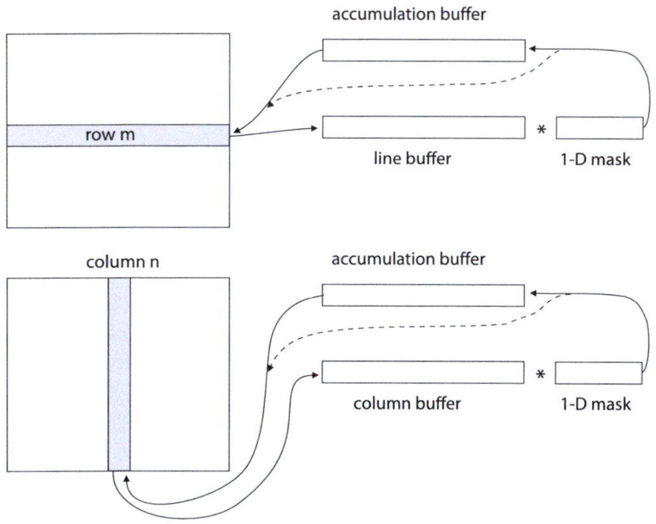

Figure 10.4: *"In-place" calculation of a separable neighborhood operation.*

the operation for the line m, the line $m + R$ is first read into one of the line memories. The line memories then contain $2R + 1$ image lines from the line $m - R$ to the line $m + R$. Therefore, all rows are available to calculate the neighborhood operation with a $(2R + 1) \times (2R + 1)$ mask for the entire row m.

To continue the calculation for row $m+1$, row $m+R+1$ is loaded into the row memory, in which row $m - R$ was located, which is no longer required. The $2R + 1$ line memory now contains the lines $m - R + 1$ to $m + R + 1$. The number of the row memory to which a row must be written is simply the row number m modulo the number of row memories, $m \bmod (2R + 1)$.

Since an entire line is available for the calculation, further optimizations are possible. The neighbourhood operation can be vectorized. The accumulation of the results can be processed simultaneously for all points of a line with one element of the mask after the other.

It is often useful to have an additional memory in which the intermediate results are accumulated. This is essential if the calculations require a higher accuracy than the input data provides. At the end of the calculation of a line, the final result is then rounded and scaled and written back to line m of the input data.

Few line memories fit well into the primary cache of a modern CPU if the lines are not too long. When the calculation progresses to the next line, only one of the line memories needs to be replaced. Therefore, most of the data remains in the cache. For all calculations performed with a point, it only needs to be loaded into the cache once.

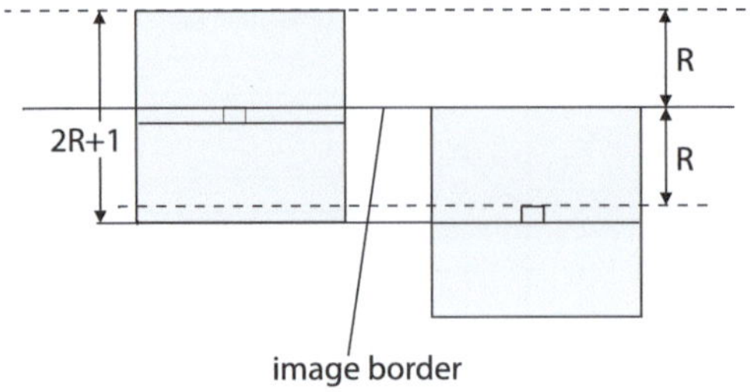

Figure 10.5: Neighborhood operator at the border of the image. A $(2R + 1) \times (2R + 1)$ *mask is shown in two positions. Once with the center in a border line, so that it protrudes R lines beyond the image border and once with its edge directly at the image border.*

Separable Neighborhood Operators

The method described above can also be used for separable masks, but a second method is also conceivable (Fig. 10.4). There is now a two-stage calculation for 2D data. First, the horizontal operation is performed. To do this, line m is read into a line memory. Now a vectorized neighbourhood operation can also be performed. If the neighborhood operation requires more precise arithmetic, the intermediate results are again accumulated in a second line memory with higher precision. If not, the accumulation can be performed directly in the line m of the input data. After the horizontal operation, the vertical operation is performed in the same way. The only difference is that column after column is read into the buffer (Fig. 10.4).

The advantage of this procedure is that it can easily be extended to higher-dimensional data. For each dimension, a line must be copied in the corresponding direction into the buffer in which the operation is then executed. This has the significant advantage that the operation in different directions requires an adapted copy operator in each case, but that the neighborhood operation can then be carried out in the same way for each direction. However, it has the disadvantage that more data must be copied. To perform a neighborhood operation on D-dimensional data, each data point must be copied D times back and forth into the buffer. In addition, copying is not effective in all those directions in which non-contiguously stored data must be accessed.

10.2.3 ♯Neighborhood Operators at the Border

If a neighborhood operator is performed with a $(2R + 1) \times (2R + 1)$ mask, then a band of R rows and columns around the image is required to perform the operation up to the border (Fig. 10.5). The image also shows that there is a border of R rows or columns at the edge, which is influenced by how the pixels outside the image are calculated. There are three possible strategies for extending an image beyond the border.

Periodic Border Extension
One could argue that this is the theoretically correct method, since the discrete Fourier transform expects a spatially limited multidimensional signal to continue periodically in all directions through the entire space (Section 5.2.4).

Addition of Zeros
However, we do not know what is outside the image. It therefore seems more appropriate to accept the lack of knowledge and simply supplement the image with zeros. Of course, this method has the disadvantage that the edges of the image appear as sharp edges, which will certainly falsify many neighborhood operations.

Signal Extrapolation beyond the Border
With this method, the data points close to the edge are used to extrapolate the values beyond the edge in one way or another. The simplest method is to use the data points directly at the edge for the values outside the edge. This method has the disadvantage that the data points at the edge are given too much weight. Linear extrapolation could therefore also be considered. This means that the first derivative and the value itself remain constant at the edge. However, every extrapolation is associated with a high degree of uncertainty, as we cannot know what the image looks like outside the recorded area.

None of the three proposed methods is ideal, each introduces errors in one way or another into the results of the neighborhood operation at the border, which extend into the image with a width R (Fig. 10.5). Therefore, there is only one sure way to avoid errors. You simply have to make sure that no important content in the data is too close to the border. However, it should be noted that with cascaded neighbor operations, this border area becomes wider and wider.

10.3 Linear Shift-invariant Filter

In this section, we turn to the most important class of neighborhood operators, the linear shift invariant filters. We describe the arithmetic operation behind them (Section 10.3.1) and discuss the most important properties in detail. The point response (Section 10.3.4) and the transfer function (Section 10.3.5) are of central importance for understanding.

10.3.1 Discrete Convolution

First, we concentrate on the question of how gray values of pixels in small neighborhoods can be linked together. A basic way of linking the pixels in a window is by an operation that multiplies each pixel in the area of the filter mask by the corresponding weighting factor of the mask, which is only position-dependent, adds the products and writes the sum to the position of the central pixel:

$$\begin{aligned} g'_{mn} &= \sum_{m'=-r}^{r} \sum_{n'=-r}^{r} h_{m'n'} g_{m-m',n-n'} \\ &= \sum_{m''=-r}^{r} \sum_{n''=-r}^{r} h_{-m'',-n''} g_{m+m'',n+n''}. \end{aligned} \tag{10.5}$$

In Section 6.3.4 the *discrete convolution* was defined in Eq. (6.40) as:

$$g'_{mn} = \sum_{m'=0}^{M-1} \sum_{n'=0}^{N-1} h_{m'n'} g_{m-m',n-n'}. \tag{10.6}$$

The definitions in Eqs. (10.5) and (10.6) are equivalent if we consider the periodicity in spatial domain as given by Eq. (6.26). From Eq. (6.26) we conclude that negative indices can be replaced by positive ones with the relation

$$g_{-n} = g_{N-n}, \quad g_{-n,-m} = g_{N-n,M-m}. \tag{10.7}$$

The restriction of the sum in Eq. (10.5) takes into account the fact that the elements of the matrix $\boldsymbol{H}$ outside the few points in the $(2R+1)\times(2R+1)$ filter mask are zero. The representation of a filtering by a filter mask instead of a matrix $\boldsymbol{H}$ is much more practical and gives a much better impression of how the filter operation works. The following 3×3 filter mask and the $M\times N$ matrix $\boldsymbol{H}$ are equivalent:

$$\begin{bmatrix} 0 & -1 & -2 \\ 1 & 0_\bullet & -1 \\ 2 & 1 & 0 \end{bmatrix} \equiv \begin{bmatrix} 0_\bullet & -1 & 0 & \ldots & 0 & 1 \\ 1 & 0 & 0 & \ldots & 0 & 2 \\ 0 & 0 & 0 & \ldots & 0 & 0 \\ \vdots & \vdots & \vdots & \ddots & \vdots & \vdots \\ 0 & 0 & 0 & \ldots & 0 & 0 \\ -1 & -2 & 0 & \ldots & 0 & 0 \end{bmatrix}. \tag{10.8}$$

D-dimensional filter operations can be written in a simplified way using vector indexing:

$$g'_{\boldsymbol{n}} = \sum_{\boldsymbol{n}'=-\boldsymbol{R}}^{\boldsymbol{R}} h_{-\boldsymbol{n}'} g_{\boldsymbol{n}+\boldsymbol{n}'} \tag{10.9}$$

with $\boldsymbol{n} = [n_1, n_2, \ldots, n_D]$, $\boldsymbol{R} = [R_1, R_2, \ldots, R_D]$, where $g_{\boldsymbol{n}}$ is an element of a D-dimensional signal $g_{n_1,n_2,\ldots,n_D}$. The notation for the sums in this equation is an abbreviation for

$$\sum_{\boldsymbol{n}'=-\boldsymbol{R}}^{\boldsymbol{R}} = \sum_{n'_1=-R_1}^{R_1} \sum_{n'_2=-R_2}^{R_2} \cdots \sum_{n'_D=-R_D}^{R_D}. \tag{10.10}$$

With vectorial indexing, expressions such as convolution sums for arbitrary-dimensional signals can be written easily.

10.3.2 Symmetries

With regard to symmetries, we can distinguish between two classes of filters: Filters with even and odd symmetry with the condition that applies in one or more directions:

$$h_{-m,n} = \pm h_{mn} \quad \text{or} \quad h_{m,-n} = \pm h_{mn}. \tag{10.11}$$

The characters + and − stand for filters with *even* or *odd* symmetry. With this definition, we can simplify the calculation of one-dimensional filters according to Eq. (10.5) to:

$$\begin{array}{ll} \text{gerade:} & g'_{mn} = h_0 g_{m,n} + \sum\limits_{n'=1}^{r} h_{n'}(g_{m,n-n'} + g_{m,n+n'}) \\ \text{ungerade:} & g'_{mn} = \sum\limits_{n'=1}^{r} h_{n'}(g_{m,n-n'} - g_{m,n+n'}). \end{array} \tag{10.12}$$

The sums only run over half the mask and exclude the central pixel. This must be treated separately as it has no symmetry partner. For filters with odd symmetry, it is omitted as it is zero after Eq. (10.11).

In the two-dimensional case, the equations are more complicated [157], as the symmetry must be considered separately in each direction. However, these equations are hardly needed, as it is much better to decompose large multidimensional convolution masks into a combination of 1D convolution masks. Then the equations are much simpler and the convolutions can be calculated much faster.

10.3.3 Calculation of Convolution Operations

Convolution is such an important operation that it is instructive to study how it works in detail. First of all, the minus signs in front of the indices m' and n' in Eq. (10.5), which stand for the mask and the image respectively, may be confusing. They mean that we mirror either the mask or the image at its center of symmetry before placing the mask over the image. (We learn the reason for this mirroring in Section 10.3.4). If we want to calculate the result of the convolution at the point $[m, n]^T$, we center the mirrored mask at this point, perform the convolution and write the result back to the position $[m, n]^T$ (Fig. 10.6). This operation is performed for all pixels of the image.

Close to the edge of the image, when the area of the filter mask extends beyond the edges of the image, we run into problems because we are missing some pixels. This problem has already been discussed in general for neighborhood operations in Section 10.2.3. Here are some specific additions for convolution operations. The correct way to solve this problem according to our assumptions in Section 6.3.4, especially in equation Eq. (6.26), is to take into account that finite image matrices must be considered as periodically repeating structures. So when we reach the left edge of the image, we use the

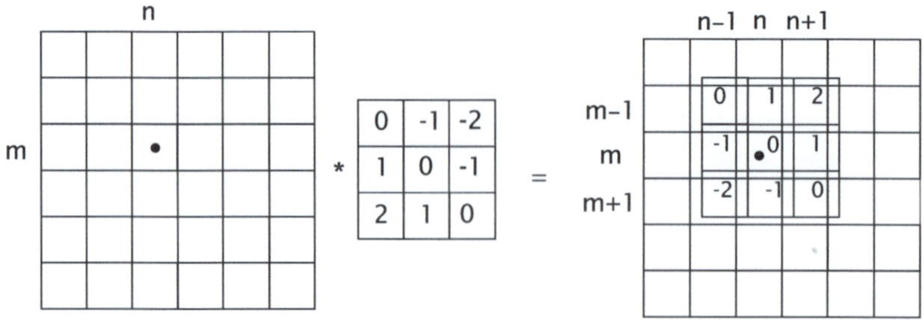

Figure 10.6: *Illustration of discrete convolution with a 3×3 filter mask.*

missing points from the right edge of the image. This process is called *cyclic convolution*. Only this type of convolution corresponds — strictly speaking — to a multiplication in Fourier space (Section 6.3).

The number of elements in a mask grows strongly with its size and dimension. A D-dimensional mask with a linear size of R contains R^D elements. The higher the dimension, the faster the number of elements increases with the mask size. In multidimensional signals, even small neighborhoods contain hundreds to thousands of elements. It is therefore a challenge to find efficient calculation methods for filter operations. An optimal result would be to find a calculation method in which the number of calculation operations depends only linearly on the mask size ($O(R^1)$), or is even independent of the mask size ($O(R^0)$). Is this conceivable?

10.3.4 Linearity and Point Spread Function

An operator is linear if the *superposition principle* applies.

Definition 10.3 (Superposition principle) *Let* $\boldsymbol{G}$ *and* $\boldsymbol{G}'$ *be two D-dimensional complex-valued signals, a and b two complex-valued scalars and* $\mathcal{H}$ *an operator. The operator is linear if and only if*

$$\mathcal{H}(a\boldsymbol{G} + b\boldsymbol{G}') = a\mathcal{H}\boldsymbol{G} + b\mathcal{H}\boldsymbol{G}'. \tag{10.13}$$

We can extend Def. 10.3 to the superposition of many signals:

$$\mathcal{H}\left(\sum_k a_k \boldsymbol{G}_k\right) = \sum_k a_k \mathcal{H}\boldsymbol{G}_k. \tag{10.14}$$

The superposition principle states that we can break down complex signals into simpler components. We can then apply a linear operator to these and reassemble the results.

The property of *shift invariance* already mentioned in Section 10.1.2 (also known as *translation invariance* or *homogeneity*) we will take another closer look. This term means that the response of an operator does not explicitly depend on the position in a signal, but only on the position difference. If we shift a signal, the result of a shift-invariant operator is also shifted. We can formulate this property more elegantly by defining a *shift operator* ${}^{n}\mathcal{S}$ as

$$ {}^{n}\mathcal{S} g_{n'} = g_{n'-n}. \tag{10.15}$$

Then we can define a shift invariant operator as follows:

Definition 10.4 (Shift invariance) *An operator is shift invariant if and only if it can be interchanged with the shift operator* $\mathcal{S}$:

$$\mathcal{H}\,{}^{mn}\mathcal{S} = {}^{mn}\mathcal{S}\mathcal{H}. \tag{10.16}$$

From the definition of the convolution operation in Eqs. (10.5) and (10.9) we can immediately conclude that it is both linear and shift invariant. Such an operator is therefore called a *linear shift invariant operator* or *LSI operator* for short. For *time series*, an operator with the same property is also called linear and time-invariant (*LTI*). It should be noted that the shift operator ${}^{mn}\mathcal{S}$ itself is also an LSI operator.

The mode of action of convolution operations is easy to understand due to linearity and shift invariance. As discussed in Section 6.3.1, we can think of each discrete image (signal) as being composed of individual pixels or *base images* ${}^{mn}\boldsymbol{P}$ Eq. (6.7):

$$\boldsymbol{G} = \sum_{m=0}^{M-1}\sum_{n=0}^{N-1} G_{mn}\,{}^{mn}\boldsymbol{P}. \tag{10.17}$$

Linearity means that we can apply an operator to these base images and then add the resulting images. Shift invariance means that the result of the convolution operation is the same for each of the base images except for one shift.

The result of a filter operation for any image is therefore completely determined by the response to a base image. The response of a convolution operation to a point image therefore has a special meaning. It is called *point spread function* (*PSF*), sometimes also *point response*. For time series, it is also called *impulse response*. The point spread function of a convolution operator is identical to the convolution mask and describes it completely in the spatial domain:

$$p'_{mn} = \sum_{m'=-r}^{r}\sum_{n'=-r}^{r} h_{-m',-n'}\,{}^{00}p_{m+m',n+n'} = h_{m,n}. \tag{10.18}$$

The point response shows the effect of the convolution from a different but equivalent perspective. The convolution sum in Eq. (10.5) says that each point is composed of a linear combination of neighboring points. The point response says that each pixel spreads into the neighborhood as described by the point response.

Extension 10.2: Convolution only linear, shift invariant neighborhood operator?

The interesting question that remains to be answered is whether there are other linear, shift-invariant operators apart from convolution. We can rule this out if we show that it follows from these properties that the neighborhood operation must be a convolution.

From our considerations here, we are already familiar with the *PSF* or *impulse response* of continuous or discrete operators. Now we introduce the formal definition of the PSF for an operator $\mathcal{H}$ on a $M \times N$-dimensional vector space:

$$\boldsymbol{H} = \mathcal{H}\,{}^{00}\boldsymbol{P}. \tag{10.19}$$

Now we can use the linearity Eq. (10.14) and shift invariance Eq. (10.16) of the operator $\mathcal{H}$ and the definition of the impulse response Eq. (10.19) to calculate the result of the operator on any image $\boldsymbol{G}$ in space:

$$
\begin{aligned}
(\mathcal{H}\boldsymbol{G})_{mn} &= \left[\mathcal{H}\left[\sum_{m'=0}^{M-1}\sum_{n'=0}^{N-1} g_{m'n'}\,{}^{m'n'}\boldsymbol{P}\right]\right]_{mn} && \text{with Eq. (10.14)}\\
&= \left[\sum_{m'=0}^{M-1}\sum_{n'=0}^{N-1} g_{m'n'}\,\mathcal{H}\,{}^{m'n'}\boldsymbol{P}\right]_{mn} && \text{linearity}\\
&= \left[\sum_{m'=0}^{M-1}\sum_{n'=0}^{N-1} g_{m'n'}\,\mathcal{H}\,{}^{m'n'}\mathcal{S}\,{}^{00}\boldsymbol{P}\right]_{mn} && \text{with Eq. (10.15)}\\
&= \left[\sum_{m'=0}^{M-1}\sum_{n'=0}^{N-1} g_{m'n'}\,{}^{m'n'}\mathcal{S}\,\mathcal{H}\,{}^{00}\boldsymbol{P}\right]_{mn} &&\\
&= \left[\sum_{m'=0}^{M-1}\sum_{n'=0}^{N-1} g_{m'n'}\,{}^{m'n'}\mathcal{S}\,\boldsymbol{H}\right]_{mn} && \text{with Eq. (10.19)}\\
&= \sum_{m'=0}^{M-1}\sum_{n'=0}^{N-1} g_{m'n'}h_{m-m',n-n'} && \text{with Eq. (10.15)}\\
&= \sum_{m''=0}^{M-1}\sum_{n''=0}^{N-1} g_{m-m'',n-n''}h_{m'',n''} && \begin{aligned} m'' &= m-m'\\ n'' &= n-n' \end{aligned}\quad.
\end{aligned}
$$

These calculations prove that a linear shift-invariant operator is always a convolution operation in space. There is no other operator type that is both linear and shift invariant.

10.3.5 Transfer Function and Eigenfunctions

In Section 6.3 we discussed that an image (signal) can also be represented in Fourier space. This representation has a special meaning for LSI operators, since the convolution is reduced to a multiplication in Fourier space according to the *convolution theorem* (Theorem B.4, p. 630):

$$G * H \quad \circ\!\!-\!\!\bullet \quad \prod_{d=1}^{D} N_d \, \tilde{G}\tilde{H} \tag{10.20}$$

The factors N_d result from the definition of the discrete Fourier transform according to Eq. (6.46b). In the following, we therefore include the factor N_d in the definition of the transfer function and replace $(N_1 N_2 \ldots N_D)\tilde{\boldsymbol{h}}$ with $\tilde{\boldsymbol{h}}$ in all future formulas for the transfer function.

The Fourier transform of the convolution mask or PSF is called *transfer function* (*TF*) of a filter. Like the point response, it clearly describes the effect of a filter operation. For each wavenumber vector, it specifies the factor by which the corresponding periodic structure is multiplied by the filter operation. This factor is a *complex number* (Section 6.3.1). Therefore, a periodic structure not only undergoes a change in amplitude, but also a phase shift:

$$\begin{aligned} \tilde{g}'_{u,v} = \tilde{h}_{u,v}\tilde{g}_{u,v} &= r_h \exp(\mathrm{i}\varphi_h)\, r_g \exp(\mathrm{i}\varphi_g) \\ &= r_h r_g \exp[\mathrm{i}(\varphi_h + \varphi_g)]. \end{aligned} \tag{10.21}$$

In the second part of the equation, the complex numbers are represented by magnitude and phase.

The symmetry of filter masks (Section 10.3.2) considerably simplifies the calculation of the transfer function. Symmetrical terms can be summarized in the Fourier transform of the PSF. For a 1D filter with even or odd symmetry, the definition of the 1D Fourier transform according to Eq. (6.46b) then results in

$$\begin{aligned} \tilde{h}_{\tilde{n}} &= \sum_{n'=-R}^{R} h_{n'} \exp\left(-\frac{2\pi \mathrm{i} n \tilde{n}}{N}\right) \quad \text{(with} \quad h_{-n'} = \pm h_{n'}) \\ &= h_0 + \sum_{n'=1}^{R} h_{n'} \left(\exp\left(-\frac{2\pi \mathrm{i} n \tilde{n}}{N}\right) \pm \exp\left(\frac{2\pi \mathrm{i} n \tilde{n}}{N}\right)\right). \end{aligned} \tag{10.22}$$

Equation (10.22) can be further simplified by introducing a continuous scaled wavenumber

$$\underset{\sim}{k} = 2\tilde{n}/N, \quad \text{with} \quad -N/2 \leq \tilde{n} < N/2. \tag{10.23}$$

This scaled wavenumber $\underset{\sim}{k}$ lies in the interval $[-1, 1[$. A wavenumber at the edge of the interval corresponds to the maximum wavenumber that satisfies the sampling theorem (Section 5.2.3). Finally, we use the Euler equation $\exp(\mathrm{i}x) = \cos x + \mathrm{i} \sin x$. Equation (10.22) simplifies for 1D filters with even and odd symmetry to

$$\boxed{\begin{aligned} \text{even:} \quad & \tilde{h}(\underset{\sim}{k}) = h_0 + 2\sum_{\tilde{n}=1}^{R} h_{\tilde{n}} \cos(\tilde{n}\pi \underset{\sim}{k}) \\ \text{odd:} \quad & \tilde{h}(\underset{\sim}{k}) = -2\mathrm{i}\sum_{\tilde{n}=1}^{R} h_{\tilde{n}} \sin(\tilde{n}\pi \underset{\sim}{k}). \end{aligned}} \tag{10.24}$$

The transfer function of a $(2R+1) \times (2R+1)$ mask with even horizontal and vertical symmetry results in a similar way to

$$\begin{aligned} \tilde{h}(\tilde{\boldsymbol{\kappa}}) &= h_{00} \\ &+ 2\sum_{n'=1}^{R} h_{0n'} \cos(n'\pi\tilde{\kappa}_1) + 2\sum_{m'=1}^{R} h_{m'0} \cos(m'\pi\tilde{\kappa}_2) \\ &+ 4\sum_{m'=1}^{R}\sum_{n'=1}^{R} h_{m'n'} \cos(n'\pi\tilde{\kappa}_1)\cos(m'\pi\tilde{\kappa}_2). \end{aligned} \tag{10.25}$$

Analogous equations apply to other symmetry combinations. All these equations are very useful as they establish the relationship between the coefficients of a *filter mask* and its *transfer function.* They form the basis for the analysis of all filters for special image processing tasks, which are dealt with in the Chapters 11 to 16.

10.3.6 Further Properties

In this section, we discuss some other properties of convolution operators that are useful for image and signal processing.

Inverse Convolution

Can we reverse filter operations? This question is important because image changes such as blurring due to movement or poorly focused optics can also be regarded as filter operations (Section 3.5.1). If an inverse operator exists and if we know the point response, the undisturbed original can be reconstructed. The problem of inverting a filter operation is called *deconvolution* or *inverse filtering*.

If we look at the filter operation in Fourier space, we immediately recognize that we can only reconstruct those wavenumbers for which the transfer function of the filter does not disappear. In practice, we are much more limited because of the quantization (Section 5.6) and because of the additional noise in the image. If a wavenumber is attenuated below a critical value, which depends on the noise and quantization, the image is no longer recoverable. It is clear that these conditions considerably reduce the success of inverse filtering. The problem of inverse filtering is discussed in more detail in Section 17.5.

Eigenfunctions

Next, we are interested in the question of whether there are special types of images $\boldsymbol{E}$ that, with the exception of multiplication by a scalar, are not changed by a linear shift invariant operator. It is clear that such images must have a special meaning for linear shift invariant operators. In mathematical terms, this means

$$\mathcal{H}\boldsymbol{E} = \lambda\boldsymbol{E}. \tag{10.26}$$

A vector (image) that fulfills this condition is called *eigenvector* (*eigenimage*) or *characteristic vector* of the operator. The scaling factor λ is an *eigenvalue* or *characteristic value* of the operator.

To find out the eigenimages of linear shift invariant operators, we consider the shift operator $\mathcal{S}$. It is obvious that for real images there is only one simple eigenimage, namely a constant image. For complex images, however, there is a whole set of eigenimages. This becomes clear when we consider the shift property of the *complex exponential function*

$$ {}^{uv}w_{mn} = \exp\left(\frac{2\pi \mathrm{i} m u}{M}\right)\exp\left(\frac{2\pi \mathrm{i} n v}{N}\right), \tag{10.27} $$

which is given by

$$ {}^{kl}\mathcal{S}\,{}^{uv}\mathbf{W} = \exp\left(-\frac{2\pi \mathrm{i} k u}{M}\right)\exp\left(-\frac{2\pi \mathrm{i} l v}{N}\right){}^{uv}\mathbf{W}. \tag{10.28} $$

The last equation states directly that the complex exponential function ${}^{uv}\mathbf{W}$ is an eigenfunction of the displacement operator. The eigenvalues are complex phase factors that depend on the wavenumber indices (u, v) and on the displacement (k, l). If the shift is one wavelength, i. e. $(k, l) = (M/u, N/v)$, the phase factor is reduced to one, as expected.

Now we want to know whether every linear shift invariant operator has such a set of eigenimages. In fact, all LSI operators have the same set of eigenimages. With the help of the *convolution theorem* (Section 6.3, Theorem B.4, p. 630) we can prove this statement. It states that a convolution is a pointwise multiplication in Fourier space. Therefore, each element of the image representation in Fourier space $\tilde{g}_{uv}$ is multiplied by the complex scalar $\tilde{h}_{uv}$. Each point in the Fourier space represents a basic image of the Fourier space, namely the complex exponential function ${}^{uv}\mathbf{W}$ from Eq. (10.28), multiplied by the scalar $\tilde{g}_{uv}$. This function is therefore an eigenfunction of every convolution operator. The eigenvalues are therefore the elements of the transfer function $\tilde{h}_{uv}$. To summarize, we can therefore write

$$ \mathcal{H}(\tilde{g}_{uv}\,{}^{uv}\mathbf{W}) = \tilde{h}_{uv}\tilde{g}_{uv}\,{}^{uv}\mathbf{W}. \tag{10.29} $$

The fact that the eigenfunctions of the LSI operators are the basis functions of the Fourier space explains why convolution in Fourier space can be expressed by a simple multiplication and emphasizes the central importance of the Fourier transform for signal and image processing.

Property 10.1 (Commutativity) *LSI operators are* commutative*:*

$$ \mathcal{H}\mathcal{H}' = \mathcal{H}'\mathcal{H}. \tag{10.30} $$

The order in which we apply convolution operators to an image is therefore irrelevant. This property is easy to prove in Fourier space, as the operations there are reduced to complex multiplications.

Property 10.2 (Associativity) *LSI operators are* associative*:*

$$\mathcal{H}'\mathcal{H}'' = \mathcal{H}. \tag{10.31}$$

Since LSI operations are associative, we can combine individual operators into a complex operator. Similarly, we can try to decompose a given complex operator into simpler operators. This is essential for an effective implementation of convolution operators. As an example, let's consider the operator

$$\begin{bmatrix} 1 & 4 & 6 & 4 & 1 \\ 4 & 16 & 24 & 16 & 4 \\ 6 & 24 & 36 & 24 & 6 \\ 4 & 16 & 24 & 16 & 4 \\ 1 & 4 & 6 & 4 & 1 \end{bmatrix}. \tag{10.32}$$

With this convolution mask, we need 25 multiplications and 24 additions per pixel. However, it is easy to see that we can split this mask into a one-dimensional horizontal and a one-dimensional vertical mask:

$$\begin{bmatrix} 1 & 4 & 6 & 4 & 1 \\ 4 & 16 & 24 & 16 & 4 \\ 6 & 24 & 36 & 24 & 6 \\ 4 & 16 & 24 & 16 & 4 \\ 1 & 4 & 6 & 4 & 1 \end{bmatrix} = [1\,4\,6\,4\,1] * \begin{bmatrix} 1 \\ 4 \\ 6 \\ 4 \\ 1 \end{bmatrix}. \tag{10.33}$$

If we perform the convolutions with the smaller masks one after the other, only 10 multiplications and 8 additions per pixel are required. Filter masks that can be decomposed into one-dimensional masks along the axes are called *separable masks*. We denote one-dimensional operators with an index for the axis. Then we can specify a separable operator $\mathcal{B}$ in three-dimensional space as follows:

$$\mathcal{B} = \mathcal{B}_z\mathcal{B}_y\mathcal{B}_x. \tag{10.34}$$

In the case of one-dimensional masks that point in orthogonal directions, the convolution is reduced to an outer product. The higher the dimension of the space, the more efficient separable filters are. As an example, let's consider a $9 \times 9 \times 9$ filter mask. A direct implementation would cost 729 multiplications and 728 additions per pixel, while a separable mask of the same size requires only 27 multiplications and 24 additions, i. e. only about 1/30 of the operations.

Property 10.3 (Distributivity over the addition) *LSI operators are* distributive *over the addition:*

$$\mathcal{H}' + \mathcal{H}'' = \mathcal{H}. \tag{10.35}$$

Since LSI operators are elements of the same vector space in which they are applied, we can define an addition of operators as an element-wise addition of vectors. Because of this property, operator additions and subtractions can also be integrated into the general operator notation that we introduced in Section 10.1.5.

10.3.7 ♯Error Propagation with Filtering.

In practice, filters are applied to measurement data that contains noise. It is therefore important to know how the statistical parameters of the filtered data result from those of the original data. This question is generally solved in Appendix C.3.3. The *covariance matrix* of a linear combination $\boldsymbol{g}' = \boldsymbol{M}\boldsymbol{g}$ of a random vector $\boldsymbol{g}$ results according to Eq. (C.28) to

$$\mathrm{cov}(\boldsymbol{g}') = \boldsymbol{M}\,\mathrm{cov}(\boldsymbol{g})\boldsymbol{M}^T. \tag{10.36}$$

We only need to apply this result to the special case of convolution. We will initially restrict our considerations to 1D signals. We assume that the covariance matrix of the signal is homogeneous, i. e.depending only on the distance of the points and not their position. Then the variance is equal to σ^2 for all elements. Similarly, the values on the secondary diagonals are all equal and the covariance matrix takes the simple form

$$\mathrm{cov}(\boldsymbol{g}) = \begin{bmatrix} \sigma_0 & \sigma_1 & \sigma_2 & \ldots & \ldots \\ \sigma_{-1} & \sigma_0 & \sigma_1 & \sigma_2 & \ldots \\ \sigma_{-2} & \sigma_{-1} & \sigma_0 & \sigma_1 & \ldots \\ \vdots & \sigma_{-2} & \sigma_{-1} & \sigma_0 & \ldots \\ \vdots & \vdots & \vdots & \vdots & \ddots \end{bmatrix}, \tag{10.37}$$

where the index indicates the distance between the points and $\sigma_0 = \sigma^2$. As a rule, the covariances become smaller with increasing pixel spacing. Often we are only dealing with a finite number of covariances σ_p not equal to zero. For statistically uncorrelated pixels, only $\sigma_0 = \sigma^2$ is non-zero.

Since the linear combination in $\boldsymbol{M}$ is now described by a convolution, it has the same form as the covariance matrix. For a filter with 3 coefficients, $\boldsymbol{M}$ results in

$$\boldsymbol{M} = \begin{bmatrix} h_0 & h_{-1} & 0 & 0 & 0 & \ldots \\ h_1 & h_0 & h_{-1} & 0 & 0 & \ldots \\ 0 & h_1 & h_0 & h_{-1} & 0 & \ldots \\ 0 & 0 & h_1 & h_0 & h_{-1} & \ldots \\ 0 & 0 & 0 & h_1 & h_0 & \ldots \\ \vdots & \vdots & \vdots & \vdots & \vdots & \ddots \end{bmatrix}. \tag{10.38}$$

Apart from boundary effects, the matrix multiplications in Eq. (10.36) are reduced to convolutions. To do this, we introduce the autocovariance vector $\boldsymbol{\sigma} = [\ldots, \sigma_{-1}, \sigma_0, \sigma_1, \ldots]^T$ and can then write Eq. (10.36)

$$\boldsymbol{\sigma}' = {}^{-}\boldsymbol{h} * \boldsymbol{\sigma} * \boldsymbol{h} = \boldsymbol{\sigma} * {}^{-}\boldsymbol{h} * \boldsymbol{h} = \boldsymbol{\sigma} * (\boldsymbol{h} \star \boldsymbol{h}), \tag{10.39}$$

where $^-\boldsymbol{h}$ is the mirrored convolution mask: $^-h_n = h_{-n}$. In the last step, we replaced the convolution with a *correlation*. The convolution of σ with $\boldsymbol{h} \star \boldsymbol{h}$ can be replaced by a correlation, since the autocorrelation function of a real function is a function with even symmetry.

For uncorrelated measurement data, the autocovariance vector is a delta function and the autocovariance vector of the noise of the filtered vector is reduced to

$$\sigma' = \sigma^2(\boldsymbol{h} \star \boldsymbol{h}). \tag{10.40}$$

For a filter with R coefficients, $2R - 1$ values of the autocovariance vector have now become non-zero. This means that pixels with a distance of up to $R - 1$ are now correlated with each other in the filtered signal.

Since the error propagation during convolution can be described by correlations, we can also directly specify the change in the noise spectrum, i. e. the power spectrum of the noise, by a convolution operation. To do this, we only need to transform Eq. (10.39) and observe the correlation theorem (> R15). Then we obtain

$$\boxed{\sigma' = \sigma * (\boldsymbol{h} \star \boldsymbol{h}) \quad \circ\!\!-\!\!\bullet \quad \tilde{\sigma}'(\kappa) = \tilde{\sigma}(\kappa)\left|\tilde{h}(\kappa)\right|^2.} \tag{10.41}$$

This means that the noise spectrum must be multiplied by the absolute value squared of the transfer function of the convolution operation. With Eqs. (10.39) and (10.41) we have everything we need to determine the changes in the statistical parameters of a signal (variance, autocovariance matrix and power spectrum of the noise) by means of a filter operation. It follows directly from Eq. (10.41) that Eq. (10.39) applies not only to 1D signals, but also to signals of any dimension.

10.4 ♯Recursive Filters

10.4.1 ♯Introduction

As the convolution is a computationally intensive operation, it makes sense to ask whether the results of the convolution operation at neighboring points can be included in the calculation of the convolution for the next point. In this way, it could be possible to calculate a convolution with fewer arithmetic operations. Unfortunately, such a filter operation, a *recursive filter*, is much more difficult to understand and apply, especially in higher dimensions.

To gain a first insight, let's look at a very simple example. The simplest one-dimensional recursive filter we can think of has the form

$$g'_n = \alpha g'_{n-1} + (1 - \alpha) g_n. \tag{10.42}$$

The filter takes the fraction α from the previous calculation and the fraction $1 - \alpha$ from the current pixel. In contrast to non-recursive filters, recursive filters therefore work in a specific direction, in our example from left to right.

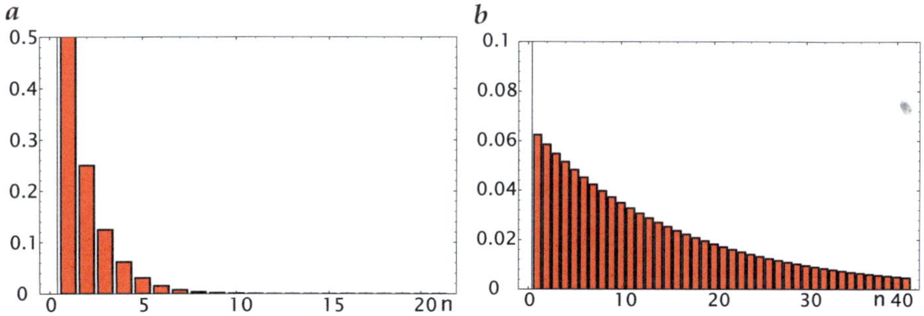

Figure 10.7: *Point response of the recursive filter* $g'_n = \alpha g'_{n-1} + (1-\alpha) g_n$ *for* ***a*** $\alpha = 1/2$ *and* ***b*** $\alpha = 15/16$.

For a time series, a preferred direction seems natural, as the current value can only depend on past values. A filter that depends only on previous values of the signal is called *causal filter*. For spatial signals, however, there is no preferred direction, and thus causal filters make no sense. We must therefore find ways to construct filters with even and odd symmetry, as required for image processing, from recursive filters.

With recursive filters, the *point response* or *impulse response* is not equal to the filter mask, but must be calculated by recursion. From (Eq. 10.42) we can calculate the point response (Section 10.3.4) of the filter as a response to the *discrete delta function*

$$\delta_n = \begin{cases} 1 & n = 0 \\ 0 & n \neq 0 \end{cases}. \tag{10.43}$$

Repeated application of Eq. (10.42) leads to

$$g'_{-1} = 0, \quad g'_0 = 1 - \alpha, \quad g'_1 = (1-\alpha)\alpha, \quad \ldots, \quad g'_n = (1-\alpha)\alpha^n. \tag{10.44}$$

This equation shows three general properties of recursive filters:

- Firstly, the point response is infinitely extended despite a finite filter mask (Fig. 10.7). For $|\alpha| < 1$ it decreases exponentially, but never becomes exactly zero. In contrast, non-recursive convolution filters always have a finitely extended point response that is as long as the filter mask. Therefore, these two types of filters are sometimes called *finite impulse response filters* (*FIR filter*) or *infinite impulse response filters* (*IIR filter*).
- FIR filters are always *stable*. A stable filter has a finite impulse response. This is equivalent to the fact that the filter result is finite for every finite signal. In contrast, the point response of an IIR filter can be infinite. The stability of a recursive filter depends on the filter coefficients. The filter in Eq. (10.42) is unstable for $|\alpha| > 1$, as the point response then diverges. In the simple case of Eq. (10.42), it is trivial to recognize the instability of a filter. In general, however, this is much more difficult, especially for higher-dimensional recursive filters.

- Each recursive filter can be replaced by a non-recursive filter with an infinitely large mask that corresponds to the point response of the recursive filter. The reverse is not possible. For all non-recursive filters with a finite filter mask, there is, e. g. no corresponding recursive filter.

10.4.2 ♯Transfer Function and Filter Stability

After this introductory example, we can look at recursive filters in a more general way. With recursive filters, the results of previous calculations are always included in the current calculation. This makes them directional. We will restrict ourselves here to 1D filters. The general formula of a filter running from left to right is

$$g'_n = -\sum_{n''=1}^{S} a_{n''} g'_{n-n''} + \sum_{n'=-R}^{R} h_{n'} g_{n-n'} \tag{10.45}$$

The neighborhood of the non-recursive part (coefficients h) is symmetrical around the central point. The recursive part (coefficients a) only uses previously calculated values. Such a filter is called *causal*.

If we place the recursive part of the filter on the left-hand side of the equation, we see that a recursive filter is equivalent to the following difference equation, which is also known as *ARMA*(S,R) process (*autoregressive moving average process*):

$$\sum_{n''=0}^{S} a_{n''} g'_{n-n''} = \sum_{n'=-R}^{R} h_{n'} g_{n-n'} \quad \text{with} \quad a_0 = 1. \tag{10.46}$$

The transfer function of such a filter with a recursive and non-recursive part can be calculated by applying the *discrete Fourier transform* (Section 6.3.2) using the *shift theorem* (Theorem B.3, p. 629). The calculation results:

$$\tilde{g}'(k) \sum_{n''=0}^{S} a_{n''} \exp(-2\pi \mathrm{i} n'' k) = \tilde{g}(k) \sum_{n'=-R}^{R} h_{n'} \exp(-2\pi \mathrm{i} n' k). \tag{10.47}$$

Therefore, the *transfer function* results in

$$\tilde{h}(k) = \frac{\tilde{g}'(k)}{\tilde{g}(k)} = \frac{\sum\limits_{n'=-R}^{R} h_{n'} \exp(-2\pi \mathrm{i} n' k)}{\sum\limits_{n''=0}^{S} a_{n''} \exp(-2\pi \mathrm{i} n'' k)}. \tag{10.48}$$

The properties of the transfer function are determined by the zeros of the numerator and denominator. A zero in the non-recursive part leads to a zero of the transfer function, i. e. a disappearance of the corresponding wavenumber, while a zero in the recursive part leads to a pole, i. e. an infinitely large transfer function.

It is not possible to determine the zeros and thus analyze the transfer function in the form of Eq. (10.48). It requires an extension similar to that from real to complex numbers when introducing the Fourier transform (Section 6.3.2). Both the denominator and the numerator contain polynomials with the *complex exponential function* $\exp(2\pi ik)$ of the form

$$\sum_{n=0}^{S} a_n \left(\exp(-2\pi ik)\right)^n . \tag{10.49}$$

These have a value of one and therefore lie on the unit circle in the complex plane. However, the zeros of the polynomial do not have to lie on the unit circle, but can be any complex number. It therefore makes sense to expand the polynomials so that they fill the entire complex plane. This is possible using the expression $z = r\exp(2\pi ik)$, which describes a circle with the radius r in the complex plane.

With this extension, we obtain polynomials of the complex number z. We can then apply the fundamental theorem of algebra, which states that every polynomial of degree N can be decomposed into N factors containing the zeros:

$$\sum_{n=0}^{N} a_n z^n = a_N z^N \prod_{n=1}^{N} \left(1 - r_n z^{-1}\right). \tag{10.50}$$

With the help of Eq. (10.50), the recursive and non-recursive parts of the polynomials in the transfer function can be broken down into the following products:

$$\begin{aligned}
\sum_{n=0}^{S} a_n z^{-n} &= z^{-S}\sum_{n'=0}^{S} a_{S-n'} z^{n'} &= \prod_{n=1}^{S}\left(1 - d_n z^{-1}\right),\\
\sum_{n=-R}^{R} h_n z^{-n} &= z^{-R}\sum_{n'=0}^{2R} h_{R-n'} z^{n'} &= h_{-R} z^{R}\prod_{n=1}^{2R}\left(1 - c_n z^{-1}\right).
\end{aligned} \tag{10.51}$$

We can therefore write the transfer function with $z = r\exp(2\pi ik)$ as

$$\boxed{\tilde{h}(z) = h_{-R} z^{R} \frac{\displaystyle\prod_{n'=1}^{2R}(1 - c_{n'} z^{-1})}{\displaystyle\prod_{n''=1}^{S}(1 - d_{n''} z^{-1})}.} \tag{10.52}$$

Each of the factors $c_{n'}$ and $d_{n''}$ is a zero of the corresponding polynomial ($z = c_{n'}$ or $z = d_{n''}$).

Now let's take a closer look at the transfer function. The decomposition of the transfer function into a product is a great advantage, as we can consider each factor as a separate filter and have thus decomposed each arbitrarily complicated recursive filter into many simple filters that can be executed one after the other. These factors have the form

$$f_n(\tilde{k}) = 1 - d_n \exp(-2\pi i \tilde{k}). \tag{10.53}$$

Since the impulse response of the filter is real, the transfer function must be Hermitian, i. e. $f(-k) = f^*(k)$ (Section 6.3.4). This can only be the case if the zeros d_n are real or if there are pairs of conjugate complex zeros, resulting in the following transfer function:

$$\begin{aligned} f(\tilde{k}) &= (1 - d \exp(-2\pi i \tilde{k})(1 - d^* \exp(-2\pi i \tilde{k}) \\ &= 1 - (d + d^* \exp(-2\pi i \tilde{k}) + dd^*) \exp(-4\pi i \tilde{k}). \end{aligned} \tag{10.54}$$

From this we can immediately conclude that there can only be two basic types of one-dimensional recursive filters, the *relaxation filter* and the *resonance filter*. Any one-dimensional recursive filter can be composed of both. They are described in detail in the Sections 10.4.4 and 10.4.5.

Extension 10.3: z-transform

The integration of the factor r into the extended transfer function leads to an extension of the discrete Fourier transform, the *z-transform*, which is defined by

$$\tilde{g}(z) = \sum_{n=-\infty}^{\infty} g_n z^{-n} \tag{10.55}$$

The z-transform of g_n can be regarded as the Fourier transform of the extended sequence $g_n r^{-n}$ [200]. The z-transform is the key to understanding one-dimensional recursive filters. It is the discrete counterpart of the *Laplace transform*. A detailed description of the z-transform can be found in Oppenheim and Schafer [234], Oppenheim et al. [235] and Poularikas [245]; the two-dimensional z-transform is treated in Lim [200].

Extension 10.4: Higher dimensional recursive filters

Recursive filters can also be written for higher-dimensional signals with equations similar to those in Eq. (10.45). Similarly, the equations for the transfer function and the z-transform are similar to those in Eq. (10.48). However, it is generally not possible to factorize the z-transform of higher-dimensional recursive filters as in Eq. (10.52) [200]. From Eq. (10.52) we can immediately conclude that it is possible to factorize *separable filter*, since the higher-dimensional polynomials are then products of 1D polynomials. Due to these inherent mathematical problems, we will restrict ourselves to recursive 1D filters in the following.

10.4.3 ♯Symmetrical Recursive Filters

Causal recursive filters are useful for real-time processing of time series. They are not very useful for processing spatial signals. There is no "before" and "after" for spatial data. In addition, there is the signal-dependent shift associated with causal recursive filters. With a single recursive filter, it is impossible to construct a so-called *zero-phase filter* with a real even transfer function. We must therefore combine several recursive filters. This combination should result in either a filter with even symmetry and real transfer function, which is suitable for smoothing, or a filter with odd symmetry and purely imaginary transfer function, which is suitable for change detection (Section 6.3.4).

We start with a causal 1D filter whose transfer function we decompose into the real and imaginary parts:

$$ {}^{+}\tilde{h}(\underset{\sim}{\kappa}) = a(\underset{\sim}{\kappa}) + \mathrm{i}b(\underset{\sim}{\kappa}). \tag{10.56} $$

The superscript "+" is intended to indicate that the filter is executed in the positive direction. The transfer function of the same filter, but running in the opposite direction, has a similar transfer function. Because of the direction reversal, $\underset{\sim}{\kappa}$ must be replaced by $-\underset{\sim}{\kappa}$. Since the transfer function of a real PSF is Hermitian (Section 6.3.4), it must be the case that $a(-\underset{\sim}{\kappa}) = a(\underset{\sim}{\kappa})$ and $b(-\underset{\sim}{\kappa}) = -b(\underset{\sim}{\kappa})$, and we obtain

$$ {}^{-}\tilde{h}(\underset{\sim}{\kappa}) = a(\underset{\sim}{\kappa}) - \mathrm{i}b(\underset{\sim}{\kappa}). \tag{10.57} $$

If the direction of the filter is reversed, only the imaginary part of the transfer function changes its sign.

We now have three options for combining the two transfer functions Eqs. (10.56) and (10.57) so that we obtain either a purely real or purely imaginary transfer function:

$$ \begin{array}{lrcl} \text{Addition:} & {}^{e}\tilde{h}(\underset{\sim}{\kappa}) & = & \frac{1}{2}\left({}^{+}\tilde{h}(\underset{\sim}{\kappa}) + {}^{-}\tilde{h}(\underset{\sim}{\kappa})\right) = a(\underset{\sim}{\kappa}), \\ \text{Subtraction:} & {}^{o}\tilde{h}(\underset{\sim}{\kappa}) & = & \frac{1}{2}\left({}^{+}\tilde{h}(\underset{\sim}{\kappa}) - {}^{-}\tilde{h}(\underset{\sim}{\kappa})\right) = \mathrm{i}b(\underset{\sim}{\kappa}), \\ \text{Multiplication:} & \tilde{h}(\underset{\sim}{\kappa}) & = & {}^{+}\tilde{h}(\underset{\sim}{\kappa})\,{}^{-}\tilde{h}(\underset{\sim}{\kappa}) = a^2(\underset{\sim}{\kappa}) + b^2(\underset{\sim}{\kappa}). \end{array} \tag{10.58} $$

Addition and multiplication (one after the other) of the filter running to the right and left result in a filter with a real transfer function and even symmetry. Subtraction results in a filter with odd symmetry and a purely imaginary transfer function.

10.4.4 ♯Relaxation Filter

The simple recursive filter discussed in Section 10.4.1

$$ g'_n = a_1 g'_{n\mp 1} + h_0 g_n \quad \text{with} \quad a_1 = \alpha,\ h_0 = (1-\alpha) \tag{10.59} $$

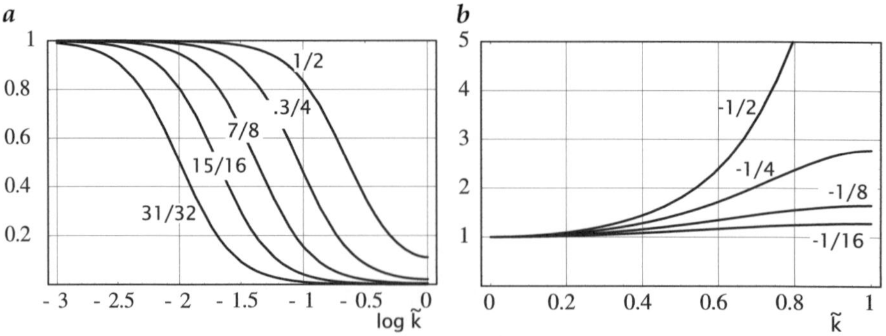

Figure 10.8: *Transfer function of the relaxation filter $g'_n = \alpha g'_{n\mp1} + (1-\alpha)g_n$ applied consecutively in the forward and reverse direction for **a** positive and **b** negative values of α, as indicated.*

with the point spread function

$$
{}^{\pm}u_{\pm n} = \begin{cases} (1-\alpha)\alpha^n & n \geq 0 \\ 0 & \text{otherwise} \end{cases} \tag{10.60}
$$

is a *relaxation filter*. The transfer function of the filter, which runs in the forward or backward direction, results directly from Eq. (10.48) with Eq. (10.59) to

$$
{}^{\pm}\tilde{u}(\tilde{k}) = \frac{1-\alpha}{1-\alpha\exp(\mp\pi\mathrm{i}\tilde{k})} \quad \text{with} \quad \alpha \in \mathbb{R}. \tag{10.61}
$$

The transfer function Eq. (10.61) is complex and can be divided into real and imaginary parts:

$$
{}^{\pm}\tilde{u}(\tilde{k}) = \frac{1-\alpha}{1-2\alpha\cos\pi\tilde{k}+\alpha^2}\left[(1-\alpha\cos\pi\tilde{k}) \mp \mathrm{i}\alpha\sin\pi\tilde{k}\right]. \tag{10.62}
$$

According to Eq. (10.58), we can calculate the transfer function $\tilde{u}$ for the resulting symmetrical filter if we run the relaxation filter forwards and backwards one after the other:

$$
\boxed{\tilde{u}(\tilde{k}) = {}^{+}\tilde{u}(\tilde{k})\,{}^{-}\tilde{u}(\tilde{k}) = \frac{(1-\alpha)^2}{1-2\alpha\cos\pi\tilde{k}+\alpha^2} = \frac{1}{(1+\beta)-\beta\cos\pi\tilde{k}}} \tag{10.63}
$$

with

$$
\beta = \frac{2\alpha}{(1-\alpha)^2} \quad \text{and} \quad \alpha = \frac{1+\beta-\sqrt{1+2\beta}}{\beta}.
$$

According to Eq. (10.60), the relaxation filter is stable for $|\alpha| < 1$. The constant β must therefore lie in the interval $]-1/2, \infty[$. The transfer function is one

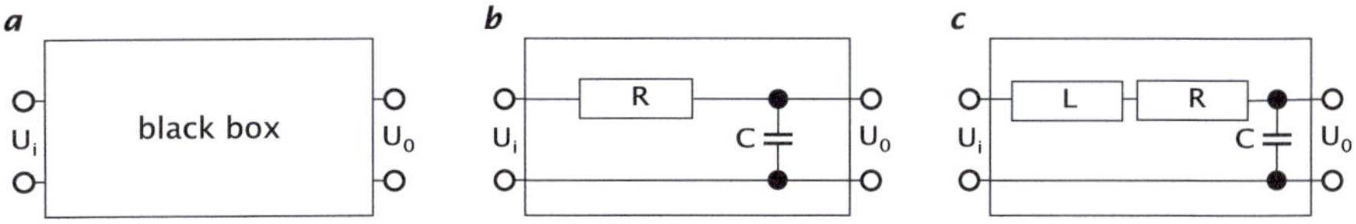

Figure 10.9: *Analog filters for time series:* ***a*** *Black box model: A signal U_i is applied to the input of an unknown system and the output signal U_o is measured.* ***b*** *A resistor-capacitor circuit as a simple example of an analog low-pass filter.* ***c*** *Damped resonant circuit consisting of an inductance L, a resistor R and a capacitance C.*

for wavenumber 0 and can be approximated for small wavenumbers with a Taylor series expansion by

$$\tilde{u}(\underset{\sim}{k}) \approx 1 - \frac{\alpha}{(1-\alpha)^2}(\pi \underset{\sim}{k})^2 + \frac{\alpha((1+10\alpha+\alpha^2)}{12(1-\alpha^2)^2}(\pi \underset{\sim}{k})^4. \tag{10.64}$$

For positive α, the filter is a smoothing filter (low-pass, Fig. 10.8a). The strength of the smoothing can be set using the parameter α. If α approaches 1, the smoothing distance becomes infinite. For negative α, the filter amplifies high wavenumbers (Fig. 10.8b).

Extension 10.5: Analog low-pass filter with resistor and capacitor

The relaxation filter is the discrete analog to the first-order differential equation

$$\frac{\mathrm{d}y}{\mathrm{d}t} + \tau y = 0, \tag{10.65}$$

which corresponds to a discrete relaxation process with a response time $\tau = -\Delta t / \ln \alpha$, where Δ is the time sampling of the discrete signal.
An example of this is a simple circuit consisting of a series connection of a resistor R and a capacitor with capacitance C (Fig. 10.9b). The differential equation for this filter results from Kirchhoff's current sum rule. The current through the resistor with the voltage difference $U_i - U_o$ must be equal to the current flowing into the capacitor. As this is proportional to the change in the potential U_o over time, we obtain the following first-order differential equation

$$\frac{U_i - U_o}{R} = C\frac{\mathrm{d}U_o}{\mathrm{d}t} \tag{10.66}$$

with the time constant $\tau = RC$.

10.4.5 ♯Resonance Filter

The second basic type of a recursive filter, which resulted from the discussion of the transfer function in Section 10.4.2, has a pair of conjugate complex zeros $r\exp(\pm \mathrm{i}\pi \underset{\sim}{k}_0)$. Therefore, the transfer function for a forward and backward

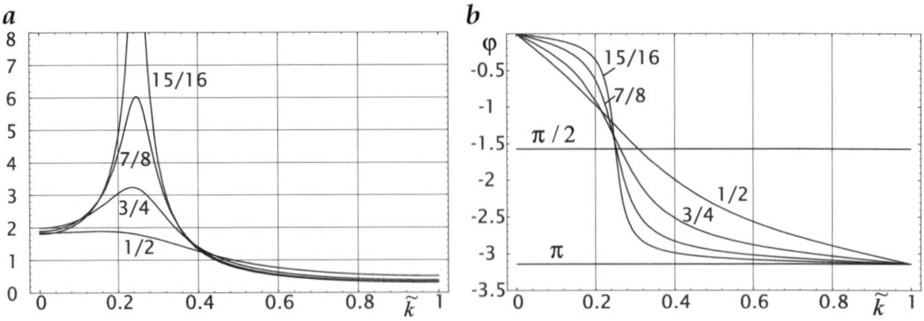

Figure 10.10: ***a*** *magnitude and* **b** *phase shift of the transfer function of the resonant filter according to Eq. (10.67) for* $\underset{\sim}{k}_0 = 1/4$ *and values for r as specified.*

filter is

$$
\begin{aligned}
{}^{\pm}\tilde{o}(\underset{\sim}{k}) &= \frac{1}{(1 - r\exp(\mathrm{i}\pi\underset{\sim}{k}_0)\exp(\mp\mathrm{i}\pi\underset{\sim}{k}))(1 - r\exp(-\mathrm{i}\pi\underset{\sim}{k}_0)\exp(\mp\mathrm{i}\pi\underset{\sim}{k}))} \\
&= \frac{1}{1 - 2r\cos(\pi\underset{\sim}{k}_0)\exp(\mp\mathrm{i}\pi\underset{\sim}{k}) + r^2\exp(\mp 2\mathrm{i}\pi\underset{\sim}{k})}.
\end{aligned}
\qquad (10.67)
$$

The second line shows that the recursive filter has the coefficients $h_0 = 1$, $a_1 = -2r\cos(\pi\underset{\sim}{k}_0)$ and $a_2 = r^2$:

$$
g'_n = g_n + 2r\cos(\pi\underset{\sim}{k}_0)g'_{n\mp1} - r^2 g'_{n\mp2}. \qquad (10.68)
$$

From the transfer function Eq. (10.67) we conclude that the filter is a *bandpass filter* with a passband wavenumber of $\pm\underset{\sim}{k}_0$ (Fig. 10.10). For $r = 1$, the transfer function has two poles at $\underset{\sim}{k} = \pm\underset{\sim}{k}_0$.

According to Oppenheim et al. [235], the point answer results in

$$
o_{\pm n} = \begin{cases} \dfrac{r^n}{\sin\pi\underset{\sim}{k}_0}\sin[(n+1)\pi\underset{\sim}{k}_0] & n \geq 0 \\ 0 & n < 0 \end{cases}. \qquad (10.69)
$$

This means that the filter behaves like a damped oscillator. The parameter $\underset{\sim}{k}_0$ specifies the wavenumber of the oscillation and the parameter r the damping (Fig. 10.11). The filter is only stable if $r \leq 1$.

If we run this filter forwards and backwards one after the other, the result is a filter with a real transfer function $\tilde{o}(\underset{\sim}{k}) = {}^{+}\tilde{o}(\underset{\sim}{k})\,{}^{-}\tilde{o}(\underset{\sim}{k})$:

$$
\tilde{o}(\underset{\sim}{k}) = \frac{1}{(1 - 2r\cos[\pi(\underset{\sim}{k} - \underset{\sim}{k}_0)] + r^2)\,(1 - 2r\cos[\pi(\underset{\sim}{k} + \underset{\sim}{k}_0)] + r^2)}. \qquad (10.70)
$$

We can normalize the transfer function so that the maximum value at the passband wavenumber becomes one by setting the non-recursive filter coefficient $h_0 = (1 - r^2)\sin(\pi\underset{\sim}{k}_0)$. This results in the following modified recursion

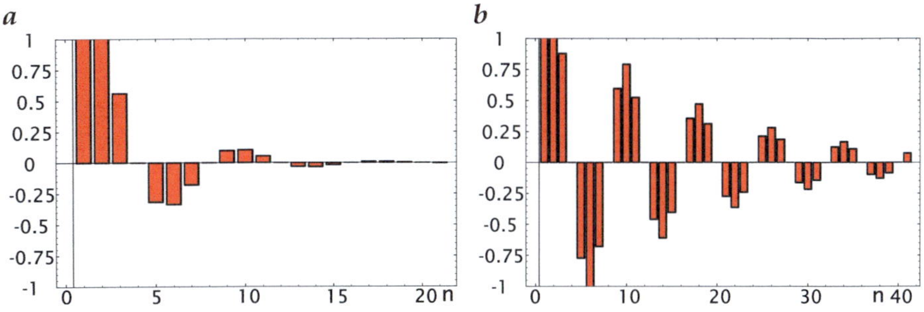

Figure 10.11: Point response of the recursive resonance filter according to Eq. (10.68) for **a** $\tilde{k}_0 = 1/4$, $r = 3/4$ *and* **b** $\tilde{k}_0 = 1/4$, $r = 15/16$.

equation:

$$g'_n = (1 - r^2)\sin(\pi \tilde{k}_0) g_n + 2r\cos(\pi \tilde{k}_0) g'_{n\mp 1} - r^2 g'_{n\mp 2}. \tag{10.71}$$

Example 10.2: Filter with resonant wavenumber $\tilde{k}_0 = 1/2$

For reasons of symmetry, the filter coefficients become particularly simple when the resonance wavenumber $\tilde{k}_0 = 1/2$. Then only two filter coefficients are not equal to zero:

$$g'_n = (1 - r^2) g_n - r^2 g'_{n\mp 2} = g_n - r^2(g_n + g'_{n\mp 2}), \tag{10.72}$$

and the transfer function simplifies to

$$\tilde{o}(\tilde{k}) = \tilde{o}(1 - \tilde{k}) = \frac{(1 - r^2)^2}{1 + r^4 + 2r^2\cos(2\pi \tilde{k})}. \tag{10.73}$$

At $\tilde{k} = 1/2$ the transfer function reaches the maximum value of 1 and at $\tilde{k} = 0$ and $\tilde{k} = 1$ the minimum value $[(1 - r^2)/(1 + r^2)]^2$.

Extension 10.6: Harmonic oscillator

The resonant filter is the discrete analog to a damped harmonic *oscillator* like the *LRC* circuit in Fig. 10.9c, which can be described with the following second-order differential equation:

$$\frac{\mathrm{d}^2 y}{\mathrm{d}t^2} + 2\tau \frac{\mathrm{d}y}{\mathrm{d}t} + \omega_0^2 y = 0. \tag{10.74}$$

The angular frequency ω_0 and the damping constant τ of the oscillator are linked to the parameters of the discrete resonant filter, r and $\tilde{k}_0$, by the following relationships [157]:

$$r = \exp(-\Delta t/\tau) \quad \text{and} \quad \tilde{k}_0 = \omega_0 \Delta t/\pi, \tag{10.75}$$

where Δ indicates the time sampling of the discrete signal.

10.4.6 ♯LSI Filters and System Theory

The last example of the damped oscillator illustrates that there is a close relationship between discrete filter operations and analog physical systems. Digital filters thus reflect real physical processes. They model how the corresponding system would react to a given input signal g. In fact, we have already used this equivalence when discussing image generation in Chapter 3. Image acquisition with a homogeneous optical system is characterized by a point response. This means that image acquisition can be described by a convolution.

Other physical systems such as electrical circuits and oscillators of all kinds can also be described in this way. Such systems are called *linear time-invariant systems* (*LTI*) for temporal processes and *linear shift-invariant systems* (*LSI*) for spatial processes.

This relationship is very helpful for image processing, as we can describe both image generation and image processing as a convolution operation. In addition, the observed images can come from a physical process that can be modeled by a linear displacement invariant system. The method used to find out how the system works can be illustrated using the black box model (Fig. 10.9a).

This term stands for systems whose internal structure or — in physical terms — whose laws we do not know. Such systems can be investigated by measuring the output signal with suitable input signals. Many experiments are carried out in this way to find out how systems function in technology and nature and to set up models for them. However, it must be carefully checked whether the system is linear. If this is not the case, then linear system theory no longer applies and a point response or transfer function no longer adequately describes the system. It is then much more difficult to find out the internal structure of an unknown system through the relationship between the input and output signal.

10.5 Nonlinear Neighborhood Operators

In Section 10.1.4 we discussed other possible neighborhood operators in addition to the linear and shift-invariant convolution operators ("weights and sums"). In this section, we will discuss a few more examples, the nonlinear but still shift invariant rank-order filters ("Sort and Select", Section 10.5.1) and nonlinear combinations of point and neighborhood operators (Section 10.5.2).

10.5.1 Rank-order Filter

Thinking about how pixels can be combined has led us to the powerful concept of linear shift invariant systems. However, we have not yet learned everything we need for this type of image processing operation. There is another class of operations that work on a completely different principle.

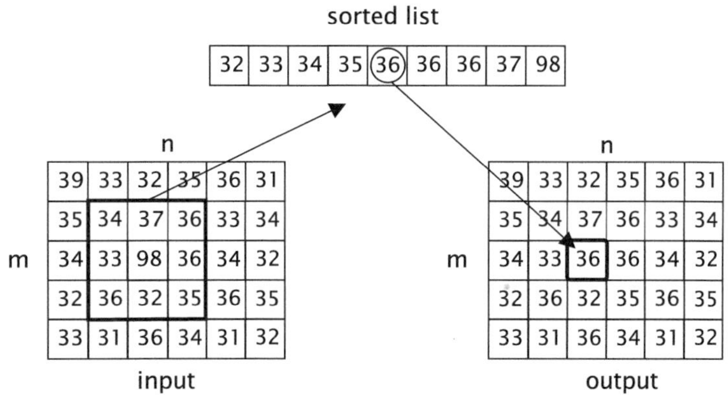

Figure 10.12: *Illustration of the principle of a rank-order filter with a* 3×3 *median filter.*

A convolution with a filter mask can be characterized by weighting and adding. The operation for combining adjacent pixels, which we are now considering, is characterized by comparing and selecting. Such filters are called *rank-order filters*. They sort all gray values of the pixels that lie within the filter mask in ascending order. This sorting process is characteristic of all rank-order filters. They differ only in the position in the list from which the gray value is taken and written back to the central pixel. The filter operation that selects the value in the middle position is called *median filter*. Fig. 10.12 illustrates how a median filter works. The filters that select the first or last value in the sorted list are called *minimum-* or *maximum filter*.

The median filter is a non-linear operator. For the sake of simplicity, we consider a one-dimensional case with a three-element median filter. It is easy to find two vectors for which the median filter is non-linear. First, we apply the median filter to the sum of two signals and obtain

$$\mathcal{M}([\cdots\ 0\ 1\ 0\ 0\ \cdots] + [\cdots\ 0\ 0\ 1\ 0\ \cdots]) = [\cdots\ 0\ 1\ 1\ 0\ \cdots].$$

We then apply the median filter to the two components first and then add the two results together:

$$\mathcal{M}[\cdots\ 0\ 1\ 0\ 0\ \cdots] + \mathcal{M}[\cdots\ 0\ 0\ 1\ 0\ \cdots] = [\cdots\ 0\ 0\ 0\ 0\ \cdots].$$

The two calculation methods lead to different results. This proves that the median filter is non-linear.

There are some significant differences between convolution filters and rank-order filters. The most important is that rank-order filters belong to the class of *nonlinear filters*. Their basic properties are therefore much more difficult to understand. Since rank-order filters do not perform arithmetic operations but select pixels, we will not encounter rounding problems. These filters map a discrete set of gray values onto themselves.

10.5.2 Non-linear Composite Filters

There is another very simple way to generate non-linear filters, namely by combining linear filters and non-linear point operations, as we have seen in Section 7.2 and 7.3. This may sound surprising, but there are actually several completely different methods of combining point operations and linear filter operations into non-linear operators.

Non-linear linking of filter operations
The results of two or more filter operations can be linked non-linearly. The simplest method for this is to multiply them together at each data point.

Weighting of the data
Not every data point has to have the same meaning. Some may have a higher error, others may carry completely incorrect information. This basic fact leads to concepts of including data points with a weighting factor in a neighborhood operation.

Non-linear filter combination
It is conceivable that a different neighborhood operation is performed at each data point, depending on the local structure of the data. This can be done either discretely or continuously. With the discrete method, a filter is selected from a set of several filters at each data point. With the continuous method, one or more parameters decide how a set of filters is combined with each other. These approaches are also known as adaptive filtering.

These simple ways of combining neighborhood and point operations have become very important in image and signal processing. In the remainder of this book, we will become familiar with a whole series of examples of this, in particular weighted averaging (Section 11.6.2), controllable averaging (Section 11.6.3), the structure tensor (Section 13.3) and inhomogeneous and anisotropic diffusion filters (Sections 17.3.1 and 17.3.2).

10.6 ♯References for Further Reading

The classical concepts of discrete signal processing, especially recursive filters, and the z-transform are covered in a number of standard textbooks, for example Oppenheim and Schafer [234] or Proakis and Manolakis [250]. A treatise on two-dimensional recursive filters can be found in Lim [200]. Multidimensional and nonlinear system theory is treated by Unbehauen [336]. Non-linear filters, in particular median filters, are described in detail by Pitas and Venetsanopoulos [242] and Arce [7].

11 Averaging and Noise Reduction

Contents

11.1 Introduction

In this chapter, we deal with neighborhood operations to perform the elementary task of averaging. This operation is one of the building blocks for more complex feature extraction operators discussed in Chapters 13–16.

B. Jähne, *Digital Image Processing and Image Acquisition*,
https://doi.org/10.1007/978-3-032-04806-6_11

In the simplest case, objects are identified as *regions* of constant values. By averaging, we obtain representative averages of the values within the object. Of course, this approach involves a simple model of the image content and only works if the object of interest is characterized by constant values that are clearly different from the background and/or other objects. However, this condition is rarely fulfilled in the real world. In general, objects have varying intensities either as an intrinsic property or due to the image generation process. Typical cases are *noise*, an *uneven illumination* or an *inhomogeneous background*.

In more complex applications, it is impossible to distinguish objects from the background based on a single property. In such cases, it may be necessary to calculate more than one feature image for the same image. We obtain a multi-channel image or *vectorial feature image*. We encounter the same situation when there is more than one image of a scene, for example with color images or all types of multispectral images. Therefore, we must also apply averaging to multi-channel images. Image sequences require an extension of averaging to the time coordinate for spatio-temporal averaging.

11.2 Properties of Averaging Filters

Convolution forms the basis for a large class of averaging or smoothing filters. These filters have a number of common properties, which are discussed in this section.

11.2.1 Zero Shift

A smoothing operator must not change the object position. Any displacement by an operator during image preprocessing would lead to errors in the subsequent position determination and possibly in the calculation of other geometric properties of an object. To avoid this, the transfer function of a filter must be real. A filter with this property is referred to as *zero-phase filter*, as it does not cause a phase shift in any of the periodic components of an image. A real transfer function implies a symmetric filter mask (Section 6.3). A D-dimensional symmetric convolution mask is defined by

$$\begin{aligned} &\text{1D:} \quad h_{-n} = h_n \\ &\text{2D:} \quad h_{-m,n} = h_{m,n}, \; h_{m,-n} = h_{m,n} \\ &\text{3D:} \quad h_{-l,m,n} = h_{l,m,n}, \; h_{l,-m,n} = h_{l,m,n}, \; h_{l,m,-n} = h_{l,m,n}. \end{aligned} \tag{11.1}$$

The symmetry relationships make it considerably easier to calculate the transfer functions, as only the cosine term of the complex exponent of the Fourier transform remains in the equations (Section 10.3.5). The transfer function for symmetrical 1D masks with an odd number of coefficients

$(2R+1)$ is

$$\tilde{h}(\tilde{k}) = h_0 + 2\sum_{\tilde{n}=1}^{R} h_{\tilde{n}} \cos(\tilde{n}\pi\tilde{k}). \tag{11.2}$$

The transfer function of a one-dimensional symmetric mask with an even number of coefficients $(2R)$ is given by

$$\tilde{h}(\tilde{k}) = 2\sum_{\tilde{n}=1}^{R} h_{\tilde{n}} \cos((\tilde{n}-1/2)\pi\tilde{k}). \tag{11.3}$$

Note that the wavenumbers are half-integer $(1/2, 3/2, ...)$, since for reasons of symmetry the result of the convolution with an even mask lies on the intermediate grid.

The equations for higher-dimensional symmetrical masks with an odd number of coefficients in all directions are considerably more complex [157]. However, we do not consider them here, as it makes much more sense to compose multidimensional smoothing filters from one-dimensional ones. This has two advantages. Firstly, these filters can be calculated much faster (Section 10.2.1) and secondly, the equations for the transfer function are much simpler.

11.2.2 Preservation of the Mean Value

The mean value should be retained with a smoothing operator. Therefore, the transfer function for the wavenumber zero is one and the sum of all coefficients of the mask is also one:

$$\sum_{n} h_n = 1 \quad \circ\!\!-\!\!\bullet \quad \tilde{h}(\mathbf{0}) = 1 \tag{11.4}$$

11.2.3 Monotonically Decreasing Transfer Function

Intuitively, we expect any smoothing operator to attenuate finer structures more than coarser ones. In other words, a smoothing operator should not make a certain structure disappear while smaller structures still remain in the image. In mathematical terms, this means that the transfer function must decrease monotonically with the wavenumber.

$$\tilde{h}(\tilde{k}_2) \le \tilde{h}(\tilde{k}_1) \quad \text{if} \quad \tilde{k}_2 > \tilde{k}_1. \tag{11.5}$$

We can specify a stricter condition according to which the transfer function for the highest wavenumber is zero. For a 1D smoothing filter with an odd number of coefficients, the following conditions result from Eq. (11.2) for $\tilde{k} = 1$:

$$\sum_{n=-R}^{R} (-1)^n h_n = 0 \quad \circ\!\!-\!\!\bullet \quad \tilde{h}(1) = 0 \tag{11.6}$$

Together with the monotonicity condition and the preservation of the mean value, this means that the transfer function for each smoothing operator should decrease monotonically from one to zero. Restricting the transfer function for smoothing filters to the interval $[0, 1]$ means that no component of the signal is amplified. Repeated application of the same smoothing operation therefore causes all structures with wavenumbers greater than zero to disappear from the signal until only the mean value remains at the end. Due to the monotonic decay of the transfer function, small-scale components disappear more quickly.

11.2.4 Isotropy

In most applications, the smoothing should be the same in all directions to avoid directional dependencies. The filter mask and transfer function should therefore be isotropic. Then the filter mask depends only on the distance from the central pixel and the transfer function only on the magnitude of the wavenumber:

$$h(\boldsymbol{x}) = h(|\boldsymbol{x}|) \quad \text{and} \quad \tilde{h}(\boldsymbol{\underset{\sim}{k}}) = \tilde{h}(|\boldsymbol{\underset{\sim}{k}}|). \tag{11.7}$$

In discrete space, the first condition can only be fulfilled approximately. An important goal for the optimization of smoothing filters is therefore the construction of discrete masks with the lowest possible anisotropy.

11.2.5 Characteristics for Smoothing Filters

Smoothing filters reduce the resolution in signals and reduce noise. This can be characterized by a number of parameters.

Half-width in Spatial Domain
The PSF or impulse response (Section 10.3.4) directly indicates how a point-like structure is "smeared". A descriptive quantity would therefore be the *half-width* of the point response. This is the distance over which the PSF drops to half. We denote this quantity by $\Delta x_{1/2}$.

Half-width in Fourier Domain
Since smoothing also leads to a drop in the transfer function, it is also possible to specify the wavenumber at which the transfer function is halved. This quantity is designated analogously with $\Delta \underset{\sim}{k}_{1/2}$.

Standard Deviation in Spatial Domain
To obtain more of an integral measure, the standard deviation of the filter mask can be calculated. Since smoothing filter masks have an even symmetry, the standard deviation for a mask with $2R + 1$ coefficients can be computed using a sum of only R coefficients:

$$\sigma_x = \left(\frac{2}{2R+1} \sum_{r=1}^{R} h_r r^2 \right)^{1/2}. \tag{11.8}$$

Standard Deviation in Fourier Domain
Similarly, a standard deviation can be calculated from the real-valued transfer function. Due to the even symmetry, the standard deviation results from integrals over only the positive wavenumbers $\underset{\sim}{k}$:

$$\sigma_{\underset{\sim}{k}} = \left(\int_0^1 \tilde{h}(\underset{\sim}{k})\underset{\sim}{k}^2 \, \mathrm{d}\underset{\sim}{k} \Bigg/ \int_0^1 \tilde{h}(\underset{\sim}{k}) \, \mathrm{d}\underset{\sim}{k} \right)^{1/2}. \tag{11.9}$$

11.3 Box Filters

11.3.1 Introduction

Smoothing filters must average the pixels in a small area. The simplest method is to add up all the pixels within the filter mask and divide the sum by the number of pixels. Such a simple filter is called *box filter* or *rectangular filter*. Box filters can be used to illustrate how a good filter is constructed. As an introduction, let's look at a 1×3 filter:

$$^3\boldsymbol{R} = \begin{bmatrix} 1 & 1 & 1 \end{bmatrix}/3. \tag{11.10}$$

The factor 1/3 scales the result of the convolution sum. This is the only way to preserve the mean value (Section 11.2.2) and thus the gray value of a pixel in a region with constant gray values. We first apply this mask to a vertical edge:

$$\begin{matrix} & \vdots & \vdots & \vdots & \vdots & \\ \cdots & 0 & 0 & 1 & 1 & \cdots \\ \cdots & 0 & 0 & 1 & 1 & \cdots \\ \cdots & 0 & 0 & 1 & 1 & \cdots \\ & \vdots & \vdots & \vdots & \vdots & \end{matrix} * \frac{1}{3}\begin{bmatrix} 1 & 1 & 1 \end{bmatrix} = \begin{matrix} & \vdots & \vdots & \vdots & \vdots & \\ \cdots & 0 & 1/3 & 2/3 & 1 & \cdots \\ \cdots & 0 & 1/3 & 2/3 & 1 & \cdots \\ \cdots & 0 & 1/3 & 2/3 & 1 & \cdots \\ & \vdots & \vdots & \vdots & \vdots & \end{matrix}$$

As we would expect from a smoothing operation, the sharp edge becomes a gradual, ramp-like transition from zero to one. Smoothing filters attenuate structures with high wavenumbers. We first test this on a vertical structure with a wavelength of three pixels:

$$\begin{matrix} \vdots & \vdots & \vdots & \vdots & \vdots & \vdots & \\ 1 & -2 & 1 & 1 & -2 & 1 & \cdots \\ 1 & -2 & 1 & 1 & -2 & 1 & \cdots \\ 1 & -2 & 1 & 1 & -2 & 1 & \cdots \\ \vdots & \vdots & \vdots & \vdots & \vdots & \vdots & \end{matrix} * \frac{1}{3}\begin{bmatrix} 1 & 1 & 1 \end{bmatrix} = \begin{matrix} \vdots & \vdots & \vdots & \vdots & \vdots & \vdots & \\ 0 & 0 & 0 & 0 & 0 & 0 & \cdots \\ 0 & 0 & 0 & 0 & 0 & 0 & \cdots \\ 0 & 0 & 0 & 0 & 0 & 0 & \cdots \\ \vdots & \vdots & \vdots & \vdots & \vdots & \vdots & \end{matrix}$$

The 1×3 rectangular filter therefore completely filters out a structure with a wavelength of three pixels. As discussed in Section 11.2.3, we expect

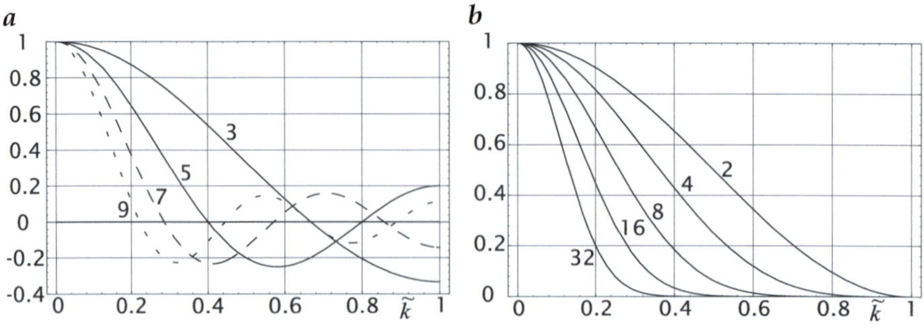

***Figure 11.1:** Transfer functions of one-dimensional smoothing filters: **a** Rectangular filter with 3, 5, 7 and 9 coefficients; **b** Binomial filter $\mathcal{B}^R$ with R = 2, 4, 8, 16 and 32.*

a good smoothing filter to filter out all structures with a wavenumber above a certain cut-off wavenumber. However, this is not the case with the 1×3 box mask under consideration. For a structure with a wavelength of two pixels, it only attenuates the amplitude to 1/3:

$$\begin{array}{ccccccc} & \vdots & \vdots & \vdots & \vdots & \\ \cdots & 1 & -1 & 1 & -1 & \cdots \\ \cdots & 1 & -1 & 1 & -1 & \cdots \\ \cdots & 1 & -1 & 1 & -1 & \cdots \\ & \vdots & \vdots & \vdots & \vdots & \end{array} * \frac{1}{3}\left[\begin{array}{ccc} 1 & 1 & 1 \end{array}\right] = \begin{array}{cccccc} & \vdots & \vdots & \vdots & \vdots & \\ \cdots & -1/3 & 1/3 & -1/3 & 1/3 & \cdots \\ \cdots & -1/3 & 1/3 & -1/3 & 1/3 & \cdots \\ \cdots & -1/3 & 1/3 & -1/3 & 1/3 & \cdots \\ & \vdots & \vdots & \vdots & \vdots & \end{array}$$

11.3.2 1D Box Filters

After the introductory considerations, we will now discuss the properties of box filters quantitatively by calculating the transfer function. We start with 1D filters. Following the considerations in Section 11.2.1, we can use Eq. (11.2) to calculate the transfer function of the box filter with the mask Eq. (11.10). Only the coefficients $h_0 = h_1 = 1/3$ are not equal to zero, and the transfer function is therefore reduced to

$$ {}^3\tilde{r}(\underset{\sim}{k}) = \frac{1}{3} + \frac{2}{3}\cos(\pi \underset{\sim}{k}). \tag{11.11}$$

The transfer function shows Fig. 11.1a. This proves our exemplary calculations at the beginning of this section. The transfer function has a zero point at $\underset{\sim}{k} = 2/3$. This corresponds to a wavenumber that is scanned three times per wavelength. The smallest possible wavelength ($\underset{\sim}{k} = 1$), which is scanned twice per wavelength, is only attenuated in amplitude by a factor of 1/3. For $\underset{\sim}{k} > 2/3$ the transfer function is negative. This means that the minima and maxima are swapped, which corresponds to a phase shift of 180 °These considerations show that the 1×3 box filter is not a good smoothing filter. The amplitude attenuation does not increase monotonically with the wavenumber, but oscillates, and structures with the highest wavenumbers are not attenuated strongly enough.

Larger box filters

$$ {}^{R}\boldsymbol{R} = \underbrace{[1\,1\,\ldots\,1]}_{R\text{-mal}}/R \tag{11.12} $$

with R coefficients and the transfer function

$$ \boxed{{}^{R}\tilde{r}(\tilde{k}) = \frac{\sin(\pi R \tilde{k}/2)}{R \sin(\pi \tilde{k}/2)}} \tag{11.13} $$

do not show any improved behavior. The oscillations are even more pronounced and the attenuation is only proportional to the wavenumber. For large filter masks, the discrete mask with R coefficients can be approximated by a continuous rectangular function of width R. Therefore, the transfer function approximates a sinc function (> R13) for $\tilde{k} \ll 1$:

$$ {}^{R}\tilde{r}(\tilde{k}) \approx \frac{\sin(\pi R \tilde{k}/2)}{\pi R \tilde{k}/2} = \operatorname{sinc}(R\tilde{k}/2). \tag{11.14} $$

11.3.3 2D Box Filters

Now we turn to the two-dimensional rectangular filters. To simplify the calculations, we use the separability of the filter and split it into vertical and horizontal 1D components:

$$ {}^{3}\boldsymbol{R} = {}^{3}\boldsymbol{R}_x * {}^{3}\boldsymbol{R}_y = \begin{bmatrix} 1 & 1 & 1 \\ 1 & 1 & 1 \\ 1 & 1 & 1 \end{bmatrix} \Big/ 9 = \begin{bmatrix} 1 & 1 & 1 \end{bmatrix} / 3 * \begin{bmatrix} 1 \\ 1 \\ 1 \end{bmatrix} \Big/ 3. $$

We already know the transfer function of the one-dimensional filter from Eq. (11.11). For the vertical filter, $\tilde{k}_x$ must be replaced by $\tilde{k}_y$. Since the convolution in spatial space corresponds to a multiplication in wavenumber space, the transfer function of $\boldsymbol{R}$ results without further calculation:

$$ {}^{3}\tilde{r} = {}^{3}\tilde{r}_x {}^{3}\tilde{r}_y = \left[\frac{1}{3} + \frac{2}{3}\cos(\pi \tilde{k}_x)\right]\left[\frac{1}{3} + \frac{2}{3}\cos(\pi \tilde{k}_y)\right]. \tag{11.15} $$

From Eq. (11.15) and Fig. 11.2a we can conclude that 2D rectangular filters are also poor low-pass filters. As with one-dimensional filters, larger 2D box filters are no better. Fig. 11.2b shows an example of a 5×5 mask.

In addition to the disadvantages of one-dimensional filters already discussed, the transfer function is not isotropic, i. e. the smoothing depends on the direction. The biggest difference is between the axial and diagonal direction (Fig. 11.2c and d). In the axial direction, the envelope of the transfer

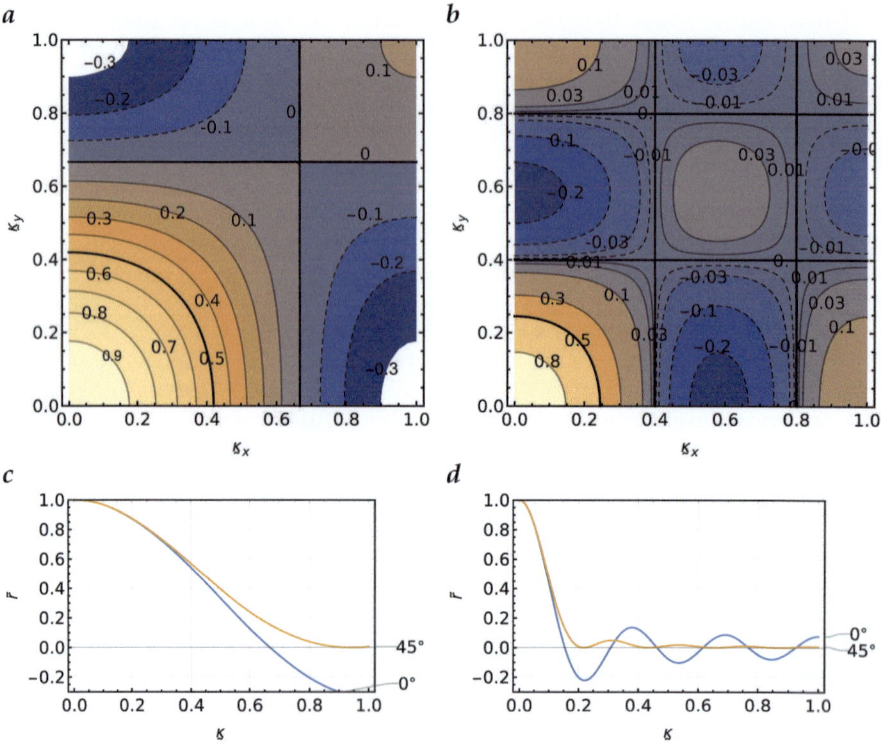

Figure 11.2: Transfer functions of two-dimensional box filters; 2D contour plots for ***a*** *2D-3 × 3 box filter and* ***b*** *5 × 5 box filter; 1D sections in axial (0°) and diagonal (45°) direction for* ***c*** *a 3 × 3 box filter and* ***d*** *a 13 × 13 box filter.*

function only decreases with the wavelength, but in the diagonal direction it decreases with the square of the wavelengths because the transfer functions multiply in the x and y directions:

$$\begin{aligned} &\text{Axial direction} && {}^{R}\tilde{r}(\tilde{k}) = \frac{\sin(\pi R \tilde{k}/2)}{R \sin(\pi \tilde{k}/2)}, \\ &\text{Diagonal direction} && {}^{R}\tilde{r}(\tilde{k}) = \frac{\sin^2(\pi R \tilde{k}/\sqrt{2})}{R^2 \sin^2(\pi \tilde{k}/\sqrt{2})}. \end{aligned} \tag{11.16}$$

If we apply a rectangular filter to any image, the disadvantages discussed have an effect, but are difficult to quantify (Fig. 11.6). However, they become immediately quantifiable if we use a carefully constructed *test image*.

The image in Fig. 11.3 consists of concentric rings, where the gray value changes sinusoidally in the radial direction, calculated with

$$g(\boldsymbol{x}) = g_0 \sin\left(\frac{\pi \tilde{k}_m |\boldsymbol{x}|^2}{r_m}\right)\left[\frac{1}{2}\tanh\left(\frac{r_m - |\boldsymbol{x}|}{w}\right) + \frac{1}{2}\right]. \tag{11.17}$$

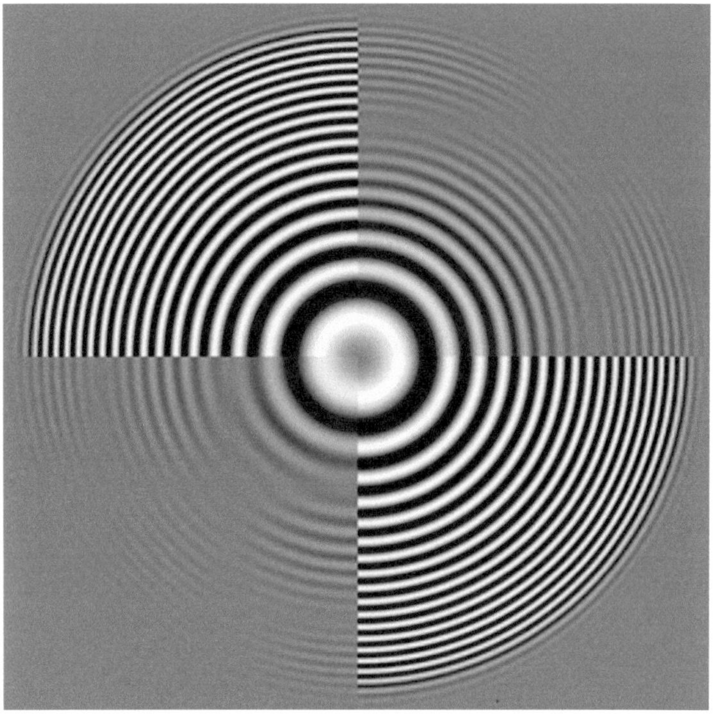

Figure 11.3: *Smoothing of a test image with concentric rings with* 5×5 *and* 9×9 *box filters in the upper right and lower left quadrant, respectively. The maximum wavenumber* $\tilde{k}$ *at the edge of the pattern is 0.6.*

Where r_m is the maximum radius of the pattern, $\tilde{k}_m$ is the maximum wavenumber at the edge and w is the width of the drop to zero at the maximum radius. The wavenumber increases linearly from zero in the center to $\tilde{k}_m$ at the edge ($\tilde{k} = \tilde{k}_m x / r_m$), as can be calculated by derivation. In optics, the pattern is known as *Fresnel zone plate*. In the *ring test pattern*, structures occur in all directions and all wavenumbers up to a selectable maximum. Furthermore, the wavenumbers are represented in the image as in two-dimensional Fourier space. Therefore, we can immediately see the transfer function, i. e. the change in amplitudes and the phase shift, when we apply a filter to the ring test pattern. If we convolve this image with a 5×5 or a 9×9 box filter, the deviations from an isotropic transfer function are clearly visible. Certain wavenumbers disappear completely, and in some areas gray value maxima change to minima or vice versa (Fig. 11.3). This is caused by a phase shift of $180°$ due to negative values in the transfer function.

We can gain an important insight from this: We must not evaluate the properties of a filter operation according to its effect on arbitrary images with our eye. The quantitative extraction of image features requires a quantitative analysis of the filter properties. This includes a thorough analysis of the transfer function and the application of the filters to carefully selected test images.

Now we return to the question of why rectangular filters are such poor smoothing filters. We can try to construct a better smoothing filter directly in wavenumber space. An ideal filter would cut off all wavenumbers above a certain threshold wavenumber. So we could use this ideal transfer function (a rectangular function) and calculate the filter mask with an inverse Fourier transform. Here, however, we run into two problems that are understandable without explicit calculation: Firstly, the inverse Fourier transform of a rectangular function is a sinc function, which means that the coefficients of the filter mask only decrease proportionally to the distance from the central pixel. We would therefore have to use very large filter masks. The filter also has the unfavorable property of reacting to an edge with strong overshoot.

11.3.4 Fast Computation of Box Filters

Despite all these disadvantages, rectangular filters also have a decisive advantage. According to the following equation, a filter of size $R = 2R' + 1$ requires only three arithmetic operations per pixel, regardless of the filter size, as if it were a recursive filter:

$$g'_m = g'_{m-1} + \frac{1}{2R' + 1}(g_{m+R'} - g_{m-R'-1}). \tag{11.18}$$

This recursion becomes clear when we compare the calculation of the convolution on neighboring pixels. If the rectangular mask is shifted one position to the right, the same weighting factor applies to all pixels except the last and the first. So we can simply use the result of the previous convolution, (g'_{m-1}), subtract the first pixel ($g_{m-R'-1}$) that has just moved out of the mask and add the gray value ($g_{m+R'}$) of the pixel that has just moved into the mask. In this way, the calculation is not dependent on the size of the rectangular filter. Only one addition, one subtraction and one multiplication are necessary to perform the filtering.

11.3.5 Noise Suppression through Box Filters

Rectangular filters perform a moving average over all pixels within the mask. All pixels are given the same weighting factor. If the noise of the individual pixels is statistically independent, as is the case with raw data from an image sensor, then the variance of the noise is reduced by the number of averages. The noise reduction factor ζ_R for a D-dimensional rectangular filter of size R is therefore

$$\zeta_R = \frac{\sigma'}{\sigma} = \frac{1}{R^{D/2}}. \tag{11.19}$$

With a two-dimensional $R \times R$ rectangular filter, the standard deviation of the noise is reduced by $1/R$.

Example 11.1: Optimal 1D smoothing filter with 3 coefficients

As a starting point for the design of an optimal smoothing filter, we check what results for a filter with three coefficients if all the requirements for an ideal smoothing filter set out in Section 11.2 are met. This results in the following equations for the three coefficients h_{-1}, h_0 and h_1:

Condition	Reference	Equation
Zero shift	Section 11.2.1	$h_{-1} = h_1$
Preservation of mean	Section 11.2.2	$h_{-1} + h_0 + h_1 = 1$
Monotonous decrease to zero	Section 11.2.3	$-h_{-1} + h_0 + -h_1 = 0$

For three unknowns, there are three equations and therefore only one filter that fulfills all three conditions. Adding the last two equations results in $h_0 = 1/2$. Since $h_{-1} = h_1$ applies, inserting it into the second or third equation yields $h_{-1} = h_1 = 1/4$ and the filter has the coefficients

$$[1\ 2\ 1]/4 = [1\ 1]/2 * [1\ 1]/2.$$

11.4 Binomial Filters

11.4.1 Basics and Basic Filter

From our experience with rectangular filters, we conclude that filter design is not a simple optimization problem. If we choose a small rectangular filter mask, we obtain a poor transfer function. If we start with an ideal transfer function that falls off sharply from a given wavenumber, this results in large filter masks and overshooting filter responses. The reason for this behavior is a fundamental relationship between smoothness and compactness of Fourier transform pairs (Section 6.3.4). An edge is a discontinuity of the signal and leads to an impulse in the first derivative. The Fourier transform of a pulse is evenly distributed over the entire Fourier space (> R13). If we consider the integral property of the Fourier transform (> R12), the integration of the derivative in space means a division by $\boldsymbol{\kappa}$ in Fourier space. Then we know without further calculation that in the one-dimensional case, the envelope of the Fourier transform of a function that has a discontinuity in local space does not decrease faster than k^{-1} in wavenumber space. This is exactly what we have found for the rectangular function whose Fourier transform is the sinc function (> R13).

Taking this basic fact into account leads us to better smoothing filters. The requirement to avoid discontinuities in the filter mask inevitably means that the filter masks must fall to zero at the edge.

11.4.2 1D Binomial Filters

We will now discuss a class of smoothing filters that fulfills this criterion and at the same time allows very efficient computation. Furthermore, these filters are excellent examples of how more complex filters can be composed of simple components. The simplest and most elementary smoothing mask we can imagine is

$$\boldsymbol{B} = [1\ 1]/2. \tag{11.20}$$

It averages the gray values of two adjacent pixels. If we apply this mask R times in succession, the following filter mask or operator equation results:

$$\underbrace{[1\ 1] * [1\ 1] * \ldots * [1\ 1]}_{R-\text{mal}}/2^R, \quad \mathcal{B}^R = \underbrace{\mathcal{B}\mathcal{B}\ldots\mathcal{B}}_{R-\text{mal}}. \tag{11.21}$$

Examples of the resulting filter masks are:

$$\begin{array}{ll} \boldsymbol{B}^2 = [1\ 2\ 1]/4 & \boldsymbol{B}^3 = [1\ 3\ 3\ 1]/8 \\ \boldsymbol{B}^4 = [1\ 4\ 6\ 4\ 1]/16 & \boldsymbol{B}^8 = [1\ 8\ 28\ 56\ 70\ 56\ 28\ 8\ 1]/256. \end{array} \tag{11.22}$$

Due to the *symmetry condition*, we are particularly interested in the masks with an odd number of coefficients. The masks contain the values of the discrete *binomial distribution*. In fact, the iterative composition of the mask by successive convolution with the 1/2 [1 1] mask corresponds to the calculation scheme of the *Pascal's triangle*:

R	f		σ^2
0	1	1	0
1	1/2	1 1	1/4
2	1/4	1 2 1	1/2
3	1/8	1 3 3 1	3/4
4	1/16	1 4 6 4 1	1
5	1/32	1 5 10 10 5 1	5/4
6	1/64	1 6 15 20 15 6 1	3/2
7	1/128	1 7 21 35 35 21 7 1	7/4
8	1/256	1 8 28 56 70 56 28 8 1	2

(11.23)

Where R is the order of the binomial, f is the scaling factor 2^{-R} and σ^2 is the variance, i. e. the effective width of the mask.

Calculating the transfer function of a binomial mask is also easy, as we only need to know the transfer function of $\mathcal{B}$. Then the transfer function of $\mathcal{B}^R$ results as the R-th power:

$$\boxed{\tilde{b}^R(\underset{\sim}{k}) = \cos^R(\pi \underset{\sim}{k}/2).} \tag{11.24}$$

The following approximation applies for small wavenumbers:

$$\tilde{b}^R(\underset{\sim}{k}) = 1 - \frac{R}{8}(\pi \underset{\sim}{k})^2 + \left(\frac{3R^2 - 2R}{384}\right)(\pi \underset{\sim}{k})^4 + O(\underset{\sim}{k}^6). \tag{11.25}$$

As we can see from the graphical representation (Fig. 11.1b), binomial filters are much better smoothing filters than rectangular filters. The transfer function decreases monotonically and approaches zero for the largest wavenumbers. The smallest symmetrical mask, $\mathcal{B}^2$, has a half-width of $\tilde{k}/2$. This is a periodic structure that is sampled just four times per wavelength.

For larger masks, both the transfer function and the filter mask approach the Gaussian distribution with an equivalent variance. Larger binomial masks result in smaller half-widths and thus a smaller cut-off frequency due to the *uncertainty relation* (Section 6.3.4).

11.4.3 Higher Dimensional Binomial Filters

Two-dimensional binomial filters result from the convolution of a horizontal with a vertical 1D binomial filter:

$$\mathcal{B}^R = \mathcal{B}_x^R \mathcal{B}_y^R. \tag{11.26}$$

The simplest mask of this type is a 3×3 binomial mask ($R = 2$):

$$\boldsymbol{B}^2 = \begin{bmatrix} 1 & 2 & 1 \end{bmatrix} /4 * \begin{bmatrix} 1 \\ 2 \\ 1 \end{bmatrix} \Big/ 4 = \begin{bmatrix} 1 & 2 & 1 \\ 2 & 4 & 2 \\ 1 & 2 & 1 \end{bmatrix} \Big/ 16. \tag{11.27}$$

The transfer function of this 2D binomial filter $\mathcal{B}^R$ with $(R+1) \times (R+1)$ coefficients can be easily derived from the transfer functions of the one-dimensional filters Eq. (11.24):

$$\tilde{b}^R = \tilde{b}_y^R \tilde{b}_x^R = \cos^R(\pi \tilde{k}_y/2) \cos^R(\pi \tilde{k}_x/2). \tag{11.28}$$

The same applies to a 3D filter:

$$\tilde{b}^R = \tilde{b}_z^R \tilde{b}_y^R \tilde{b}_x^R = \cos^R(\pi \tilde{k}_z/2) \cos^R(\pi \tilde{k}_y/2) \cos^R(\pi \tilde{k}_x/2). \tag{11.29}$$

Fig. 11.4 shows the transfer functions of $\mathcal{B}^2$ and $\mathcal{B}^4$. Even the small 3×3 filter shows remarkably small deviations from isotropic smoothing. Larger deviations from the circular contour lines are only perceived for higher wavenumbers when the transfer function has decreased to 0.3 (Fig. 11.4a). We can show this property by a Taylor series expansion in $\tilde{k}$ using cylindrical coordinates $\tilde{\boldsymbol{k}} = [\tilde{k}, \phi]^T$:

$$\tilde{b}^R \approx 1 - \frac{R}{8}(\pi \tilde{k})^2 + \frac{2R^2 - R}{256}(\pi \tilde{k})^4 - \frac{R \cos 4\phi}{768}(\pi \tilde{k})^4. \tag{11.30}$$

Only the second-order term is isotropic. One of the fourth-order terms, on the other hand, contains an anisotropic part that increases the transfer function in the direction of the diagonal (Fig. 11.4a). A larger filter becomes less anisotropic, as the isotropic fourth-order term increases quadratically with R, while the anisotropic fourth-order term only increases linearly with R. The 5×5 filter (Fig. 11.4b) is already significantly less anisotropic. The low anisotropy of the binomial filter is also illustrated by the application to the ring test image in Fig. 11.5.

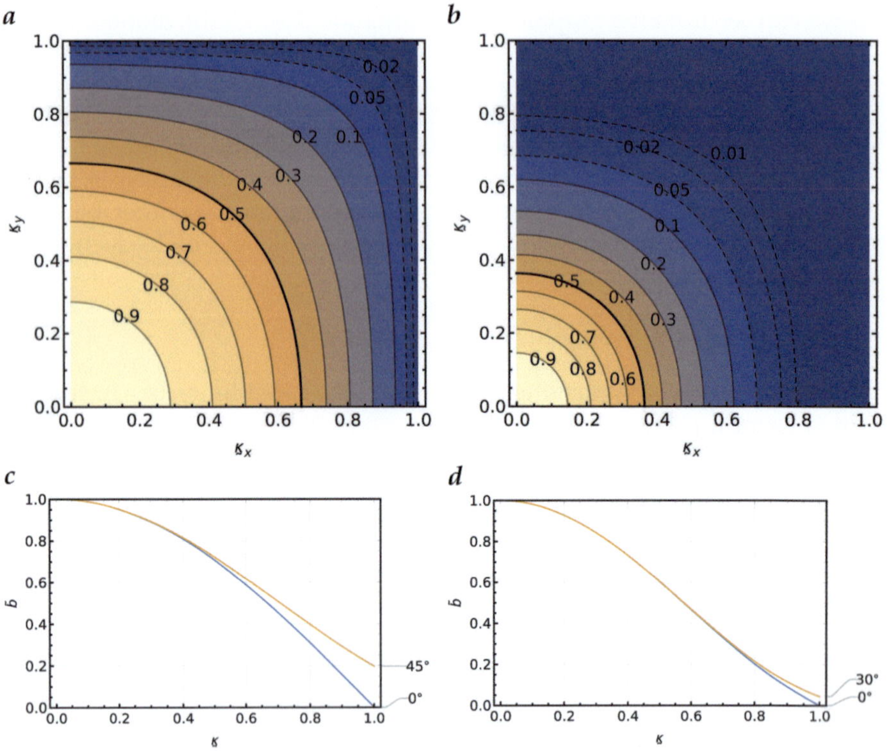

Figure 11.4: *Transfer function of 2D binomial filters shown as a contour plot for* **a** $\mathcal{B}^1$ *and* **b** $\mathcal{B}^4$. *For reasons of symmetry, only a quadrant is shown. Illustration of isotropy using sections:* **c** $\mathcal{B}^1$ *in axial and diagonal direction and* **d** $\mathcal{B}^1_{hex}$ *on a hexagonal grid at 0° and 30°.*

Extension 11.1: Binomial filter on a hexagonal grid

On a hexagonal grid, there are three equivalent directions in which 1D filters can run, 0°, 60° and 120°. The interesting question is: How isotropic is a binomial filter on a hexagonal grid? The transfer function is somewhat more complicated due to the non-orthogonal directions and is as follows

$$\tilde{b}_0^R \tilde{b}_{60}^R \tilde{b}_{120}^R = \cos^R(\pi \tilde{k}_x/2)\cos^R(\pi(\sqrt{3}\tilde{k}_y + \tilde{k}_x)/8)\cos^R(\pi(\sqrt{3}\tilde{k}_y - \tilde{k}_x)/8). \quad (11.31)$$

A Taylor series expansion for small $\tilde{k}$ results in

$$\tilde{b}_0^R \tilde{b}_{60}^R \tilde{b}_{120}^R \approx 1 - \frac{3R}{16}(\pi \tilde{k})^2 + \frac{9R^2 - 3R}{512}(\pi \tilde{k})^4 - \frac{R\cos 6\phi}{30720}(\pi \tilde{k})^6. \quad (11.32)$$

The result is astonishing. An anisotropic term with $\cos 6\phi$ only appears with the 6th order term and is a good 40 times smaller than with the square lattice. Obviously, filters on a hexagonal grid are much more isotropic than on a square grid (Fig. 11.4c and d).

Figure 11.5: Smoothing of a test image with concentric rings with a $\mathcal{B}^4$ and a $\mathcal{B}^{16}$ binomial filter in the upper right and lower left quadrant. The maximum wavenumber $\tilde{k}$ at the edge of the pattern is 0.6.

Fig. 11.6c and d show the effect of smoothing with two different binomial filters. The edges become blurred. Fine structures such as the branches of the trees are lost. Smoothing is a method for *noise reduction*. Binomial filters can considerably reduce the noise level of *mean-free, normally distributed noise* (Section C.4.2), but only at the expense of detail resolution (Fig. 11.7a and c). *Binary noise* or *impulse noise*, in which isolated, randomly distributed gray values have incorrect values (for example due to transmission errors), is only poorly suppressed by linear filters. Such images become blurred and the errors caused by binary noise are not eliminated, but only distributed.

11.4.4 Fast Computation

Unfortunately, there is no direct trick like with the rectangular filters to calculate the binomial filters with constant calculation effort regardless of the mask size (Section 11.3.4). However, binomial filters are separable. This means that a $(R+1)^D$ filter mask $D(R+1)$ requires multiplications and DR additions. If we break down the 1D binomial masks into the elementary smoothing masks $1/2\,[1\ 1]$ and apply them R times in each direction, we can save all but one multiplication. Since only powers of 2 occur, these can be computed much more efficiently by shift operations.

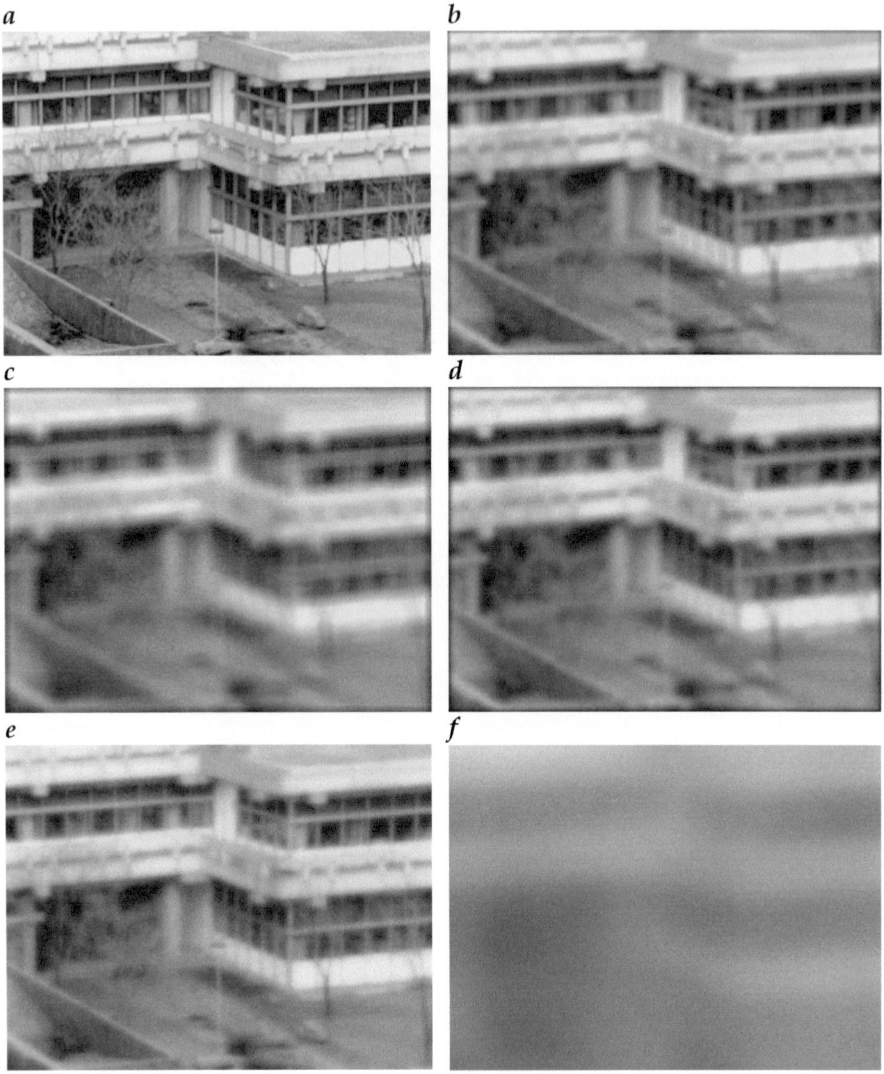

Figure 11.6: *Application of smoothing filters:* ***a*** *Original image;* ***b*** 5×5 *box filter;* ***c*** 9×9 *box filter;* ***d*** 17×17 *binomial filter* ($\mathcal{B}^{16}$)*;* ***e*** *and* ***f*** *each a set of recursive filters Eq. (11.42) running in horizontal and vertical direction (**e** $\alpha = 1/2$; **f** $\alpha = 1/16$).*

a b c d e f

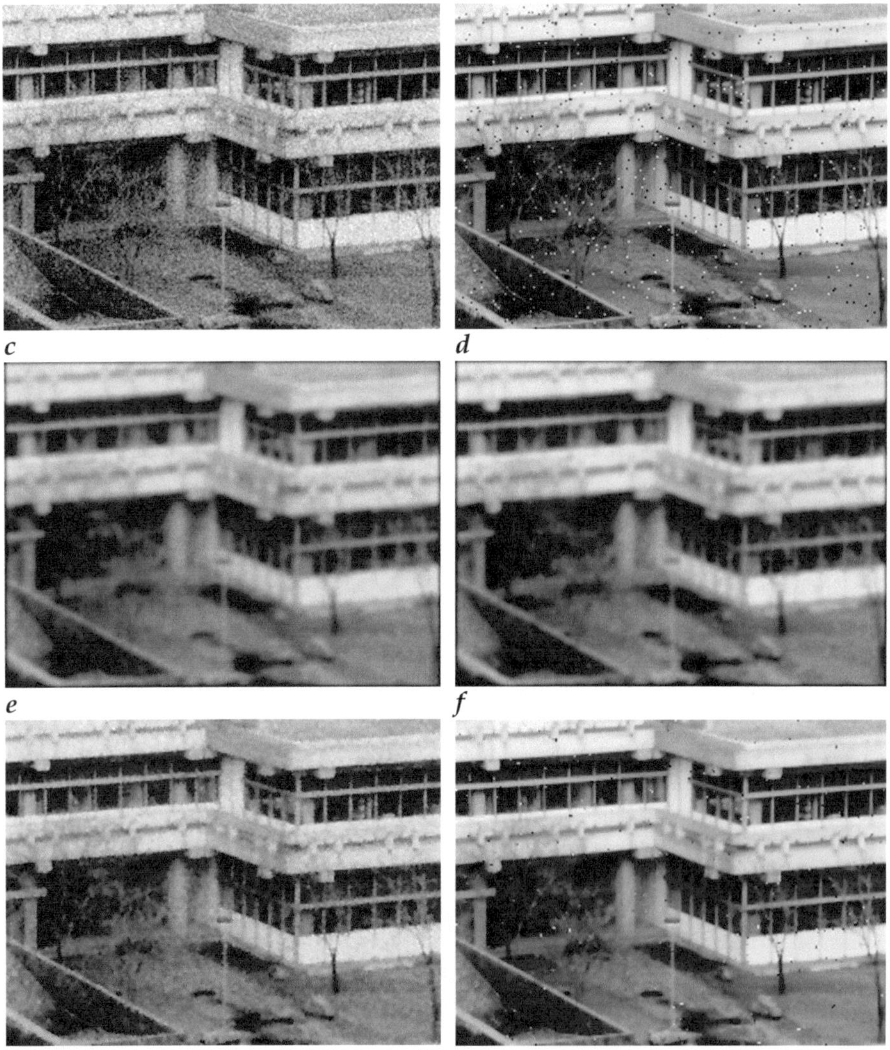

Figure 11.7: noise reduction *with smoothing filters: Image from Fig. 11.6a with* ***a*** *Gaussian noise,* ***b*** *with binary noise;* ***c*** *and* ***d*** *image* ***a*** *and* ***b*** *each filtered with a* 9×9 *binomial filter* ($\mathcal{B}^8$)*;* ***e*** *and* ***f*** *image* ***a*** *and* ***b*** *each filtered with a* 3×3 *median filter (Section 11.6.1).*

11.5 Efficient Large-scale Averaging

Despite the efficient calculation of binomial smoothing filters $\mathcal{B}^R$ by cascaded convolution with $\mathcal{B}$, the computational effort for smoothing masks increases significantly with the distance over which smoothing is performed. This is due to the fact that the standard deviation of the filters according to Eq. (C.45) is proportional to the square root of R:

$$\sigma = \sqrt{R/4}. \tag{11.33}$$

We consider a smoothing operation with an isotropic convolution mask with a variance of $\sigma^2 = 3$. This corresponds to smoothing over a circle with the radius $\sigma \approx 1.73$ and the area $\sigma^2\pi$ which contains slightly more than just 9 pixels. After Eq. (11.33) we have to apply the binomial operator $\mathcal{B}^{12}$ for this smoothing. Even with efficient implementation using separability, this requires 24 (36) additions and 2 (3) shift operations for each pixel of a 2D (3D) image. If we want to perform smoothing over twice the distance ($\sigma^2 = 12$, radius ≈ 3.5, $\mathcal{B}^{48}$), the number of additions quadruples to 96 (144) per pixel in 2D (3D) space.

11.5.1 Multistep Averaging

The problem of slow large-scale averaging is due to the small distance of the pixels that are averaged in the elementary mask $\boldsymbol{B} = 1/2\,[1\ 1]$. To overcome this problem, we can use the same basic averaging process, only with pixels further apart. In doing so, we increase the standard deviation for smoothing according to the step size. In two dimensions, e. g. the following masks along the diagonal ($\sigma \cdot \sqrt{2}$) can be used:

$$\boldsymbol{B}^2_{x+y} = \frac{1}{4}\begin{bmatrix} 1 & 0 & 0 \\ 0 & 2 & 0 \\ 0 & 0 & 1 \end{bmatrix}, \quad \boldsymbol{B}^2_{x-y} = \frac{1}{4}\begin{bmatrix} 0 & 0 & 1 \\ 0 & 2 & 0 \\ 1 & 0 & 0 \end{bmatrix}. \tag{11.34}$$

Possible masks with double step width along the axes ($\sigma \cdot 2$) are

$$\boldsymbol{B}^2_{2x} = \frac{1}{4}\,[1\,0\,2\,0\,1], \quad \boldsymbol{B}^2_{2y} = \frac{1}{4}\begin{bmatrix} 1 \\ 0 \\ 2 \\ 0 \\ 1 \end{bmatrix}, \quad \boldsymbol{B}^2_{2z} = \frac{1}{4}\begin{bmatrix} 1 \\ 0 \\ 2 \\ 0 \\ 1 \end{bmatrix}_z. \tag{11.35}$$

The indices of these masks denote the number of steps along the specified coordinate axes between two pixels to be averaged. $\boldsymbol{B}^2_{x+y}$ averages the gray values of two adjacent pixels in the direction of the main diagonal. $\boldsymbol{B}^2_{2x}$ calculates the mean value from the values of a pixel and its next but one

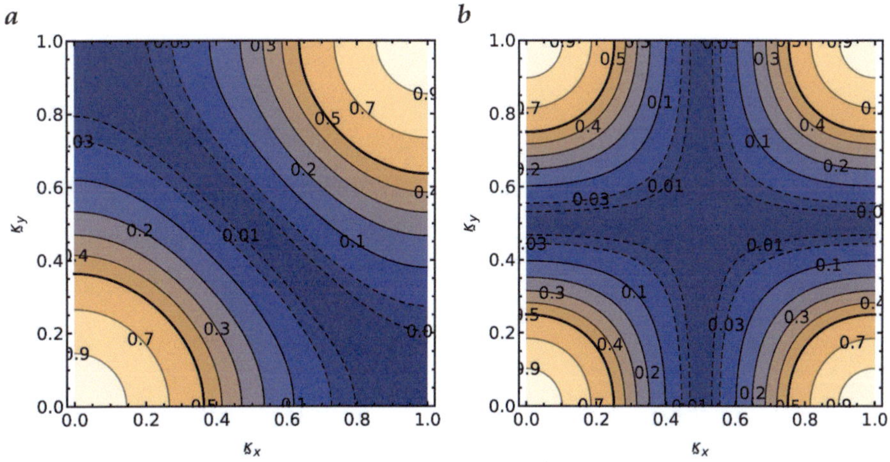

Figure 11.8: *Transfer functions of binomial masks with larger step sizes **a** in diagonal direction $\mathcal{B}^2_{x+y}\mathcal{B}^2_{x-y}$ and **b** with double step size in the axial direction $\mathcal{B}^2_{2x}\mathcal{B}^2_{2y}$.*

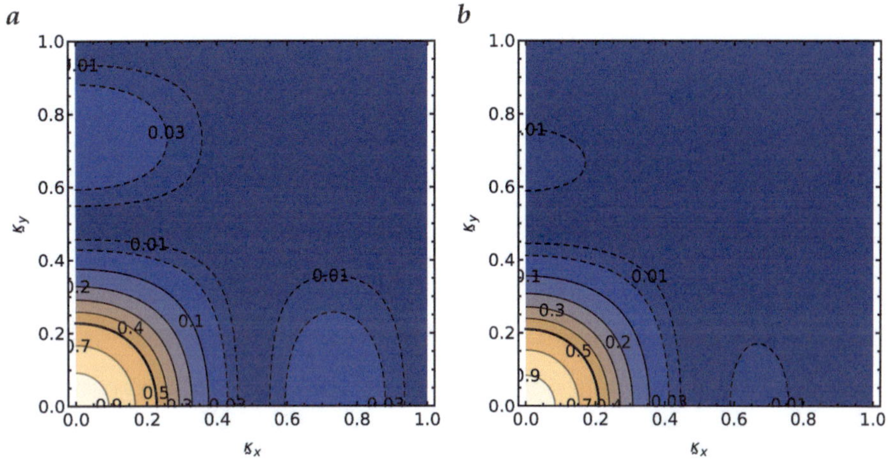

Figure 11.9: *Transfer functions of nested multi-step binomial filters: **a** $\mathcal{B}^2_2\mathcal{B}^2_1$ and **b** $\mathcal{B}^4_2\mathcal{B}^4_1$.*

neighbor in the x direction. The standard deviation of these filters is proportional to the distance between the pixels. Multi-step masks along the axes are the simplest. They have the additional advantage that they can be easily extended to higher-dimensional images.

The problem with these filters is that they perform sampling. This means that they are no longer smoothing filters for large wavenumbers. For example, if we use a symmetric 2D $\mathcal{B}^2_{2x}\mathcal{B}^2_{2y}$ filter, we are actually working on a grid with double the grid constant in the spatial domain. This means that the reciprocal grid in wavenumber space has half the grid width and the

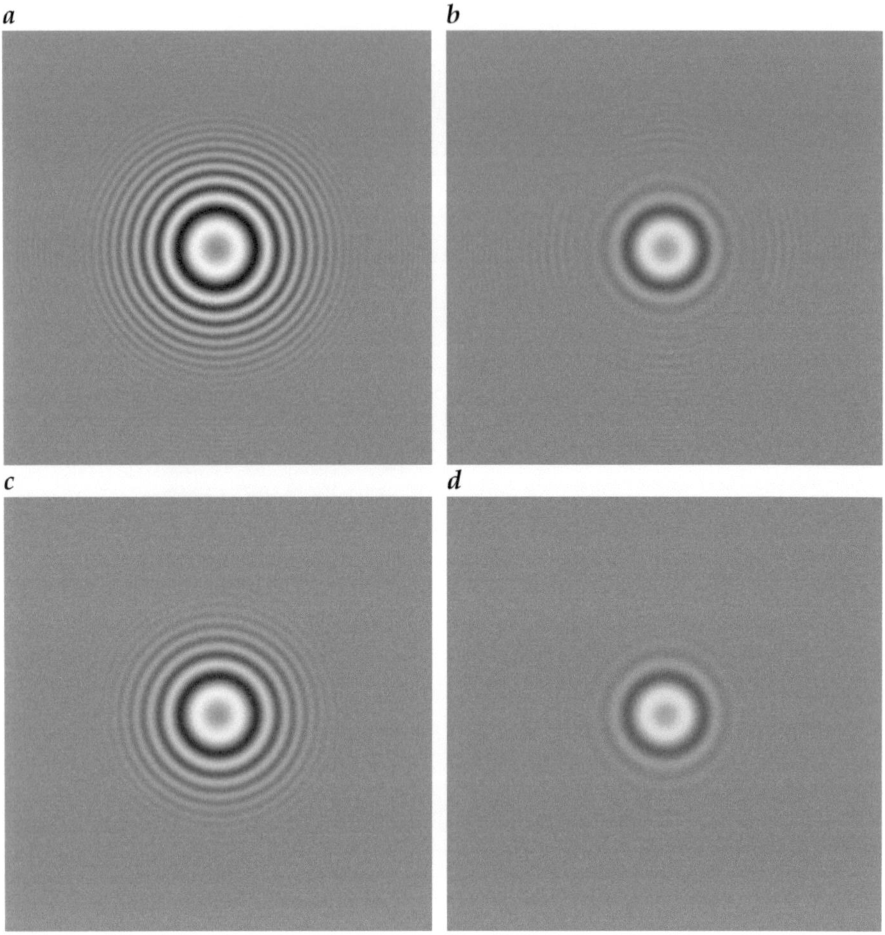

Figure 11.10: *Cascaded multi-step averaging with step width doubling according to Eq. (11.36), applied to the ring test pattern:* ***a*** $\mathcal{B}_2^2\mathcal{B}_1^2$, ***b*** $\mathcal{B}_4^2\mathcal{B}_2^2\mathcal{B}_1^2$, ***c*** $\mathcal{B}_2^4\mathcal{B}_1^4$ *and* ***d*** $\mathcal{B}_4^4\mathcal{B}_2^4\mathcal{B}_1^4$.

transfer function is repeated once in both directions (Fig. 11.8). In general, the zeros of the transfer functions of masks with a larger step size reflect this reciprocal grid because the binomial masks at the grid lines are zero. For a convolution with two neighboring pixels in the direction of the two diagonals, the reciprocal grid is rotated by 45°The grid constant of this grid is smaller than that of the original grid by a factor of $\sqrt{2}$.

If these filters were applied individually, they would be useless for smoothing. However, if we use them in cascade, starting with directly adjacent pixels, the zeros of the transfer functions, which are different for each pixel pitch, cause the transfer function to approach zero for large wavenumber ranges.

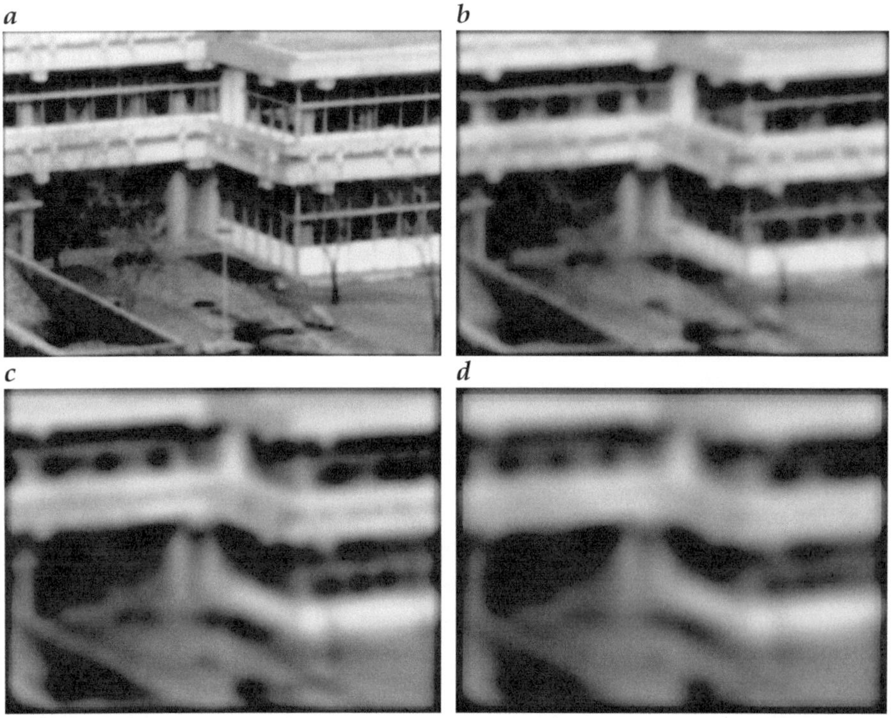

Figure 11.11: *Cascaded multi-step averaging with step size doubling according to Eq. (11.36), applied to the image Fig. 11.6a with **a** one, **b** two, **c** three and **d** four steps with the $\mathbf{B}^4$ filter.*

Cascaded multi-step binomial filtering provides a significant speed gain for large scale smoothing. For normal separable binomial filters, the number of calculations is proportional to σ^2 ($O(\sigma^2)$). For multi-step binomial filtering, the effort only depends logarithmically on σ ($O(\mathrm{ld}\sigma^2)$) if the step size is recursively doubled when cascading the filter operations:

$$\underbrace{\mathcal{B}^R_{2^{S-1}x}\cdots\mathcal{B}^R_{8x}\mathcal{B}^R_{4x}\mathcal{B}^R_{2x}\mathcal{B}^R_{x}}_{S\text{–mal}}. \tag{11.36}$$

Such a mask has the standard deviation

$$\sigma^2 = \underbrace{R/4 + R + 4R + \ldots + 4^{S-1}R}_{S\text{–mal}} = \frac{R}{12}(4^S - 1) \tag{11.37}$$

and the transfer function

$$\prod_{s=0}^{S-1}\cos^R(2^{s-1}\pi\tilde{k}). \tag{11.38}$$

Then only RS additions are required for S steps, while the standard deviation grows exponentially with approximately $\sqrt{R/12}\cdot 2^S$ grows.

With the parameter R we can adjust the degree of isotropy and the degree of residual inhomogeneities in the transfer function. We achieve a very efficient implementation with $R = 2$ ($\boldsymbol{B}^2 = 1/4[1\ 2\ 1]$ in each direction). However, the remaining side maxima at high wavenumbers with amplitudes of up to 0.08 still represent significant perturbations (Fig. 11.9a, b, Fig. 11.10a, b).

The next larger mask ($R = 4$, $\boldsymbol{B}^4 = 1/16[1\ 4\ 6\ 4\ 1]$ in each direction) significantly reduces the anisotropy (Fig. 11.9; Fig. 11.10c and d). Such interference can be further suppressed with even larger masks. Fig. 11.11 shows the first four steps of multistep averaging with the mask $\boldsymbol{B}^4$. You can see how quickly the smoothing progresses by doubling the step size.

11.5.2 ♯Recursive Averaging

The recursive filters introduced in Section 10.4 represent a completely different approach to large-scale averaging. Recursion essentially gives a convolution filter an infinite point response. The fundamental advantage of recursive filters is their ease of adjustment, as shown using the example of a simple low-pass filter in Section 10.4.4. In this section, the focus is on the design of smoothing filters that fulfill the criteria discussed in Section 11.2, in particular the freedom from displacement. This condition is not met by causal recursive filters.

Essentially, recursive filters work in the same way as non-recursive filters. In principle, we can replace any recursive filter with a non-recursive filter whose filter mask is identical to the point response of the recursive filter. The problem is the construction of recursive filters, i. e. the determination of the filter coefficients for the desired transfer function. While the theory of one-dimensional recursive filters is part of the standard knowledge of digital signal processing (see for example Oppenheim and Schafer [234], Oppenheim et al. [235]), the design of two-dimensional filters is still not sufficiently understood. The main reason for this are fundamental differences in the mathematics of one- and higher-dimensional z-transformations and polynomials [200].

Despite these theoretical problems, recursive filters can be successfully applied in digital image processing. To avoid problems in construction, we will only use recursive filters that are easy to understand and combine them into more complex filters, similar to how we composed binomial filters from the elementary smoothing mask $1/2\,[1\ 1]$. In this way, we obtain a class of recursive filters that are not necessarily optimal from a filter design point of view, but are suitable for practical applications.

In the first step, we combine causal recursive filters into symmetric filters. The starting point is a general one-dimensional recursive filter with the following transfer function:

$$^{+}\tilde{c} = \tilde{a}(\underset{\sim}{\kappa}) + \mathrm{i}\tilde{b}(\underset{\sim}{\kappa}). \tag{11.39}$$

The index + indicates the direction of movement of the filter in the positive coordinate direction. The transfer function of the same filter in the opposite

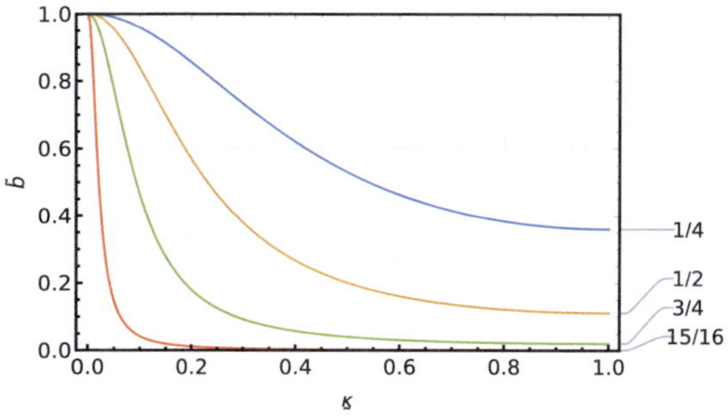

Figure 11.12: *Transfer function of the recursive low-pass filter Eq. (11.45) for different values α (1/4, 1/2, 3/4 and 15/16).*

direction is

$$^{-}\tilde{c} = \tilde{a}(\tilde{k}) - \mathrm{i}\tilde{b}(\tilde{k}). \tag{11.40}$$

Only the sign of the imaginary part of the transfer function changes, as it corresponds to the odd part of the point response, while the real part corresponds to the even part. We now have two options for combining the forward and backward filters into symmetrical smoothing filters:

$$\begin{array}{lll} \text{Addition} & \frac{1}{2}\left[{}^{+}\tilde{c} + {}^{-}\tilde{c}\right] & = \tilde{a}(\tilde{k}) \\ \text{Multiplication} & {}^{+}\tilde{c}\,{}^{-}\tilde{c} & = \tilde{a}^2(\tilde{k}) + \tilde{b}^2(\tilde{k}). \end{array} \tag{11.41}$$

Both combinations provide real transfer functions, i. e. straight filters without shifting, which are therefore suitable for smoothing.

As an elementary recursive smoothing filter, we use the low-pass filter with two elements that we have already studied in Section 10.4.4:

$$^{\pm}\mathcal{U}_x : \; g'_{mn} = g'_{m,n\mp1} + \alpha(g_{mn} - g'_{m,n\mp1}) \quad \text{with} \quad 0 \le \alpha \le 1 \tag{11.42}$$

with the *point spread function* (*PSF*)

$$({}^{\pm}\boldsymbol{U}_x)_{m,n} = \begin{cases} \alpha(1-\alpha)^n & n > 0, m = 0 \\ 0 & \text{otherwise.} \end{cases} \tag{11.43}$$

The transfer function of this filter can be easily calculated by considering that the Fourier transform of Eq. (11.43) forms a *geometric series*:

$$^{\pm}\tilde{u}_x(\tilde{k}) \approx \frac{\alpha}{1 + (1-\alpha)\exp(\mp\mathrm{i}\pi\tilde{k})}. \tag{11.44}$$

This relationship only applies approximately, as we have to break off the infinite sum of Eq. (11.43) at $n = N - 1$ due to the limited image size.

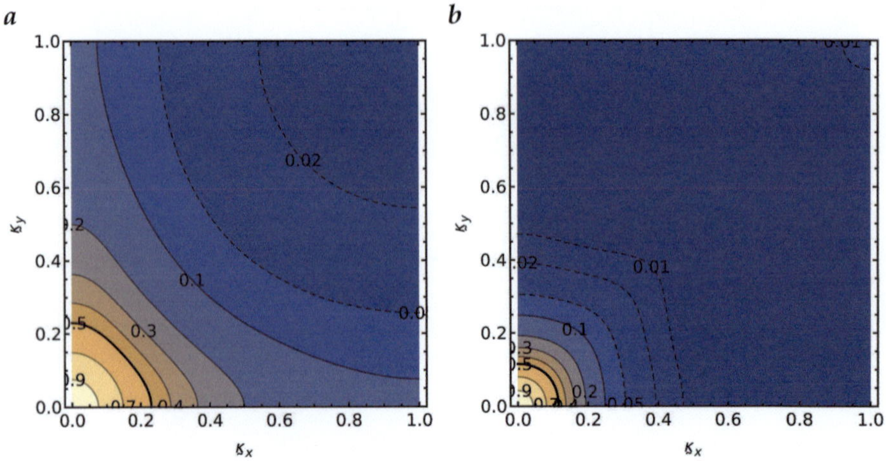

Figure 11.13: *Transfer functions of two-dimensional recursive low-pass filters:* **a** $\mathcal{U}$ *and* **b** $\mathcal{U}'$*, each with* $\alpha = 1/2$*.*

Successive filtering with one filter to the left and one to the right corresponds to a multiplication of the transfer functions:

$$\tilde{u}_x(\tilde{k}) = {}^{+}\tilde{u}_x(\tilde{k})\ {}^{-}\tilde{u}_x(\tilde{k}) \approx \frac{\alpha^2}{\alpha^2 + 2(1-\alpha)(1-\cos(\pi\tilde{k}))}. \tag{11.45}$$

The transfer function (Fig. 11.12) shows the characteristic expected for low-pass filters. For $\tilde{k} = 0$, $\tilde{A}_x(\tilde{k}) = 1$ applies; for small $\tilde{k}$, the transfer function decreases proportionally to $\tilde{k}^2$,

$$\tilde{u}_x \approx 1 - \frac{1-\alpha}{\alpha^2}(\pi\tilde{k})^2 \quad \text{for} \quad \tilde{k} \ll 1, \tag{11.46}$$

and has a half-wavenumber $\tilde{k}_c$ ($\tilde{A}_x(\tilde{k}_c) = 1/2$) of

$$\tilde{k}_c \approx \frac{1}{\pi} \arcsin \frac{\alpha}{\sqrt{2(1-\alpha)}} \approx \frac{\alpha}{\sqrt{2}\pi}, \tag{11.47}$$

where the last approximation only applies to $\alpha \ll 1$. For the largest wavenumber $\tilde{k} = 1$, the transfer function drops to

$$\tilde{u}_x(1) \approx \frac{\alpha^2}{4(1-\alpha) + \alpha^2}. \tag{11.48}$$

In contrast to the binomial filters, it is not exactly zero, but sufficiently small even for moderately small values of α (Fig. 11.12).

Two-dimensional filters can be composed of one-dimensional filters that run in horizontal and vertical directions:

$$\mathcal{U} = \mathcal{U}_x\mathcal{U}_y = {}^{+}\mathcal{U}_x{}^{-}\mathcal{U}_x{}^{+}\mathcal{U}_y{}^{-}\mathcal{U}_y. \tag{11.49}$$

This filter (Fig. 11.13a) has a considerably lower isotropy than binomial filters (Fig. 11.4). The smoothing in the axial direction is significantly lower than in the other directions. However, recursive filters have the great advantage that the computational effort does not depend on the degree of smoothing. Even with the simple first-order recursive filter, we can freely set the degree of smoothing by selecting the filter parameter α Eq. (11.47). The isotropy of the recursive filters can be further improved by running additional filters along the diagonals (Fig. 11.13b):

$$\mathcal{U}' = \mathcal{U}_x \mathcal{U}_y \mathcal{U}_{x-y} \mathcal{U}_{x+y}. \tag{11.50}$$

The indices $x - y$ and $x + y$ denote the main and secondary diagonals respectively.

Finally, some considerations on the computational effort follow. As already mentioned, the computational effort for recursive filters — unlike for non-recursive filters — does not depend on the limiting wavenumber. If $\alpha = 2^{-l}$ in Eq. (11.42), the filter can be calculated without a single multiplication:

$$g'_{mn} = \left[g'_{m,n\pm1} \cdot 2^l - g'_{m,n\pm1} + g_{mn} \right] \cdot 2^{-l} \quad \text{for} \quad l > 1. \tag{11.51}$$

The two-dimensional filter $\mathcal{U}$ then only requires 8 additions and shift operations per pixel, while the filter $\mathcal{U}'$, which runs in four directions, requires twice as many operations. This makes the recursive filter more effective than multi-step averaging with binomial masks (Section 11.5.1), but has a significantly poorer isotropy.

11.6 Context-dependent Averaging

Linear smoothing filters blur edges. Even worse: if the mask of a smoothing operator runs over an object edge, it contains pixels from the object and background. The filter produces a nonsensical result at this point. The same applies if a certain number of pixels have incorrect values due to transmission errors, for example. The question is therefore whether smoothing is possible that does not go over object edges or that omits certain pixels. Of course, such a procedure can only be applied if we have already detected the edges or the faulty pixels.

In this section, we discuss three types of nonlinear smoothing filters: the classical median filters (Section 11.6.1), the weighted averaging (Section 11.6.2), also known as normalized convolution, and the averaging controlled by the local image properties (Section 11.6.3).

11.6.1 Median Filter

Linear filters suppress Gaussian noise effectively, but binary noise only very poorly (Fig. 11.7). By weighting and summing with linear filters, we assume that each pixel carries useful information. However, pixels that are disturbed by transmission errors have lost their original gray value. Linear smoothing does not eliminate this misinformation, but transfers it to neighboring pixels. The only correct procedure for such interference is therefore to detect and eliminate it.

This is exactly what an *rank-order filter* (Section 10.5.1) does. The gray values within the mask are sorted according to their size and a pixel is selected. The *median filter* selects the middle value. Since binary noise completely changes the gray value, it is very unlikely that the incorrect gray value is the same as the mean gray value in the immediate vicinity. In this way, the mean gray value of the surroundings is used to restore the gray value of the disturbed pixel.

The following examples illustrate how a horizontal 3-median filter $\mathcal{M}$ works:

$$\begin{aligned}
\mathcal{M}[\cdots\ 1\ 2\ 3\ 7\ 8\ 9\ \cdots] &= [\cdots\ 1\ 2\ 3\ 7\ 8\ 9\ \cdots],\\
\mathcal{M}[\cdots\ 1\ 2\ 102\ 4\ 5\ 6\ \cdots] &= [\cdots\ 1\ 2\ 4\ 5\ 5\ 6\ \cdots],\\
\mathcal{M}[\cdots\ 0\ 0\ 0\ 9\ 9\ 9\ \cdots] &= [\cdots\ 0\ 0\ 0\ 9\ 9\ 9\ \cdots].
\end{aligned}$$

As expected, outliers are eliminated. The two other gray value structures — a monotonically increasing ramp and an edge between two areas of constant gray values — are retained. A median filter thus effectively eliminates binary noise without significantly affecting the sharpness of the image. However, it is less suitable for suppressing normally distributed noise (Fig. 11.7e and f).

The most important deterministic properties of a one-dimensional $(2N+1)$ median filter can be formulated using the following terms:

- A *constant neighborhood* is a region with $N + 1$ equal values.
- An *edge* is an area with monotonically increasing or decreasing values between two constant neighborhoods.
- A *pulse* is an area of at most N points that are bounded on the right and left by neighborhoods with only lower or only higher values with at least N values each.
- A *root* or a *fix point* is a signal that does not change under a median filter operation.

With these definitions, the deterministic properties of a median filter are as follows:

- Constant neighborhoods and edges are fix points.
- Impulses are eliminated.

If the median filter is applied repeatedly, what remains is an image that consists only of constant image areas and edges. If only individual pixels are disturbed, a 3-pixel median filter ($N = 1$) is sufficient to eliminate them. If groups of disturbed pixels occur, larger median filters must be used.

The statistical properties of median filters can be illustrated with an image containing only constant neighborhoods, edges and impulses. The power spectrum of the impulses is flat (*white noise*). As the median filter eliminates impulses, the power spectrum decreases for all wavenumbers. However, the contribution of the edges to a specific wavenumber is not removed. This example also underlines the non-linear nature of median filters.

11.6.2 Weighted Averaging

Image data, like other experimental data, are characterized by statistical uncertainties (Appendix C.1) and systematic errors that must be taken into account in further processing. As an introduction, we consider the averaging of a set of N independent data points g_n with the standard deviation σ_n. From elementary statistics, we know that for correct averaging, each data point g_n must be weighted with the reciprocal of the variance $w_n = 1/\sigma_n^2$. We then obtain an estimate of the mean with

$$\overline{g} = \sum_{n=1}^{N} g_n/\sigma_n^2 \Bigg/ \sum_{n=1}^{N} 1/\sigma_n^2 \tag{11.52}$$

with the standard deviation

$$\sigma_{\overline{g}}^2 = 1 \Bigg/ \sum_{n=1}^{N} 1/\sigma_n^2 \,. \tag{11.53}$$

The weight of an individual data point for the calculation of the mean value in Eq. (11.52) is the higher, the lower its statistical error is.

The application of *weighted averaging* to image processing is known as *normalized convolution* [115]. Smoothing is now extended to local neighborhoods. Each pixel is included in the convolution sum with a weighting factor assigned to it. Accordingly, two images are required for the normalized convolution; one is the image to be processed, the other contains the weighting factors. Analogous to Eqs. (11.52) and (11.53), the normalized convolution is defined as follows:

$$\boxed{\boldsymbol{G}' = \frac{\boldsymbol{H} * (\boldsymbol{W} \cdot \boldsymbol{G})}{\boldsymbol{H} * \boldsymbol{W}},} \tag{11.54}$$

where $\boldsymbol{H}$ is an arbitrary convolution mask, $\boldsymbol{G}$ is the image to be processed and $\boldsymbol{W}$ is the image containing the weighting factors. Essentially, the normalized convolution with the mask $\boldsymbol{H}$ transforms the image $\boldsymbol{G}$ into a new image $\boldsymbol{G}'$ and the weighting image $\boldsymbol{W}$ into a new weighting image $\boldsymbol{W}' = \boldsymbol{H} * \boldsymbol{W}$; both can then be sent for further processing.

In this sense, normalized convolution is nothing special or complicated, but merely the appropriate consideration of pixels with spatially varying statistical errors. The standard convolution can be seen as a special case of the normalized convolution. In this case, all pixels are assigned the same weighting factor and the weighting image is omitted because it is constant.

The flexibility of normalized convolution lies in the choice of the weighting image, which is not necessarily associated with an error. It can also be

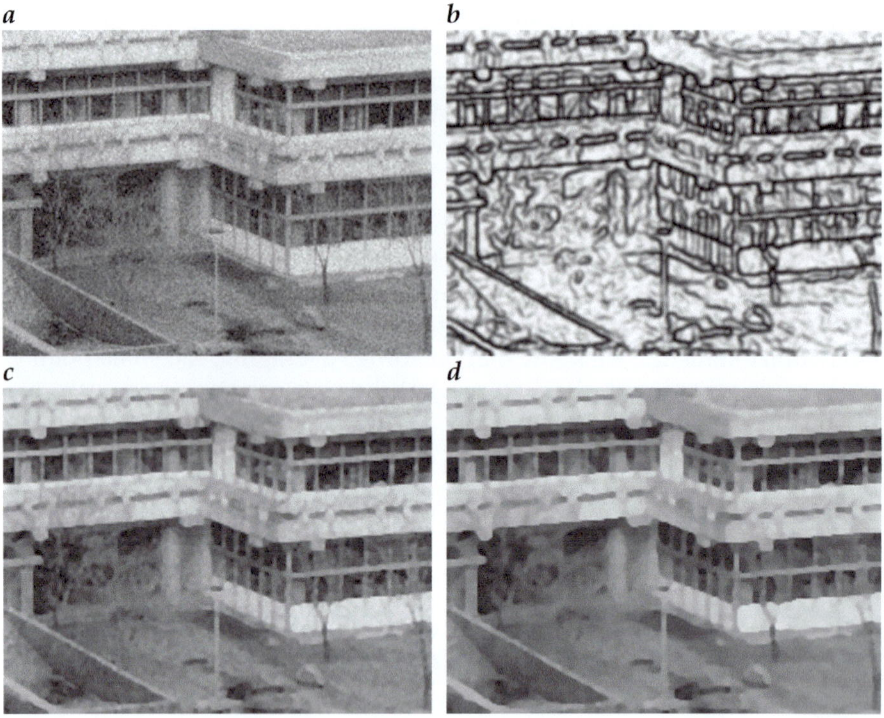

Figure 11.14: Weighted averaging using edge strength to prevent smoothing at edges: ***a*** *Image of Fig. 11.6a with additive normally distributed noise;* ***b*** *weighting image after five cascaded weighted averages; image after* ***c*** *two and* ***d*** *five cascaded weighted averages using a* $\mathcal{B}^2$ *binomial mask (compare with Fig. 11.7).*

used to select and/or enhance pixels with certain properties. This makes the normalized convolution a versatile non-linear operator. As an example, Fig. 11.14 shows the weighted averaging of a noisy image with a weighting image that reduces the smoothing at edges.

11.6.3 Steerable Averaging

The idea of *steerable filter* is to make the convolution mask dependent on the local image structure. This general concept is not limited to smoothing, but can be applied to all types of convolution processes. The basic idea is explained below. Adjustable filters have some freely adjustable parameters that can control the filtering. These can be very different properties such as the degree of smoothing, the direction of smoothing or both. It is easy to write down a filter mask with adjustable parameters.

For example, we have done this for recursive filters in Eq. (11.42), where the parameter α determines the degree of smoothing. However, it is not efficient in terms of computational effort to calculate an image with masks that are different at each pixel. The advantage of being able to separate the masks is lost.

Alternatively, a basis of a few filters can be sought with which a set of filtered images can be calculated. These images are then interpolated using adjustable parameters. In operator notation, this looks as follows:

$$\boxed{\mathcal{H}(\alpha) = \sum_{p=1}^{P} f_p(\alpha)\mathcal{H}_p.} \qquad (11.55)$$

Where $\mathcal{H}_p$ is the p-th filter and $f_p(\alpha)$ is a scalar interpolation function of the control parameter α. Two problems need to be solved if we use adjustable filters. First, it is not clear whether such a filter basis H_p exists at all. Then the relationship between the adjustable parameters α and the interpolation function f_p must be found. Once the first problem has been solved, the solution to the second usually takes care of itself.

As an example, a directional smoothing filter with the following transfer function is to be constructed:

$$\tilde{h}_{\theta_0}(k,\theta) = 1 - f(k)\cos^2(\theta - \theta_0). \qquad (11.56)$$

In this equation, cylindrical coordinates (k, θ) are used in Fourier space. The filter in Eq. (11.56) is a *polar separable* filter with an arbitrary radial function $f(k)$. This radial component represents an arbitrary isotropic smoothing filter. The adjustability of the angle is made possible by the function $\cos^2(\theta - \theta_0)$. Structures that are oriented in the direction θ_0 remain in the image, while those that are perpendicular to θ_0 are completely filtered out. The half-width of the angular distribution of this simple directional smoothing filter is $\pm 45°$.

We decompose the cosine function in Eq. (11.56) into trigonometric functions that depend either only on θ or on θ_0. This then results in

$$\tilde{h}_{\theta_0}(k,\theta) = 1 - \frac{1}{2} f(k)\left[1 + \cos(2\theta_0)\cos(2\theta) + \sin(2\theta_0)\sin(2\theta)\right] \qquad (11.57)$$

with the basic filters

$$\tilde{h}_1 = 1 - \frac{1}{2} f(k), \quad \tilde{h}_2 = -\frac{1}{2} f(k)\cos(2\theta), \quad \tilde{h}_3 = -\frac{1}{2} f(k)\sin(2\theta) \qquad (11.58)$$

and the interpolation functions

$$f_1(\theta_0) = 1, \quad f_2(\theta_0) = \cos(2\theta_0), \quad f_3(\theta_0) = \sin(2\theta_0). \qquad (11.59)$$

Three basic filters are therefore required. The filter $\tilde{h}_1$ is an isotropic smoothing filter, while the other two directional filters are with the main directions 0° and 45°.

Although the equations for this adjustable directional smoothing filter are simple, it is not easy to implement polar separable base filters as they

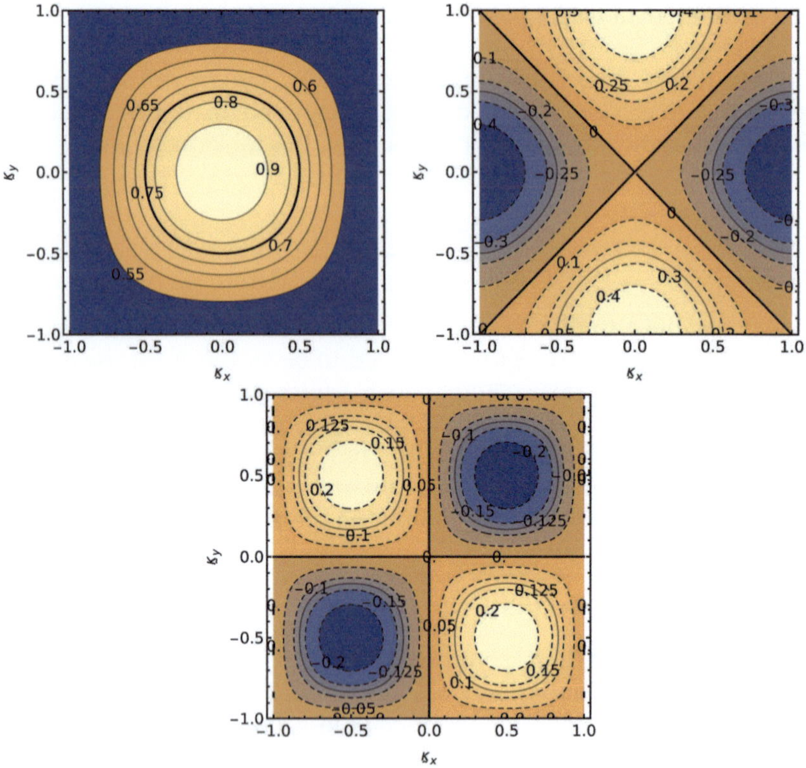

Figure 11.15: Transfer function of the three base filters for directional smoothing according to Eq. (11.58).

are not separable in the Cartesian coordinate system and therefore require careful optimization.

Nevertheless, it is possible to implement an adjustable directional smoothing filter with 3×3 base filters can be implemented. Due to the symmetries imposed by the transfer functions, we do not have much choice for the filter coefficients and obtain the following three base filters:

$$
\boldsymbol{H}_1 = \frac{1}{32}\begin{bmatrix} 1 & 2 & 1 \\ 2 & 20 & 2 \\ 1 & 2 & 1 \end{bmatrix}, \quad \boldsymbol{H}_2 = \frac{1}{32}\begin{bmatrix} 0 & -4 & 0 \\ 4 & 0 & 4 \\ 0 & -4 & 0 \end{bmatrix}, \quad \boldsymbol{H}_3 = \frac{1}{32}\begin{bmatrix} -2 & 0 & 2 \\ 0 & 0 & 0 \\ 2 & 0 & -2 \end{bmatrix}
$$

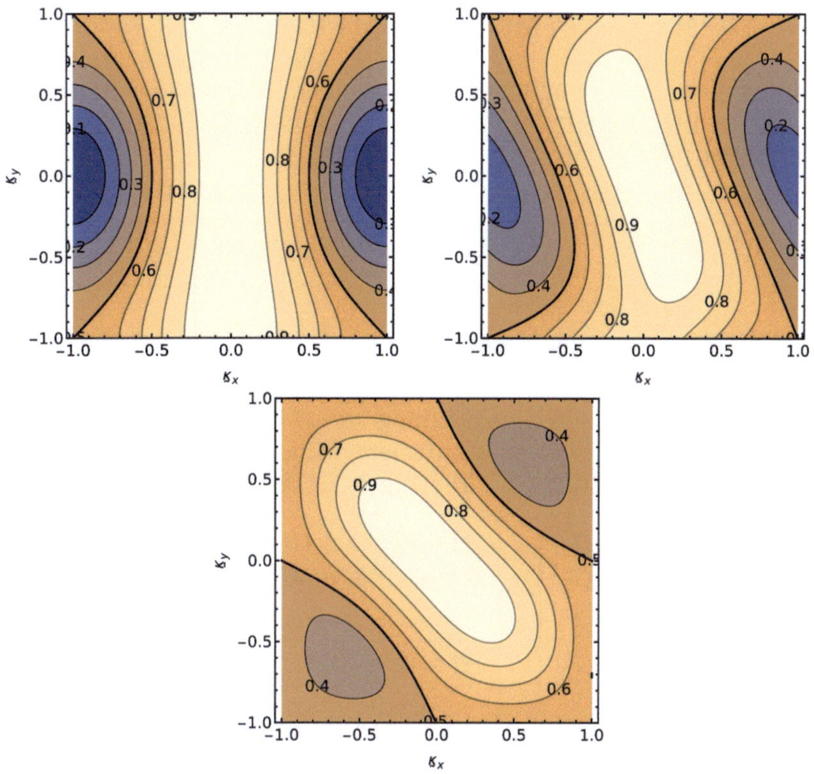

__Figure 11.16:__ Transfer function for controllable directional smoothing in 0°, 22.5° and 45° to the x-axis according to Eq. (11.57) with the base Eq. (11.60).

with the transfer functions (Fig. 11.15)

$$
\begin{aligned}
\tilde{h}_1 &= \frac{1}{2} + \frac{1}{2}\cos^2(\pi \underset{\sim}{k}_1/2)\cos^2(\pi \underset{\sim}{k}_2/2) && \approx 1 - \frac{\pi^2 \underset{\sim}{k}^2}{8}, \\
\tilde{h}_2 &= \frac{1}{4}\left(\cos(\pi \underset{\sim}{k}_1) - \cos(\pi \underset{\sim}{k}_2)\right) && \approx \frac{\pi^2 \underset{\sim}{k}^2}{8}\cos(2\theta), \\
\tilde{h}_3 &= \frac{1}{8}\left(\cos(\pi(\underset{\sim}{k}_1 + \underset{\sim}{k}_2)) - \cos(\pi(\underset{\sim}{k}_1 - \underset{\sim}{k}_2))\right) && \approx \frac{\pi^2 \underset{\sim}{k}^2}{8}\sin(2\theta).
\end{aligned} \tag{11.60}
$$

Fig. 11.16 shows that this simple implementation works well up to moderate wavenumbers. At high wavenumbers ($\underset{\sim}{k} > 0.5$), however, the filter is less directional as it smooths structures in all directions.

11.7 ♯Averaging in Multi-channel Images

At first glance, the averaging of multi-channel images seems to be nothing special, because we simply apply the smoothing mask to each of the P channels separately:

$$\mathbf{G}' = \begin{bmatrix} G'_1 \\ G'_2 \\ \vdots \\ G'_p \end{bmatrix} = \boldsymbol{H} * \mathbf{G} = \begin{bmatrix} \boldsymbol{H} & * & G_1 \\ \boldsymbol{H} & * & G_2 \\ & \vdots & \\ \boldsymbol{H} & * & G_p \end{bmatrix}. \tag{11.61}$$

This simple concept can also be extended to normalized convolution (Section 11.6.2). If the same smoothing kernel is used for all components, a common weighting image is sufficient, which can be appended as the $(P+1)$-th component of the multi-component image:

$$\begin{bmatrix} G'_1 \\ G'_2 \\ \vdots \\ G'_P \\ W' \end{bmatrix} = \begin{bmatrix} (\boldsymbol{H} * (W \cdot G_1))/(\boldsymbol{H} * W) \\ (\boldsymbol{H} * (W \cdot G_2))/(\boldsymbol{H} * W) \\ \vdots \\ (\boldsymbol{H} * (W \cdot G_P))/(\boldsymbol{H} * W) \\ \boldsymbol{H} * W \end{bmatrix}. \tag{11.62}$$

If multi-channel images contain cyclical features that can be mapped to angular coordinates, we are dealing with an interesting special case. Variables with this property are, for example, the direction of an edge or the phase of a periodic signal. These quantities are cyclical and can only be poorly represented in Cartesian coordinates and therefore cannot be averaged in this representation. Let's imagine two angles of +175° and –179°. The average angle is 178°, since –179° = 360° – 179° = 181° is close to 175° and not (175° –179°) / 2 = –2°, as would result from a direct averaging.

Cyclic properties such as angles are therefore better represented as unit vectors of the form $\hat{\boldsymbol{n}}_\theta = [\cos\theta, \sin\theta]^T$. This allows them to be averaged correctly, as Fig. 11.17 shows. The averaged vector points in the correct direction, but its magnitude is generally less than one:

$$(\hat{\boldsymbol{n}}_{\theta_1} + \hat{\boldsymbol{n}}_{\theta_2})/2 = \begin{bmatrix} \cos[(\theta_1 + \theta_2)/2] \\ \sin[(\theta_1 + \theta_2)/2] \end{bmatrix} \cos[(\theta_2 - \theta_1)/2]. \tag{11.63}$$

For an angle difference of 180° the averaged vector has a value of zero. The decrease in its magnitude is interpreted intuitively: The greater the dispersion of the angle, the less certain the averaged value is. In fact, the vector sum disappears if all directions are equally probable, while it increases if the dispersion of the angle is small.

These considerations can also be extended to a weighted smoothing for cyclic variables. To do this, we set the vector length equal to the coefficient

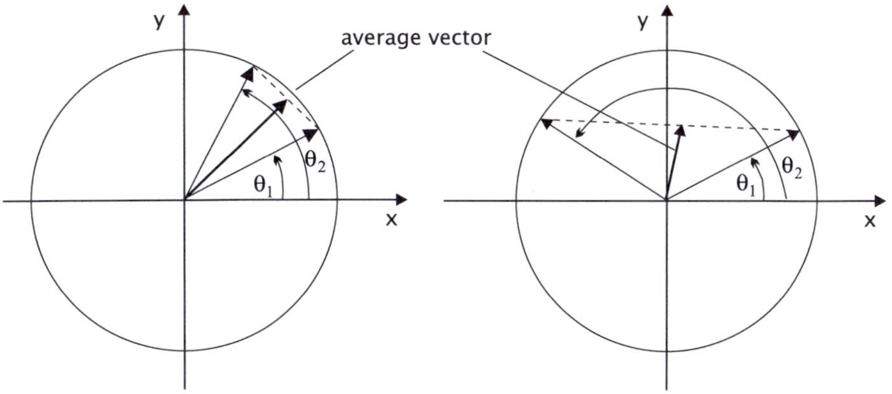

Figure 11.17: Averaging of a cyclic variable represented by the unit vector $\hat{\boldsymbol{n}}_\theta = [\cos\theta, \sin\theta]^T$. *The averaged vector* $(\hat{\boldsymbol{n}}_{\theta_1} + \hat{\boldsymbol{n}}_{\theta_2})/2$ *points in the correct direction* $(\theta_1 + \theta_2)/2$, *but its magnitude decreases with the angle difference.*

of determination of the variable represented by the angle of the vector. Short vectors then contribute little, long ones correspondingly more to the averaging. This is a very attractive form of weighted convolution because — in contrast to normalized convolution (Section 11.6.2) — it does not require any complex division. Of course, this method only works with properties that can be adequately mapped to an angle.

Finally, we consider a measure to characterize the scattering in the direction of the vector. Fig. 11.17 makes it clear that the sum vector is only slightly lower than the sum of the vector lengths if the scattering is low. Accordingly, we can use

$$c = \frac{|\boldsymbol{H} * \mathbf{G}|}{|\mathbf{G}|} \tag{11.64}$$

define a coherence measure, where $\boldsymbol{H}$ is an arbitrary smoothing convolution operator. This measure is one if all vectors of the neighborhood covered by the convolution operator point in the same direction and zero if they are uniformly distributed. This definition of a coherence measure applies not only in two-channel but also in higher-dimensional vector spaces. In one-dimensional vector spaces (scalar images), the coherence measure is of course always one.

11.8 ♯References for Further Reading

The journal articles by Simonds [303] and Wells III [352] deal with fast algorithms for large Gaussian masks. Readers interested in the general principles of efficient algorithms should refer to the textbooks by Aho et al. [5] or Sedgewick [298]. The textbook by Blahut [21] deals exclusively with fast algorithms for digital signal processing. The classic design techniques for filters, especially for recursive filters, are covered in the textbooks on digital signal processing, e. g. Proakis and Manolakis [250] or Oppenheim and Schafer [234]. The textbook by Lim [200] on 2D signal processing also includes the design of recursive 2D filters.

A detailed description of the deterministic and statistical properties of median filters and other non-linear filters can be found in Pitas and Venetsanopoulos [242] and Arce [7]. The monograph by Granlund and Knutsson [115] on signal processing for machine vision also discusses weighted averaging (normalized convolution, Section 11.6.2). Adjustable filters (Section 11.6.3) were introduced by the journal articles of Freeman and Adelson [93] and Simoncelli et al. [302].

12 Discontinuities: Edges, Lines, Surfaces and Corners

Contents

B. Jähne, *Digital Image Processing and Image Acquisition*,
https://doi.org/10.1007/978-3-032-04806-6_12

12.1 Introduction

Objects can be recognized in images if they differ from the background in at least one channel of the image signal. There is then a discontinuity at the border of the object or an edge. This is limited to a small area in the image. Therefore, *edge detection* is possible with a neighborhood operator that shows changes and suppresses areas of constant values. In this way, a feature image is generated in which *edges* are bright and all other areas are dark.

However, it is not necessary that there is a gray value edge at the edge of an object. If the difference in contrast between the object and the background is too small, the edge of an object cannot be detected. Conversely, gray value edges can occur not only at the edge of a 3D object (a so-called *depth edge*), but also within an object. This is obviously the case if the object is patterned. But even if the object has a constant optical property, such as e. g. a Lambertian surface with a constant reflection coefficient (Section 2.6.4), edges can occur within an object when three-dimensional objects are projected onto an image plane (Fig. 2.2). Shadows also create edges. Detecting an edge is therefore only the first step, interpreting the edge is the second.

In this chapter, we only deal with edge detection. An ideal edge represents a discontinuity of the spatial gray value function $g(\boldsymbol{x})$ in the image plane. This simple idea is not entirely correct, as an edge in an image is not infinitely steep. In Section 3.5.4 we have already learned that the limited resolution of an image widens a step edge to approximately the minimum resolvable wavelength.

Nevertheless, an edge is characterized by a high gradient. Edge detection is therefore always based on derivatives in one form or another. In discrete images, these must be replaced by discrete differences. The errors associated with this approximation must be carefully considered. Firstly, edge detection can be anisotropic, i. e. edges are not detected equally well in all directions. Secondly, the edge direction may have an error. We will therefore carry out this analysis for all edge detectors.

In a two-dimensional image, edges are characterized by the fact that the gray values in a neighbourhood only change perpendicular to the edge, but not parallel to the edge. If the direction of an edge changes abruptly, there is a *corner*. In higher-dimensional data, the situation is much more complex. In three-dimensional *volume images*, the edges of objects are *surfaces*, while edges represent abrupt changes in the direction of the surfaces. Three edges come together at corners. Our imagination fails with even higher dimensional signals. In four dimensions, objects are bounded by three-dimensional hypervolumes, in hypersurfaces hypervolumes converge and so on.

In this chapter, as with smoothing filters in Section 11.2, we first look at the basic properties of edge filters (Section 12.2 and 12.3). Then we get to know the two basic classes of edge filters: Gradient-based (Section 12.4) and edge detection by zero-crossings with second derivatives (Section 12.5). Then

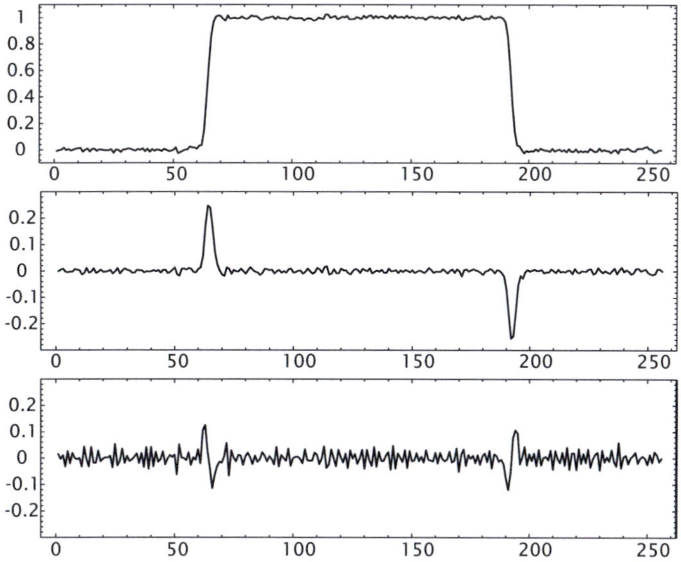

Figure 12.1: *One-dimensional noisy gray value edges and their first and second derivatives.*

we turn to optimized edge detectors. In image processing, both the most accurate derivative filters (Section 12.6) and the most noise-insensitive edge detectors (Section 12.7) are required. While in scalar images it is clear what an edge is, edges in multi-channel or vectorial images can mean discontinuities in only one or more components. How we can handle this uniformly and consistently will be discussed in Section 12.8.

12.2 Differential Description of Signal Changes

Smoothing filters suppress high wavenumbers. The detection of edges, on the other hand, requires a filter operation that amplifies gray value changes and suppresses areas of constant gray values. Derivative operators are suitable for this (Fig. 12.1). The first derivative has extreme values at edges (maximum positive or negative slopes), the second has zero crossings (no curvature) where edges have their steepest fall or rise. Both criteria can be used to determine edges.

In higher dimensions, the differential description of signal changes is much more difficult. First we look at 2D images. Here we can consider edges, corners, lines and local extreme values as relevant features for image processing. At an *edge* there is a large change in the signal perpendicular to the direction of the edge, but in the direction of the edge the signal change is small. However, if the curvature perpendicular to the direction of the gradient becomes high, an edge turns into an *edge*. A *line* is characterized by the fact that both the gradient and curvature in the direction of the line are small, while now instead of the gradient, as with the edge, the curvature perpendicular to the line is high. At a *local extreme value* the first-order derivatives are zero, but the curvatures in all directions are but large.

In three dimensions, i. e. with *volume images*, the situation is even more complicated. Now there is *surface* with a strong first-order signal change perpendicular to the surface and small slopes and curvatures in the two directions that lie in the surface. At an edge, there are small signal changes only in the direction of the edge and at a corner the signal changes in all directions.

Because of this rich set of differential features to describe local signal changes, it is worth getting to know the basic properties of differential operators in multidimensional signals before we design suitable operators to detect these features in images.

12.2.1 First Derivative and the Gradient

In wavenumber space, a derivative filter of order p means a multiplication of the image spectrum by $(2\pi ik)^p$ (Section 6.3, ≻ R12):

$$\frac{\partial}{\partial x_d} \quad \circ\!\!-\!\!\bullet \quad 2\pi \mathrm{i}\kappa_d, \qquad \frac{\partial^2}{\partial x_d^2} \quad \circ\!\!-\!\!\bullet \quad -4\pi^2\,\kappa_d^2. \tag{12.1}$$

The partial derivatives of a D-dimensional signal $g(\boldsymbol{x})$ in all directions form the D-dimensional *gradient vector*:

$$\nabla = \left[\frac{\partial}{\partial x_1}, \frac{\partial}{\partial x_2}, \ldots, \frac{\partial}{\partial x_D}\right]^T \quad \circ\!\!-\!\!\bullet \quad 2\pi \mathrm{i}\boldsymbol{\kappa}. \tag{12.2}$$

When the coordinate system is rotated, the gradient is transformed like any other vector by multiplication with an orthogonal rotation matrix $\boldsymbol{R}$ (Section 3.2.2):

$$\nabla' = \boldsymbol{R}\nabla. \tag{12.3}$$

The first derivative in a specific direction, the so-called *directional derivative* [29], results from the scalar product between the gradient vector and a vector $\hat{\boldsymbol{n}} = [\cos\theta,\ \sin\theta]^T$ with magnitude one, which points in the corresponding direction:

$$\frac{\partial}{\partial \hat{\boldsymbol{n}}} = \hat{\boldsymbol{n}}^T\,\nabla = [\cos\theta, \sin\theta]\begin{bmatrix}\dfrac{\partial}{\partial x_1}\\[2ex] \dfrac{\partial}{\partial x_2}\end{bmatrix} \quad \circ\!\!-\!\!\bullet \quad 2\pi \mathrm{i}\hat{\boldsymbol{n}}^T\,\boldsymbol{\kappa}. \tag{12.4}$$

The magnitude of the gradient vector,

$$|\nabla| = \|\nabla\|_2 = \left(\nabla^T\nabla\right)^{1/2} = \left(\sum_{d=1}^{D}\left(\frac{\partial}{\partial x_w}\right)^2\right)^{1/2}, \tag{12.5}$$

is invariant under a rotation of the coordinate system. If we rotate the coordinate system so that the gradient vector is parallel to the rotated x'-axis, then all other components of the gradient vector disappear and the directional derivative in the direction of the x'-axis is maximum and equal to the magnitude of the gradient.

12.2.2 Second Derivative and Curvature

Second-order differential operators detect *curvature*. All possible combinations of the second-order partial derivatives of a D-dimensional signal form a $D \times D$ matrix, the *Hessian matrix*:

$$\boldsymbol{H} = \begin{bmatrix} \dfrac{\partial^2}{\partial x_1^2} & \dfrac{\partial^2}{\partial x_1 x_2} & \cdots & \dfrac{\partial^2}{\partial x_1 x_D} \\ \dfrac{\partial^2}{\partial x_1 x_2} & \dfrac{\partial^2}{\partial x_2^2} & \cdots & \dfrac{\partial^2}{\partial x_2 x_D} \\ \vdots & \vdots & \ddots & \vdots \\ \dfrac{\partial^2}{\partial x_1 x_D} & \dfrac{\partial^2}{\partial x_2 x_D} & \cdots & \dfrac{\partial^2}{\partial x_D^2} \end{bmatrix} \quad \circ\!\!-\!\!\bullet \quad -4\pi^2 \boldsymbol{\kappa}\boldsymbol{\kappa}^T. \tag{12.6}$$

When the coordinate system is rotated, the Hessian matrix is transformed by multiplication with an orthogonal rotation matrix $\boldsymbol{R}$ from the left and right:

$$\boldsymbol{H}' = \boldsymbol{R}\nabla\boldsymbol{R}^T. \tag{12.7}$$

The Hessian matrix is symmetrical. Symmetrical matrices can be diagonalized by a suitable coordinate transformation (rotation) $\boldsymbol{R}$ (Appendix C.3.3). In this system, the so-called *principle coordinate system*, the Hessian matrix looks as follows:

$$\boldsymbol{H}' = \begin{bmatrix} \dfrac{\partial^2}{\partial x_1'^2} & 0 & \cdots & 0 \\ 0 & \dfrac{\partial^2}{\partial x_2'^2} & \cdots & 0 \\ \vdots & \vdots & \ddots & \vdots \\ 0 & 0 & \cdots & \dfrac{\partial^2}{\partial x_D'^2} \end{bmatrix}. \tag{12.8}$$

For the gradient, only one component was not equal to zero in the principle coordinate system; this is not the case for the curvature. They can all be non-zero.

The trace of this matrix, i. e. the sum of the diagonal elements, is called *Laplace operator* and is denoted by Δ:

$$\Delta = \operatorname{trace} \boldsymbol{H} = \sum_{d=1}^{D} \frac{\partial^2}{\partial x_d^2} \quad \circ\!\!-\!\!\bullet \quad -4\pi^2 \sum_{d=1}^{D} k_d^2 = -4\pi^2 \kappa^2. \tag{12.9}$$

As the Laplace operator is the trace of the Hessian matrix, it does not change when the coordinate system is rotated.

12.3 General Properties of Edge Filters

In the following Sections 12.3.1–12.3.5 we discuss the general properties of the filters for edge detection. This discussion is similar to the one about the general properties of the smoothing filters in Sections 11.2.1–11.2.4.

12.3.1 Zero Shift

No shifts in the object position may occur during edge detection. This is the most important property of a derivative operator. With smoothing filters, this is achieved using a real transfer function and a symmetrical convolution mask (Section 11.2.1). A real transfer function is not useful for a first-order derivative filter, as extremes should be mapped to zero crossings and the largest slopes to extreme values. This means a phase shift by 90° and therefore an imaginary transfer function. An imaginary transfer function in turn implies an antisymmetric filter mask, which is defined by

$$h_{-n} = -h_n. \tag{12.10}$$

This equation shows that the central coefficient for a convolution mask with an odd number of coefficients is is zero.

Second-order derivative filters detect curvature. Extreme values of the spatial gray value curve should correspond to extreme values of the curvature. Therefore, a second-order derivative filter, like a smoothing filter, should be symmetrical. Then all the properties discussed in Section 11.2.1 for symmetrical filters also apply to second-order derivative filters.

12.3.2 Suppression of the Mean Value

Derivative filters of any order must not show a response to constant values or an offset in the signal. This condition implies that the sum of the coefficients and the transfer function must apply:

$$\sum_n h_n = 0 \quad \circ\!\!-\!\!\bullet \quad \hat{h}(\mathbf{0}) = 0. \tag{12.11}$$

12.3.3 Symmetry Properties

Let's take a closer look at the symmetry properties, as they form the basis of a more efficient convolution due to a multiplications and simplify the calculation of the transfer functions. For a first-order derivative operator, the requirement of zero shift (Section 12.3.1) generally results in a 1D mask of odd symmetry with $2R+1$ or $2R$ coefficients:

$$\begin{aligned} 2R+1: &\quad [h_R, \ldots, h_1, 0, -h_1, \ldots, -h_R], \\ 2R: &\quad [h_{R-1/2}, \ldots, h_{1/2}, -h_{1/2}, \ldots, -h_{R-1/2}]. \end{aligned} \tag{12.12}$$

Therefore, the calculation of the convolution can be simplified as

$$\begin{aligned} 2R+1: &\quad g'_n = \sum_{n'=1}^{R} h_{n'}(g_{n-n'} - g_{n+n'}), \\ 2R: &\quad g'_n = \sum_{n'=1}^{R} h_{n'-1/2}\left(g_{n-(n'-1/2)} - g_{n+n'-1/2}\right). \end{aligned} \tag{12.13}$$

With $2R$ coefficients, the positions are described from the perspective of the grid shifted by half a pixel pitch, to which the results are written. This means that only R multiplications are required. However, the number of additions remains unchanged at $2R-1$.

The calculation of the transfer functions is also considerably simplified by the symmetry relationships, as only the sine functions of the complex exponent of the Fourier transforms remain in the equations. The transfer function of a one-dimensional mask of odd symmetry is therefore as follows:

$$\begin{aligned} 2R+1: &\quad \tilde{g}(\underset{\sim}{k}) = 2\mathrm{i}\sum_{\tilde{n}=1}^{R} h_{\tilde{n}} \sin(\tilde{n}\pi\underset{\sim}{k}), \\ 2R: &\quad \tilde{g}(\underset{\sim}{k}) = 2\mathrm{i}\sum_{\tilde{n}=1}^{R} h_{\tilde{n}} \sin[(\tilde{n}-1/2)\pi\underset{\sim}{k}]. \end{aligned} \tag{12.14}$$

For second-order derivative filters, we can use all equations derived in Section 11.2.1 for the smoothing filters, since symmetry in the derivative direction must be given.

12.3.4 Correct, Non-selective Derivation

Intuitively, we expect that each derivative operator amplifies finer structures more than coarser ones, since the transfer function for a pth order derivative operator increases with k^p. Therefore, we could argue that the transfer function of a good discrete derivative operator should approximate the ideal transfer function, the right part in Eq. (12.1), as well as possible.

However, this condition is too strong a restriction. Suppose we first apply a smoothing filter to an image before using a derivative filter. We will still recognize the combined operation as a derivative; the mean value is suppressed, and the operator only reacts to spatial changes. The strict condition is only necessary if it is not just a matter of detecting edges or signal changes, but if a derivative is required directly.

Therefore, it is sufficient in signal processing to require the ideal transfer function in Eq. (12.1) only for *small* wavenumbers. This is easily possible if we combine the ideal derivative filter with a smoothing filter

$$\tilde{h}(\underset{\sim}{\boldsymbol{\kappa}}) = (\pi \mathrm{i} \underset{\sim}{\kappa}_d)^p \tilde{b}(|\underset{\sim}{\boldsymbol{\kappa}}|), \tag{12.15}$$

which is subject to the following conditions:

$$\tilde{b}(0) = 1, \quad \boldsymbol{\nabla}_k \tilde{b}(\underset{\sim}{\boldsymbol{\kappa}})\Big|_{\underset{\sim}{\boldsymbol{\kappa}}=\mathbf{0}} = \mathbf{0}, \quad \tilde{b}(\underset{\sim}{\boldsymbol{\kappa}}) \geq 0, \quad \text{decreasing monotonously.} \tag{12.16}$$

It is important to add here that if several different derivatives are used, including the signal itself, these must all have the *same* smoothing factor $\tilde{b}(\underset{\sim}{\boldsymbol{\kappa}})$. Otherwise, errors will occur when the results of these filters are subsequently combined.

The requirement that a derivative filter should have the correct ideal transfer function for small wavenumbers also results in conditions for the coefficients of the filter masks. For a first-order derivative operator in the direction d, this means that when applied to a signal with a slope of one in this direction, the value one must also result. The means

$$\frac{\partial}{\partial x_d} : \qquad \sum_n n_d h_n = -1. \tag{12.17}$$

For a second-order derivative filter, the following applies accordingly by applying it to a signal with a curvature of one

$$\frac{\partial^2}{\partial x_d^2} : \qquad \sum_n n_d^2 h_n = 2 \quad \text{and} \quad \sum_n n_d h_n = 0. \tag{12.18}$$

Since second-order derivative filters should not react to a constant gradient in the direction of the derivative, the second condition was added.

12.3.5 Isotropy

For good edge detection, it is important that the operator response does not depend on the direction of the edge. If this is the case, we speak of an isotropic edge detector. The isotropy of an edge detector can best be investigated using its transfer function. The most general form for an isotropic derivative operator of p-th order is given by using the same approach as in Eq. (12.15), only additionally requiring that the smoothing term may only depend on the magnitude of the wavenumber:

$$\tilde{h}(\underset{\sim}{\boldsymbol{\kappa}}) = (\pi \mathrm{i} \underset{\sim}{\kappa}_d)^p \tilde{b}(|\underset{\sim}{\boldsymbol{\kappa}}|). \tag{12.19}$$

Otherwise, the same requirements apply to $\tilde{b}(|\underset{\sim}{k}|)$ as in Eq. (12.16). All requirements for derivative filters are clearly summarized in Appendix D (> R25 and > R26).

Example 12.1: 3×1 **derivative filter**

This filter has only 3 coefficients $[h_{-1}, h_0, h_1]$. What restrictions on the coefficients result from all the above conditions? The following applies to the first derivative:

$$\begin{array}{ll} \text{Symmetry} & h_{-1} = -h_1, \quad h_0 = 0 \\ \text{Suppression of mean value} & h_{-1} + h_0 + h_1 = 0 \\ \text{Correct derivation} & -h_{-1} + h_1 = -1 \end{array} .$$

These are four conditions for three coefficients. Since the conditions do not contradict each other, there is a unique solution with $[1, 0, -1]/2$.
The following applies to the second derivative:

$$\begin{array}{ll} \text{Symmetry} & h_{-1} = h_1 \\ \text{Suppression of mean value} & h_{-1} + h_0 + h_1 = 0 \\ \text{Correct derivation} & h_{-1} + h_1 = 2 \quad \text{and} \quad -h_{-1} + h_1 = 0 \end{array} .$$

Again, there are four non-contradictory conditions and a unique solution with $[1, -2, 1]$. With the 3×1 filter, there are therefore no possibilities for optimization. Filters with larger masks must therefore be used.

12.4 Gradient-based Edge Detection

12.4.1 Principle

In relation to first derivatives, an edge represents an extreme value (Fig. 12.1). Edge detection with first-order derivatives therefore means searching for the largest changes, i. e. maxima in the magnitude of the gradient vector ∇ Eq. (12.2). We must therefore calculate partial first-order derivatives in all directions. In representation-invariant operator notation (Section 10.1.5), we can write this for two- and three-dimensional images as

$$\mathcal{D} = \begin{bmatrix} \mathcal{D}_x \\ \mathcal{D}_y \end{bmatrix} \quad \text{or} \quad \mathcal{D} = \begin{bmatrix} \mathcal{D}_x \\ \mathcal{D}_y \\ \mathcal{D}_z \end{bmatrix} . \tag{12.20}$$

Since the gradient is a vector, its magnitude Eq. (12.5) is invariant under a rotation of the coordinate system. This is a necessary condition for isotropic edge detection. The calculation of the gradient magnitude can be performed in two dimensions using the operator equation

$$|\mathcal{D}| = \left[\mathcal{D}_x \cdot \mathcal{D}_x + \mathcal{D}_y \cdot \mathcal{D}_y\right]^{1/2} . \tag{12.21}$$

The symbol · indicates the point-by-point multiplication of the two images resulting from the filter operations $\mathcal{D}_x$ and $\mathcal{D}_y$ (Section 10.1.5). The square root is also calculated point by point in the spatial domain. The application of the operator $|\mathcal{D}|$ to the image $\mathbf{G}$ means the following chain of operations according to the operator equation in Eq. (12.21):

1. Filtering the image $\mathbf{G}$ independently of each other with the operators $\mathcal{D}_x$ and $\mathcal{D}_y$,
2. Calculation of the square of the two resulting images,
3. their addition and
4. Calculation of the square root of the sum image.

Example 12.2: Does a simplified calculation of the gradient make sense?

Calculating the amount of the gradient may seem complex due to the many calculation steps. For this reason, it is often approximated by

$$|\mathcal{D}| \approx |\mathcal{D}_x| + \left|\mathcal{D}_y\right|. \tag{12.22}$$

However, this approximation is anisotropic even for small wavenumbers. Edges in the diagonal direction are detected more sensitively by a factor of $\sqrt{2}$ than edges in the axial direction. However, the calculation of the magnitude of the gradient vector can be quickly performed as *dyadic point operation* with a *lookup table* (Section 7.4.2).

12.4.2 Error in Magnitude and Direction

The fundamental problem with all edge detectors is that derivative operators can only be approximated on a discrete grid. Two types of errors can occur (Fig. 12.2). Firstly, the edge detection can be anisotropic, i. e. the calculation of the magnitude of the gradient depends on the direction of the edge. Secondly, the calculated edge direction can deviate from the actual direction. It makes sense to introduce an error measure for both types of error. We will specify these error measures in the description of the various edge filters.

The magnitude of the two-dimensional gradient is given by

$$\left|\tilde{\boldsymbol{d}}(\underset{\sim}{\boldsymbol{\kappa}})\right| = \left(\left|\tilde{d}_x(\underset{\sim}{\boldsymbol{\kappa}})\right|^2 + \left|\tilde{d}_y(\underset{\sim}{\boldsymbol{\kappa}})\right|^2\right)^{1/2}, \tag{12.23}$$

where $\tilde{\boldsymbol{d}}(\underset{\sim}{\boldsymbol{\kappa}})$ is the vectorial transfer function of the gradient operator. The anisotropy of the calculation of the magnitude can then be specified as a deviation from the magnitude of the gradient in the x direction. This is given by

$$e_m(\underset{\sim}{\boldsymbol{\kappa}}) = \left|\tilde{\boldsymbol{d}}(\underset{\sim}{\boldsymbol{\kappa}})\right| - \left|\tilde{d}_x(\underset{\sim}{\boldsymbol{\kappa}})\right|. \tag{12.24}$$

This error measure can also be used for higher-dimensional signals.

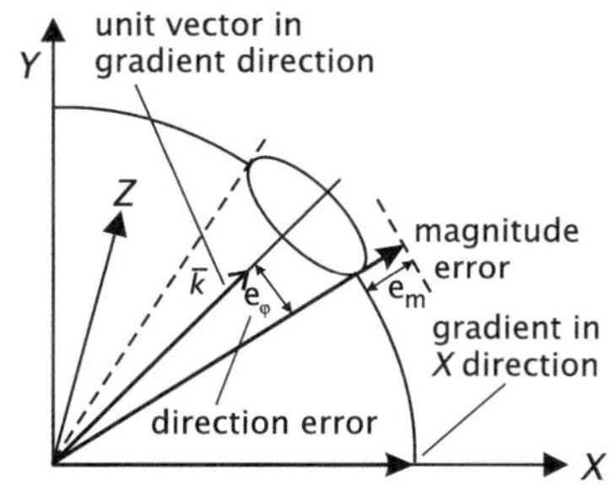

Figure 12.2: *Illustration of the magnitude and direction error of the gradient vector.*

The error in the direction of the gradient can be calculated in a similar way. From the components of the 2D gradient, we calculate its direction, which is given by the angle ϕ', as

$$\phi' = \arctan \frac{\tilde{d}_y(\underset{\sim}{k}, \phi)}{\tilde{d}_x(\underset{\sim}{k}, \phi)}. \tag{12.25}$$

This results in the angular error

$$e_\phi(\underset{\sim}{k}, \phi) = \arctan \frac{\tilde{d}_y(\underset{\sim}{k}, \phi)}{\tilde{d}_x(\underset{\sim}{k}, \phi)} - \phi. \tag{12.26}$$

For higher-dimensional signals, the angular deviation is a vector quantity. However, we can specify a scalar direction error by calculating the scalar product between a unit vector in the true gradient direction and the calculated gradient vector $\tilde{\boldsymbol{d}}(k)$ (Fig. 12.2):

$$\cos e_\phi = \frac{\hat{\boldsymbol{k}}^T \tilde{\boldsymbol{d}}(\underset{\sim}{\boldsymbol{k}})}{\mathrm{i}\,\left|\tilde{d}(\underset{\sim}{\boldsymbol{k}})\right|} \quad \text{with} \quad \hat{\boldsymbol{k}} = \frac{\boldsymbol{\kappa}}{|\boldsymbol{\kappa}|}. \tag{12.27}$$

In contrast to the angular error for the 2D gradient in Eq. (12.26), the angular error according to Eq. (12.27) only assumes positive values. It cannot indicate the direction of the deviation.

Extension 12.1: Error for derivative operators with isotropic smoothing

In Section 12.3.5 we combined an ideal derivative filter with an isotropic smoothing filter and claimed that it is a correct derivative operator only in the limiting case of small wavenumbers. Here we now investigate whether this approach leads to a gradient-based edge detector that calculates the magnitude of the gradient independent of the direction and the angle of the edge correctly. To do this, we only have to insert Eq. (12.19) with $p = 1$ into Eq. (12.24) and Eq. (12.26) and obtain

$$e_m(\boldsymbol{\kappa}) = \left(\pi^2 \underset{\sim}{k}^2 \cos^2\phi\, \tilde{b}(\underset{\sim}{k})^2 + \pi^2 \underset{\sim}{k}^2 \sin^2\phi\, \tilde{b}(\underset{\sim}{k})^2\right)^{1/2} - \pi \underset{\sim}{k} \tilde{b}(\underset{\sim}{k}) = 0 \tag{12.28}$$

and

$$e_\phi(\kappa, \phi) = \arctan \frac{\pi \underset{\sim}{\kappa} \sin \phi \tilde{b}(\underset{\sim}{\kappa})}{\pi \underset{\sim}{\kappa} \cos \phi \tilde{b}(\underset{\sim}{\kappa})} - \phi = 0. \tag{12.29}$$

For the angular error in dimensions higher than 2, calculate with Eq. (12.27)

$$\cos e_\phi = \frac{1}{k} \left[\underset{\sim}{\kappa} \cos \phi, \underset{\sim}{\kappa} \sin \phi \right] \left[\begin{array}{c} \mathrm{i}\underset{\sim}{\kappa} \cos \phi \, \tilde{b}(\underset{\sim}{\kappa}) \\ \mathrm{i}\underset{\sim}{\kappa} \sin \phi \, \tilde{b}(\underset{\sim}{\kappa}) \end{array} \right] \frac{1}{\mathrm{i}\pi \underset{\sim}{\kappa} \tilde{b}(\underset{\sim}{\kappa})} = 1. \tag{12.30}$$

This proves that the approach of a gradient operator with isotropic smoothing according to Eq. 12.19 can actually detect edges in all directions equally and without angular errors.

There is a wide range of edge detectors. We will examine some of them carefully in the Sections 12.4.3–12.6.

12.4.3 Discrete First-order Differences

First-order differences are the simplest approach to calculating the gradient. We can approximate the first partial derivative in the x-direction, $\partial g(x, y)/\partial x$, by:

$$\begin{array}{lll} \text{Backward difference} & \dfrac{\partial g(x,y)}{\partial x} \approx \dfrac{g(x,y) - g(x - \Delta x, y)}{\Delta x}, & \\ \text{Forward difference} & \dfrac{\partial g(x,y)}{\partial x} \approx \dfrac{g(x + \Delta x, y) - g(x,y)}{\Delta x} & \text{or} \\ \text{Symmetrical difference} & \dfrac{\partial g(x,y)}{\partial x} \approx \dfrac{g(x + \Delta x, y) - g(x - \Delta x, y)}{2\Delta x}. & \end{array} \tag{12.31}$$

These approximations correspond to the filter masks

$$\begin{array}{llll} \text{Backward} & {}^{-}\boldsymbol{D}_x & = & [1_\bullet \; -1] \\ \text{Forward} & {}^{+}\boldsymbol{D}_x & = & [1 \; -1_\bullet] \\ \text{Symmetric} & \boldsymbol{D}_{2x} & = & 1/2 \, [1 \; 0 \; -1] . \end{array} \tag{12.32}$$

The index $\bullet$ denotes the point at which the result is written back. The symmetry required in Section 12.3.3 only shows the last mask. We can also consider the two-element masks that correspond to the backward or forward gradient as odd or antisymmetric masks if the result is not stored at the position of the right or left pixel, but between the two pixels. This corresponds to a shift of the grid by half a pixel distance. The transfer function for the backward gradient is then

$${}^{-}\tilde{d}_x = \exp(\mathrm{i}\pi \underset{\sim}{\kappa}_x/2) \left[1 - \exp(-\mathrm{i}\pi \underset{\sim}{\kappa}_x)\right] = 2\mathrm{i} \sin(\pi \underset{\sim}{\kappa}_x/2), \tag{12.33}$$

where the first term results from the shift by half a grid spacing.

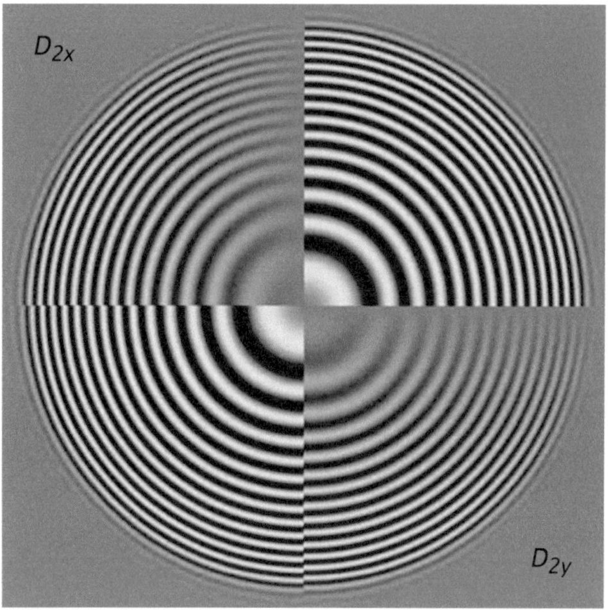

Figure 12.3: Application of the symmetrical difference filters $\mathcal{D}_{2x}$ and $\mathcal{D}_{2y}$ to the test image from Fig. 11.5.

Using Eq. (12.14), the transfer function of the symmetric derivative operator is reduced to

$$\tilde{d}_{2x} = \mathrm{i}\sin(\pi\tilde{k}_x) = \mathrm{i}\sin(\pi\tilde{k}\cos\phi). \qquad (12.34)$$

This operator can also be thought of as being composed of the backward difference ${}^{-}\boldsymbol{D}_x$ and the elementary binomial smoothing filter $\boldsymbol{B}_x$ (Section 11.4.2):

$$\boldsymbol{D}_{2x} = {}^{-}\boldsymbol{D}_x\,\boldsymbol{B}_x = [1_\bullet\ -1] * 1/2\ [1\ 1_\bullet] = 1/2\,[1\ 0\ -1]\,.$$

Corresponding equations apply for difference filters in other directions. For example, the transfer function of the symmetrical difference filter in the y direction is given by

$$\tilde{d}_{2y} = \mathrm{i}\sin(\pi\tilde{k}_y) = \mathrm{i}\sin(\pi\tilde{k}\sin\phi). \qquad (12.35)$$

When applying $\mathcal{D}_{2x}$ and $\mathcal{D}_{2y}$ to the ring test pattern in Fig. 12.3, the directional dependence and phase shift of 90° of these filters becomes clear. The horizontal derivative $\mathcal{D}_{2x}$ sees no horizontally oriented edges, but vertical and diagonal ones (Fig. 12.4c), while the vertical derivative sees no vertical edges (Fig. 12.4c). In both images, a medium gray value means no change, lighter positive and darker gray values negative slopes (cf. Section 5.6.3). The magnitude of the gradient and the sum of the magnitudes of $\mathcal{D}_{2x}$ and $\mathcal{D}_{2y}$ are shown in Fig. 12.4e and f.

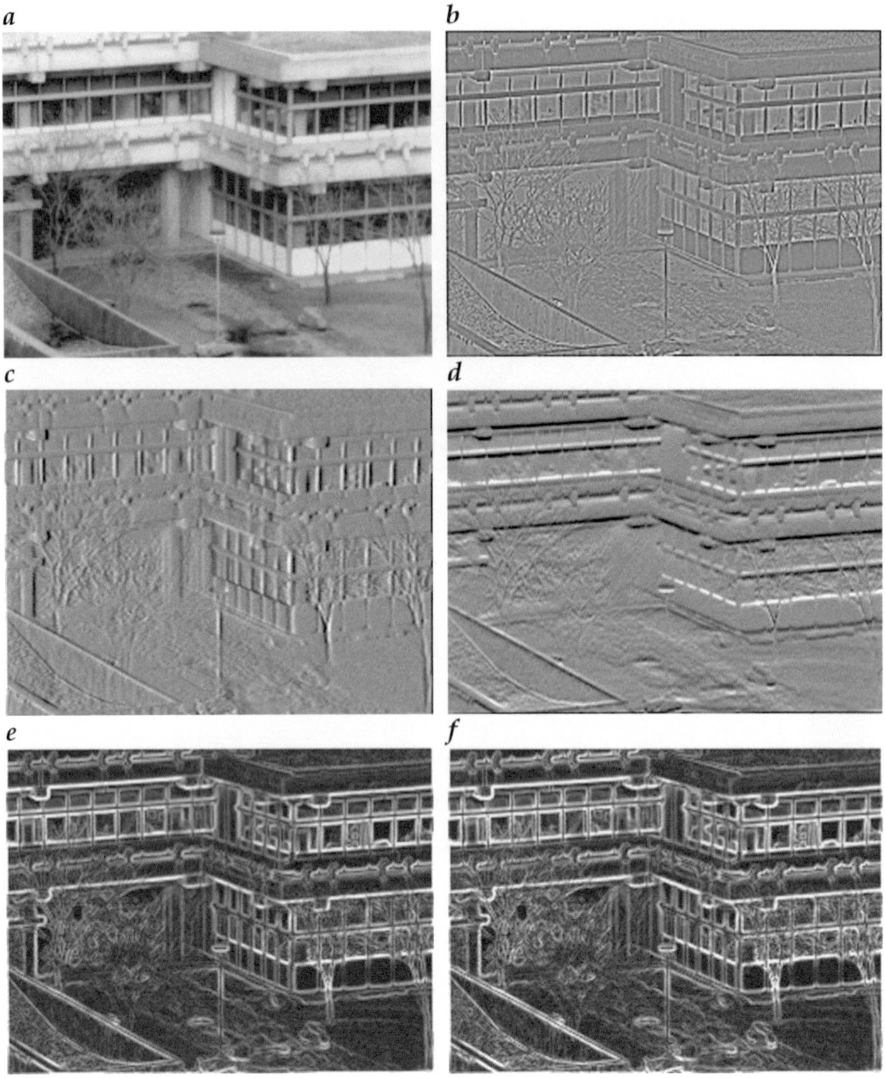

Figure 12.4: Detection of edges with derivation operators: ***a*** *Original image,* ***b*** *Laplacian operator $\mathcal{L}$,* ***c*** *horizontal derivative $\mathcal{D}_{2x}$,* ***d*** *vertical derivative $\mathcal{D}_{2y}$,* ***e*** *Magnitude of the gradient $(\mathcal{D}_{2x} \cdot \mathcal{D}_{2x} + \mathcal{D}_{2y} \cdot \mathcal{D}_{2y})^{1/2}$ and* ***f*** *Sum of the amounts of* ***c*** *and* ***d*** *according to Eq. (12.22).*

The transfer functions shown in Fig. 12.5 show that the simple discrete difference filters deviate from the ideal linearly increasing transfer function. For the difference of neighboring pixels ($\mathcal{D}_x$), the transfer function still increases monotonically, while for the symmetrical difference $\mathcal{D}_{2x} = \mathcal{D}_x\mathcal{B}_x$ the transfer function drops to zero due to the additional smoothing filter for the highest wavenumber $\tilde{k} = 1$.

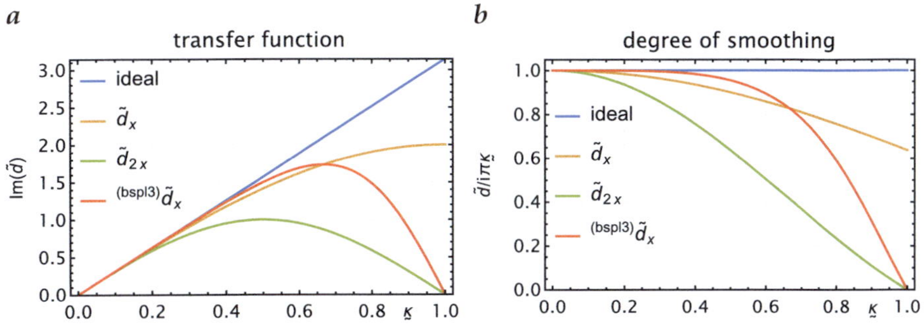

Figure 12.5: ***a*** *Imaginary part of the transfer functions of simple one-dimensional derivative filters as specified and their* ***b*** *degree of smoothing relative to the ideal derivative filter with the transfer function* $\mathrm{i}\pi\tilde{k}$.

These simple gradient filters are only a poor approximation for edge detectors. From Eqs. (12.34) and (12.35) we can deduce that the magnitude and direction of the gradient can be derived from

$$|\tilde{\boldsymbol{d}}| = \left(\sin^2(\pi\tilde{k}\cos\phi) + \sin^2(\pi\tilde{k}\sin\phi)\right)^{1/2} \tag{12.36}$$

and

$$\phi' = \arctan\frac{\sin(\pi\tilde{k}\sin\phi)}{\sin(\pi\tilde{k}\cos\phi)} \tag{12.37}$$

where the wavenumber was written in polar coordinates (k, ϕ). The resulting errors are shown as contour plots in Fig. 12.6 as a function of the magnitude of the wavenumber and the angle ϕ to the x-axis.

The magnitude of the gradient decreases rapidly from the correct value. A Taylor series of Eq. (12.36) in $\tilde{k}$ results for the error of the magnitude

$$e_m(\tilde{k}, \phi) \approx \frac{(\pi\tilde{k})^3}{12}\sin^2 2\phi + O(\tilde{k}^5). \tag{12.38}$$

The amount is greater in the diagonal direction. The filter is therefore anisotropic. The error in the direction of the gradient is also large. While the error in the direction of the axes and diagonals is zero, it reaches values of around ± 10° (Fig. 12.6b) in the directions in between at $\tilde{k} = 0.5$. A Taylor series of Eq. (12.37) in $\tilde{k}$ results for small $\tilde{k}$ approximately in the angular error

$$e_\phi(\tilde{k}, \phi) \approx \frac{(\pi\tilde{k})^2}{24}\sin 4\phi + O(\tilde{k}^4). \tag{12.39}$$

From this equation we see that the angular error for $\phi = n\pi/4$ with $n \in \mathbb{Z}$, i. e. for $\phi = 0°, 45°, 90°, \ldots$ is zero.

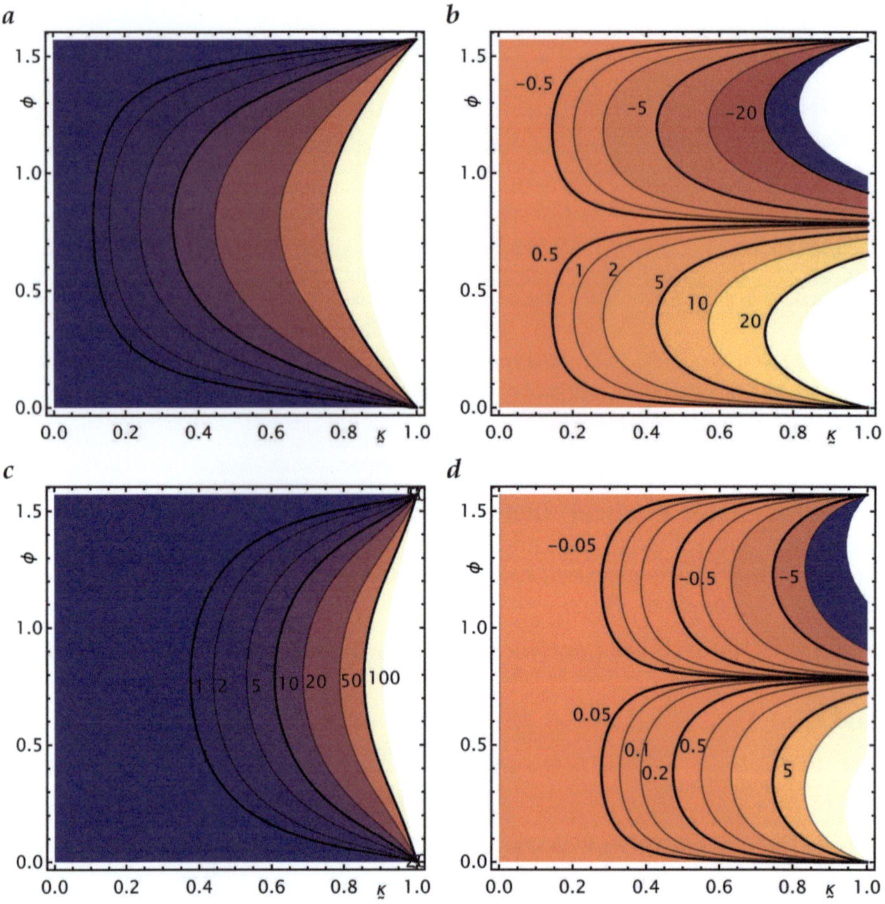

Figure 12.6: ***a*** *Relative anisotropy in magnitude and* ***b*** *error in the direction of the gradient for the symmetric gradient filter* $\left[\mathcal{D}_{2x}, \mathcal{D}_{2y}\right]^T$. ***c*** *Relative anisotropy in magnitude and* ***d*** *error in the direction of the gradient for the cubic B-spline derivative operator according to Eq. (12.43). The parameters are the magnitude of the wavenumber* κ *(0 to 1) and the angle* ϕ *to the x-axis (0 to* $\pi/2$*). The relative error in the magnitude is given as a percentage, the angular error in degrees.*

12.4.4 ♯Spline-based Edge Detection

The cubic B-spline transformation, which was used in Section 5.5.5 for interpolation, has led to a continuous representation of an image in which the first and second order derivatives are also continuous:

$$g_3(x) = \sum_n c_n \beta_3(x - n), \tag{12.40}$$

where $\beta_3(x)$ is the cubic B-spline function defined in Eq. (5.41). Based on this continuous representation, the calculation of the spatial derivative of $g_3(x)$

is simple:

$$\frac{\partial g_3(x)}{\partial x} = \sum_n c_n \frac{\partial \beta_3(x-n)}{\partial x}. \tag{12.41}$$

For a discrete derivative filter, we only need the derivatives at the grid points. Fig. 5.11a shows that the cubic B-spline function is extended over a maximum of 5 grid points. The maximum of the spline function lies at the central grid point. The derivative is therefore zero at this point, as well as at the two outer grid points. The derivative is therefore only non-zero at the direct right and left neighbors of the central grid point. At the grid point x_m it is therefore reduced to

$$\left.\frac{\partial g_3(x)}{\partial x}\right|_{x_m} = (c_{m+1} - c_{m-1})/2. \tag{12.42}$$

This means that the calculation of the first-order derivative based on the cubic B-spline transformation is actually an efficient solution. We first perform the cubic B-spline transformation in the direction of the derivative to be calculated (Section 5.5.5) and then apply the simple $\mathcal{D}_{2x}$ operator. The transfer function then results in

$$\tilde{d}_x = \mathrm{i}\frac{\sin(\pi \underset{\sim}{k}_x)}{2/3 + 1/3\cos(\pi \underset{\sim}{k}_x)} = \mathrm{i}\pi \underset{\sim}{k}_x - \mathrm{i}\frac{\pi^5 \underset{\sim}{k}_x^5}{180} + O(\underset{\sim}{k}_x^7). \tag{12.43}$$

The errors of magnitude and direction of a gradient vector based on the B-spline derivative filter are shown in Fig. 12.6c and d. They are more than ten times smaller than for simple derivative filters (Fig. 12.6a and b).

This also results from Taylor series for the error of the magnitude of the gradient and for the angular error:

$$e_m(\underset{\sim}{k}, \phi) \approx -\frac{(\pi \underset{\sim}{k})^5}{240}\sin^2 2\phi + O(\underset{\sim}{k}^7) \tag{12.44}$$

$$e_\phi(\underset{\sim}{k}, \phi) \approx \frac{(\pi \underset{\sim}{k})^4}{720}\sin 4\phi + O(\underset{\sim}{k}^6). \tag{12.45}$$

The errors now only appear in the terms with $\underset{\sim}{k}^5$ and $\underset{\sim}{k}^4$. Compare Eqs. (12.44) and (12.45) with Eqs. (12.36) and (12.37) and with Eqs. (12.70) and (12.71).

12.5 Edge Detection through Zero Crossings

12.5.1 Principle

Edges are *zero crossings* in the second derivative (Fig. 12.1). Therefore, the sum of the partial second derivatives in all directions after Eq. (12.9) forms a linear isotropic edge detector with the transfer function $-(\pi \underset{\sim}{k})^2$, which is known as *Laplace operator* Δ. However, it is also clear from Fig. 12.1 that not every zero crossing represents an edge. Only if there are signal

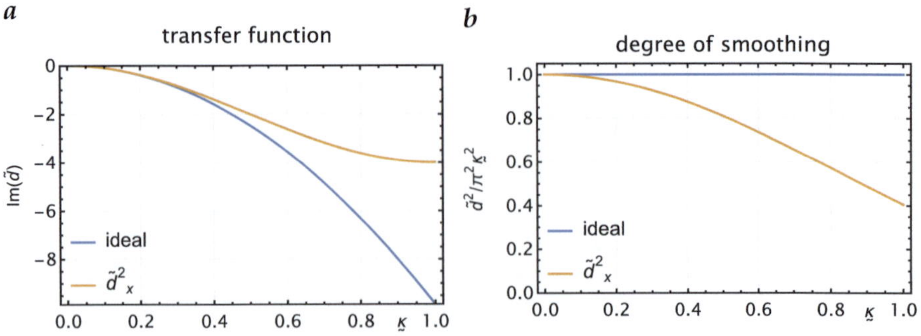

Figure 12.7: ***a*** *Transfer functions of simple one-dimensional second-order derivative filters as specified and their* ***b*** *degree of smoothing relative to the ideal derivative filter with the transfer function* $-(\pi \tilde{k})^2$

peaks immediately before and after the zero crossing that are significantly higher than the noise level is there a significant edge. We can also see from Fig. 12.1 that edge detection with the Laplacian operator is significantly more susceptible to noise than gradient-based edge detection.

12.5.2 Laplace Filter

We arrive at second-order difference operators directly by applying the first-order operators twice:

$$\mathcal{D}_x^2 = {}^{-}\mathcal{D}_x \, {}^{+}\mathcal{D}_x. \tag{12.46}$$

In local space, this corresponds to the convolution masks

$$[1_\bullet \; -1] * [1 \; -1_\bullet] = [1 \; -2 \; 1]\,. \tag{12.47}$$

Although the convolution mask does not contain a smoothing term, the transfer function is monotonic like the first-order difference, $\boldsymbol{D}_x$, but does not fall off as much as the second-order derivative (Fig. 12.7). Structures with the maximum wavenumber $\tilde{k} = 1$ are amplified by a factor of 4, but by less than half than with the second derivative.

The discrete Laplacian operator $\mathcal{L} = \mathcal{D}_x^2 + \mathcal{D}_y^2$ for 2D images has the filter mask

$$\boldsymbol{L} = \begin{bmatrix} 1 & -2 & 1 \end{bmatrix} + \begin{bmatrix} 1 \\ -2 \\ 1 \end{bmatrix} = \begin{bmatrix} 0 & 1 & 0 \\ 1 & -4 & 1 \\ 0 & 1 & 0 \end{bmatrix} \tag{12.48}$$

and the transfer function:

$$\tilde{l}(\tilde{\boldsymbol{k}}) = -4\sin^2(\pi \tilde{k}_x/2) - 4\sin^2(\pi \tilde{k}_y/2). \tag{12.49}$$

Like other discrete approximations of operators, the discretized Laplace operator is only isotropic for small wavenumbers (Fig. 12.8a):

$$\tilde{l}(\underset{\sim}{k}, \phi) = -(\pi \underset{\sim}{k})^2 + \frac{3}{48}(\pi \underset{\sim}{k})^4 + \frac{1}{48}\cos 4\phi (\pi \underset{\sim}{k})^4 + O(\underset{\sim}{k}^6). \quad (12.50)$$

There are many other ways to construct a discrete approximation for the Laplace operator. The use of a binomial mask is interesting. With Eq. (11.28) we can approximate all binomial masks for sufficiently small wavenumbers by the following equation:

$$\tilde{b}^{2R}(\underset{\sim}{k}) \approx 1 - \frac{R}{4}(\underset{\sim}{k}\pi)^2 + O(\underset{\sim}{k}^4). \quad (12.51)$$

From this we conclude that every operator $\mathcal{B}^p - \mathcal{I}$ forms a Laplace operator for small wavenumbers. For example

$$\boldsymbol{L}' = 4(\boldsymbol{B}^2 - \boldsymbol{I}) = \frac{1}{4}\begin{bmatrix} 1 & 2 & 1 \\ 2 & 4 & 2 \\ 1 & 2 & 1 \end{bmatrix} - \begin{bmatrix} 0 & 0 & 0 \\ 0 & 4 & 0 \\ 0 & 0 & 0 \end{bmatrix} = \frac{1}{4}\begin{bmatrix} 1 & 2 & 1 \\ 2 & -12 & 2 \\ 1 & 2 & 1 \end{bmatrix} \quad (12.52)$$

with the transfer function

$$\tilde{l}'(\underset{\sim}{\boldsymbol{k}}) = 4\cos^2(\pi \underset{\sim}{k}_x/2)\cos^2(\pi \underset{\sim}{k}_y/2) - 4 \quad (12.53)$$

is another realization of a discrete Laplace operator. For small wavenumbers, it can be approximated by

$$\tilde{l}'(\underset{\sim}{k}, \phi) \approx -(\pi \underset{\sim}{k})^2 + \frac{3}{32}(\pi \underset{\sim}{k})^4 - \frac{1}{96}\cos 4\phi (\pi \underset{\sim}{k})^4 + O(\underset{\sim}{k}^6). \quad (12.54)$$

For large wavenumbers, the transfer functions of both Laplacian operators $\mathcal{L}$ and $\mathcal{L}'$ show considerable deviations from an ideal Laplacian operator $-(\pi \tilde{k})^2$ (Fig. 12.8). However, $\mathcal{L}'$ has a significantly lower anisotropy than $\mathcal{L}$.

When comparing the transfer function of the two Laplacian operators in Eq. (12.50) and Eq. (12.54), it is noticeable that the anisotropy terms with $\cos 4\phi(\pi \underset{\sim}{k})^4$ have a reversed sign and that of $\mathcal{L}$ is twice as high as that of $\mathcal{L}'$. Therefore, this term can be eliminated in the combined Laplacian operator:

$$\boldsymbol{L}'' = (2\boldsymbol{L}' - \boldsymbol{L})/3 = \frac{1}{6}\begin{bmatrix} 1 & 2 & 1 \\ 2 & -12 & 2 \\ 1 & 2 & 1 \end{bmatrix} - \frac{1}{6}\begin{bmatrix} 0 & 2 & 0 \\ 2 & -8 & 2 \\ 0 & 2 & 0 \end{bmatrix} = \frac{1}{6}\begin{bmatrix} 1 & 4 & 1 \\ 4 & -20 & 4 \\ 1 & 4 & 1 \end{bmatrix}. \quad (12.55)$$

This has the transfer function

$$\tilde{l}''(\underset{\sim}{\boldsymbol{k}}) = \left(8\cos^2(\pi \underset{\sim}{k}_x/2)\cos^2(\pi \underset{\sim}{k}_y/2) - 4\sin^2(\pi \underset{\sim}{k}_x/2) - 4\sin^2(\pi \underset{\sim}{k}_y/2) - 8\right)/3, \quad (12.56)$$

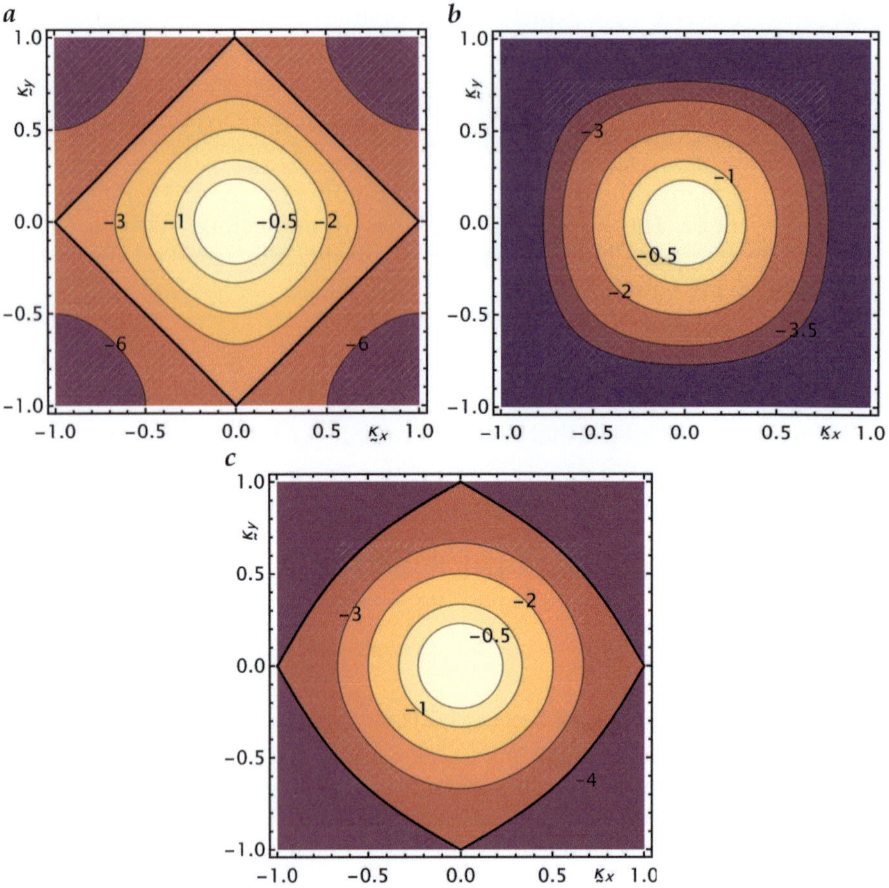

Figure 12.8: *Transfer functions of discrete Laplace operators and their anisotropy:* ***a*** $\mathcal{L}$ *Eq. (12.48),* ***b*** $\mathcal{L}'$ *Eq. (12.52) and* ***c*** $(2\mathcal{L}' + \mathcal{L})/3$ *Eq. (12.55).*

which can be approximated for small wavenumbers by

$$\tilde{l}''(\underset{\sim}{k}, \phi) \approx -(\pi \underset{\sim}{k})^2 + \frac{(\pi \underset{\sim}{k})^4}{12} + \frac{(\pi \underset{\sim}{k})^6}{1440}(\cos 4\phi - 5) + O(\underset{\sim}{k}^8). \tag{12.57}$$

As expected, an anisotropic term with $\cos 4\phi$ only results in the 6th order. The two-dimensional transfer function clearly shows that this Laplacian operator is more isotropic (Fig. 12.8c).

12.6 ♯Optimized Derivative Filters

We have already learned that for edge detection it is not necessary to match a discrete derivative filter as well as possible to the ideal derivative (Section 12.3.5 and 12.4.2). However, for other applications, such as motion analysis (Chapter 14) or when a differential equation system is needed to solve an image processing operation (Chapter 17), such operators are required.

Therefore, in this section we will use the example of first-order derivative filters to show that they can be flexibly optimized using the *least squares* method. The basic idea is to apply a 1D filter with R coefficients and odd symmetry in the derivative direction d and to vary the coefficients in such a way that the deviation of the transfer function from the ideal transfer function $\mathrm{i}\pi\underset{\sim}{k}_d$ is minimized. This means that the *objective function* of the optimization approach is

$$\tilde{t}(\underset{\sim}{k}_d) = \mathrm{i}\pi\underset{\sim}{k}_d \tag{12.58}$$

and the transfer function of the 1D filter of size $2R+1$ with R unknown coefficients is

$${}^R\tilde{d}(\underset{\sim}{k}_d) = -\mathrm{i}\sum_{\tilde{n}=1}^{R} 2d_{\tilde{n}} \sin(\tilde{n}\pi\underset{\sim}{k}_d). \tag{12.59}$$

The coefficients are now determined in such a way that the sum of the quadratic deviations between ${}^R\tilde{d}(\underset{\sim}{k}_d)$ and $\tilde{t}(\underset{\sim}{k}_d)$ is minimal:

$$\int\limits_0^1 w(\underset{\sim}{k}_d)\left|{}^R\tilde{d}(\underset{\sim}{k}_d) - \tilde{t}(\underset{\sim}{k}_d)\right|^2 \mathrm{d}\underset{\sim}{k}_d. \tag{12.60}$$

The wavenumber-dependent function $w(\underset{\sim}{k}_d)$ determines how strongly the different wavenumbers are weighted.

As an additional useful constraint, we can require the transfer function to be equal to $\mathrm{i}\pi\underset{\sim}{k}$ for small wavenumbers. should be the same. This constraint reduces the number of degrees of freedom for the filter optimization from R coefficients to $R-1$. Instead of Eq. (12.59), the following approach results:

$${}^R\tilde{d} = -\mathrm{i}\sin(\pi\underset{\sim}{k}_d) - \mathrm{i}\sum_{v=\tilde{n}}^{R} 2d_{\tilde{n}}\left(\sin(\tilde{n}\pi\underset{\sim}{k}_d) - \tilde{n}\sin(\pi\underset{\sim}{k}_d)\right) \tag{12.61}$$

and

$$d_1 = 1 - \sum_{\tilde{n}=2}^{R} \tilde{n} d_{\tilde{n}}. \tag{12.62}$$

Derivative filters can be further improved by compensating for the drop in the transfer function at high wavenumbers using a forward and backward relaxation filter (Section 10.4.4, Fig. 10.8b). This results in the following approach for the transfer function:

$${}^{(R,\beta)}\tilde{d} = \frac{-\mathrm{i}\sin(\pi\underset{\sim}{k}) - \mathrm{i}\sum\limits_{\tilde{n}=2}^{R} 2d_{\tilde{n}}\left(\sin(\tilde{n}\pi\underset{\sim}{k}_d) - \tilde{n}\sin(\pi\underset{\sim}{k}_d)\right)}{1+\beta-\beta\cos(\pi\underset{\sim}{k}_d)} \tag{12.63}$$

with the additional parameter β.

Filters optimized in this way can have significantly lower errors than the derivative filter based on cubic B-spline interpolation discussed in Section 12.4.4. A detailed discussion of the design of optimal derivative filters with filter coefficient tables can be found in Jähne et al. [166].

12.7 Regularized Edge Detection

12.7.1 Principle

The edge detectors discussed so far are useful for calculating the derivative, but less so for detecting edges in noisy images. Due to the small mask size, the transfer function is high for large wavenumbers. However, there is often more noise than useful signal. Optimal edge detectors must therefore be set to the wavenumber range at which the edge signal is at a maximum compared to the noise. So we have to develop filters that derive in one direction or determine differences, but also smooth as evenly as possible in all directions for noise suppression.

Smoothing is particularly effective in higher-dimensional signals, as smoothing in all directions perpendicular to the gradient does not smear the edge. A difference filter that includes smoothing is called a *regularized edge detector*, as it provides a robust solution to the ill-posed problem of determination of derivatives from discrete signals.

12.7.2 Regularized 2 × 2-Edge Detectors

The smallest regularized 2D gradient operator has the 2×2 masks

$$\boldsymbol{D}_x\boldsymbol{B}_y = \frac{1}{2}\begin{bmatrix} 1 & -1 \\ 1 & -1 \end{bmatrix} \quad \text{and} \quad \boldsymbol{D}_y\boldsymbol{B}_x = \frac{1}{2}\begin{bmatrix} 1 & 1 \\ -1 & -1 \end{bmatrix} \tag{12.64}$$

and the transfer functions

$$\begin{aligned} \tilde{d}_x\tilde{b}_y(\tilde{\boldsymbol{k}}) &= 2\mathrm{i}\sin(\pi\tilde{k}_x/2)\cos(\pi\tilde{k}_y/2) \\ \tilde{d}_y\tilde{b}_x(\tilde{\boldsymbol{k}}) &= 2\mathrm{i}\sin(\pi\tilde{k}_y/2)\cos(\pi\tilde{k}_x/2). \end{aligned} \tag{12.65}$$

There are no optimization options for these small masks. The filters $\boldsymbol{D}_x = [1\ -1]$ and $\boldsymbol{D}_y = [1\ -1]^T$ are not suitable for forming a gradient operator, as $\boldsymbol{D}_x$ and $\boldsymbol{D}_y$ each shift the convolution result by half a pixel distance in the x or y direction.

The errors in the magnitude and direction of the gradient are for small wavenumbers

$$e_m(\tilde{k}, \phi) \approx -\frac{(\pi\tilde{k})^3}{24}\sin^2 2\phi + O(\tilde{k}^5) \tag{12.66}$$

and

$$e_\phi(\tilde{k}, \phi) \approx -\frac{(\pi\tilde{k})^2}{48}\sin 4\phi + O(\tilde{k}^4). \tag{12.67}$$

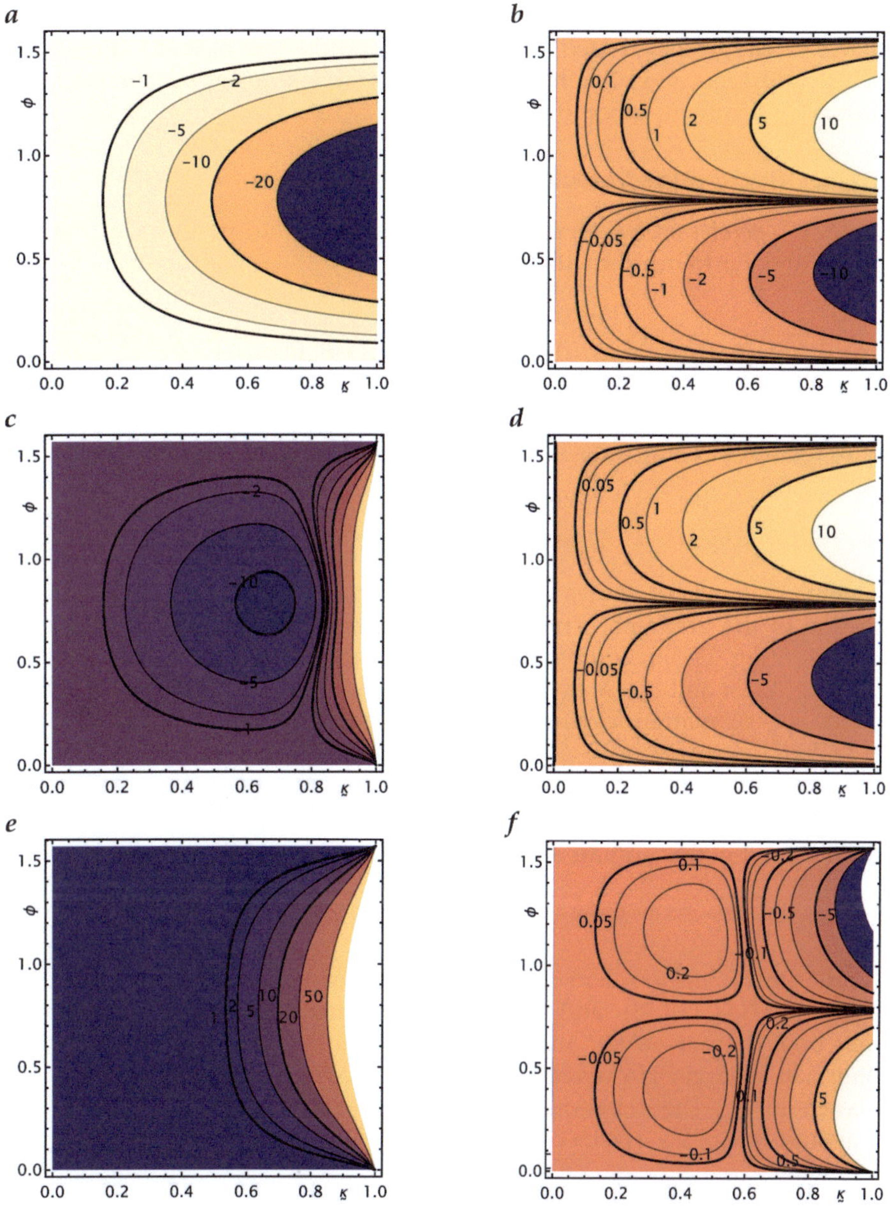

Figure 12.9: *Relative anisotropy of the magnitude and error of the direction of the gradient for three regularized gradient operators:* ***a*** *and* ***b*** 2×2 *edge detectors after Eq. (12.64),* ***c*** *and* ***d*** 3×3 *Sobel operator according to Eq. (12.68) and* ***e*** *and* ***f*** *optimized* 3×3 *Scharr operator according to Eq. (12.81). The parameters are the magnitude of the wavenumber (0 to 1) and the angle to the x-axis (0 to* $\pi/2$*). The relative error of the magnitude is given in percent and the angular error in degrees. The absolute value error is shown on the left and the angular error on the right.*

The errors are significantly smaller (for small wavenumbers by about half) than with the gradient based on the symmetric difference operator (Figs. 12.6a and b and 12.9a and b), although the anisotropic terms appear in the same order in the wavenumber in Eqs. (12.38) and (12.39).

12.7.3 Sobel Edge Detectors

The *Sobel operator* uses difference filters that average the image in the direction perpendicular to the derivation direction:

$$ {}^2\mathbf{S}_x = \mathbf{D}_x\mathbf{B}_x\mathbf{B}_y^2 = \frac{1}{8}\begin{bmatrix} 1 & 0 & -1 \\ 2 & 0 & -2 \\ 1 & 0 & -1 \end{bmatrix}, \quad {}^2\mathbf{S}_y = \mathbf{D}_y\mathbf{B}_y\mathbf{B}_x^2 = \frac{1}{8}\begin{bmatrix} 1 & 2 & 1 \\ 0 & 0 & 0 \\ -1 & -2 & -1 \end{bmatrix} \tag{12.68} $$

with the transfer functions

$$ \begin{aligned} \tilde{d}_x\tilde{b}_x\tilde{b}_y^2(\tilde{\mathbf{k}}) &= 2\mathrm{i}\sin(\pi\tilde{k}_x/2)\cos(\pi\tilde{k}_x/2)\cos^2(\pi\tilde{k}_y/2) \\ \tilde{d}_y\tilde{b}_y\tilde{b}_x^2(\tilde{\mathbf{k}}) &= 2\mathrm{i}\sin(\pi\tilde{k}_y/2)\cos(\pi\tilde{k}_y/2)\cos^2(\pi\tilde{k}_x/2). \end{aligned} \tag{12.69} $$

Fig. 12.9c and d shows the resulting errors of the magnitude and direction of the gradient. The improvement over the symmetric difference operator (Fig. 12.6) is comparable to that of the regularized2 $\times$ 2 edge detector (Fig. 12.9a and b). A Taylor series expansion provides the same results as an approximation for small wavenumbers (compare Eqs. (12.66) and (12.67)):

$$ e_m(\tilde{k},\phi) \approx -\frac{(\pi\tilde{k})^3}{24}\sin^2 2\phi + O(\tilde{k}^5) \tag{12.70} $$

for the error in the amount and

$$ e_\phi(\tilde{k},\phi) \approx -\frac{(\pi\tilde{k})^2}{48}\sin 4\phi + O(\tilde{k}^4) \tag{12.71} $$

for the error of the gradient direction. A comparison with the corresponding equations Eqs. (12.38) and (12.39) for the simple derivative filters shows that both the anisotropy and the angular error of the Sobel operator are only half as large. However, the error still increases with the square of the wavenumber. The directional error of the Sobel operator is up to 5° for a wavenumber of 0.5, which cannot be tolerated for most applications.

12.7.4 Derivatives of the Gaussian function

A long-known class of regularized derivative filters are filters that are based on the derivative of smoothing filters based on the Gaussian function. For instance, Canny [39] used such a filter for optimal edge detection (*Canny filter*). For discrete data, these operators are best approximated by derivatives of the binomial smoothing filter (Section 11.4) as

$$ {}^{(B,R)}\mathcal{D}_d = \mathcal{D}_{2d}\mathcal{B}^R \tag{12.72} $$

with a non-square $(2R+3)\times(2R+1)^{D-1}$ D-dimensional filter mask and the transfer function

$$^{(B,R)}\tilde{d}_d(\underset{\sim}{\boldsymbol{k}}) = \mathrm{i}\sin(\pi\underset{\sim}{k}_d)\prod_{d=1}^{D}\cos^{2R}(\pi\underset{\sim}{k}_d/2). \tag{12.73}$$

Surprisingly, these filters turn out to be a poor choice, as the anisotropy is the same as with the symmetric difference filter. This can be seen immediately for the direction of the gradient. The smoothing terms are the same for both directions and are therefore shortened to Eq. (12.26). The remaining terms are the same as for the symmetrical difference filter.

Similarly, the same ansiotropy results for the Sobel-type R^D-difference operators

$$^{R}\mathcal{S}_d = \mathcal{D}_d\mathcal{B}_d^{R-1}\prod_{d'\neq d}\mathcal{B}_{d'}^{R} \tag{12.74}$$

with a $(2R+1)^D$ large D-dimensional filter mask and the transfer function

$$^{R}\tilde{s}_d(\underset{\sim}{k}) = \mathrm{i}\tan(\pi\underset{\sim}{k}_d/2)\prod_{d=1}^{D}\cos^{2R}(\pi\underset{\sim}{k}_d/2) \tag{12.75}$$

as with the 3×3 Sobel operator.

12.7.5 LoG and DoG Filter

Laplacian filters tend to significantly increase noise in images because the transfer function is proportional to the square of the wavenumber. A better edge detector is obtained if we first smooth the image and then apply the Laplacian filter. This gives us a kind of controllable edge detection. Such a filter is called in the literature *Laplace of Gaussian* filter (abbreviated *LoG* filter) or as *Marr-Hildreth operator* [216].

In the discrete case, an LoG filter is approximated by first smoothing the image with a binomial mask and then filtering it with the discrete Laplacian operator. This gives us the operator $\mathcal{L}\mathcal{B}^p$ with the following transfer function:

$$\tilde{l}\tilde{b}^p(\underset{\sim}{\boldsymbol{k}}) = -4\left[\sin^2(\pi\underset{\sim}{k}_x/2)+\sin^2(\pi\underset{\sim}{k}_y/2)\right]\cos^p(\pi\underset{\sim}{k}_x/2)\cos^p(\pi\underset{\sim}{k}_y/2). \tag{12.76}$$

For small wavenumbers, this transfer function can be approximated to

$$\tilde{l}\tilde{b}^p(\underset{\sim}{k},\phi) \approx -(\pi\underset{\sim}{k})^2+\left[\frac{1}{16}+\frac{1}{8}p+\frac{1}{48}\cos(4\phi)\right](\pi\underset{\sim}{k})^4 \tag{12.77}$$

In Section 12.5.2 we saw that a Laplace filter can be better described with operators of type $\mathcal{B}^p-\mathcal{I}$. If we additionally smooth, this approximation leads to a filter type called *difference of Gaussian* filter or abbreviated *DoG* filter:

$$4(\mathcal{B}^q-\mathcal{I})\mathcal{B}^p = 4(\mathcal{B}^{p+q}-\mathcal{B}^p). \tag{12.78}$$

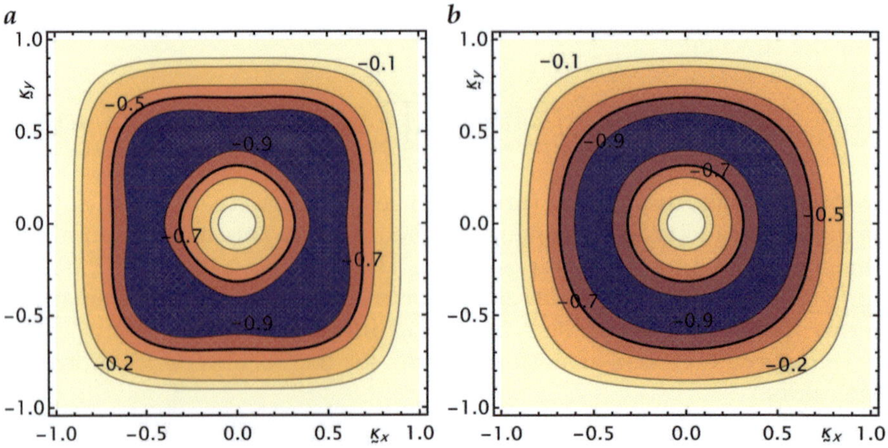

*Figure 12.10: Contour plot of the transfer functions **a** of the LoG filter $\mathcal{L}\mathcal{B}^2$ and **b** of the DoG filter $4(\mathcal{B}^4 - \mathcal{B}^2)$.*

This DoG filter has the transfer function

$$\begin{aligned} 4(\tilde{b}^{p+q} - \tilde{b}^p)(\boldsymbol{\kappa}) &= 4\cos^{p+q}(\pi\tilde{k}_x/2)\cos^{p+q}(\pi\tilde{k}_y/2) \\ &- 4\cos^{p}(\pi\tilde{k}_x/2)\cos^{p}(\pi\tilde{k}_y/2), \end{aligned} \tag{12.79}$$

which can be approximated for small wavenumbers to

$$4(\tilde{b}^{p+q} - \tilde{b}^p)(\tilde{k},\phi) \approx -\frac{q}{2}(\pi\tilde{k})^2 - \frac{1}{192}\left[3q - 12pq - 6q^2 - q\cos(4\phi)\right](\pi\tilde{k})^4 \tag{12.80}$$

The transfer functions of the LoG and DoG filters are compared in Fig. 12.10. Obviously, the DoG filter shows smaller deviations from an isotropic filter.

A filter with even less deviation can be constructed if we compare Eqs. (12.77) and (12.80). The anisotropic terms $\cos 4\phi$ have different signs. They can therefore be easily compensated for by mixing LoG and DoG operators according to the formula 2/3DoG + 1/3LoG. This corresponds to the operator $(8/3\mathcal{B}^2 - 8/3\mathcal{I} - 1/3\mathcal{L})\mathcal{B}^p$.

Filters that are similar to LoG and DoG filters are also found in the human visual system [215].

12.7.6 $^\sharp$Optimized Regularized Edge Detectors

We can easily derive an optimized Sobel operator with a minimum error of the direction of the gradient. The comparison of Eqs. (12.37) and (12.71) shows that the two filters have angular errors in opposite directions. It appears that the Sobel operator smoothes too much perpendicular to the derivative direction, while the symmetric difference operator smooths too little, i. e. not at all. It can therefore be assumed that a minimum angular

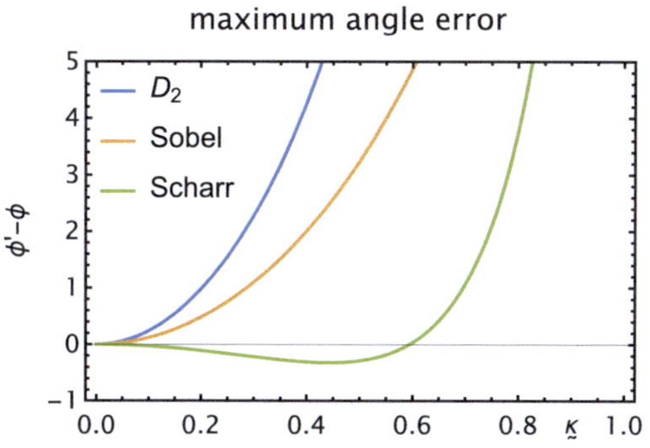

Figure 12.11: *Comparison of the maximum angular error for the determination of an edge direction as a function of the wavenumber for derivative filters as specified.*

error can be achieved by a suitable combination of the two operators, i.e. by setting the transverse smoothing.

Cross-smoothing can be reduced by increasing the central coefficient. In fact, Scharr et al. [287] used a non-linear optimization method to show that the following operators are optimal with regard to the angular error of the gradient (Fig. 12.9f):

$$
\begin{aligned}
{}^{2\mathrm{opt}}\boldsymbol{S}_x &= 1/4\boldsymbol{D}_x\boldsymbol{B}_x(3\boldsymbol{B}_y^2+\boldsymbol{I}) = \frac{1}{32}\begin{bmatrix} 3 & 0 & -3 \\ 10 & 0 & -10 \\ 3 & 0 & -3 \end{bmatrix} \\
{}^{2\mathrm{opt}}\boldsymbol{S}_y &= 1/4\boldsymbol{D}_y\boldsymbol{B}_y(3\boldsymbol{B}_x^2+\boldsymbol{I}) = \frac{1}{32}\begin{bmatrix} 3 & 10 & 3 \\ 0 & 0 & 0 \\ -3 & -10 & -3 \end{bmatrix}
\end{aligned} \tag{12.81}
$$

with the transfer function

$$
\begin{aligned}
{}^{2\mathrm{opt}}\tilde{s}_x(\underset{\sim}{\boldsymbol{k}}) &= \mathrm{i}\sin(\pi\underset{\sim}{k}_x/2)\cos(\pi\underset{\sim}{k}_x/2)(3\cos^2(\pi\underset{\sim}{k}_y/2)+1)/2 \\
{}^{2\mathrm{opt}}\tilde{s}_y(\underset{\sim}{\boldsymbol{k}}) &= \mathrm{i}\sin(\pi\underset{\sim}{k}_y/2)\cos(\pi\underset{\sim}{k}_y/2)(3\cos^2(\pi\underset{\sim}{k}_x/2)+1)/2.
\end{aligned} \tag{12.82}
$$

This filter has also been included in the OpenCV library under the name *Scharr filter*. The maximum angular error of this 3×3 filter mask is shown in Fig. 12.11. It is less than 0.3° for wavenumbers up to 0.6 and thus more than an order of magnitude smaller than that of the *Sobel filter* ${}^2\mathcal{S}$ or the symmetrical difference filter $\mathcal{D}_2$. Similar optimizations are possible for derivative operators with larger masks [166].

12.8 ♯Edges in Multichannel Images

Analyzing edges in multichannel images is much more difficult than averaging, which was discussed in Section 11.7. The basic problem lies in the fact that the different channels may contain conflicting data about edges. The gradient in channel A can point in a different direction than that in channel B. The simple addition of the gradients of all channels is therefore not useful:

$$\sum_{p=1}^{P} \nabla g_p(\boldsymbol{x}). \tag{12.83}$$

It is even possible for the gradients of two channels to point in opposite directions so that they cancel each other out. cancel each other out. This case cannot then be distinguished from constant areas in both channels.

A better measure of the resulting edge strength is therefore the sum of the squared amounts of the gradients of all channels:

$$\sum_{p=1}^{P} |\nabla g_p|^2 = \sum_{p=1}^{P}\sum_{d=1}^{D}\left(\frac{\partial g_p}{\partial x_d}\right)^2. \tag{12.84}$$

Although this expression gives a useful estimate of the resulting edge thickness, it does not solve the problem of contrary edge directions. edge directions. An analysis of how the edges are distributed in the P channels is possible with the following symmetric $D \times D$ matrix $\boldsymbol{D}$, where D is the dimension of the multichannel image:

$$\boldsymbol{D} = \boldsymbol{J}^T \boldsymbol{J}. \tag{12.85}$$

$\boldsymbol{J}$ is the *Jacobi matrix* defined as follows:

$$\boldsymbol{J} = \begin{bmatrix} \dfrac{\partial g_1}{\partial x_1} & \dfrac{\partial g_1}{\partial x_2} & \cdots & \dfrac{\partial g_1}{\partial x_D} \\ \dfrac{\partial g_2}{\partial x_1} & \dfrac{\partial g_2}{\partial x_2} & \cdots & \dfrac{\partial g_2}{\partial x_D} \\ \vdots & & \ddots & \vdots \\ \dfrac{\partial g_P}{\partial x_1} & \dfrac{\partial g_P}{\partial x_2} & \cdots & \dfrac{\partial g_P}{\partial x_D} \end{bmatrix}. \tag{12.86}$$

The elements of the matrix $\boldsymbol{D}$ are therefore

$$D_{kl} = \sum_{p=1}^{P} \frac{\partial g_p}{\partial x_k}\frac{\partial g_p}{\partial x_l}. \tag{12.87}$$

Since $\boldsymbol{D}$ is a symmetrical matrix, we can diagonalize and write it using a coordinate transformation:

$$\boldsymbol{D}' = \begin{bmatrix} \sum_{P} \left(\frac{\partial g_p}{\partial x'_1}\right)^2 & 0 & \cdots & 0 \\ 0 & \sum_{P} \left(\frac{\partial g_p}{\partial x'_2}\right)^2 & \ddots & 0 \\ 0 & \ddots & \ddots & 0 \\ 0 & \cdots & \cdots & \sum_{P} \left(\frac{\partial g_p}{\partial x'_D}\right)^2 \end{bmatrix}. \tag{12.88}$$

In the case of an ideal edge, only one of the diagonal terms is not equal to zero. The corresponding coordinate direction is perpendicular to the discontinuity. The diagonal terms are zero in all other directions. This case can therefore be recognized by the fact that the rank of $\boldsymbol{D}$ is one.

However, if the edges in the different channels happen to point in all directions, all terms are non-zero. In this way, it is in principle possible to distinguish random gray value changes due to noise from coherent edges. The trace of the matrix $\boldsymbol{D}$

$$\text{trace}(\boldsymbol{D}) = \sum_{d=1}^{D} D_{dd} = \sum_{d=1}^{D} \sum_{p=1}^{P} \left(\frac{\partial g_p}{\partial x_d}\right)^2 \tag{12.89}$$

leads to the measure of edge strength already defined in Eq. (12.84). This is independent of the orientation of the edge, as the trace of a symmetrical matrix is invariant when the coordinate system is rotated.

12.9 ♯References for Further Reading

There is a bewildering wealth of work on edge detection in the literature. Only a few carefully selected references are given here. The development of edge detection based on first-order difference filters can be traced through a few key publications. Canny [39] developed an optimal edge detector based on derivatives of the Gaussian function, Deriche [57] presented a fast recursive implementation of the Canny detector, Lanser and Eckstein [193] improved the isotropy of Deriche's recursive filter and Jähne et al. [166] developed a general nonlinear optimization strategy for edge filters with optimal isotropy.

Edge detection based on second-order difference filters (zero crossings) was strongly influenced by biological vision systems. Fundamental work can be found in Marr and Hildreth [216] and Marr [215]. For more advanced work towards a universal system for neighborhood operators, see the work of Koenderink and van Doorn [188] and Danielsson et al. [50].

13 Local Orientation, Local Wavenumber and Phase

Contents

13.1 Introduction

In the last two chapters we have dealt with neighborhood operations for averaging and edge detection. Thus, we have only studied the simplest structures in a local neighborhood, namely constant regions and edges (discontinuities). However, patterns can also be present in a local neighborhood. In this chapter, we consider simple patterns in a local neighborhood, which

B. Jähne, *Digital Image Processing and Image Acquisition*,
https://doi.org/10.1007/978-3-032-04806-6_13

*Figure 13.1: Object recognition through differences **a** of the gray value, **b** of the orientation or **c** of the size of a pattern.*

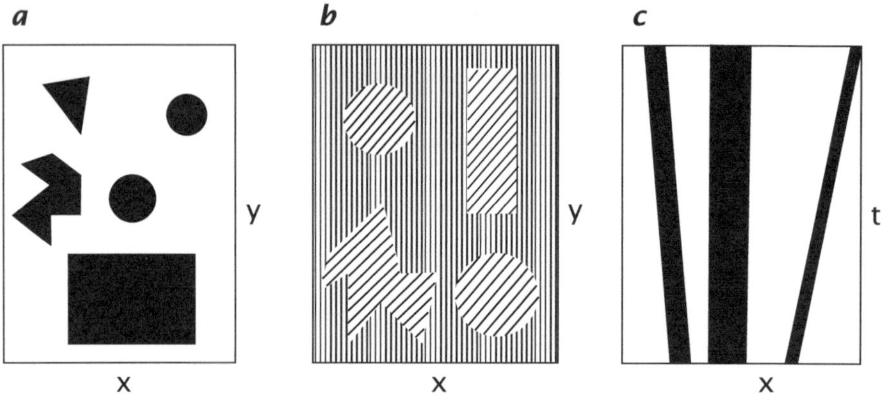

*Figure 13.2: Three interpretations of oriented local structures in 2D images: **a** Object edge; **b** Orientation of patterns; **c** Motion as orientation in a 2D place/time image.*

we will refer to as a simple neighborhood. As an introduction, we ask ourselves which types of patterns are suitable for distinguishing objects from the background with our visual system.

As Fig. 13.1 shows, we can recognize objects even if they do not differ from the background by their mean gray value, but only by the orientation or size of a pattern. To perform this task with a digital image processing system, we need operators that determine the orientation and size of patterns and thus convert a gray value image into a feature image. In a feature image, we can then easily separate patterns that differ in orientation or size.

We refer to local environments that can be described by an orientation as *simple neighborhoods*. The development of suitable operators to capture orientation and structure size is an important and necessary prerequisite for the analysis of *complex structures*. The meaning of one and the same local structure can be very different, as Fig. 13.2 shows for 2D images:

- In the simplest case, the object and background differ by the gray value (Fig. 13.2a). Then a gray value change in a local environment means that an *object edge* is present. An analysis of the orientation results in the edge orientation.

- In Fig. 13.2b the objects differ from the background by the orientation of their *pattern* or their *texture*. Local spatial structures do not represent an edge, but characterize the texture. We deal with texture analysis in Chapter 16.
- In image sequences, local structures in space-time are determined by movement. This is shown in Fig. 13.2c for a two-dimensional space-time image. Motion is an important feature that identifies objects like any other. It will be discussed in detail in Chapter 14.

Although the three examples refer to completely different image data, they have in common that the local structure is characterized by orientation, i. e. the local gray values only change in one direction. In this sense, the concept of orientation is just an extension of the concept of edges.

In this chapter, we first discuss the basic mathematical properties of simple neighborhoods in which the gray value structures only change in one direction (Section 13.2). This lays the foundation for the second step, which is to work out the image processing operators that make it possible to determine the orientation of these structures and at the same time determine whether the gray values actually only change in one direction. We will see that this requires tensor operators (Section 13.3). Then, in Section 13.4, we examine methods with which we can determine the characteristic structure size in a neighborhood. The decisive quantity here will be the phase of the signal, from which it is possible to determine a local wavenumber. In this section, we will also become familiar with methods that can be used to determine both the local orientation and the local wavenumber.

13.2 Properties of Simple Neighborhoods

13.2.1 Representation in the Spatial Domain

Local environments are best described mathematically using continuous rather than discrete functions. This approach has two significant advantages. First, it makes it easier to formulate concepts and investigate their properties analytically. As long as the sampled image satisfies the *sampling theorem* (Section 5.2.3), all obtained results remain valid as it is an exact representation of the continuous gray value function. Second, we can distinguish between errors related to the chosen approach and those due to the discretization.

A local environment with ideal local orientation is characterized by the fact that the gray value only changes in one direction. Since the gray values are constant along lines, local orientation is also referred to as *linear symmetry* [19]. The term *simple neighborhood* was coined by Granlund and Knutsson [115]. If we orient an axis of the coordinate system along the direction in which the gray values change, the gray values become a one-dimensional

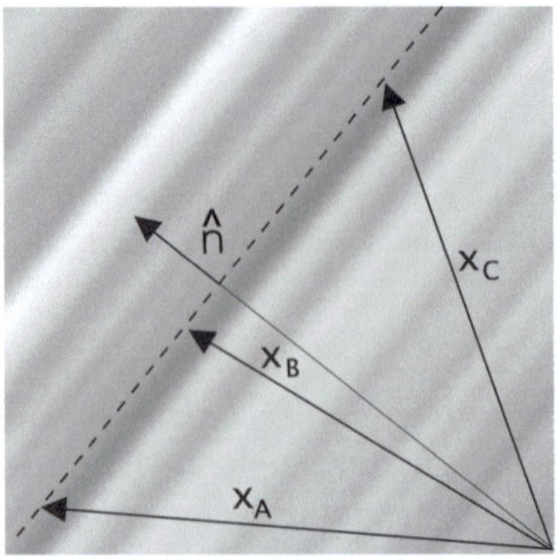

Figure 13.3: Illustration of a linearly symmetrical or simple neighborhood. The gray values only depend on one coordinate in the direction of the unit vector $\hat{\boldsymbol{n}}$.

function with only one coordinate. In general, we denote the direction of the local orientation with a unit vector $\hat{\boldsymbol{n}}$, which is perpendicular to the lines of the constant gray values. Then a simple neighborhood is mathematically represented as follows:

$$g(\boldsymbol{x}) = g(\boldsymbol{x}^T\hat{\boldsymbol{n}}), \tag{13.1}$$

where we write the scalar product as $\boldsymbol{x}^T\hat{\boldsymbol{n}}$ for the sake of simplicity. Equation Eq. (13.1) also applies to images of higher dimensions. The projection of the vector $\boldsymbol{x}$ onto the unit vector $\hat{\boldsymbol{n}}$ means that the gray values only depend on one scalar quantity, the coordinate in the direction of $\hat{\boldsymbol{n}}$ (Fig. 13.3). The correctness of this representation can be easily verified by calculating the gradient:

$$\nabla g(\boldsymbol{x}^T\hat{\boldsymbol{n}}) = \begin{bmatrix} \dfrac{\partial g(\boldsymbol{x}^T\hat{\boldsymbol{n}})}{\partial x_1} \\ \vdots \\ \dfrac{\partial g(\boldsymbol{x}^T\hat{\boldsymbol{n}})}{\partial x_D} \end{bmatrix} = \begin{bmatrix} \hat{n}_1 \dot{g}(\boldsymbol{x}^T\hat{\boldsymbol{n}}) \\ \vdots \\ \hat{n}_D \dot{g}(\boldsymbol{x}^T\hat{\boldsymbol{n}}) \end{bmatrix} = \hat{\boldsymbol{n}}\dot{g}(\boldsymbol{x}^T\hat{\boldsymbol{n}}). \tag{13.2}$$

We use $\dot{g}$ to denote the derivative of g with respect to the scalar quantity $\boldsymbol{x}^T\hat{\boldsymbol{n}}$. In the hyperplane perpendicular to the gradient, the values are locally constant. Equation Eq. (13.2) proves that the gradient is in the direction of $\hat{\boldsymbol{n}}$.

13.2.2 Representation in Fourier Space

A simple neighborhood also has a special form in Fourier space. To derive it, we assume in a first step that the entire image is described by Eq. (13.1) and therefore $\hat{\boldsymbol{n}}$ is independent of the position. Then we can conclude — based on the fact that a simple environment is constant in all directions except $\hat{\boldsymbol{n}}$ —

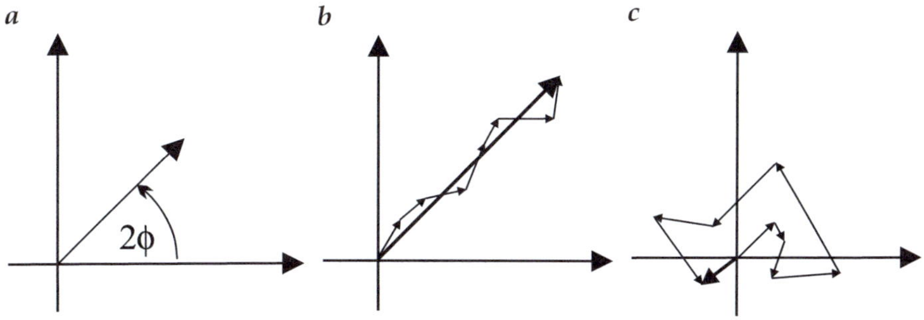

Figure 13.4: Vector representation of the local orientation: ***a*** *Orientation vector;* ***b*** *Averaging of the orientation vectors of a region with homogeneous orientation;* ***c*** *the same for a region with randomly distributed orientation.*

that the Fourier transform is restricted to one line. The direction of the line is given by $\hat{\boldsymbol{n}}$:

$$g(\boldsymbol{x}^T\hat{\boldsymbol{n}}) \quad \circ\!\!-\!\!\bullet \quad \hat{g}(\boldsymbol{\kappa})\delta(\boldsymbol{\kappa} - \hat{\boldsymbol{n}}(\boldsymbol{\kappa}^T\hat{\boldsymbol{n}})) = \hat{g}(\kappa_\parallel), \tag{13.3}$$

where $\kappa_\parallel$ is the coordinate in Fourier space in the direction of $\hat{\boldsymbol{n}}$. The argument in the δ function is only zero if the wavenumber vector $\boldsymbol{\kappa}$ is parallel to $\hat{\boldsymbol{n}}$. In a second step, we now limit Eq. (13.3) to a local environment by multiplying $g(\boldsymbol{x}^T\hat{\boldsymbol{n}})$ in the local space by a window function $w(\boldsymbol{x} - \boldsymbol{x}_0)$. We therefore select a local environment around $\boldsymbol{x}_0$ whose size and shape is determined by the window function. With a window function that gradually approaches zero, the influence of the pixels as a function of their distance from the central pixel disappears. A multiplication in local space corresponds to a convolution in Fourier space (Section 6.3). This results in

$$\boxed{w(\boldsymbol{x} - \boldsymbol{x}_0) \cdot g(\boldsymbol{x}^T\hat{\boldsymbol{n}}) \quad \circ\!\!-\!\!\bullet \quad \hat{w}(\boldsymbol{\kappa}) * \hat{g}(\boldsymbol{\kappa})\delta(\boldsymbol{\kappa} - \hat{\boldsymbol{n}}(\boldsymbol{\kappa}^T\hat{\boldsymbol{n}})),} \tag{13.4}$$

where $\hat{w}(\boldsymbol{\kappa})$ is the Fourier transform of the window function.

The limitation to a local environment therefore smears the line in the Fourier space into a sausage-like shape. Due to the reciprocity of the scales between the two spaces, their thickness is inversely proportional to the size of the window. Based on this elementary relationship, we can already qualitatively conclude that the accuracy of the determination of the orientation is directly related to the ratio of the window size to the wavelength of the smallest structures in the window.

13.2.3 Vector Representation of Local Neighborhoods

In order to correctly represent simple neighborhoods, we must first distinguish between the terms *orientation* and *direction*. The direction is defined over the entire angular range of 2π (360°). Two vectors that point in opposite directions, i. e. are rotated by 180° against each other, are considered to be different. For example, the gradient vector always points in the direction of increasing gray values. For a light object on a dark background, this means that the gradient at the edge points in the direction of the object.

In contrast, when describing the direction of a local environment, an angle range of 360° makes no sense. We cannot distinguish a pattern rotated by 180° from one that is not rotated, as the pattern still has the same direction. This means that the direction of a simple neighborhood is different from that of a gradient. While gradients pointing in opposite directions represent inconsistent information for object edges, they are consistent information for the direction of a simple neighborhood.

To distinguish the two types of direction, we speak of *orientation* in all cases where only an angular range of 180° is required. Of course, the orientation is still a *cyclic* quantity. If you increase the orientation beyond 180° it starts again at 0°. Therefore, a correct representation of the orientation requires a doubling of the angle.

After the basic discussion of the representation of the orientation, we ask about the appropriate representation of a simple neighborhood. Obviously, a simple scalar with the doubled orientation angle is not sufficient. We also need a coefficient of determination that describes how well the environment corresponds to a simple neighborhood. Both pieces of information, the direction and the coefficient of determination, can be combined to form a vector: The magnitude of the vector is the coefficient of determination and its direction is the doubled orientation angle (Fig. 13.4a). This vector representation of the orientation has two decisive advantages:

Firstly, it is better suited for further processing than a separate representation with two scalar quantities. Vectors are summed by appending them together so that the resulting sum vector extends from the starting point of the first vector to the end point of the last vector (Fig. 13.4b). The weight of an individual vector in the vector sum is determined by its length. The coefficient of determination of the orientation measurement is thus adequately taken into account. The vector representation of the orientation also has favorable averaging properties. In a region of homogeneous orientation, the vectors result in a large vector, i. e. a reliable estimate of the orientation (Fig. 13.4b). In a region with randomly distributed orientation, the resulting vector remains small because there is no significant local orientation (Fig. 13.4c).

On the other hand, it is difficult to represent orientation as a gray value image. While orientation is a cyclic quantity, the gray value representation has an unnatural jump between the smallest and the largest angle. This jump in orientation images does not give a useful impression of the distribution of the orientation. However, the orientation vector can be displayed well as a color image.

Intuitively, we assign the coefficient of determination of the brightness and the orientation angle to the color tone. Our attention is then drawn to the bright parts of the image where we can easily distinguish colors. The darker the image, the more difficult it is to visually distinguish the individual colors. In this way, our visual impression matches the certainty of the orientation determination in the image.

13.3 First-order Tensor Representation

13.3.1 The Structure Tensor

The vector representation discussed in Section 13.2.3 is not yet complete; it is only sufficient for representing the orientation of simple neighborhoods. However, it is not possible to distinguish between neighborhoods with constant gray values and those with isotropic orientation distribution, such as uncorrelated noise. In both cases, an orientation vector with a magnitude of zero results.

It is therefore obvious that an adequate representation of the changes in a local environment is of a complex nature. Such a representation should be able to determine the preferred direction of the gray value changes (given by a unit vector $\hat{\boldsymbol{n}}$) and distinguish constant environments from those without local orientation.

With the following optimization strategy for determining the orientation of a simple neighborhood, we arrive at a suitable representation. The optimal orientation is defined as the orientation that shows the least deviation from the gradient direction. A suitable measure for the deviation must equally consider gradients pointing in opposite directions. This criterion is fulfilled by the squared scalar product of the gradient vector $\boldsymbol{\nabla} g$ and the unit vector $\hat{\boldsymbol{n}}$, which reflects the local orientation:

$$(\boldsymbol{\nabla} g^T \hat{\boldsymbol{n}})^2 = |\boldsymbol{\nabla} g|^2 \cos^2\left(\angle(\boldsymbol{\nabla} g, \hat{\boldsymbol{n}})\right). \tag{13.5}$$

This quantity is proportional to the squared cosine of the angle between the gradient vector and the orientation vector and is therefore maximum if $\boldsymbol{\nabla} g$ and $\hat{\boldsymbol{n}}$ are parallel or antiparallel and zero if they are perpendicular to each other. In a D-dimensional local environment, the following expression must therefore be maximized:

$$\int w(\boldsymbol{x} - \boldsymbol{x}')\left(\boldsymbol{\nabla} g(\boldsymbol{x}')^T \hat{\boldsymbol{n}}\right)^2 \mathrm{d}^D x', \tag{13.6}$$

where the window function w determines the size and shape of the environment around a point $\boldsymbol{x}$ in which the orientation is averaged. The maximization problem must be solved for each point $\boldsymbol{x}$. Equation Eq. (13.6) can be written as follows:

$$\hat{\boldsymbol{n}}^T \boldsymbol{J} \hat{\boldsymbol{n}} \rightarrow \text{maximum} \tag{13.7}$$

with

$$\boldsymbol{J} = \int w(\boldsymbol{x} - \boldsymbol{x}')\left(\boldsymbol{\nabla} g(\boldsymbol{x}') \boldsymbol{\nabla} g(\boldsymbol{x}')^T\right) \mathrm{d}^D x',$$

where $\nabla g \nabla g^T$ denotes the outer or Cartesian product. The components of this symmetric $D \times D$ tensor, the *structure tensor*, are

$$J_{pq}(\boldsymbol{x}) = \int_{-\infty}^{\infty} w(\boldsymbol{x} - \boldsymbol{x}') \left(\frac{\partial g(\boldsymbol{x}')}{\partial x'_p} \frac{\partial g(\boldsymbol{x}')}{\partial x'_q} \right) \mathrm{d}^D x'. \tag{13.8}$$

These equations prove that a tensor is an adequate first-order representation of a local neighborhood. The addition "first order" has a double meaning. On the one hand, only first-order derivatives are involved. On the other hand, only simple neighborhoods can be described. More complex structures such as those with multiple orientations cannot be distinguished.

The complexity of Eqs. (13.7) and (13.8) somewhat hides the simple meaning. The tensor is symmetrical and can be brought into a diagonal form by rotating the coordinate system. Then Eq. (13.7) simplifies in the two-dimensional case to

$$J = \left[\hat{n}'_1, \hat{n}'_2 \right] \begin{bmatrix} J'_{11} & 0 \\ 0 & J'_{22} \end{bmatrix} \begin{bmatrix} \hat{n}'_1 \\ \hat{n}'_2 \end{bmatrix} \quad \rightarrow \quad \text{maximum.} \tag{13.9}$$

A unit vector $\hat{\boldsymbol{n}}' = [\cos\theta \;\; \sin\theta]$ in the direction θ results in the values

$$J = J'_{11} \cos^2\theta + J'_{22} \sin^2\theta.$$

Without limiting the generality, we can assume that $J'_{11} > J'_{22}$. Then it is obvious that the unit vector $\hat{\boldsymbol{n}}' = [1 \; 0]^T$ maximizes the expression in Eq. (13.9) with the maximum value J'_{11}.

This approach not only leads to a tensor representation for local environments, but also shows a way to determine the orientation. Essentially, we solve the so-called *eigenvalue problem*. Eigenvalues λ_d and eigenvectors $\boldsymbol{e}_d$ of a $D \times D$ matrix $\boldsymbol{J}$ are defined by the following equation:

$$\boldsymbol{J}\boldsymbol{e}_d = \lambda_d \boldsymbol{e}_d. \tag{13.10}$$

An eigenvector $\boldsymbol{e}_d$ of $\boldsymbol{J}$ is therefore a vector that is not rotated by multiplication with $\boldsymbol{J}$, but only multiplied by a scalar factor, the eigenvalue λ_d. This makes it clear that the coordinate system in which the structure tensor is diagonal Eq. (13.9) is spanned by the eigenvectors. For our further discussion, it is important that we are aware of the following basic properties of eigenvalues and eigenvectors of a symmetric matrix:

1. Eigenvalues are always real and
2. eigenvectors form an orthogonal basis.

According to the *maximization problem* formulated here, the eigenvector of the maximum eigenvalue gives the orientation of the local environment.

Table 13.1: Classification of the eigenvalue of the structure tensor in 2D images.

Condition	Rank(J)	Explanation
$\lambda_1 = \lambda_2 = 0$	0	Both eigenvalues are zero. The mean squared magnitude of the gradient ($\lambda_1 + \lambda_2$) is zero. The local environment is constant.
$\lambda_1 > 0, \lambda_2 = 0$	1	An eigenvalue is zero. The values therefore do not change in the direction of the associated eigenvector. The local environment is a simple neighborhood with ideal orientation.
$\lambda_1 > 0, \lambda_2 > 0$	2	Both eigenvalues are not equal to zero. The gray values change in all directions. In the special case $\lambda_1 = \lambda_2$ we speak of an isotropic gray value structure, as it changes uniformly in all directions.

13.3.2 Classification of the Eigenvalues

The power of the tensor approach becomes clear when we classify the eigenvalues of the structure tensor. The classification criterion is the number of eigenvalues that are zero. If an eigenvalue is zero, this means that the gray values do not change in the direction of the corresponding eigenvector.

The number of eigenvalues with the value zero is closely related to the rank of a matrix, which is defined as the dimension of the subspace for which $\boldsymbol{J}\boldsymbol{x} \neq \boldsymbol{0}$. The space for which $\boldsymbol{J}\boldsymbol{x} = \boldsymbol{0}$ is called *null space*. The dimension of the null space is the dimension of the row and column vectors of the matrix minus their rank; it is equal to the number of null eigenvalues. Which cases occur in two or three dimensions and what they mean are summarized in Tables 13.1 and 13.2.

In practice, it is not checked whether the eigenvalues are zero, but only whether they are below a critical limit, which is determined by the noise level in the image. We learned how to measure the temporal noise of cameras in Section 4.5.4. The application is complicated by the strong gray value dependence of the noise. It can therefore be useful to make the noise independent of the gray value using a simple non-linear gray value transformation (*noise equivalization*, see Section 7.2.4).

13.3.3 Orientation Vector

Here we solve the eigenvalue problem to determine the orientation vector. In two dimensions, the orientation angle is determined by rotating the structure tensor on the principal axis system. To do this, we rotate the coordinate system counterclockwise by an unknown angle θ. The coordinates of a

Table 13.2: *Classification of the eigenvalue of the structure tensor in 3D (volume) images.*

Condition	Rank(J)	Explanation
$\lambda_1 = \lambda_2 = \lambda_3 = 0$	0	The gray values do not change; there is a constant environment.
$\lambda_1 > 0, \lambda_2 = \lambda_3 = 0$	1	The gray values only change in one direction, which is determined by the eigenvector of the non-zero eigenvalue. In the environment we have boundaries between two objects or a layered texture. In a space-time image, this means the constant movement of a spatially oriented pattern (planar wave).
$\lambda_1 > 0, \lambda_2 > 0, \lambda_3 = 0$	2	The gray values change in two directions and are constant in a third. The eigenvector of the eigenvalue with the value zero indicates the direction of the constant gray values.
$\lambda_1 > 0, \lambda_2 > 0, \lambda_3 > 0$	3	The gray values change in all three directions.

vector in the rotated system are then

$$x' = Rx \quad \text{or} \quad \begin{bmatrix} x' \\ y' \end{bmatrix} = \begin{bmatrix} \cos\theta & \sin\theta \\ -\sin\theta & \cos\theta \end{bmatrix} \begin{bmatrix} x \\ y \end{bmatrix}$$

and for the structure tensor the result is $J' = RJR^T$

$$\begin{bmatrix} \lambda_1 & 0 \\ 0 & \lambda_2 \end{bmatrix} = \begin{bmatrix} \cos\theta & \sin\theta \\ -\sin\theta & \cos\theta \end{bmatrix} \begin{bmatrix} J_{11} & J_{12} \\ J_{12} & J_{22} \end{bmatrix} \begin{bmatrix} \cos\theta & -\sin\theta \\ \sin\theta & \cos\theta \end{bmatrix}.$$

Taking into account the trigonometric identities $\sin 2\theta = 2\sin\theta\cos\theta$ and $\cos 2\theta = \cos^2\theta - \sin^2\theta$, performing the matrix multiplications results in the following expression:

$$\begin{bmatrix} \lambda_1 & 0 \\ 0 & \lambda_2 \end{bmatrix} =$$

$$\begin{bmatrix} \cos\theta & \sin\theta \\ -\sin\theta & \cos\theta \end{bmatrix} \begin{bmatrix} J_{11}\cos\theta + J_{12}\sin\theta & -J_{11}\sin\theta + J_{12}\cos\theta \\ J_{22}\sin\theta + J_{12}\cos\theta & J_{22}\cos\theta - J_{12}\sin\theta \end{bmatrix} =$$

$$\begin{bmatrix} J_{11}\cos^2\theta + J_{22}\sin^2\theta - J_{12}\sin 2\theta & 1/2(J_{22}-J_{11})\sin 2\theta + J_{12}\cos 2\theta \\ 1/2(J_{22}-J_{11})\sin 2\theta + J_{12}\cos 2\theta & J_{11}\sin^2\theta + J_{22}\cos^2\theta + J_{12}\sin 2\theta \end{bmatrix}. \tag{13.11}$$

Now we can compare the matrix coefficients on the left and right sides of the equation. Since the matrices are symmetrical, we have three equations with the three unknowns θ, λ_1 and λ_2. Although the system of equations is non-linear, it can easily be solved for the angle θ. A comparison of the non-diagonal elements on both sides initially results in

$$1/2(J_{22} - J_{11}) \sin 2\theta + J_{12} \cos 2\theta = 0 \tag{13.12}$$

and after elementary transformations the orientation angle is

$$\boxed{\tan 2\theta = \frac{2J_{12}}{J_{11} - J_{22}}.} \tag{13.13}$$

Without us having specified any conditions, the expected angle doubling for the orientation has occurred. Since $\tan 2\theta$ results from a quotient, we can consider the dividend as the y component and the divisor as the x component of a vector. From this we can form the vectorial orientation operator $\boldsymbol{o}$ as introduced by Granlund [114]:

$$\boxed{\boldsymbol{o} = \begin{bmatrix} J_{11} - J_{22} \\ 2J_{12} \end{bmatrix}. \quad \text{orientation vector}} \tag{13.14}$$

The argument of *orientation vector* gives the orientation angle and the magnitude a degree of certainty for local orientation.

The result of Eq. (13.14) is remarkable because the calculation of the components of the orientation vector from those of the orientation tensor only requires a subtraction and a multiplication by two. Since these components of the orientation vector are all we need for the further processing steps, this makes solving the eigenvalue problem in two dimensions trivial.

13.3.4 Coherence

The orientation vector reduces a local structure to a local orientation. Only two of the three independent components of the symmetric tensor are used. If we cannot determine an oriented structure in a neighborhood, we do not know whether we are dealing with constant gray values or distributed orientations. This information is contained in the unused sum of the diagonal elements of the tensor, $J_{11} + J_{22}$, which provides the mean squared magnitude of the gradient. From this we conclude that a good structure operator must also take the third component into account. The following equation represents a suitable linear combination:

$$\begin{bmatrix} J_{11} + J_{22} \\ J_{11} - J_{22} \\ 2J_{12} \end{bmatrix}. \tag{13.15}$$

This structure operator contains the two components of the orientation vector and additionally the mean squared magnitude of the gradient, which is a rotation invariant parameter. If we compare the magnitude squared of the gradient with the magnitude of the orientation vector, we can distinguish between a range of constant gray values and an isotropic gray value structure without a preferred direction. In the first case, both squared values are zero, in the second only the magnitude of the orientation vector. If we have a perfectly oriented pattern, both values are equal. The ratio between these quantities seems to be a good *coherence measure* c_c for local orientation:

$$c_c = \frac{|o|}{|\nabla g|^2} = \frac{\sqrt{(J_{11} - J_{22})^2 + 4J_{12}^2}}{J_{11} + J_{22}} = \frac{\lambda_1 - \lambda_2}{\lambda_1 + \lambda_2} \quad \text{coherence measure.} \tag{13.16}$$

The coherence c_c varies between zero and one. It is one for an ideal orientation ($\lambda_2 = 0, \lambda_1 > 0$), but zero for an isotropic gray value structure ($\lambda_1 = \lambda_2 > 0$).

13.3.5 Color Coding of the 2D Structure Tensor

At the end of Section 13.2.3 we discussed a color representation of the orientation vector. The question arises whether it is also possible to represent the structure tensor as a color image. A symmetric 2D tensor has three independent pieces of information that fit well with the three degrees of freedom we have available for color, for example intensity, hue and saturation.

For a color representation of the structure tensor, we only need two small changes compared to the orientation vector. First, the magnitude squared of the gradient is mapped to the intensity instead of the length of the orientation vector. Then the coherence measure Eq. (13.16) is used for the saturation. In the color representation of the orientation vector, the saturation is always one. Its angle is displayed as a hue.

A slight modification of this color representation is useful for practical applications. The squared value of the gradient shows too large variations to be displayed in the small dynamic range of a screen with only 256 brightness levels. A corresponding normalization is therefore necessary. The basic idea of this normalization is to compare the squared amount of the gradient with the noise level. If the gradient is clearly above the noise level, it is regarded as significant information. This suggests the following normalization of the luminance I:

$$I = \frac{J_{11} + J_{22}}{(J_{11} + J_{22}) + \gamma\sigma_n^2}, \tag{13.17}$$

where σ_n is an estimate of the standard deviation of the noise level. This normalization causes a rapid transition of the illuminance from one if the magnitude of the gradient is greater than σ_n to zero if the gradient is less than σ_n. The factor γ is used to optimize the display.

In summary, we can therefore state that the following three parameters of the structure tensor best describe a simple local neighborhood:

$$\boxed{s = \begin{bmatrix} I\left(|\nabla g|^2\right) \\ |o| / |\nabla g|^2 \\ \theta \end{bmatrix}}. \tag{13.18}$$

The first two parameters, the normalized magnitude of the gray value gradient and the coherence measure, are rotation invariant and have values between 0 and 1, while the third, the orientation vector, specifies the orientation of the local neighborhood. With the help of the definition of the orientation vector, a simple relationship for the eigenvalues can also be specified from the non-linear system of equations (13.11) to

$$\lambda_{1,2} = \left(|\nabla g|^2 \pm |o|\right)/2, \tag{13.19}$$

Example 13.1: Structural tensor with ideal local structures

To illustrate the structure tensor, we calculate the values of all parameters for a series of example structures. We assume either ideally oriented in different directions or an isotropic local structure. The trace of the structure tensor, i.e. the averaged magnitude squared of the gradient, has the same value $s = J_{11} + J_{22}$. This results in the values in the following table:

θ	J_{11}	J_{22}	J_{12}	o_x	o_y	$\|o\|$	c_c	λ_1	λ_2
0°	s	0	0	s	0	s	1	s	0
45°	$s/2$	$s/2$	$s/2$	0	s	s	1	s	0
90°	0	s	0	$-s$	0	s	1	s	0
—	$s/2$	$s/2$	0	0	0	0	0	$s/2$	$s/2$

13.3.6 Implementation

The structure tensor (Section 13.3.1) can be calculated directly from a combination of *linear convolution* and *nonlinear point operations*. The partial derivatives in Eqs. (13.8) and (13.28) are approximated by discrete derivative operators. The integration weighted with the window function is replaced by a convolution with a smoothing filter. If we denote the discrete partial *derivation operator* in the direction of the coordinate p by $\mathcal{D}_p$ and the isotropic *smoothing operator* by $\mathcal{B}$, we can calculate the local structure of a gray value image with the following structure tensor operator:

$$\mathcal{J}_{pq} = \mathcal{B}(\mathcal{D}_p \cdot \mathcal{D}_q). \tag{13.20}$$

Pixel-by-pixel multiplication is denoted by $\cdot$ to distinguish it from the sequential application of convolution operators. Equation (13.20) literally means: The component $\mathcal{J}_{pq}$ of the tensor is calculated by separately convolving the image with $\mathcal{D}_p$ and $\mathcal{D}_q$, multiplying the two convolved images pixel by pixel and smoothing the resulting image with $\mathcal{B}$.

These operators apply to images of any dimension $D \geq 2$. In a D-dimensional image, the structure tensor $D(D+1)/2$ has independent components, three in 2D images, six in 3D images and ten in 4D images. These components are best stored in a multi-channel image with $D(D+1)/2$ components.

The smoothing operators require the largest number of operations. Therefore, we first need a fast smoothing algorithm for an efficient implementation. The basis for this is the general observation that higher order properties always show a lower resolution than the properties from which they were calculated. This means that a structure tensor can be stored on a coarser grid and thus in a smaller image. A convenient and appropriate subsampling is to store only every second pixel in every second line. Using this method, we automatically arrive at multi-grid data structures, which we will discuss in detail in Chapter 15. Multi-step averaging was discussed in Section 11.5.1.

In addition to saving memory space, saving higher-order properties on coarser scales has the further important advantage that fewer pixels have to be processed in each subsequent step. A linear size reduction to half leads to a reduction of the pixels and the calculation steps to 1/4 in the 2D case and to 1/8 in the 3D case.

Figure 13.5 shows the steps for calculating a structure tensor and the sizes calculated from it using the ring test pattern. This test pattern is particularly suitable for orientation analysis, as all orientations and wavenumbers are contained in one image. The color representations of the orientation analysis in Fig. 13.8 are of course much more intuitive. The colors are selected so that perpendicular structures are shown in complementary colors. Horizontal structures are shown in red and vertical structures in green.

Further application examples are shown in Fig. 13.6. Here, the lettering "IMAGE" was made visible either by different gray values or differently oriented patterns. In the color representation (Fig. 13.6c) of the structure tensor, corners with vertically abutting edges appear white. Individual pixels with deviating gray values, which are barely visible in the original image (Fig. 13.6d), are immediately noticeable as white dots in the blue background in the color representation of the structure tensor (Fig. 13.6f).

13.3.7 Error Analysis

The accuracy of the orientation angle depends on the implementation of the derivative filter. The direct implementation of the algorithm using the symmetric difference operator (Section 12.4.3) or the *Sobel operator* (Section 12.7.3) results in a surprisingly high angular error (Fig. 13.7a) of more than 7° at a wavenumber $\tilde{k} = 0.7$. The error depends on both the wavenumber and the orientation of the local structure. The high error and the structure of the error pattern result from the transfer function of the derivative filter, as this

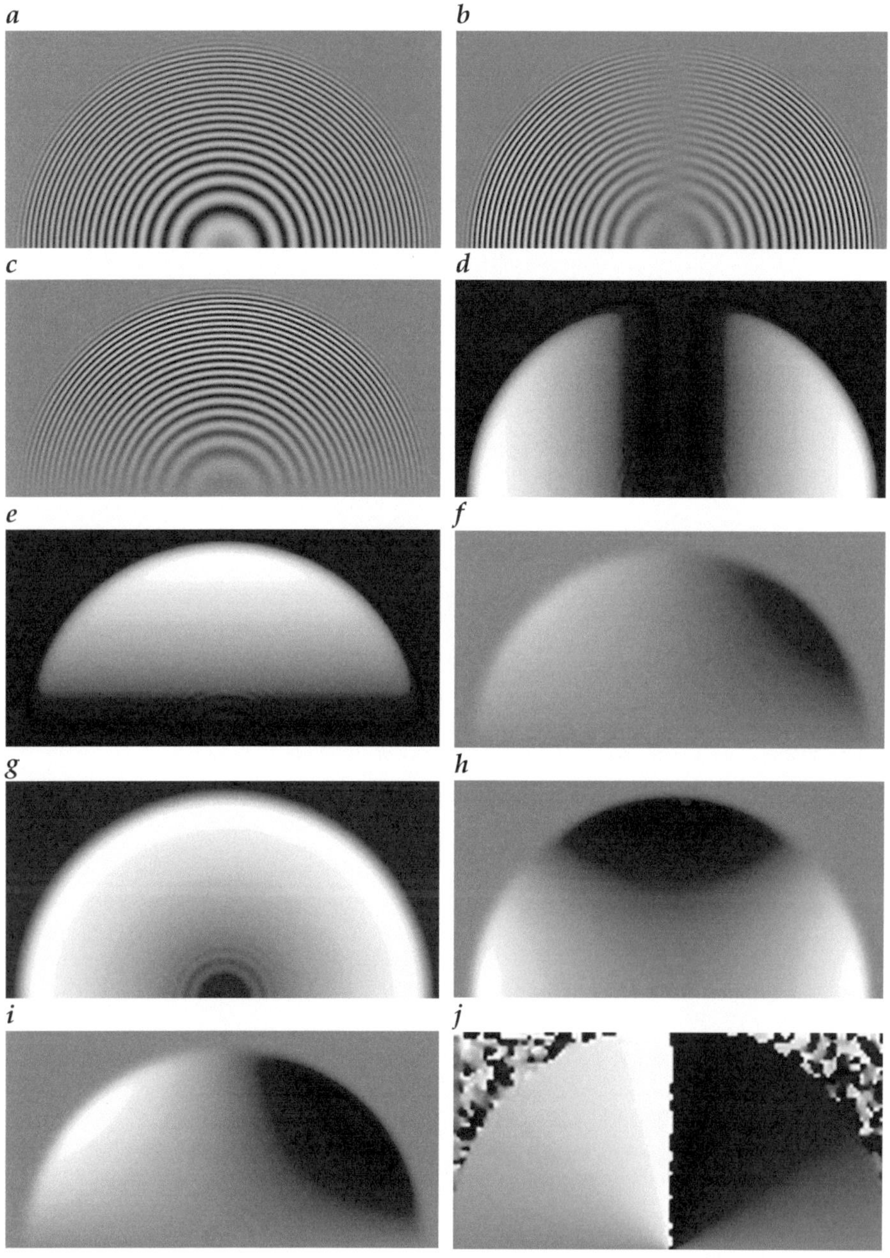

Figure 13.5: *Steps for calculating the structure tensor:* ***a*** *Original image;* ***b*** *horizontal derivative* $\mathcal{D}_x$; ***c*** *vertical derivative* $\mathcal{D}_y$; ***d – f*** *averaged components of the structure tensor:* $J_{11} = \mathcal{B}(\mathcal{D}_x \cdot \mathcal{D}_x)$, $J_{22} = \mathcal{B}(\mathcal{D}_y \cdot \mathcal{D}_y)$, $J_{12} = \mathcal{B}(\mathcal{D}_x \cdot \mathcal{D}_y)$; ***g*** *squared amount of the gradient* $J_{11} + J_{22}$; ***h*** *x-component of the orientation vector* $J_{11} - J_{22}$, ***i*** *y-component of the orientation vector* $2J_{12}$ *and* ***j*** *orientation angle of* $[-\pi/2, \pi/2]$, *mapped to the gray value range* $[0, 255]$.

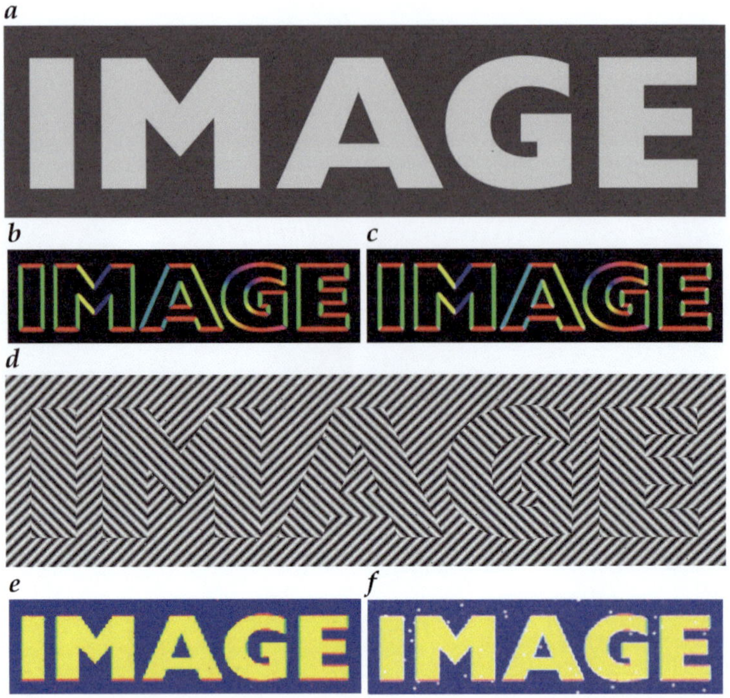

Figure 13.6: *Orientation analysis on **a** edges and **d** oriented patterns: **b** and **e** a color representation of the orientation vector and **c** and **f** of the structure tensor.*

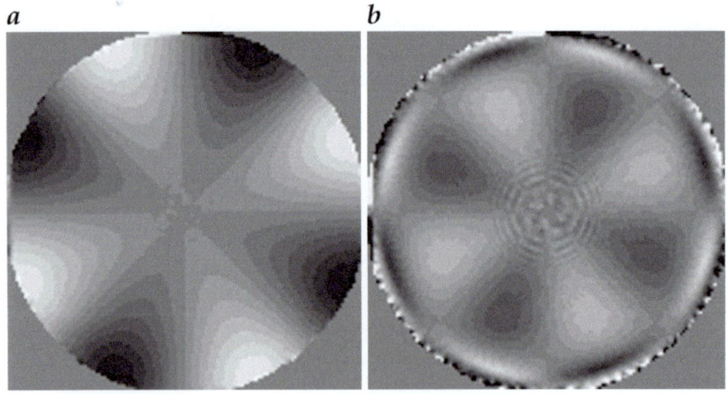

Figure 13.7: *Systematic errors in the determination of the orientation angle when using different derivative operators with the ring test pattern, which has wavenumbers of maximum $\underset{\sim}{k} = 0.7$. Error maps when using **a** of the Sobel operator (angular range ±7° in 16 gray value steps) and **b** of the optimized Scharr operator with ten times smaller angular range ±0.7° in 16 gray value steps.*

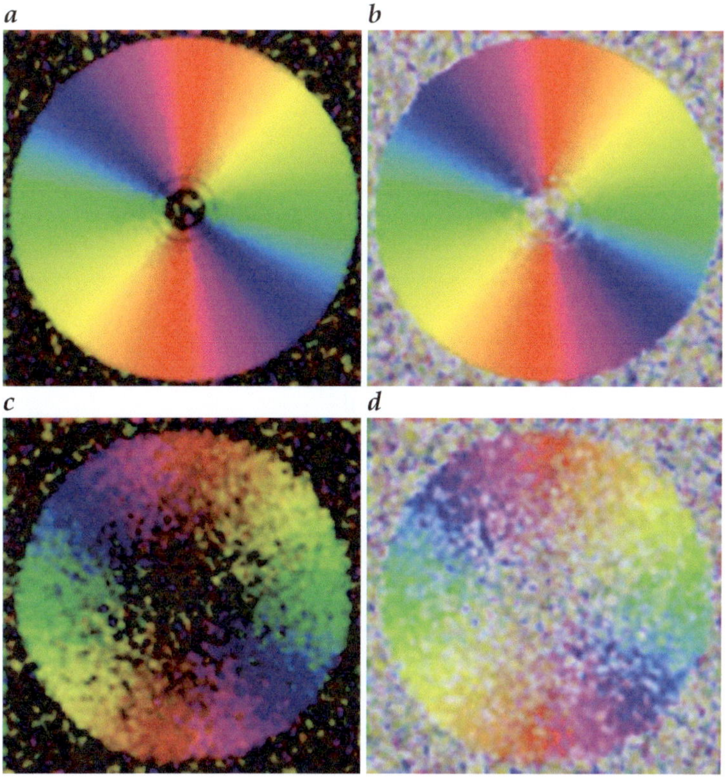

Figure 13.8: *Color representation of the orientation analysis with noisy ring test patterns:* ***a*** *and* ***b*** *with an amplitude of the ring pattern of 100 and normally distributed noise with a standard deviation of 10,* ***c*** *and* ***d*** *with an amplitude of the ring pattern of 50 and normally distributed noise with a standard deviation of 50,* ***a*** *and* ***c*** *show a color representation of the orientation vector (Section 13.2.3),* ***b*** *and* ***d*** *of the structure tensor (Section 13.3.5).*

shows significant deviations from the transfer function of an ideal derivative filter at high wavenumbers (Section 12.3). According to Eq. (13.13), the orientation angle depends on the ratio of the derivatives. Along the axes, one of the derivatives is zero and therefore no error occurs there. Along the diagonals, the derivatives are the same in the x and y directions; therefore, the error is canceled out in the ratio of the derivatives.

The error of the orientation angle can be significantly reduced by using better derivative filters. Fig. 13.7 compares the error in estimating the orientation using the example of the *Sobel operator* (Section 12.7.3) with the optimized *Scharr operator* (Section 12.7.6). By using this optimized 3×3 gradient filters of the same size, the angular error is reduced by slightly more than one order of magnitude to less than $0.5°$. The explanations in Sections 12.4–12.7 on the various derivative filters allow the computational effort and accuracy to be weighed up.

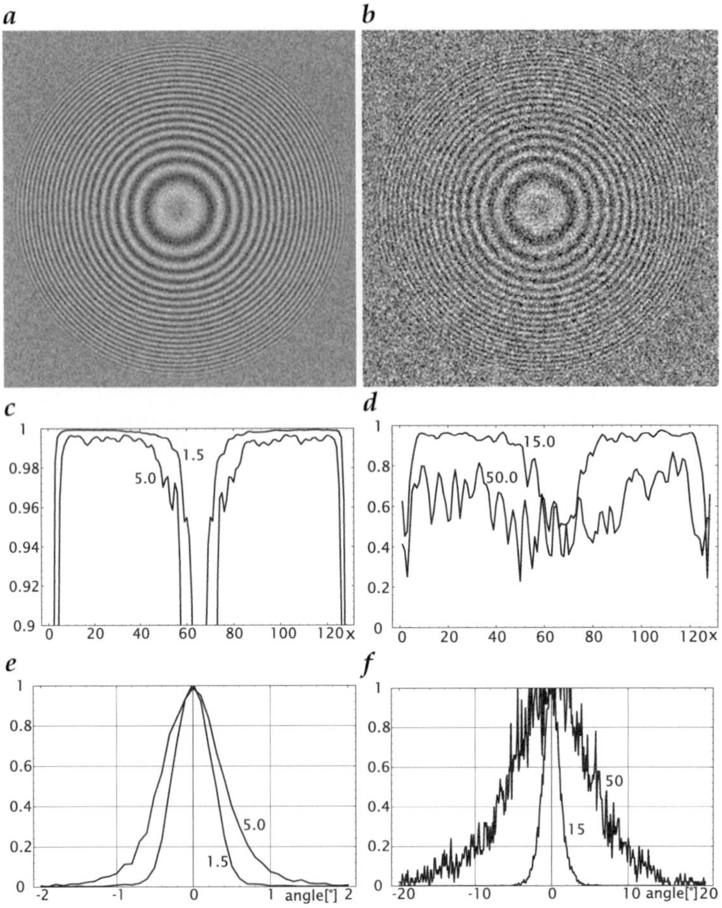

Figure 13.9: *Orientation analysis with a noisy ring test pattern using the optimized Sobel operator: Ring pattern with an amplitude of 50 and normally distributed noise with a standard deviation of* ***a*** *15 and* ***b*** *50.* ***c*** *and* ***d*** *radial section through the coherence measure of the ring pattern with standard deviations of the normally distributed noise of 1.5 and 5 and 15 and 50, respectively;* ***e*** *and* ***f*** *histograms of the angular error for the same conditions.*

An important property of every image processing algorithm is its *robustness*. This term refers to the algorithm's insensitivity to noise or other disturbances. Two questions are important in this context: Firstly, how large is the statistical error of the estimated properties in a noisy image? Secondly, are the values still correct?

To answer the first question, the laws of statistics are used for error propagation. They state that noise only makes the estimation more uncertain, but not more erroneous. The mean value is still correct provided that we make a sufficiently large number of estimates. In a noisy image, however, an operator can also deliver erroneous results, so that the mean value deviates significantly from the correct value. In the worst case — to answer the second question — an algorithm can even become unstable and produce nonsensical results.

Figures 13.8 and 13.9 illustrate that determining the orientation with the structure tensor is a particularly robust algorithm. Even with a low signal-to-noise ratio of one, the orientation is still estimated correctly without systematic deviation when using the optimized derivative operator. As the noise increases, the *coherence* (Section 13.3.4) decreases (Fig. 13.9c and d). The statistical error of the orientation angle estimate increases, which can be recognized by the increasing broadening of the distribution of the angular errors (Fig. 13.9e and f). Nevertheless, even if the standard deviation of the noise is as large as the amplitude of the ring pattern, and thus the *signal-to-noise ratio* is just one, the half-width of the distribution of the angular error is only about 5°. If you look at the original image in Fig. 13.9b, it is difficult to see the ring pattern in the finer rings. This makes orientation analysis with the structure tensor one of the few examples where an image processing operation is at least as good as the human visual system.

13.3.8 ♯Other Tensor Representation

In this section, we examine an alternative approach to describing local image structures using tensors. This approach is based on observations in Fourier space. We start by considering a structure ideally oriented according to Eq. (13.1) in wavenumber space. To easily calculate the Fourier transform of Eq. (13.1), we rotate the x_1 axis of the coordinate system in the direction of $\hat{\boldsymbol{n}}$. Then the gray value function is constant in the x_2 direction and the Fourier transform is reduced to a δ line in the direction of $\hat{\boldsymbol{n}}$ (≻ R13).

Determining the local orientation in Fourier space sounds promising, because all we have to calculate is the orientation of the line where the spectral densities are non-zero. Bigün and Granlund [19] suggest the following steps:

- A small neighborhood is selected from the image using a window function.
- The image multiplied by the window function is transformed. The smaller the selected window, the more blurred the calculated energy spectrum (*blur relation*, Theorem B.7, p. 633 and Section 13.2.2). Even an ideal local orientation results in a band-shaped distribution of spectral energy.
- By determining a regression line through the spectral density distribution, the direction of the local orientation can then be determined from the slope of the line.

The decisive step in this procedure is the calculation of the regression line. Since this problem is generally overdetermined, it cannot be solved exactly; we can only minimize the error measure. This is an example of a discrete inverse problem (Appendix C.6), which we solve here with the

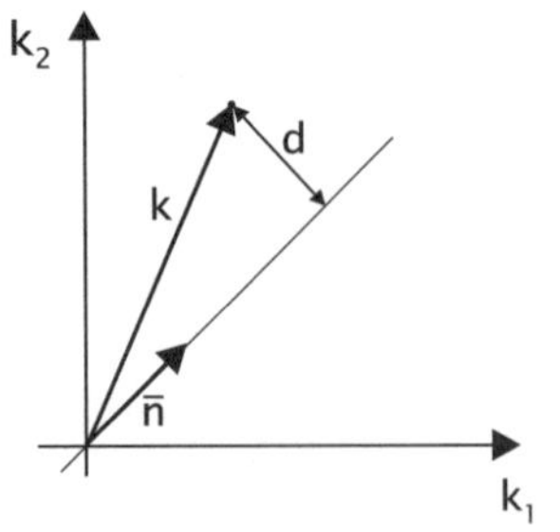

Figure 13.10: Distance of a point in wavenumber space from the straight line in the direction of the unit vector $\hat{\boldsymbol{n}}$.

standard error measure, the magnitude squared of the deviations, the L_2 norm Eq. (C.87).

For the optimal straight line, we minimize the sum of the squares of the distances of all given points to the straight line:

$$\int_{-\infty}^{\infty} d^2(\boldsymbol{\kappa}, \hat{\boldsymbol{n}}) |\tilde{g}(\boldsymbol{\kappa})|^2 \mathrm{d}^D k \rightarrow \text{minimum}. \tag{13.21}$$

The distance function is abbreviated as $d(\boldsymbol{\kappa}, \hat{\boldsymbol{n}})$. The integral runs over the entire wavenumber space; the wavenumbers are weighted with the spectral density $|\tilde{g}(\boldsymbol{\kappa})|^2$. Equation Eq. (13.21) is not limited to two dimensions, but is generally valid for local orientation or linear symmetry in a D-dimensional space.

The distance vector $\boldsymbol{d}$ can be derived from Fig. 13.10:

$$\boldsymbol{d} = \boldsymbol{\kappa} - (\boldsymbol{\kappa}^T \hat{\boldsymbol{n}})\hat{\boldsymbol{n}}. \tag{13.22}$$

The square of the distance then results in

$$|\boldsymbol{d}|^2 = |\boldsymbol{\kappa} - (\boldsymbol{\kappa}^T \hat{\boldsymbol{n}})\hat{\boldsymbol{n}}|^2 = |\boldsymbol{\kappa}|^2 - (\boldsymbol{\kappa}^T \hat{\boldsymbol{n}})^2. \tag{13.23}$$

To express the distance more clearly as a function of the vector $\hat{\boldsymbol{n}}$, we rewrite it as follows:

$$|\boldsymbol{d}|^2 = \hat{\boldsymbol{n}}^T (\boldsymbol{I}(\boldsymbol{\kappa}^T \boldsymbol{\kappa}) - (\boldsymbol{\kappa}\boldsymbol{\kappa}^T))\hat{\boldsymbol{n}}, \tag{13.24}$$

where $\boldsymbol{I}$ is the unit matrix. If we substitute this expression into Eq. (13.21), we get

$$\hat{\boldsymbol{n}}^T \boldsymbol{J}' \hat{\boldsymbol{n}} \rightarrow \text{minimum}, \tag{13.25}$$

where $\boldsymbol{J}'$ is a symmetrical tensor with the diagonal elements

$$J'_{pp} = \sum_{q \neq p} \int_{-\infty}^{\infty} k_q^2 |\tilde{g}(\boldsymbol{\kappa})|^2 \mathrm{d}^D k \tag{13.26}$$

Table 13.3: Eigenvalue classification of the inertia tensor in 3D images.

Condition	explanation
Ideal local orientation	The body of rotation is a line. When rotating around this line, the inertia disappears. It follows that the eigenvector of the eigenvalue with the value zero coincides with the direction of the line. The other eigenvector is orthogonal to this line and the associated eigenvalue is not equal to zero. This results in the axis of rotation with maximum inertia.
Isotropic gray value structure	In this case, the rotational body is a kind of flat, isotropic disk without a preferred direction. Both eigenvalues are equal and the inertia is the same for rotations around all axes. There is no minimum.
Constant gray values	The body of revolution degenerates to a point at the origin of the wavenumber space. The inertia is zero for rotations around each axis and both eigenvalues disappear.

and the secondary diagonal elements

$$J'_{pq} = -\int_{-\infty}^{\infty} k_p k_q |\tilde{g}(\boldsymbol{\kappa})|^2 \mathrm{d}^D k, \quad p \neq q. \tag{13.27}$$

The tensor $\boldsymbol{J}'$ is very similar to an important physical quantity, the *inertia tensor*. If we replace the wavenumber coordinates with spatial coordinates and the spectral density $|\tilde{g}(\boldsymbol{\kappa})|^2$ with the specific density ρ, Eqs. (13.21) and (13.25) are the equations for calculating the inertia of a body rotating around the axis $\bar{\boldsymbol{n}}$.

With this analogy, we can reformulate the problem of determining local orientation. We need to find the axis around which the body formed by the spectral density in Fourier space rotates with minimum inertia. The rotating body can take different forms. The shape of the rotating body is related to the different solutions we obtain for the eigenvalues of the inertia tensor and thus for the solution of the local orientation problem (Table 13.3).

We have thus derived the inertia tensor approach in Fourier space. Now we will show how the coefficients of the inertia tensor can be calculated in space. The integrals in Eqs. (13.26) and (13.27) contain expressions of the form

$$k_q^2 |\tilde{g}(\boldsymbol{\kappa})|^2 = |\mathrm{i}k_q \tilde{g}(\boldsymbol{\kappa})|^2$$

and

$$k_p k_q |\tilde{g}(\boldsymbol{\kappa})|^2 = [\mathrm{i}k_p \tilde{g}(\boldsymbol{\kappa})]^* [\mathrm{i}k_q \tilde{g}(\boldsymbol{\kappa})].$$

If we integrate these expressions, we obtain an *inner* or *scalar product* of the functions $\mathrm{i}k_p \tilde{g}(\boldsymbol{\kappa})$. Since this product is preserved during the Fourier transformation (> R12), we can calculate the corresponding integrals just as well in the spatial domain. The multiplication of $\tilde{g}(\boldsymbol{\kappa})$ by $\mathrm{i}k_p$ in wavenumber space corresponds to the first derivative in the direction x_p in local space:

$$
\begin{aligned}
J'_{pp}(\boldsymbol{x}) &= \sum_{q\neq p}\int_{-\infty}^{\infty} w(\boldsymbol{x}-\boldsymbol{x}')\left(\frac{\partial g}{\partial x_q}\right)^2 \mathrm{d}^D x' \\
J'_{pq}(\boldsymbol{x}) &= -\int_{-\infty}^{\infty} w(\boldsymbol{x}-\boldsymbol{x}')\frac{\partial g}{\partial x_p}\frac{\partial g}{\partial x_q}\mathrm{d}^D x'.
\end{aligned}
\tag{13.28}
$$

In Eq. 13.28 we have already taken into account the weighting with the window function w to select a local environment.

The window in Section 13.3.1 Eq. (13.8) is closely related to the inertia tensor:

$$
\boldsymbol{J}' = \mathrm{trace}(\boldsymbol{J})\boldsymbol{I} - \boldsymbol{J}. \tag{13.29}
$$

From this relationship we can see that both matrices have the same eigenvectors. The eigenvalues λ_p are related to each other via the following equation:

$$
\lambda_p = \sum_{q=1}^{n}\lambda_q - \lambda'_p, \quad \lambda'_p = \sum_{q=1}^{n}\lambda_q - \lambda_p. \tag{13.30}
$$

This means that the eigenvalue analysis can be performed with either of the two matrices. For the inertia tensor, the direction of the local orientation is given by the minimum eigenvalue, but for the structure tensor it is given by the maximum eigenvalue.

Extension 13.1: Method of directional derivatives

In their article on analyzing oriented patterns, Kass and Witkin [172] chose a — at first glance — completely different method. In fact, however, it is equivalent to the tensor method, as we will now show. The authors started from *directed derivatives* with a *DoG* filter (Section 12.7.5). The resulting operator can be written in operator notation:

$$
\mathcal{R}(\theta) = [\cos\theta \;\; \sin\theta]\begin{bmatrix} \mathcal{D}_x(\mathcal{B}_1 - \mathcal{B}_2) \\ \mathcal{D}_y(\mathcal{B}_1 - \mathcal{B}_2) \end{bmatrix} = [\cos\theta \;\; \sin\Theta]\begin{bmatrix} \mathcal{R}_x \\ \mathcal{R}_y \end{bmatrix},
$$

where $\mathcal{B}_1$ and $\mathcal{B}_2$ denote two Gaussian smoothing masks with different variances. The direction in which this directional derivative is maximum in terms of quadratic deviations indicates the direction of the normal vector perpendicular to lines of constant gray values. This approach results in the following expression for the variance of the directional derivative:

$$
\mathcal{V}(\theta) = \mathcal{B}(\mathcal{R}(\theta)\cdot\mathcal{R}(\theta)). \tag{13.31}
$$

The directional derivative is squared and then smoothed with a binomial mask. The associated inertia tensor has the following form:

$$
\begin{bmatrix} \mathcal{B}(\mathcal{R}_y\cdot\mathcal{R}_y) & -\mathcal{B}(\mathcal{R}_x\cdot\mathcal{R}_y) \\ -\mathcal{B}(\mathcal{R}_x\cdot\mathcal{R}_y) & \mathcal{B}(\mathcal{R}_x\cdot\mathcal{R}_x) \end{bmatrix}. \tag{13.32}
$$

The approach of Kass and Witkin [172] is therefore identical to the structural tensor method. They only use a special type of derivative filter.

Extension 13.2: Many names for the same method

Shortly after the already cited work of Bigün and Granlund [19], Harris and Stevens [121] presented a combined edge/corner detector, which is often cited under the name *Harris operator*.
The year before, Förstner and Gülch [98] also presented a fast operator for corner and point detection, obviously without knowing the work of Harris and Bigün. The basic considerations and derivations of Förstner are very similar to those presented here.
Without knowing about the earlier work of Bigün and Granlund [19] and the contemporaneous investigations of Knutsson [186], Rao and Schunck [257] and Rao [256] proposed the same structure tensor (called moment tensor by them).
Obviously, the period at the end of the 1980s was ripe for independently developing similar nonlinear and tensor-based operators, which are still among the best local operators for spatial structure analysis of small neighborhoods.

13.4 Local Wavenumber and Phase

13.4.1 Introduction

So far in this chapter, we have discussed in detail the analysis of simple neighborhoods in terms of their orientation. In this section, we will look at another elementary property of simple neighborhoods. We must ask ourselves not only in which directions the gray values change, but also how quickly they change. This question leads us to the size of the *local wavenumber*. The key to determining it is the *phase* of the signal. As an introduction, we will discuss the one-dimensional periodic signal as a simple example

$$g(x) = g_0 \cos(2\pi\kappa x). \tag{13.33}$$

The argument of the cosine function is referred to as the phase of the periodic signal:

$$\phi(x) = 2\pi\kappa x. \tag{13.34}$$

The equation shows that the phase is a linear function of the position x and the wavenumber κ. We therefore obtain the wavenumber of the periodic signal by forming the first-order spatial derivative of the phase signal:

$$\boxed{\kappa = \frac{1}{2\pi}\frac{\partial\phi(x)}{\partial x}. \quad \text{local wavenumber}} \tag{13.35}$$

These simple considerations once again emphasize the importance of phase in image processing, as we have already discussed in Section 6.3.5.

Figure 13.11: Application of a Hilbert filter to the ring test sample; left upper quadrant: in horizontal direction; right lower quadrant: in vertical direction.

We will discuss several approaches to determine the phase of a signal: the Hilbert transform (Section 13.4.2) and the associated analytic signal (Section 13.4.3), its extension to higher dimensions with the Riesz transform and the monogenic signal (Section 13.4.4), and the quadrature filters (Section 13.4.5). Then we introduce efficient techniques to calculate the local wavenumber from the phase gradients (Section 13.4.6). Finally, we consider polar separable quadrature filters as an alternative to determine the local orientation and the wavenumber simultaneously (Section 13.4.7).

13.4.2 Hilbert Transform and Hilbert Filter

To explain the principle of calculating the phase of a signal, let's take the simple periodic signal from the previous section as an example again. Let's assume that there is an operator for delaying the signal by a phase difference of 90°. This would transform the signal $g_e(x) = g_0 \cos(kx)$ into the signal $g_o(x) = -g_0 \sin(kx)$ (Fig. 13.11). Using both signals, we can compute the phase of $g(x)$ with

$$\phi(g(x)) = \arctan\left(\frac{-g_o(x)}{g_e(x)}\right). \tag{13.36}$$

Since only the ratio of $g_o(x)$ to $g_e(x)$ is included in Eq. (13.36), the phase is actually independent of the amplitude. If we take the signs of the two functions $g_o(x)$ and $g_e(x)$ into account, the phase can be calculated over the entire range of 360°.

All we need to determine the phase of a signal is a linear operator that shifts the phase by 90°. This operator is called *Hilbert filter* $\boldsymbol{H}$ or *Hilbert operator* $\mathcal{H}$. It has the following transfer function:

$$\tilde{h}(k) = \begin{cases} \mathrm{i} & k > 0 \\ 0 & k = 0 \\ -\mathrm{i} & k < 0 \end{cases}. \tag{13.37}$$

As the amplitude is not changed, the magnitude of the transfer function is one. The fact that the Hilbert filter should generate a real signal again means that its purely imaginary transfer function must have an odd symmetry. Therefore, positive wavenumbers are shifted by 90° ($\pi/2$) and negative ones by -90° (-$\pi/2$). An exception is the wavenumber zero, for which the transfer function is also zero. This can be illustrated as follows: A signal with wavenumber zero is a constant and can be considered as a cosine function with infinite wavenumber, sampled at phase zero. The filtered signal is therefore the corresponding sine function, which disappears at phase zero.

Because of the discontinuity of the transfer function of the Hilbert filter at the origin, its point response is infinitely extended:

$$h(x) = -\frac{1}{\pi x}. \tag{13.38}$$

We can write the convolution with Eq. (13.38) as follows:

$$g_h(x) = \frac{1}{\pi} \int_{-\infty}^{\infty} \frac{g(x')}{x' - x} \mathrm{d}x'. \tag{13.39}$$

This integral transformation is referred to as *Hilbert transform* [245].

Since the convolution mask of the ideal Hilbert filter is infinitely extended, it is impossible to construct an exact discrete Hilbert filter for arbitrary signals with a small convolution mask. Therefore, we need to restrict the class of signals to which the filter can be applied. To do this, we consider the following approach to effectively implement a Hilbert filter.

Firstly, the filter should shift the phase exactly by $\pi/2$. This requirement results from the fact that we cannot tolerate errors in the phase as it contains the information about the position. A wavenumber-dependent phase shift would cause wavenumber-dependent errors. All convolution kernels of odd symmetry have the desired property. On the other hand, the requirement for the magnitude of one can be restricted if the Hilbert filter is applied to bandpass-filtered signals, for example a Laplacian pyramid. The Hilbert filter must then only have a magnitude of one in the passband of the filter used. With this approach, we avoid the discontinuities of the transfer function at the wavenumber of zero and obtain a convolution kernel of finite size.

Hilbert filters can be optimized like first-order derivative filters (Section 12.6) using the *least squares* method (Fig. 13.12). Due to the odd symme-

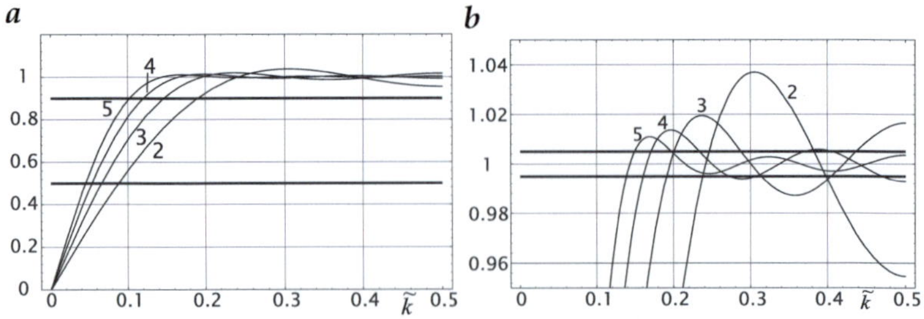

Figure 13.12: ***a*** *Transfer functions of a family of Hilbert operators optimized by the* least squares *method according to Eq. (13.40) with R=2–5 filter coefficients.* ***b*** *The enlarged section of* ***a*** *illustrates the deviation from an ideal Hilbert filter. As the filters are symmetrical around* $\tilde{k}$*=0.5, only the wavenumber range 0 to 0.5 is shown.*

try, we use the following transfer function:

$$\tilde{h}(\tilde{k}) = 2\mathrm{i}\sum_{\tilde{n}=1}^{R} h_{\tilde{n}} \sin\left((2\tilde{n}-1)\pi\tilde{k}\right). \tag{13.40}$$

Since only sine functions with odd wavenumbers occur, the transfer function is also symmetrical around $\tilde{k} = 1/2$. This results in the filter mask

$$[h_R, 0, \cdots, h_2, 0, h_1, 0, -h_1, 0, -h_2, \cdots, 0, -h_R] \tag{13.41}$$

with $4R-1$ coefficients, where $2R-1$ are zero. The filter with $R=4$ (the mask contains 15 coefficients) with

$$\boldsymbol{h} = \{0,6208, 0, 1683, 0, 0630, 0, 0191\} \tag{13.42}$$

for example, has an amplitude error that is only slightly greater than 1% in the wavenumber range of [0.16, 0.84] and no phase error due to the design. Convolution with this mask requires 4 multiplications and 7 additions or subtractions.

13.4.3 ♯The Analytical Signal

A real-valued signal and its Hilbert transform can be combined to form a complex-valued signal:

$$g_a = g - \mathrm{i}g_h. \tag{13.43}$$

This complex-valued signal is referred to as the *analytical function* or the *analytical signal*. According to Eq. (13.43), the analytical filter has the point response

$$a(x) = 1 + \frac{\mathrm{i}}{\pi x} \tag{13.44}$$

and the transfer function

$$\tilde{a}(k) = \begin{cases} 2 & k > 0 \\ 1 & k = 0 \\ 0 & k < 0 \end{cases}. \tag{13.45}$$

All negative wavenumbers are therefore suppressed. Although the transfer function of the analytical filter is real, it generates a complex signal due to its asymmetry. With a real-valued signal, no information is lost if we suppress the negative wavenumbers. They can be reconstructed because the Fourier transform of a real signal is Hermitian (Section 6.3.4). The analytical signal can be considered as another representation of a real signal with two important properties. The magnitude of the analytical signal gives the *local amplitude*

$$|\mathcal{A}|^2 = \mathcal{I} \cdot \mathcal{I} + \mathcal{H} \cdot \mathcal{H} \tag{13.46}$$

and the argument the *local phase*

$$\arg(\mathcal{A}) = \arctan\left(\frac{-\mathcal{H}}{\mathcal{I}}\right) \tag{13.47}$$

with $\mathcal{A}$ for the analytical operator and $\mathcal{H}$ for the Hilbert operator. We obtain the original signal and its Hilbert transform from the analytical signal with Eq. (13.43):

$$\begin{array}{lcl} g(x) & = & (g_a(x) + g_a^*(x))/2 \\ g_h(x) & = & \mathrm{i}(g_a(x) - g_a^*(x))/2. \end{array} \tag{13.48}$$

The concept of the analytic signal also makes it easy to extend the idea of the local phase to several dimensions. The transfer function of the analytic operator uses only the positive wavenumbers, i. e. only one half of the Fourier space. With multiple dimensions, we have more than one way to divide the Fourier space into two half-spaces. Instead of the wavenumber, we can use the scalar product between the wavenumber vector $\boldsymbol{\kappa}$ and any unit vector $\hat{\boldsymbol{n}}$ and then suppress the half-space for which the scalar product $\boldsymbol{\kappa}^T\hat{\boldsymbol{n}}$ is negative:

$$\tilde{a}(\boldsymbol{\kappa}) = \begin{cases} 2 & \boldsymbol{\kappa}^T\hat{\boldsymbol{n}} > 0 \\ 1 & \boldsymbol{\kappa}^T\hat{\boldsymbol{n}} = 0 \\ 0 & \boldsymbol{\kappa}^T\hat{\boldsymbol{n}} < 0 \end{cases}. \tag{13.49}$$

The unit vector $\hat{\boldsymbol{n}}$ indicates the direction in which the Hilbert filter is applied. The definition Eq. (13.49) of the transfer function of the analytical signal implies that the Hilbert operator can only be applied to directionally filtered signals. This results from the following considerations: For one-dimensional signals, we have seen that a discrete Hilbert filter is not suitable for small wavenumbers (Fig. 13.12). For multiple dimensions, this means that a Hilbert filter does not work well when $\tilde{\boldsymbol{\kappa}}^T\hat{\boldsymbol{n}} \ll 1$. All wavenumbers that are almost orthogonal to the direction of the Hilbert filter must therefore not be present if errors are to be avoided.

This fact makes the application of the Hilbert filter and thus the determination of the local phase for higher-dimensional signals considerably more complex. It is not sufficient to use bandpass-filtered images such as, e. g. a Laplacian pyramid (Section 15.3.4). In addition, the bandpass-filtered images must be further broken down into directional components. We need at least one component for each spatial direction.

13.4.4 ♯The Monogenic Signal

The extension of the Hilbert transform from a 1D signal to higher-dimensional signals is not satisfactory, as it can only be applied to directionally filtered signals. The Hilbert transform does not work for wavenumbers near the separation plane (Fig. 13.11). Therefore, it is necessary to look for an isotropic extension of the Hilbert transform. Obviously, no scalar transformation can be both isotropic and odd for multidimensional signals.

With a vectorial transformation, however, both conditions can be fulfilled. This results in the *monogenic signal*, which was introduced into image processing by Felsberg and Sommer [79]. The monogenic signal is composed of the original signal and the *Riesz transform*. The transfer function of the Riesz transform is

$$\tilde{\boldsymbol{h}}(\boldsymbol{\kappa}) = \mathrm{i}\frac{\boldsymbol{\kappa}}{|\boldsymbol{\kappa}|}. \tag{13.50}$$

The magnitude of the vector $\tilde{\boldsymbol{h}}$ is one for all $\boldsymbol{\kappa}$. Therefore, the Riesz transformation is isotropic. It is also odd, since

$$\tilde{\boldsymbol{h}}(-\boldsymbol{\kappa}) = -\tilde{\boldsymbol{h}}(\boldsymbol{\kappa}). \tag{13.51}$$

The Riesz transformation can be applied to signals of any dimension. In the 1D case, it is reduced to the Hilbert transform.

For 2D signals, the transfer function of the Riesz transform can be written using polar coordinates as

$$\tilde{\boldsymbol{h}}(\boldsymbol{\kappa}) = \mathrm{i}\,[\cos\theta,\ \sin\theta]^T. \tag{13.52}$$

The transfer function is similar to that for the gradient operator (Section 12.2.1, Eq. (12.2)). The difference is the division with the magnitude of the wavenumber.

The convolution mask or PSF of the Riesz transform is

$$\boldsymbol{h}(\boldsymbol{x}) = -\frac{\boldsymbol{x}}{2\pi\,|\boldsymbol{x}|^3}. \tag{13.53}$$

The original signal and the signal convolved with the Riesz transform can be combined to form the monogenic 3D signal for a 2D signal:

$$g_m(\boldsymbol{x}) = [p, q_1, q_2]^T \quad \text{with} \quad p = g, q_1 = h_1 * g, q_2 = h_2 * g. \tag{13.54}$$

As with the analytical signal Eq. (13.46), the local amplitude of the monogenic signal results as the magnitude of the signal to

$$\left|g_m\right|^2 = p^2 + q_1^2 + q_2^2. \quad (13.55)$$

Not only the *local phase* ϕ can be determined from the monogenic signal as from the *analytical signal*. The monogenic signal also allows the *local orientation* θ to be determined using the following relationships:

$$p = a\cos\phi, \quad q_1 = a\sin\phi\cos\theta, \quad q_2 = a\sin\phi\sin\theta. \quad (13.56)$$

Therefore, we can conclude that the monogenic signal combines a determination of local orientation and phase. This is of great importance for image processing, since the two most important properties of a local environment, the local orientation and the wavenumber, which we can calculate from the derivative of the phase (Section 13.4.6), can be determined together in a unified concept.

13.4.5 Quadrature Filter

Quadrature filters are an alternative way in which we can obtain a signal pair that differs only by a phase shift of 90° ($\pi/2$). The complex form of the quadrature filter is the easiest to explain. Essentially, the transfer function of a quadrature filter for $\boldsymbol{\kappa}\hat{\boldsymbol{n}} < 0$ is just as zero as that of the analytical filter. However, the magnitude of the transfer function is not one, but can be any real-valued function $h(\boldsymbol{\kappa})$:

$$\tilde{q}(\boldsymbol{\kappa}) = \begin{cases} 2h(\boldsymbol{\kappa}) & \boldsymbol{\kappa}^T\hat{\boldsymbol{n}} > 0 \\ 0 & \text{otherwise.} \end{cases} \quad (13.57)$$

The quadrature filter therefore also transforms a real-valued signal into an analytical signal. In contrast to the analytical operator, however, a wavenumber weighting is performed. We can derive the real quadrature filter pair from the complex form of the quadrature filter by taking into account that it represents the parts of even and odd symmetry of Eq. (13.57). Thus

$$\begin{aligned} \tilde{q}_e(\boldsymbol{\kappa}) &= (\tilde{q}(\boldsymbol{\kappa}) + \tilde{q}(-\boldsymbol{\kappa}))/2, \\ \tilde{q}_o(\boldsymbol{\kappa}) &= (\tilde{q}(\boldsymbol{\kappa}) - \tilde{q}(-\boldsymbol{\kappa}))/2. \end{aligned} \quad (13.58)$$

Even and odd parts of the quadrature filter pair have a phase shift of 90° and can therefore also be used to calculate the local phase.

Quadrature filters can also be designed on the basis of the monogenic signal (Section 13.4.4). These quadrature filters then have one more component than the dimension of the signal. The transfer function can be written as

$$\tilde{\boldsymbol{q}}(\boldsymbol{\kappa}) = \left[\tilde{q}_e(\boldsymbol{\kappa}),\ \mathrm{i}\boldsymbol{\kappa}\tilde{q}_e(\boldsymbol{\kappa})/\left|\boldsymbol{\kappa}\right|\right]^T. \quad (13.59)$$

The best-known quadrature filter pair is the *Gabor filter*. It is a bandpass filter that selects a wavenumber range around the center $\boldsymbol{\kappa}_0$ with the weighting of a Gaussian bell curve. The complex transfer function of the D-dimensional Gabor filter has the following form:

$$
{}^{(\text{Gabor})}\tilde{q}(\boldsymbol{\kappa}) = \begin{cases} (2\pi)^{D/2}\,\sigma_x^D \exp\left(-2\pi^2\sigma_x^2|\boldsymbol{\kappa}-\boldsymbol{\kappa}_0|^2\right) & \boldsymbol{\kappa}^T\boldsymbol{\kappa}_0 > 0 \\ 0 & \text{else.} \end{cases} \tag{13.60}
$$

If $|\boldsymbol{\kappa}_0|\sigma_x \gg 1$, Eq. 13.60 simplifies to

$$
{}^{(\text{Gabor})}\tilde{q}(\boldsymbol{\kappa}) = (2\pi)^{D/2}\,\sigma_x^D \exp\left(-2\pi^2\sigma_x^2|\boldsymbol{\kappa}-\boldsymbol{\kappa}_0)|^2\right). \tag{13.61}
$$

If we use the relationships specified in Eq. (13.58), the transfer functions for the even and odd components result in

$$
\begin{aligned}
{}^{(\text{Gabor})}\tilde{q}_e(\boldsymbol{\kappa}) &= \frac{1}{2}\left[\exp\left(-|\boldsymbol{\kappa}-\boldsymbol{\kappa}_0|^2\sigma_x^2/2\right) + \exp\left(-|\boldsymbol{\kappa}+\boldsymbol{\kappa}_0|^2\sigma_x^2/2\right)\right], \\
{}^{(\text{Gabor})}\tilde{q}_o(\boldsymbol{\kappa}) &= \frac{1}{2}\left[\exp\left(-|\boldsymbol{\kappa}-\boldsymbol{\kappa}_0|^2\sigma_x^2/2\right) - \exp\left(-|\boldsymbol{\kappa}+\boldsymbol{\kappa}_0|^2\sigma_x^2/2\right)\right].
\end{aligned} \tag{13.62}
$$

The point response of these filters can easily be calculated using the *shift theorem* (Theorem B.3, p. 629, > R12):

$$
\begin{aligned}
{}^{(\text{Gabor})}Q_e(\boldsymbol{x}) &= \cos(2\pi\boldsymbol{\kappa}_0\boldsymbol{x})\exp\left(-\frac{|\boldsymbol{x}|^2}{2\sigma_x^2}\right), \\
{}^{(\text{Gabor})}Q_o(\boldsymbol{x}) &= \mathrm{i}\sin(2\pi\boldsymbol{\kappa}_0\boldsymbol{x})\exp\left(-\frac{|\boldsymbol{x}|^2}{2\sigma_x^2}\right).
\end{aligned} \tag{13.63}
$$

By combining them into a complex filter mask, this becomes

$$
{}^{(\text{Gabor})}Q(\boldsymbol{x}) = \exp(2\pi\mathrm{i}\boldsymbol{\kappa}_0\boldsymbol{x})\exp\left(-\frac{|\boldsymbol{x}|^2}{2\sigma_x^2}\right). \tag{13.64}
$$

Example 13.2: Applying quadrature filters to image sequences

Gabor filters are often used for image analysis in spatial or wavenumber space. Fig. 13.13 shows an application [266, 267]. An image of small-scale wind-generated water surface waves was decomposed by a series of Gabor filters. The center wavelength $\boldsymbol{\kappa}_0$ was placed in x-direction, parallel to the wind direction. The center wavelengths of the filters were set to 1.2, 2.4 and 4.8 cm and the bandwidths were set proportional to the center wavenumber. The left column of the images in Fig. 13.13 shows the filtering with the straight Gabor filter, the right column shows the local amplitude, which is a measure of the energy of the waves. The filtered images show that waves with different wavelengths are partially coupled: In areas where the large wavelength waves

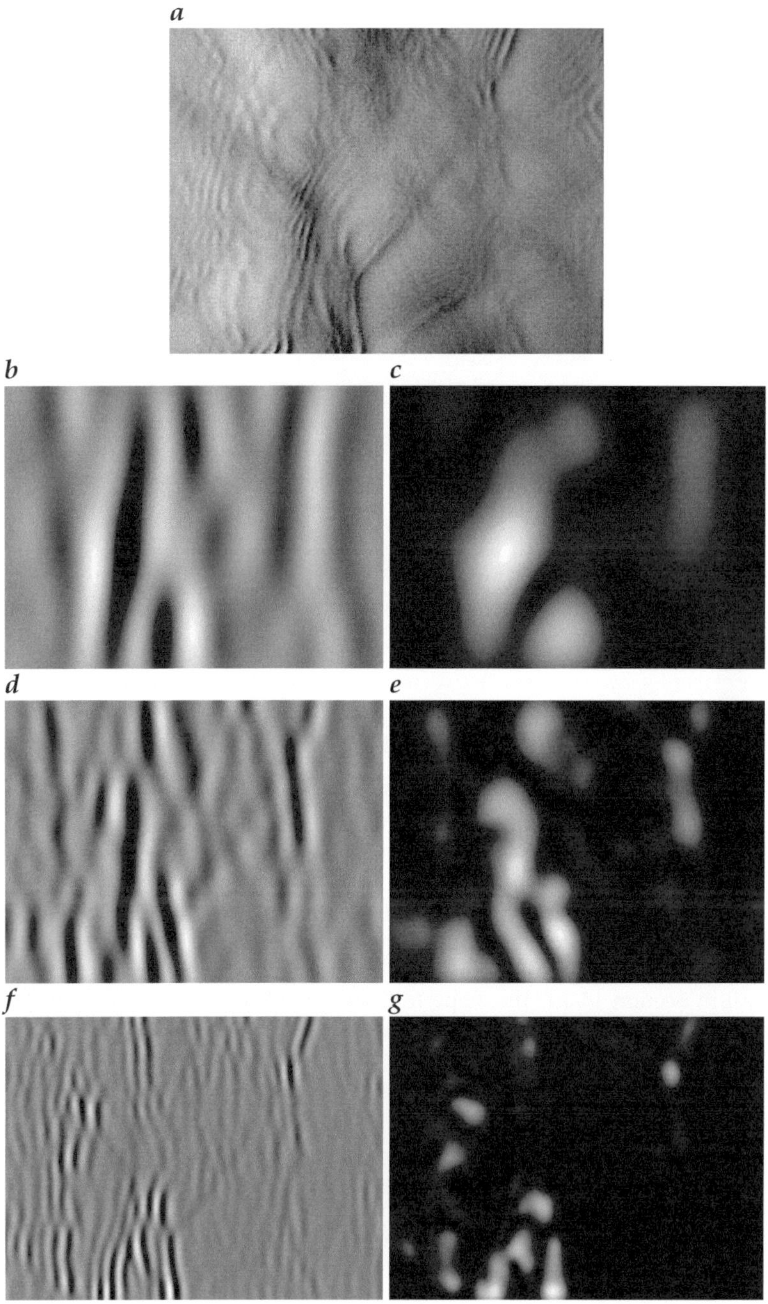

Figure 13.13: *Analysis of an image (**a**, 40 cm × 30 cm) of wind-generated water surface waves. The brightness is proportional to the slope of the waves in the wind direction. Straight part (**b**, **d**, **f**) and magnitude squared (energy, **c**, **e**, **g**) of the Gabor-filtered image with centroid wavelengths of 48, 24 and 12 mm, respectively.*

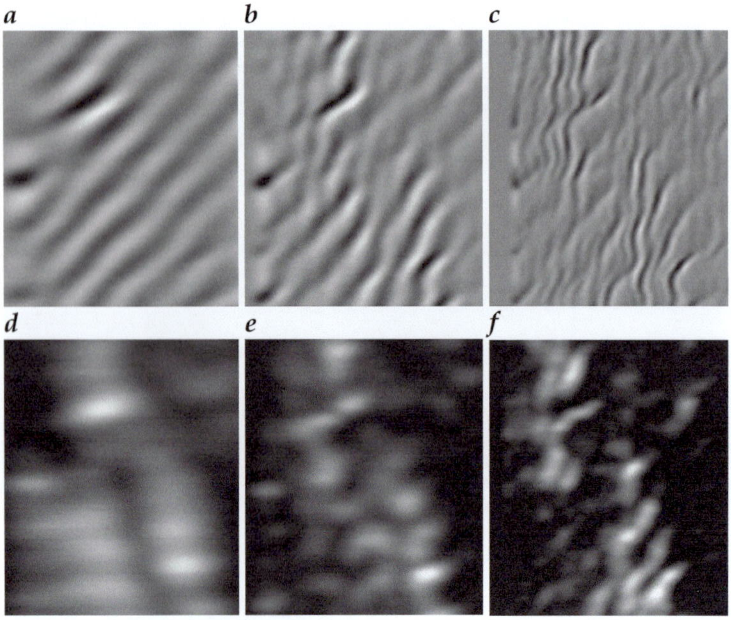

*Figure 13.14: Analysis of a 5 s long space-time slice in wind direction through an image sequence of wind-generated water surface waves. The time axis is vertical. Even part (wave amplitude, **a–c**) and magnitude squared (energy, **d–f**) of Gabor-filtered images with centroid wavelengths at 48, 24 and 12 mm, respectively.*

have large amplitudes, the small-scale waves (capillary waves) also have large amplitudes. The energy of the waves is not evenly distributed over the water surface.

Extending this analysis to image sequences gives a direct insight into the complex non-linear wave-wave interactions. In Fig. 13.14 the temporal evolution of a line of the image from Fig. 13.13 is shown. As we will discuss in more detail in Section 14.3.1, the slope to the time axis of the structures in these place/time images is proportional to the speed of the moving objects. This shows that the small-scale waves are modulated by large waves and that the *group velocity* (velocity of the wave energy) of the small waves is lower than the phase velocity of the capillary waves.

13.4.6 ♯Local Wavenumber Estimation

To determine the local wavenumber, we only need to calculate the first spatial derivative of the phase signal (Section 13.4.1, Eq. (13.35)). The phase results from

$$\phi(x) = \arctan\left(\frac{-g_o(x)}{g_e(x)}\right), \tag{13.65}$$

where g_e and g_o denote the signals filtered with the even and odd parts of the quadrature filter or from the original signal and the Hilbert transform.

However, due to the discontinuities inherent in the phase signal, it is not advisable to calculate the partial derivatives directly from these equations. A phase calculated with the arc tangent is limited to the interval $[-\pi, \pi[$ and inevitably leads to a jump in the phase signal between π and $-\pi$. As Fleet [83], Fleet and Jepson [85] showed, this problem can be avoided by calculating the phase gradient directly from the gradients of $g_e(\boldsymbol{x})$ and $g_o(\boldsymbol{x})$. The result is:

$$\boldsymbol{\kappa} = \frac{1}{2\pi}\boldsymbol{\nabla}\phi(\boldsymbol{x}) = \frac{1}{2\pi}\boldsymbol{\nabla}\arctan\left(\frac{-g_o(\boldsymbol{x})}{g_e(\boldsymbol{x})}\right) = \frac{1}{2\pi}\frac{g_o\,\boldsymbol{\nabla}g_e - g_e\,\boldsymbol{\nabla}g_o}{g_e^2 + g_o^2}. \tag{13.66}$$

This formulation of the phase gradient also eliminates the need to calculate trigonometric functions and is therefore significantly faster.

13.4.7 ♯Polar Separable Quadrature Filter Sets

Quadrature filters open another way of analyzing simple neighborhoods to determine both the *local orientation* and the *local wavenumber*. Historically, this technique, which emerged from the work of Granlund [114], was the first for analyzing local structures. The inertial and structural tensor techniques came later [19, 172, 256, 257].

The quadrature filter method is based on the idea of extracting structures in a certain range of wavenumbers and directions. To determine the local orientation, we need to apply a series of directional filters, where each filter is sensitive to structures of a different orientation. We then compare the filter responses and obtain a maximum filter response from the directional filter whose direction best matches the local orientation. Similarly, a series of quadrature filters can be used to determine the local wavenumber.

If we obtain a pronounced maximum from one of the filters, but only low responses from the other filters, the neighborhood contains a locally oriented pattern. If the filter responses are comparable, the neighborhood consists of a distribution of differently oriented patterns. So far, the concept seems simple; however, a few problems still need to be solved. What properties do the directional filters have to fulfill for an exact determination of the local orientation, if this is possible at all? In order to maximize computational speed, we need to apply the smallest possible number of filters necessary to interpolate the angle of local orientation. But what is the minimum number?

This section is based on work by Granlund [114], Knutsson [185] and Knutsson et al. [187]. They were summarized in the monograph by Granlund and Knutsson [115]. While the mentioned authors have formulated the quadrature filter set technique for several dimensions, we will only consider the two-dimensional case here.

For an appropriate set of directional filters, each filter should be a rotated copy of the next filter. This requirement implies that the transfer function of the filters can be separated into an angular part $d(\phi)$ and a wavenumber part $r(\kappa)$. Such a filter is called *polar separable*:

$$\tilde{q}(\kappa, \phi) = \tilde{r}(\kappa)\tilde{d}(\phi), \tag{13.67}$$

where $\kappa = \sqrt{\kappa_1^2 + \kappa_2^2}$ is the magnitude and $\phi = \arctan(\kappa_2/\kappa_1)$ is the argument of the wavenumber. For a set of directional filters, only the angular part of the transfer function is important, as the radial part must be the same for all filters, but can be of any shape. The opposite applies to a set of filters for determining the local wavenumber.

Knutsson [185] proposed the following basic quadrature filter:

$$\begin{aligned} \tilde{r}(k) &= \exp\left[-\frac{(\ln\kappa - \ln\kappa_0)^2}{(B/2)^2 \ln 2}\right] \\ \tilde{d}(\phi) &= \begin{cases} \cos^{2l}(\phi - \phi_k) & |\phi - \phi_k| < \pi/2 \\ 0 & \text{otherwise.} \end{cases} \end{aligned} \tag{13.68}$$

In this equation, the constant κ_0 denotes the maximum wavenumber; B determines the half-width of the wavenumber in octaves and l the angular resolution of the filter. With a logarithmic wavenumber scale, the filter has the form of a Gaussian function. Therefore, the radial, wavenumber-dependent part has a *lognormal* (logarithmically normally distributed) form.

Furthermore, the complex notation for quadrature filters is used (Section 13.4.5). The filter direction is ϕ_k; the unit vector in this direction is $\hat{\boldsymbol{n}}_k = \left[\cos\phi_k, \sin\phi_k\right]$. The filter is continuous, as the cosine function in the division plane is zero for both half-spaces ($|\phi - \phi_k| = \pi/2$ or $\boldsymbol{\kappa}^T\hat{\boldsymbol{n}}_k = 0$). If we use the unit vector $\hat{\boldsymbol{n}}_k$ in the direction of the filter, we can also write the angle part as follows:

$$\tilde{d}(\boldsymbol{\kappa}) = \begin{cases} (\boldsymbol{\kappa}^T\hat{\boldsymbol{n}}_k)^{2l} & (\boldsymbol{\kappa}^T\hat{\boldsymbol{n}}_k) > 0 \\ 0 & \text{otherwise.} \end{cases} \tag{13.69}$$

In the even and odd filter of the real quadrature filter pair, the radial parts are the same, only the angular parts differ:

$$\begin{aligned} \tilde{d}_+(\phi) &= \cos^{2l}(\phi - \phi_k) \\ \tilde{d}_-(\phi) &= \mathrm{i}\cos^{2l}(\phi - \phi_k)\,\mathrm{signum}(\cos(\phi - \phi_k)). \end{aligned} \tag{13.70}$$

Fig. 13.15 shows the radial and angular part of the transfer functions for different maximum wavenumbers κ_0 and filter directions ϕ_k. We obtain a set of directional filters by selecting correspondingly different ϕ_k:

$$\phi_k = \frac{\pi k}{K} \quad k = 0, 1, \cdots, K-1. \tag{13.71}$$

Knutsson uses four filters with 45° increments in the directions 22.5°, 67.5°, 112.5° and 157.5°. These directions have the advantage that only *one* filter core needs to be constructed. The cores for the filters of the other directions are obtained by mirroring the axes and diagonals. These filters are

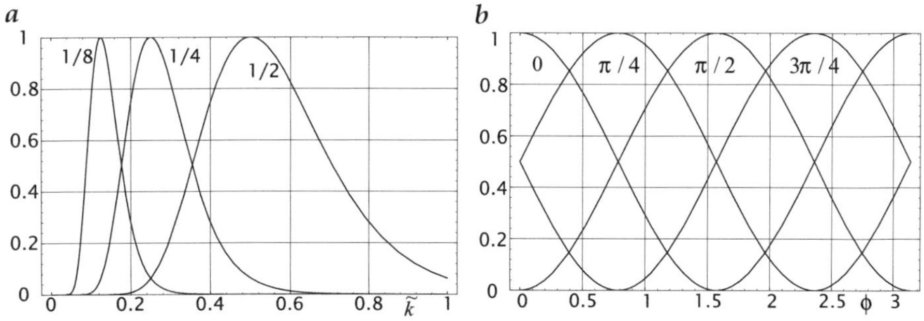

Figure 13.15: ***a*** *radial (wavenumber) and* ***b*** *angular part of a quadrature filter according to Eq. (13.68) with $B = 1$ and $l = 1$ with different maximum wavenumbers κ and in different directions.*

constructed in wavenumber space. The filter coefficients are obtained by inverse Fourier transformation. If we choose a small filter mask, we cut off some of the filter coefficients that are not equal to zero. This results in deviations from the ideal transfer function. For this reason, Knutsson has optimized the coefficients of the filter kernel so that the ideal transfer function is approximated as well as possible. According to Knutsson, this requires a filter mask of at least the size 15×15.

Determination of the Orientation Vector

The local orientation can be calculated from the responses of the four quadrature filters by vector addition. The idea of the approach is simple. We assign an orientation vector to the individual directional filters. The magnitude of the vector corresponds to the response of the quadrature filter. The direction of the vector is given by twice the angle of the filter direction (Section 13.3.3). In this representation, each filter response shows how exactly the direction of the examined pattern corresponds to the direction of the filter. An estimate of the orientation vector is then given by the vector sum of the orientation vectors belonging to the individual filters.

If we use a representation with complex numbers for the orientation vector, we can write the filter response of the filter in the direction ϕ_k as

$$Q_{\phi_k} = |Q| \exp(2i\phi_k) \tag{13.72}$$

and the orientation vector as vector sum

$$O = \sum_{k=0}^{K-1} Q_{\phi_k}. \tag{13.73}$$

The example in Fig. 13.16, in which the response of all four filters is the same, shows why it is necessary to double the angles compared to those of

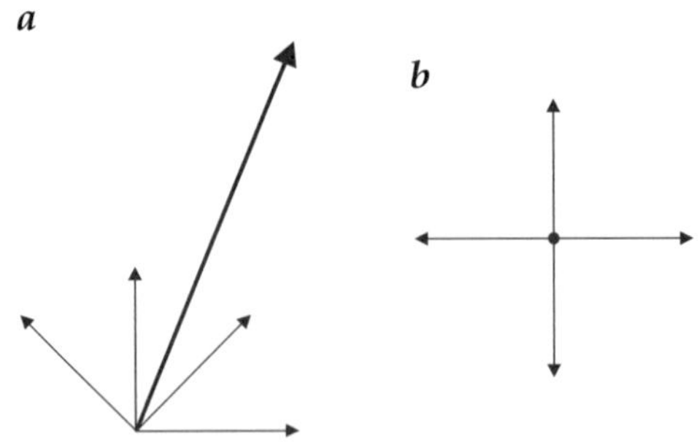

*Figure 13.16: Calculation of the local orientation by vector addition of the four filter responses. Shown is an example with an isotropic environment in relation to the orientation: All four filter responses are the same. The angles of the vectors correspond to the filter directions in **a** and the doubled filter direction angles in **b**.*

the corresponding orientation filters. There is therefore no local orientation and the environment contains structures in all directions. The vector sum of all filters is zero if we double the orientation angle (Fig. 13.16b), but not if we omit this step (Fig. 13.16a).

Extension 13.3: Calculation of the orientation vector with ideal orientation

Here, we will go into more detail about how the local orientation can be calculated exactly if there is an ideal orientation in any direction ϕ_0. At the same time, we see how many filters are required as a minimum. The calculations can be simplified if we only consider the angle-dependent terms, as the filter responses show the same wavenumber dependence.
Using Eq. (13.68), Eq. (13.71) and Eq. (13.72), we can formulate the angular part of the filter response of the k-th filter as follows:

$$\tilde{d}_k(\phi_0) = \exp(2\pi i k/K)\cos^{2l}(\phi_0 - \pi k/K).$$

Under the condition $K > l + 1$, the sum of the K filter responses is [115]

$$\sum_{k=0}^{K-1} \tilde{d}_k = \frac{K}{2^{2l}} \binom{2l}{l+1} \exp\left(i2\phi_0\right), \tag{13.74}$$

a vector in the complex plane where the angle of the local orientation is doubled. From $l > 0$ and $K > l + 1$ we conclude that at least $K = 3$ directional filters are necessary. This condition can be understood intuitively. If we only have two filters ($K = 2$), their orientation vectors and the sum vector lie on a straight line (Fig. 13.17a), and it is not possible to determine the orientation. The sum vector can only assume all directions if there are three or more vectors (Fig. 13.17b and c).

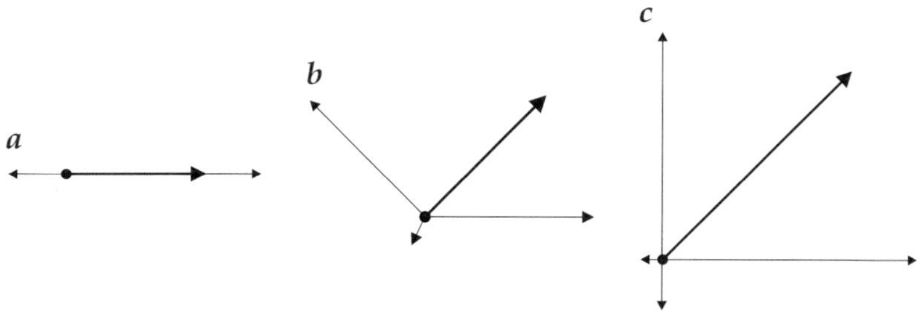

Figure 13.17: *Vector addition of the filter responses of K directional filters to determine the local orientation;* ***a*** *K = 2;* ***b*** *K = 3;* ***c*** *K = 4; the sum vector is drawn with a thicker line.*

It can also be shown that the sum over the transfer functions of the K filters for $K > l$ results in an isotropic function:

$$\sum_{k=0}^{K-1} \cos^{2l}(\phi - \pi k/K) = \frac{K}{2^{2l}}\binom{2l}{l} \frac{K}{2^{2l}} \frac{(2l)!}{l!^2}. \tag{13.75}$$

In other words: the sum of all filter responses results in an *orientation invariant* response, there is no preferred direction. This is the deeper reason why the local orientation can be easily and accurately determined with a very limited number of filters and a simple linear procedure like a vector addition.

Determination of the Local Wave Number

The *lognormal* form of the radial part of the quadrature filter enables direct determination of the *local wavenumber* of a narrowband signal. According to Eq. (13.68), we can write the radial part of the transfer function of the quadrature filter as

$$\tilde{r}_l(k) = \exp\left[-\frac{(\ln k - \ln k_l)^2}{2\sigma^2 \ln 2}\right]. \tag{13.76}$$

We examine the ratio of the output at two different radial center frequencies k_1 and k_2 and obtain:

$$\frac{\tilde{r}_2}{\tilde{r}_1} = \left(\frac{k}{\sqrt{k_1 k_2}}\right)^{\ln(k_2/k_1)/(\sigma^2 \ln 2)}. \tag{13.77}$$

In general, the ratio of two different radial filters is directly proportional to the local wavenumber. The ratio is particularly simple if the exponent of the last expression is one. This is the case, for example, when $k_2/k_1 = 2$ and $\sigma = 1$. Then

$$\frac{\tilde{r}_2}{\tilde{r}_1} = \frac{k}{\sqrt{k_1 k_2}}. \tag{13.78}$$

Determination of the Structur Tensor

In the last section of this chapter, we will relate the quadrature filter set technique discussed in Section 13.4.5 to the tensor technique (Section 13.3). The structure tensor can be calculated from the answers of these filters. Granlund and Knutsson [115] present a general equation for calculating the structure tensor from the quadrature filter responses:

$$J(\boldsymbol{x}) = \sum_{k=0}^{K-1} Q_k g(\boldsymbol{x}) \left(\alpha \hat{\boldsymbol{n}}_k \hat{\boldsymbol{n}}_k^T - \beta \boldsymbol{I} \right), \tag{13.79}$$

where $Q_k g(\boldsymbol{x})$ is the amplitude of the k-th quadrature filter, $\boldsymbol{I}$ is the unit matrix and $\alpha = 4/3$, $\beta = 1/3$ in two dimensions.

We show this relationship using a quadrature with three filters. These point in the directions 0°, 60° and 120°. This results in the unit direction vectors being

$$\begin{aligned} \hat{\boldsymbol{n}}_0 &= [1, 0]^T \\ \hat{\boldsymbol{n}}_1 &= \left[1/2,\ \sqrt{3}/2\right]^T \\ \hat{\boldsymbol{n}}_2 &= \left[-1/2,\ \sqrt{3}/2\right]^T. \end{aligned} \tag{13.80}$$

With these values for $\hat{\boldsymbol{n}}_k$, Eq. (13.79) can be written as follows:

$$\begin{aligned} J(\boldsymbol{x}) &= Q_0 g(\boldsymbol{x}) \begin{bmatrix} 1 & 0 \\ 0 & -1/3 \end{bmatrix} \\ &+ Q_1 g(\boldsymbol{x}) \begin{bmatrix} 0 & 1/\sqrt{3} \\ 1/\sqrt{3} & 2/3 \end{bmatrix} \\ &+ Q_2 g(\boldsymbol{x}) \begin{bmatrix} 0 & -1/\sqrt{3} \\ -1/\sqrt{3} & 2/3 \end{bmatrix}. \end{aligned} \tag{13.81}$$

The matrices provide the contribution of the individual quadrature filters to the associated elements of the structure tensor. For an isotropically oriented pattern, the response of all quadrature filters is the same. If we denote it by $q(\boldsymbol{x})$, Eq. (13.81) gives the correct structure tensor for an isotropically oriented pattern:

$$J(\boldsymbol{x}) = \begin{bmatrix} q(\boldsymbol{x}) & 0 \\ 0 & q(\boldsymbol{x}) \end{bmatrix}. \tag{13.82}$$

Conversely, the answer for a directional pattern is $q(\boldsymbol{x}) \cos^2(\phi_0 - \phi_k)$, and we get

$$J(\boldsymbol{x}) = q(\boldsymbol{x}) \begin{bmatrix} \cos^2(\phi_0) & \sin(2\phi_0)/2 \\ \sin(2\phi_0)/2 & \sin^2(\phi_0) \end{bmatrix}. \tag{13.83}$$

Again, this is the correct form of the structure tensor for an ideally oriented structure in the direction ϕ_0. (This can be shown, for example, by checking that the determinant of the matrix is zero and by calculating the orientation angle according to Eq. (13.13)).

There is only one subtle but important difference between the quadrature filter technique and the structure tensor method. The quadrature filter technique does not require averaging to calculate the components of the structure tensor. However, averaging is an important element of this method, without which the coherence measure (see Eq. (13.16) in Section 13.3.4) would always be one.

13.5 ♯References for Further Reading

The quadrature filter approach (Section 13.4.5) is described in detail in the monograph by Granlund and Knutsson [115], the inertia tensor method (Section 13.3.8) in an article by Bigün and Granlund [19]. Poularikas [245, Chapters 7 and 20] deals with the mathematics of the Hilbert transform and related techniques. The extension of the analytical signal to higher dimensions (Section 13.4.4) was published by Felsberg and Sommer [78, 80]. More on the mathematics of the monogenic signal and geometric algebra for computer vision can be found in the monograph by Sommer [310].

14 Motion

Contents

14.1 Introduction and Basics

For a long time, motion analysis was a special field of research that had little to do with general image processing. There were two reasons for this separation. Firstly, the techniques for analyzing motion in image sequences were different. Secondly, the large storage space and computing power required

B. Jähne, *Digital Image Processing and Image Acquisition*,
https://doi.org/10.1007/978-3-032-04806-6_14

to process image sequences limited motion analysis to a few specialized institutes that could afford the expensive equipment. The author still remembers the first work in his working group on image sequence analysis in the early 1990s, which was only possible because an expensive digital real-time storage system from the Fraunhofer Institute IITB in Karlsruhe (now IOSB) could be used to digitize video image sequences [266, 267].

Today, it is hard to imagine the limitations of that time. The rapid progress, especially in computer hardware in terms of storage capacity and processing speed, but also in algorithms, now allows image sequences to be recorded, stored and analyzed not only on standard PCs and workstations but also on mobile devices.

The integration of motion analysis into general image processing can be compared to the transition from photography of individual images to film. This makes it possible to view motion as just another feature used to identify and characterize objects and to understand scenes. All dynamic processes that cause temporal changes and that can be made visible are potential tasks for image sequence analysis. This means that more than movement can be analyzed. Motion analysis is therefore an adequate tool for analyzing the underlying dynamic processes.

This results in many possibilities for scientific, technical and general applications. Here are just a few examples: the investigation of flow and transport processes, biological growth processes both in the molecular area and in the entire ecosystem, changes in the daily or annual rhythm, process sequences in industry, the analysis of traffic scenes and the control of autonomous vehicles and robots.

Motion analysis can be approached conceptually in two ways. One possibility is to look at only two images and determine the movement by determining the positional displacement of the objects from the first to the second image. We will examine the fundamental problems associated with this in more detail in this introductory section. In the following Section 14.2 we will examine correlation methods for determining the displacement vector.

The second approach is based on understanding an image sequence as a three-dimensional data structure with two spatial and one temporal coordinate. If the *sampling theorem* (Section 5.2.3) is also satisfied in the temporal direction, we can consider the image sequence analysis in a continuous space-time domain and the corresponding wavenumber-frequency space (Section 14.2). The transition from two discrete images to a continuous image sequence can be compared to the transition from a stereo camera (Section 8.5.1) to a light field image (Section 8.5.3).

With this powerful concept, it is possible to systematically compare the confusing abundance of approaches to image sequence analysis. The gray value structures in the temporal direction have a different characteristic than in the spatial direction and therefore look unusual. Therefore, in Section 14.3 we will first deal fundamentally with the space-time and wavenumber-frequency domains for image sequences. Based on this, we can then discuss

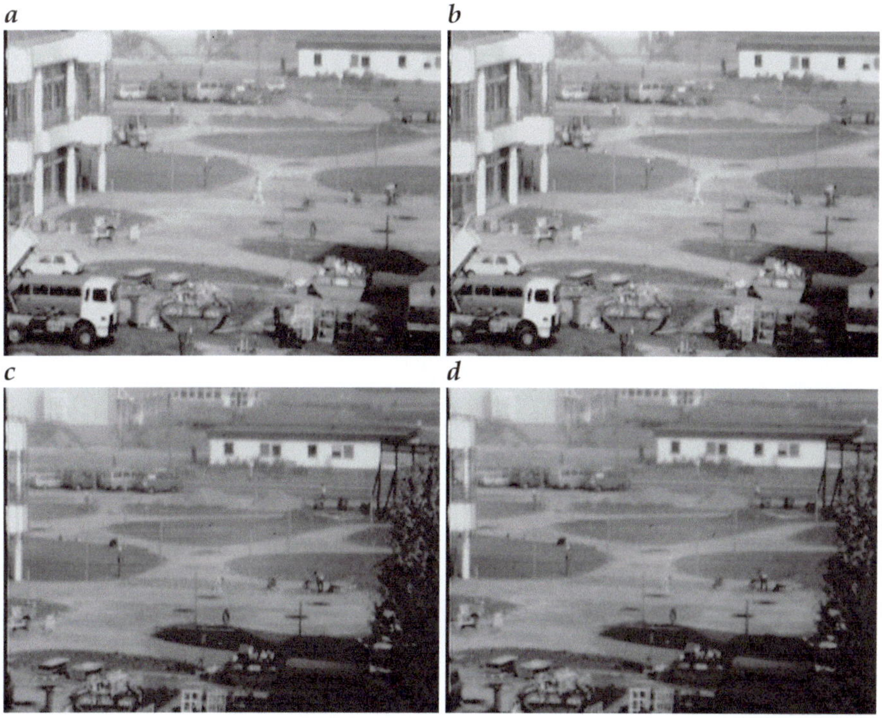

Figure 14.1: *Two pairs of pictures of the construction site of the new head clinic at Heidelberg University (**a** / **b** and **c** / **d**). What has changed from the left to the right picture?*

elementary methods for motion estimation, such as differential (Section 14.4), tensor (Section 14.5) and phase methods (Section 14.6). It will come as no surprise that derivative operators in the spatial and temporal direction, which we discussed in detail in Chapter 12, will play a central role.

14.1.1 Motion and Gray Value Changes

Intuitively, we associate movement with change. So let's start the discussion of motion analysis by looking at the differences between two images in a sequence. In Fig. 14.1a and b we see a pair of images of a construction site at the University of Heidelberg. There are differences between the left and right images, but these are not immediately obvious. However, if we subtract the two images from each other, the differences become immediately visible (Fig. 14.3a). At the bottom left of the image, a truck has moved, while the car behind it is obviously parked. In the center of the image, we see the outline of a pedestrian, which is barely visible in the original images. The bright spots in the upper part of the picture are cyclists on a cycle path in the background. From the greater displacement of the double contours in the difference image, we can estimate that they are moving faster than the

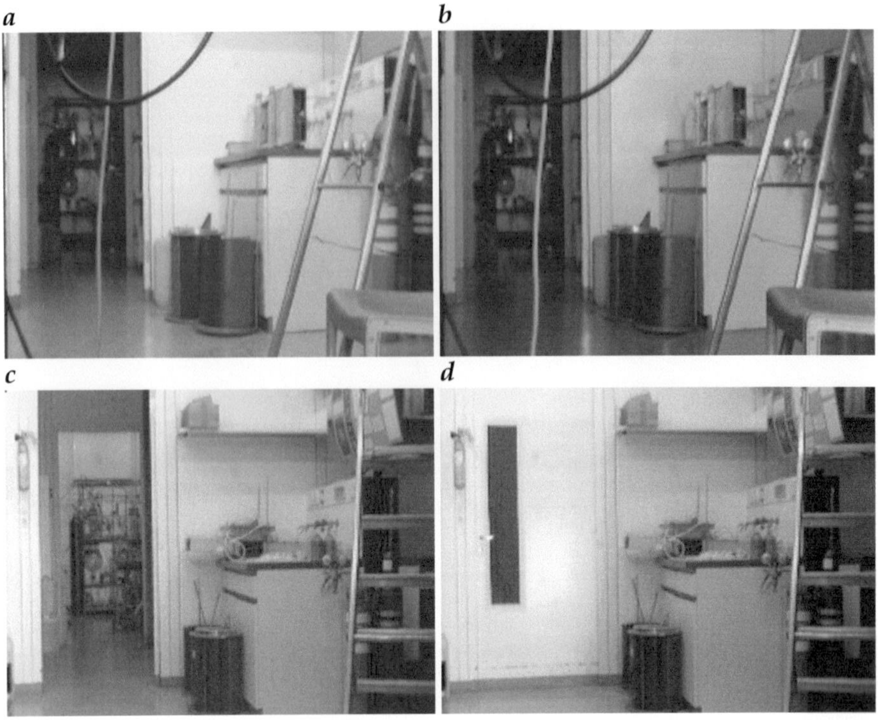

Figure 14.2: *Two pairs of images from a laboratory (**a** / **b** and **c** / **d**). What are the differences between the left and right images?*

pedestrian. Even such a qualitative description shows that motion analysis helps us considerably to understand a scene. Without the moving cyclists, for example, it would be much harder to recognize the cycle path.

Fig. 14.1c and d seem to show the same scene. However, we can already recognize the changes between the two images. If we look at the edges, we can see that the images are slightly shifted horizontally in relation to each other. What has happened? Obviously the camera was panned. In the difference image Fig. 14.3b, all object edges appear as bright lines. However, in places where the spatial gray value changes in the original image are small. The difference image is dark in most regions. From this it can be concluded that movement is only detected where gray values change spatially. This simple observation illustrates the central role of spatial gray value changes for motion detection.

So far, we can summarize our experience in the statement that movement can result in temporal gray value changes. Unfortunately, the reverse conclusion, that all temporal gray value changes are due to motion, is not valid. At first glance, the images in Fig. 14.2a and b look identical. However, the difference image in Fig. 14.4a shows that some parts in the left image are brighter than in the right image. Obviously, the lighting conditions have

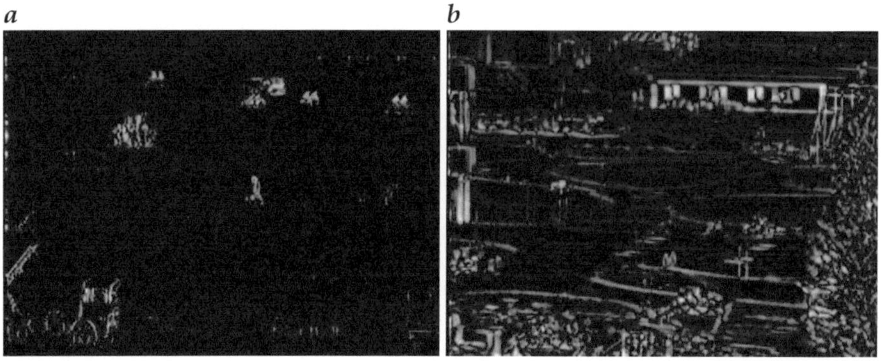

Figure 14.3: *Difference image* ***a*** *of the images* ***a*** *and* ***b*** *in Fig. 14.1,* ***b*** *of the images* ***c*** *and* ***d*** *in Fig. 14.1.*

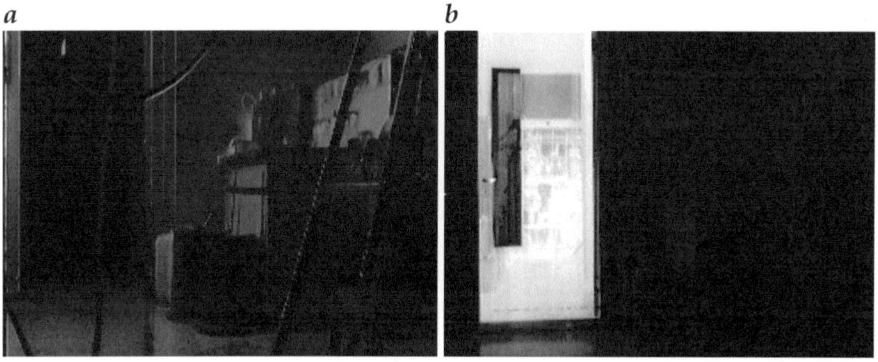

Figure 14.4: *Difference image* ***a*** *of the images* ***a*** *and* ***b*** *in Fig. 14.2,* ***b*** *of the images* ***c*** *and* ***d*** *in Fig. 14.2.*

changed. In fact, a lamp outside the image section was switched off before the image in Fig. 14.2b was taken. But where is this lamp? The difference image shows that not all areas are equally bright. Surfaces facing the camera are approximately equally bright in both images, while surfaces facing to the left are significantly brighter in the left image. From this we can conclude that the lamp is on the left outside the image section.

Another pair of images (Fig. 14.2c and d) shows a much more complicated scene, although the lighting has not been changed, only the laboratory door has been closed. Of course, we see strong gray value differences in the area of the door. However, closing the door also causes a change in the lighting conditions. As more light is now reflected into the area below the door, we also see gray value differences in the objects located there (Fig. 14.4b).

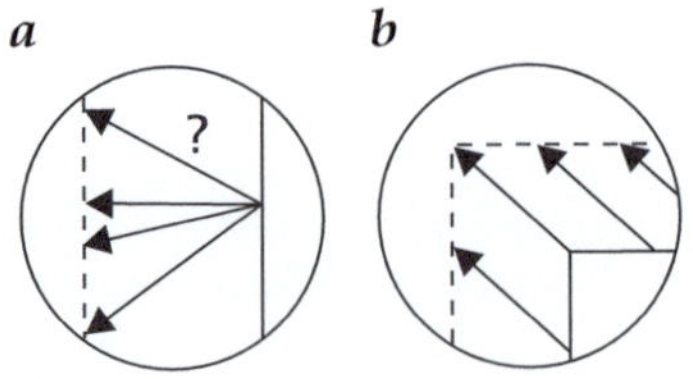

Figure 14.5: *Local operators provide only sparse information about movement. Illustration of the aperture problem in motion analysis:* ***a*** *ambiguity of the displacement vectors at an edge;* ***b*** *Uniqueness of the displacement vector at a corner.*

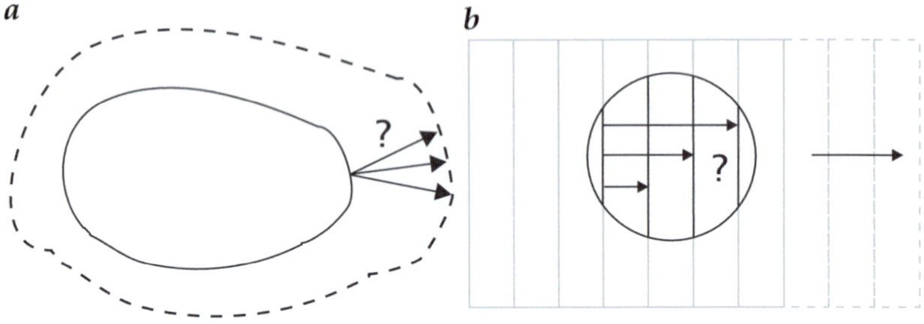

Figure 14.6: *The correspondence problem with* ***a*** *a deformable two-dimensional solid and* ***b*** *a regular lattice.*

14.1.2 The Aperture Problem

So far we have learned that motion analysis is closely related to spatial and temporal gray value differences. Both quantities can be obtained quite easily with local operators, which form the spatial and temporal derivatives. However, such a local operator "sees" only a small section of the observed object according to the size of its mask. This can be illustrated by placing an aperture on the image.

Fig. 14.5a shows an edge that has been moved from the position of the solid line to the position of the dashed line. The movement from one position to the next can be described by a *displacement vector*, abbreviated *DV*. In this case, we cannot determine the displacement uniquely, since the displacement vector can point from a point on the edge in the first image to any point on the displaced edge in the next image (Fig. 14.5a). We can only determine the component of the DV perpendicular to the edge, while the component parallel to the edge remains unknown. This ambiguity is referred to as the *aperture problem*. An unambiguous determination of the DV is only possible if the operator mask includes the corner of an object (Fig. 14.5b). The example shows that local operators only provide sparse information about motion.

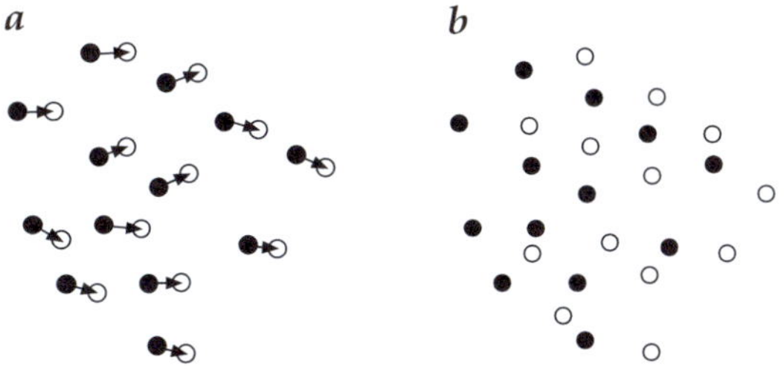

Figure 14.7: The correspondence problem with indistinguishable particles: ***a*** *the mean particle distance is greater than the displacement vectors;* ***b*** *the opposite case. The filled and open circles represent the particles in the first and second images, respectively.*

14.1.3 The Correspondence Problem

The reason for the aperture problem lies in the fact that we cannot find a certain point of an edge in the following image of an image sequence, since we have no way of distinguishing between different points of an edge. In this sense, we can understand the aperture problem as a special case of the more general *correspondence problem*. In general, this means that we cannot determine uniquely correlated points in two consecutive images of a sequence. In this section we look at further examples of the correspondence problem.

Fig. 14.6a shows a two-dimensional deformable object, e. g. a color blob that gradually spreads out. It is obvious that a clear determination of the displacement vectors is not possible even at the edge of the blob. Nothing at all can be said about the movement inside, as there are no recognizable features whose movement could be tracked.

Initially, one might assume that the correspondence problem does not occur for rigid objects with pronounced gray value differences. A grid as an example of a periodic texture (see Fig. 14.6b) shows that this is not true. As long as we consider the shift of the grid with a local operator and thus only in a section of the object, the shift by a multiple of the mesh size is not recognizable in principle. Only when we see the entire grid with its edge does the shift become clearly determinable.

Another variant of the correspondence problem occurs in images with many objects that are indistinguishable in shape, e. g. small particles that are introduced into a flow field to measure the flow velocity. In this case, it is not possible to determine the corresponding particle in the subsequent image (Fig. 14.7b). A solution to the problem arises if we take the successive images at such short time intervals that the mean displacement vector is significantly smaller than the mean particle distance. With this additional information, we can search for the nearest neighbor of a particle in the following image. However, such an approach is never error-free, as the distance of particles is statistically distributed.

From these simple examples, the fundamental problem of motion analysis becomes clear. At a higher level of abstraction, we can see that the *physical* or *real correspondence* of real objects does not have to match the *visual correspondence* in the image. The problem has two faces. First, we can find a visual correspondence without a physical correspondence, for example with objects or periodic object textures that are indistinguishable from one another. Secondly, a physical correspondence does not necessarily mean that a visual correspondence also exists. This is the case when objects have no recognizable features or when we cannot determine the visual correspondence due to changes in lighting.

14.2 Correlation Method

14.2.1 Basics

The correlation method has its origin in an analysis of the shift between two consecutive images. In order to find a characteristic feature from the first image in the second, we use a small section from the first image $g(\boldsymbol{x}, t_1) = g_1$ and search for it in the second image $g(\boldsymbol{x}, t_2) = g_2$ within a certain displacement range. We are thus looking for the position of the optimal similarity between the two images.

But when do we consider two features to be equal? The similarity measure should be insensitive to illumination changes. So we consider two spatial patterns to be equal if they differ only by a constant factor α that reflects the differences in illumination. In the language of vector spaces, this means that the two feature vectors g_1 and g_2 are parallel. This is exactly the case when equality occurs in the *Cauchy-Schwarz inequality*:

$$\left| \int\limits_{-\infty}^{\infty} g_1(\boldsymbol{x}) g_2(\boldsymbol{x} - \boldsymbol{s}) \mathrm{d}^2 x \right|^2 \le \int\limits_{-\infty}^{\infty} g_1^2(\boldsymbol{x}) \mathrm{d}^2 x \int\limits_{-\infty}^{\infty} g_2^2(\boldsymbol{x} - \boldsymbol{s}) \mathrm{d}^2 x. \tag{14.1}$$

In other words, we need to maximize the *cross-correlation coefficient*:

$$r(\boldsymbol{s}) = \frac{\int\limits_{-\infty}^{\infty} g_1(\boldsymbol{x}) g_2(\boldsymbol{x} - \boldsymbol{s}) \mathrm{d}^2 x}{\left(\int\limits_{-\infty}^{\infty} g_1^2(\boldsymbol{x}) \mathrm{d}^2 x \int\limits_{-\infty}^{\infty} g_2^2(\boldsymbol{x} - \boldsymbol{s}) \mathrm{d}^2 x \right)^{1/2}}. \tag{14.2}$$

The cross-correlation coefficient is a useful measure of similarity. It is zero for completely dissimilar (orthogonal) patterns and reaches a maximum value of one for equal features. The cross-correlation coefficient can be performed as a combination of convolution and point operations. The first step is to introduce a window function w into the definition of the cross-correlation coefficient. The size of the window determines the spatial resolution with which the shift can be determined. This window is moved over the image to calculate the local cross-correlation coefficient. Equation (14.2) then becomes

$$r(\boldsymbol{x}, \boldsymbol{s}) = \frac{\int\limits_{-\infty}^{\infty} w(\boldsymbol{x}-\boldsymbol{x}') g_1(\boldsymbol{x}') g_2(\boldsymbol{x}'-\boldsymbol{s}) \mathrm{d}^2 x'}{\left(\int\limits_{-\infty}^{\infty} w(\boldsymbol{x}-\boldsymbol{x}') g_1^2(\boldsymbol{x}') \mathrm{d}^2 x' \int\limits_{-\infty}^{\infty} w(\boldsymbol{x}-\boldsymbol{x}') g_2^2(\boldsymbol{x}'-\boldsymbol{s}) \mathrm{d}^2 x' \right)^{1/2}}. \tag{14.3}$$

or in a more compact notation for weighted averaging with an overline:

$$\boxed{r(\boldsymbol{x}, \boldsymbol{s}) = \frac{\overline{g_1(\boldsymbol{x}) g_2(\boldsymbol{x}-\boldsymbol{s})}}{\left[\overline{g_1^2(\boldsymbol{x})}\;\overline{g_2^2(\boldsymbol{x}-\boldsymbol{s})}\right]^{1/2}} \rightarrow \text{maximum.}} \tag{14.4}$$

The resulting cross-correlation coefficient is a four-dimensional function that depends on the position in the image $\boldsymbol{x}$ and the shift $\boldsymbol{s}$.

14.2.2 ♯Fast Iterative Maximum Search

The correlation method in the form presented so far is a very computationally expensive operation. We can speed it up considerably if we restrict the calculation to a quick approach to finding the position of the maximum of r, since this is all we are interested in.

One way to calculate the position of the maximum directly is to approximate the cross-correlation function by a Taylor series. We therefore develop the cross-correlation coefficient at the position of the maximum $\breve{\boldsymbol{s}}$ into a second-order Taylor series:

$$\begin{aligned} r(\boldsymbol{s}) &\approx r(\breve{\boldsymbol{s}}) + \tfrac{1}{2}\breve{r}_{xx}(\breve{\boldsymbol{s}})(s_1-\breve{s}_1)^2 + \tfrac{1}{2}\breve{r}_{yy}(\breve{\boldsymbol{s}})(s_2-\breve{s}_2)^2 \\ &\quad + \breve{r}_{xy}(\breve{\boldsymbol{s}})(s_1-\breve{s}_1)(s_2-\breve{s}_2) \\ &= r(\breve{\boldsymbol{s}}) + \tfrac{1}{2}(\boldsymbol{s}-\breve{\boldsymbol{s}})^T \boldsymbol{H}(\breve{\boldsymbol{s}})(\boldsymbol{s}-\breve{\boldsymbol{s}}), \end{aligned} \tag{14.5}$$

where $\boldsymbol{H}$ is the *Hessian matrix* defined in Eq. (12.6). We do not know the position of the maximum correlation coefficient. Therefore, we assume that the second-order derivatives are constant near the maximum and calculate the maximum from the position of the previous iteration $\boldsymbol{s}^{(i)}$. If we have no other information, we set our initial guess to zero: $\boldsymbol{s}^{(0)} = \boldsymbol{0}$. As long as we have not yet found the position of the maximum correlation coefficient, the first derivative at $\boldsymbol{s}^{(i)}$ will not yet disappear and can be calculated from Eq. (14.5):

$$\nabla r(\boldsymbol{s}^{(i)}) = \boldsymbol{H}(\boldsymbol{s}^{(i)})(\boldsymbol{s}^{(i)} - \breve{\boldsymbol{s}}). \tag{14.6}$$

Assuming that the Hessian matrix is invertible, we obtain the following iteration:

$$s^{(i+1)} = s^{(i)} - H^{-1}(s^{(i)})\nabla r(s^{(i)}) \quad \text{with} \quad s^{(0)} = \mathbf{0}. \tag{14.7}$$

This type of iteration is known as *Newton-Raphson iteration* [248]. To calculate the shift, we therefore only need to calculate the first and second derivatives of the cross-correlation coefficient.

14.2.3 ♯Rating

The correlation approach allows changes in light intensity between the two images to the extent that the illumination within a selected window can be globally different. This makes techniques based on the correlation method very useful for a special case of motion analysis, namely the processing of stereo images where small intensity variations always occur between the left and right images due to the two different cameras used.

In fact, the fast search for maxima described in the previous section is a standard approach for determining the disparity of stereo images. Quam [252] uses this method together with a coarse-to-fine strategy and Nishihara [232] in a modified version with the sign of the Laplacian-of-Gaussian filter as a feature. Nishihara reports an accuracy of about 0.1 pixels for small shifts. Gelles et al. [105] can determine movements in cells with a precision of about 0.02 pixels using the correlation method. However, they use a more complex approach by calculating the centroid of the cross-correlation function. The model-adapted approach of Diehl and Burkhardt [61] can be regarded as an extended correlation method, as it also allows for rotation and other forms of movement. Due to the non-linear parameters, an iterative approach is required.

The correlation method differs from the motion estimation methods discussed below in that it is based on the comparison of only two images. Even if we extend the correlation method to more than two images through multiple correlations, it remains an approach with a discrete time step that lacks the elegance of the other methods that can be formulated in continuous space. Another disadvantage of the correlation method is the high computational cost.

14.3 Motion in Space-time Images

14.3.1 Orientation in Space-time

The discussion in the previous Sections 14.1–14.2 has shown that the analysis of motion from only two consecutive images is fraught with serious problems. The question is whether these problems can be at least partially overcome if we extend the analysis to an entire image sequence. With a pair of images, we only have a "snapshot" of the motion field, so we do not know how the movement continues over time. This means, for example, that we

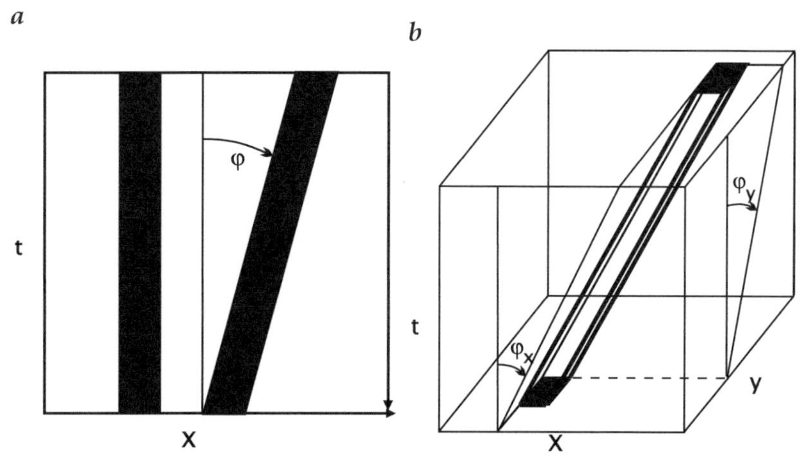

*Figure 14.8: Space-time images: **a** two-dimensional space-time image with a space and a time coordinate; **b** three-dimensional space-time image.*

have no information about acceleration and cannot observe how parts of an object appear and disappear again.

In this section, we consider the basics of image sequence analysis in a multidimensional space that is spanned over one time coordinate and one to three space coordinates. We refer to this space as *space-time domain* or ***xt*** space.

In a *space-time image*, an image point or pixel becomes a *voxel*, i.e. it represents the gray value in a small volume element with the extents Δx, Δy and Δt. When capturing real 3D data, we reach the limits of our visual imagination (see the discussion in Section 8.2). This results in the great need for suitable representations of such data to visualize properties of interest.

We first consider a simple example with one space and one time coordinate (Fig. 14.8a). A stationary one-dimensional object shows vertically oriented gray value structures. If an object moves, it shifts from image to image and appears as an inclined gray value structure. The velocity u is directly linked to the orientation in place/time images; in the simple case of a 2D place/time image, it is given by

$$u = -\tan\varphi, \tag{14.8}$$

where φ is the angle between the t-axis and the direction in which the gray values are constant. The minus sign in Eq. (14.8) comes from the fact that counterclockwise angles are counted positively. The extension to two spatial dimensions is simple; it is shown in Fig. 14.8b. The speed results in

$$\boldsymbol{u} = -\begin{bmatrix} \tan\varphi_x \\ \tan\varphi_y \end{bmatrix}. \tag{14.9}$$

Figure 14.9: *Traffic scene on the Hanauer Landstraße in Frankfurt/Main as an example of a space-time image, depicted as an image cube. The time axis approaches the viewer from the depth. On the right-hand side of the cube, a yt slice can be seen for the x coordinate marked with a vertical white line in the xy image, while the upper side shows a xt slice marked with the horizontal white line (from Jähne [156]).*

The angles φ_x and φ_y are defined as the angles between the projections onto the xt and yt planes of a vector in the direction of the constant gray values and the t axis.

A practical image example shows Fig. 14.9. A three-dimensional space-time image can be thought of as a stack of consecutive images and represented as an *image cube* (Fig. 14.9). On each visible face of the cube, we display a cross-section through the image stack in the corresponding direction. Thus, the top side shows an xt slice and the right side a yt slice. The slices originate from the row and the column, which are each marked by white lines in the last image of the time series on the front side of the cube.

The movement on the roadway under consideration is almost vertical, so that the yt cross-section can be viewed as a 2D time/space image. The movement is immediately recognizable. If the cars stop at a traffic light, the structures are oriented horizontally. Phases with increasing and constant speed can be easily distinguished from each other.

We can summarize the following: Motion appears in space-time images as *orientation*. This fundamental fact forms the basis of movement analysis in space-time. The conceptual difference to the approach with only two consecutive images is that velocity in space-time images can be measured *directly as orientation* and not as discrete *displacement*. The difference between these two concepts is greater than it first appears. We can now formulate algorithms for motion estimation in continuous $\boldsymbol{x}t$ space and consider *analytically* before applying a suitable discretization strategy. This makes it possible to clearly distinguish between fundamental deficiencies of an approach and errors due to discretization.

The presented motion analysis approach has much in common with the problem of reconstructing 3D images from projections (Section 9.3). In fact, we can think of motion analysis as a geometric determination of velocity by observing transparent three-dimensional space-time images from different angles. At the observation angle that corresponds to a certain speed, all individual images must lie directly on top of each other in the time axis. The speed then results from the angle between the viewing direction and the time axis. If we only look at the edge of a moving object, it appears as an inclined plane in the space-time image. We can therefore change our viewing direction along the edge at will and still get the edges in the entire image stack to coincide. Thus we again encounter the *aperture problem* considered in Section 14.1.2, this time from a different perspective.

14.3.2 Motion in Fourier Space

The introduction of the space-time domain has the significant advantage that we can also analyze motion in the corresponding Fourier space, the *wavenumber-frequency space* or $\boldsymbol{\kappa}\nu$ space for short. As an introduction, let us consider the example of a 3D image sequence in which all objects move at a constant speed. Such a sequence $g(\boldsymbol{x}, t)$ can be described by

$$g(\boldsymbol{x}, t) = g(\boldsymbol{x} - \boldsymbol{u}t). \tag{14.10}$$

The Fourier transform of this sequence is

$$\tilde{g}(\boldsymbol{\kappa}, \nu) = \int_t \int_x g(\boldsymbol{x} - \boldsymbol{u}t) \exp[-2\pi \mathrm{i}(\boldsymbol{\kappa}\boldsymbol{x} - \nu t)] \mathrm{d}^2 x \mathrm{d}t. \tag{14.11}$$

With

$$\boldsymbol{x}' = \boldsymbol{x} - \boldsymbol{u}t$$

we get

$$\tilde{g}(\boldsymbol{\kappa}, \nu) = \int_t \left[\int_{x'} g(\boldsymbol{x}') \exp(-2\pi \mathrm{i}\boldsymbol{\kappa}\boldsymbol{x}') \right] \exp(-2\pi \mathrm{i}\boldsymbol{\kappa}\boldsymbol{u}t) \exp(2\pi \mathrm{i}\nu t) \mathrm{d}^2 x' \mathrm{d}t.$$

The inner integral contains the spatial coordinates and results in the spatial Fourier transform $\tilde{g}(\boldsymbol{\kappa})$ of the image $g(\boldsymbol{x}')$. The outer integral over the time coordinate is reduced to a δ function, resulting in a total:

$$\tilde{g}(\boldsymbol{\kappa}, \nu) = \tilde{g}(\boldsymbol{\kappa})\delta(\boldsymbol{\kappa}\boldsymbol{u} - \nu). \tag{14.12}$$

This equation states that an object moving with the speed $\boldsymbol{u}$ only occupies a two-dimensional subspace of the three-dimensional $\boldsymbol{\kappa}\nu$ space. The equation for the plane results directly from the argument of the δ function in Eq. (14.12):

$$\nu = \boldsymbol{\kappa}\boldsymbol{u}. \tag{14.13}$$

This plane intersects the k_1k_2 plane normal to the direction of velocity, as the scalar product $\boldsymbol{\kappa}\boldsymbol{u}$ disappears in this direction. The gradient of the plane, a two-component vector, gives the velocity:

$$\nabla_\kappa \nu = \nabla_\kappa(\boldsymbol{\kappa}\boldsymbol{u}) = \boldsymbol{u}.$$

The index κ in the gradient operator indicates that the partial derivatives are calculated in relation to $\boldsymbol{\kappa}$.

These considerations show — at least in principle — how to determine the velocity in an image sequence with constant velocity. Calculate the Fourier transform of the sequence and then determine the slope of the plane on which the spectrum of the sequence is located. This works best if the scene contains small structures, i. e. high wavenumbers distributed in many directions. However, the slope of the plane cannot be determined unambiguously if the spectrum on the plane becomes a line, which happens with spatially oriented gray value structures. Then we only obtain the component of the slope of the plane in the direction of the spatial local orientation from the line in Fourier space. This brings us back to the *aperture problem* (Section 14.1.2) in the $\kappa\nu$ space.

14.3.3 Optical Flow

The examples discussed in Section 14.1.1 have shown that the motion and gray value changes are not equivalent. We want to quantify the relationship between the two in this section. Two terms are important in this context: the *motion field* and the *optical flow*. The motion field of an image is the projection of the movements in the 3D scene onto the image plane. This variable is to be extracted from an image sequence. The optical flow is defined as the "flow of gray values" in the image plane. This is what we observe. The optical flow and the field of motion are only the same if the illumination is not changing with time. Although this equivalence initially seems plausible, on closer inspection it actually only applies in very few cases. The fundamental questions are therefore how significant the deviations are and whether we can assume the equivalence of optical flow and motion field in practical application.

Two classic examples in which the projected field of motion and the optical flow are not the same have been described by Horn [141]. The first is a rotating sphere with an arbitrary uniform surface. Such a sphere can rotate about any axis through its center of gravity without resulting in an optical flow field. The opposite example is the same sphere at rest and illuminated by a moving light source. Now the field of motion is zero, but the changes in the gray values caused by the moving light source result in an optical flow field.

At this point, we must first clarify the different notations for motion in image sequences, as there is considerable confusion in the literature due to the many different terms. The *optical flow* or *image flow* is the visible motion in the image plane. It has the dimension of a velocity and we denote it by $\boldsymbol{f} = [f_1, f_2]^T$. If the optical flow is determined from two consecutive images, it appears as a *displacement vector* (*DV*) from the features of the first to those of the second image. If there is a displacement vector at each image point, this is referred to as a *displacement vector field* (*DVF*) $\boldsymbol{s} = [s_1, s_2]^T$. We obtain an approximation of the optical flow by dividing the DVF by the time interval between the two images. It is important to note that the optical flow is a term belonging to continuous space and the DVF is its discrete counterpart. The field of motion $\boldsymbol{u} = [u_1, u_2]^T = [u, v]^T$ in the image plane is the projection of the three-dimensional physical field of motion through the optics.

The concept of optical flow originates from *hydrodynamics*. Applied to images, the motion causes gray values (corresponding to an optical signal) to "flow" across the image plane, just as volume elements flow in liquids or gases. The continuity equation plays an important role in hydrodynamics. It expresses the fact that mass is conserved in a flow, i. e. fluid particles cannot disappear in a flow. The question now is whether a similar continuity equation can also be formulated for gray values and under what conditions gray values are preserved.

In hydrodynamics, the continuity equation for the density ϱ of the fluid results in

$$\frac{\partial \varrho}{\partial t} + \nabla(\boldsymbol{u}\varrho) = \frac{\partial \varrho}{\partial t} + \boldsymbol{u}^T\nabla\varrho + \varrho\nabla\boldsymbol{u} = 0. \tag{14.14}$$

This equation, which applies to two- and three-dimensional flows, describes the conservation of mass in a fluid in differential form. The change in density over time is compensated for by the divergence of the flow density $\boldsymbol{u}\varrho$. By integrating the continuity equation over an arbitrary volume element, we can also write the equation in integral form:

$$\int_V \left(\frac{\partial \varrho}{\partial t} + \nabla(\boldsymbol{u}\varrho)\right) \mathrm{d}V = \frac{\partial}{\partial t}\int_V \varrho \,\mathrm{d}V + \oint_A \varrho\boldsymbol{u}\,\mathrm{d}\boldsymbol{a} = 0. \tag{14.15}$$

The second volume integral was replaced by a surface integral around the volume using Gauss's integral theorem. $\mathrm{d}\boldsymbol{a}$ is a vector normal to the surface element $\mathrm{d}A$. The integral form of the continuity equation states that the change in mass over time is caused by a net flux into the volume, integrated over the entire surface of the volume.

How can we now derive a similar continuity equation for the optical flow $\boldsymbol{f}$, the *brightness change constraint equation* (*BCCE*) or *optical flow constraint* (*OFC*)? Can we simply replace the velocity $\boldsymbol{u}$ with the optical flow $\boldsymbol{f}$? The quantity analogous to the density ϱ would be the irradiance E or the gray

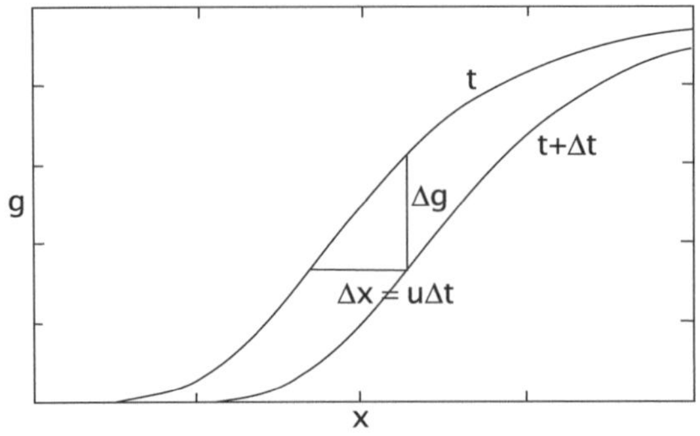

Figure 14.10: Illustration of the differential method for determining the movement in the one-dimensional case.

value g. However, we should be careful and take a closer look at the terms in Eq. (14.14). The left-hand term $\boldsymbol{f}^T\nabla g$ describes the change in brightness over time due to a moving gray value gradient.

However, the second term with the divergence of the velocity field $g\nabla\boldsymbol{f}$ appears questionable. It would cause a temporal change even in a region with a constant irradiance if the divergence of the flow field is not equal to zero. Such a case exists, for example, when an object moves away from the camera. However, the irradiance in the image plane remains constant if the object illumination does not change. The total captured radiation decreases with the square of the distance of the object. However, as the projected area of the object on the image plane decreases in the same proportion, the irradiance does not change.

We therefore omit the last part of the continuity equation for the optical flow and obtain

$$\boxed{\frac{\partial g}{\partial t} + \boldsymbol{f}^T\nabla g = 0.} \tag{14.16}$$

In the one-dimensional case, the continuity of the optical flow takes the following simple form:

$$\frac{\partial g}{\partial t} + f\frac{\partial g}{\partial x} = 0, \tag{14.17}$$

from which we can directly calculate the one-dimensional velocity, provided that the spatial derivative does not disappear:

$$\boxed{f = -\frac{\partial g}{\partial t}\bigg/\frac{\partial g}{\partial x}.} \tag{14.18}$$

The velocity is therefore the ratio of the temporal to the spatial derivative. As Fig. 14.10 shows, this basic relationship can also be derived geometrically. In the time interval Δt, a gray value is shifted by the distance $\Delta x = f\Delta t$. The gray value changes by $g(x, t + \Delta t) - g(x, t)$. The gray value changes can also be expressed by the gradient of the gray value edge:

$$g(x, t + \Delta t) - g(x, t) = -\frac{\partial g(x, t)}{\partial x}\Delta x = -\frac{\partial g(x, t)}{\partial x}u\Delta t. \tag{14.19}$$

From this we obtain the continuity equation for the optical flow Eq. (14.17) through the limit transition $\Delta t \to 0$.

The continuity equation of the optical flow Eq. (14.16) can only be a first approximation after the considerations in the introductory Section 14.1.1 about motion and gray value changes. The reason lies in the nature of the reflection of opaque surfaces, which depends on the viewing direction, the surface normal and the directions of the incident light. Each object receives radiation not only directly from light sources, but also from all other objects in the scene that can be seen from the object.

The radiant flux density emanating from the surface of an object therefore depends on the position of all other objects in a scene. In the quest for photorealistic computer-generated images, such problems are studied in detail in computer graphics. A major step towards this goal was a method called *radiosity*, which solves exactly the above described relationships of the radiance flux density of objects [87]. A general expression for the object radiance flux density — the now famous *rendering equation* — goes back to the work of Kajiya [171]. In image sequence processing, it is in principle necessary to reverse this equation and thereby determine the surface reflectivity from the measured object radiant flux density. The surface reflectivity is a property that is independent of the surface orientation and the position of other objects and is therefore ideal for motion analysis. However, such an approach is unrealistic as it requires a reconstruction of the three-dimensional scene before the inversion of the rendering equation can even be tackled.

Since there is no generally valid continuity equation for the optical flow, it is important that we compare possible additional terms with those in the standard continuity equation. All other terms essentially depend on the rate of change of a number of quantities, but not on the brightness gradients. If the gray value gradient is large, the influence of additional terms becomes small. From this we can conclude that the determination of the speed is most reliable for steep gray value edges, while it can be strongly disturbed in regions with only small gray value gradients. This conclusion aligns with the findings of Verri and Poggio [341, 342], who emphasize that the distinction between optical flow and motion fields is insignificant for practical applications.

Another observation is also important. It is certainly true that the historical approach to determine the displacement vector from only two consecutive images is not robust. In general, we cannot distinguish whether gray value changes come from a displacement vector or from any other source. However, in space-time images, the determination of the optical flow becomes more robust. We will investigate this using two examples.

First, we show that gray value changes caused by global illumination differences can be distinguished from those caused by motion. Fig. 14.11

Figure 14.11: *Static scene with lighting changes.* ***a*** *xt cross-section on the upper bright line (sky in* ***b****);* ***b*** *first image of the sequence;* ***c*** *xt-cross-section on the lower bright line (roof area in* ***b****). The time axis spans 3,4 h and runs from top to bottom (from Jähne [156]).*

shows an image sequence of a static scene recorded at a speed of five frames per minute. The two space-time slices (Fig. 14.11a and c), whose location is marked with two white lines in the first image of the sequence (Fig. 14.11b), span a period of approximately 3.4 h. The upper line runs through the skyscraper and the sky. We can see from the area of the sky that clouds and sections with direct sunlight have alternated. The lower line crosses several

a

b

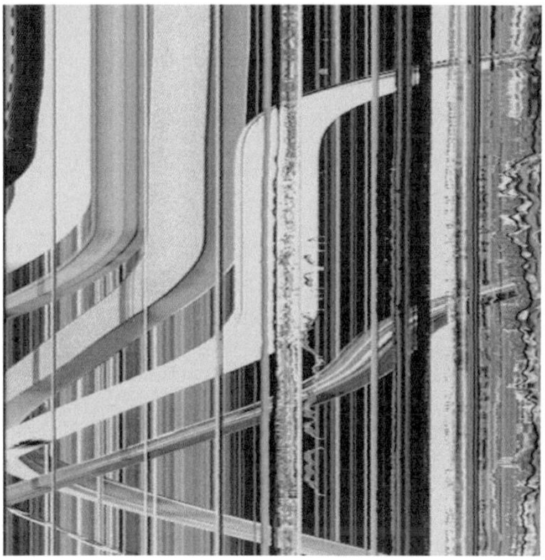

Figure 14.12: *Traffic scene at the city limits of Hanau:* ***a*** *last image of the sequence;* ***b*** *xt cross-section in the area of the line marked in* ***a****. The time axis spans 20.5 s and runs from top to bottom (from Jähne [156]).*

roofs with skylights and a gable wall. In both *xt* sections, the changes in illumination can be seen as horizontal stripes that appear to be transparently superimposed on the vertical stripes that characterize a static scene. Since a horizontal pattern means an object moving at infinite speed, these patterns can be eliminated by directional filtering, for example, without disturbing the motion analysis.

The second example shows that it is still possible to determine motion in space-time images when overlays occur and the local lighting of an object changes because it is rotating. In Fig. 14.12 you can see a traffic scene on the outskirts of Hanau. In the last image of the sequence (Fig. 14.12a), you can see that a road junction has been observed through the branches of a tree in the right foreground. A road runs horizontally from left to right, with the intersection controlled by a traffic light on the left. The space-time slice (Fig. 14.12b) comes from the image sequence in the area of the horizontal line in Fig. 14.12a. It shows various overlays: The car tracks disappear under the static vertical pattern of tree branches, lampposts and traffic lights. In addition, we see that the time trace of the truck shows significant gray value changes as it turned at the intersection and therefore the lighting conditions changed as it moved through the scene. Nevertheless, the temporal trace is continuous and promises a reliable speed estimation.

We can conclude that it is best to stick with the standard continuity equation Eq. (14.16) for motion estimation and use it as the basis for the methods developed in this section. Because of the enormous variety of additional terms, this approach still seems to be the most reasonable and broadly applicable, as it contains the fundamental continuity condition.

14.4 Differential Methods in Space-time Images

14.4.1 Basics

Differential methods are the classical approach for determining motion from two consecutive images. In this chapter, we discuss how these techniques can be applied to space-time images. The continuity equation for the optical flow Eq. (14.16) is the starting point:

$$\frac{\partial g}{\partial t} + \boldsymbol{f}^T \boldsymbol{\nabla} g = 0. \tag{14.20}$$

This single scalar equation contains D unknown vector components in D-dimensional space. We therefore cannot uniquely determine the optical flow $\boldsymbol{f} = [f_1, f_2]^T$ even in two-dimensional space. The scalar product $\boldsymbol{f}^T \boldsymbol{\nabla} g$ is equal to the magnitude of the gray value gradient multiplied by the component of $\boldsymbol{f}$ that runs in the direction of the gradient, i. e. perpendicular to the local gray value edge:

$$\boldsymbol{f}^T \boldsymbol{\nabla} g = f_\perp |\boldsymbol{\nabla} g|.$$

Only the component perpendicular to the edge can therefore be determined from the optical flow. This is referred to as the *aperture problem*, which we have already discussed qualitatively in Section 14.1.2. The optical flow perpendicular to the edge results in

$$f_\perp = -\frac{\partial g}{\partial t} / |\boldsymbol{\nabla} g| \, . \tag{14.21}$$

Accordingly, it is not possible to determine the complete vector via first-order derivatives at a *individual* point in the time/space image.

14.4.2 Least Squares Method

Instead of a single point, we can use a neighborhood to determine the optical flow. We assume that the optical flux in the neighborhood is constant and discuss under which conditions an unambiguous determination of the optical flux is possible. Now we still have the two unknowns $\boldsymbol{f} = [f_1, f_2]^T$, but at many points the continuity equation Eq. (14.20) for the optical flow. We therefore obtain an overdetermined system of equations. So there is no exact solution, but only one that minimizes an error norm (Appendix C.6.2). Using the *least squares* method, we look for a solution that minimizes Eq. (14.20) within a local neighborhood. Therefore, the convolution integral

$$\|e\|_2^2 = \int_{-\infty}^{\infty} w(\boldsymbol{x}-\boldsymbol{x}', t-t')\left(f_1\dot{g}_x(\boldsymbol{x}') + f_2\dot{g}_y(\boldsymbol{x}') + \dot{g}_t(\boldsymbol{x}')\right)^2 \mathrm{d}^2x'\mathrm{d}t' \tag{14.22}$$

is to be minimized as an error norm. Note that $\boldsymbol{f} = [f_1, f_2]^T$ is considered constant within the neighborhood. Like $\|e\|$, it naturally depends on $\boldsymbol{x}$. In order to obtain a compact notation, we omit the explicit dependence of $\dot{g}_x$, $\dot{g}_y$ and $\dot{g}_t$ with $\boldsymbol{x}'$ in the following equations. We use $\dot{g}_p$ as the abbreviation for the partial derivative $\partial g/\partial p$. In the integral, the square of the remaining deviation from the continuity equation is averaged over the region that depends on the window function w. To further simplify the following equations, we abbreviate this weighted averaging procedure with

$$\boxed{\|e\|_2^2 = \overline{\left(f_1\dot{g}_x + f_2\dot{g}_y + \dot{g}_t\right)^2} \rightarrow \text{minimum.}} \tag{14.23}$$

The window function, w, determines the size of the neighborhood. This makes the method flexible. Although optional, the averaging can be extended in the time direction. If we choose a rectangular environment with the same weighting factor for all points, this corresponds to a simple block matching technique and averaging with a *box filter*. However, since we have learned about the poor smoothing properties of rectangular filters in (Section 11.3), a weighting function that slowly decreases towards the edge is a better solution. In the continuous case, a Gaussian function is a good choice; for discrete signals, an *binomial filter* (Section 11.4) makes sense.

Equation (14.23) can be solved by setting the partial derivatives to zero:

$$\begin{aligned} \frac{\partial \|e\|_2^2}{\partial f_1} &= \overline{2\dot{g}_x\left(f_1\dot{g}_x + f_2\dot{g}_y + \dot{g}_t\right)} \stackrel{!}{=} 0, \\ \frac{\partial \|e\|_2^2}{\partial f_2} &= \overline{2\dot{g}_y\left(f_1\dot{g}_x + f_2\dot{g}_y + \dot{g}_t\right)} \stackrel{!}{=} 0. \end{aligned} \tag{14.24}$$

A linear system of equations follows from these conditions

$$\boxed{\begin{bmatrix} \overline{\dot{g}_x\dot{g}_x} & \overline{\dot{g}_x\dot{g}_y} \\ \overline{\dot{g}_x\dot{g}_y} & \overline{\dot{g}_y\dot{g}_y} \end{bmatrix}\begin{bmatrix} f_1 \\ f_2 \end{bmatrix} = -\begin{bmatrix} \overline{\dot{g}_x\dot{g}_t} \\ \overline{\dot{g}_y\dot{g}_t} \end{bmatrix},} \tag{14.25}$$

or in compact matrix notation

$$\boldsymbol{M}\boldsymbol{f} = \boldsymbol{d}. \tag{14.26}$$

The expressions $\overline{\dot{g}_p \dot{g}_q}$ represent a regularized estimate composed of convolutions and nonlinear point operations. In operator notation, we can formulate them by

$$\mathcal{B}(\mathcal{D}_p \cdot \mathcal{D}_q), \tag{14.27}$$

where $\mathcal{D}_p$ is a suitable first-order derivative operator in direction p (Chapter 12) and $\mathcal{B}$ is a smoothing filter (Chapter 11).

This results in the following sequence of image processing operators:

1. Applying the derivative operators $\mathcal{D}_p$ and $\mathcal{D}_q$ to the image. An image with a first-order derivative in the direction p or q is obtained.
2. Point-by-point multiplication of the two derivative images.
3. Convolution of the resulting image with the smoothing mask $\mathcal{B}$. The choice of smoothing mask determines the size and shape of the window function in Eq. (14.27).

As the pointwise multiplication is a non-linear operation, it must not be interchanged with the smoothing in the sequence.

The linear system of equations Eq. (14.25) can be solved if the matrix is invertible. This is the case if the determinant of the matrix is not zero:

$$\det \boldsymbol{M} = \overline{\dot{g}_x \dot{g}_x}\, \overline{\dot{g}_y \dot{g}_y} - \overline{\dot{g}_x \dot{g}_y}^2 \neq 0. \tag{14.28}$$

From this inequality we can conclude that two conditions must be met:

1. Not all partial derivatives $\dot{g}_x$ and $\dot{g}_y$ may be zero. In other words, the neighborhood must not be a surface with constant gray values.
2. The gradients in the neighborhood must not all point in the same direction. If this were the case, we could express $\dot{g}_y$ by $\dot{g}_x$ except for a constant factor ($\dot{g}_y = c\dot{g}_x$) and the determinant of $\boldsymbol{M}$ in Eq. (14.28) would be zero.

The solution for the optical flow $\boldsymbol{f}$ can be written directly, as it is easy to invert the 2×2 matrix $\boldsymbol{M}$:

$$\boldsymbol{M}^{-1} = \frac{1}{\det \boldsymbol{M}} \begin{bmatrix} \overline{\dot{g}_y \dot{g}_y} & -\overline{\dot{g}_x \dot{g}_y} \\ -\overline{\dot{g}_x \dot{g}_y} & \overline{\dot{g}_x \dot{g}_x} \end{bmatrix} \quad \text{if} \quad \det \boldsymbol{M} \neq 0. \tag{14.29}$$

With $\boldsymbol{f} = \boldsymbol{M}^{-1}\boldsymbol{d}$ we then get

$$\begin{bmatrix} f_1 \\ f_2 \end{bmatrix} = -\frac{1}{\det \boldsymbol{M}} \begin{bmatrix} \overline{\dot{g}_x \dot{g}_t}\, \overline{\dot{g}_y \dot{g}_y} - \overline{\dot{g}_y \dot{g}_t}\, \overline{\dot{g}_x \dot{g}_y} \\ \overline{\dot{g}_y \dot{g}_t}\, \overline{\dot{g}_x \dot{g}_x} - \overline{\dot{g}_x \dot{g}_t}\, \overline{\dot{g}_x \dot{g}_y} \end{bmatrix}. \tag{14.30}$$

The solution still looks quite complicated. It can be considerably simplified by the fact that $\boldsymbol{M}$ is a symmetric matrix. Every symmetric matrix can be brought into diagonal form by rotating the coordinate system in the so-called *principle axis system* (Appendix C.3.3). The matrix $\boldsymbol{M}$ is then reduced to

$$\boldsymbol{M}' = \begin{bmatrix} \overline{\dot{g}_{x'}\dot{g}_{x'}} & 0 \\ 0 & \overline{\dot{g}_{y'}\dot{g}_{y'}} \end{bmatrix}, \tag{14.31}$$

the determinate simplifies to $\det \boldsymbol{M}' = \overline{\dot{g}_{x'}\dot{g}_{x'}}\ \overline{\dot{g}_{y'}\dot{g}_{y'}}$, and the optical flow results in

$$[f_{1'}, f_{2'}]^T = -\left[\frac{\overline{\dot{g}_{x'}\dot{g}_t}}{\overline{\dot{g}_{x'}\dot{g}_{x'}}}, \frac{\overline{\dot{g}_{y'}\dot{g}_t}}{\overline{\dot{g}_{y'}\dot{g}_{y'}}}\right]^T. \tag{14.32}$$

This equation reflects quantitatively the qualitative discussion of the *aperture problem* that we discussed in Section 14.1.2. The principal axes are oriented along the directions of minimum and maximum root mean square values of the spatial gray value changes, which are perpendicular to each other. Since the matrix $\boldsymbol{M}'$ is diagonal, both values are not correlated with each other. Now we can distinguish three cases:

1. $\overline{\dot{g}_{x'}\dot{g}_{x'}} > 0$, $\overline{\dot{g}_{y'}\dot{g}_{y'}} > 0$: The gray values change in all directions; both components of the optical flow can be determined.
2. $\overline{\dot{g}_{x'}\dot{g}_{x'}} > 0$, $\overline{\dot{g}_{y'}\dot{g}_{y'}} = 0$: The gray values only change in one direction (perpendicular to an edge). Then only the component of the optical flow in the x' direction can be determined (aperture problem). The component of the optical flow parallel to the edge remains undetermined.
3. $\overline{\dot{g}_{x'}\dot{g}_{x'}} = \overline{\dot{g}_{y'}\dot{g}_{y'}} = 0$: The gray values do not change in any direction. In this case, there is a region with constant gray values and both components of the optical flow remain undetermined.

It should be noted that only the matrix $\boldsymbol{M}$ determines the solution of the least squares method. This matrix does not contain any temporal derivatives, only spatial ones. This means that the spatial derivatives and thus the spatial structures of the image completely determine whether and how precisely the optical flow can be calculated.

14.4.3 ♯Error Analysis

Noise can systematically distort the determination of the optical flow. Here we show how we can analyze the influence of noise in a very general way. We assume that an image sequence consists of a structure moving at a constant velocity $\boldsymbol{u}$, overlaid with zero-mean isotropic noise:

$$g'(\boldsymbol{x}, t) = g(\boldsymbol{x} - \boldsymbol{u}t) + n(\boldsymbol{x}, t). \tag{14.33}$$

This is a very general approach, as we make no assumption about the shape of the gray value structure. The expression $g(\boldsymbol{x} - \boldsymbol{u}t)$ only states that an arbitrary spatial gray value structure moves with a constant velocity $\boldsymbol{u}$. In this way, we have reduced a general function $g(x_1, x_2, t)$ with three parameters to a function with only two parameters: $g(x_1 - u_1 t, x_2 - u_2 t)$. We also assume that the partial derivatives of the noise function are not correlated with each other and with the partial derivatives of the image function. We therefore use the conditions

$$\overline{n} = 0, \quad \overline{n_p n_q} = \sigma_n^2 \delta_{p-q}, \quad \overline{g_p n_q} = 0, \tag{14.34}$$

and the partial derivative are

$$\nabla g' = \nabla g + \nabla n \quad \dot{g}'_t = -\boldsymbol{u}\nabla g + \partial_t n_t. \tag{14.35}$$

These conditions result in the following value for the optical flow:

$$f = \boldsymbol{u}(\overline{\nabla g \nabla g^T} + \overline{\nabla n \nabla n^T})^{-1}\overline{\nabla g \nabla g^T}. \tag{14.36}$$

The key to understanding this matrix equation lies in the observation that the noise matrix $\overline{\nabla n \nabla n^T}$ is diagonal in any coordinate system because of the conditions given by Eq. (14.34). Therefore, we can transform the equation into the principal axis system in which $\nabla g \nabla g^T$ is a diagonal matrix. Then we obtain

$$f = \boldsymbol{u}\begin{bmatrix} \overline{\dot{g}_{x'}{}^2} + \sigma_n^2 & 0 \\ 0 & \overline{\dot{g}_{y'}{}^2} + \sigma_n^2 \end{bmatrix}^{-1} \begin{bmatrix} \overline{\dot{g}_{x'}{}^2} & 0 \\ 0 & \overline{\dot{g}_{y'}{}^2} \end{bmatrix}.$$

As long as the noise is not zero, the inverse of the first matrix always exists and we get

$$f = \boldsymbol{u}\begin{bmatrix} \dfrac{\overline{\dot{g}_{x'}{}^2}}{\overline{\dot{g}_{x'}{}^2} + \sigma_n^2} & 0 \\ 0 & \dfrac{\overline{\dot{g}_{y'}{}^2}}{\overline{\dot{g}_{y'}{}^2} + \sigma_n^2} \end{bmatrix}. \tag{14.37}$$

This equation shows that the determination of the optical flow is distorted towards smaller values. If the variance of the noise is as large as the magnitude squared of the gradient, then the estimated values are only half as large as the true values. This behavior makes the differential method a non-robust method, as it provides incorrect values when the noise is high.

If the noise is negligible, then the optical flow is estimated correctly. This result contradicts the widespread statement that differential methods do not provide accurate results if the gray value curve cannot be well approximated by a first-order Taylor series expansion (see, e. g. [304]). For example, Kearney et al. [177] performs an error analysis of the gradient method and comes to the conclusion that it provides incorrect values as soon as second-order spatial derivatives cannot be neglected.

These contradictory statements are clarified when we consider the additional errors caused by an inadequate discretization of the partial derivatives (see the discussion on optimal derivative filters in Section 12.4). The error of the optical flow is directly related to the errors in the direction of the discrete gradient (see also the discussion on orientation analysis in Section 13.3.6). Therefore, carefully optimized regularized gradient operators, as we discussed in Section 12.7.6, are essential for the accurate determination of the optical flow.

14.4.4 ♯Other Differential Methods

Second-order Differential Methods

The differential first-order method has the fundamental problem that with the continuity of the optical flow there is only one condition for the two unknown components of the optical flow $\boldsymbol{f} = [f_1, f_2]^T$ (Section 14.4.1). So far, we have only been able to make up for this deficit by considering the velocity in a neighborhood as constant and thus using it to determine the optical flow (Section 14.4.2).

Another approach is to use more than one property image or multi-channel images. This gives us two or more independent conditions at one point and enables us to determine both components of the optical flow locally. The crucial point, however, is that additional information is really introduced. It does not help if the new condition is correlated with existing ones.

In this way, we arrive at an important generalization of the differential method. We can apply any preprocessing to image sequences or extract any features and apply the methods discussed so far to them. If the continuity of the optical flow is preserved in the original image, this also applies to any derived feature image. We can apply non-linear point operations as well as neighborhood operations.

We first discuss the method of Girosi et al. [107]. He applied the continuity of optical flow to two feature images, namely the horizontal and vertical spatial derivatives:

$$\begin{aligned} \boldsymbol{f}\nabla\dot{g}_x + \ddot{g}_{xt} &= 0 \\ \boldsymbol{f}\nabla\dot{g}_y + \ddot{g}_{yt} &= 0. \end{aligned} \tag{14.38}$$

This results in a second-order differential method with the solution

$$\boxed{\boldsymbol{f} = -\boldsymbol{H}^{-1}\nabla\dot{g}_t \quad \text{if} \quad \det\boldsymbol{H} \neq 0,} \tag{14.39}$$

where $\boldsymbol{H}$ is the *Hessian matrix* as defined in Eq. (12.6).

If we also add the normal continuity equation for the optical flow, we obtain the following overdetermined system of equations with three equations and two unknowns:

$$\begin{bmatrix} \dot{g}_x & \dot{g}_y \\ \ddot{g}_{xx} & \ddot{g}_{xy} \\ \ddot{g}_{xy} & \ddot{g}_{yy} \end{bmatrix} \begin{bmatrix} f_1 \\ f_2 \end{bmatrix} = - \begin{bmatrix} \dot{g}_t \\ \ddot{g}_{xt} \\ \ddot{g}_{yt} \end{bmatrix}. \tag{14.40}$$

In this context, images taken with different sensors are also a promising method. For example, Markandey and Flinchbaugh [213] used multispectral images taken in the visible and infrared. Image sequence analysis of scenes illuminated with light sources from different directions have been investigated by Woodham [356]. This approach is particularly interesting because it allows direct reflections from shiny surfaces to be detected, thus eliminating an important source of errors.

Differential Geometric Modeling

The discussion in the previous sections has shown that the spatial structure of the gray values determines the motion. This fundamental fact is not adequately taken into account in the previous approach, as it relies only on first-order spatial derivatives. Second-order differential methods allow a direct, local solution under the condition that the Hessian matrix can be inverted Eq. (14.39). In this section, we consider the differential methods from the point of view of *differential geometry*. We assume that the gray value structures in two consecutive images differ only by a locally constant shift $\boldsymbol{s}$:

$$g(\boldsymbol{x} - 1/2\boldsymbol{s}, t_1) = g(\boldsymbol{x} + 1/2\boldsymbol{s}, t_2) . \tag{14.41}$$

This approach contains a different formulation of the continuity equation under the assumption that locally only a translation of the image takes place, neglecting any rotations or deformations of surface elements. We therefore assume that the velocity field does not change in a small neighborhood. For symmetry reasons, we distribute the displacement evenly over the two images. Assuming that the displacement vector $\boldsymbol{s}$ and the size of the surface element are small, we can develop the gray values at the point $\boldsymbol{x} = \boldsymbol{0}$ into a Taylor series. First, we consider a first-order evolution, i. e. we approximate the gray value distribution with a *plane*:

$$g(\boldsymbol{x} \pm 1/2\boldsymbol{s}) = g_0 + \nabla g \cdot (\boldsymbol{x} \pm 1/2\boldsymbol{s}) . \tag{14.42}$$

The planes of both images differ only in the shift $\boldsymbol{s}$. We sort the terms in Eq. (14.42) by increasing power of $\boldsymbol{x}$ in order to be able to perform a coefficient comparison:

$$g(\boldsymbol{x} \pm 1/2\boldsymbol{s}) = \underbrace{g_0 \pm 1/2\nabla g\, \boldsymbol{s}}_{\text{offset}} + \underbrace{\nabla g}_{\text{slope}}\, \boldsymbol{x}. \tag{14.43}$$

The first two terms contain the offset and the third the gradient of the plane. Now we can derive the offset $\boldsymbol{s} = (p, q)^T$ from the condition that both planes must be identical. Therefore, the two coefficients must also be identical and we obtain two equations:

$$\begin{array}{rcl} g_0(t_1) - g_0(t_2) & = & 1/2\,(\nabla g(t_1) + \nabla g(t_2))\,\boldsymbol{s}, \\ \nabla g(t_1) & = & \nabla g(t_2). \end{array} \tag{14.44}$$

The second equation states that the gradient in both images must be equal. Otherwise, a plane fit of the spatial gray values would not be a meaningful representation either. The first equation corresponds to the continuity of the optical flow Eq. (14.16). In Eq. (14.44) only the temporal derivative is already discretely expressed as the difference of the mean gray values of both images. Therefore, the gradient is replaced by the mean gradient in both images. In addition, we use the displacement vector field (DVF) $\boldsymbol{s}$ instead of the optical flow $\boldsymbol{f}$. As expected, a plane fit of the gray value distribution yields nothing new. We can still only determine the velocity component in the direction of the gray value gradient. Therefore, we have to develop Eq. (14.41) into a second-order Taylor series and obtain

$$
\begin{aligned}
g(\boldsymbol{x} \pm 1/2\boldsymbol{s}) &= g_0 \\
&+ \dot{g}_x \cdot (x \pm 1/2s_1) + \dot{g}_y \cdot (y \pm 1/2s_2) \\
&+ 1/2\ddot{g}_{xx} \cdot (x \pm 1/2s_1)^2 + 1/2\ddot{g}_{yy} \cdot (y \pm 1/2s_2)^2 \\
&+ \ddot{g}_{xy} \cdot (x \pm 1/2s_1)(y \pm 1/2s_2).
\end{aligned}
$$

Nagel [225] carried out a similar modeling of the gray value geometry by expanding it into a second-order Taylor series. However, he obtained complex non-linear equations that are only easy to solve under certain conditions. He called these conditions *gray value corners* and *gray value extrema*. The reason for the different results lies in the solution approach. Nagel compared the Taylor series of two images using the method of least squares, whereas here a direct comparison of coefficients is carried out.

A comparison of the coefficients of the second-order fit results in a total of six equations. The quadratic terms result in three equations, which state that all second-order spatial derivatives must match in both images:

$$
\begin{aligned}
\ddot{g}_{xx}(t_1) &= \ddot{g}_{xx}(t_2), \\
\ddot{g}_{yy}(t_1) &= \ddot{g}_{yy}(t_2), \\
\ddot{g}_{xy}(t_1) &= \ddot{g}_{xy}(t_2).
\end{aligned}
$$

If this is not the case, either the gray value distribution cannot be adequately described by a second-order Taylor series, or the requirement of a constant shift in the neighborhood is not met. The comparison of the coefficients of the zero-order and first-order terms results in the following three equations:

$$
\begin{aligned}
-(g_0(t_2) - g_0(t_1)) &= \tfrac{1}{2}\left(\dot{g}_x(t_1) + \dot{g}_x(t_2)\right)s_1 \\
&\quad +\tfrac{1}{2}\left(\dot{g}_y(t_1) + \dot{g}_y(t_2)\right)s_2, \\
-(\dot{g}_x(t_2) - \dot{g}_x(t_1)) &= \ddot{g}_{xx}s_1 + \ddot{g}_{xy}s_2, \\
-(\dot{g}_y(t_2) - \dot{g}_y(t_1)) &= \ddot{g}_{yy}s_2 + \ddot{g}_{xy}s_1.
\end{aligned}
\tag{14.45}
$$

Surprisingly, the coefficient comparison for the zero-order terms (offset) provides the same result as the plane fit Eq. (14.44). This means that the DVF is calculated correctly by a simple plane fit, even if the gray value distribution is no longer represented by a plane, but by a second-order polynomial.

The other two equations can be written as a simple linear system of equations with two unknowns:

$$\begin{bmatrix} \ddot{g}_{xx} & \ddot{g}_{xy} \\ \ddot{g}_{xy} & \ddot{g}_{yy} \end{bmatrix} \begin{bmatrix} s_1 \\ s_2 \end{bmatrix} = - \begin{bmatrix} \dot{g}_x(t_2) - \dot{g}_x(t_1) \\ \dot{g}_y(t_2) - \dot{g}_y(t_1) \end{bmatrix}. \tag{14.46}$$

The 2×2 matrix on the left-hand side can be easily inverted if $\ddot{g}_{xx}\ddot{g}_{yy} - (\ddot{g}_{xy})^2$ does not vanish. For this reason, it is possible to estimate the displacement between two images from a local neighborhood if we consider the *curvature* of the gray value distribution. So far, we have not discussed the conditions that a gray value distribution must fulfill for Eq. (14.45) to be reversible. These conditions are met if there is either a *gray value extremum* or a *grey value corner*. As already mentioned, these terms were coined by Nagel [225]. In a gray value extremum (and also in a saddle point), the two principal curvatures are not equal to zero. Then Eq. (14.46) is solvable. At a gray value corner, only one principal curvature is zero, but not the gradient in this direction. Therefore, the first and second equation from Eq. (14.45) can be used to determine both components of the optical flow vector.

No smoothing is required for the differential geometric method as second order derivatives are only used at one point. Nevertheless, smoothing of the image is often applied for a more stable estimation of the derivatives. Since convolution operations are commutative, this smoothing can also be performed after the derivatives have been calculated.

The difference of the first-order spatial derivatives in the images at time t_2 and t_1 in Eq. (14.46) is a discrete approximation for a temporal derivative, which can be replaced by a temporal derivative operator. Then the displacement vector must also be replaced by the optical flow vector. This finally gives us the following continuous formulation of the differential geometric method:

$$\begin{bmatrix} \overline{\ddot{g}_{xx}} & \overline{\ddot{g}_{xy}} \\ \overline{\ddot{g}_{xy}} & \overline{\ddot{g}_{yy}} \end{bmatrix} \begin{bmatrix} f_1 \\ f_2 \end{bmatrix} = - \begin{bmatrix} \overline{\ddot{g}_{xt}} \\ \overline{\ddot{g}_{yt}} \end{bmatrix}. \tag{14.47}$$

Space-time Energy Models

Gabor-like quadrature filters (Section 13.4.5) are often used to explain the motion vision of humans and animals. They are the basis for *space-time energy models* or *motion energy models* [2, 3, 129]. These terms are easily misunderstood. What is meant is not the kinetic energy of moving objects, but the energy (squared amplitude) of a signal at a sensor in a certain interval $\kappa\nu$. We will now compare this approach with the previously discussed differential method.

One of the simplest models for one-dimensional motion analysis uses only three quadrature filters. This set of directional filters detects objects moving to the right, to the left, or standing still. We denote the squared magnitude of these quadrature filters as $\mathcal{R}$, $\mathcal{L}$ and $\mathcal{S}$. Then we obtain an estimate of the one-dimensional optical flow using the operator [2, 3]:

$$\mathcal{U} = \frac{\mathcal{R} - \mathcal{L}}{\mathcal{S}}. \tag{14.48}$$

We find an interesting connection between this approach and the differential method if we also consider the latter (Section 14.4.2) as a method for energy extraction. We perform the comparison here for the analysis of one-dimensional motion, i. e. in a 2D place/time image. In this case, the solution of the differential method in operator notation according to Eq. (14.32) can be expressed as

$$\mathcal{U}' = -\frac{\mathcal{B}_{xt}\,(\mathcal{D}_t \cdot \mathcal{D}_x)}{\mathcal{B}_{xt}\,(\mathcal{D}_x \cdot \mathcal{D}_x)}. \tag{14.49}$$

We reformulate this equation with a slight modification in which the operators are smoothed with the binomial mask $\mathcal{B}_{xt}$ before we apply the derivative operators (Section 12.7):

$$\mathcal{U}' = -\frac{\mathcal{B}_{xt}\,[(\mathcal{D}_t\mathcal{B}_{xt}) \cdot (\mathcal{D}_x\mathcal{B}_{xt})]}{\mathcal{B}_{xt}\,[(\mathcal{D}_x\mathcal{B}_{xt}) \cdot (\mathcal{D}_x\mathcal{B}_{xt})]}. \tag{14.50}$$

Smoothing with $\mathcal{B}_{xt}$ is nothing more than a regularization of the derivative operator. The indices xt mean that the smoothing is performed along the temporal and spatial axes. Let's use the operator identity

$$\mathcal{A}\mathcal{B} = \frac{1}{4}\left[(\mathcal{A} + \mathcal{B})^2 - (\mathcal{A} - \mathcal{B})^2\right] \tag{14.51}$$

and the abbreviations

$$\mathcal{R}' = (\mathcal{D}_x + \mathcal{D}_t)\mathcal{B}_{xt}, \quad \mathcal{L}' = (\mathcal{D}_x - \mathcal{D}_t)\mathcal{B}_{xt}, \quad \mathcal{S}' = 2\mathcal{D}_x\mathcal{B}_{xt},. \tag{14.52}$$

Then we can rewrite Eq. (14.50) and get an expression very similar to Eq. (14.48):

$$\mathcal{U}' = \frac{\mathcal{B}_{xt}(\mathcal{R}' \cdot \mathcal{R}' - \mathcal{L}' \cdot \mathcal{L}')}{\mathcal{B}_{xt}(\mathcal{S}' \cdot \mathcal{S}')}. \tag{14.53}$$

The filters $\mathcal{R}'$, $\mathcal{L}'$ and $\mathcal{S}'$ are derivatives of binomial filters. The transfer functions show that objects are selected that move to the right or left or are at rest (Fig. 14.13). These filters are *no* quadrature filters. However, squaring the filter responses and further smoothing with $\mathcal{B}_{xt}$ leads approximately to a phase-independent detection of the squared amplitude as with a quadrature filter under certain conditions. Let us imagine a fine-scale periodic structure.

The derivative filters will preserve these structures but remove the mean gray value. Therefore, squaring will result in a mean gray value whose magnitude is half the amplitude and a fast spatial gray value oscillation with twice the wavenumber (half the wavelength).

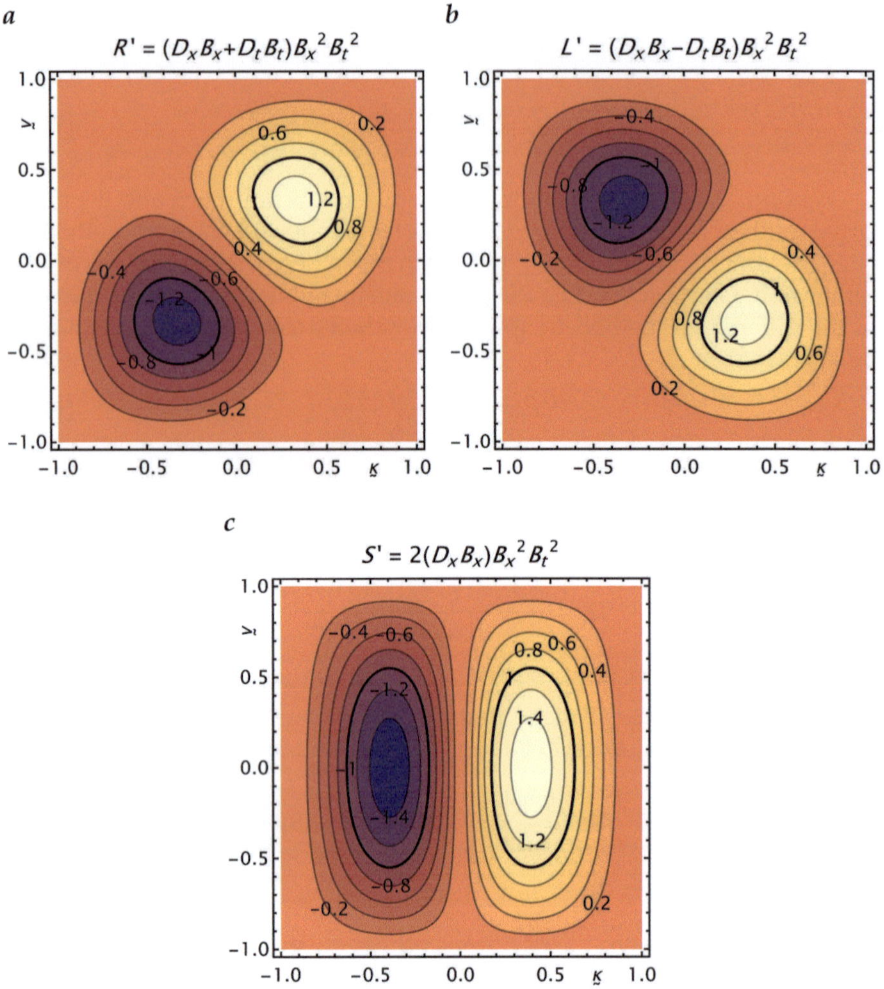

Figure 14.13: *Transfer function of the convolution operators in Eq. (14.52) for detecting objects that move to the right or left or are stationary in wavenumber-frequency space:* ***a*** $\mathcal{R}'$, ***b*** $\mathcal{L}'$ *and* ***c*** $\mathcal{S}'$. *The following convolution masks were used for the operators: for* $\mathcal{D}_d$ *the mask* $\mathbf{D}_d\mathbf{B}_d$ *and for* $\mathcal{B}_d$ *the mask* $\mathbf{B}_d^2$.

If the subsequent smoothing removes these oscillations, the result is a phase-independent response to the filter, as with a quadrature filter. Unlike with quadrature filters, we can only achieve these results in regions where the scales of the structures are fine enough so that the double wavelength can be removed with the smoothing filter.

14.5 Tensor Method

The tensor method for analyzing local orientation has already been discussed in detail in Section 13.3. Since motion in 3D place/time images is represented as orientation, we only need to extend the tensor method to three dimensions. First, we will revisit the optimization criterion used for the tensor approach in Section 14.5.1 to distinguish this technique from the differential method (Section 14.4).

14.5.1 Optimization Strategy

In Section 13.3.1 we found that the optimal orientation is given by the fact that it has minimal deviations from the direction of the gradient. We have introduced the squared scalar product between the gradient vector and a unit vector that reflects the optimal orientation:

$$(\boldsymbol{\nabla} g^T \hat{\boldsymbol{n}})^2 = |\boldsymbol{\nabla} g|^2 \cos^2\left(\angle(\boldsymbol{\nabla} g, \hat{\boldsymbol{n}})\right). \tag{14.54}$$

This measure can be used in vector spaces of any dimension. Therefore, in the space-time space the space-time gradient is used

$$\boldsymbol{\nabla}_{xt} g = \left[\boldsymbol{\nabla}^T g, \frac{\partial g}{\partial t}\right]^T = \left[\dot{g}_x, \dot{g}_y, \ldots, \dot{g}_t\right]^T \tag{14.55}$$

and write

$$(\boldsymbol{\nabla}_{xt} g^T \hat{\boldsymbol{n}})^2 = |\boldsymbol{\nabla}_{xt} g|^2 \cos^2\left(\angle(\boldsymbol{\nabla}_{xt} g, \hat{\boldsymbol{n}})\right). \tag{14.56}$$

In the 2D orientation analysis, we have maximized the expression

$$\int w(\boldsymbol{x} - \boldsymbol{x}') \left(\boldsymbol{\nabla} g(\boldsymbol{x}')^T \hat{\boldsymbol{n}}\right)^2 \mathrm{d}^D x' = \overline{(\boldsymbol{\nabla} g\, \hat{\boldsymbol{n}})^2}, \tag{14.57}$$

to find the optimal orientation. For the analysis of motion in space-time, we are not interested in the direction of maximum gray value changes, but in the minimum gray value change. We specify this orientation with the unit vector $\hat{\boldsymbol{e}}_3 = [e_{31}, e_{32}, e_{33}]^T$. According to the considerations in Section 14.3.1 Eq. (14.9), the optical flow results from this 3D vector as

$$\boldsymbol{f} = \frac{1}{e_{33}} \begin{bmatrix} e_{31} \\ e_{32} \end{bmatrix}. \tag{14.58}$$

By analogy with Eq. (14.57) we therefore minimize

$$\int w(\boldsymbol{x} - \boldsymbol{x}', t - t') \left(\boldsymbol{\nabla}_{xt} g(\boldsymbol{x}', t')^T \hat{\boldsymbol{e}}_3\right)^2 \mathrm{d}^D x' \mathrm{d}t' \tag{14.59}$$

or written more compactly

$$\boxed{\overline{(\boldsymbol{\nabla}_{xt} g^T \hat{\boldsymbol{e}}_3)^2} \rightarrow \text{minimum.}} \tag{14.60}$$

The window function w is also extended in the time direction and determines the size and shape of the neighborhood around the point $[\boldsymbol{x}, t]^T$, over which the orientation is averaged. Equation (14.60) must be compared with the corresponding expression Eq. (14.23), which is minimized in the differential method:

$$\overline{(\boldsymbol{f}\nabla g + \dot{g}_t)^2}. \tag{14.61}$$

The optimization strategies between Eqs. (14.60) and (14.61) are slightly different. Both are approaches in the sense that the optical flow is determined in such a way that the deviations from the continuity equation are minimized using the method of *least squares*. However, the optical flow is determined directly on the one hand and a 3D unit vector on the other. The latter amounts to a *total least squares* method [340]. This method is better adapted to the problem, as it considers not only the time derivatives as error-prone as in Eq. (14.61), but all components of the space-time gradient.

In analogy to the discussion in Section 13.3.1, we can conclude that the determination of the optical flow in space-time images is equivalent to the problem of finding the *eigenvector* $\hat{\boldsymbol{e}}_3$ to the smallest eigenvalue λ_3 of the structure tensor

$$\boldsymbol{J} = \begin{bmatrix} \overline{\dot{g}_x\dot{g}_x} & \overline{\dot{g}_x\dot{g}_y} & \overline{\dot{g}_x\dot{g}_t} \\ \overline{\dot{g}_x\dot{g}_y} & \overline{\dot{g}_y\dot{g}_y} & \overline{\dot{g}_y\dot{g}_t} \\ \overline{\dot{g}_x\dot{g}_t} & \overline{\dot{g}_y\dot{g}_t} & \overline{\dot{g}_t\dot{g}_t} \end{bmatrix}, \tag{14.62}$$

where $\overline{g_p g_q}$ with $p, q \in \{x, y, t\}$ is given by

$$\overline{\dot{g}_p\dot{g}_q(\boldsymbol{x}, t)} = \int w(\boldsymbol{x} - \boldsymbol{x}', t - t')\dot{g}_p(\boldsymbol{x}', t')\dot{g}_q(\boldsymbol{x}', t')\mathrm{d}^2x'\mathrm{d}t'. \tag{14.63}$$

At this point, we can compare the tensor method with the differential first-order technique. While the tensor method performs an eigenvalue analysis of a symmetric tensor with six regularized products of spatial and temporal partial derivatives, the differential method uses only five of these terms. The term $\overline{g_t g_t}$ is not used. We will see in the next section that this additional term allows the tensor method to directly determine whether there is a constant velocity in a neighborhood at all. This is not directly possible with the differential method.

14.5.2 Eigenvalue Analysis

Unfortunately, the *eigenvalue analysis* of a symmetric 3×3 tensor is not as simple as for a symmetric 2×2 tensor. In two dimensions, we were able to solve the eigenvalue problem directly. Thus, in Section 13.3.3 we transformed the three independent components of the symmetric 2×2 tensor into the three parameters orientation, determinacy and coherence measure.

The symmetric 3×3 tensor now contains six independent components, and we need to find a corresponding number of parameters that adequately describe the local structure of the space-time image. Again, it is helpful to decompose these six parameters into rotation-variant and rotation-invariant parameters.

As already mentioned, the solution of the eigenvalue problem cannot be simply stated, but requires a suitable numerical method. We will not deal with this problem because, although it is not simple, it is a standard problem in numerical mathematics for which there are a number of efficient solutions [109, 248]. So we assume that we have solved the eigenvalue problem and that a set of three orthonormal eigenvectors and three eigenvalues are available. With the solution of the eigenvalue problem, we essentially have a principal axis coordinate system in which the structure tensor lies diagonally and which contains the eigenvalues as diagonal elements:

$$J' = \begin{bmatrix} \lambda_1 & 0 & 0 \\ 0 & \lambda_2 & 0 \\ 0 & 0 & \lambda_3 \end{bmatrix}. \tag{14.64}$$

Without limiting the generality, we have sorted the eigenvalues by size:

$$\lambda_1 \geq \lambda_2 \geq \lambda_3 \geq 0. \tag{14.65}$$

The principle axis coordinate system is spanned by the three *eigenvectors*. Rotation in this coordinate system requires three independent parameters, as discussed in Section 3.2.2. Three of the six parameters are therefore used to describe the orientation in space-time. This information is contained in the three orthonormal eigenvectors.

The remaining parameters are the three rotation-invariant *eigenvalues*. We will now show how the different classes of local structures in space-time images can be distinguished by the three eigenvalues. This approach also helps us to an efficient implementation of tensor-based motion analysis.

We can distinguish four classes of neighborhoods in a space-time image. They correspond to ranks 0 to 3 of the symmetric tensor:

Constant Gray Value
All elements and eigenvalues of the tensor are zero:

$$\lambda_1 = \lambda_2 = \lambda_3 = 0. \tag{14.66}$$

The rank of the tensor is also zero. Therefore, the optical flow is also zero. This condition is easy to recognize. The sum of the eigenvalues must be below a critical limit, which is determined by the noise level in the image sequence. Since the sum of the eigenvalues is equal to the trace of the tensor, we do not need an eigenvalue analysis to check this condition:

$$\text{trace}(J) = \sum_{p=1}^{3} \overline{\dot{g}_p \dot{g}_p} < \gamma, \tag{14.67}$$

where γ is a suitable measure for the noise level of the image sequence. For all points for which the condition Eq. (14.67) is fulfilled, the eigenvalue analysis can be omitted.

Spatial Orientation and Constant Motion
In this case, two eigenvalues are zero, as the gray values only change in one direction:

$$\lambda_1 > 0 \quad \text{and} \quad \lambda_2 = \lambda_3 = 0. \tag{14.68}$$

The rank of the tensor is one. The spatial gray value structure shows a *linear symmetry*. This condition can also be easily checked without performing an eigenvalue analysis, since the determinant of the upper 2×2 partial tensor must be below a threshold γ^2:

$$\overline{\dot{g}_x \dot{g}_x}\,\overline{\dot{g}_y \dot{g}_y} - \overline{\dot{g}_x \dot{g}_y}^2 < \gamma^2. \tag{14.69}$$

The eigenvector $\hat{\boldsymbol{e}}_1$ belonging to the only non-zero eigenvalue points in the direction of the maximum gray value change. It provides both the spatial orientation and the velocity in this direction. It should be noted that due to the *aperture problem* only the *normal velocity*, i. e. the velocity in the direction of the spatial gradient, can be calculated (Section 14.1.2). The spatial orientation is given by the two spatial coordinates of the eigenvector $\hat{\boldsymbol{e}}_1$. As the normal optical flow points in this direction, it results from

$$\boldsymbol{f}_\perp = -\frac{e_{1t}}{e_{1x}^2 + e_{1y}^2}\begin{bmatrix} e_{1x} \\ e_{1y} \end{bmatrix}, \tag{14.70}$$

and its amount to

$$|\boldsymbol{f}_\perp| = \sqrt{\frac{e_{1t}^2}{e_{1x}^2 + e_{1y}^2}} = \sqrt{\frac{e_{1t}^2}{1 - e_{1t}^2}}. \tag{14.71}$$

Distributed Spatial Structures and Constant Motion
In this case, only one eigenvalue is zero:

$$\lambda_1, \lambda_2 > 0 \quad \text{and} \quad \lambda_3 = 0. \tag{14.72}$$

As the movement is constant, the principle axis coordinate system moves with the scene. The eigenvector $\hat{\boldsymbol{e}}_3$ at eigenvalue zero points in the direction of movement. This results in the optical flow

$$\boldsymbol{f} = \frac{1}{e_{3t}}\begin{bmatrix} e_{3x} \\ e_{3y} \end{bmatrix} \tag{14.73}$$

and its amount

$$|\boldsymbol{f}| = \sqrt{\frac{e_{3x}^2 + e_{3y}^2}{e_{3t}^2}} = \sqrt{\frac{1 - e_{3t}^2}{e_{3t}^2}}. \tag{14.74}$$

Distributed Spatial Structures and Non-constant Motion
All three eigenvalues are greater than zero, and the rank of the tensor is three:

$$\lambda_1, \lambda_2, \lambda_3 > 0. \tag{14.75}$$

Meaningful calculation of the optical flow is not possible in this case.

After this detailed classification, we turn to the question of which three rotation-invariant parameters independent of the velocity and the spatial orientation of the gray value parameters can be extracted from the structure tensor to describe the structure.

Certainty Measure
The first parameter is a *certainty measure*, which is a measure of the gray value changes. There are two possibilities to choose either the mean square of the spatial gradient (trace of the upper 2×2 partial tensor) or the mean square of the space-time gradient. For practical reasons, the mean square of the spatial gradient is preferable, as the spatial gradient does not change in a sequence when the velocity increases.

The mean square of the space-time gradient, on the other hand, increases with increasing speed, as larger temporal gradients are added. Therefore, surprisingly, the mean square of the spatial gradient is the better coefficient of determination:

$$c_c = \overline{\dot{g}_x \dot{g}_x} + \overline{\dot{g}_y \dot{g}_y}. \tag{14.76}$$

Spatial Coherency Measure
As a second measure, we use the coherence known from the analysis of local neighborhoods (Section 13.3.4) and refer to it as *spatial coherence measure*:

$$c_s = \frac{(\overline{\dot{g}_x \dot{g}_x} - \overline{\dot{g}_y \dot{g}_y})^2 + 4\overline{\dot{g}_x \dot{g}_y}^2}{(\overline{\dot{g}_x \dot{g}_x} + \overline{\dot{g}_y \dot{g}_y})^2}. \tag{14.77}$$

Its value is between zero and one and indicates whether only the normal optical flow or both components of the optical flow can be determined.

Total Coherency Measure
Finally, we need an additional measure, the *total coherence measure*, which tells us whether we are dealing with a local neighborhood with constant velocity. It should be independent of the spatial coherence. The following measure, which uses the largest and smallest eigenvalues, fulfills this condition:

$$c_t = \left(\frac{\lambda_1 - \lambda_3}{\lambda_1 + \lambda_3}\right)^2. \tag{14.78}$$

The total coherence measure is one as soon as the eigenvalue λ_3 is zero. The two remaining eigenvalues can then take on any other value. The total coherence approaches zero if all three eigenvalues are equal. In contrast to the other two measures c_c and c_s, total coherence requires an eigenvalue analysis, as the smallest and largest eigenvalues are necessary for the calculation.

The following special case must be taken into account when interpreting the measure: The total coherence measure is also one for a spatially oriented pattern and a non-constant motion. However, this case can be recognized from the fact that both the spatial and the total coherence are then one, but only one eigenvalue is zero. Another simple criterion is that the eigenvector of the eigenvalue zero then lies in the xy-plane. This results in $e_{33} = 0$, and we obtain an infinite value for the optical flow vector after Eq. (14.73).

14.6 ♯Phase Method

14.6.1 ♯Basics

With the exception of the computationally expensive correlation method, all other methods that calculate the optical flow are more or less sensitive to changes in illumination over time. This raises the question of whether we can use another feature instead of the gray value itself that is less sensitive to illumination changes but still contains the essential image information. Fleet and Jepson [85] and Fleet [83] suggest using the phase to calculate the optical flow. We have already discussed the critical role of the phase in Sections 6.3.5 and 13.4.1. In Section 6.3.5 it was shown that the phase of the Fourier transform of a signal carries the decisive information. An image can still be recognized if the amplitude information is lost, but not if the phase is missing [200]. In our context, it is important to note that global illumination only influences the amplitude of a signal, but not its phase.

As an introduction to the phase method, we consider a planar 1D wave with wavenumber k and frequency ν, which moves forward with a phase velocity of $u = \nu/k$:

$$g(x,t) = g_0 \exp[-2\pi\mathrm{i}(\phi(x,t))] = g_0 \exp[-2\pi\mathrm{i}(kx - \nu t)]. \tag{14.79}$$

The position and thus also the displacement is determined by the *phase*. The phase depends on both the spatial and temporal coordinates. For a plane wave, the phase varies linearly in time and space:

$$\phi(x,t) = 2\pi(kx - \nu t) = 2\pi(kx - ukt), \tag{14.80}$$

where k is the wavenumber and ν is the frequency of the wave. The calculation of the temporal and spatial derivatives of the phase, i. e. the gradient in space-time, provides both the wavenumber and the frequency of the moving periodic structures:

$$\nabla_{xt}\phi = \begin{bmatrix} \phi_x \\ \phi_t \end{bmatrix} = 2\pi \begin{bmatrix} k \\ -\nu \end{bmatrix}. \tag{14.81}$$

The speed is then the ratio of the frequency to the wavenumber:

$$u = \frac{\nu}{k} = -\phi_t \big/ \phi_x\,. \tag{14.82}$$

This formula is very similar to the optical flow based estimation Eq. (14.18). In both cases, the velocity is expressed as the ratio of the temporal and spatial derivatives.

The direct calculation of the partial derivatives from the phase signal is not advisable due to the discontinuity inherent in the phase signal (restriction to the main interval $[-\pi, \pi[$). As already discussed in Section 13.4.6, it is possible to calculate the phase gradients directly from the result of a quadrature filter pair. If we denote the quadrature filter pair by $g_e(x,t)$ and $g_o(x,t)$, the place/time phase gradient results from (see Eq. 13.66)

$$\nabla_{xt}\phi(x,t) = \frac{g_e(x,t)\,\nabla_{xt}g_o(x,t) - g_o(x,t)\,\nabla_{xt}g_e(x,t)}{g_e^2(x,t) + g_o^2(x,t)}. \tag{14.83}$$

With Eq. 14.82 the optical flux calculated from the phase

$$\boxed{f = -\frac{g_e\,(\dot{g}_o)_t - g_o\,(\dot{g}_e)_t}{g_e\,(\dot{g}_o)_x - g_o\,(\dot{g}_e)_x}.} \tag{14.84}$$

14.6.2 ♯Evaluation and Comparison

At first, the phase method seems to offer nothing new. However, replacing the gray value with the phase is a significant improvement because the phase is much less dependent on the illumination than the gray value itself. If we use only the phase signal, the amplitude of the gray value variations can change without affecting velocity estimates at all.

So far, we have only considered an ideal periodic gray value structure. In general, however, images are composed of gray value structures with different wavenumbers, from which we cannot obtain meaningful phase estimates. So we have to break down the image into individual wavenumber ranges. This implies that the phase method is not suitable for two-dimensional displacements, but is essentially a 1D concept that measures motion of a linearly oriented structure, for example a plane wave, in the direction of its gray value gradients.

From this fact, Fleet and Jepson [84] derive a new paradigm of motion analysis. The image is decomposed using directional filters, and the *normal velocity* is determined in each of the components. The two-dimensional motion field is then composed of these normal velocities. This approach has the advantage that the composition into a complete motion field is moved to a second processing step, which can be adapted to the form of motion occurring in the image. Therefore, more complex cases such as the overlapping of transparent objects can also be handled with this approach.

Fleet and Jepson [84] use a set of *Gabor filters* (Section 13.4.5) with an angular resolution of 30° and a bandwidth of 0.8 octaves for directional bandpass decomposition. Alternatively, a bandpass decomposition and *Hilbert filter* (Section 13.4.2) can be used. The motivation for this idea stems from the fact that the decomposition with a set of Gabor filters, as proposed by Fleet and

Jepson, does not allow a simple reconstruction of the original image. The Gabor decomposed sub-images do not add up exactly to the original image. Rather, the resulting transfer function has a considerable ripple, as Riemer [266] shows.

A bandpass decomposition with, for example, a Laplacian pyramid [36, 37] does not have this disadvantage (Section 15.3.4). In addition, it is more efficient in terms of computational effort. On the other hand, we face the problem that we do not obtain a directional decomposition. Jähne [151, 153] shows how the concept of the Laplace pyramid can be efficiently extended to a (see also Section 15.3.6). Each plane of the pyramid is further decomposed into two or four directional components that add directly to the corresponding isotropically filtered pyramid planes (see also Section 15.3.6).

14.6.3 #From Normal Flow to 2D Flow

DSince the phase method only provides the optical flow perpendicular to the direction of the filtered structures, we need a technique to determine the two-dimensional optical flow from the normal flow. The basic relationship between the two fluxes is as follows. We assume that $\boldsymbol{f}_\perp$ is a normal flow vector. It is the result of the projection of the 2D flow vector $\boldsymbol{f}$ in the direction of the normal flow. Then we can write:

$$f_\perp = \hat{\boldsymbol{f}}_\perp \boldsymbol{f}, \tag{14.85}$$

where $\hat{\boldsymbol{f}}_\perp$ is a unit vector in the direction of the normal flow. From Eq. (14.85) it is obvious that we can calculate the unknown two-dimensional optical flow using the *least squares* method if we have more than two estimates of the normal flow in different directions. This approach yields the following system of linear equations in a similar way as in Section 14.4.2:

$$\begin{bmatrix} \overline{\hat{f}_{\perp x}\hat{f}_{\perp x}} & \overline{\hat{f}_{\perp x}\hat{f}_{\perp y}} \\ \overline{\hat{f}_{\perp x}\hat{f}_{\perp y}} & \overline{\hat{f}_{\perp y}\hat{f}_{\perp y}} \end{bmatrix} \begin{bmatrix} f_x \\ f_y \end{bmatrix} = \begin{bmatrix} \overline{\hat{f}_{\perp x} f_\perp} \\ \overline{\hat{f}_{\perp y} f_\perp} \end{bmatrix} \tag{14.86}$$

with

$$\overline{\hat{f}_{\perp p}\hat{f}_{\perp q}} = \int w(\boldsymbol{x}-\boldsymbol{x}', t-t')\hat{f}_{\perp p}\hat{f}_{\perp q}\,\mathrm{d}^2x'\mathrm{d}t' \tag{14.87}$$

and

$$\overline{\hat{f}_{\perp p}\hat{f}_{\perp}} = \int w(\boldsymbol{x}-\boldsymbol{x}', t-t')\hat{f}_{\perp p} f_\perp\,\mathrm{d}^2x'\mathrm{d}t'. \tag{14.88}$$

14.7 #References for Further Reading

The following monographs on motion analysis are available: Singh [304], Fleet [83] and Jähne [156]. The articles by Beauchemin and Barron [13],

Haußecker and Spies [127] and Jähne et al. [167] also provide a good overview. The second also includes the estimation of higher order motion fields. Readers interested in the visual detection of motion in biological systems are referred to the monograph by Smith and Snowden [306]. The extension of motion analysis to parameters of dynamic processes and illumination changes is described in Haußecker and Fleet [126] and Garbe et al. [101]. Methods for analyzing complex motion fields are dealt with by Jähne et al. [168].

15 Scale Spaces, Multigrid

Contents

15.1 Scales in Signal Processing

15.1.1 Introduction

The neighborhood operations discussed in Chapters 10 to 14 can only be the starting point for image analysis, as they only extract local features in the order of a few pixels at most. It is clear that images also contain large-scale information. To extract them, we need correspondingly larger filter masks. However, the computational effort increases considerably when using large filter masks. If we use a non-separable mask of size R^D in a D-dimensional image, the number of arithmetic operations is proportional to R^D. Doubling the size therefore leads to an increase in the number of arithmetic operations by a factor of four and eight for two- and three-dimensional images respectively. For scales that are larger by a factor of 10, the number of operations

B. Jähne, *Digital Image Processing and Image Acquisition*,
https://doi.org/10.1007/978-3-032-04806-6_15

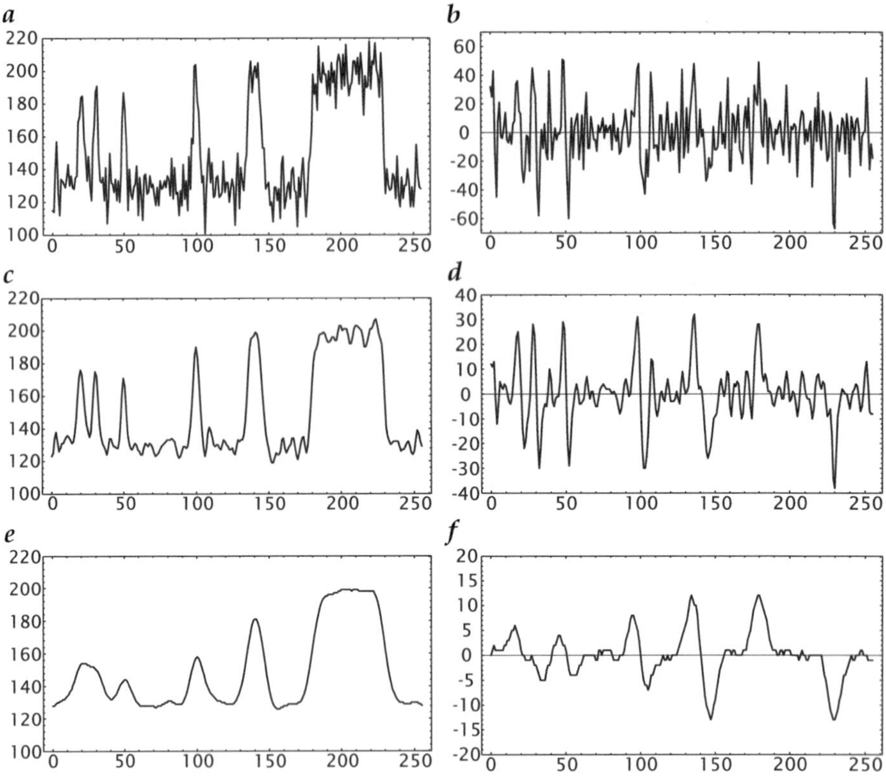

Figure 15.1: *Lines and edges with* **a** *high,* **c** *medium and* **e** *low resolution.* ***b***, ***d*** *and* ***f*** *Difference of neighboring points for the detection of edges in* ***a***, **c** *or* **e**.

increases by a factor of 100 for two-dimensional images and by a factor of 1000 for three-dimensional images.

The sharp increase in computational effort is only the superficial expression of a much deeper problem. This is illustrated by a simple task, the detection of edges and lines at different resolutions. To do this, we use the same image line, but smooth it to different degrees (Fig. 15.1). We then define the associated scale as the distance over which the image was smoothed and analyze the gray value differences of adjacent points.

First, we examine gray value differences at high resolution with a step size of only one pixel (Fig. 15.1a, b). At this fine scale, the gray value differences are dominated by the background noise in the image line. This makes it difficult to determine gray value changes due to contrast differences between the objects and the background. This problem is due to *scale mismatch*: The gray values change over larger distances, which the operators that are supposed to detect them do not.

However, if we look at the image at low resolution (Fig. 15.1e, f), we can see that the difference in brightness to the background is significantly smaller in some cases. Therefore, the lines cannot be optimally detected in this case either. In addition, the two closely adjacent lines in the left part merge into one object. However, at a resolution that corresponds to the width of the lines, they can be optimally detected (Fig. 15.1c, d). Here the noise is significantly lower than with the fine scale (Fig. 15.1a), and the contrast between line and background is preserved in contrast to Fig. 15.1e.

This example shows that the detection of certain features in an image depends on the correct scale. This in turn, of course, depends on the characteristic sizes in the object to be detected. For optimal processing, an image must therefore be available in different scales, which requires a representation in several resolution levels.

In this chapter, we first look at the relationship between the spatial and wavenumber representation of images from this perspective (Section 15.1.2). Then we introduce the *scale space* as a concept with a continuous scale parameter, discuss how it can be generated, and describe its basic properties (Section 15.2). The scale space is very helpful to understand the influence of scales on image processing. However, because it increases the dimension of the data by one, it requires enormously more storage capacity and is therefore theoretically interesting, but not suitable for effective image processing.

However, the basic approaches and properties of the scale space are the basis for efficient *multigrid representations* (Section 15.3). These are one of the most important concepts of modern image processing, as they allow very effective solutions for many even complex image processing algorithms. In this chapter, we will discuss the *Gaussian pyramid* (Section 15.3.2), the *Laplacian pyramid* (Section 15.3.4) and directional decompositions on the Laplacian pyramid (Section 15.3.6).

15.1.2 Spatial and Wavenumber Representation

In Chapter 6, we analyzed the representation of images in spatial and wavenumber space in detail. In this section, we will look at both forms from the point of view of generating a scale space representation of an image.

If we represent an image on a grid in spatial domain, we have no information about the wavenumbers in a pixel. We know the position with the accuracy of the lattice constant Δx, but the local wavenumber at this position can be in the full range of possible wavenumbers of $\pm M\Delta k/2 = 1/(2\Delta x)$.

The exact opposite is true for the wavenumber representation. Each pixel represents a wavenumber in the highest possible wavenumber resolution for a given image size. However, any information about the position in space is lost, since a point in the wavenumber space represents a periodic structure that is distributed over the entire image.

These considerations show that the representation of an image in either spatial or wavenumber space represents two opposing extremes. Either we have an optimal spatial resolution or an optimal wavenumber resolution, but the opposite information is no longer available. What we need for an image representation in scale space is a kind of combination of both representations that allows a separation into different wavenumber ranges (scales) and still maintains the spatial resolution as far as possible.

15.1.3 Local Fourier Transform

One way to achieve a combined space/wavenumber representation is the *local Fourier transform*. As the name suggests, the Fourier transform is not applied to the entire image, but only to a region that is created when the image is multiplied by a window function $w(\boldsymbol{x})$. The window function has a maximum at $\boldsymbol{x} = \boldsymbol{0}$ and decreases monotonically with increasing $|\boldsymbol{x}|$ towards zero. The maximum of the window function is positioned at each point $\boldsymbol{x}$ and a local Fourier transform is calculated for each point:

$$\tilde{g}(\boldsymbol{x}, \boldsymbol{\kappa}_0) = \int_{-\infty}^{\infty} g(\boldsymbol{x}')w(\boldsymbol{x}' - \boldsymbol{x}) \exp\left(-2\pi \mathrm{i} \boldsymbol{\kappa}_0 \boldsymbol{x}'\right) \mathrm{d}x'^2. \tag{15.1}$$

The integral in Eq. (15.1) looks almost like a convolution integral (Eq. 6.39), > R12). To convert it to a convolution integral, we use that $w(-\boldsymbol{x}) = w(\boldsymbol{x})$ and change the second part of Eq. (15.1):

$$w(\boldsymbol{x}' - \boldsymbol{x}) \exp\left(-2\pi \mathrm{i} \boldsymbol{\kappa}_0 \boldsymbol{x}'\right) = w(\boldsymbol{x} - \boldsymbol{x}') \exp\left(2\pi \mathrm{i} \boldsymbol{\kappa}_0 (\boldsymbol{x} - \boldsymbol{x}') \exp\left(-2\pi \mathrm{i} \boldsymbol{\kappa}_0 \boldsymbol{x}\right)\right). \tag{15.2}$$

Then we can write Eq. (15.1) as a convolution:

$$\tilde{g}(\boldsymbol{x}, \boldsymbol{\kappa}_0) = \left[g(\boldsymbol{x}) * w(\boldsymbol{x}) \exp\left(2\pi \mathrm{i} \boldsymbol{\kappa}_0 \boldsymbol{x}\right)\right] \exp\left(-2\pi \mathrm{i} \boldsymbol{\kappa}_0 \boldsymbol{x}\right). \tag{15.3}$$

The local Fourier transform corresponds to a convolution with a complex convolution kernel $w(\boldsymbol{x})\exp(2\pi \mathrm{i} \boldsymbol{\kappa}_0 \boldsymbol{x})$, with the exception of a phase factor $\exp(-2\pi \mathrm{i} \boldsymbol{\kappa}_0 \boldsymbol{x})$. Using the *shift theorem* (Theorem B.3, p. 629, > R12), the transfer function of the convolution kernel can be calculated as

$$w(\boldsymbol{x}) \exp\left(2\pi \mathrm{i} \boldsymbol{\kappa}_0 \boldsymbol{x}\right) \circ\!\!-\!\!\bullet\ \tilde{w}(\boldsymbol{\kappa} - \boldsymbol{\kappa}_0). \tag{15.4}$$

This means that the convolution is a *bandpass filter* with a maximum at the wavenumber $\boldsymbol{\kappa}_0$. The width of the bandpass is inversely proportional to the width of the window function. In this way, the spatial and wavenumber resolution are related to each other. As an example, let us consider a Gaussian window function and the corresponding transfer function (> R12, > R13)

$$\exp\left(-\frac{x^2}{2\sigma_x^2}\right) \circ\!\!-\!\!\bullet\ \frac{1}{\sqrt{2\pi}\sigma_x} \exp\left(-2\pi^2 k^2 \sigma_x^2\right). \tag{15.5}$$

Therefore, the product of the variances in spatial and wavenumber space ($\sigma_k^2 = 1/(4\pi\sigma_x^2)$) is a constant: $\sigma_x^2\sigma_k^2 = 1/(4\pi)$). This represents the classical *uncertainty principle* (Theorem B.7, p. 633). It states that the product of the variances of each Fourier transform pair is greater than or equal to $1/(4\pi)$. Since the Gaussian window function reaches the theoretical minimum, it is optimal; a better wavenumber resolution cannot be achieved with a given local resolution.

15.2 ♯Scale Spaces

As we saw in the example of the local Fourier transform in the last section, the introduction of a characteristic *scale* leads to a new coordinate in the representation of image data. The new scale parameter next to the spatial resolution is denoted by ξ and characterizes the current resolution of the image data. The data structure consisting of a sequence of images with different resolutions is referred to as *scale space*. We write $g(\boldsymbol{x}, \xi)$ for the scale space of the image $g(\boldsymbol{x})$.

In the next section (Section 15.2.1), we first discuss how we can create a sequence of images that become increasingly blurred. Obviously, we need to smooth an image more and more using continued smoothing operators, as we learned in Chapter 11. So far, however, we have only become familiar with these smoothing operators as operators on discrete images and the step size of the smoothing is also discrete. Here, however, we want to smooth continuous images continuously. This is possible by imagining smoothing as a process that can be described by a partial differential equation. This process is known in physics as diffusion. This allows us to create a scale space (Section 15.2.1).

We then discuss the general properties of scale spaces in Section 15.2.2. We discover that smoothing becomes slower and slower the further the smoothing progresses. We therefore look at methods for accelerating the smoothing process in Section 15.2.3.

Then in Section 15.2.4 we discuss the particularly interesting question of how an image changes with increasing smoothing. This is referred to as a differential scale space. Finally, we are also interested in the transition from a continuous to a discrete scale space (Section 15.2.5), as we can only perform practical calculations in such a space. In other words, this corresponds to the question of how to sample a scale space so that no artifacts occur and the continuous scale space can be reconstructed from the sampled scale space.

15.2.1 ♯Generation by Diffusion

The creation of a scale space requires a process that can blur images in a controlled manner. Diffusion is a transport process that leads to the equalization of concentration differences [47, 170]. In physics, diffusion processes in the transport of heat, matter and momentum (in flows, the momentum density $\rho\boldsymbol{u}$ is transported) lead to an increasingly far-reaching spatial equalization of concentration. If we identify time with the scale parameter ξ, the diffusion process creates a scale space.

To apply a diffusion process to a multidimensional signal with D dimensions, we consider the gray value g as the concentration of a chemical substance. The elementary law of diffusion states that the flux density $\boldsymbol{j}$ is opposite to and proportional to the direction of the concentration gradient $\boldsymbol{\nabla} g$:

$$\boldsymbol{j} = -D\boldsymbol{\nabla} g, \tag{15.6}$$

where the constant D is referred to as *diffusion coefficient*. The continuity equation

$$\frac{\partial g}{\partial t} + \boldsymbol{\nabla}\boldsymbol{j} = 0 \tag{15.7}$$

results in the instationary diffusion equation

$$\boxed{\frac{\partial g}{\partial t} = \boldsymbol{\nabla}(D\boldsymbol{\nabla} g) \quad \text{diffusion equation.}} \tag{15.8}$$

In the case of a homogeneous diffusion process in which D is independent of the position, the equation is reduced to

$$\frac{\partial g}{\partial t} = D\Delta g, \quad \text{with} \quad \Delta = \sum_{d=1}^{D} \frac{\partial^2}{\partial x_d^2}, \tag{15.9}$$

where Δ is the *Laplace operator*. D in the summation here denotes the dimension of the space (not the diffusion constant!) We can easily show that the general solution of this equation is equivalent to a convolution with a smoothing mask. Finally, we perform a Fourier transform in space and obtain with the help of Theorem B.5, p. 631

$$\frac{\partial \tilde{g}(\boldsymbol{\kappa})}{\partial t} = -4\pi^2 D|\boldsymbol{\kappa}|^2 \tilde{g}(\boldsymbol{\kappa}) \tag{15.10}$$

and thus a reduction of the equation to a linear differential equation of the first order with the general solution

$$\tilde{g}(\boldsymbol{\kappa}, t) = \exp(-4\pi^2 D|\boldsymbol{\kappa}|^2 t)\tilde{g}(\boldsymbol{\kappa}, 0), \tag{15.11}$$

where $\tilde{g}(\boldsymbol{\kappa}, 0)$ is the Fourier-transformed image at time zero.

The multiplication of an image in Fourier space with the Gaussian function in Eq. (15.11) corresponds to a convolution with the same function, but with reciprocal width (Theorem B.4, p. 630, > R12 and > R14). Therefore

$$g(\boldsymbol{x}, t) = \frac{1}{[2\pi\sigma^2(t)]^{D/2}} \exp\left(-\frac{|\boldsymbol{x}|^2}{2\sigma^2(t)}\right) * g(\boldsymbol{x}, 0). \tag{15.12}$$

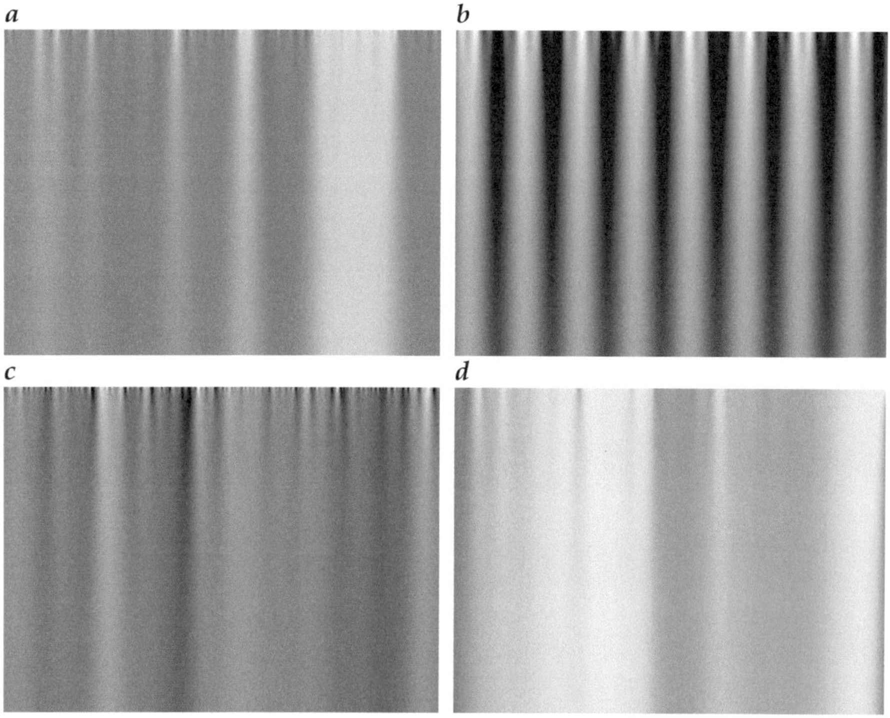

Figure 15.2: *Scale spaces of some one-dimensional signals:* ***a*** *edges and lines,* ***b*** *superposition of two periodic patterns,* ***c*** *a random signal and* ***d*** *line 10 from the image in Fig. 11.6a. The vertical coordinate is the scale parameter* ξ*.*

D in the exponent denotes the dimension of the space (not the diffusion constant!) and

$$\sigma(t) = \sqrt{2Dt}. \tag{15.13}$$

The last equation shows that the degree of smoothing expressed by the standard deviation σ only increases over time with the square root. We therefore set the scale parameter ξ equal to the square of the standard deviation:

$$\xi = 2Dt. \tag{15.14}$$

It is important to note that this formulation of the scale space applies to images of any dimension, i. e. it can also be extended to image sequences. The scale parameter is not identical to time, although we have used a physical diffusion process that progresses in time to derive it. If we calculate a scale space representation of an image sequence, it is useful to scale the time coordinate with a characteristic speed u_0 so that it has the same dimension as the spatial coordinates:

$$t' = u_0 t. \tag{15.15}$$

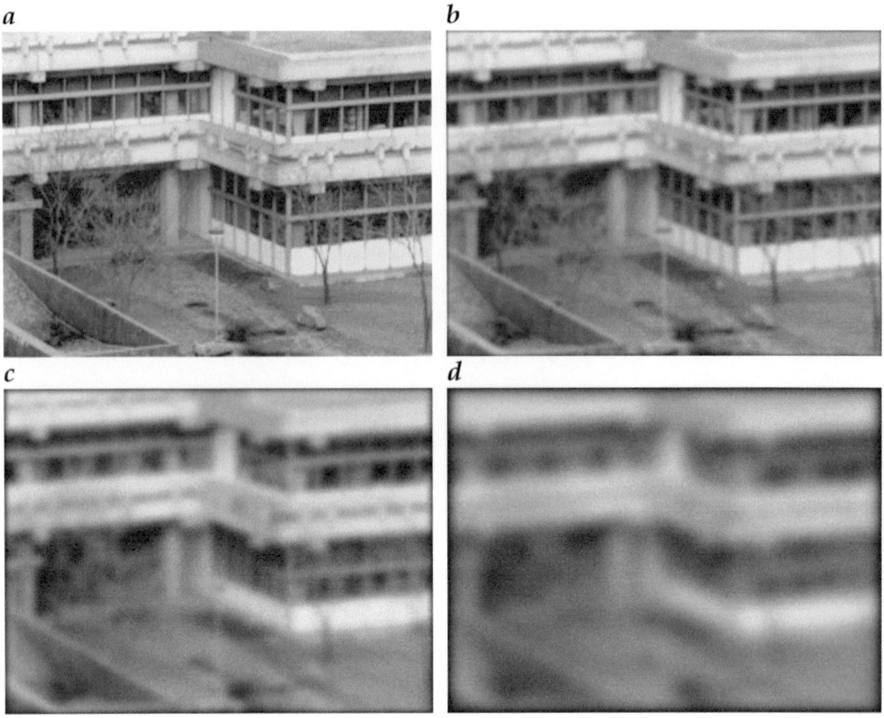

Figure 15.3: Scale space representation of a two-dimensional image: ***a*** *original resolution and a scale parameter ξ of* ***b*** *1,* ***c*** *2 and* ***d*** *4.*

We add this coordinate to the spatial coordinates and obtain a new coordinate vector

$$\boldsymbol{x} = [x_1, x_2, u_0 t]^T \quad \text{or} \quad \boldsymbol{x} = [x_1, x_2, x_3, u_0 t]^T. \tag{15.16}$$

We can also extend the wavenumber vector by a scaled frequency:

$$\boldsymbol{\kappa} = [k_1, k_2, \nu/u_0]^T \quad \text{or} \quad \boldsymbol{\kappa} = [k_1, k_2, k_3, \nu/u_0]^T. \tag{15.17}$$

With Eqs. (15.16) and (15.17), all previously derived equations, for example Eqs. (15.11) and (15.12), can also be applied to scale spaces of image sequences. For discrete spaces, the scaling of the time axis is automatically determined by the spatial and temporal sampling intervals: $u_0 = \Delta x/\Delta t$.

Fig. 15.2 illustrates the scale space of some characteristic one-dimensional signals: noisy edges and lines, a periodic pattern, a random signal and an image line. These examples illustrate the general properties of scale spaces. As the scale parameter ξ increases, the signals become blurred and more and more details are lost.

This property can be derived most easily from the transfer function of the scale space representation in Eq. (15.11). The transfer function is always positive and decreases monotonically with increasing scale parameter ξ for all wavenumbers. This means that no structures are amplified, but all structures are attenuated with increasing ξ, smaller structures always faster than coarser ones. In the limiting case $\xi \to \infty$, the scale space converges to an image with the mean value as a constant gray value. A certain feature only exists over a certain scale range. In Fig. 15.2a we can see that edges and lines disappear and two objects merge into one. For two-dimensional images, a continuous representation of the scale space results in a three-dimensional data structure. Therefore, Fig. 15.3 shows individual images for different scale parameters ξ.

15.2.2 ♯General Properties

In this section, we discuss some general properties of scale spaces. More precisely, we want to know which conditions must be fulfilled by a filter kernel that generates a scale space. Two basic requirements are relevant. First, no new details may be added as the scale parameter increases. From the point of view of information theory, we can say that the information content in the signal should continuously decrease with the scale parameter. The second property is related to the general principle of *scale invariance*. This essentially means that given any scale parameter in the scale space, we can start smoothing the signal to get the same scale space. We will only discuss some basic ideas about these elementary properties here, but will not give any proofs. For a detailed treatment of scale space theory, please refer to the monograph by Lindeberg [201].

According to Eq. (15.12), the linear, homogeneous and isotropic diffusion process has the convolution kernel

$$\boldsymbol{B}(\boldsymbol{x}, \xi) = \frac{1}{2\pi\xi} \exp\left(-\frac{|\boldsymbol{x}|^2}{2\xi}\right) \tag{15.18}$$

and after Eq. (15.11) the transfer function

$$\tilde{b}(\boldsymbol{\kappa}, \xi) = \exp(-4\pi^2|\boldsymbol{\kappa}|^2\xi/2). \tag{15.19}$$

In these equations, we have replaced the explicit time dependence with the scale parameter ξ using Eq. (15.14). We can represent the operator that generates the scale space in an independent way by

$$\mathcal{B}(\xi). \tag{15.20}$$

The property of the scale space that the information decreases with ξ can be formulated mathematically in different ways. We use the *minimum-maximum principle* here. It states that local extreme values must not be amplified. This means that the gray value must not increase at a local maximum and must not decrease at a minimum. This is an immediately understandable property for a diffusion process. If we imagine a heat transport process, a hot spot

*Figure 15.4: Scale space of a 1D signal with different wavenumbers, calculated with **a** a Gaussian filter and **b** a box filter. The scale parameter runs from top to bottom.*

cannot get hotter and a cool spot cannot get cooler. The Gaussian kernel Eq. (15.18) fulfills the minimum-maximum principle.

The second important property of the scale space concerns scale invariance. We want to start the process with an arbitrary scale parameter and still obtain the same scale space. We can formulate this property as follows:

$$\mathcal{B}(\xi_2)\mathcal{B}(\xi_1) = \mathcal{B}(\xi_1 + \xi_2). \tag{15.21}$$

This equation states that smoothing the scale space with the scale ξ_1 by an operator with the scale ξ_2 is equivalent to applying the scale space operator with the scale $\xi_1 + \xi_2$ to the original image. Alternatively, we can say that the representation at a coarser resolution ξ_2 can be calculated from a representation with a finer resolution ξ_1 by applying

$$\mathcal{B}(\xi_2) = \mathcal{B}(\xi_2 - \xi_1)\mathcal{B}(\xi_1) \quad \text{with} \quad \xi_2 > \xi_1. \tag{15.22}$$

From Eqs. (15.18) and (15.19), Eqs. (15.21) and (15.22) can be easily verified. In mathematics, the properties Eqs. (15.21) and (15.22) are referred to as *semigroup property*.

We can ask ourselves which kernels exist for generating scale spaces that fulfill both the minimum-maximum principle and the semigroup property. The answer may be surprising. The Gaussian kernel is the *only* convolution kernel that fulfills both criteria and is additionally isotropic and homogeneous [201]. This property gives the Gaussian convolution kernel and its discrete counterpart, the binomial kernel, a special significance for image processing. We have already looked at binomial filters in Section 11.4.

It is always helpful to discuss a counterexample. The simplest smoothing kernel for a D-dimensional image — known as the *running mean* — is the box filter

$$\boldsymbol{R}(\boldsymbol{x}, \xi) = \frac{1}{\xi^D} \prod_{d=1}^{D} \Pi\left(\frac{x_d}{\xi}\right) \tag{15.23}$$

with the transfer function

$$\tilde{r}(\boldsymbol{\kappa}, \xi) = \prod_{d=1}^{D} \frac{\sin(k_d \xi/2)}{k_d \xi/2}. \tag{15.24}$$

This kernel fulfills neither the minimum-maximum principle nor the half-group property. Fig. 15.4 compares scale spaces of a periodic signal with varying wavenumbers that were generated with a Gaussian and a rectangular filter. In Fig. 15.4b it becomes clear that the rectangular filter does not fulfill the minimum-maximum principle, as structures disappear but then reappear.

15.2.3 ♯Accelerated Scale Spaces

Despite the mathematical brilliance of generating scale spaces with a Gaussian convolution kernel, a significant disadvantage of this approach must be taken into account. The standard deviation of the smoothing only increases with the square root of time Eq. (15.13). Therefore, the scale parameter ξ is only proportional to the square of the standard deviation. The result is a non-linear scale coordinate. While smoothing occurs quickly for fine scales, it becomes increasingly slower for larger scales.

There is a simple solution to this problem. We need a diffusion process in which the diffusion constant increases with time. First, we discuss a diffusion coefficient that increases linearly with time. This approach results in the following diffusion equation:

$$\frac{\partial g}{\partial t} = D_0 t \Delta g. \tag{15.25}$$

A spatial Fourier transformation results in

$$\frac{\partial \tilde{g}(\boldsymbol{\kappa})}{\partial t} = -4\pi^2 D_0 t |\boldsymbol{\kappa}|^2 \tilde{g}(\boldsymbol{\kappa}). \tag{15.26}$$

This equation has the general solution

$$\tilde{g}(\boldsymbol{\kappa}, t) = \exp(-2\pi^2 D_0 t^2 |\boldsymbol{\kappa}|^2) \tilde{g}(\boldsymbol{\kappa}, 0), \tag{15.27}$$

which is equivalent to a convolution in local space. Therefore

$$g(\boldsymbol{x}, t) = \frac{1}{2\pi D_0 t^2} \exp\left(-\frac{|\boldsymbol{x}|^2}{2D_0 t^2}\right) * g(\boldsymbol{x}, 0). \tag{15.28}$$

From these equations, we can write the convolution kernel and the transfer function in the same form as in Eqs. (15.18) and (15.19) with the only exception being for the scale parameter:

$$\xi_q = D_0 t^2. \tag{15.29}$$

Now the standard deviation for smoothing the diffusion process, which increases linearly over time, is proportional to time. Since the scale parameter ξ_q is proportional to the square of time, we refer to this scale space as the *quadratic scale space*. This modified scale space still fulfills the minimum-maximum principle and the semigroup property.

For even faster smoothing, we can construct an *exponential scale space*, i. e. a scale space in which the logarithm of the scale parameter increases linearly with time. To do this, we use a diffusion coefficient that grows exponentially with time:

$$\frac{\partial g}{\partial t} = D_0 \exp(t/\tau)\Delta g. \tag{15.30}$$

Again, we obtain a convolution kernel and a transfer function as in Eqs. (15.18) and (15.19), but this time with the scale parameter

$$\xi_l = 2D_0\tau \exp(t/\tau). \tag{15.31}$$

15.2.4 ♯Differential Scale Spaces

The interest in *differential scale spaces* results from the fact that we want to select optimal scales for the processing of image properties. In a differential scale space, the change of the image is represented with the scale. We use the transfer function of the scale space kernel from Eq. (15.19), which also applies to quadratic and logarithmic scale spaces. The general solution for the scale space can be written in Fourier space as

$$\tilde{g}(\boldsymbol{\kappa}, \xi) = \exp(-2\pi^2|\boldsymbol{\kappa}|^2\xi)\tilde{g}(\boldsymbol{\kappa}, 0). \tag{15.32}$$

Differentiating this signal according to the scale parameter ξ results in

$$\frac{\partial \tilde{g}(\boldsymbol{\kappa}, \xi)}{\partial \xi} = -2\pi^2|\boldsymbol{\kappa}|^2 \exp(-2\pi^2|\boldsymbol{\kappa}|^2\xi)\tilde{g}(\boldsymbol{\kappa}, 0) = -2\pi^2|\boldsymbol{\kappa}|^2\tilde{g}(\boldsymbol{\kappa}, \xi). \tag{15.33}$$

Multiplication by $-|\boldsymbol{\kappa}|^2$ is equivalent to a second-order spatial derivative, the *Laplace operator* (≻ R12). So we can write in spatial space:

$$\frac{\partial g(\boldsymbol{x}, \xi)}{\partial \xi} = \frac{1}{2}\Delta g(\boldsymbol{x}, \xi). \tag{15.34}$$

Equations Eqs. (15.33) and (15.34) show a fundamental property of the differential scale space. This is equivalent to a second-order derivative with the Laplace operator and therefore leads to an isotropic *bandpass decomposition* of the image. The transfer function at the scale ξ is

$$-2\pi^2|\boldsymbol{\kappa}|^2 \exp(-2\pi^2|\boldsymbol{\kappa}|^2\xi). \tag{15.35}$$

For small wavenumbers, the transfer function is proportional to $-|\boldsymbol{\kappa}|^2$. It reaches a maximum at

$$k^2_{\max} = \frac{2}{\xi} \tag{15.36}$$

and then falls exponentially.

15.2.5 ♯Discrete Scale Spaces

The construction of *discrete scale spaces* requires a discretization of the diffusion equation. We start with a discretization of the one-dimensional diffusion equation:

$$\frac{\partial g(x,\xi)}{\partial \xi} = D\frac{\partial^2 g(x,\xi)}{\partial x^2}. \tag{15.37}$$

The derivatives are replaced by discrete differences as follows:

$$\begin{aligned} \frac{\partial g(x,\xi)}{\partial \xi} &= \frac{g(x,\xi+\Delta\xi)-g(x,\xi)}{\Delta\xi} \\ \frac{\partial^2 g(x,\xi)}{\partial x^2} &= \frac{g(x+\Delta x,\xi)-2g(x,\xi)+g(x-\Delta x,\xi)}{\Delta x^2}. \end{aligned} \tag{15.38}$$

This leads to the following iteration scheme for calculating a discrete scale space with $\epsilon = D\Delta\xi/\Delta x^2$:

$$g(x,\xi+\Delta\xi) = \epsilon g(x+\Delta x,\xi) + (1-2\epsilon)g(x,\xi) + \epsilon g(x-\Delta x,\xi) \tag{15.39}$$

or in discrete coordinates ($\xi \to i, x \to n$):

$${}^{i+1}g_n = \epsilon\, {}^{i}g_{n+1} + (1-2\epsilon)\, {}^{i}g_n + \epsilon\, {}^{i}g_{n-1}. \tag{15.40}$$

Lindeberg [201] shows that this iteration leads to a discrete scale space that satisfies the minimum-maximum principle and the semigroup property if

$$\epsilon \leq 1/4. \tag{15.41}$$

The limiting case of $\epsilon = 1/4$ results in the particularly simple iteration

$${}^{i+1}g_n = 1/4\, {}^{i}g_{n+1} + 1/2\, {}^{i}g_n + 1/4\, {}^{i}g_{n-1}. \tag{15.42}$$

Here, each step of the scale space calculation is determined by a spatial smoothing of the signal with the mask $\boldsymbol{B}^2 = [1\ 2\ 1]/4$. We can also formulate the general operator for generating a scale space from Eq. (15.39) with the convolution operator $\mathcal{B}$. In the operator notation introduced in Section 10.1.5, an iteration step for scale space generation results as

$$(1-4\epsilon)\mathcal{I} + 4\epsilon\mathcal{B}^2 \quad \text{with} \quad \epsilon \leq 1/4, \tag{15.43}$$

where $\mathcal{I}$ denotes the identity operator.

This expression is important because it can be extended directly for higher dimensions by replacing $\mathcal{B}^2$ with a corresponding higher-dimensional smoothing operator. The convolution mask $\boldsymbol{B}^2$ is the simplest mask from the class of binomial smoothing filters without having to change the grid for symmetry reasons. Smoothing filters were discussed in detail in Section 11.4.

15.3 Multigrid Structures

15.3.1 Introduction

If you want to process signals at different scales, this can be done most effectively in a *multigrid representation*. The underlying idea is simple. While the representation of fine scales requires full resolution, a lower resolution is sufficient for coarse structures. This leads to a sequence of images that become smaller and smaller from level to level. In the following two sections, we will analyze the *Gaussian pyramid* (Section 15.3.2) and the *Laplacian pyramid* (Section 15.3.4).

Gaussian and Laplacian pyramids are examples of multigrid data structures that were introduced to image processing in the early 1980s. They offer the possibility to perform neighborhood operators simultaneously on all scales without significant additional effort. As a result, they have considerably accelerated many image processing algorithms.

15.3.2 Gaussian Pyramid

If we want to reduce the size of an image, we cannot simply sample the image by taking every second pixel in every second line, for example. If we were to do this, the sampling theorem would be disregarded (Section 5.2.3). For example, a structure that is sampled three times per wavelength in the original image would only be sampled one and a half times in the next level. This would result in an incorrect pattern, as we showed in Section 5.1.

We must therefore ensure that all structures sampled less than four times per wavelength are suppressed by a suitable smoothing filter. For the generation of a scale space, this means that size reduction must go hand in hand with appropriate smoothing.

In general, the requirement for the smoothing filter can be formulated as follows:

$$\tilde{b}(\underset{\sim}{\boldsymbol{\kappa}}) = 0 \quad \forall \underset{\sim}{\kappa}_p \geq \frac{1}{r_p}, \tag{15.44}$$

where r_p is the sampling rate in the direction of the p-th coordinate.

The combined smoothing and size reduction when calculating the $(q+1)$-th pyramid plane from the q-th plane can be expressed with a single operator if we use the following notation:

$$\boxed{\boldsymbol{G}^{(0)} = \boldsymbol{G}, \quad \boldsymbol{G}^{(q+1)} = \mathcal{B}_{\downarrow 2}\boldsymbol{G}^{(q)}.} \tag{15.45}$$

The number after the $\downarrow$ in the index indicates the sampling rate. The zeroth plane is the original image.

The series of images obtained by applying this operator several times is called *Gaussian pyramid* is designated. From one layer to the next, the resolution decreases by half and the size of the image is reduced accordingly.

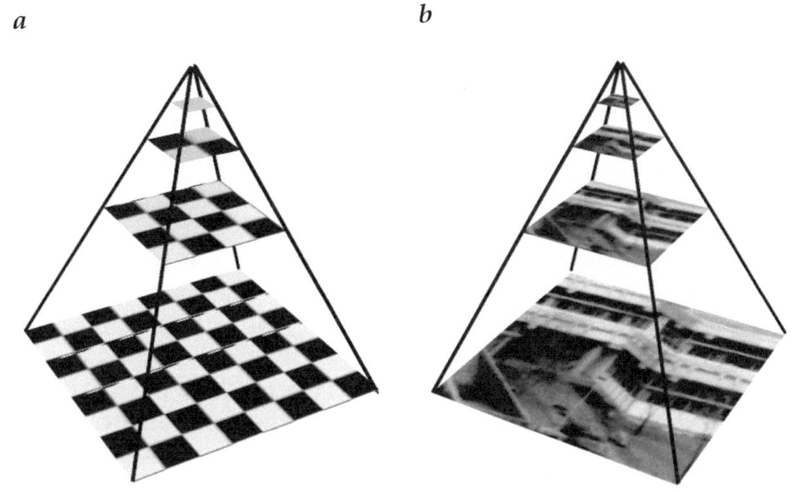

Figure 15.5: *Gaussian pyramid:* ***a*** *schematic representation; the squares of the chessboard correspond to pixels;* ***b*** *example.*

We can imagine this series of images as shown in Fig. 15.5, arranged in the shape of a pyramid.

The pyramid has a surprisingly low memory requirement. If we consider the formation of a pyramid from a D-dimensional image with M pixels in each coordinate direction, the total number of pixels at a sampling rate of two

$$M^D\left(1+\frac{1}{2^D}+\frac{1}{2^{2D}}+\ldots\right)< M^D\frac{2^D}{2^D-1}. \tag{15.46}$$

With a two-dimensional image, the entire pyramid only requires 1/3 more memory than the original image, with a three-dimensional image only 1/7 more. The calculation of the pyramid is just as effective. The same smoothing filter is applied to each level of the pyramid. This means that the calculation of the entire pyramid requires only 4/3 of the arithmetic operations for a two-dimensional image and 8/7 of those for a three-dimensional image.

The pyramid brings large scales into the range of local neighborhood operations with small masks. In addition, these operations are efficient to calculate. Once the pyramid is calculated, we can perform neighborhood operations with large scales in the upper levels of the pyramid. These are much more efficient than with finer scales due to the smaller image size.

The Gaussian pyramid represents a series of low-pass filtered images in which the cut-off wavelength decreases by half (one octave) from level to level. This means that only increasingly coarser details remain in the image (Fig. 15.5). Only a few pyramid levels are required to span the entire range of wavenumbers. From a $N \times N$ image, we can calculate a pyramid with a maximum of $\mathrm{ld}\,N + 1$ levels. The smallest image consists of only one pixel.

15.3.3 Multigrid Averaging

The Gaussian pyramid can also be understood as a fast *multigrid method* for cascaded smoothing. In this sense, we compare such a method here with the multi-step averaging introduced in Section 11.5.1.

The Gaussian pyramid means that we gradually apply the same mask with increasingly larger step sizes to correspondingly coarser grids. This means that the last operation before using a mask with a larger step size only has to calculate the convolution for those grid points that are taken into account by the subsequent coarser grid operator. This sampling procedure is marked with a special syntax in the index. $\mathcal{O}_{\downarrow 2}$ means, for example: Apply the operator in the all direction and advance the mask noy by one but by two pixels in all direction. This means that the result of the filter operator only has half as many pixels in all directions as the input image.

With multigrid averaging, the number of calculations is essentially independent of the standard deviation of the smoothing mask. We consider the cascading of the binomial masks

$$\underbrace{\mathcal{B}^R_{\downarrow 2} \cdots \mathcal{B}^R_{\downarrow 2}\mathcal{B}^R_{\downarrow 2}}_{S\text{times}}.$$

If $\mathcal{B}^R_{\downarrow 2}$ requires P operations per pixel with a D-dimensional signal, then the following number of operations are needed for the operator sequence

$$P\sum_{s=1}^{S}\frac{1}{2^{D(s-1)}} = P\left(1+\frac{1}{2^{D(S-1)}}\right) < 2P.$$

The variance of the mask of the operator sequence is the same as for the multi-step method

$$\sigma^2 = \frac{R}{12}(4^S - 1). \tag{15.47}$$

Smoothing of any degree therefore requires no more than twice as many operations as the first step. As with a multi-step binomial filter, the standard deviation increases by a factor of two from step to step. Furthermore, since $\tilde{b}^R(\tilde{k}) = 0 \quad \forall \tilde{k} \geq 1/2$, the same filters can be used as for multistep smoothing (Section 11.5.1).

Extension 15.1: Optimal and efficient smoothing filters for generating the Gaussian pyramid

To generate a Gaussian pyramid, Burt [37] used a separable 5 × 5 smoothing filter that approximated a Gaussian distribution as well as possible and he arrived at the following one-dimensional convolution mask

$$\begin{array}{c} \boldsymbol{B}_{\text{Burt}} = (4\boldsymbol{B}^4 + \boldsymbol{B}^2)/5 = [1\ 5\ 8\ 5\ 1]/20 \\ \circ\!\!-\!\!\bullet \\ b_{\text{Burt}}(\tilde{k}) = \left(4\cos^4(\pi\tilde{k}/2) + \cos^2(\pi\tilde{k}/2)\right)/5, \end{array} \tag{15.48}$$

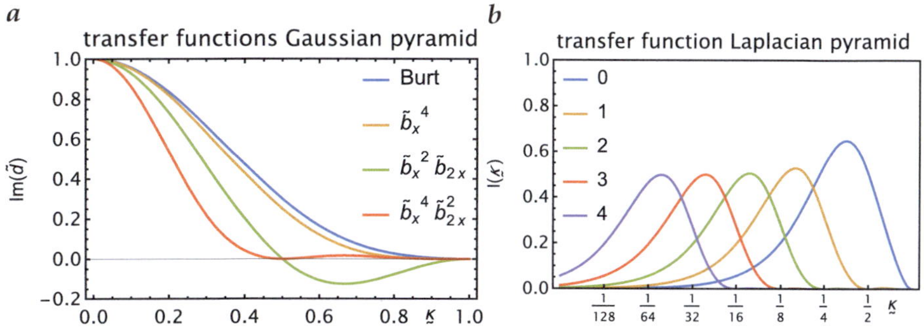

***Figure 15.6:** **a** Transfer functions of filters for smoothing for a Gaussian pyramid as indicated. **b** Transfer functions of the first five levels of a Laplacian pyramid, shown with a logarithmic wavenumber axis. The smoothing operator $\mathcal{B}^2\mathcal{B}^4_{\downarrow 2}$ was used.*

which was applied in the horizontal and vertical direction. The transfer function of this mask is very similar to the binomial mask $\boldsymbol{B}^4$. However, both filters do not fulfill the sampling condition for generating a Gaussian pyramid that the transfer function is zero for all wavenumbers $\tilde{k} \geq 1/2$ (Fig. 15.6a).
If you want the transfer function to be zero for wavenumbers $\tilde{k} = 1/2$, the filter must have a partial filter with double step size (Section 11.5.1). With a 5×5 filter, there is then only one possibility:

$$\begin{array}{c} \boldsymbol{B}^2\boldsymbol{B}_2 = [1\ 2\ 1]/4 * [1\ 0\ 1]/2 = [1\ 2\ 2\ 2\ 1]/8 \\ \circ\!\!-\!\!\bullet \\ \cos^2(\pi\tilde{k}/2)\cos(\pi\tilde{k}). \end{array} \tag{15.49}$$

This filter is better than Burt's filter. At wavenumbers $\tilde{k} \geq 0.5$, however, the structures are sometimes insufficiently attenuated and the transfer function becomes negative. It is much better to take the 5×5 $\boldsymbol{B}^4_{\downarrow 2}$ smoothing filter and then to smooth the signal at the next stage with a 3×3 $\boldsymbol{B}^2$ smoothing filter. This is very effective as it is a smaller filter that only needs to be applied to a quarter of the data points in two dimensions. This filter combination corresponds to the filter operator $\mathcal{B}^2\mathcal{B}^4_{\downarrow 2}$. It removes all structures with wavenumbers above $\tilde{k} \geq 1/2$ very effectively (Fig. 15.6a).
In this one-dimensional consideration, no attention has yet been paid to optimal isotropy. This requires further optimization steps.

15.3.4 Laplacian Pyramid

From the Gaussian pyramid, we can calculate another type of pyramid: the *Laplacian pyramid*, which leads to a sequence of bandpass-filtered images. In contrast to the Fourier transform, the Laplacian pyramid only leads to a coarse wavenumber decomposition without directional decomposition. All wavenumbers within the range of approximately one octave (factor 2) are contained in one plane of the pyramid regardless of their direction. Due to the coarser wavenumber resolution, a good spatial resolution is maintained.

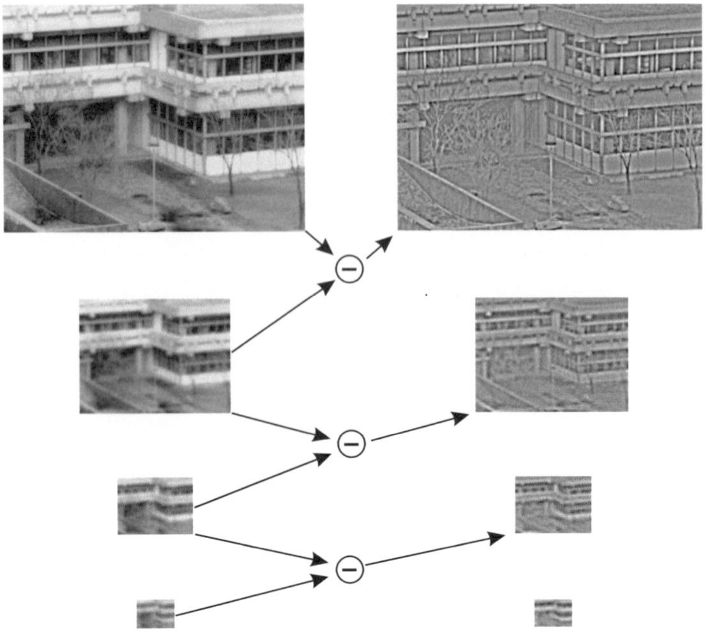

Figure 15.7: Construction of the Laplacian pyramid (right) from a Gaussian pyramid (left) by calculating the difference between two consecutive images of the Gaussian pyramid.

Each plane of the pyramid contains only matching scales, which are scanned several times (two to six times) per wavelength. In this way, the Laplacian pyramid represents an efficient data structure that is well adapted to the limits of the product between wavenumber and spatial resolution set by the *uncertainty relation* (Section 15.1.3 and Theorem B.7, p. 633,).

To achieve this, two levels of the Gaussian pyramid must be subtracted from each other. To do this, the image in the coarser plane must first be expanded. This operation is performed by the *expansion operator* $\uparrow_2$. As with the reducing smoothing operator Eq. (15.45), the degree of expansion is indicated by the number after the $\uparrow$ sign in the index. The expansion is considerably more difficult than the size reduction, as the missing information must be interpolated. To increase the size by a factor of two in all directions, every second pixel in each line must first be interpolated and then every second line. The interpolation was discussed in detail in Section 5.5. With the introduced notation, we can write the generation of the q-th plane of the Laplacian pyramid as follows:

$$\boxed{\boldsymbol{L}^{(q)} = \boldsymbol{G}^{(q)} - \uparrow_2 \boldsymbol{G}^{(q+1)}, \quad \boldsymbol{L}^{(Q)} = \boldsymbol{G}^{(Q)}.} \tag{15.50}$$

The Laplacian pyramid is an effective scheme for a *bandpass decomposition* of an image. The central wavenumber is halved from level to level. The last image of the Laplacian pyramid, $\boldsymbol{L}^{(Q)}$, is a low-pass filtered image $\boldsymbol{G}^{(Q)}$ containing only the coarsest structures.

The Laplacian pyramid has the significant advantage that the original image can be quickly restored from the image sequence of the Laplacian pyramid by recursively expanding and summing the images. The corresponding recursion is the inversion of the recursion in Eq. (15.50). For a Laplacian pyramid with $p+1$ levels, the level q (counting starts with zero!) is the $(q+1)$th level of the Gaussian pyramid. The $(q-1)$-th plane of the Gaussian pyramid can then be reconstructed with

$$\boxed{\boldsymbol{G}^{(Q)} = \boldsymbol{L}^{(Q)}, \quad \boldsymbol{G}^{(q-1)} = \boldsymbol{L}^{(q-1)} + \uparrow_2 \boldsymbol{G}^{q}.} \tag{15.51}$$

Note that this is merely a reversal of the construction scheme of the Laplacian pyramid. This means that an erroneous interpolation algorithm for the expansion of the image only affects the Laplacian pyramid, but not the reconstruction of the Gaussian pyramid from the Laplacian pyramid, since the same algorithm is used.

The recursion in Eq. (15.51) is repeated with lower levels until the level 0, i. e. the original image, is reached again. As Fig. 15.7 illustrates, increasingly finer details become visible during the reconstruction process. Due to the possibility of progressive reconstruction of details, the Laplacian pyramid was also used for image compression. Today, however, more efficient compression schemes based on a wavelet transform are available, which are based on a similar principle as the Laplacian pyramid [1, 42].

15.3.5 Multigrid Neighborhood Operators

The Laplacian pyramid is ideal for carrying out any filter operations simultaneously on all scales. A particular advantage is that the gray value changes change at approximately the same spatial rate on all levels because the wavenumber distribution in relation to the wavelength range is approximately the same on each level.

This can be seen from the similarity of the transfer functions at all stages in Fig. 15.6b. A logarithmic wavenumber axis was used in this figure. The transfer functions of the individual stages are each shifted by a factor in the wavenumber. With the logarithmic axis, this means a constant shift from stage to stage.

A further advantage results from the reduction in noise due to the bandpass filtering in each stage of the Laplacian pyramid. No further filter operation is required for multiscale edge detection if you want to detect edges through zero crossings, as each stage of the Laplacian pyramid corresponds to the application of a Laplacian filter (Section 12.5).

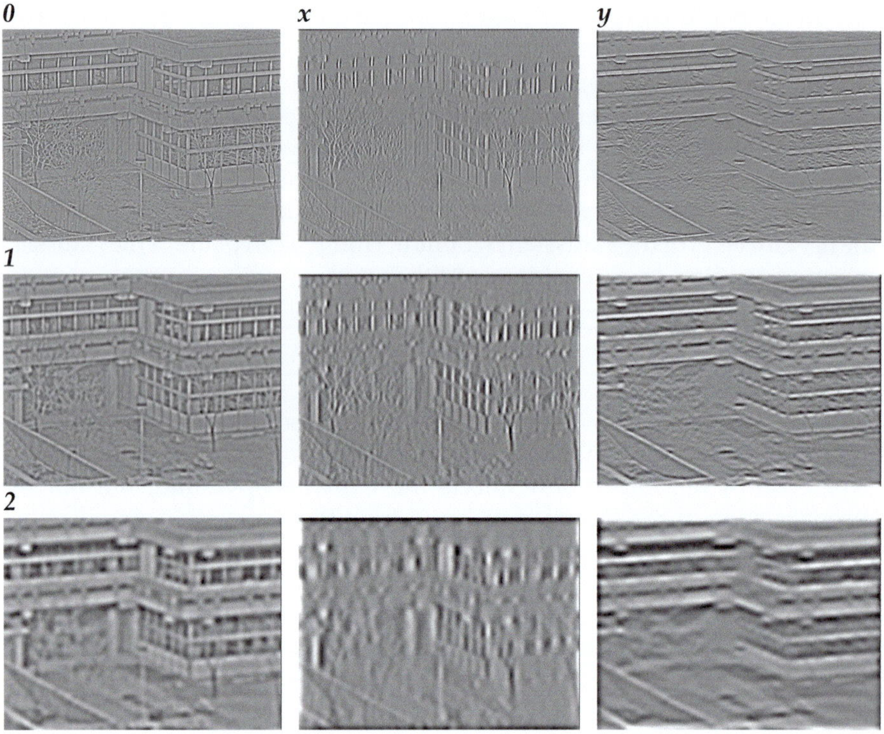

Figure 15.8: The first three levels of the directional decomposition on a pyramid of the image Fig. 15.3a: the rows show the levels 0, 1 and 2, the columns respectively $\mathbf{L}$, $\mathbf{L}_x$, $\mathbf{L}_y$ *according to Eqs. (15.53) and (15.54).*

15.3.6 ♯Directio-pyramidal Decomposition

For multidimensional signals, directional decomposition is just as important as scale decomposition. Directional filters are necessary for this. Ideally, the directional components should add up to the total signal, just like the individual planes in the Laplacian pyramid. A combined decomposition into a pyramid and directional components is known as *directio-pyramidal decomposition* [151].

In general, such a decomposition is a difficult filter design problem. Therefore, we illustrate this decomposition here only with a simple scheme with two directional components on each pyramid level.

Smoothing is performed with a separable smoothing filter. One filter ($\mathcal{B}_x$) only smoothes in the x direction, the other ($\mathcal{B}_y$) in the y direction. Then the next higher level of the Gaussian pyramid results as in Eq. (15.45) as

$$\mathbf{G}^{(q+1)} = \downarrow_2 \mathcal{B}_x \mathcal{B}_y \mathbf{G}^{(q)}. \tag{15.52}$$

The Laplacian pyramid is

$$L^{(q)} = G^{(q)} - \uparrow_2 G^{(q+1)}. \tag{15.53}$$

The two directional components are

$$\begin{aligned} L_x^{(q)} &= 1/2(G^{(q)} - \uparrow_2 G^{(q+1)} - (\mathcal{B}_x - \mathcal{B}_y)G^{(q)}), \\ L_y^{(q)} &= 1/2(G^{(q)} - \uparrow_2 G^{(q+1)} + (\mathcal{B}_x - \mathcal{B}_y)G^{(q)}). \end{aligned} \tag{15.54}$$

From Eq. (15.54) it is immediately clear that L_x and L_y add up to the isotropic Laplacian pyramid: $L = L_x + L_y$. Example images with the first 3 levels of a pyramidal directional decomposition are shown in Fig. 15.8.

15.4 ♯References for Further Reading

Multi-grid image processing was developed in the early 1980s. A nice compilation of the early work can be found in Rosenfeld [271]. The equivalent development for time series, multirate signal processing, is described in the textbook by Fliege [86]. The mathematics of multigrid methods is dealt with in the monograph by Trottenberg et al. [331].

The monograph by Lindeberg [201] deals in detail with linear scale space theory. Non-linear scale spaces including inhomogeneous and anisotropic diffusion are treated by Weickert [350].

Current developments in scale-space analysis can be followed in the proceedings of the international scale-space conferences held every two years: 1997 [322], 1999 [231], 2001 [179], 2003 [116], 2005 [180], and from 2007 as Scale-Space and Variational Methods (SSVM), most recently in 2025 [30, 31].

16 Texture and Complex Features

Contents

16.1 Introduction

In Chapters 11 to 13 we discussed smoothing, edge detection, and simple neighborhoods. Now we will use and extend these important building blocks together with the multiscale processing introduced in Chapter 15 to analyze complex patterns, called *texture* in image processing. Such patterns make the difference between an artificial object world, whose surfaces are only characterized by color and reflection properties, and real images.

We can easily recognize and distinguish patterns with our visual system (Fig. 16.1). Much more difficult is the characterization and differentiation of the more "diffuse" quantities of a texture with precisely defined parameters that allow a computer to perform this task.

In this chapter, we will systematically examine operators for analyzing and distinguishing patterns. With the help of these operators, even complex patterns can be described with a small number of characteristic variables. In doing so, we reduce the problem of pattern recognition to the simple differentiation of gray values.

How can we define a texture? An arbitrary pattern that extends over a large area of an image is certainly not recognized as a texture. A fundamental property of texture is therefore the small elementary pattern that is repeated periodically or quasiperiodically in space, like the pattern on a wallpaper. It is therefore sufficient to describe the small basic pattern and the repetition rules. The latter define the characteristic size of the texture.

B. Jähne, *Digital Image Processing and Image Acquisition*,
https://doi.org/10.1007/978-3-032-04806-6_16

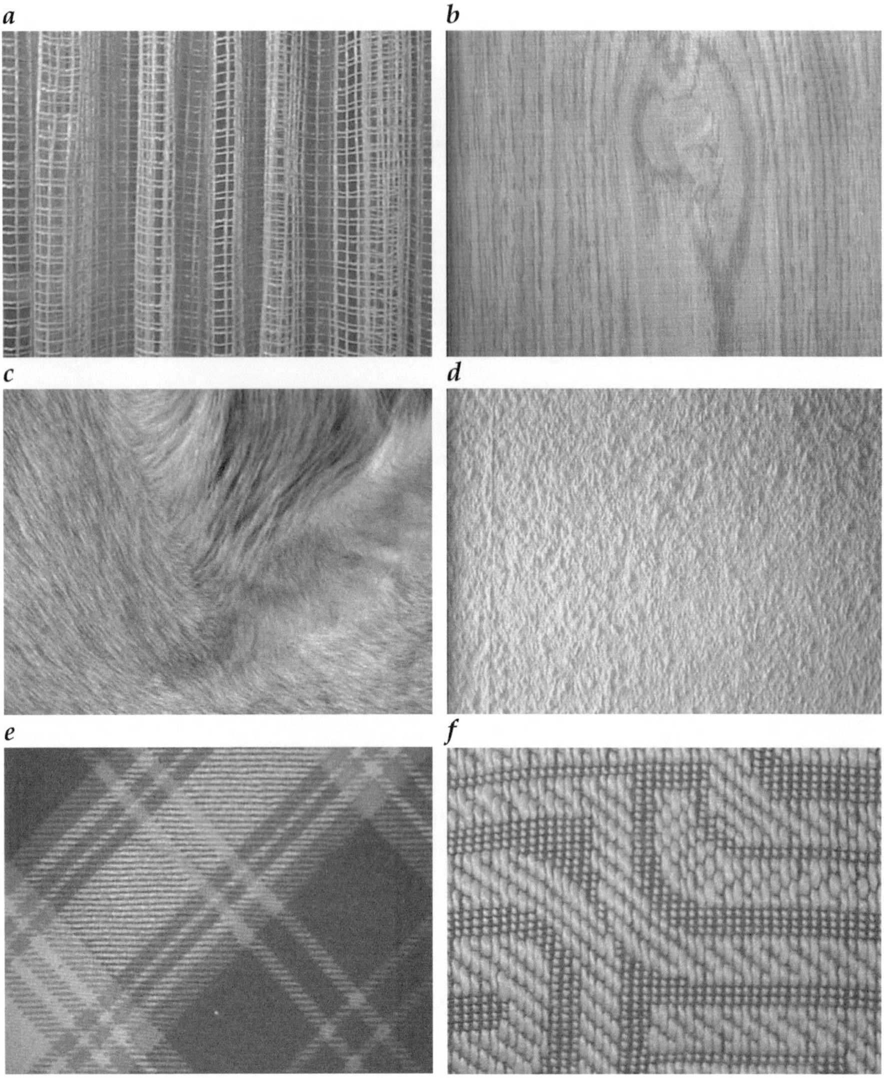

Figure 16.1: *Texture examples:* ***a*** *curtain;* ***b*** *wood;* ***c*** *dog fur;* ***d*** *woodchip wallpaper;* ***e*** *and* ***f*** *fabrics.*

Texture analysis can be compared to the analysis of solid state structures, a field of research that we find in solid state physics, chemistry and mineralogy. A solid-state physicist is concerned with the repetition patterns and distribution of atoms in the elementary cell. Texture analysis is complicated by the fact that both the pattern and the periodic repetition can be subject to significant random variations (Fig. 16.1).

Figure 16.2: *Hierarchical organization of a pattern, shown using the example of the curtain from Fig. 16.1a by recording in different resolutions.*

Textures can be structured *hierarchical*, i. e. they can look very different at different magnifications. A nice example is the curtain shown in Fig. 16.1a. At the highest resolution, our attention is on the individual fabric thread (Fig. 16.2a), and the characteristic size is the thickness of the threads, which also have a local orientation. At the next coarser resolution, we recognize the meshes of the fabric (Fig. 16.2b) with the mesh diameter as the characteristic variable. In this plane, the local orientation is evenly distributed in all directions. At even lower magnification, we no longer recognize the individual meshes, but only the folds of the curtain (Fig. 16.2c). They are also characterized by their own characteristic size, the fold density and their orientation. These considerations underline the importance of *multi-scale texture analysis*. The multi-scale data structures discussed in the previous Chapter 15 are therefore essential for texture analysis.

Due to the task at hand, the texture parameters can generally be divided into two classes. The first class contains the rotation and size invariant parameters, while the second class contains the rotation and size variant parameters. Let us imagine a typical industrial or scientific application in which it is important to recognize randomly oriented objects in an image. We are interested in distinguishing the objects from each other, not the orientation of the individual objects. Texture parameters that depend on orientation are therefore of no interest. We can use them, but only if the objects have a characteristic shape that allows us to determine their orientation. Similar arguments apply to size-invariant properties. If the objects of interest are at different distances from the camera, the texture parameters used to recognize them should also be size invariant. If they are not, the object recognition depends on the distance to the camera. However, if the texture properties change with the magnification, as in the example of the curtain in Fig. 16.1a, there are no size-invariant texture properties at all. This makes the use of patterns for object characterization at different distances a difficult task.

In the previous examples, we were more interested in the objects themselves than their orientation in space. Surface orientation is a key feature in another image processing task, namely the reconstruction of a three-dimensional scene. If we know that the surface of an object has a regular pattern, we can use the orientation and size of the pattern to analyze the orientation of the surface in space. To do this, we need the characteristic sizes and orientations of the texture.

Texture analysis is one of the areas of image processing that still lacks basic knowledge. Accordingly, we find many different empirical and semi-empirical approaches in the literature. We will not deal with these approaches here, but with a simple method that builds complex texture operators from elementary operators.

For texture analysis, we only need four fundamental texture operators:

- *mean value*,
- *variance*,
- *orientation*, and
- *scale* (local wavenumber).

These operators are used at different levels of the image processing hierarchy. For example, if we have calculated the local orientation and size, mean and variance operators can be applied again, but this time not to calculate the mean and variance of the gray values, but of the local orientation and the local size.

The four elementary texture operators can be divided into the two classes mentioned above. Mean and variance are invariant to rotation and size, while the orientation and size operators naturally determine the orientation and size respectively. This means that we have already completed the important class separation of the parameters. The importance of this approach lies in the simplicity and orthogonality of the parameter set as well as in the possibility of applying it hierarchically.

16.2 First-order Statistics

16.2.1 Basics

All texture parameters that are based on first-order statistics, i. e. local gray value histograms, are invariant to a permutation of the pixels. They therefore depend neither on the orientation nor the size of the objects. The latter, however, only as long as fine structures do not disappear at coarser resolutions. Therefore, we can consider this class of texture parameters as invariant to rotation and scale.

However, this invariance of first-order statistics to pixel permutations also has a serious disadvantage. Textures that have different spatial patterns with the same distribution of gray values cannot be distinguished. Here is a simple example. A texture with equally wide white and black stripes and a texture with a black/white checkerboard pattern have the same bimodal gray value distribution with a completely different spatial structure of the patterns.

Therefore, many textures cannot be differentiated using parameters calculated from local histograms. Additional classes of texture parameters must be used for a better differentiation.

16.2.2 Local Variance

A statistical distribution of a variable such as gray or color values can be characterized by variance and higher moments in addition to the mean value (Section C.2.2). For texture analysis, these parameters must be averaged in a local neighborhood. This leads us to a new operator for estimating the *local variance*. Local higher moments of a distribution can also be calculated in a similar way. Here we only show the procedure using the example of the variance.

In the simplest case, we can select a mask M and calculate the parameters from the pixels in this window alone. The *variance operator* then results in

$$v_n = \frac{1}{Q-1} \sum_{\boldsymbol{n}' \in M} \left(g_{n-n'} - \overline{g}_n \right)^2 . \tag{16.1}$$

The sum runs over Q pixels of the window M. The vector indexing introduced in Section 10.3.1 was used. The expression $\overline{g}_n$ denotes the mean value of the gray values calculated over the same window M at the point $\boldsymbol{n}$:

$$\overline{g}_n = \frac{1}{Q} \sum_{\boldsymbol{n}' \in M} g_{n-n'} . \tag{16.2}$$

It is important to note that the variance operator is not linear. Nevertheless, it is similar to the general form of a neighborhood operation, the convolution. The combination of Eqs. (16.1) and (16.2) makes it clear that the variance operator consists of a linear convolution and non-linear point operations:

$$v_n = \frac{1}{Q-1} \left[\sum_{\boldsymbol{n}' \in M} g_{n-n'}^2 - \left(\frac{1}{Q} \sum_{\boldsymbol{n}' \in M} g_{n-n'} \right)^2 \right] , \tag{16.3}$$

or in operator notation:

$$\mathcal{V} = \mathcal{R}(\mathcal{I} \cdot \mathcal{I}) - (\mathcal{R} \cdot \mathcal{R}). \tag{16.4}$$

The operator $\mathcal{R}$ denotes a smoothing over all pixels with a rectangular filter of the size of the window W. The operator $\mathcal{I}$ is the identity operator. The operator $\mathcal{I} \cdot \mathcal{I}$ thus performs a non-linear point operation, namely the squaring of the gray values of each pixel. Finally, the variance operator subtracts the square of a smoothed gray value from the smoothed squared gray values. From the discussion of smoothing in Section 11.3, we know that a rectangular filter is not a good smoothing filter. We therefore obtain a better variance operator if we replace the rectangular filter $\mathcal{R}$ with a binomial filter $\mathcal{B}$:

$$\mathcal{V} = \mathcal{B}(\mathcal{I} \cdot \mathcal{I}) - (\mathcal{B} \cdot \mathcal{B}). \tag{16.5}$$

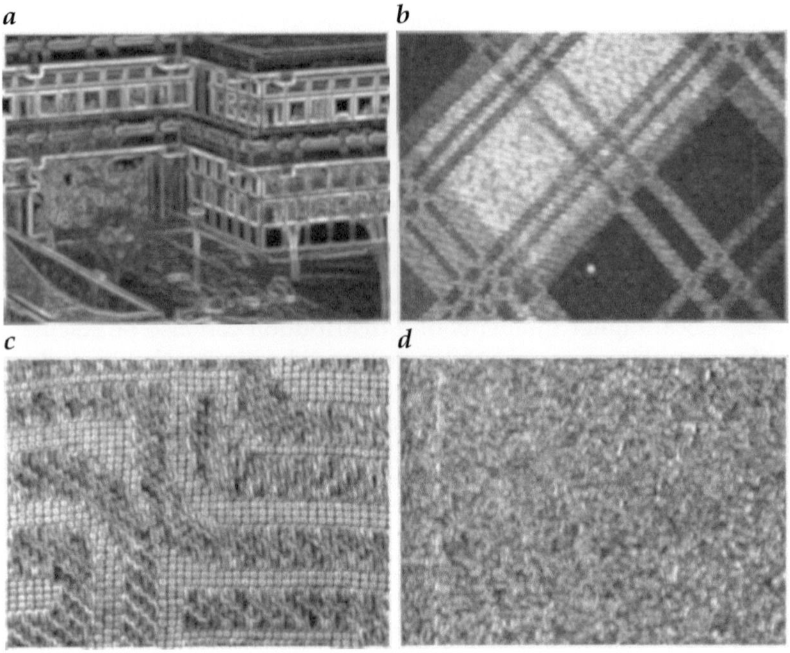

*Figure 16.3: Variance operator applied to **a** Fig. 11.6a; **b** Fig. 16.1e; **c** Fig. 16.1f; **d** Fig. 16.1d.*

The variance operator is isotropic and also size-invariant as long as the window is larger than the largest structure in the textures and as long as no fine structures of the texture disappear when objects are further away from the camera. From this it can be assumed that a size-invariant texture operator only exists if the texture itself is size-invariant.

Fig. 16.3 shows the application of the variance operator Eq. (16.5) with $\mathcal{B}^{16}$ to different images. In Fig. 16.3a, the variance operator is an isotropic edge detector, since the original image contains areas with more or less uniform gray values. The other three examples in Fig. 16.3 show variance images of patterned surfaces. The variance operator can distinguish the areas with the fine horizontal stripes in Fig. 16.1e from the more uniform areas. They appear as uniform bright areas in the variance image (Fig. 16.3b). However, the variance operator cannot distinguish between the two textures in Fig. 16.3c. As the resolution is still significantly smaller than the repeat length of the texture, the variance operator cannot provide a uniform mean value of the variance in the pattern. For the woodchip wallpaper (Fig. 16.3d), the variance operator does not provide a uniform answer, as the structures show too strong fluctuations.

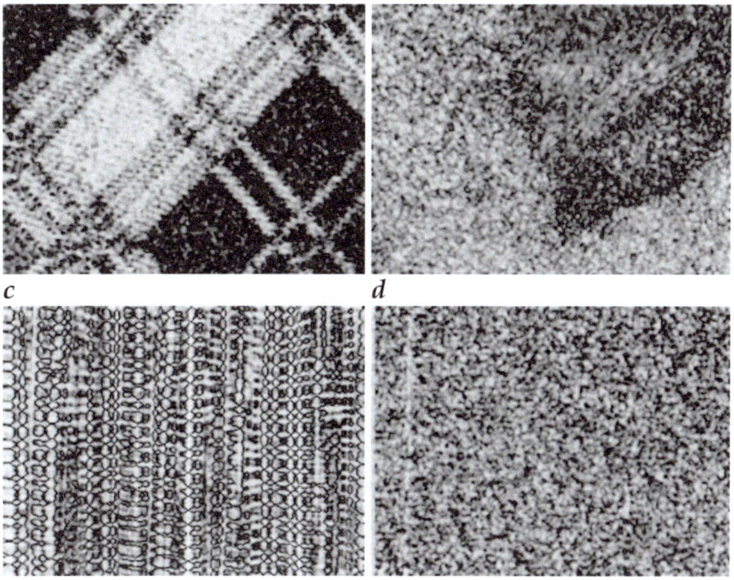

Figure 16.4: *Coherence of local orientation:* ***a*** *fabric with regions of horizontal stripes (Fig. 16.1e),* ***b*** *dog fur (Fig. 16.1c),* ***c*** *curtain (Fig. 16.1a) and* ***d*** *woodchip wallpaper (Fig. 16.1d).*

16.2.3 Higher Moments

In addition to the variance, we can also use the higher moments of the gray value distribution (see Section C.2.2 for a detailed description). Let us illustrate the importance of this approach with two examples of quite different gray value distributions, a normal distribution and a bimodal distribution:

$$p(g) = \frac{1}{\sqrt{2\pi}\sigma} \exp\left(-\frac{g-\overline{g}}{2\sigma^2}\right), \quad p'(g) = \frac{1}{2}\left(\delta(\overline{g}+\sigma) + \delta(\overline{g}-\sigma)\right).$$

Both distributions have the same mean value and the same variance, but differ in moments of higher order.

16.3 Rotation and Size Variant Features

16.3.1 Local Orientation

Since local orientation has already been discussed in detail in Chapter 13, we only discuss a few examples here to illustrate the importance of local orientation for texture analysis.

Fig. 16.4 shows the coherence measure for local orientation defined in Section 13.3.4. This measure is one for an ideally oriented texture in which

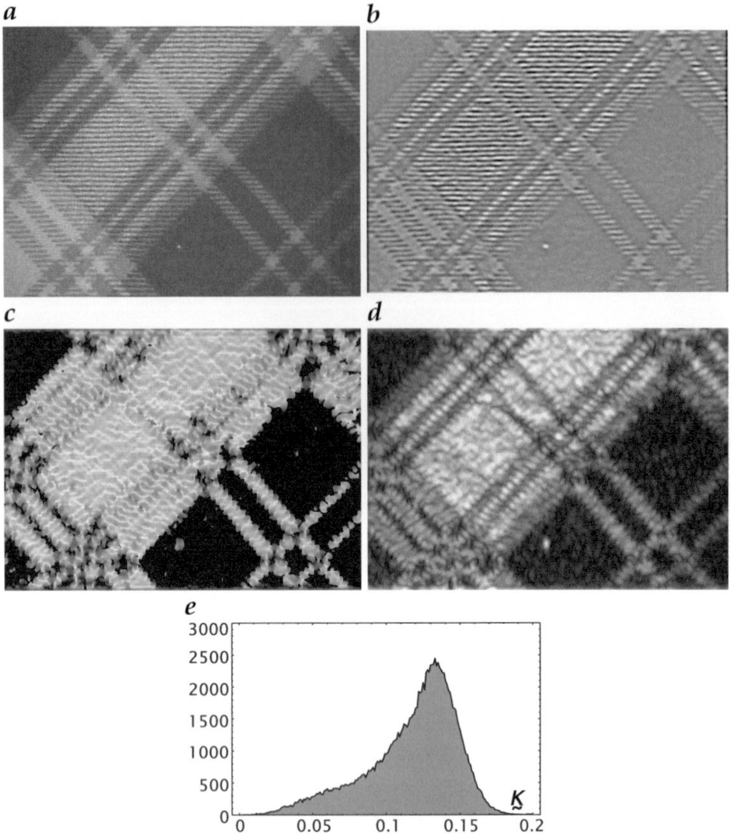

Figure 16.5: Determination of the characteristic values of a fabric sample by calculating the local wavenumber: ***a*** *original pattern,* **b** *filtered with a directional bandpass using levels one and two of the vertical component of a directional pyramidal decomposition* **c** *local wavenumber estimation (all structures below a certain threshold are mapped to black)* ***d*** *amplitude of the local wavenumber and* **e** *Histogram of the local wavenumber distribution (unit: number of periods per pixel).*

the gray values only change in one direction and zero for a distributed gray value structure. The coherence measure approaches one in areas with horizontal stripes of the fabric (Fig. 16.4a) and in the dense areas of dog fur (Fig. 16.4b).

In the curtain (Fig. 16.1a), the coherence is high along the individual threads, but not where two threads cross and in most parts of the background visible through the stitches (Fig. 16.4c). In the case of woodchip wallpaper (Fig. 16.1d), the coherence is uneven, as this texture has no predominant local orientation (Fig. 16.4d).

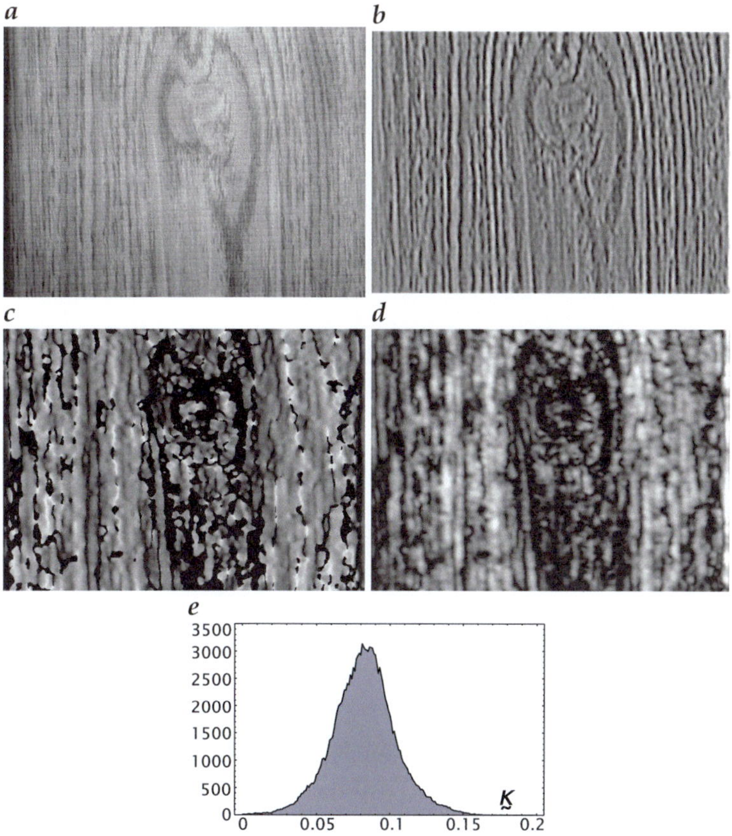

Figure 16.6: *Like Fig. 16.5, but applied to a wood grain.*

16.3.2 Local Wavenumber

In Section 13.4 we discussed the calculation of the *local wavenumber* from a *quadrature filter pair* using either a *Hilbert filter* (Section 13.4.2) or a quadrature filter (Section 13.4.5). In this section, we will use these techniques to calculate the characteristic size of a texture by first using a directio-pyramidal decomposition (Section 15.3.6) as a bandpass filter and then using Hilbert filters.

In the piece of fabric in Fig. 16.5a, stripes running horizontally to each other can be seen in certain areas. This image is first bandpass filtered using levels one and two of the vertical component of a directional pyramidal decomposition of the image (Fig. 16.5b). Fig. 16.5c shows the approximation of the local wavenumber (component in vertical direction). The areas in which the amplitude of the associated structure is not significantly higher than the noise level are hidden (Fig. 16.5d). A local wavenumber could be calculated in all areas with horizontal stripes. The histogram in Fig. 16.5e shows that

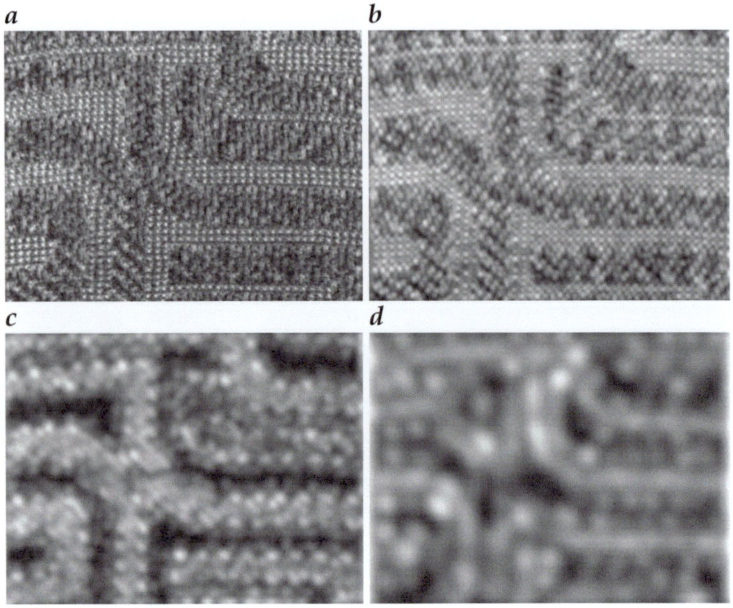

Figure 16.7: Application of the variance operator to the levels 0 to 3 of the Laplacian pyramid of the image from Fig. 16.1f.

the most frequent local wavenumber is about 0.133. This corresponds to a structure that is scanned about 7.5 times per wavelength. The elongated distribution towards small wavenumbers is noteworthy. The texture must therefore contain a second, larger structure. This is indeed the case with the narrow diagonal stripes.

Fig. 16.6 shows the same analysis for a wood texture. This time the pattern is more random. Nevertheless, it is possible to determine the local wavenumber. It is important to mask out the areas where the bandpass filtered image has no significant amplitudes. If this masking is not performed, the estimation of the local wavenumber is considerably disturbed. With masking, we obtain a narrow distribution of the local wavenumber with a maximum at a wavenumber of 0.085.

16.3.3 Pyramidal Texture Analysis

The Laplacian pyramid is an alternative to the local wavenumber operator as it performs a bandpass decomposition of an image. This decomposition does not directly calculate a local wavenumber, but we get a series of images that show the texture at different scales of the pattern.

The variance operator takes a very simple form with a Laplacian pyramid, since the mean gray value — except for the lowest resolution level — is zero:

$$\mathcal{V} = \mathcal{B}(\mathcal{L}^{(p)} \cdot \mathcal{L}^{(p)}). \tag{16.6}$$

Figure 16.7 shows how the different textures of the image in Fig. 16.1f appear in different levels of the Laplacian pyramid. At the finest resolutions in the zero and one levels of the pyramid, the variance is dominated by the texture itself. Thus, in Fig. 16.7a and b, the variance around the point-like stitches in one of the two patterns appears particularly pronounced. In the second level of the Laplacian pyramid (Fig. 16.7c), the stitches have disappeared due to smoothing; the variance becomes small in this pattern, while it is still significant in areas with larger vertical and diagonal stitches. Finally, the third level (Fig. 16.7d) is too coarse for both patterns and is dominated by the edges between the areas of the two patterns that have different mean gray values.

The Laplacian pyramid is very well suited for the analysis of hierarchically organized textures that show different characteristics at different magnifications, such as the curtain discussed in Section 16.1. In this way, we can apply such operators as local variance and local orientation in each level of the pyramid. The simultaneous application of variance and local orientation at different scales provides many properties that make it possible to distinguish even complex hierarchically organized patterns. It should be noted that applying these operations to all levels of the pyramid in 2D images only increases the number of calculations by a factor of 4/3.

16.4 ♯References for Further Reading

The textbooks by Jain [146, Section 9.11] and Pratt [246, Chapter 17] also deal with texture analysis. Further references for texture analysis are the monograph by Rao [256], the handbook by Jähne et al. [165, Vol. 2, Chapter 12] and the Workshop Proceedings published by Burkhardt [34].

17 Regularization and Modelling

Contents

B. Jähne, *Digital Image Processing and Image Acquisition*,
https://doi.org/10.1007/978-3-032-04806-6_17

17.1 Introduction

17.1.1 Local Analysis and Global Knowledge

All image processing operations that we have considered so far from Chapter 10 to 16 were neighborhood operations, even though we have succeeded in extending this effective class of operators to large scales using multiscale approaches (Chapter 15). Nevertheless, the feature images obtained in this way are incomplete in several respects.

On the one hand, the image quality can be insufficient, so that desired features can only be extracted incompletely. One might think that this should not happen if the image extraction methods described in the first part of this book (Chapter 2 to 9) are used optimally.

Unfortunately, things are not quite that simple in some cases. When working on the edge of what is technically possible in research applications, the *signal-to-noise ratio* (Section 4.5.3) is often the biggest problem. When processing images of scenes where the lighting conditions cannot be controlled, it can happen that objects cannot be completely distinguished from the background despite careful selection of the wavelength ranges and recording parameters. When determining motion (Section 14.1.1), we have learned that a change in gray value is not necessarily a clear indication of motion. The lighting may also have changed.

Unintentional disturbances can also occur during image acquisition. An image can be blurred due to movement or an incorrectly set distance.

In both 3D reconstruction using stereoscopy (Section 8.5.1) and motion detection (Section 14.1), we have learned that neither the depth information nor the field of motion can be determined locally across the entire area. Where there are no gray value changes, we can determine neither depth nor motion. On straight edges, it is only partially possible to determine motion. The velocity component parallel to the edge remains unknown.

The question now arises as to what possibilities there are to complete the missing features. This will not be possible without additional knowledge. The only question is how we can link such knowledge with the incompletely determined local information.

In this chapter, we discuss very general approaches to linking local with global information that do not require explicit models of the object, but use flexible conditions to account for global information. The underlying idea is as follows: On the one hand, a model should match the data contained in the image as closely as possible. This requirement is called *similarity condition*. On the other hand, the modeled data should fulfill some global conditions. In the simplest case, this could be a *smoothness condition*.

In general, it will not be possible to obtain exact solutions. Since real image data always contain a certain degree of uncertainty, an exact fit to the data does not make sense either. Rather, we expect a certain deviation of the calculated model values from the measured image data, which we can compare with the expected standard deviation of the noise of the data.

Therefore, we arrive at a *global optimization approach*. The two conditions must be combined in a suitable way to find a solution with a minimum error for a given error norm.

This approach can be applied to a wide range of image processing tasks. These include tasks as diverse as

- *restoration* of images that were subjected to distortions during image acquisition (Chapter 3),
- calculation of *depth images* from *stereo images* or any other triangulation-based sensor for depth image acquisition (Chapter 8.5.1),
- calculation of *depth images* or *surface normals* by various method (Chapter 8 and 9),
- calculation of *motion* or *displacement vector fields* from image sequences (Chapter 14),
- division of an image into regions (*segmentation*) and
- calculation of object boundaries (*active contours* or *snakes*).

Most of the calculated features are scalar fields. However, some of them, such as motion fields or surface normals, are vector fields. Therefore, we will extend the methods discussed here to vector fields.

But before we start, it is useful to have some general thoughts about the purpose and limitations of modeling (Section 17.1.2). After discussing the general approach of variational image modeling in Section 17.2, in Section 17.2.5 we turn to the important question of how to adequately account for discontinuities in global smoothness conditions. The variational approach leads to partial differential equations that are equivalent to transport equations with diffusion and reaction terms. This is an equivalent approach to variational methods, which allows a particularly comprehensible approach to how global smoothness conditions can be modified (Section 17.3). We then discuss a third approach, discrete network models, which looks at modeling from a broader perspective (Section 17.4).

In the second part of this chapter, we turn to two further examples. In Section 17.5 we show how image distortions can be restored by *inverse filtering*. With the Hough transform, we show an example of model-based segmentation of object edges (Section 17.6).

17.1.2 Purpose and Limits of Models

The term *model* reflects the fact that all natural phenomena can only be described correctly up to a certain limit. One of the most important principles in all scientific disciplines is the search for the simplest and most general description of a fact that reflects the observations with the smallest possible deviation. A few fundamental laws of physics describe a variety of phenomena quantitatively in this way.

In the same sense, models are useful and valid approximations for image processing tasks. However, models must be used with appropriate caution.

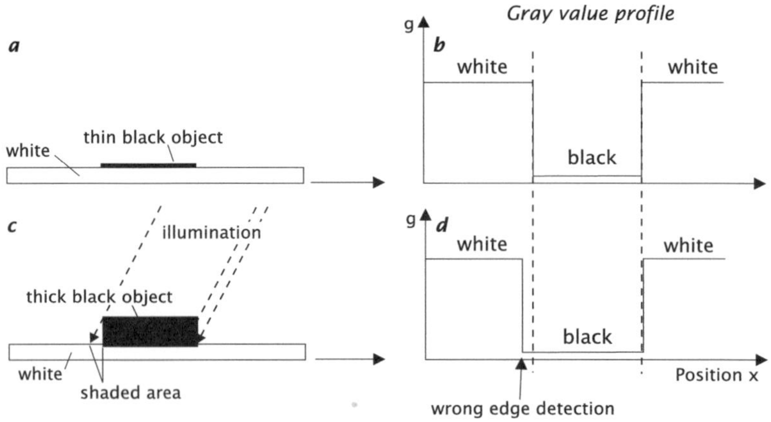

Figure 17.1: Demonstration of a systematic error that cannot be derived from the perceived image. ***a*** *and* **c** *Drawing of the object and the lighting conditions.* **b** *and* **d** *resulting gray value profiles of the images* ***a*** *and* **d** *respectively* **c** *.*

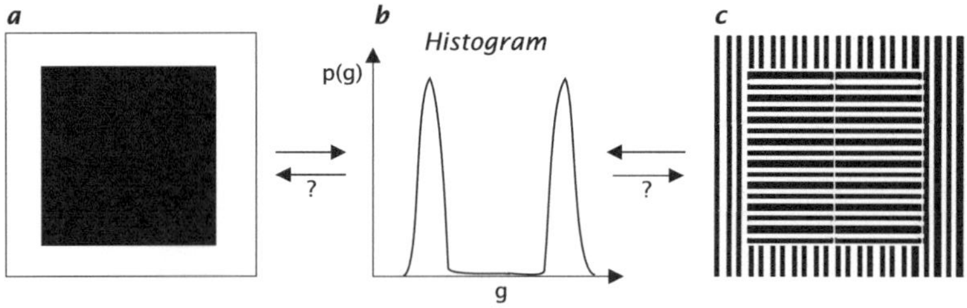

Figure 17.2: Systematic deviation from a model assumption (black object, white background) that cannot be derived from the image histogram.

Even if the data appear to match the model assumptions perfectly, there is no guarantee that the model assumptions are correct.

An illustrative example is shown in Fig. 17.1. The model assumes a flat, black object that lies in front of a white background and is uniformly illuminated (Fig. 17.1a). The object can be clearly recognized by the low gray values in the image, and the transitions between the high and low values indicate the edges of the object.

However, if the black object has a non-negligible thickness and the scene is illuminated by parallel, oblique light (Fig. 17.1c), we get exactly the same profile type as with the assumptions in Fig. 17.1a. We therefore see no deviations from the model assumptions in the image. However, only the right edge is still detected correctly. The left edge is shifted to the left due to the shaded area. This makes the object appear too large.

Figure 17.2 shows another example. A black, flat object fills half of the image on a white background. The histogram, which represents the distribution of the gray values, shows a bimodal shape with two peaks of the same height. From this we conclude that there are essentially only two gray values in the image, the lower one belonging to the black object and the upper one to the white background, and each fills half of the image.

However, this does not mean that every bimodal histogram originates from an image in which a black object on a white background fills half the image. Many other interpretations are possible. For example, there may also be a white object on a black background. The same bimodal histogram also results in an image in which the object and background are striped in black and white. In this case, a segmentation that identifies all pixels below a certain threshold as belonging to the object and the others to the background would not extract the object, but the black stripes. Such a simple method only works if the model fulfills the assumption that both the object and background have a constant value.

These two examples show that even in simple cases we can end up in situations where the model assumptions seem to be correct — as can be deduced from certain parameters of the image, such as the histogram — but in fact they are wrong. While the wrong model assumptions are relatively easy to recognize in simple cases, this is much more difficult, if not impossible, in more complex situations and models.

17.2 Variational Methods

As discussed in the introduction (Section 17.1.1), a mathematically sound approach to image modeling requires a model function or objective function and an error functional. The latter measures the residual deviations between the measured data and the calculated model values.

In image segmentation, a piecewise smooth function $f(\boldsymbol{x})$ would be a suitable objective function. The free parameters of this model are the gray values in the individual regions and the boundary curves between the regions. These two parameters should be varied in such a way that the deviations between the model function $f(\boldsymbol{x})$ and the image data $g(\boldsymbol{x})$ are minimized.

Despite the generality of the approach, the global specifications of this segmentation model are still quite rigid. It would fail with inhomogeneous illumination. Smoothness constraints are a more general approach. They state that we expect only slow changes in the values. This allows us to solve segmentation problems with inhomogeneous illumination. However, we must be able to recognize discontinuities in order to weaken the smoothness condition at these points.

Such general global conditions can be formulated and solved in a very general way with *variational calculus*. Before we apply the calculus of variations to image modeling, it makes sense to start with a simple example from physics.

17.2.1 Simple Temporal Variational Problem

The calculus of variations is widely used in the natural sciences. It is particularly at home in physics. All fundamental concepts of theoretical physics can be described as extremal principles. Probably the best known is the *Hamilton's principle*, from which the Lagrange equation of *theoretical mechanics* results [108].

To illustrate this, we will discuss the motion of a point of mass. Without external forces, it moves at a constant speed. The higher its mass, the higher the forces required to change its speed. Therefore, its mass acts against changes in velocity when the point of mass moves through a potential field $V(x,t)$ that changes in space and time and exerts the force $F = V_x(x,t)$ on it. Hamilton's principle states that the movement follows a curve for which the following integral becomes extremal:

$$\int_{t_1}^{t_2} \left(\frac{1}{2} m\, \dot{x}_t^2 - V(x,t)\right) \mathrm{d}t. \tag{17.1}$$

The time derivative of x is denoted by $\dot{x}_t$ in Eq. (17.1). The function in the integral is called *Lagrange function* $L(x, \dot{x}_t, t)$. The Lagrange function depends on the spatial position x and the time t via the potential $V(x,t)$ and on the kinetic energy $m\,\dot{x}_t^2/2$ of the mass point via the time derivative of the position, i. e. the velocity.

The above integral equation is solved with the *Euler-Lagrange equation*:

$$\frac{\partial L}{\partial x} - \frac{\mathrm{d}}{\mathrm{d}t}\frac{\partial L}{\partial \dot{x}_t} = 0 \quad \text{or short} \quad \dot{L}_x - \ddot{L}_{\dot{x}_t t} = 0. \tag{17.2}$$

This equation can be used to convert the integral equation Eq. (17.1) for a given Lagrange function into a differential equation.

To illustrate this, we calculate the motion of a mass point in the harmonic potential well $V(x) = \epsilon x^2/2$. The Lagrange function of this system is

$$L(x, \dot{x}_t, t) = T - V = \frac{1}{2} m (\dot{x}_t)^2 - \frac{1}{2}\epsilon x^2. \tag{17.3}$$

The derivatives of the Lagrange function result in

$$\frac{\partial L}{\partial x} = -\epsilon x, \qquad \frac{\partial L}{\partial \dot{x}_t} = m\,\dot{x}_t, \qquad \frac{\mathrm{d}}{\mathrm{d}t}\frac{\partial L}{\partial x_t} = m\,\ddot{x}_{tt}. \tag{17.4}$$

From the Euler-Lagrange equation Eq. (17.2) we obtain the simple second-order differential equation

$$m\,\ddot{x}_{tt} + \epsilon x = 0, \tag{17.5}$$

which describes a harmonic oscillation of the mass point in the potential well with the angular frequency $\omega = \sqrt{\epsilon/m}$.

17.2.2 Spatio-temporal Variation Problems

For image processing, we need to formulate variation problems for variables that vary in space and time. The path of the mass point $x(t)$, a scalar function, must be replaced by a spatial function $f(\boldsymbol{x})$, i. e. by a scalar function f of the multidimensional vector variable $f(\boldsymbol{x})$. For image sequences, one of the components of $\boldsymbol{x}$ is the time t.

Consequently, the Lagrange function now depends on the vector variable $\boldsymbol{x}$. Furthermore, it is not only a function of $f(\boldsymbol{x})$ and $\boldsymbol{x}$, but also of additional variables that depend on spatial partial derivatives of f. They are necessary as soon as we require f to depend on f at a point in the neighborhood. To summarize, the following general equation for the *error functional* $\varepsilon(f)$ as a variational integral for f results in

$$\varepsilon(f) = \int_{\Omega} L(f, \nabla f, \boldsymbol{x})\, \mathrm{d}x^D \rightarrow \text{minimum}. \tag{17.6}$$

The integral is calculated over a certain area of the space $\Omega \in \mathbb{R}^D$. Equation Eq. (17.6) already contains the knowledge that the extreme value is a minimum. This results from the fact that f should show a minimum deviation from the given functions at certain points with additional conditions.

The corresponding Euler-Lagrange equation is:

$$\dot{L}_f - \sum_{d=1}^{D} \ddot{L}_{f_{x_d} x_d} = 0. \tag{17.7}$$

The variational approach can also be extended to vectorial properties such as the velocity with image sequences. Then the Lagrange function depends on the vectorial property $\boldsymbol{f} = [f_1, f_2, \ldots, f_D]^T$, the partial derivatives of each component f_d of the property in all directions $(\dot{f}_p)_{x_d}$ and explicitly on the coordinate $\boldsymbol{x}$:

$$\boxed{\varepsilon(\boldsymbol{f}) = \int_{\Omega} L(\boldsymbol{f}, \nabla f_1, \ldots, \nabla f_D, \boldsymbol{x})\, \mathrm{d}x^D \rightarrow \text{minimum.}} \tag{17.8}$$

From this equation we obtain an Euler-Lagrange equation for each component f_d of the vectorial property:

$$\boxed{\dot{L}_{f_d} - \sum_{d=1}^{D} \ddot{L}_{(\dot{f}_d)_{x_d} x_d} = 0.} \tag{17.9}$$

17.2.3 Similarity Conditions

The similarity term is used to determine the equality between the modeled and the measured feature. In the simplest case, in segmentation, the measured feature is the gray value itself, and the similarity term S results in

$$L(f, \boldsymbol{x}) = S(f, \boldsymbol{x}) = \|f(\boldsymbol{x}) - g(\boldsymbol{x})\|_n. \tag{17.10}$$

This formulation of the similarity term in the Lagrange function means that the deviation between the modeled feature and the measured image should be minimal according to the L_n norm (Appendix C.6.2). The most commonly used norm is the L_2 norm, which results in the well-known *least squares* (LS) method.

More complex similarity terms are required to model the features derived from the image data. We show this with two examples.

Example 17.1: Similarity term for image restoration

For a linear *restoration*, we assume that the original image $f(\boldsymbol{x})$ was distorted by a convolution operation with the point response $h(\boldsymbol{x})$ (for more details see Section 17.5). Therefore, the measured image $g(\boldsymbol{x})$ results in

$$g(\boldsymbol{x}) = h(\boldsymbol{x}) * f(\boldsymbol{x}). \tag{17.11}$$

A minimal deviation between the measured and reconstructed image results from the similarity term

$$S(f, \boldsymbol{x}) = \|h(\boldsymbol{x}) * f(\boldsymbol{x}) - g(\boldsymbol{x})\|_n. \tag{17.12}$$

Example 17.2: Similarity term for motion determination

In Section 14.4.2 we discussed that the optical flow should fulfill the continuity equation (14.16):

$$\boldsymbol{f}(\boldsymbol{x}, t)^T \boldsymbol{\nabla} g(\boldsymbol{x}, t) + \dot{g}_t(\boldsymbol{x}, t) = 0. \tag{17.13}$$

In Eq. (14.22) we used the method of least squares to minimize the deviation from the continuity equation. More generally, the L_n-norm yields the following similarity term:

$$S(\boldsymbol{f}, \boldsymbol{x}, t) = \|\boldsymbol{f}^T \boldsymbol{\nabla} g + \dot{g}_t\|_n. \tag{17.14}$$

This equation in turn says nothing other than that the continuity equation for the optical flow Eq. (14.16) should be fulfilled as well as possible. It should be noted that the equality now also explicitly depends on time, as the minimization problem is extended from images to space-time images.

Extension 17.1: Similarity condition alone is not sufficient

Using the example of motion determination, we learn that the similarity conditions alone are not of much use in a variational approach. With Eq. (17.14), the Lagrange function depends only on the optical flow $\boldsymbol{f}$. To calculate the Euler-Lagrange equations, we only need to consider the partial derivatives of the similarity term Eq. (17.14) according to the components of the optical flow, $\partial L/\partial f_p$:

$$\dot{L}_{f_p} = 2\left(\boldsymbol{f}^T \nabla g + \dot{g}_t\right)\dot{g}_{x_p}. \tag{17.15}$$

By inserting Eq. (17.15) into Eq. (17.9) we get

$$\left(\boldsymbol{f}^T \nabla g + \dot{g}_t\right)\dot{g}_x = 0 \quad \text{and} \quad \left(\boldsymbol{f}^T \nabla g + \dot{g}_t\right)\dot{g}_y = 0 \tag{17.16}$$

or, written as a vector equation,

$$\left(\boldsymbol{f}^T \nabla g + \dot{g}_t\right)\nabla g = \boldsymbol{0}. \tag{17.17}$$

These equations initially mean that the optical flow cannot be determined if the spatial gradient of ∇g is a zero vector. Otherwise, there is only the condition that the continuity of the optical flow must be maintained locally, but there are no restrictions on the temporal or spatial variation of the optical flow. This example impressively demonstrates the limits of local similarity conditions. Real progress can only be achieved if global conditions are taken into account.

Therefore, it is necessary to add another term to the Lagrange function, which must also depend on the derivatives of f:

$$\boxed{L(f, \nabla f, \boldsymbol{x}) = S(f, \boldsymbol{x}) + R(f, \nabla f, \boldsymbol{x}).} \tag{17.18}$$

17.2.4 Global Smoothness Conditions

One of the basic global regularizers is smoothness (Appendix C.6.6). For many image processing problems, it makes sense to require that a quantity to be modeled changes only slowly in space (and in time). This is the case with segmentation, for example. An object is characterized by the fact that it is a continuous region with constant or only slowly changing features. Similarly, the depth of a surface or the velocity field of a moving object is continuous, at least in most points.

For these reasons, we now look for suitable *smoothness conditions* to add to the Lagrangian function in order to enforce a spatially smooth solution in addition to minimizing the deviation from the data. Such a term requires spatial partial derivatives of the modeled features. The simplest term, containing only first-order derivatives, is for a scalar feature f in a 2D image:

$$R\left(\dot{f}_x, \dot{f}_y\right) = \alpha^2\left(\dot{f}_x^2 + \dot{f}_y^2\right) = \alpha^2|\nabla f|^2. \tag{17.19}$$

For a vectorial characteristic $\boldsymbol{f} = [f_1, f_2]^T$ this results in

$$R(\nabla f_1, \nabla f_2) = \alpha^2 \left(|\nabla f_1|^2 + |\nabla f_2|^2\right). \tag{17.20}$$

In this additional term, the partial derivatives appear as a squared sum. This means that we can evaluate the *smoothness term* with the same norm (L_2-norm, sum of *least squares*, see Appendix C.6.2) as the similarity term. In addition, all partial derivatives are equally weighted in this formula. The factor α^2 stands for the relative weight of the smoothness term compared to the similarity term.

The complete Lagrange function L for the determination of motion with similarity and smoothness term then results in

$$\boxed{L(\boldsymbol{f}, \nabla f_1, \nabla f_2, \boldsymbol{x}) = \left(\boldsymbol{f}^T \nabla g + \dot{g}_t\right)^2 + \alpha^2 \left(|\nabla f_1|^2 + |\nabla f_2|^2\right).} \tag{17.21}$$

By inserting this Lagrange function into the Euler-Lagrange equation Eq. (17.9) we obtain the following system of differential equations:

$$\begin{array}{rcll} \left(\nabla g^T \boldsymbol{f} + \dot{g}_t\right) g_x & - & \alpha^2 \left((\ddot{f_1})_{xx} + (\ddot{f_1})_{yy}\right) & = & 0, \\ \left(\nabla g^T \boldsymbol{f} + \dot{g}_t\right) g_y & - & \alpha^2 \left((\ddot{f_2})_{xx} + (\ddot{f_2})_{yy}\right) & = & 0. \end{array} \tag{17.22}$$

Summarized in a vector equation, the result is

$$\boxed{\underbrace{\left(\nabla g^T \boldsymbol{f} + \dot{g}_t\right) \nabla g}_{\text{similarity}} \quad - \quad \underbrace{\alpha^2 \Delta \boldsymbol{f}}_{\text{smoothness}} \quad = \boldsymbol{0}.} \tag{17.23}$$

It is easy to deduce how the optical flow results from this formula. First, we imagine that the intensity changes strongly in one direction. The similarity term then dominates over the smoothness term, and the velocity is calculated according to the local optical flow. If, on the other hand, the change in intensity is small, the smoothness term dominates. The local velocity is calculated so that it corresponds as closely as possible to the velocity in the neighborhood. In other words, the flux vectors are interpolated from the surrounding flux vectors.

This process is further illustrated by an extreme example. Consider a bright object of constant brightness moving in front of a dark background. The similarity term disappears completely within the object, while at edges only the velocity component perpendicular to the edge can be calculated from this term. This is an old and well-known problem in physics: How do we calculate the potential function $\Delta \boldsymbol{f} = \boldsymbol{0}$ without sources and sinks given boundary conditions at the edges of the object?

The equation $\Delta \boldsymbol{f} = \boldsymbol{0}$ is the so-called *Laplace equation*. In areas for which the similarity term is zero, we can immediately deduce the form of the solution. Since the second-order derivatives are zero, the first-order spatial derivatives are constant. This results in an optical flow $\boldsymbol{f}$ that changes linearly in space.

17.2.5 Controlled Smoothness

After discussing the basic properties of the smoothness conditions, we now turn to the question of how we can appropriately handle spatial and temporal discontinuities with the variational approach. In a segmentation problem, the modeled feature is discontinuous at the object edges. The same is true for the optical flow. However, the smoothness condition formulated so far does not allow for discontinuities. We have set up a global smoothness condition and therefore obtain a globally smooth feature field even if the data has discontinuities. We must therefore develop methods to recognize these in the data and take them adequately into account in the modeling.

We will first discuss the basic possibilities for varying the minimum problem within the chosen framework. To do this, we rewrite the integral equation Eq. (17.6) for the minimum problem by applying the knowledge about the meaning of the Lagrange function from the previous section:

$$\int_{\Omega} \big(\underbrace{S(f)}_{\text{similarity}} + \underbrace{R(\nabla f)\big)}_{\text{smoothness}} \, \mathrm{d}^D x \rightarrow \text{minimum.} \qquad (17.24)$$

Two approaches are possible to take discontinuities into account:

Limiting Integration Area Ω
The integration range is one of the options for solving the discontinuities in feature f. If there are discontinuities within the integration area, this results in incorrect values. We must therefore find algorithms that recognize the edges in f and, as a consequence, restrict the integration area to the segmented areas. This is obviously a difficult iterative procedure. First, the edges in the image itself do not necessarily coincide with the edges of the feature f. Secondly, in many cases, especially in motion analysis, there is only a patchy feature field, so that segmentation into individual regions is not possible.

Modifying Smoothness Term
Modifying the smoothness term is another way of solving the discontinuity problem. At points where a discontinuity is suspected, the smoothness condition can be softened or disappear completely. This means that discontinuities are permitted. This is also an iterative algorithm. The smoothness term must contain a control function that switches off the smoothness condition under certain conditions. This property is called *controlled smoothness* [323].

In the following, we first discuss two methods for modifying the integration surface. The modification of the smoothness term is discussed in detail in Section 17.3.

Integration along Closed Zero Crossings
Hildreth [134] uses the Laplacian-filtered image and limits all further calcula-

a *b*

Figure 17.3: *Two images of the Hamburg cab scene, which was recorded by the Department of Computer Science at the University of Hamburg and has since become one of the best-known early test sequences for image sequence analysis.*

tions to the zero crossings, as these represent gray value edges (Section 12.3), i. e. the features that we can use to calculate the velocity component normal to the edge. The great advantage of this approach is that by preselecting promising features, the required calculations are considerably reduced.

By selecting the zero crossings, the smoothness condition is restricted to a specific contour line. This seems helpful because a zero crossing is likely to surround or lie within an object, but not leave the object edges to the outside. However, this is not necessarily the case. If there is a zero crossing within an object, the velocity along the contour should not show any discontinuities. If we choose a contour instead of a surface for the smoothness condition, the integral changes from a surface integral to a line integral along a contour s:

$$\oint \left\{ \left(\hat{\boldsymbol{n}}^T \boldsymbol{f} - f_\perp \right)^2 + \alpha^2 \left[\left((\dot{f}_1)_s \right)^2 + \left((\dot{f}_2)_s \right)^2 \right] \right\} \mathrm{d}s \to \text{minimum}. \qquad (17.25)$$

$\hat{\boldsymbol{n}}$ is a unit vector normal to the edge and $f_\perp$ is the velocity normal to the edge.

The derivatives of the velocity are calculated along the contour line. The component $f_\perp$ normal to the edge is derived directly from the similarity term, while the velocity term parallel to the edge must be derived from the smoothness condition along the entire edge.

Despite the elegance of this method, it also has a serious disadvantage. It is not certain that zero crossings include only one object. Therefore, we cannot assume that the optical flow field is continuous along the zero crossings. Since only edges are used to calculate the optical flow field, only one component of the displacement vector can be calculated locally. In this way, all features that occur either as gray value maxima or gray value corners and allow an unambiguous determination of a displacement vector are neglected.

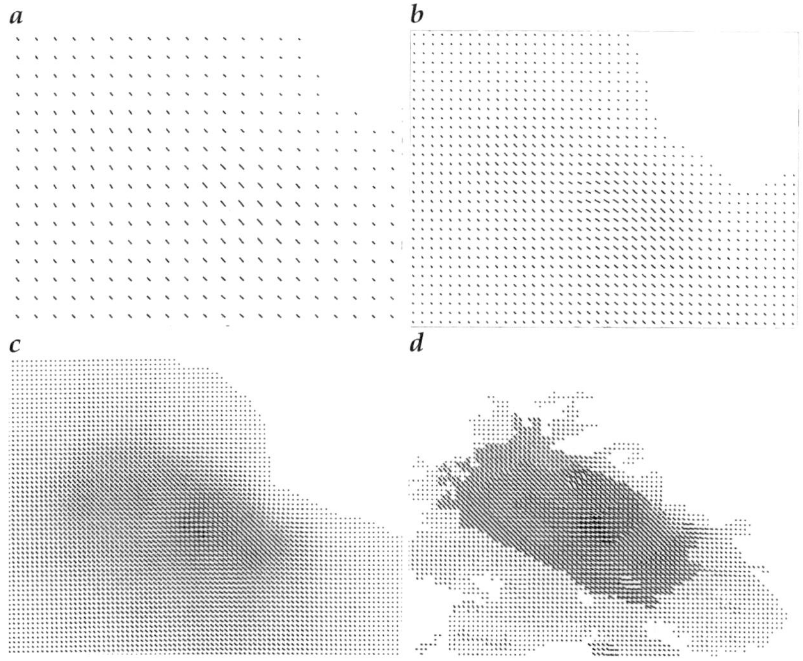

Figure 17.4: *Determination of the optical flow in the cab scene (Fig. 17.3) using the dynamic pyramid method:* ***a*** *to* ***c*** *Three levels of the optical flow field using a global smoothness condition;* ***d*** *Final result of the optical flow using the region-oriented smoothness condition. Kindly provided by M. Schmidt and J. Dengler, German Cancer Research Center, Heidelberg.*

Limiting integration to segmented regions

A region-oriented approach does not omit such points, but still tries to limit the smoothness within objects. Again, *zero crossings* is used to divide the image into regions. Region-limited smoothness merely disregards the continuity condition at the region boundaries. The simplest approach for this type of condition is to limit the integration surfaces to the different regions of an image and evaluate the surfaces separately.

As expected, a smoothness condition restricted to regions causes discontinuities at the boundaries (Fig. 17.4d) in clear contrast to the globally smoothed optical flow in Fig. 17.4c. We can immediately recognize the cab at the boundaries of the optical flow.

However, we can also see that the car is divided into further sub-regions with different optical flow, as can be seen from the cab sign on the roof and the rear and side windows. The small regions in particular show an optical flow that differs significantly from that in larger regions. Thus, a simple region-limited smoothness condition does not reflect the fact that separate regions can exist within objects. The optical flow can be smooth across these boundaries.

17.2.6 ♯Elasticity Models

At this point, it is useful to consider an analogous physical problem that gives us further insight into how the similarity and smoothness terms balance each other. In a physical model, these two terms correspond to two types of forces.

We use the example of determining motion again and think of the images as being painted on an *elastic membrane*. The motion will move the membrane from image to image. Specifically, non-uniform motion will slightly expand or contract the membrane. The similarity term acts as an external force that tries to move the membrane in the direction of the local *displacement vector* (*DV*). The internal elastic forces of the membrane try to distribute these displacements as evenly as possible over the entire membrane and thus cause a smooth *displacement vector field* (*DVF*).

Let us first take a closer look at the external forces. It does not make sense to rigidly predetermine the distortions at the points where we can calculate the DV of the estimated displacement. Rather, the less precisely the calculated displacement can be determined, the greater the deviations from the calculated displacement we will allow. Physically, this corresponds to a pair of springs whose spring constant is proportional to the accuracy with which the DV can be determined. The zero point of the spring system corresponds to the calculated displacement vector. As the membrane is two-dimensional, two pairs of springs are required at each point. The direction of the springs is determined according to the *local orientation* (Section 13.3). At an edge, only the displacement perpendicular to the edge can be calculated (*aperture problem*, Section 14.1.2). Consequently, only one pair of springs is attached perpendicular to the edge here; the displacement parallel to the edge is free.

The external spring forces counteract the internal elastic forces of the membrane, which try to compensate for the different displacements and minimize the deformation of the membrane. From this point of view, we can now understand the Euler-Lagrange equation for the optical flow Eq. (17.23) as follows:

$$\underbrace{\left(\boldsymbol{\nabla} g^T \boldsymbol{f} + \dot{g}_t\right)\boldsymbol{\nabla} g}_{\text{external force}} - \underbrace{\alpha^2 \Delta \boldsymbol{f}}_{\text{internal force}} = \mathbf{0}. \qquad (17.26)$$

Here α^2 represents the *elasticity constant*. The external forces are balanced by the internal forces. The second partial derivatives of the DVF appear with the internal forces, as a constant gradient of the DVF does not generate any internal forces.

All elasticity properties of the membrane are summarized in the constant α. A deeper insight into the inner structure of the membrane can be obtained by looking at the Lagrangian function Eq. (17.23)

$$L(\boldsymbol{f}, \boldsymbol{\nabla} f_1, \boldsymbol{\nabla} f_2, \boldsymbol{x}) = \underbrace{\left(\boldsymbol{\nabla} g^T \boldsymbol{f} + \dot{g}_t\right)^2}_{-V,\ \text{potential}} + \underbrace{\alpha^2 \left(|\boldsymbol{\nabla} f_1|^2 + |\boldsymbol{\nabla} f_2|^2\right)}_{T,\ \text{deformation energy}}. \qquad (17.27)$$

The Lagrangian function is composed of the *potential* of the external force, which results from the continuity of the optical flow, and the *deformation energy*, which is related to the internal forces. This energy appears here instead of the kinetic energy in the classical example of the Lagrangian function for a mass point (Section 17.2.1), because the minimum problem is not formulated in a time integral but in an area integral.

The term with the deformation energy can be divided into several terms that correspond to the different types of deformation:

$$T = \frac{1}{2}\left[\underbrace{\left((\dot{f}_1)_x + (\dot{f}_2)_y\right)^2}_{\text{dilatation}} + \underbrace{\left((\dot{f}_1)_x - (\dot{f}_2)_y\right)^2 + \left((\dot{f}_1)_y + (\dot{f}_2)_x\right)^2}_{\text{shear}} + \underbrace{\left((\dot{f}_1)_y - (\dot{f}_2)_x\right)^2}_{\text{rotation}}\right]. \quad (17.28)$$

This shows that the elasticity properties of the membrane are optimally adapted to the movement kinematics. Each possible deformation due to the different modes of 2D movement on the image plane is given equal weight. The DVF is therefore calculated in such a way that its divergence, *rotation* and shear are equally minimized.

Physically, this membrane does not make sense. The differential equation for a real physical membrane looks somewhat different [81]:

$$\boldsymbol{f} - (\lambda + \mu)\boldsymbol{\nabla}(\boldsymbol{\nabla u}) - \mu\Delta\boldsymbol{u} = \boldsymbol{0}. \quad (17.29)$$

The elasticity of a physical membrane is described by the two constants λ and μ. $\lambda = -\mu$ is not possible, so that only the additional term with $\nabla(\nabla \boldsymbol{u})$ compared to the model membrane for the DVF can never disappear with a physical membrane. If there is no transverse contraction, λ can at best be zero.

In the membrane model, only the deflection is continuous, but not its first derivative. These discontinuities occur exactly at the points where the external forces act. This can be seen directly from Eq. (17.23). A locally applied external force corresponds to a δ *distribution* (> R13) in the similarity term. After integration of Eq. (17.23), the DVF has discontinuities in the first-order derivatives.

These considerations call into question the smoothness conditions considered so far, because we know that the motion of plane surface elements has no discontinuities in the first derivative. A smoothness of the first-order derivatives can be enforced if we include second-order derivatives in the smoothness term Eq. (17.23) or the deformation energy Eq. (17.27). From a physical point of view, this turns a membrane into a thin *elastic plate* that cannot be kinked like a membrane.

17.3 Diffusion Models: Continuous and Discrete

In this section, we look at continuous modeling from a new perspective. The quadratic error functional for the motion determination Eq. (17.23)

$$\left(\nabla g^T f + g_t\right)\nabla g - \alpha^2 \Delta f = \mathbf{0} \tag{17.30}$$

can be regarded as the stationary solution of a *diffusion-reaction system* with homogeneous diffusion, if we consider the constant α^2 as the diffusion constant D:

$$\frac{\partial f}{\partial t} = D\Delta f - \left(\nabla g^T f + g_t\right)\nabla g. \tag{17.31}$$

In this equation, compared to the transient partial differential equation for homogeneous diffusion (see Eq. (15.9) in Section 15.2.1), an additional source term is added, which represents the similarity term. The source strength is proportional to the deviation from the continuity equation for the optical flow. Therefore, this term modifies the values of f to fit the continuity equation as closely as possible.

After this introductory example, we can formulate the relationship between the error functional of the variational approach and a diffusion-response system in a more general way. The *Euler-Lagrange equation*

$$\sum_{d=1}^{D} \ddot{L}_{f_{x_d} x_d} - \dot{L}_f = 0, \tag{17.32}$$

which minimizes the error functional

$$\varepsilon(f) = \int_{\Omega} L\left(f, f_{x_1}, \ldots, f_{x_D}, \boldsymbol{x}\right) \mathrm{d}x^D \tag{17.33}$$

for the scalar time/space function $f(\boldsymbol{x}), \boldsymbol{x} \in \Omega$, can be regarded as the equilibrium solution of the diffusion-reaction system

$$f_t = \sum_{d=1}^{D} \ddot{L}_{f_{x_d} x_d} - \dot{L}_f. \tag{17.34}$$

In the following, we will analyze the aspect of the modeling that we have so far only briefly considered in Section 17.2.5, namely the local modification of the smoothness term, in more detail. In the language of diffusion models, this means a locally varying diffusion constant in the first term on the right-hand side of Eq. (17.34). From the discussion above, we know that for each approach for a locally varying diffusion constant, there is a corresponding error functional for the variation approach.

In Section 15.2.1 we have considered the homogeneous diffusion process that generates a multiscale representation of an image, the linear *scale space*. If now the smoothness term is made dependent on local properties of the image, such as, e. g. a function of the gradient, then the corresponding inhomogeneous diffusion process leads to the generation of a nonlinear scale space. In terms of modeling, the interesting point here is that image segmentation can be achieved without using a similarity term.

17.3.1 Inhomogeneous Diffusion

The simplest approach for a spatially varying smoothness term that takes discontinuities into account is to reduce the diffusion coefficient at edges. Then one speaks of *inhomogeneous diffusion*. This makes the diffusion constant dependent on the edge strength, which is given by the magnitude of the gradient:

$$D(f) = D(|\nabla f|^2). \tag{17.35}$$

With a locally varying diffusion constant, we obtain the diffusion-reaction system

$$\dot{f}_t = \nabla\left(D(|\nabla f|^2)\nabla f\right) - L_f. \tag{17.36}$$

It would not be correct to write $D(|\nabla f|^2)\,\Delta f$, as we can see from the derivation of the transient diffusion equation in Section 15.2.1. With Eq. (17.36), the smoothness term (regularization term) R in the Lagrange function is

$$R = R(|\nabla f|^2), \tag{17.37}$$

where the diffusion constant is the derivative of the function R: $D = \dot{R}$. We can easily convince ourselves of this by substituting Eq. (17.37) into Eq. (17.32).

Perona and Malik [239] use the following dependence on the magnitude of the gradient:

$$D(|\nabla f|) = D_0 \frac{\lambda^2}{|\nabla f|^2 + \lambda^2}, \tag{17.38}$$

where λ is an adjustable parameter. For small gradients ($|\nabla f| \ll \lambda$), the diffusion constant approaches D_0, while for large gradients with $|\nabla f| \gg \lambda$ it approaches zero. The parameter λ can therefore be understood as a measure of the amount of the gradient from which it can be considered reliable compared to the noise in the image.

As simple and straightforward as this idea may seem, it also has its problems. Depending on the function we apply to the relationship between D and $|\nabla f|$, the diffusion process can become unstable and lead to a splitting of the edges. A safe way to avoid this problem is to use a regularized gradient calculated from a smoothed version of the image [349, 350]. One possible approach is

$$D = D_0\left[1 - \exp\left(-\frac{c_m}{(|\nabla(B^R * f)(\boldsymbol{x})|/\lambda)^m}\right)\right]. \tag{17.39}$$

This equation states that the diffusion constant is constant for small gradients. However, if the magnitude of the gradient exceeds a certain value, the diffusion constant quickly drops to zero. The higher the exponent m, the steeper the transition. With the values $m = 4$ and $c_4 = 3.31488$ used by Weickert [349], the diffusion coefficient drops from 1 at $|\nabla f|/\lambda = 1$ to about 0.15 at $|\nabla f|/\lambda = 2$.

It should be noted that a regularized gradient was selected in Eq. (17.39). It is not calculated directly from the image $f(x)$, but after smoothing the image with the binomial smoothing mask $\mathcal{B}^p$. A carefully implemented regularized gradient operator (Section 12.7) is sufficient to stabilize the inhomogeneous smoothing process and avoid the splitting of edges, so that no pre-smoothing is necessary.

A simple explicit discretization of the inhomogeneous diffusion uses the regularized derivative operators discussed in Section 12.7. In the first step, a gradient image is calculated with the vector operator

$$\begin{bmatrix} \mathcal{D}_1 \\ \mathcal{D}_2 \end{bmatrix}. \tag{17.40}$$

In the second step, the gradient image is multiplied point by point with a control operator C, which computes the diffusion constant according to Eq. (17.38) or Eq. (17.39) with $D_0 = 1$:

$$\begin{bmatrix} C \cdot \mathcal{D}_1 \\ C \cdot \mathcal{D}_2 \end{bmatrix}. \tag{17.41}$$

It is one in constant regions and drops to small values at edges. In the third step, the gradient operator is applied a second time:

$$[\mathcal{D}_1, \mathcal{D}_2] \begin{bmatrix} C \cdot \mathcal{D}_1 \\ C \cdot \mathcal{D}_2 \end{bmatrix} = \mathcal{D}_1(C \cdot \mathcal{D}_1) + \mathcal{D}_2(C \cdot \mathcal{D}_2). \tag{17.42}$$

Weickert [350] uses a more sophisticated implicit solution scheme. However, this is significantly more computationally expensive and less isotropic than the explicit scheme in Eq. (17.42) when used with the gradient operators presented in Section 12.7.6, which are carefully optimized for isotropy.

An even simpler, but only approximate, implementation of inhomogeneous diffusion uses binomial smoothing filters (Section 11.4) and the operator

$$\mathcal{I} + C \cdot (\mathcal{B} - \mathcal{I}). \tag{17.43}$$

Where $\mathcal{I}$ is the identity operator and the control operator C has values between 0 and 1.

Figure 17.5 shows the application of inhomogeneous diffusion to the segmentation of a noisy image. The test image contains a triangle and a rectangle. Normal smoothing can significantly suppress the noise, but

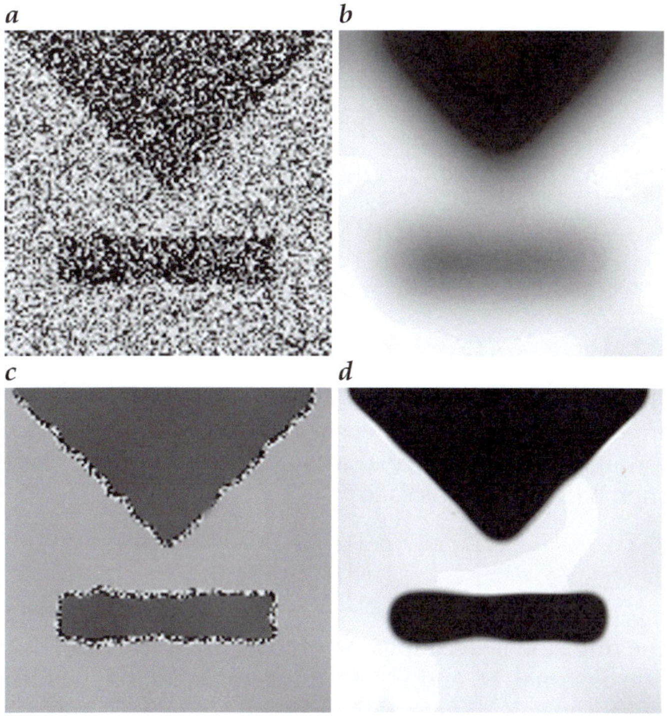

Figure 17.5: ***a*** *Original image, smoothed via* ***b*** *linear diffusion,* ***c*** *inhomogeneous but isotropic diffusion and* ***d*** *anisotropic diffusion. From Weickert [349].*

the edges become blurred (Fig. 17.5b). With inhomogeneous diffusion, the edges remain sharp and we still achieve an almost perfect segmentation of the triangle and rectangle (Fig. 17.5c). The only disadvantage is that the edges themselves remain noisy, as the smoothing was suppressed there.

17.3.2 Anisotropic Diffusion

As shown in the last example, inhomogeneous diffusion has the decisive disadvantage that the diffusion does not pass over edges; these remain noisy in all directions. However, edges only become blurred if the diffusion runs perpendicular to them, but not if the diffusion runs parallel. In this case, the edges are even stabilized.

A method in which the diffusion depends on the edge direction is called *anisotropic diffusion*. With this approach, the flow no longer runs parallel to the gradient, so that the diffusion can no longer be described by a scalar diffusion constant. We now need a *diffusion tensor* because the diffusion is no longer the same in all directions and therefore the current density j no longer has to be parallel to the gradient ∇f:

$$j = -D\nabla f = -\begin{bmatrix} D_{11} & D_{12} \\ D_{12} & D_{22} \end{bmatrix} \begin{bmatrix} \dot{f}_1 \\ \dot{f}_2 \end{bmatrix}. \tag{17.44}$$

With a diffusion tensor, we obtain the diffusion-reaction system

$$f_t = \nabla\left(D(\nabla f \nabla f^T)\nabla f\right) - L_f, \tag{17.45}$$

and the corresponding smoothness term in the Lagrange function is

$$R = \operatorname{trace} R\left(\nabla f \nabla f^T\right) \tag{17.46}$$

with $D = \dot{R}$.

The properties of the diffusion tensor are best understood if the symmetric tensor is brought into its principal axis system by rotating the coordinate system. Then Eq. (17.44) reduces to

$$j' = -\begin{bmatrix} D_1' & 0 \\ 0 & D_2' \end{bmatrix} \begin{bmatrix} \dot{f}_1' \\ \dot{f}_2' \end{bmatrix} = -\begin{bmatrix} D_1'\dot{f}_1' \\ D_2'\dot{f}_2' \end{bmatrix}. \tag{17.47}$$

Now the diffusion is decoupled in both axial directions. The two coefficients on the diagonal, D_1' and D_2', are the *eigenvalues* of the diffusion tensor. In analogy to isotropic diffusion, the general solution of anisotropic diffusion in local space with $\sigma_1'(t) = \sqrt{2D_1't}$ and $\sigma_2'(t) = \sqrt{2D_2't}$ can be written as follows:

$$f(x,t) = \frac{1}{2\pi\sigma_1'(t)\sigma_2'(t)} \exp\left(-\frac{x'^2}{2\sigma_1'(t)}\right) * \exp\left(-\frac{y'^2}{2\sigma_2'(t)}\right) * f(x,0) \tag{17.48}$$

The equation states that the anisotropic diffusion is equivalent to a cascaded convolution with two one-dimensional Gaussian convolution masks with different standard deviation, which are aligned in the direction of the main axes of the diffusion tensor. If one of the two eigenvalues of the diffusion tensor is significantly larger than the other, the diffusion only takes place in the direction of the corresponding eigenvector. The gray values are therefore only smoothed in this direction. The spatial smearing is — as with any diffusion process — proportional to the square root of the diffusion constant Eq. (15.13).

Using this property of anisotropic diffusion, it is easy to implement a diffusion process that mainly smoothes along the edges, but not perpendicular to them. With the following approach, only smoothing across edges is prevented [349]:

$$\begin{aligned} D_1' &= 1 - \exp\left(-\frac{c_m}{(|\nabla(B^r * f)(x)|/\lambda)^m}\right) \\ D_2' &= 1. \end{aligned} \tag{17.49}$$

As Scharr and Weickert [286] have shown, an effective and exact implementation is possible with regularized first-order derivative operators that have been optimized for minimum anisotropy, as in the case of inhomogeneous diffusion:

$$[\mathcal{D}_1, \mathcal{D}_2] \begin{bmatrix} C_{11} & C_{12} \\ C_{12} & C_{22} \end{bmatrix} \begin{bmatrix} \mathcal{D}_1 \\ \mathcal{D}_2 \end{bmatrix} = \mathcal{D}_1(C_{11} \cdot \mathcal{D}_1 + C_{12} \cdot \mathcal{D}_2) + \mathcal{D}_2(C_{12} \cdot \mathcal{D}_1 + C_{22} \cdot \mathcal{D}_2). \tag{17.50}$$

with

$$\begin{bmatrix} C_{11} & C_{12} \\ C_{12} & C_{22} \end{bmatrix} = \begin{bmatrix} \cos\theta & \sin\theta \\ -\sin\theta & \cos\theta \end{bmatrix} \begin{bmatrix} C_1' & 0 \\ 0 & C_2' \end{bmatrix} \begin{bmatrix} \cos\theta & -\sin\theta \\ \sin\theta & \cos\theta \end{bmatrix}.$$

C_{pq}, C_1' and C_2' are control images with values between 0 and 1, with which the diffusion is aligned parallel to edges. C_1' and C_2' are calculated directly from Eq. (17.49), the direction of the edge, the angle θ, can be calculated, e. g. with the structure tensor (Section 13.3).

In contrast to inhomogeneous diffusion, the edges are also smoothed in anisotropic diffusion (Fig. 17.5d). However, smoothing along the edges has the disadvantage that the corners become blurred as with linear diffusion. This does not happen with inhomogeneous diffusion (Fig. 17.5c).

17.4 Network Models

In this section, we discuss the *network models* originating from electrical engineering. Their advantage is that they directly represent a discrete model that can be transferred directly to image data. The presentation in this section largely follows the work of Harris [122, 123]. The study of network models has become topical since network structures can be mapped directly on parallel computer systems, such as the *Connection Machine* of the Massachusetts Institute of Technology (MIT) [123], or on highly integrated analog circuits (VLSI) [223].

17.4.1 One-dimensional Networks

We first consider the simpler one-dimensional case. In a chain of nodes with the voltage U_n, these are connected to their neighbors by a resistor R (Fig. 17.6). These connections enforce continuity of the potential. We can apply an external potential U_{0n} to each node U_n via a resistor S_n. If the potential is only specified at a single point in the network, the entire network receives this potential. If we apply a different potential at a second point in the network, this results in a linear change in potential between the two points. Thus, the link with resistors causes the *smoothness*, while the local potential specification U_{0n} enforces the similarity.

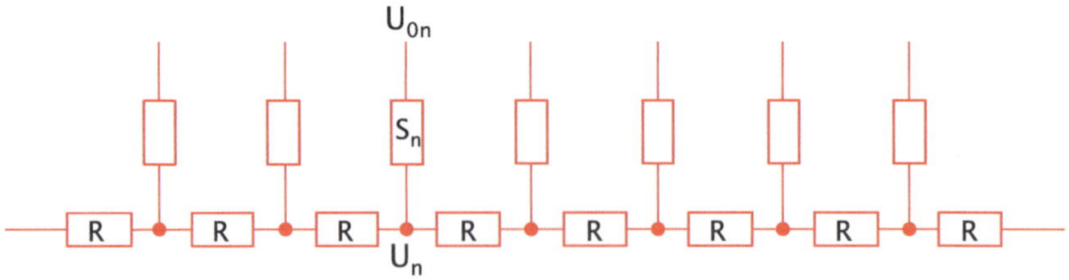

Figure 17.6: Simple discrete network model for smooth one-dimensional modeling; according to Harris [123].

Similar to the elasticity model (Section 17.2.6), where the *similarity condition* is not specified rigidly but via springs, in the network model the potential U_{0n} is not specified directly at the nodes U_n but via a resistance S_n on the network (Fig. 17.6). The more uncertain the similarity term, i. e. the default potential, is at the corresponding point, the greater this resistance is chosen.

Several types of boundary conditions are possible: You can set the edge of the resistance chain to a potential and thus specify a fixed value at the edge (zero-order boundary condition). Another possibility is to specify no connection at the edge. This corresponds to setting the first-order spatial derivative to zero. The potential is then determined by the nearest potential input.

The difference equation for the network model results from the *Kirchhoff's current sum rule* that the sum of all incoming and outgoing currents must be zero at each node of the network. With the definitions from Fig. 17.6, the following results for the node n of the network

$$\frac{U_n - U_{0n}}{S_n} + \frac{U_n - U_{n-1}}{R} + \frac{U_n - U_{n+1}}{R} = 0. \tag{17.51}$$

The two right-hand fractions together form the discrete second-order derivative operator $\mathcal{D}_x^2$ (see Section 12.5.2), so that we can write Eq. (17.51) as follows:

$$\frac{1}{S}(U - U_0) - \frac{1}{R}\ddot{U}_{xx} = 0. \tag{17.52}$$

This equation is the one-dimensional discrete form of the continuous equation Eq. (17.23), which we repeat for the one-dimensional case for better comparison:

$$(\dot{g}_x)^2\left(f + \frac{\dot{g}_t}{\dot{g}_x}\right) - \alpha^2 \ddot{f}_{xx} = 0. \tag{17.53}$$

Now we can quantify the analogy between the differential equation Eq. (17.53) and the network model Eq. (17.52). The specification of the potential U_0 corresponds to the calculation of the local velocity with $-\dot{g}_t/\dot{g}_x$. The similarity term $(\dot{g}_x)^2$ corresponds to the coupling coefficient $1/S$ and the smoothness term is weighted with the coupling coefficient $1/R$ instead of α^2.

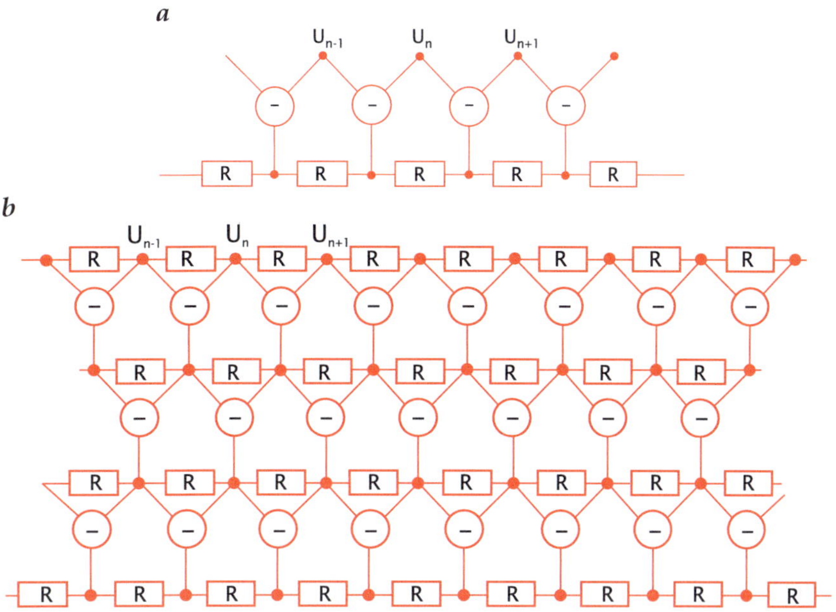

Figure 17.7: ***a*** *Simple discrete network models for modeling a scalar feature where* ***a*** *the first derivative is smooth;* ***b*** *The signal and all derivatives up to third order are smooth; according to Harris [123].*

17.4.2 Generalized Networks

Now we turn to the question of how a continuity of the first derivative can be integrated into the network model. For this purpose, Harris [122] uses an active subtraction module that forms the difference between two signals. In the module, all three connections are both inputs and outputs. The potential at one output is set according to the potential at any two other inputs. Active electronic components are required to implement such a subtraction module.

Fig. 17.7a shows the integration of this subtraction module into a network. It calculates the potential difference between two neighboring nodes. This difference is then applied to the resistor network instead of the potential itself. This gives us a network in which the first derivatives remain continuous.

This model can easily be extended so that the continuity of derivatives of any order is given. To do this, only several levels with subtraction modules need to be connected on top of each other (Fig. 17.7b).

17.4.3 Networks with Discontinuities

The network model makes it possible to integrate discontinuities in a simple way. In the network model with the continuity requirement only on the DVF itself (Fig. 17.6), the resistance must simply be omitted or made sufficiently large at a discontinuity point. A potential jump is then possible

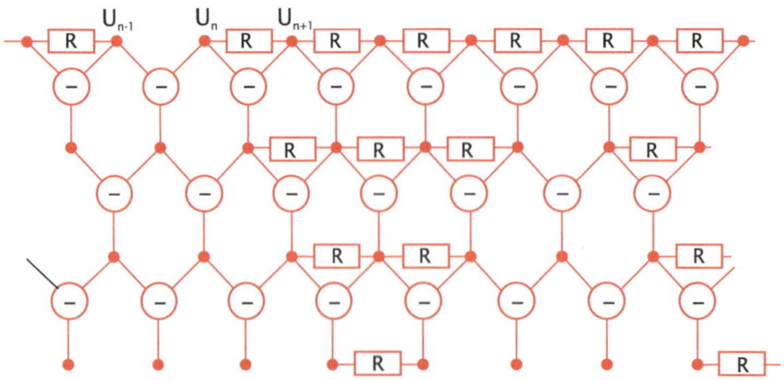

Figure 17.8: Generalized one-dimensional network with one discontinuity in the DVF and one in the first derivative of the DVF, as marked.

between the corresponding nodes in the network. In the sense of controlled smoothness (Section 17.2.5), we can also imagine a model with non-linear resistors in which the resistance increases with the voltage. This provides a mechanism with which discontinuities are automatically generated. These considerations clearly show how flexible and illustrative network models are.

In the generalized network model, the integration of discontinuities is no more difficult. Here, a discontinuity can be introduced at each level of the network, i. e. at the measured variable itself or at each derivation level, by removing a resistance in the corresponding level. However, you cannot remove the resistor in question alone, but must also remove all resistors from lower-lying nodes that lead to the discontinuity point (Fig. 17.8). Otherwise, the continuity in the higher derivatives remains and causes continuity in the lower derivatives as well.

17.4.4 Two-dimensional networks

The network model can also be applied to higher-dimensional problems. For a two-dimensional model with first-order continuity, we build a two-dimensional network of resistors. The construction of generalized two-dimensional networks with higher-order continuity conditions is more complex. At each stage, the continuity of several partial derivatives has to be taken into account. For the first spatial derivative there are two partial derivatives, one horizontal and one vertical. For each of these directions, a separate plane with subtraction modules is constructed as in Fig. 17.7, so that the smoothness condition is fulfilled. For further details, the reader is referred to the original literature [123].

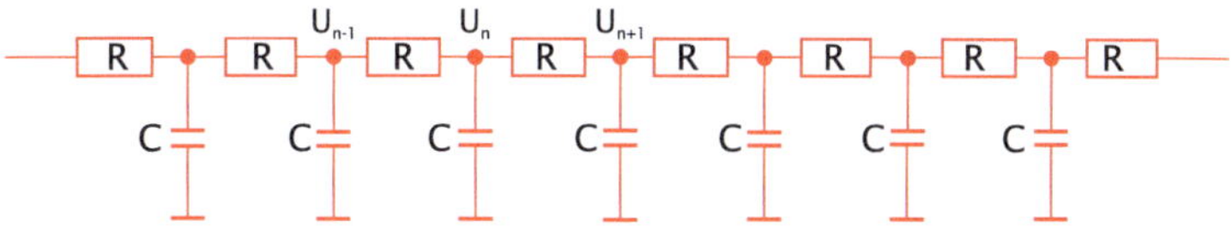

Figure 17.9: *One-dimensional network with capacitors for simulating iterative solutions.*

17.4.5 Multi-grid Networks

One of the most important problems with large systems of equations, which we discuss here, is to find fast iterative solution methods. We can also model this aspect with networks. Iteration brings a time constant into play, which can be easily simulated by an additional capacitor at each node (Fig. 17.9). The static behavior of the network and thus the solution of the equation system is not changed by the installation of the capacitors.

The typical starting point of the iteration is that the features are only known at isolated points with strong gray value edges. The question is how many iterations we need until this information is transferred to distant points from which we have no displacement information. To answer this question, we derive the difference equation of the resistor-capacitor chain (Fig. 17.9). This in turn results from Kirchhoff's current sum law. In addition, we only need to know that the current flowing into a capacitor is proportional to its capacitance C and the time derivative of the voltage $\partial U/\partial t$. This results in

$$\frac{U_{n-1} - U_n}{R} + \frac{U_{n+1} - U_n}{R} - C\frac{\partial U_n}{\partial t} = 0 \tag{17.54}$$

or

$$\frac{\partial U_n}{\partial t} = \frac{(\Delta x)^2}{RC}\frac{\partial^2 U_n}{\partial x^2}. \tag{17.55}$$

In the second equation, Δx is the spatial distance between neighboring points of the chain so that a spatial derivative can be formulated. $\tau = RC$ is the time constant of a single resistor-capacitor circuit.

Equation Eq. (17.55) is the one-dimensional formulation of a very well-known equation used across many fields of science, the one-dimensional *transport* or *diffusion equation*, which we have already described in detail in Sections 15.2.1 and 17.3. Without explicitly solving Eq. (17.55), we can answer the question of the time constant required to transport the information a certain distance. We assume a spatially periodically varying potential with the wavelength λ, whose amplitude decreases exponentially with a time constant τ_λ that depends on the wavelength λ (cf. Section 15.2.1):

$$U(x) = U_0(x)\exp(-t/\tau)\exp(\mathrm{i}kx). \tag{17.56}$$

If we use this approach in Eq. (17.55), then we get

$$\tau_\lambda = \frac{\tau}{(\Delta x\, k)^2} = \frac{\tau}{4\pi^2(\Delta x)^2}\lambda^2. \tag{17.57}$$

We have thus obtained the essential answer to the question of the convergence of the iteration: The convergence time grows quadratically with the wavelength of the structure we want to smooth. It therefore takes four times as long to equilibrate values that are twice as far apart. Let us arbitrarily assume that one iteration step is necessary to bring neighboring nodes into equilibrium. Then we need 100 iteration steps for nodes that are 10 pixels apart. With only a few known values, the convergence of the method is therefore far too slow.

This can be remedied with multigrid methods, which we discussed in Section 15.3. They are an efficient tool for accelerating the convergence of the iteration. The iteration of the system of equations is started at a rough level. The calculated values are then close together. For a pyramid with only six levels, the distances shrink to 1/32. We can therefore calculate the large structures of the solution with a convergence rate that is a factor of 1000 faster than on the original image. At this stage, we do not capture the small-scale variations. However, we can use the coarse solution as a starting point for further iterations at the next finer resolution level.

In this way, we can refine the solution from level to level and obtain a solution with the full resolution at the lowest level of the pyramid. The calculations in the higher levels of the pyramid do not cost much computing time, as the number of pixels in all levels of the pyramid together is only 1/3 higher than in the lowest level. The calculation of a motion field from the cab scene (Fig. 17.3) using this method is shown in Fig. 17.4.

17.5 Inverse Filtering for Restoration

In this section, we investigate a special class of inverse problems that occurs very frequently in image processing and show how to solve very large inverse problems by fast iterative methods.

17.5.1 Image Restoration

Systems for generating images are not perfect due to the physical limits of optical systems. Therefore, images are never identical to their original and, in particular, have a limited resolution. For many technical and scientific applications, however, it is crucial to improve the resolution. When optical means are no longer sufficient, image processing methods can help. The causes of a reduction in image quality include operating errors or design faults in imaging systems. These include blurring due to incorrect focusing, movement of objects or the camera, a mechanically unstable optical system or errors in the design of optical systems. Such errors are more common than you might think. A famous example is the defect in the optics of the Hubble Space Telescope. An error in the test procedures for measuring the primary mirror led to a significant residual aberration of the telescope.

The correction of known and unknown image distortions is called *restoration*. The question arises as to whether distortions are reversible, and if so, to what extent. It is clear that information that is no longer contained in a disturbed image can no longer be restored. To illustrate this point, let us assume the extreme case that only the mean gray value of an image has been preserved. Then we have no way of reconstructing the image content. On the other hand, images contain a lot of redundant information, so we can hope that an image defect only partially destroys important information, even if we can no longer "see" it directly.

In Sections 3.5 and 5.2.1 we have seen that an optical system including digitization is a *linear displacement invariant system* and can therefore be well described by a *point response* and a *transfer function*.

The first task will be to describe the perturbation of the image as accurately as possible. This is possible either by a theoretical analysis of the image acquisition system or experimentally by using suitable test images. If this is not possible, the disturbed image remains the only source of information.

17.5.2 Survey of Image Distortions

Due to the diverse possibilities of image generation (Chapter 3), there are also many reasons for image distortions. Image sharpness is limited by a number of lens errors (aberrations). However, even with a perfect optical system, sharpness is limited by the diffraction of electromagnetic waves at the input aperture (aperture) of the optical system. In addition to these inherent image distortions in an optical system, *defocusing* often causes an image to become blurred. Other causes of image blurring are unwanted movements and vibrations of the camera system during the exposure time. Systems with a small angle of view or long *focal length* are particularly sensitive to this type of interference. Objects that move by more than one pixel in the image plane during the exposure time also cause blurring.

Since defocusing and *lens aberrations* are directly related to the optical system, we discuss them together in this section. The effect of a blur or aberration is described by the *point response* (*PSF*) $h(\boldsymbol{x})$ or the *optical transfer function* (*OTF*) $\tilde{h}(\boldsymbol{\kappa})$; see Section 3.5. The relationship between the object $g(\boldsymbol{x})$ and the image $g'(\boldsymbol{x})$ is thus given by

$$g'(\boldsymbol{x}) = (h * g)(\boldsymbol{x}) \quad \circ\!\!-\!\!\bullet \quad \tilde{g}'(\boldsymbol{\kappa}) = \tilde{h}(\boldsymbol{\kappa})\tilde{g}(\boldsymbol{\kappa}). \tag{17.58}$$

Aberrations are generally more difficult to handle, as they increase strongly with the distance from the optical axis, so they are not actually shift invariant and therefore cannot be described by a spatially invariant PSF. On the other hand, aberrations only change slowly and continuously with the position in the image. As long as the resulting blur is limited to a range in which the aberration can be considered constant, it can still be treated with the theory of linear shift invariant systems. The only difference is that PSF and OTF vary with the position in the image.

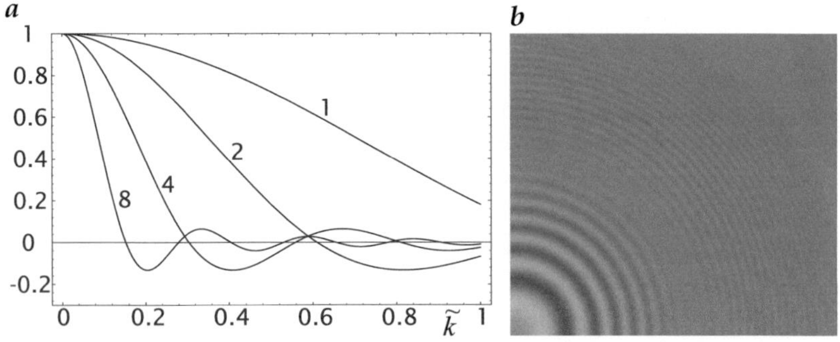

Figure 17.10: ***a*** *Transfer functions for circular blurring. The parameters for the different curves are the radius of the blur circle;* ***b*** *defocused image of the ring test pattern.*

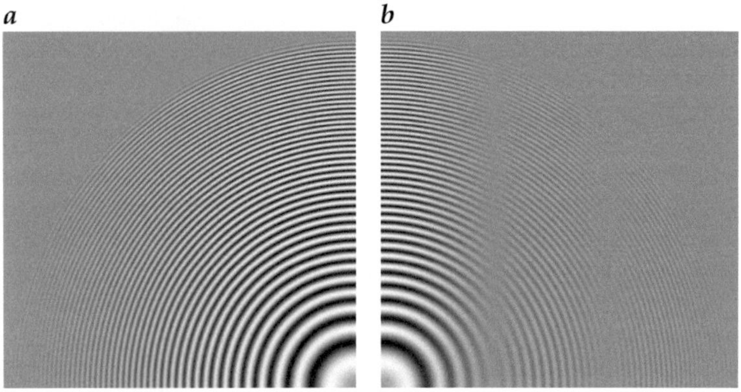

Figure 17.11: *Simulation of motion blur with the ring pattern:* ***a*** *small and* ***b*** *large horizontal motion blur.*

The point response has the shape of the aperture if defocusing is the dominant cause of the blur. Most apertures can be approximately described by a circular disk. The Fourier transform of a circular disk with the radius r is a Bessel function (> R13):

$$\frac{1}{\pi r^2}\Pi\left(\frac{|x|}{2r}\right) \circ\!\!-\!\!\bullet \frac{J_1(2\pi|\kappa|r)}{\pi|\kappa|r}. \qquad (17.59)$$

This Bessel function (Fig. 17.10a) has a series of zeros and thus completely eliminates certain wavenumbers. This effect is shown in Fig. 17.10b using a defocused image of the ring test pattern.

While blurring due to defocus and aberrations of the optical system tends to be isotropic, blurring effects due to motion are usually one-dimensional blurs in the direction of motion (Fig. 17.11b). In the simplest case, the motion is constant during the exposure time, so the PSF of the motion blur is a one-dimensional rectangular function. Without limiting the generality, we first assume that the movement runs along the x-axis. Then (> R12, > R13) is

$$h_{blur}(x) = \frac{1}{u\Delta t}\Pi\left(\frac{x}{u\Delta t}\right) \circ\!\!-\!\!\bullet\; \tilde{h}_{blur}(\kappa) = \operatorname{sinc}(\kappa u \Delta t), \tag{17.60}$$

where u is the amount of velocity and Δt is the exposure time. This results in the blur length $\Delta x = u\Delta t$.

If the velocity $\boldsymbol{u}$ is oriented in a different direction, Eq. (17.60) can be generalized as follows:

$$h_{blur}(\boldsymbol{x}) = \frac{1}{|\boldsymbol{u}|\Delta t}\Pi\left(\frac{\boldsymbol{x}\hat{\boldsymbol{u}}}{|\boldsymbol{u}|\Delta t}\right)\delta(\boldsymbol{u}\boldsymbol{x}) \circ\!\!-\!\!\bullet\; \tilde{h}_{blur}(\boldsymbol{\kappa}) = \operatorname{sinc}(\boldsymbol{\kappa}\boldsymbol{u}\Delta t), \tag{17.61}$$

where $\hat{\boldsymbol{u}} = \boldsymbol{u}/|\boldsymbol{u}|$ is a unit vector in the direction of the motion blur.

17.5.3 Deconvolution

Defocusing, motion blur and 3D imaging techniques such as focus series and confocal laser scanning microscopy (Section 9.2.3) have in common that the object function $g(\boldsymbol{x})$ is convolved with a point response. Therefore, the basic procedures for reconstructing or restoring the object function are the same. Essentially, it is a *deconvolution* or *inverse filtering*, as the effects of the convolution must be undone with the PSF. Using the relationships in Eq. (17.58), inverse filtering is in principle a simple process. The effect of the convolution operator $\mathcal{H}$ is reversed by applying the inverse operator $\mathcal{H}^{-1}$. In Fourier space, we can then write as follows:

$$\tilde{\boldsymbol{G}}_R = \frac{\tilde{\boldsymbol{G}}'}{\tilde{\boldsymbol{H}}'} = \tilde{\boldsymbol{H}}^{-1} \cdot \tilde{\boldsymbol{G}}'. \tag{17.62}$$

The reconstructed image $\boldsymbol{G}_R$ then results from an inverse Fourier transformation:

$$\boldsymbol{G}_R = \mathcal{F}^{-1}\tilde{\boldsymbol{H}}^{-1} \cdot \mathcal{F}\boldsymbol{G}. \tag{17.63}$$

The reconstruction process therefore consists of the following steps: The Fourier-transformed image $\mathcal{F}\boldsymbol{G}'$ is multiplied by the inverse OTF $\tilde{\boldsymbol{H}}^{-1}$ and then back-transformed into the spatial domain. The inverse filtering can also be performed in the spatial domain by convolution with a mask given by the back-transformed inverse OTF:

$$\boldsymbol{G}_R = (\mathcal{F}^{-1}\tilde{\boldsymbol{H}}^{-1}) * \boldsymbol{G}'. \tag{17.64}$$

At first glance, inverse filtering appears simple. However, a closer look reveals that Eqs. (17.63) and (17.64) are not applicable in most cases. The reason for this is that the OTF is often zero in wide areas. This is the case, for example, for the OTF of motion blur Eq. (17.61) and defocus Eq. (17.59). In these areas, the inverse OTF becomes infinite.

Not only the zeros of the OTF are problematic, but also all areas in which the OTF becomes small, as noise has a major influence here. For a quantitative analysis, we use the following simple image generation model:

$$\boldsymbol{G}' = \boldsymbol{H} * \boldsymbol{G} + \boldsymbol{N} \quad \circ\!\!-\!\!\bullet \quad \tilde{g}' = \tilde{h} \cdot \tilde{g} + \tilde{n}. \tag{17.65}$$

Here $\boldsymbol{N}$ is an image with additive noise and $\tilde{n}$ is the associated Fourier transform, the noise spectrum. Equation Eq. (17.65) states that noise *after* the disturbance is added to the image. With this model, the inverse filtering according to Eq. (17.62) results under the condition that $\tilde{h}$ is not equal to zero everywhere:

$$\tilde{g}_R = \tilde{h}^{-1} \cdot \tilde{g}' = \tilde{g} + \tilde{h}^{-1} \cdot \tilde{n}. \tag{17.66}$$

This means that the restored image is the original image $\tilde{\boldsymbol{G}}$ with the noise amplified by $\tilde{h}^{-1}$.

If $\tilde{h}$ approaches zero, $\tilde{h}^{-1}$ and the noise level become very high. From Eqs. (17.65) and (17.66) it can also be seen that the signal-to-noise ratio has not improved, but has remained the same, since the noise level and the useful information in the image are multiplied by the same factor.

In conclusion, we can state that inverse filtering does not improve the image quality in the slightest. In fact, it is clear that every linear technique behaves in this way. With linear techniques, we can only amplify the structures attenuated by the interference to the extent that the noise level no longer exceeds an unacceptable threshold.

As an example of inverse filtering, consider the 3D reconstruction from a microscopic *focus series*. A focus series is an image stack of microscope images in which the focused plane is gradually adjusted. Due to the shallow depth of field, only objects in a thin layer appear in focus in each image (Section 3.4.4). This is why the image stack immediately produces a three-dimensional image. However, the 3D objects in it are disturbed by the 3D point spread function of the optical image.

Certain object structures are therefore completely filtered out, and blurred objects are superimposed on those in focus. Inverse filtering can be used to try to reduce these effects.

Obviously, precise knowledge of the PSF is essential for a good reconstruction. In Section 3.5.1 we have calculated the 3D-*PSF* of optical systems neglecting lens aberrations and resolution limits due to diffraction. Now we have to take into account that highly magnified microscopic images have diffraction-limited resolution.

The diffraction-limited 3D PSF was calculated by Erhardt et al. [73]. Essentially, the resolution limit changes the double cone of the PSF (Fig. 8.1) only in the vicinity of the focused plane. Here, a point is not mapped onto a point, but onto a diffraction disk. As a result, the *OTF* drops for higher wavenumbers in the $k_x k_y$ plane. As a first approximation, we can consider the resolution limited by diffraction as an additional low-pass filter. The PSF of the geometric optics is convolved with this additional low-pass filter, the OTF is multiplied with its TF.

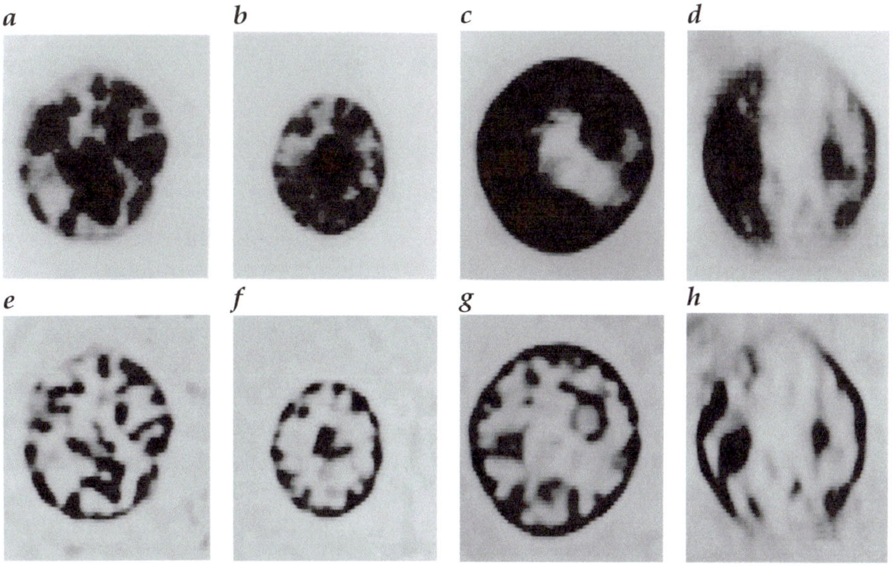

Figure 17.12: *3D reconstruction of a focal series of a cell nucleus taken with conventional microscopy. Top row:* ***a–c*** *selected original images;* ***d*** *xz cross-section perpendicular to the image plane. Bottom row:* ***e–h*** *reconstructions of the overlying images; kindly provided by Dr. Schmitt and Prof. Dr. Komitowski, German Cancer Research Center, Heidelberg.*

The simplest approach for an optimal reconstruction is to limit the inverse OTF to the wavenumber components that are not attenuated below a critical limit. This limit depends on the noise in the image. In this way, the true inverse OTF is replaced by an *effective inverse OTF*, which approaches zero again in the wavenumber ranges that cannot be reconstructed.

Fig. 17.12 shows the result of such a reconstruction procedure using the example of a $64 \times 64 \times 64$ focal series recorded from the nucleus of a cancer cell from a rat liver. The resolution is $0.22\,\mu$m in all directions. The images verify the theoretical observations. The reconstruction considerably improves the resolution in the xy-image plane, while the resolution in the z-direction remains — as expected — significantly worse. Structures that only change in the z-direction have been completely filtered out of the image stack by the PSF of the optical image and can therefore no longer be reconstructed.

17.5.4 ♯Iterative inverse Filterung

Iterative techniques are an interesting variant of inverse filtering, as they allow control over the degree of reconstruction. Let $\mathcal{H}$ be the blur operator. In addition, we introduce the new operator $\mathcal{H}' = \mathcal{I} - \mathcal{H}$. Then the inverse operator

$$\mathcal{H}^{-1} = \frac{\mathcal{I}}{\mathcal{I} - \mathcal{H}'} \tag{17.67}$$

can be approximated by a Taylor series:

$$\mathcal{H}^{-1} = \mathcal{I} + \mathcal{H}' + \mathcal{H}'^2 + \mathcal{H}'^3 + \ldots. \tag{17.68}$$

This means explicitly for the OTF in Fourier space:

$$\tilde{h}^{-1}(\boldsymbol{\kappa}) = 1 + \tilde{h}' + \tilde{h}'^2 + \tilde{h}'^3 + \ldots. \tag{17.69}$$

To understand how the iteration works, we first consider periodic structures and one that is only slightly weakened, i. e. $\tilde{h}$ is only slightly smaller than one. This means that $\tilde{h}'$ is small and the iteration converges quickly.

The other extreme would be when the periodic structure has almost disappeared. Then $\tilde{h}'$ is almost one. Accordingly, the amplitude of the periodic structure increases by the same amount with each iteration step (linear convergence). This procedure has the decisive advantage that we can stop the iteration as soon as the noise becomes visible.

A direct implementation of the iteration scheme makes little sense, as the convolution masks become larger and larger with increasing exponents and therefore the computational effort increases from iteration step to iteration step. A much more effective iteration scheme, the *Van Cittert iteration*, uses the Horner scheme to quickly calculate a polynomial:

$$\boldsymbol{G}_0 = \boldsymbol{G}', \quad \boldsymbol{G}_{k+1} = \boldsymbol{G}' + (\boldsymbol{I} - \boldsymbol{H}) * \boldsymbol{G}_k. \tag{17.70}$$

The convergence of the iteration can easily be examined in Fourier space. From Eq. (17.70) we obtain

$$\tilde{g}_k(\boldsymbol{\kappa}) = \tilde{g}'(\boldsymbol{\kappa}) \sum_{i=0}^{k} (1 - \tilde{h}(\boldsymbol{\kappa}))^i. \tag{17.71}$$

This equation represents a geometric series with the initial value $a_0 = \tilde{g}'$ and the factor $q = 1 - \tilde{h}$. The series only converges if $|q| = |1 - \tilde{h}| < 1$. Then the sum is given by

$$\tilde{g}_k(\boldsymbol{\kappa}) = a_0 \frac{1 - q^k}{1 - q} = \tilde{g}'(\boldsymbol{\kappa}) \frac{1 - |1 - \tilde{h}(\boldsymbol{\kappa})|^k}{\tilde{h}(\boldsymbol{\kappa})} \tag{17.72}$$

and converges to the correct value $\tilde{g}'/\tilde{h}$. Unfortunately, the condition for convergence is not fulfilled for all those transfer functions that have negative values. Therefore, the Van Cittert iteration cannot be applied for motion blur and defocus.

However, a small modification of the iteration process allows it to be used for negative transfer functions as well. The simple trick is to apply the transfer function twice. The transfer function $\tilde{h}^2$ of the cascaded filter $\boldsymbol{H} * \boldsymbol{H}$ is then always positive. The modified iteration scheme is

$$\boldsymbol{G}_0 = \boldsymbol{H} * \boldsymbol{G}', \quad \boldsymbol{G}_{k+1} = \boldsymbol{H} * \boldsymbol{G}' + (\boldsymbol{I} - \boldsymbol{H} * \boldsymbol{H}) * \boldsymbol{G}_k. \tag{17.73}$$

With $a_0 = \tilde{h}\tilde{g}'$ and $q = 1 - \tilde{h}^2$ the iteration again converges to the correct value

$$\lim_{k\to\infty} \tilde{g}_k(\boldsymbol{\kappa}) = \lim_{k\to\infty} \tilde{h}\tilde{g}' \frac{1 - |1 - \tilde{h}^2|^k}{\tilde{h}^2} = \frac{\tilde{g}'}{\tilde{h}}, \quad \text{if} \quad |1 - \tilde{h}^2| < 1. \tag{17.74}$$

17.6 Model-based Segmentation

17.6.1 Introduction

Simple segmentation methods only use local information. In Section 1.6 (Fig. 1.16) it was about the remarkable property of the human eye to recognize objects even if they are not fully represented. It is obvious that information provided by local neighborhood operators is not sufficient for this task. What is needed is specific knowledge about the geometric shape of the objects, which must be compared with the local information.

These thoughts lead us to *model-based segmentation*, which can be applied if the basic shape of the objects contained in the image is known. We only consider the case of straight object edges here.

17.6.2 Parameter Space, Hough Transform

The method we are discussing here recognizes straight edges, even if they are interrupted by noise or only partially visible. To explain the method, we first assume a segmented image that contains straight edge pieces of this type. The fact that such an edge point lies on a straight line results in meaningful conditions that can be used to determine the position of the edge. All points $[x_n, y_n]^T$ of a straight line must fulfill the following condition:

$$y_n = a_0 + a_1 x_n, \tag{17.75}$$

where a_0 and a_1 are the intercept and slope of the line. We can also read Eq. (17.75) as a condition for the parameters a_0 and a_1:

$$a_1 = \frac{y_n}{x_n} - \frac{1}{x_n} a_0. \tag{17.76}$$

This is again the equation for a straight line in a new space that is spanned by the parameters a_0 and a_1. In this space, the straight line has the offset y_n/x_n and a gradient of $-1/x_n$. With only one point on an edge, we can no longer freely choose a_0 and a_1, but the two parameters must fulfill equation Eq. (17.76).

The space spanned by the model parameters a_0 and a_1 is called *model space* or *parameter space*. Each point reduces the model space to a straight line. We can therefore draw a straight line in the model space for each point in the data space (Fig. 17.13). If all points in the data space lie on a straight edge segment, all associated straight lines in the model space meet at the point that defines the parameters a_0 and a_1 of the edge segment in the data space. Since a line segment contains many points, we obtain a solid estimate of the two line parameters. A line segment in the data space is mapped to a point in the model space in this way. The transformation from the data space to the model space via a model equation is called *Hough transform*. More on the geometric illustration of the mathematics in Appendix C.6.4.

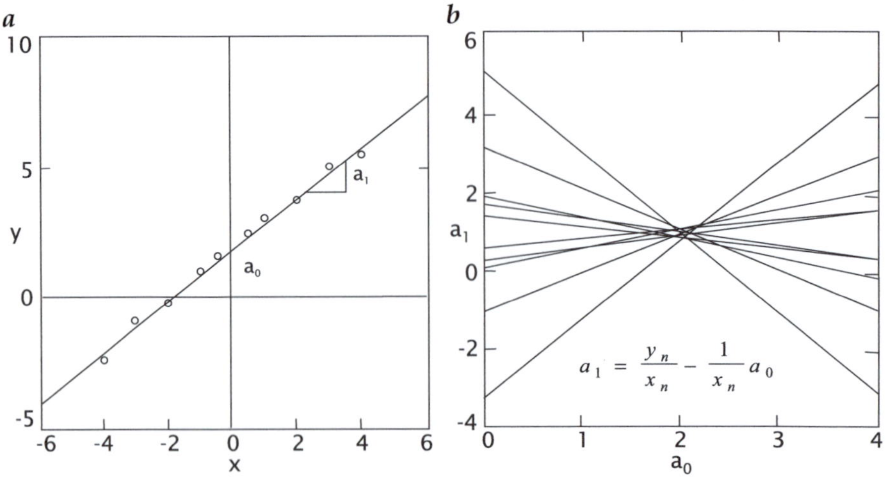

*Figure 17.13: Hough transform of straight lines: the data space $[x, y]^T$ (**a**) is mapped to the model space $[a_0, a_1]^T$ (**b**). Points on a straight line in the data space lead to an intersection of a corresponding number of straight lines in the model space.*

The linear equation Eq. (17.75) is not used for practical applications, as the gradient of a straight line can become infinite and is therefore unusable for a discrete model space. Another option is to use the slope angle of the line and the distance of the line from the origin of the coordinate system. With these two parameters, we can write the equation of the line as

$$\hat{\boldsymbol{n}}\boldsymbol{x} = d \quad \text{or} \quad x \cos\theta + y \sin\theta = d, \tag{17.77}$$

where $\hat{\boldsymbol{n}}$ is a vector normal to the line and θ is the angle of this vector to the x-axis of the image coordinate system.

The disadvantage of the Hough transform for line detection is the high computational effort. A straight line must be calculated in the parameter space for each image point, and each point in the model space through which the straight line runs must be incremented.

17.6.3 ♯Fast Orientation-based Hough transform

A significant acceleration of the Hough transform can be achieved by using additional information from the first steps of image processing. The analysis of the local neighborhoods with the *structure tensor method* not only allows the detection of edges, but also provides the gradient of the edges. Therefore, we have two pieces of information for each pixel if it lies on an edge: the point through which the edge passes and its orientation. This completely describes a straight line. Accordingly, each point on a straight line in image space no longer corresponds to a straight line — as discussed in Section 17.6.2 — but to a single point in parameter space. The one-to-one correspondence

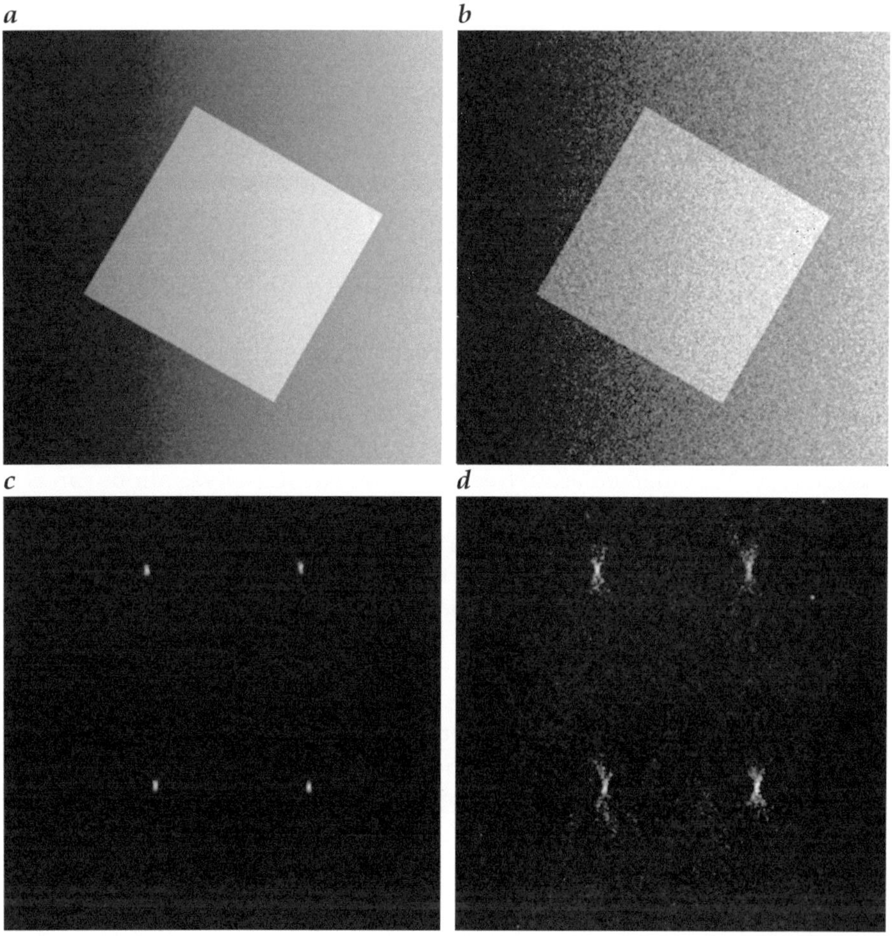

Figure 17.14: *Orientation-based fast Hough transform:* ***a*** *and* ***b*** *unevenly illuminated noisy squares;* ***c*** *and* ***d*** *Hough model space with the distance d (horizontal axis) and the angle θ (vertical axis) of the straight line according to Eq. (17.77) for* ***a*** *and* ***b****, respectively.*

speeds up the calculation of the Hough transform considerably. For each pixel, we simply need to add *one* point to the parameter space.

The application of this Hough transform to a difficult segmentation task can be seen in Fig. 17.14. Fig. 17.14a, b shows unevenly illuminated noisy images of a square. To extract its edges, no segmentation of the edges is required with the orientation-based Hough transform. We only need to determine the components of the structure tensor using the techniques described in Section 13.3.6. Then for each pixel θ and d are calculated according to Eq. (17.77). We use the length of the orientation vector as the weighting factor for the contribution of a point to the parameter space. In this way, the points are entered into the parameter space according to the reliability measure for the local orientation and thus the edge strength.

In the Hough parameter space (Fig. 17.14c and d), four clusters appear corresponding to the four sides of the square. The clusters appear in pairs, as two sides are parallel to each other and only differ in the distance to the image center. It is worth noting how well the technique works even at high noise levels.

17.7 #References for Further Reading

The topic of this chapter requires the methods of matrix algebra to a large extent. The monograph by Golub and van Loan [109] gives an excellent overview. Variational methods in image processing (Section 17.2) are discussed in detail by Jähne et al. [165, Vol. 2, Chapter 16] and Schnörr and Weickert [293]. A mathematical treatise on the calculus of variations can also be found in Courant and Hilbert [46] and Scherzer et al. [288]. The use of the membrane model (Section 17.2.6) was first published by Broit [28], who used it for image registration in computer tomography. It was later extended by Dengler [56] and used for image sequence analysis. Today, elasticity models are widely used in image processing for tasks as diverse as edge modeling and tracking (active contours) [173], reconstruction of 3D objects [325] and reconstruction of surfaces [324]. Anisotropic diffusion (Section 17.3) and nonlinear scale spaces are still an active subject of research. An excellent overview of this topic is provided by Weickert [350]. Optimal filters for fast anisotropic diffusion are investigated by Scharr [284] and Scharr and Uttenweiler [285].

18 Morphology and Shape Analysis

Contents

B. Jähne, *Digital Image Processing and Image Acquisition*,
https://doi.org/10.1007/978-3-032-04806-6_18

18.1 Introduction

Segmentation is used to extract objects from images, i. e. to identify which pixels belong to which objects. Now we can perform the next step and analyze the *shape* of the objects. In this chapter, we first discuss a class of neighborhood operations, the morphological operators for binary images, which can be used to modify and analyze the shape of objects (Section 18.2 to 18.4).

All operations that we discussed in Chapters 11 to 16 for extracting properties from images return images as a result. However, it is clear that the shape of objects can be stored much more compactly. The complete information about the shape of an object is, e. g. contained in its edge pixels.

In Section 18.5 we therefore deal with the question of which data structures we can use to represent segmented objects. The following data structures for binary objects are covered: *run-length encoding* (Section 18.5.1), *tree structures* (Section 18.5.2) and *directional chains* (Section 18.5.3). Two other object representations, the *moments* and the *Fourier descriptors*, have their own sections dedicated to their importance (Sections 18.6 and 18.7).

A compact representation of the shape of objects would be of little use if it would require a lot of effort to calculate it and if it would be difficult to determine shape parameters directly from the representation. Therefore, in Section 18.8 we also deal with the question of how shape parameters can be calculated from the individual object representations.

Parameters that describe the shape of an object are used to compare them with sample objects or to categorize them into classes of different shapes. In this context, the important question arises as to how these parameters can be made invariant with respect to different geometric transformations. Objects can be viewed from different distances and directions. Therefore, it is of interest to find shape parameters that are size or rotation invariant or even invariant with respect to an affine or projective mapping.

18.2 Neighborhood operations with binary images

18.2.1 Binary Convolution

Operators that relate pixels in a small neighborhood have proven to be versatile and powerful tools for extracting features from scalar and vector images (Chapter 10). The result of such an operation in binary images can only be zero or one. So neighborhood operators in binary images will edit the object shape by adding pixels to an object or deleting pixels from an object. In the Sections 10.3 and 10.5.1 we discussed the two basic operations for linking neighboring pixels in gray value images: convolution ("weigh and sum") and nonlinear rank order filtering ("sort and select").

With binary images, we do not have much choice in terms of the type of operations to be performed. Pixels can only be combined with the logical

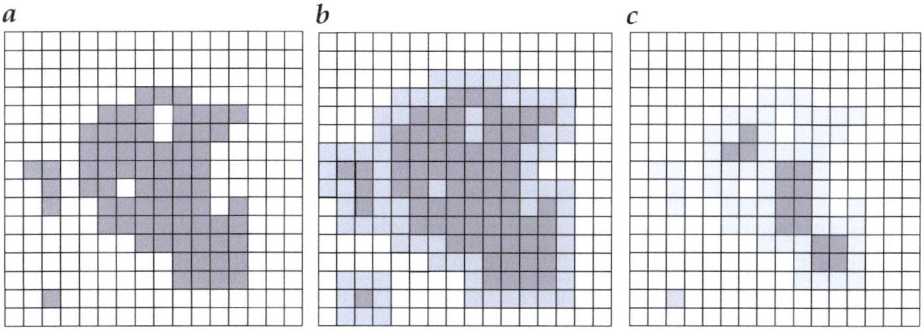

Figure 18.1: ***b*** *Dilation and* **c** *Erosion of a binary object in* ***a*** *with a* 3×3 *mask. The added (dilation) or removed (erosion) pixels are shown in a lighter color.*

operators of Boolean algebra. We can introduce a *binary convolution* by replacing the multiplication of the image pixels with the mask pixels with an *and operation* and the summation with an *or operation*:

$$\boxed{g'_n = \bigvee_{n'=-R}^{R} m_{n'} \wedge g_{n+n'} \quad \text{dilation.}} \tag{18.1}$$

The characters $\wedge$ and $\vee$ stand for the logical AND operation and the OR operation respectively. As with a sum character, all binary pixel values contained in a multidimensional mask are linked with each other using OR operations with the large OR character. In Eq. (18.1) and all corresponding following ones, the vectorial indexing introduced in Section 10.3.1 is used so that the equations are valid for arbitrary-dimensional signals. The binary image $\boldsymbol{G}$ is convolved in Eq. (18.1) with a D-dimensional mask $\boldsymbol{M}$. Note that, unlike the convolution, the mask is not mirrored (Section 10.3.4).

What does this operation do? Let's assume that all coefficients of the mask are set to one. If there are one or more object pixels, i. e. ones, within the mask, the result of the operation is one, otherwise zero (Fig. 18.1). The object is therefore expanded; small holes or cracks are filled and the contours become smoother (Fig. 18.2b). The operator defined in Eq. (18.1) is called *dilation operator*. Interestingly, we have the same effect if we apply a *rank-order filter* to binary images (see Section 10.5.1).

Let's look at the *maximum operator*. The maximum will be one if one or more ones occur within the mask. This leads to the same result as the binary convolution operation in Eq. (18.1). The *minimum operator* has the opposite effect. Now the result is only one if the mask is completely inside the object. In this way, the object is eroded. Objects that are smaller than the mask disappear completely, those that are connected with a narrow bridge are separated (Fig. 18.1c). The *erosion* of an object can also be performed with binary convolution:

$$\boxed{g'_n = \bigwedge_{n'=-R}^{R} m_{n'} \wedge g_{n+n'} \quad \text{erosion.}} \tag{18.2}$$

By transferring the concept of neighborhood operations for gray value images to binary images, we have obtained an important tool for editing the shape of objects. In Fig. 18.1 we have already seen that these operations are suitable for filling small holes and cracks or eliminating small objects. The size of the mask determines the effect of the operator. Therefore, the mask is often referred to as *structural element*. For example, an erosion operation acts like a mesh with holes in the shape of the mask. All objects that fit through the holes slip through and disappear from the image. An object only remains in the image if the mask is completely covered by object pixels at least at one point; otherwise it disappears. An operator that influences the shape of objects is called *morphological operator*. The name comes from morphology, which describes the shape of objects in biology and the geosciences.

18.2.2 Set Operations with Pixels

We have chosen a rather unconventional way to introduce morphological operations. Normally, these operations are defined as set operations with pixels. For this purpose, we consider $\boldsymbol{G}$ as the set of all non-zero pixels of the image matrix. $\boldsymbol{M}$ is the set of non-zero mask pixels. With $\boldsymbol{M}_p$ we denote the mask shifted with its reference point (generally, but not necessarily, the center) to the image point p. The erosion is then defined as

$$\boxed{\boldsymbol{G} \ominus \boldsymbol{M} = \{p : \boldsymbol{M}_p \subseteq \boldsymbol{G}\} \quad \text{erosion}} \tag{18.3}$$

and the dilation as

$$\boxed{\boldsymbol{G} \oplus \boldsymbol{M} = \{p : \boldsymbol{M}_p \cap \boldsymbol{G} \neq \emptyset\} \quad \text{dilation.}} \tag{18.4}$$

These definitions are equivalent to Eqs. (18.1) and (18.2). We can now express the erosion of the pixel set $\boldsymbol{G}$ by the pixel set $\boldsymbol{M}$ as the set of all pixels p for which $\boldsymbol{M}_p$ is completely contained in $\boldsymbol{G}$. In contrast, the dilation of $\boldsymbol{G}$ by $\boldsymbol{M}$ is the set of all pixels for which the intersection of $\boldsymbol{G}$ and $\boldsymbol{M}_p$ is not the empty set. Since this theoretical approach leads to more compact and descriptive formulas, we will use it from now on. However, the equations Eqs. (18.1) and (18.2) are still important for the implementation of morphological operators with logical operations.

Erosion and dilation operators can be regarded as elementary morphological operators from which more complex operators can be generated. We will examine their properties in detail in the next section.

18.3 General Properties

Morphological operators share most, if not all, of the properties of linear convolution operators (Section 10.3). The properties we discuss below are not limited to 2D images, but apply to N-dimensional image data in general.

18.3.1 Shift Invariance

The *shift invariance* results directly from the definition of the erosion and dilation operators as convolutions with binary data in Eqs. (18.1) and (18.2). If we use the shift operator ${}^{mn}\mathcal{S}$ defined in Eq. (10.15), we can formulate the shift invariance of any morphological operator $\mathcal{M}$ in operator notation as follows:

$$\mathcal{M}\left({}^{mn}\mathcal{S}G\right) = {}^{mn}\mathcal{S}\left(\mathcal{M}G\right). \tag{18.5}$$

18.3.2 Superposition Principle

For gray value images, the *superposition principle* is defined as follows:

$$\mathcal{H}(aG + bG') = a\mathcal{H}G + b\mathcal{H}G'. \tag{18.6}$$

In binary images, the factors a and b have no meaning, and the sum of two gray value images corresponds to the union or logical *or* of two binary images. If the superposition principle for morphological operations $\mathcal{M}$ applies to binary images, it has the following form:

$$\mathcal{M}(G \cup G') = (\mathcal{M}G) \cup (\mathcal{M}G') \quad \text{or} \quad \mathcal{M}(G \vee G') = (\mathcal{M}G) \vee (\mathcal{M}G'). \tag{18.7}$$

The operation $G \vee G'$ stands for a pointwise *or* of the elements of the matrices G and G'. In general, morphological operators are not additive in the sense of Eq. (18.7). While the dilation operation fulfills the superposition principle, this does not apply to erosion. The erosion of the union of two objects is a superset of the union of two eroded objects:

$$\begin{array}{rcl} (G \cup G') \ominus M & \supseteq & (G \ominus M) \cup (G' \ominus M) \\ (G \cup G') \oplus M & = & (G \oplus M) \cup (G' \oplus M). \end{array} \tag{18.8}$$

18.3.3 Commutativity and Associativity

Morphological operators are generally not *commutative*:

$$M_1 \oplus M_2 = M_2 \oplus M_1, \text{ but } M_1 \ominus M_2 \neq M_2 \ominus M_1. \tag{18.9}$$

We see that the erosion is not commutative if we consider the special case that $M_1 \supset M_2$. Then the erosion of M_2 by M_1 results in the empty set. However, if erosion and dilation masks are applied consecutively to the same image G, they are commutative:

$$\boxed{\begin{array}{lll} (G \ominus M_1) \ominus M_2 & = G \ominus (M_1 \oplus M_2) & = (G \ominus M_2) \ominus M_1 \\ (G \oplus M_1) \oplus M_2 & = G \oplus (M_1 \oplus M_2) & = (G \oplus M_2) \oplus M_1. \end{array}} \tag{18.10}$$

These equations are important for the implementation of morphological operations. The cascaded operation with k structure elements $\boldsymbol{M}_1, \boldsymbol{M}_2, \ldots, \boldsymbol{M}_k$ is equivalent to the operation with the structure element $\boldsymbol{M} = \boldsymbol{M}_1 \oplus \boldsymbol{M}_2 \oplus \ldots \oplus \boldsymbol{M}_k$ (*associativity*). We can therefore decompose large structural elements in the same way as we do with linear shift invariant operators. An important example is the construction of separable structural elements from horizontal and vertical elements $\boldsymbol{M} = \boldsymbol{M}_x \oplus \boldsymbol{M}_y$. A less trivial example is the construction of large one-dimensional structural elements from elements containing many zeros:

$$
\begin{gathered}
[1\,1\,1] \oplus [1\,0\,1] = [1\,1\,1\,1\,1] \\
[1\,1\,1\,1\,1] \oplus [1\,0\,0\,0\,1] = [1\,1\,1\,1\,1\,1\,1\,1\,1] \\
[1\,1\,1\,1\,1\,1\,1\,1\,1] \oplus [1\,0\,0\,0\,0\,0\,0\,0\,1] \\
= [1\,1\,1\,1\,1\,1\,1\,1\,1\,1\,1\,1\,1\,1\,1\,1\,1] .
\end{gathered}
\tag{18.11}
$$

In this way, we can assemble large exponentially growing structural elements with a minimum number of logical operations, as we have constructed large convolution masks by cascading in Section 11.5. However, isotropic, i. e. circular, structural elements are more difficult to obtain. The problem lies in the fact that the dilation of horizontal and vertical structural elements always leads to a rectangular structural element, but not to a circular mask. However, such a mask can be approximated by one-dimensional structural elements that run in several directions and not just along the axes. It should be noted that multi-step masks can also be efficiently cascaded to form large masks.

18.3.4 Monotony

Erosion and dilation are *monotonic* operations:

$$
\begin{array}{lcl}
G_1 \subseteq G_2 & \leadsto & G_1 \oplus \boldsymbol{M} \subseteq G_2 \oplus \boldsymbol{M} \\
G_1 \subseteq G_2 & \leadsto & G_1 \ominus \boldsymbol{M} \subseteq G_2 \ominus \boldsymbol{M}.
\end{array}
\tag{18.12}
$$

The monotony property means that the subset relations are invariant with regard to erosion and dilation.

18.3.5 Distributivity and Duality

Linear shift invariant operators are distributive with respect to addition. For erosion and dilation, a distinction must be made between the union and intersection of two images G_1 and G_2:

$$
\begin{array}{lcl}
(G_1 \cap G_2) \oplus \boldsymbol{M} & \subseteq & (G_1 \oplus \boldsymbol{M}) \cap (G_2 \oplus \boldsymbol{M}), \\
(G_1 \cap G_2) \ominus \boldsymbol{M} & = & (G_1 \ominus \boldsymbol{M}) \cap (G_2 \ominus \boldsymbol{M})
\end{array}
\tag{18.13}
$$

and

$$
\begin{array}{lcl}
(G_1 \cup G_2) \oplus \boldsymbol{M} & = & (G_1 \oplus \boldsymbol{M}) \cup (G_2 \oplus \boldsymbol{M}), \\
(G_1 \cup G_2) \ominus \boldsymbol{M} & \supseteq & (G_1 \ominus \boldsymbol{M}) \cup (G_2 \ominus \boldsymbol{M}).
\end{array}
\tag{18.14}
$$

a *b* *c* *d*

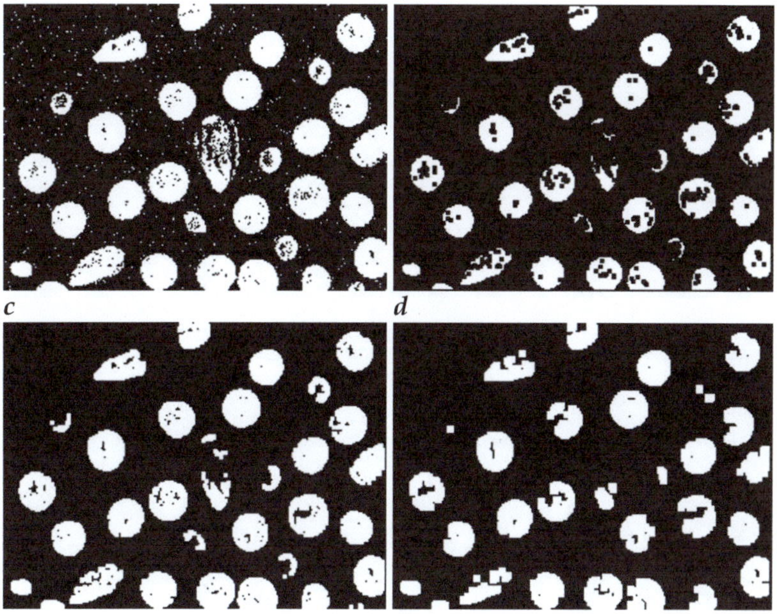

Figure 18.2: *Erosion and opening:* ***a*** *original image;* ***b*** *erosion with a* 3×3 *mask;* ***c*** *opening with a* 3×3 *mask;* ***d*** *opening with a* 5×5 *mask.*

Erosion is therefore only distributive with regard to the intersection operation, while dilation is only distributive with regard to the union operation.

Erosion and dilation are *dual operators*. Erosion with the negation of a binary image is equivalent to the negation of the dilation of the original image. The same applies when erosion and dilation are swapped:

$$\begin{aligned} \overline{G} \ominus M &= \overline{G \oplus M} \\ \overline{G} \oplus M &= \overline{G \ominus M}. \end{aligned} \tag{18.15}$$

18.4 Composite morphological operators

18.4.1 Opening and Closing

Using the elementary erosion and dilation operations, we now develop further operators for processing the object form. While the previous section focused on general and theoretical aspects of morphological operations, we will now concentrate on the application.

Erosion is used to remove small objects. However, it has the disadvantage that all objects remaining in the image become smaller (Fig. 18.2a and b). This can be avoided by subsequently dilating the image with the same structural

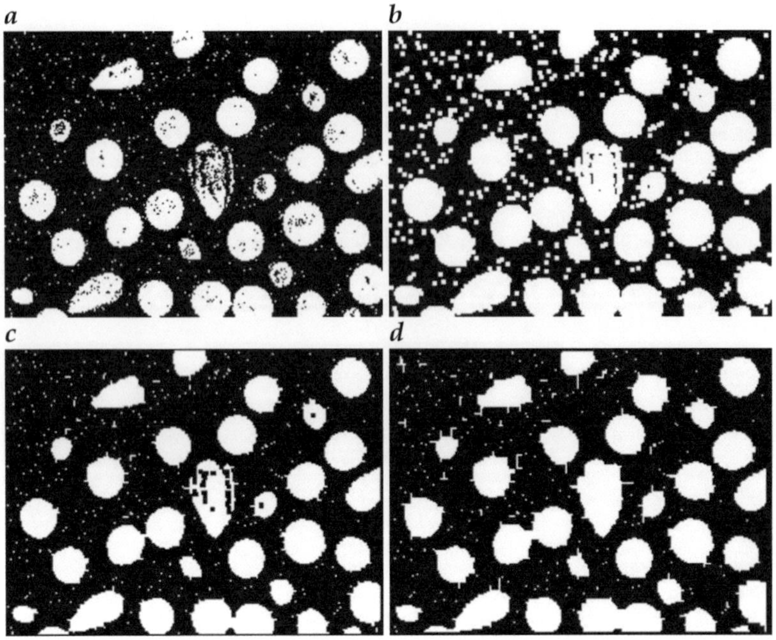

*Figure 18.3: Dilation and closing: **a** original image; **b** dilation with a 3×3 mask; **c** closing with a 3×3 mask; **d** closing with a 5×5 mask.*

element. The combination of erosion and dilation is called *opening*:

$$\boxed{G \circ M = (G \ominus M) \oplus M \quad \text{opening.}} \tag{18.16}$$

Opening sifts out all objects that do not completely contain the structural element at any point, but avoids the size reduction of all objects (Fig. 18.2c and d). It is also ideal for removing lines whose thickness is less than the diameter of the structural element. The object boundaries also become smoother.

In contrast, dilation enlarges objects and fills small holes and cracks (Fig. 18.3a and b). The general enlargement of the objects due to the size of the structural elements can be compensated by subsequent erosion (Fig. 18.3c and d). This combination of dilation and erosion is called *closing*:

$$\boxed{G \bullet M = (G \oplus M) \ominus M \quad \text{closing.}} \tag{18.17}$$

The change in the area of objects due to the different operations can be summarized by the following relationships:

$$G \ominus M \subseteq G \circ M \subseteq G \subseteq G \bullet M \subseteq G \oplus M. \tag{18.18}$$

Opening and closing are *idempotent operators*:

$$\begin{aligned} G \bullet M &= (G \bullet M) \bullet M \\ G \circ M &= (G \circ M) \circ M, \end{aligned} \tag{18.19}$$

This means that a second application of a close and an open operation with the same structural element does not cause any further changes.

18.4.2 Hit-Miss Operator

The *hit-miss operator* has its origin in the question of whether it is possible to detect objects of a specific shape. The erosion operator can only remove objects that do not completely contain the structural element at any point, and thus deletes objects of very different shapes. For the detection of a specific shape, the combination of two morphological operators is necessary. As an example, we will discuss the detection of objects with three consecutive horizontal pixels.

Let's erode the image with a 1×3 mask that corresponds to the shape of the object,

$$M_{<3} = [1\;1\;1], \tag{18.20}$$

then all objects that are smaller than the target object are removed. However, all objects that are larger than the mask are retained, i. e. for which the shifted mask is a subset of the object G ($M_p \subset G$, Fig. 18.4d). We therefore need a second operation that removes all objects that are larger than the target object.

This is possible by analyzing the background of the original image. As a second step, we can erode the background with a 3×5 mask M_2 in which all coefficients are zero except for the pixels of the background surrounding the object. This is a negative mask for the object:

$$M_{>3} = \begin{bmatrix} 1 & 1 & 1 & 1 & 1 \\ 1 & 0 & 0 & 0 & 1 \\ 1 & 1 & 1 & 1 & 1 \end{bmatrix}. \tag{18.21}$$

The eroded background then contains all pixels of the background with the shape of M_2 or larger ($M_2 \subseteq \overline{G}$, Fig. 18.4b). This corresponds to all objects that have the desired shape or are smaller. Since the first erosion leaves all objects that are equal or larger, the intersection of the image eroded with M_1 with the background eroded with M_2 provides all central pixels of the objects with three consecutive pixels (Fig. 18.4e). The *hit-miss operator* is generally defined as follows:

$$\boxed{\begin{aligned} G \otimes (M_1, M_2) &= (G \ominus M_1) \cap (\overline{G} \ominus M_2) \\ &= (G \ominus M_1) \cap \overline{(G \oplus M_2)} \qquad \text{hit-miss} \\ \text{with} \quad & M_1 \cap M_2 = \emptyset. \end{aligned}} \tag{18.22}$$

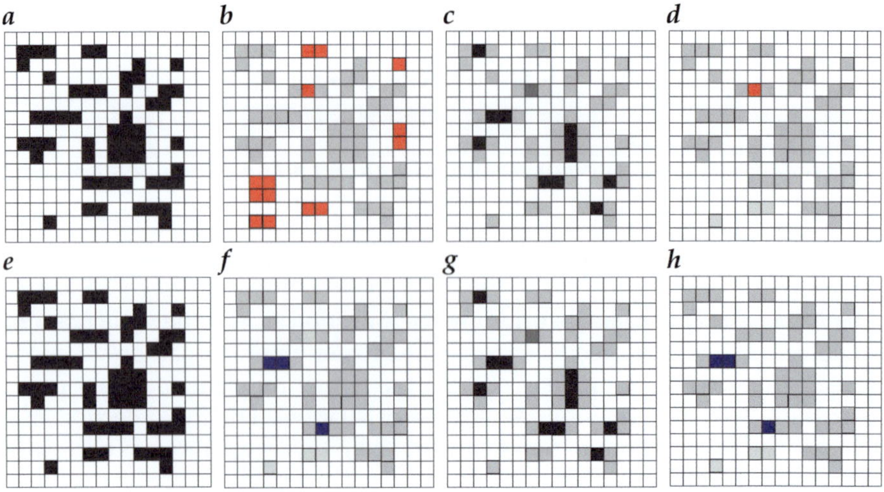

Figure 18.4: *Illustration of the processing chain of the hit-miss operator for extracting all objects that contain three (top row) or three to four (bottom row) horizontally consecutive pixels:*
***a** and **e** Original image (in all the following images, the black pixels of the original image with the value 1 are shown as light gray pixels. The dark and red pixels are the pixels with the value 1 created after the operation),*
*Top row: Processing chain for extracting objects with three consecutive horizontal pixels: **b** with a 3×5 mask Eq. (18.21) eroded background, **c** object eroded with a 1×3 mask Eq. (18.20), **d** intersection of **b** and **c** ;*
*Bottom row: Processing chain for extracting objects with 3 to 5 consecutive horizontal pixels and a 3×7 large free background: **f** with a 3×7 mask Eq. (18.23) eroded background, **g** object eroded with a 1×3 mask Eq. (18.20), **h** intersection of **f** and **g** .*

The condition $\boldsymbol{M}_1 \cap \boldsymbol{M}_2 = \emptyset$ is necessary, otherwise the hit-miss operator will result in an empty set. With the hit-miss operator, we have a flexible tool in our hands that enables the detection of objects with a specific shape. The benefits of this operator can be easily illustrated using the example of another miss mask:

$$\boldsymbol{M}_{>5} = \begin{bmatrix} 1 & 1 & 1 & 1 & 1 & 1 & 1 \\ 1 & 0 & 0 & 0 & 0 & 0 & 1 \\ 1 & 1 & 1 & 1 & 1 & 1 & 1 \end{bmatrix}. \tag{18.23}$$

Eroding the background with this mask leaves all pixels in the binary image for which the union of the mask $\boldsymbol{M}_3$ with the object is the empty set (Fig. 18.4c). This can only be the case for objects with one to five consecutive pixels. The hit-miss operator with $\boldsymbol{M}_1$ and $\boldsymbol{M}_3$ therefore returns all central pixels of the objects with 3 to 5 horizontally consecutive pixels and a 3×7 large free background around the object (Fig. 18.4h).

As the hit and miss masks of the hit-miss operator are disjoint, they can be combined into one mask. We use the following notation: The combined mask is labeled with 1 where the hit mask is one, with -1 where the miss mask is one, and with 0 if both masks are zero. Thus, the hit-miss mask for detecting objects with 3 to 5 horizontally consecutive pixels and a 3×7 free background is

$$\boldsymbol{M}_{3,5} = \begin{bmatrix} -1 & -1 & -1 & -1 & -1 & -1 & -1 \\ -1 & 0 & 1 & 1 & 1 & 0 & -1 \\ -1 & -1 & -1 & -1 & -1 & -1 & -1 \end{bmatrix}. \tag{18.24}$$

If there are no elements with 0 in a hit-miss mask, objects of exactly the size corresponding to the ones in the mask are extracted. If there are 0 elements in the mask, the 1 elements specify the minimum of the objects searched for. The union of the 1 elements with the 0 elements, on the other hand, specifies the maximum of the objects to be detected.

Another example is the following hit-miss mask; it detects isolated pixels:

$$\boldsymbol{M}_1 = \begin{bmatrix} -1 & -1 & -1 \\ -1 & 1 & -1 \\ -1 & -1 & -1 \end{bmatrix}. \tag{18.25}$$

The operation $\boldsymbol{G}/\boldsymbol{G} \otimes \boldsymbol{M}_1$ therefore removes isolated pixels from a binary image. The / symbol represents the difference set operator.

The hit-miss operator only detects certain shapes if the miss mask surrounds the hit mask. If the hit mask touches the edges of the hit-miss mask, only certain shapes at the edge of an object are detected. The hit-miss mask

$$\boldsymbol{M}_C = \begin{bmatrix} 0 & 1 & -1 \\ 1 & 1 & -1 \\ -1 & -1 & -1 \end{bmatrix} \tag{18.26}$$

detects the bottom right-hand corners of objects, for example.

18.4.3 Boundary Extraction

We can also use morphological operators to extract the edges of binary objects. This is an important operation, as the object edge is a compact yet complete representation of the geometry of an object. We can extract further shape parameters from the object edge, as we will discuss from Section 18.5.

Edge pixels have no neighbor on at least one side. An erosion operator with a mask that contains all possible neighbors therefore removes edge pixels. These masks are available for the 4-neighborhood and the 8-neighborhood:

$$\boldsymbol{M}_{b4} = \begin{bmatrix} 0 & 1 & 0 \\ 1 & 1 & 1 \\ 0 & 1 & 0 \end{bmatrix} \quad \text{and} \quad \boldsymbol{M}_{b8} = \begin{bmatrix} 1 & 1 & 1 \\ 1 & 1 & 1 \\ 1 & 1 & 1 \end{bmatrix}. \tag{18.27}$$

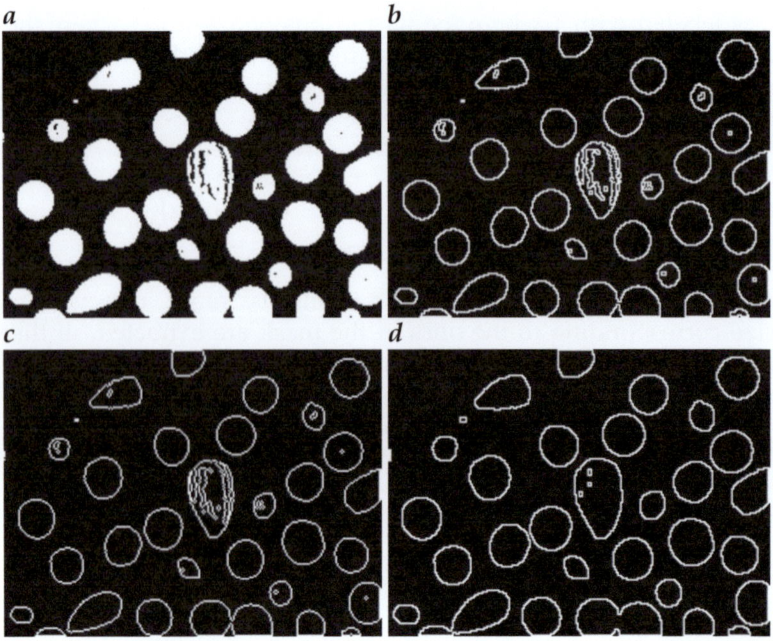

Figure 18.5: Extraction of boundaries with morphological operators: ***a*** *binary original image;* ***b*** *8-connected boundary, extracted with* $\boldsymbol{M}_{b4}$*, and* ***c*** *4-connected boundary, extracted with* $\boldsymbol{M}_{b8}$ *Eq. (18.27).* ***d*** *8-connected boundary of the background, extracted with Eq. (18.29).*

We then obtain the edge from the quantity difference (/ operator) between the object and the eroded object:

$$\begin{aligned} \partial \boldsymbol{G} &= \boldsymbol{G}/(\boldsymbol{G} \ominus \boldsymbol{M}_b) \\ &= \boldsymbol{G} \cap \overline{(\boldsymbol{G} \ominus \boldsymbol{M}_b)} \\ &= \boldsymbol{G} \cap (\overline{\boldsymbol{G}} \oplus \boldsymbol{M}_b). \end{aligned} \tag{18.28}$$

As Eq. (18.28) shows, we can also represent the boundary as the intersection of the object with the background extended by a dilation. Fig. 18.5b and c illustrates the extraction of the 4 and 8 boundary for binary objects using Eq. (18.27).

In a similar way, the boundary of the background can be determined by dilating the object and then subtracting the object (Fig. 18.5d):

$$\partial \boldsymbol{G}_B = (\boldsymbol{G} \oplus \boldsymbol{M}_b)/\boldsymbol{G}. \tag{18.29}$$

18.4.4 ♯Distance Transforms

The boundary consists of all the points that have a distance of zero to the object edge. If we apply the boundary extraction again to an object that was eroded with the mask Eq. (18.27), we obtain all points that have a distance of one to the object boundary. A recursive application of the boundary extraction procedure therefore returns the distance of all points of the object to the boundary. Such a transformation is called *distance transform* and is written as follows:

$$D = \bigcup_{n=1}^{\infty} \left[(G \ominus M_b^{n-1})/(G \ominus M_b^{n}) \cdot n \right], \tag{18.30}$$

where the operation $\cdot$ is a point-by-point multiplication of the nth boundary curve by the number n. This direct distance transformation has two serious disadvantages. Firstly, it is a slow iterative procedure. Secondly, it does not provide the desired Euclidean distance, but — depending on the selected neighborhood type — the block or checkerboard distance (see Section 6.2.3).

Fortunately, fast algorithms are available for calculating the Euclidean distance. The Euclidean distance transformation is so important because it makes morphological operations isotropic. All morphological operations suffer from the fact that the Euclidean distance is not a natural measure of a rectangular grid. The square structural elements, for example, are characterized by the checkerboard distance. Successive dilation with such a structural element mask causes, for example, the object to take on more and more the shape of a square. The Euclidean distance transformation can also be used for isotropic erosion and dilation operations. In an erosion operation with the radius r, only pixels with a distance greater than r from the edge of the object are retained. Similarly, an isotropic dilation can be realized by Euclidean distance transformation of the background and a subsequent isotropic erosion of the background.

18.5 Shape Presentation an Analysis

18.5.1 Run-length Encoding

A compact, simple and widely used representation of an image is the *run-length encoding*. It is generated with the following procedure: An image is scanned line by line. If a line contains a sequence of p identical pixels, the same number is not stored p times, but the fact that the value occurs p times is recorded (Fig. 18.6). In this way, large uniform line elements can be stored very efficiently.

The code is particularly efficient for binary images because only the two values zero and one occur. Since a sequence of zeros is always followed by a sequence of ones, there is no need to store the value of the pixel. We only need to record how often a pixel value occurs (Fig. 18.6b). The only problem is the start of the line, as a line can start with a zero or a one. This problem is solved by assuming that a line starts with a zero. However, if it actually begins with a sequence of ones, the run-length encoding starts with a zero, which indicates that there are no zeros at the beginning of the line (Fig. 18.6b).

a) ***Gray value image***

Original line (hex): 12 12 12 20 20 20 20 25 27 25 20 20 20 20 20 20

Code (hex): 82 **12** 83 **20** 2 **25 27 25** 85 **20**

b) ***Binary image***

Original line (hex): 1 1 1 1 1 0 0 0 1 1 1 0 0 1 0 0 0 0 0 1 1 1 1 1 1 1

Code (hex) 0 6 3 3 2 1 5 8

Figure 18.6: *Illustration of the run-length code for* ***a*** *a gray value image and* ***b*** *a binary image.*

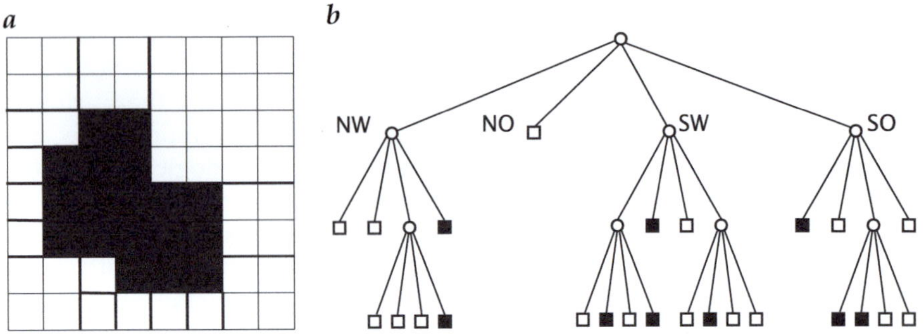

Figure 18.7: *Representation of a binary image with a region quadtree:* ***a*** *step-by-step division of the binary image into quadrants;* ***b*** *the corresponding region quadtree.*

Run-length encoding is suitable for compact image storage. It has become an integral part of some standard image formats such as the *TIFF* file format. However, run-length encoding is less suitable for the direct processing of images as it is not object-oriented. Its use is therefore focused on compact image storage, although not all image types can be successfully compressed with this scheme. Digitized grayscale images, for example, always contain noise, so the probability of a sufficiently long sequence of pixels with the same grayscale value is very low. However, large data reduction factors can be achieved for binary images and many types of computer-generated grayscale and color images.

18.5.2 Tree Structures

The run-length coding discussed in the previous section is a line-oriented representation of binary images. It therefore encodes one-dimensional data and not two-dimensional data; the two-dimensional structure is not taken into account at all. A *tree structure* (a *quadtree*), on the other hand, is based on the principle of recursive decomposition of space, as shown in Fig. 18.7 for a binary image.

First, the entire image is divided into four equally sized *quadrants*. If a quadrant does not contain a uniform region, i. e. if the quadrant is not completely filled by the object or the background, it is again divided into four sub-quadrants. The decomposition ends when only uniform quadrants occur or when the quadrants contain only one pixel.

The recursive decomposition can be represented as a data structure, which in computer science is referred to as *tree* (Fig. 18.7b). The decomposition starts at the top level of the tree, the *root*. The root corresponds to the entire binary image. It is connected via four edges to four *son nodes*, which represent the quadrants NW, NE, SW and SE from left to right. If a quadrant does not need to be further subdivided, it is shown as *end* or *leaf node* in the tree. It is designated as black if the quadrant belongs to an object and as white if it belongs to the background. Accordingly, filled or open squares are used in the representation. Nodes that are not leaf nodes are further subdivided and designated as gray. They are displayed with open circles (Fig. 18.7b).

Quadtrees can be coded, for example, using the so-called *depth-first search algorithm*, starting at the root. Only the node type needs to be saved using the symbols b (black), w (white) and g (gray).

We start the code with the value of the root node. Then we list the values of the son nodes from left to right. Each time we encounter a gray node, however, we traverse the tree first to the next lower level. This rule is applied recursively, which means that we only continue from left to right on a level of the tree when a branch has been completely traversed to the leaf. This type of tree traversal is therefore referred to as a depth-first search.

For example, the quadtree from Fig. 18.7b results in the following code:

$$ggwwgwwwbbwggwbwbbwgwbwwgbwgbbwww.$$

The code becomes easier to read if we put a left bracket whenever we go down one level in the tree and a right bracket when we go up again:

$$g(g(wwg(wwwb)b)wg(g(wbwb)bwg(wbww))g(bwg(bbww)w)).$$

However, the code is unique even without the brackets. A quadtree is a compact representation of a binary image if it contains many leaf nodes in high levels. In the worst case, however, for example in a regular checkerboard pattern, all leaf nodes are in the lowest level. The quadtree then contains as many leaf nodes as pixels and requires far more memory than the direct matrix representation of the binary image.

The region quadtree discussed here is only one of many possibilities for recursive spatial decomposition and the representation of a binary image using a tree structure. Three-dimensional binary images can be subdivided in a similar way. The 3D image is broken down into eight octants of equal size. The resulting data structure is referred to as *octree*. Quadtrees and octrees have become very important in geographical information systems and computer graphics.

Quadtrees are more suitable for encoding images than line-oriented run-length encoding. However, they are less suitable for image analysis. It

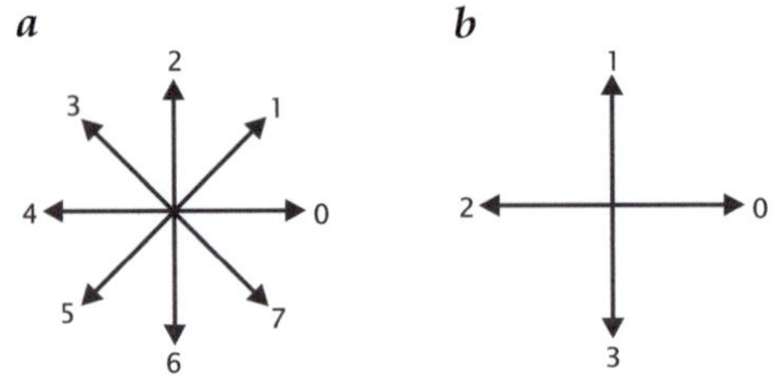

*Figure 18.8: Chain coding in **a** of an 8-bit neighborhood with 3-bit code and **b** of a 4-bit neighborhood with 2-bit code.*

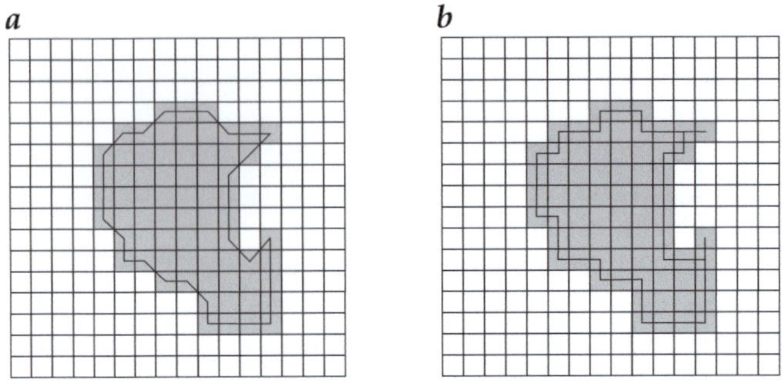

*Figure 18.9: Representation of borders with the chain code: **a** 8-neighborhood; **b** 4-neighborhood.*

is relatively difficult to carry out image analysis directly with quadtrees. Without going into further detail, this can be made plausible by the simple fact that an object shifted by one pixel in any direction leads to a completely different quad tree. Region quadtrees share with run-length coding the serious disadvantage that they are global image decomposition techniques, but do not represent objects extracted from images in a compact way.

18.5.3 Chain Code

In contrast to run-length encoding and quadtrees, the *chain code* is an object-related data structure for effectively representing the edge of a binary object on a discrete grid. Instead of storing the positions of all edge pixels, we select a starting image point and store only its coordinates. If we use an algorithm that scans the image line by line, this will be the top leftmost pixel of the object. Then we follow the edge counterclockwise.

With an 8-neighborhood, there are eight possible directions to go in, and with a 4-neighborhood, there are four possible directions to go in. These directions can be encoded with a 3-bit or a 2-bit code, as shown in Fig. 18.8a and b. Extracted edges are shown in Fig. 18.9 for an 8- and a 4-neighborhood. Chain coding has a number of obvious advantages over the matrix representation of a binary object:

First, chain coding is compact. Imagine a disk-shaped object with a diameter of R pixels. In a direct matrix representation, we need to store the *surrounding rectangle* of the object (handled in Section 18.8.4), i. e. about R^2 pixels in R^2 bits. The surrounding rectangle is the smallest rectangle enclosing the object. If we use a border of 8, the slice has about πR border points. The chain coding of the πR points can be stored in approximately $3\pi R$ bits. For objects with a diameter of more than 10 pixels, the chain code is therefore a more compact representation.

Secondly, the chain code is a *translation invariant* representation of a binary object. This property simplifies the comparison of objects. However, the chain code is neither rotation-invariant nor scale-invariant. This is a major disadvantage for object recognition. However, chain coding can be used to extract rotation-invariant parameters such as the area of an object.

Third, the chain code is a complete representation of an object or curve. Therefore, we can — at least in principle — calculate any shape feature from the chain code.

As we will discuss in Section 18.8, we can compute a number of shape parameters — including perimeter and area — more efficiently using the chain code representation than in the matrix representation of a binary image. One limitation, however, is that chain coding is a digital curve on a discrete grid and as such can only describe the edge of the object within the precision of a discrete grid.

If the object is not contiguous or has holes, more than one chain encoding is required to represent it. We must therefore store the information as to whether an edge encloses an object or a hole. Reconstructing a binary image from the chain coding is very simple. First, we can draw the outline of the object, determine the inside and outside and then use a *fill operation* for coloring.

18.6 ♯Moment-based Shape Features

18.6.1 ♯Definitions

We could continue to define parameters like the ones in the last section. However, we will get a better insight into the shape description if we take a systematic approach. In this section, we will first define *moments* for gray value and binary images and then show how this approach can be used to obtain useful shape parameters. Similarly, we will introduce Fourier descriptors in the following section.

Moments are used to describe the probability density function for gray values (Section C.2.2). Here we extend this description to two dimensions and define the moments of the gray value function $g(\boldsymbol{x})$ of an object by

$$\mu_{p,q} = \int (x_1 - \overline{x_1})^p (x_2 - \overline{x_2})^q \, g(\boldsymbol{x}) \mathrm{d}^2 x, \qquad (18.31)$$

with

$$\overline{x_i} = \int x_i g(\boldsymbol{x}) \mathrm{d}^2 x \Big/ \int g(\boldsymbol{x}) \mathrm{d}^2 x. \tag{18.32}$$

The integration is formed over the area of the object. Instead of the gray value, we can generally use any pixel-based feature with only positive values to calculate object moments. The vector $\overline{\boldsymbol{x}} = (\overline{x_1}, \overline{x_2})$ is called *center of gravity* in analogy to classical mechanics. If we consider $g(\boldsymbol{x})$ as the density $\rho(\boldsymbol{x})$ of the object, then the zero-order moment $\mu_{0,0}$ is the total mass of the object.

All moments defined in Eq. (18.31) are related to the center of gravity. Therefore, they are often referred to as *central moments*. Central moments are translation invariant and therefore well suited for describing the object shape.

For discrete binary images, the calculation of the moments is reduced to

$$\mu_{p,q} = \sum (x_1 - \overline{x_1})^p (x_2 - \overline{x_2})^q. \tag{18.33}$$

The sum includes all pixels that belong to the object. To describe the object shape, we can use moments based on either binary, gray value or feature images. The moments determined from gray value or feature images not only reflect the geometric shape of an object, but also the distribution of the features within the object. As such, they differ fundamentally from moments that originate from binary images.

18.6.2 ♯Size-invariant Moments

Shape parameters are often required that do not depend on the size of the objects. This is always the case when objects are to be compared that are viewed from different distances. If we scale an object $g(\boldsymbol{x})$ with α so that $g'(\boldsymbol{x}) = g(\boldsymbol{x}/\alpha)$, the moments are scaled as follows:

$$\mu'_{p,q} = \alpha^{p+q+2} \mu_{p,q}.$$

If we then divide the moments by the zero-order moment, $\mu_{0,0}$, we obtain *size-invariant moments*:

$$\underset{\sim}{\mu}_{p,q} = \frac{\mu_{p,q}}{\mu_{0,0}^{(p+q+2)/2}}. \tag{18.34}$$

Since the zero-order moment of a binary object describes the area of the object Eq. (18.33), the normalized moments are to be scaled with a corresponding power of the object area. Second-order moments ($p + q = 2$), for example, are scaled by the square of the area.

18.6.3 ♯Moment Tensor

The analysis of shape that goes beyond area measurement begins with second-order moments. This is because the zero-order moment provides

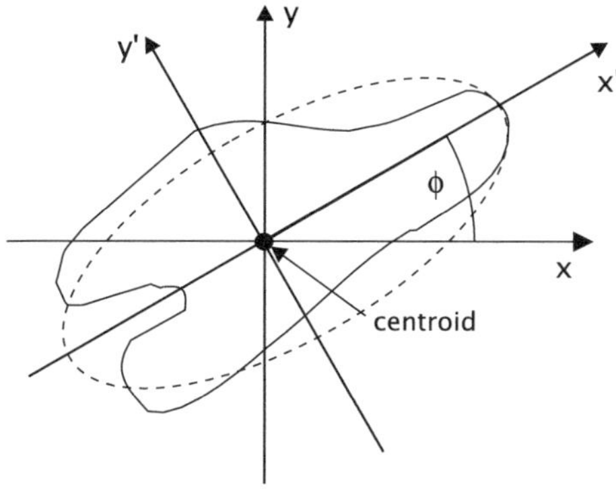

Figure 18.10: *Main axes of the inertia tensor of an object for rotation around the center of gravity.*

the area or total mass of a binary or gray value object, and the central first-order moments are zero by definition. Here too, the analogy to mechanics helps us to understand the meaning of the second-order moments $\mu_{2,0}$, $\mu_{0,2}$ and $\mu_{1,1}$. They contain terms in which the gray value function, i. e. the density of the object, is multiplied by the square of the distance from the center of gravity. We also find exactly the same terms in the inertia tensor, which was discussed in Section 13.3.8 (see Eqs. (13.26) and (13.27)). The three second-order moments form the components of the *inertia tensor* for the rotation of the object around its center of gravity:

$$J = \begin{bmatrix} \mu_{2,0} & -\mu_{1,1} \\ -\mu_{1,1} & \mu_{0,2} \end{bmatrix}. \tag{18.35}$$

Based on this analogy, we can transfer all results from Section 13.3 to the shape description with second-order moments.

The *orientation* of the object is defined as the angle between the x-axis and the axis around which the object can be rotated with minimum inertia. This is the eigenvector to the minimum eigenvalue. The object has its largest extension in this direction (Fig. 18.10). According to Eq. (13.13), the orientation angle results in

$$\theta = \frac{1}{2} \arctan \frac{2\mu_{1,1}}{\mu_{2,0} - \mu_{0,2}}. \tag{18.36}$$

As a measure for the *eccentricity* ε we can use the quantity that we have defined as a coherence measure for the local orientation Eq. (13.16):

$$\varepsilon = \frac{(\mu_{2,0} - \mu_{0,2})^2 + 4\mu_{1,1}^2}{(\mu_{2,0} + \mu_{0,2})^2}. \tag{18.37}$$

The eccentricity assumes values from 0 to 1. It is 0 for a round object and 1 for a linear object. It is therefore a better defined quantity than roundness with its non-normalized value range (Section 18.8.3).

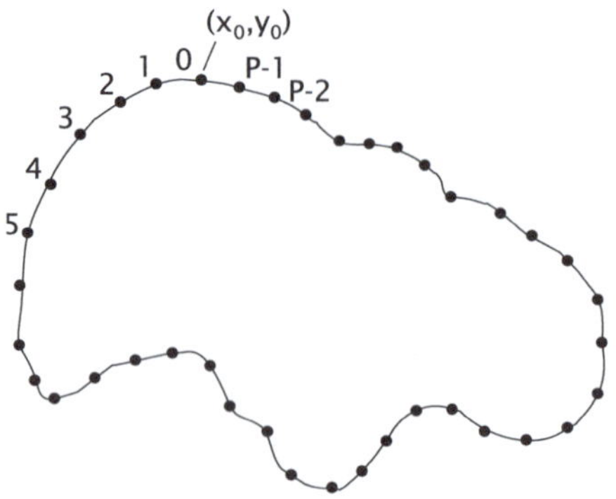

Figure 18.11: Illustration of the parameter representation of a closed curve. The parameter p is the path length from the starting point $[x_0, y_0]^T$ *in a counterclockwise direction. An equidistant sampling of the curve with P points is also shown.*

The shape description by second-order moments in the *moment tensor* essentially models the object as an *ellipse*. The combination of the three second-order moments into one tensor results in two rotation invariant terms, the trace $\mu_{2,0} + \mu_{0,2}$ of the tensor, which describes the radial distribution of features in the object, and the eccentricity Eq. (18.37), which is a measure of roundness, as well as a term that measures the orientation of the object. Moments allow a complete shape description [260]. It becomes more detailed the more moments of higher order are used.

18.7 Fourier Descriptors

18.7.1 Cartesian Fourier Descriptors

Like chain coding, Fourier descriptors only use the object edges, but do not describe a curve on a discrete grid. Fourier descriptors can be formulated for continuous or sampled curves. Consider the closed curve drawn in Fig. 18.11. We can reproduce it in parameter representation by using the path length p from a starting point $[x_0, y_0]^T$ as a parameter.

It is not easy to create a curve with evenly spaced sample points. Discrete boundary curves such as directional chains have serious disadvantages in this respect. The sampling points are not equidistant with an 8-point neighborhood, but they are with a 4-point neighborhood. Here, however, the edge is frayed because the sections of the edge curve can only run in a horizontal and vertical direction. Therefore, the circumference is too large. It is therefore not a good idea to form a continuous boundary curve from points on a regular grid. The only alternative is to extract object edges with subpixel accuracy directly from the gray value images, but this is not an easy task. Thus, the correct determination of Fourier descriptors from contours in images is still a challenging research topic.

A continuous boundary curve has the components $x(p)$ and $y(p)$. We can combine these two curves with the complex numbers $z(p) = x(p) + iy(p)$ to form a cyclic curve. If P is the circumference of the curve, then the following applies:

$$z(p + nP) = z(p) \quad n \in \mathbb{Z}. \tag{18.38}$$

A cyclic or periodic curve can be developed into a Fourier series. The coefficients of the Fourier series are determined as follows:

$$\tilde{z}_{\tilde{p}} = \frac{1}{P}\int_0^P z(p)\exp\left(\frac{-2\pi i p\tilde{p}}{P}\right)\mathrm{d}p \quad u \in \mathbb{Z}. \tag{18.39}$$

The periodic curve can be reconstructed from the Fourier coefficients:

$$z(p) = \sum_{\tilde{p}=-\infty}^{\infty} \tilde{z}_{\tilde{p}} \exp\left(\frac{2\pi i p\tilde{p}}{P}\right). \tag{18.40}$$

The coefficients $\tilde{z}_{\tilde{p}}$ are referred to as *cartesian Fourier descriptors* of the boundary curve. The geometric meaning of the descriptors is easy to recognize. The first coefficient

$$\tilde{z}_0 = \frac{1}{P}\int_0^P z(p)\mathrm{d}p = \frac{1}{P}\int_0^P x(p)\mathrm{d}p + \frac{i}{P}\int_0^P y(p)\mathrm{d}p \tag{18.41}$$

provides the center or the *centroid* of the boundary curve. The second coefficient describes a circle:

$$z_1(p) = \tilde{z}_1 \exp\left(\frac{2\pi i p}{P}\right) = r_1 \exp\left(i\varphi_1 + 2\pi i p/P\right). \tag{18.42}$$

The radius r_1 and the starting point at an angle of φ_1 are determined by $\tilde{z}_1 = r_1\exp(i\varphi_1)$. The coefficient $\tilde{z}_{-1}$ also results in a circle, which, however, is traversed in the opposite direction (with the clock hand):

$$z_{-1}(p) = r_{-1}\exp\left(i\varphi_{-1} - 2\pi i p/P\right). \tag{18.43}$$

With both complex coefficients together — a total of four parameters — an ellipse can be formed with any semi-axes a and b, any orientation ϑ and any starting angle φ_0 on the ellipse. As an example, let's consider $\varphi_1 = \varphi_{-1} = 0$. Then

$$z_1 + z_{-1} = (r_1 + r_{-1})\cdot\cos\left(\frac{2\pi p}{P}\right) + i(r_1 - r_{-1})\sin\left(\frac{2\pi p}{P}\right). \tag{18.44}$$

This is the parameter form of an ellipse whose semi-axes are parallel to the axes of the coordinate system and whose starting point is on the x-axis.

From this discussion we see that Fourier descriptors always occur in pairs. The pairing of higher order coefficients also results in an ellipse. However,

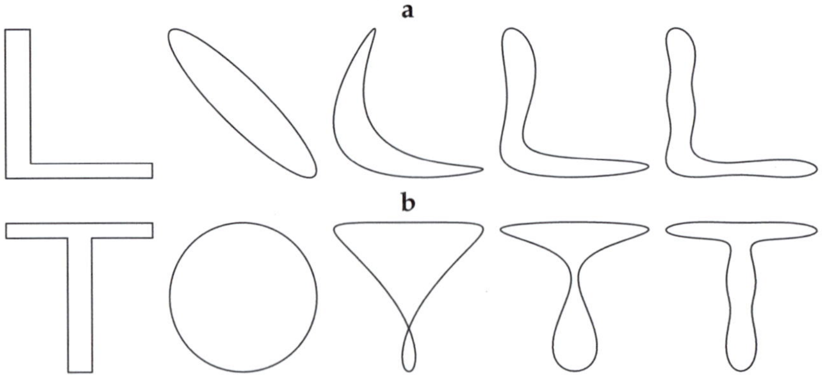

Figure 18.12: Reconstruction of the shape by an increasing number of Fourier descriptors for **a** *the letter L and* **b** *the letter T (with 2, 3, 4 and 8 Fourier descriptor pairs).*

these ellipses are traversed n times. The superposition with the base ellipse of the first pair means that the higher order Fourier descriptors contribute more and more details to the boundary curve. To illustrate this, consider the reconstruction of the letters T and L with an increasing number of Fourier descriptors (Fig. 18.12). The example shows that only a few coefficients are necessary to describe even quite complicated shapes.

Fourier descriptors can also be calculated from sampled edges z_n. If the circumference of the closed curve is P, N sampled points must be tapped at equidistant intervals P/N (Fig. 18.11). Then

$$\tilde{z}_{\tilde{n}} = \frac{1}{N}\sum_{n=0}^{N-1} z_n \exp\left(-\frac{2\pi i n\tilde{n}}{N}\right). \tag{18.45}$$

All other equations also apply to sampled edges. Sampling has only changed the Fourier series into a discrete Fourier transform with only N wavenumber coefficients. The coefficients take the values from 0 to $N-1$ or from $-N/2$ to $N/2-1$ (see also Table B.1).

18.7.2 ♯Polar Fourier Descriptors

An alternative approach to the Fourier descriptors uses a different parameterization of the boundary curve. Instead of the path length p, the angle θ is used, which the straight line through the center and the considered point on the boundary curve forms with the x-axis. We then describe the edge of the object using the *object radius* as a function of the angle. We only need a real-valued sequence $\boldsymbol{r}$ with N points at equal angular distances to describe the edge. The coefficients of the discrete Fourier transforms of this sequence,

$$\tilde{r}_{\tilde{n}} = \frac{1}{N}\sum_{n=0}^{N-1} r_n \exp\left(-\frac{2\pi i n\tilde{n}}{N}\right), \tag{18.46}$$

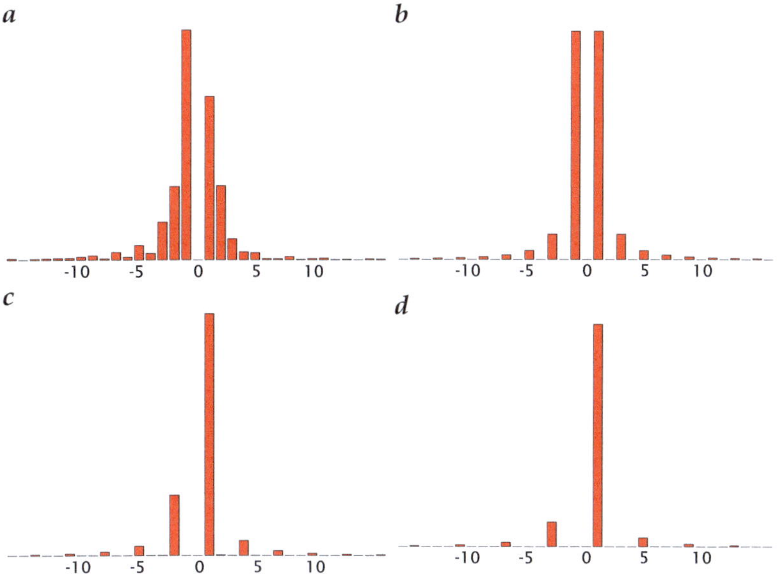

Figure 18.13: *Effect of the symmetry of an object on its Fourier descriptors **a** of the letter L, **b** of a line, **c** of a triangle and **d** of a square. The figures show the magnitudes of the Fourier descriptors from $\tilde{n} = -16$ to $\tilde{n} = 16$.*

are referred to as *polar Fourier descriptors* of the edge. Here, the first coefficient $\tilde{r}_0$ is equal to the mean radius. Polar Fourier descriptors cannot be used for all edge types. The radial edge parameterization $r(\theta)$ must be unique. Due to this important restriction, we will only consider Cartesian Fourier descriptors in the following.

18.7.3 Object Symmetries

symmetries can be easily detected in Fourier descriptors. If the contour has *m-number rotational symmetry*, only the descriptors $z_{1\pm vm}$ can be non-zero. This is shown in Fig. 18.13 using the example of the Fourier descriptors of a vertical line, a triangle and a square. If the contour is the mirror image of another contour, the Fourier descriptors are conjugate complex to each other.

Fourier descriptors can also be used for non-closed curves. To close such curves, we simply trace them forwards and then backwards again. Open curves can be easily recognized because their area is zero. From Eq. (18.48) we can then conclude that $|\tilde{z}_{-\tilde{n}}| = |\tilde{z}_{\tilde{n}}|$. If the run begins at one of the end points, even $\tilde{z}_{-\tilde{n}} = \tilde{z}_{\tilde{n}}$.

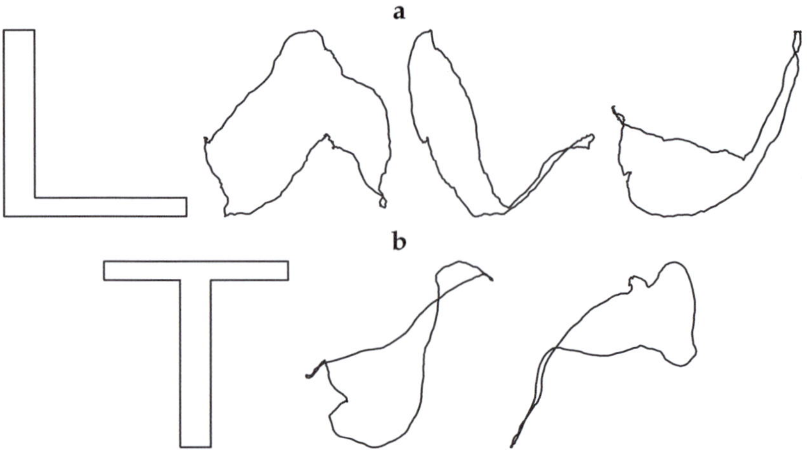

Figure 18.14: Importance of the phase for the description of the shape with Fourier descriptors: Next to the original letters (left), three random phase variations are shown with unchanged magnitude of the Fourier descriptors.

18.7.4 Invariant Shape Description

Translation Invariance
The position of the object only influences a single coefficient, namely $\tilde{z}_0$, the center point (centroid). All other coefficients are translation invariant.

Scale Invariance
If the contour is scaled with a coefficient α, all Fourier descriptors are also scaled with α. For an object with a non-zero area whose contour is traced counterclockwise, the first coefficient is always non-zero. Thus, we can relate all Fourier descriptors to $|\tilde{z}_1|$ to obtain scale invariant descriptors for the object shape. Note that the scaled descriptors still contain the full information.

Rotation Invariance
If a contour is rotated counterclockwise by the angle φ_0, the Fourier descriptor $\tilde{z}_{\tilde{n}}$ is multiplied by the phase factor $\exp(\mathrm{i}\tilde{n}\varphi_0)$ according to the shift theorem of the Fourier transform (Theorem B.3, p. 629, > R12). This simple shift property makes the construction of rotationally invariant Fourier descriptors easy. For example, we can relate the phase of all Fourier descriptors to the phase φ_1 of $\tilde{z}_1$ and subtract the phase shift $\tilde{n}\varphi_1$ from the phases of the other descriptors. The descriptors then contain the phase difference to the phase of the first descriptor and are therefore rotationally invariant except for the first one.

Thus, both Fourier descriptors (Section 18.7) and moments (Section 18.6) provide the basis for scale and rotation invariant shape parameters. The Fourier descriptors are the more versatile instrument. However, they limit the object description to the boundary curve, while moments of gray value objects also take into account the spatial distribution of the gray values in an object.

Ideally, shape parameters describe the shape of an object completely and unambiguously. This means that different shapes must not be mapped to the same feature set. The amount of Fourier descriptors provides a scale- and rotation-invariant, but incomplete shape description. Using the example of the letters T and L, Fig. 18.14 shows the different object shapes that arise when the phase is changed randomly while retaining the descriptor amounts.

Only the complete set of Fourier descriptors therefore guarantees an unambiguous shape description. Note that one degree of freedom is lost for each invariance. In the case of translation invariance, we omit the Fourier descriptor $\tilde{z}_0$ (two degrees of freedom). For the scaling invariance, we set the magnitude of the Fourier descriptor $\tilde{z}_1$ to one (one degree of freedom), and for the rotation invariance, we relate all phases to the phase of $\tilde{z}_1$ (another degree of freedom). If all three invariances are present, we are missing four degrees of freedom.

The elegance of the Fourier descriptors lies in the fact that these invariances are contained in the first two Fourier descriptors. If we normalize all other Fourier descriptors to the phase and magnitude of the second Fourier descriptor, we have a complete translation-, rotation- and scale-invariant description of the shape of an object. If we omit higher order Fourier descriptors, we can remove fine details from the shape description in a controlled manner.

A measure for shape differences is obtained if we take into account that Fourier descriptors form a complex-valued vector. The following then applies to the magnitude of the difference vector:

$$d_{zz'} = \sum_{\tilde{n}=-N/2}^{N/2-1} \left|\tilde{z}_{\tilde{n}} - \tilde{z}'_{\tilde{n}}\right|^2 . \tag{18.47}$$

Depending on which normalization we apply to the Fourier descriptors, this measure is scale and/or rotation invariant.

18.8 Elementary Shape Parameters

Having discussed the various ways of representing binary objects extracted from image data, we now turn to the question of how the shape of these objects can be described. This section is limited to elementary geometric parameters such as area and perimeter.

18.8.1 Area

One of the simplest shape parameters is the *area* A of an object. In a discrete binary image, the area is given by the number of pixels that belong to the

Table 18.1: *Calculation of the area of an object from the chain code. At the beginning, the area is set to zero, B to any value. With each step, the area and the parameter B are increased according to the value of the chain code; according to Zamperoni [360].*

4-chain code	8-chain code	Increase in area	Increase in B
0	0	$-B$	0
	1	$-(B+1/2)$	1
1	2	0	1
	3	$B+1/2$	1
2	4	B	0
	5	$B-1/2$	-1
3	6	0	-1
	7	$-(B-1/2)$	-1

object. In the matrix or pixel list representation of the object, the area calculation therefore simply means counting the pixels. At first glance, calculating the area of an object described by the chain code appears to be a complex operation. However, the opposite is the case. The area calculation from the chain code is much faster than the pixel counting, since the object edge contains only a small part of the pixels of an object and requires only two additions per edge pixel.

The algorithm works in a similar way to numerical integration. We imagine a horizontal baseline at an arbitrary vertical position in the image. Then we start the integration of the area at the top pixel of the object. We denote the distance of this point to the baseline by B. Then we follow the edge of the object and increase the area of the object according to the specifications in Table 18.1.

For example, if we move to the left (chain code 4), the area increases by B. If we move to the top left (chain code 3), the area increases by $B + 1/2$ and B must be increased by 1 because the distance between the edge pixel and the baseline has increased by 1. The area is reduced for all movements to the right. In this way, we subtract the area between the bottom edge of the object and the baseline, which was initially added too much in the area calculation during the movement to the right.

It should be noted that the edge line of the chain code lies in the *center* of the pixel. It therefore does not provide an area that is equal to the number of pixels. A line that is only one pixel wide has no area, a square of 2×2 pixels has an area of one. At the beginning, the area is set to zero.

The *area* can also be calculated directly from the Fourier descriptors using

$$A = \pi \sum_{\tilde{n}=-N/2}^{N/2-1} v|\tilde{z}_{\tilde{n}}|^2. \qquad (18.48)$$

This is a fast algorithm that requires at most as many operations as the boundary curve contains points. The Fourier descriptors have the additional advantage that we can calculate the area for a certain degree of smoothness by considering only a corresponding number of Fourier descriptors. The more Fourier descriptors we use, the more detailed the boundary curve becomes, as shown in Fig. 18.12.

18.8.2 Perimeter

The perimeter is another geometric parameter that can be easily calculated from the chain coding of the object edge. We only need to count the length of the chain coding and take into account that diagonal steps are longer by a factor of $\sqrt{2}$. The perimeter p then results in the following for an 8-neighborhood chain coding

$$p = n_e + \sqrt{2} n_o, \tag{18.49}$$

where n_e and n_o are the number of chain codes with even and odd codes respectively. The steps with odd codes are in the diagonal direction.

In contrast to the area, the perimeter is a parameter that is influenced by the noise level in the image. The higher the noise level, the more wavy and therefore longer the edge of an object becomes during segmentation. This means that caution is required when comparing edges extracted from different images. We need to make sure that the smoothness of the edges is comparable in all images.

Unfortunately, there is no simple formula to calculate the *perimeter* from Fourier descriptors, since calculating the perimeter of ellipses leads to elliptic integrals. However, a good approximation of the circumference can be obtained directly from the construction of the boundary curve with equally spaced sampling points by multiplying the number of sampling points by the mean distance between the points.

18.8.3 Circularity

Area and perimeter are the two parameters that describe the size of an object in one way or another. In order to compare objects observed from different distances, we need to use shape parameters that are independent of the size of the object in the image plane. The *roundness* is one of the simplest parameters of this kind. It is defined as follows:

$$c = \frac{p^2}{A}. \tag{18.50}$$

The roundness is a dimensionless number with a minimum value of $4\pi \approx 12.57$ for circles. It is 16 for squares and $12\sqrt{3} \approx 20.8$ for an equilateral triangle. It generally tends towards large values for elongated, long objects.

Area, perimeter and roundness are shape parameters that do not depend on the orientation of the object on the image plane. They are therefore useful for distinguishing objects independently of their orientation.

18.8.4 Bounding Box

Another simple and useful parameter for roughly describing the size of an object is the bounding box (*bounding box*). It is defined as the rectangle that is just large enough to contain all object pixels. This rectangle also provides a rough description of the shape of the object. In contrast to the area (Section 18.8.1), it is not rotation invariant. However, it becomes invariant if the object is first rotated to a standard orientation, for example via the orientation of the moment tensor (Section 18.6.3). In any case, the surrounding rectangle is a helpful feature if further object-oriented pixel processing such as extracting the object pixels is necessary.

18.9 ♯References for Further Reading

The mathematical foundations of morphological image processing, including the complete proofs of the properties mentioned in this chapter, can be found in the classic book by Serra [300]. For the practical application of morphological image processing operations, we recommend Jähne and Haußecker [161, Chapter 14] and Soille [309]. The breadth of theory and applications can be seen in conference proceedings, for example in Serra and Soille [301].

A detailed treatment of the various tree structures and their applications can be found in Samet [282, 283]. A detailed discussion of moment-based shape analysis with a focus on invariant features can be found in the monograph by Reiss [260]. Gray value-based invariant shape features are discussed by Burkhardt and Siggelkow [35].

Part III

Mathematical Foundations and Reference Material

A Notation

Contents

A.1 Principles and Problems

Due to the multidisciplinary nature of image processing — in contrast to other fields — there is no consistent and generally accepted terminology. Therefore, two problems arise:

- *Contradictory terminology:* Different disciplines often use different symbols and names for one and the same quantity.
- *Multiple meanings:* Due to the large number of terms used in image processing and related fields, the same symbol is often used for different sizes.

There is no trivial solution to this unpleasant situation. Therefore, a reasonable compromise must be found. The following guidelines were used in this book:

- *Compliance with generally accepted standards:* First and foremost, the symbols recommended by international organizations such as the International Organization for Standardization (ISO) and the International Union of Pure and Applied Chemistry (IUPAC, [145]) were used and important manuals [11, 29, 245, 248, 253, 274] and internet sources such as *Wolfram Science World* and *Wikipedia* were consulted. In addition, the notations in some important textbooks were consulted [23, 88, 234, 280]. Deviations from the commonly used symbols were only introduced for some quantities that are not uniformly designated.

B. Jähne, *Digital Image Processing and Image Acquisition*,
https://doi.org/10.1007/978-3-032-04806-6

- *Use of compact notation:* If there was a choice between different notations, the easier to understand and more compact one was generally preferred. In some cases, it also seemed useful to use more than one notation. Depending on the context, it makes more sense to write $\boldsymbol{x} = [x_1, x_2]^T$ or $\boldsymbol{x} = [x, y]^T$.
- *Use of a symbol for several quantities:* One and the same symbol can have different meanings. This is not as confusing as it seems at first glance, as the meaning is clear from the context. Symbols with multiple meanings are only used in such cases.

In order to familiarize readers approaching image processing from different disciplines with the notation used in this book, we would like to point out some important deviations from the usual usage in some disciplines.

Vectors, Scalar Product and Outer Product

In this book, vectors are always understood as column vectors and marked with bold italic lower case letters. Example:

$$\boldsymbol{x} = \begin{bmatrix} x_1 \\ x_2 \\ x_3 \end{bmatrix} = [x_1, x_2, x_3]^T. \tag{A.1}$$

This means that vectors are regarded as special cases of matrices with only one column. The superscript T means a matrix transposition. Vector operations such as the *scalar product* (*inner product*) or the *outer product* can be understood as *matrix multiplications* and written as such:

$$\text{Scalar product} \quad \boldsymbol{x}^T \boldsymbol{y} = [x_1, x_2, x_3] \begin{bmatrix} y_1 \\ y_2 \\ y_3 \end{bmatrix} = x_1 y_1 + x_2 y_2 + x_3 y_3, \tag{A.2}$$

$$\text{Outer product} \quad \boldsymbol{x}\boldsymbol{y}^T = \begin{bmatrix} x_1 \\ x_2 \\ x_3 \end{bmatrix} [y_1, y_2, y_3] = \begin{bmatrix} x_1 y_1 & x_1 y_2 & x_2 y_3 \\ x_2 y_1 & x_2 y_2 & x_2 y_3 \\ x_3 y_1 & x_3 y_2 & x_3 y_3 \end{bmatrix}. \tag{A.3}$$

Time Series and Image Matrices

According to [234] $x[n]$ is the generally used notation for elements of a *time series*. However, this is too cumbersome for multidimensional signals: $g[k][m][n]$. Therefore, the more compact notation with indices is used (x_n and $g_{k,m,n}$).

Derivatives and Partial Derivatives

For a compact notation of derivatives and partial derivatives, the point notation introduced by Newton is extended where appropriate:

$$\frac{\mathrm{d}g}{\mathrm{d}x} = \dot{g}, \quad \frac{\partial g}{\partial x} = \dot{g}_x \quad \text{and} \quad \frac{\partial^2 g}{\partial x^2} = \ddot{g}_{xx}. \tag{A.4}$$

The dot on the symbol allows you to clearly differentiate between derivatives and indices for components.

Imaginary Unit
We denote the imaginary unit by i. In electrical engineering and related fields, the letter j is often used for this.

Wavenumber
Unfortunately, there is no generally accepted notation for the *wavenumber*: A smaller number of authors [32, 245, 247] use the *circular wavenumber*, which is also common in physics

$$k = \frac{2\pi}{\lambda}, \tag{A.5}$$

while most [10, 23, 88, 110, 113, 147, 334, 339] use the wavenumber defined without the factor 2π:

$$\tilde{\nu} = \frac{1}{\lambda}, \tag{A.6}$$

which is more commonly used in spectroscopy and mathematics.

In fact, the second definition is the clearer and better choice for image processing. The continuous Fourier transform can be written more simply and without prefactors (Appendix B.1.5) and the equations look more similar to those of the discrete Fourier transform (Section 6.3). There is little agreement on the symbols used for the wavenumber. Bracewell [23] takes s in the one-dimensional case and u, v in the two-dimensional case, Jain et al. [147] ξ_1, ξ_1, Barrett and Myers [10], Easton Jr. [67] ξ, η, ζ and Goodman [113] f_X, f_Y.

This book has therefore taken the liberty of using a new symbol for the wavenumber that is as appropriate as possible and does not conflict with the notation for other variables in image processing. The wavenumber is denoted by the symbol κ and the wavenumber vector in multidimensional with

$$\boldsymbol{\kappa} = [\kappa_1, \kappa_2, \ldots]^T \quad \text{with} \quad |\boldsymbol{\kappa}| = \kappa = \frac{1}{\lambda}. \tag{A.7}$$

The discrete wavenumber is denoted by $\tilde{n}$ and the discrete wave vector by $\tilde{\boldsymbol{n}}$. This makes it immediately clear that it is a quantity in Fourier space.

Fourier Transform
For the continuous Fourier transform, it is common in mathematics and signal processing to denote the functions in space or time with small letters ($g(\boldsymbol{x})$) and in Fourier space with capital letters ($G(\boldsymbol{\kappa})$) [23, 29, 245]. This notation is not useful in image processing because continuous signals ($f(\boldsymbol{x})$) and matrices ($\boldsymbol{G}$) have to be transformed. Therefore, in this book, as is common in theoretical physics [22], quantities in Fourier space are marked with a tilde ($\tilde{g}(\boldsymbol{\kappa})$ and $\tilde{\boldsymbol{G}}$).

A.2 Lists of Symbols

A.2.1 Typography

Typography	Description
e, i, d, w	Upright symbols have a special meaning; examples: e for the base of the natural logarithm, $\mathrm{i} = \sqrt{-1}$, differential operator: $\mathrm{d}g$, $\mathrm{w} = \mathrm{e}^{2\pi\mathrm{i}}$
$a, b, \ldots$	Slanted (but not boldface): *scalar*, real-valued or integer
$\mathfrak{a}, \mathfrak{b}, \ldots$	fractal: complex-valued scalar
$\boldsymbol{g}, \boldsymbol{\kappa}, \boldsymbol{u}, \boldsymbol{x}, \ldots$	Slanted bold lower case: *vector*, i. e. a coordinate vector, a time series, a row of an image, . . .
$\boldsymbol{G}, \boldsymbol{H}, \boldsymbol{J}, \ldots$	Slanted bold capital letters: *matrix*, *tensor*, a discrete image, a 2D convolution mask, a structure tensor; also used for signals of any dimension including 1D
$\mathcal{B}, \mathcal{R}, \mathcal{F}, \ldots$	Calligraphic: representation-independent *operator*
$\mathbb{N}, \mathbb{Z}, \mathbb{R}, \mathbb{C}$	Used for sets of numbers, points or other quantities

A.2.2 Accents

Accents	Description
$\hat{\boldsymbol{k}}, \hat{\boldsymbol{n}}, \ldots$	The hat symbol denotes a *unit vector*.
$\underset{\sim}{k}, \underset{\sim}{\boldsymbol{k}}, \underset{\sim}{\boldsymbol{x}}, \ldots$	Underlining with a tilde denotes a dimensionless normalized quantity (of a dimensioned quantity).
$\breve{G}, \breve{g}(k, t), \ldots$	A breve denotes a partially Fourier-transformed variable
$\tilde{G}, \tilde{g}(k), \ldots$	A tilde denotes a quantity in *Fourier space*

A.2.3 Superscripts

Superscripts	Description
$\boldsymbol{A}^{-1}, \boldsymbol{A}^{-g}$	Inverse of the square matrix $\boldsymbol{A}$; generalized inverse of the (non-square) matrix $\boldsymbol{A}$
$\boldsymbol{A}^T, \boldsymbol{a}^T$	Transpose of a matrix or a vector
$\boldsymbol{A}^*, \boldsymbol{a}^*$	Conjugated complex matrix/vector
$\boldsymbol{a}^T\boldsymbol{b}, \langle \boldsymbol{a} \mid \boldsymbol{b} \rangle$	Scalar product of two real vectors
$\boldsymbol{a}^{*T}\boldsymbol{b}, \langle \boldsymbol{a} \mid \boldsymbol{b} \rangle$	Scalar product of two complex vectors (the first vector is conjugated)
$\mathfrak{a}^*$	Conjugate of a complex number

A.2.4 Symbols for Indices and Ranges

Indices	Description
K, L, M, N	Number of elements of a discrete image in t-, z-, y- and x-direction
k, l, m, n	Indexing the elements of a discrete image in t-, z-, y- and x-direction
$\boldsymbol{N}$	Vectorial number of elements of a D-dimensional signal in spatial domain with a D-dimensional vector
$\boldsymbol{n}$	Vectorial indexing of the elements of a D-dimensional signal in spatial domain with a D-dimensional vector
$\tilde{k}, \tilde{l}, \tilde{m}, \tilde{n}$	Indexing the elements of a discrete image in Fourier space in ν-, κ_z-, κ_y- and κ_x-direction
$\tilde{\boldsymbol{n}}$	Vectorial indexing of the elements of a D-dimensional signal in Fourier space with a D-dimensional vector
P	Number of components in a multi-channel image; dimension of the feature space, number of components, pyramid levels or measured values
Q	Number of quantization levels, number of classes for classification, number of regression parameters
R	Size of the mask of a neighborhood operator
D	Dimension of a signal
p, q, r, d	Corresponding indices for P, Q, R, D

A.2.5 Meaning of Indices

Index	Description
g_n	Element n of the vector $\boldsymbol{g}$
g_{mn}	Element m, n of the matrix $\boldsymbol{G}$
$g_{\boldsymbol{n}}$	Element $\boldsymbol{n} = [n_1, n_2, \ldots, n_D]$ of a multidimensional discrete signal $\boldsymbol{G}$
$\dot{g}_p$	Compact notation for a first-order partial derivative of a continuous function g in the direction p: $\partial g(\boldsymbol{x})/\partial x_p$
$\ddot{g}_{pq}$	Compact notation for a second-order partial derivative of a continuous function g in the directions p and q: $\partial^2 g(\boldsymbol{x})/(\partial x_p \partial x_q)$

A.2.6 Functions

Function	Description
$\cos(x)$	cosine function
$\exp(x)$	exponential function
$\mathrm{ld}(x)$	logarithm to base 2
$\ln(x)$	logarithm to the base e (natural logarithm)
$\log(x)$	logarithm to base 10
$\sin(x)$	sine function
$\mathrm{sinc}(x)$	sinc function: $\mathrm{sinc}(x) = \sin(\pi x)/(\pi x)$
$\delta(x)$	delta distribution $\delta(x) = \begin{cases} \infty & x = 0 \\ 0 & \text{otherwise} \end{cases}, \int\limits_{-\infty}^{\infty} \delta(x)\mathrm{d}x = 1$
$\Pi(x)$	box function $\Pi(x) = \begin{cases} 1 & \lvert x\rvert < 1/2 \\ 0 & \text{otherwise} \end{cases}$
$\Lambda(x)$	triangular function $\Lambda(x) = \begin{cases} 1 - \lvert x\rvert & \lvert x\rvert < 1 \\ 0 & \text{otherwise} \end{cases}$
$\Theta(x)$	unit-step function $\Theta(x) = \begin{cases} 0 & x < 0 \\ 1 & x > 0 \end{cases}$
$\mathrm{sgn}(x)$	sign function $\mathrm{sgn}(x) = \begin{cases} 0 & x < -1 \\ 1 & x > 1 \end{cases}$
$\det(\boldsymbol{G})$	Determinant of a square matrix
$\mathrm{diag}(\boldsymbol{G})$	Vector with the diagonal elements of a square matrix
$\mathrm{trace}(\boldsymbol{G})$	Trace (sum of the diagonal elements) of a square matrix
$\mathrm{cov}(\boldsymbol{g})$	Covariance matrix of a vectorial random variable
$E(g)$, $\mathrm{var}(\boldsymbol{G})$	Expected value (mean) and variance
$\mathrm{Re}\, z$	Real part of a complex number z
$\mathrm{Im}\, z$	Imaginary part of a complex number z
$\lvert z\rvert$	Magnitude of a complex number $z = re^{i\varphi}$: $\lvert z\rvert = r$
$\arg(z)$	Argument of a complex number $z = re^{i\varphi}$, which is the counter-clockwise angle of a complex number z to the real axis: $\arg z = \varphi$

A.2.7 Image Operators

Image operator	Description
$\cdot$	Point-by-point multiplication of two images
$*$	Convolution, also multidimensional
$\star$	Correlation
$\ominus, \oplus$	Morphological erosion and dilation
$\circ, \bullet$	Morphological opening and closing
$\otimes$	Morphological hit-miss operator
$\vee, \wedge$	Boolean *Or* and *And*
$\cup, \cap$	Union and intersection of sets
$\subset, \subseteq$	Set is subset of, subset of or equal to
$\circlearrowleft$	Shift operator
$\downarrow_s$	Sampling or reduction operator: only every sth element is taken
$\uparrow_s$	Expansion or interpolation operator: increase the number of sampled values by the factor s; the new points are interpolated from the original ones

A.2.8 Greek and Latin Symbols

Symbol	Definition, [unit]	Meaning
Greek characters		
α	$[\mathrm{m}^{-1}]$	absorption coefficient
β	$[\mathrm{m}^{-1}]$	scattering coefficient
γ	$[\mathrm{m}^{-1}]$	extinction coefficient, sum of absorption and scattering coefficient
$\delta(x)$, δ_n		δ distribution) and discrete delta function Eq. (B.44)
Δ	$\sum\limits_{d=1}^{D} \dfrac{\partial^2}{\partial x_d^2}$	Laplacian operator
$\Delta x, \Delta k, \ldots$	sampling distance	Specifies the distance between sampling points in spatial or Fourier space.
$\Box x, \Box k, \ldots$	sampling size	Specifies the extent of a digitized signal in spatial or Fourier space.
ϵ	[1]	specific emissivity
ϵ	$[(\mathrm{mol/l})^{-1}\,\mathrm{cm}^{-1}]$	molar absorption coefficient

continues on the next page

Symbol	Definition, [unit]	Meaning
Continued from the previous page		
κ	$1/\lambda$, [m^{-1}]	wavenumber (number of wavelengths per unit length); also in radial direction
$\boldsymbol{\kappa}$	[m^{-1}]	wavenumber vector
$\underset{\sim}{\kappa}$	$2\kappa\Delta x$	wavenumber normalized to the maximum possible wavenumber (Nyquist wavenumber)
$\boldsymbol{\nabla}$	$\left[\frac{\partial}{\partial x_1}, \ldots, \frac{\partial}{\partial x_D}\right]^T$	D dimensional gradient
λ	[m]	wavelength
ν	[s^{-1}], [Hz] (Hertz)	frequency
$\boldsymbol{\nabla}\times$		rotation operator
η	[1]	quantum efficiency of an image sensor
φ	[rad], [°]	phase shift or phase difference
ϕ	[rad], [°]	azimuth angle
Φ	[J/s], [W], [s^{-1}], [lm]	radiant or luminous flux
Φ_e, Φ_p	[W], [s^{-1}], [lm]	energy-based and photon-based radiant flux
$\rho, \rho_\parallel, \rho_\perp$	[1]	reflectivity for unpolarized, parallel or perpendicular polarized light
ρ	[kg/m^3]	density
σ_x		standard deviation of the random variable x
σ	$5.6696 \cdot 10^{-8}$Wm^{-2}K^{-4}	Stefan-Boltzmann constant
$\sigma_a, \sigma_l, \sigma_s$	[cm^2]	absorption cross-section, luminescence cross-section, scattering cross-section
τ	[1]	transmissivity
τ	[s]	time constant
θ	[rad], [°]	incidence angle, polar or zenith angle (to the z-axis)
θ_b	[rad], [°]	Brewster angle
θ_c	[rad], [°]	critical angle (for total internal reflection)
Ω	[sr] (Steradiant)	solid angle
ω	$\omega = 2\pi\nu$, [rad s^{-1}]	circular frequency
Latin characters		
A	[m^2]	surface area
A	[1]	absorbance or optical density
$a, \boldsymbol{a}$	$\boldsymbol{a} = \boldsymbol{x}_{tt} = \boldsymbol{u}_t$, [m/s^2]	acceleration
		continues on the next page

Symbol	Definition, [unit]	Meaning
Continued from the previous page		
a	[m]	diameter (aperture) of a lens
$\tilde{b}(\underset{\sim}{\boldsymbol{\kappa}})$		transfer function of a binomial mask
$\boldsymbol{B}$	[Vs/m^2]	magnetic field
$\boldsymbol{B}$		binomial mask
$\mathcal{B}$		binomial convolution operator
c	[m/s]	phase velocity of a wave
c_0	299 792 458 ms^{-1}	speed of light in a vacuum
c_g	[m/s]	group velocity of a wave
$\mathbb{C}$		set of complex numbers
d, d'	[m]	axial distance in object and image space to the corresponding focal plane
$\tilde{d}(\underset{\sim}{\boldsymbol{\kappa}})$		transfer function of $\mathcal{D}$
D	[m^2/s]	diffusion constant
$\boldsymbol{D}$		mask of a first-order difference filter
$\mathcal{D}$		first-order difference operator
e	$1.6022 \cdot 10^{-19}$ As	elementary electric charge
e	2.718281 ...	base for the natural logarithm
E	[W/m^2], [lm/m^2], [lx]	irradiance or illuminance
$\boldsymbol{E}$	[V/m]	electric field
$\hat{\boldsymbol{e}}$	[1]	eigenvector of length one of a matrix
f, f_e	[m]	(effective) focal length of an optical system
f_b, f_f	[m]	rear and front focal length
$\boldsymbol{f}$		optical flow
$\boldsymbol{f}$		feature vector
F	[N] (Newton)	force
$\boldsymbol{G}$		image matrix
$\boldsymbol{H}$		general filter mask
h	$6.6262 \cdot 10^{-34}$ Js	Planck's constant (quantum of action)
$\hbar$	$h/(2\pi)$ [Js]	
i	$\sqrt{-1}$	imaginary unit
I	[W/sr], [lm/sr]	radiant or luminous intensity
I	[A]	electric current
$\boldsymbol{I}$		unit matrix
$\mathcal{I}$		identity operator
$\boldsymbol{J}$		structure tensor, inertia tensor
k	f/a, [1]	f-number, aperture relative to focal length of an optical system

continues on the next page

Symbol	Definition, [unit]	Meaning
Continued from the previous page		
k_B	$\approx 0,86173 \cdot 10^{-4}$ eV/K	Boltzmann constant
k	$2\pi/\lambda$, [rad m^{-1}]	circular wavenumber
$\boldsymbol{k}$	[rad m^{-1}]	circular wavenumber vector
K_q	[l/mol]	quench constant
K_I	[1]	equilibrium constant of a pH indicator
L	[W/(m^{2}sr)], [1/(m^{2}sr)], [lm/(m^{2}sr)], [cd/m^2]	radiance or luminance
$\boldsymbol{L}$		mask of the Laplacian operator
$\mathcal{L}$		Laplacian operator
m	[kg]	mass
m_l, m_a	[1]	image scale (magnification) of an optical system, in lateral and axial direction
M	[W/m^2], [1/(s m^2)]	excitance, outgoing radiative flux density
M_e	[W/m^2]	energy-based excitance
M_p	[1/(s m^2)]	photon-based excitance
n	[1]	index of refraction
NA	[1]	numerical aperture of an optical system
$\hat{\boldsymbol{n}}$	[1]	surface normal, unit vector perpendicular to a surface
$\mathbb{N}$		set of natural numbers: $\{0, 1, 2, 3, \ldots\}$
$\mathbb{N}^*$		set of natural numbers without zero: $\{1, 2, 3, \ldots\}$
p	[kg m/s], [N s]	momentum
p	[N/m^2]	pressure
pH	[1]	pH value, negative logarithm of the proton concentration
q	[As], [lm s]	electric charge
Q	[Ws] = [J], [lm s], [1]	radiant energy radiometric, photometric and number of photons
r	[m]	radius
$\boldsymbol{r}_{m,n}$	$\boldsymbol{r}_{m,n} = [m\Delta x, n\Delta y]^T$	translation vector of a grid in the spatial domain
$\tilde{\boldsymbol{r}}_{\tilde{m},\tilde{n}}$	$\tilde{\boldsymbol{r}}_{\tilde{m},\tilde{n}} = [\tilde{m}/\Delta x, \tilde{n}/\Delta y]^T$	translation vector of the reciprocal lattice in Fourier space
R	Φ/s, [A/W]	responsivity of a radiation detector
$\boldsymbol{R}$		mask of the box (rectangular) filter
$\mathbb{R}$		set of real numbers
s	[A]	sensor signal

continues on the next page

Symbol	Definition, [unit]	Meaning
Continued from the previous page		
s	$[\mathrm{cm}^{-1}]$, $[\mathrm{mol/cm}^2]$	column density
T	[K]	absolute temperature
T	[s]	period of a wave or oscillation
t	[1]	transmissivity
t	[s]	time
u	[m/s]	velocity
$\boldsymbol{u}$	[m/s]	velocity vector
u, u'	[m]	diameter of the blur circle in object and image space
U	[V]	voltage, electrical potential
V	$[\mathrm{m}^3]$	volume
$V(\lambda)$	[lm/W]	spectral sensitivity of vision in bright light
$V'(\lambda)$	[lm/W]	spectral sensitivity of vision in the dark
w	$\mathrm{e}^{2\pi\mathrm{i}}$	
w_N	$\exp(2\pi\mathrm{i}/N)$	
$\boldsymbol{x}'$	$[x', y']^T, [x'_1, x'_2]^T$	image coordinates
$\boldsymbol{x}$	$[x, y, z]^T, [x_1, x_2, x_3]^T$	world coordinates
z, z'	[m]	axial distance in object and image space to the corresponding principle points (object and image distance)
$\mathbb{Z}$		set of integers $\{\ldots, -2, -1, 0, 1, 2, \ldots\}$
$\mathbb{Z}_+ = \mathbb{N}$		set of non-negative integers $\{0, 1, 2, \ldots\}$
$\mathbb{Z}_N$		set $\{0, 1, 2, \ldots, N-1\}$

B Vector Spaces and Unitary Transformations

Contents

B.1 Wave Number Space and Fourier Transform

B.1.1 Vector Spaces and Orthogonal Basis Functions

In the spatial representation of digital images, each image can be thought of as being composed of basic images in which only one pixel has the value one, while all other pixels have the value zero (Fig. 6.7). We denote such a *base image* with a value of one in row m and column n with

$$ {}^{m,n}\boldsymbol{P}: \quad {}^{m,n}p_{m',n'} = \begin{cases} 1 & m = m' \wedge n = n' \\ 0 & \text{otherwise.} \end{cases} \tag{B.1} $$

This allows any image to be assembled from the base images in Eq. (B.1):

$$ \boldsymbol{G} = \sum_{m=0}^{M-1}\sum_{n=0}^{N-1} g_{m,n}\, {}^{m,n}\boldsymbol{P}, \tag{B.2} $$

where $g_{m,n}$ denotes the gray value at position m, n.

B. Jähne, *Digital Image Processing and Image Acquisition*, https://doi.org/10.1007/978-3-032-04806-6

We can easily convince ourselves that the MN basis images ${}^{m,n}\boldsymbol{P}$ form an *orthonormal basis*. We need an *inner product*, which corresponds to the *scalar product* for vectors. The inner product of two images $\boldsymbol{G}$ and $\boldsymbol{H}$ is defined as

$$\langle \boldsymbol{G} | \boldsymbol{H} \rangle = \sum_{m=0}^{M-1} \sum_{n=0}^{N-1} g_{m,n} h_{m,n}. \tag{B.3}$$

The notation for the inner product is borrowed from quantum mechanics to distinguish it from matrix multiplication, which is denoted by $\boldsymbol{GH}$. From Eq. (B.3) we can directly derive the orthonormality relation for the basis images ${}^{m,n}\boldsymbol{P}$:

$$\sum_{m=0}^{M-1} \sum_{n=0}^{N-1} {}^{m',n'}p_{m,n} \, {}^{m'',n''}p_{m,n} = \delta_{m'-m''} \delta_{n'-n''}. \tag{B.4}$$

The equation says that the inner product of two base images is zero if two different base images are used. In contrast, the scalar product of a base image with itself is one. The MN basis images thus span a $M \times N$-dimensional vector space over the body of real numbers.

From the variety of possible other decompositions of multidimensional signals, only one other has gained outstanding importance for image processing besides the spatial representation. Its basic images are periodic spatial patterns. The transformation between the spatial and wavenumber representation is the *Fourier transform*. This requires the expansion from real to complex numbers in order to be able to represent periodic structures with amplitude and phase (Sections 2.2.1 and 6.3.1). We understand a real image as a complex image with a vanishing imaginary part.

The following sections of this appendix describe the definition and basic properties of continuous and discrete Fourier transforms and, in the second part, introduce the concept of unitary transforms and further examples of transforms that are important for image and signal processing.

B.1.2 One-dimensional Fourier Transform

First we consider the *one-dimensional Fourier transform*.

Definition B.1 (1D-FT) *Let* $g(x) : \mathbb{R} \mapsto \mathbb{C}$ *a square-integrable function, that is*

$$\int_{-\infty}^{\infty} |g(x)|^2 \, \mathrm{d}x < \infty. \tag{B.5}$$

Then the Fourier transform of $g(x)$, $\tilde{g}(k)$, *is given by*

$$\boxed{\tilde{g}(k) = \int_{-\infty}^{\infty} g(x) \exp(-2\pi \mathrm{i} kx) \, \mathrm{d}x, \quad \text{1D Fourier transform.}} \tag{B.6}$$

The Fourier transform maps the vector space of the square-integrable functions onto itself. The inverse Fourier transform of $\tilde{g}(k)$ results in the original function $g(x)$:

$$\boxed{g(x) = \int_{-\infty}^{\infty} \tilde{g}(k) \exp(2\pi \mathrm{i} kx)\, \mathrm{d}k, \quad \textit{inverse 1D Fourier transform.}} \tag{B.7}$$

We can write the Fourier transform more compactly using the abbreviation

$$\mathrm{w} = \mathrm{e}^{2\pi \mathrm{i}} \tag{B.8}$$

and by rewriting the integral as a *scalar product*:

$$\langle g(x) \,|\, h(x) \rangle = \int_{-\infty}^{\infty} g^*(x) h(x) \mathrm{d}x. \tag{B.9}$$

The symbol * denotes the conjugate complex. This results in the shorthand notation

$$\tilde{g}(k) = \left\langle \mathrm{w}^{kx} \,\middle|\, g(x) \right\rangle. \tag{B.10}$$

We can imagine the function w^t as a vector that rotates counterclockwise on the *unit circle* in the *complex plane*. The variable t indicates the number of revolutions.

Sometimes it is practical to write the Fourier transform as an operator:

$$\tilde{g} = \mathcal{F} g \quad \text{and} \quad g = \mathcal{F}^{-1} \tilde{g}. \tag{B.11}$$

A function and its transform, a *Fourier transform pair*, is specified by the notation $g(x) \circ\!\!-\!\!\bullet\, \tilde{g}(k)$.

In the *discrete Fourier transform* (*DFT*), the wavenumber is an integer that indicates how many wavelengths fit into an interval of length N.

Definition B.2 (1D DFT) *The DFT maps an ordered N-tuple of complex numbers g_n, the* complex-valued column vector

$$\boldsymbol{g} = \left[g_0, g_1, \ldots, g_{N-1}\right]^T, \tag{B.12}$$

to another vector $\tilde{\boldsymbol{g}}$ of a vector space with the same dimension N:

$$\boxed{\tilde{g}_{\tilde{n}} = \frac{1}{N} \sum_{n=0}^{N-1} g_n \exp\left(-\frac{2\pi \mathrm{i} n\tilde{n}}{N}\right), \quad 0 \le \tilde{n} < N.} \tag{B.13}$$

The inverse transformation is given by

$$\boxed{g_n = \sum_{\tilde{n}=0}^{N-1} \tilde{g}_{\tilde{n}} \exp\left(\frac{2\pi \mathrm{i} n\tilde{n}}{N}\right), \quad 0 \le n < N.} \tag{B.14}$$

As with the continuous Fourier transform, it is useful to use an abbreviation for the kernel of the DFT Eq. (B.8):

$$\mathrm{w}_N = \mathrm{w}^{1/N} = \exp\left(\frac{2\pi \mathrm{i}}{N}\right). \tag{B.15}$$

The DFT can also be understood as the scalar product of the vector $\boldsymbol{g}$ with a set of N orthonormal basis vectors:

$$\boldsymbol{b}_{\tilde{n}} = \frac{1}{\sqrt{N}} \left[\mathrm{w}_N^0, \mathrm{w}_N^{\tilde{n}}, \mathrm{w}_N^{2\tilde{n}}, \ldots, \mathrm{w}_N^{(N-1)\tilde{n}}\right]^T. \tag{B.16}$$

Then

$$\tilde{g}_{\tilde{n}} = \frac{1}{N} \sum_{n=0}^{N-1} \mathrm{w}_N^{-n\tilde{n}} g_n = \frac{1}{\sqrt{N}} \left\langle \boldsymbol{b}_{\tilde{n}} \middle| \boldsymbol{g} \right\rangle = \frac{1}{\sqrt{N}} \boldsymbol{b}_{\tilde{n}}^{*T} \boldsymbol{g}. \tag{B.17}$$

Note the second compact notation of the *scalar product* with the superscript T on the right-hand side of the equation, which includes the formation of the conjugate complex of the first vector. Equation (B.17) means that the coefficient $\tilde{g}_{\tilde{n}}$ is obtained in Fourier space by projecting the vector $\boldsymbol{g}$ onto the basis vector $\boldsymbol{b}_{\tilde{n}}$. The N basis vectors $\boldsymbol{b}_{\tilde{n}}$ are orthogonal to each other and have a magnitude of one:

$$\boldsymbol{b}_{\tilde{n}}^{*T} \boldsymbol{b}_{\tilde{n}'} = \delta_{\tilde{n}-\tilde{n}'} = \begin{cases} 1 & \tilde{n} = \tilde{n}' \\ 0 & \text{otherwise.} \end{cases} \tag{B.18}$$

Therefore, the $\boldsymbol{b}_{\tilde{n}}$ form an orthonormal basis of the vector space, i. e. each vector of the vector space can be represented as a linear combination of these basis vectors. The DFT calculates the projections of the vector $\boldsymbol{g}$ onto all basis vectors of the Fourier space, i. e. the components of $\boldsymbol{g}$ in the direction of the basis vectors.

The real and imaginary parts of the base vectors are sampled sine and cosine functions of different wavelengths (Fig. 6.9). The index ν denotes how often the wavelength of the function fits into the interval $[0, N]$. The base vector for wavenumber zero, $\boldsymbol{b}_0$, is a real, constant vector. The projection onto this vector provides the mean value of the elements of the vector $\boldsymbol{g}$ multiplied by $\sqrt{N}$.

In addition to the continuous and discrete Fourier transform, there are two other forms that are often used in the natural sciences and engineering. The *Fourier series* (*FS*) maps a function in a finite interval $[0, \Delta x]$ to an infinite sequence of complex coefficients. Conversely, the *infinite discrete Fourier transform* (*IDFT*) maps an infinite sequence of complex numbers to a finite interval $[0, 1/\Delta x]$ in Fourier space. It is therefore instructive to compare the DFT with these transformations (Table B.1).

Table B.1: *Comparison of the continuous Fourier transform (FT), the Fourier series (FS), the infinite discrete Fourier transform (IDFT) and the discrete Fourier transform (DFT) in one dimension.*

Type	Forward transform	Backward transform
FT: $x, k \in \mathbb{R}$	$\tilde{g}(k) = \int\limits_{-\infty}^{\infty} g(x) \mathrm{w}^{-kx} \mathrm{d}x$	$g(x) = \int\limits_{-\infty}^{\infty} \tilde{g}(k) \mathrm{w}^{kx} \mathrm{d}k$
FS: $x \in [0, \Delta x]$, $\tilde{n} \in \mathbb{Z}$	$\tilde{g}_{\tilde{n}} = \frac{1}{\Delta x} \int\limits_{0}^{\Delta x} g(x) \mathrm{w}^{-\tilde{n}x/\Delta x} \mathrm{d}x$	$g(x) = \sum\limits_{\tilde{n}=-\infty}^{\infty} \tilde{g}_{\tilde{n}} \mathrm{w}^{\tilde{n}x/\Delta x}$
IDFT: $n \in \mathbb{Z}$, $k \in [0, 1/\Delta x]$	$\tilde{g}(k) = \sum\limits_{n=-\infty}^{\infty} g_n \mathrm{w}^{-nk\Delta x}$	$g_n = \Delta x \int\limits_{0}^{1/\Delta x} \tilde{g}(k) \mathrm{w}^{nk\Delta x} \mathrm{d}k$
DFT: $n, \tilde{n} \in \mathbb{Z}_N$	$\tilde{g}_{\tilde{n}} = \frac{1}{N} \sum\limits_{n=0}^{N-1} g_n \mathrm{w}_N^{-n\tilde{n}}$	$g_n = \sum\limits_{\tilde{n}=0}^{N-1} \tilde{g}_{v} \mathrm{w}_N^{n\tilde{n}}$

B.1.3 Multidimensional Fourier Transform

The Fourier transform can easily be extended to *multidimensional* signals.

Definition B.3 (Multidimensionale FT) *Let $g(\boldsymbol{x}) : \mathbb{R}^W \mapsto \mathbb{C}$ a square-integrable function, that is*

$$\int\limits_{-\infty}^{\infty} \left|g(\boldsymbol{x})\right|^2 \mathrm{d}^D x = \left\langle g(\boldsymbol{x}) \,\middle|\, g(\boldsymbol{x}) \right\rangle = \left\| g(\boldsymbol{x}) \right\|_2^2 < \infty. \tag{B.19}$$

Then the Fourier transform *of $g(\boldsymbol{x})$, $\tilde{g}(\boldsymbol{\kappa})$ is given by*

$$\boxed{\tilde{g}(\boldsymbol{\kappa}) = \int\limits_{-\infty}^{\infty} g(\boldsymbol{x}) \exp\left(-2\pi \mathrm{i} \boldsymbol{\kappa}^T \boldsymbol{x}\right) \mathrm{d}^D x = \left\langle \mathrm{w}^{-\boldsymbol{x}^T \boldsymbol{\kappa}} \,\middle|\, g(\boldsymbol{x}) \right\rangle} \tag{B.20}$$

and the inverse Fourier transform *by*

$$\boxed{g(\boldsymbol{x}) = \int\limits_{-\infty}^{\infty} \tilde{g}(\boldsymbol{\kappa}) \exp\left(2\pi \mathrm{i} \boldsymbol{\kappa}^T \boldsymbol{x}\right) \mathrm{d}^D k = \left\langle \mathrm{w}^{\boldsymbol{x}^T \boldsymbol{\kappa}} \,\middle|\, \tilde{g}(\boldsymbol{\kappa}) \right\rangle.} \tag{B.21}$$

The *scalar product* in the exponent of the kernel $\boldsymbol{x}^T \boldsymbol{\kappa}$ makes the kernel of the Fourier transform separable, i. e. it can be written as the product

$$\mathrm{w}^{\boldsymbol{x}^T \boldsymbol{\kappa}} = \prod_{d=1}^{D} \mathrm{w}^{k_d x_d}. \tag{B.22}$$

The discrete multidimensional Fourier transform is discussed using the example of the 2D DFT; the extension to higher dimensions is straightforward.

Definition B.4 (2D DFT) *The 2D DFT maps complex-valued $M \times N$ matrices to complex-valued $M \times N$ matrices:*

$$\boxed{\tilde{g}_{\tilde{m}\tilde{n}} = \frac{1}{MN}\sum_{m=0}^{M-1}\sum_{n=0}^{N-1} g_{mn} \exp\left(-\frac{2\pi i m\tilde{m}}{M}\right)\exp\left(-\frac{2\pi i n\tilde{n}}{N}\right)} \tag{B.23}$$

or

$$\tilde{g}_{\tilde{m}\tilde{n}} = \frac{1}{MN}\sum_{m=0}^{M-1}\left(\sum_{n=0}^{N-1} g_{mn} w_N^{-n\tilde{n}}\right) w_M^{-m\tilde{m}}. \tag{B.24}$$

In the second line, the abbreviation introduced in Eq. (B.15) is used. Analogous to 1D DFT, a matrix is expanded into a set of NM basis matrices that span the $M \times N$-dimensional vector space over the body of complex numbers. The basis matrices have the form

$$\underbrace{\boldsymbol{B}_{\tilde{m}\tilde{n}}}_{M\times N} = \frac{1}{\sqrt{MN}}\begin{bmatrix} w^0 \\ w_M^{\tilde{m}} \\ w_M^{2\tilde{m}} \\ \vdots \\ w_M^{(M-1)\tilde{m}} \end{bmatrix}\left[w^0, w_N^{\tilde{n}}, w_N^{2\tilde{n}}, \ldots, w_N^{(N-1)\tilde{n}}\right]. \tag{B.25}$$

In this equation, the basis matrices are represented as an *outer product* of a column and row vector, which represent the basis vectors of the 1D DFT Eq. (B.16). This shows the separability of the 2D DFT.

We can also write the 2D DFT as a scalar product

$$\tilde{g}_{\tilde{m}\tilde{n}} = \frac{1}{\sqrt{MN}}\left\langle \boldsymbol{B}_{-\tilde{m},-\tilde{n}} \,|\, \boldsymbol{G} \right\rangle, \tag{B.26}$$

where the scalar product of two complex-valued matrices is defined as

$$\langle \boldsymbol{G} | \boldsymbol{H} \rangle = \sum_{m=0}^{M-1}\sum_{n=0}^{N-1} g_{mn}^* h_{mn}. \tag{B.27}$$

The inverse 2D DFT is given by

$$\boxed{g_{mn} = \sum_{\tilde{m}=0}^{M-1}\sum_{v=0}^{N-1} \tilde{g}_{\tilde{m}\tilde{n}} w_M^{m\tilde{m}} w_N^{n\tilde{n}} = \sqrt{MN}\left\langle \boldsymbol{B}_{\tilde{m}\tilde{n}} \middle| \tilde{\boldsymbol{G}} \right\rangle.} \tag{B.28}$$

B.1.4 Properties of the Fourier Transform

In this section, we summarize the most important properties of the continuous and discrete Fourier transform. We focus on those properties of the FT that are most important for image and signal processing. Together with some elementary Fourier transform pairs (> R13), these properties (> R12, > R15) form a powerful tool with which further properties of the Fourier transform and the transformation of many functions can be easily derived.

Periodicity
The core of the DFT shows a characteristic *[*periodicity]periodicity!DFT:

$$\exp\left(-\frac{2\pi \mathrm{i}(n+lN)}{N}\right) = \exp\left(-\frac{2\pi \mathrm{i}n}{N}\right), \mathrm{w}_N^{(n+lN)} = \mathrm{w}_N^n \quad \forall\, l \in \mathbb{Z}. \tag{B.29}$$

The definitions of the DFT limit the space and the Fourier space to a finite number of values. If we ignore this limitation and calculate the forward and backward transformations for all integers, we find the same periodicities directly from the definitions in Eqs. (B.26) and (B.28) in the functions in space and Fourier space:

$$\begin{array}{lll} \text{Wavenumber domain} & \tilde{g}_{\tilde{m}+kM,\tilde{n}+lN} = \tilde{g}_{\tilde{m}\tilde{n}}, & \forall\, k,l \in \mathbb{Z} \\ \text{Spatial domain} & g_{m+kM,n+lN} = g_{m,n}, & \forall\, k,l \in \mathbb{Z}. \end{array} \tag{B.30}$$

These equations describe a periodic replication in all directions in both spaces beyond the original domain of definition.

Symmetry
Four types of *symmetries* are important for the Fourier transform;

$$\begin{array}{lrcr} \text{even} & g(-\boldsymbol{x}) & = & g(\boldsymbol{x}), \\ \text{odd} & g(-\boldsymbol{x}) & = & -g(\boldsymbol{x}), \\ \text{Hermitian} & g(-\boldsymbol{x}) & = & g^*(\boldsymbol{x}), \\ \text{anti-Hermitian} & g(-\boldsymbol{x}) & = & -g^*(\boldsymbol{x}) \end{array} \tag{B.31}$$

The symbol $*$ denotes the complex conjugate. The complete symmetry relationships for the Fourier transform are as follows:

$$\begin{array}{lcl} \text{real} & \circ\!\!-\!\!\bullet & \text{Hermitian} \\ \text{imaginary} & \circ\!\!-\!\!\bullet & \text{anti-Hermitian} \\ \text{Hermitian} & \circ\!\!-\!\!\bullet & \text{real} \\ \text{anti-Hermitian} & \circ\!\!-\!\!\bullet & \text{imaginary} \\ \text{even} & \circ\!\!-\!\!\bullet & \text{even} \\ \text{odd} & \circ\!\!-\!\!\bullet & \text{odd} \\ \text{real and even} & \circ\!\!-\!\!\bullet & \text{real and even} \\ \text{real and odd} & \circ\!\!-\!\!\bullet & \text{imaginary and odd} \\ \text{imaginary and even} & \circ\!\!-\!\!\bullet & \text{imaginary and even} \\ \text{imaginary and odd} & \circ\!\!-\!\!\bullet & \text{real and odd} \end{array} \tag{B.32}$$

The DFT shows the same symmetry properties as the FT in Eqs. (B.31) and (B.32). In the definition for even and odd functions $g(-x) = \pm g(x)$, only the continuous function must be replaced by the corresponding vector ($g_{-n} = \pm g_n$) or matrix elements ($g_{-m,-n} = \pm g_{m,n}$). Due to the periodicity of the DFT, these symmetry relations can also be written as

$$g_{-m,-n} = \pm g_{m,n} \equiv g_{M-m,N-n} = \pm g_{m,n} \tag{B.33}$$

with the sign + for even functions and the sign − for odd functions. This rewriting shifts the center of symmetry from the origin to the point $[M/2, N/2]^T$.

Separability
According to Eq. (B.22), the kernel of the Fourier transform is *separable.* Therefore, the Fourier transform of separable functions is also separable:

$$\boxed{\prod_{d=1}^{D} g(x_d) \circ\!\!-\!\!\bullet \prod_{d=1}^{D} \tilde{g}(k_d).} \tag{B.34}$$

Since many significant multidimensional functions are separable, this property is important for effectively calculating the transformation of multidimensional functions from 1D transformations.

Similarity
The similarity theorem states how a Fourier transform behaves when the coordinate system is scaled. In one dimension, a function can only be scaled ($x' = ax$). In higher dimensions, a more general transformation of the coordinate system is possible using an *affine transform* ($\boldsymbol{x}' = \boldsymbol{A}\boldsymbol{x}$), i. e. the new basis vectors are a linear combination of the old ones. The rotation of the coordinate system is a special case.

Theorem B.1 (Similarity) *Let a be a non-zero real number,* $\boldsymbol{A}$ *a real, invertible matrix and* $\boldsymbol{R}$ *an orthogonal matrix that describes a rotation of the coordinate system* ($\boldsymbol{R}^{-1} = \boldsymbol{R}^T$, $\det \boldsymbol{R} = 1$)*. Then the following similarity relationships apply:*

$$\begin{array}{llcl} \textit{Scalar} & g(a\boldsymbol{x}) & \circ\!\!-\!\!\bullet & \dfrac{1}{|a|^W}\tilde{g}(\boldsymbol{\kappa}/a), \\ \textit{Affine transform} & g(\boldsymbol{A}\boldsymbol{x}) & \circ\!\!-\!\!\bullet & \dfrac{1}{\det \boldsymbol{A}}\tilde{g}((\boldsymbol{A}^T)^{-1}\boldsymbol{\kappa}), \\ \textit{Rotation} & g(\boldsymbol{R}\boldsymbol{x}) & \circ\!\!-\!\!\bullet & \tilde{g}(\boldsymbol{R}\boldsymbol{\kappa}). \end{array} \tag{B.35}$$

If a function is compressed in local space, it is stretched accordingly in Fourier space, and vice versa. A rotation of the coordinate system in local space causes the same rotation of the coordinate system in Fourier space.

The above similarity theorem cannot be used for the discrete Fourier transform, as arbitrary scaling and rotation are not possible here. A discrete function can be stretched by an integer factor K (*upsampling*) by filling the newly created points with zeros:

$$(g_{\uparrow K})_n = \begin{cases} g_{n/K} & n = 0, K, 2K, \ldots (N-1)K) \\ 0 & \text{otherwise.} \end{cases} \tag{B.36}$$

Theorem B.2 (Similarity, discrete) *Let g be a complex-valued vector with N elements and $K \in \mathbb{N}$. Then the discrete Fourier transform of the upsampled vector $g_{\uparrow K}$ with KN elements is given by*

$$g_{\uparrow K} \quad \circ\!\!-\!\!\bullet \quad \frac{1}{K}\tilde{g} \quad \text{with} \quad \tilde{g}_{kN+\tilde{n}} = \tilde{g}_{\tilde{n}}. \tag{B.37}$$

An upward sampling by the factor K therefore only causes a K-fold replication of the Fourier transform. In Equation (B.37) it should be noted that due to the periodicity of the discrete Fourier transforms discussed at the beginning of this section: $\tilde{g}_{kN+\tilde{n}} = \tilde{g}_{\tilde{n}}$.

Shift and Modulation

As a direct consequence of the linearity of the Fourier transform, we can formulate the following *shift theorem* and *modulation theorem*:

Theorem B.3 (Shift and Modulation) *Let the Fourier transform of $g(\boldsymbol{x})$ be $\tilde{g}(\boldsymbol{\kappa})$. Then $g(\boldsymbol{x} - \boldsymbol{x}_0)$ has the Fourier transform* $\exp(-2\pi i \boldsymbol{x}_0^T \boldsymbol{\kappa})\tilde{g}(\boldsymbol{\kappa})$ *and* $\exp(2\pi i \boldsymbol{\kappa}_0^T \boldsymbol{x}) g(\boldsymbol{x})$ *has the Fourier transform $\tilde{g}(\boldsymbol{\kappa} - \boldsymbol{\kappa}_0)$:*

$$\begin{array}{lcl} g(\boldsymbol{x} - \boldsymbol{x}_0) & \circ\!\!-\!\!\bullet & \exp(-2\pi i \boldsymbol{x}_0^T \boldsymbol{\kappa})\tilde{g}(\boldsymbol{\kappa}), \\ \exp(2\pi i \boldsymbol{\kappa}_0^T \boldsymbol{x}) g(\boldsymbol{x}) & \circ\!\!-\!\!\bullet & \tilde{g}(\boldsymbol{\kappa} - \boldsymbol{\kappa}_0). \end{array} \tag{B.38}$$

A shift in spatial space therefore does not change the Fourier transform in amplitude, but only in phase, whereby the phase change $-2\pi \boldsymbol{x}_0^T \boldsymbol{\kappa}$ is dependent on the wavenumber vector. A shift in Fourier space modulates a signal in spatial space with the complex exponential function $\exp(2\pi i \boldsymbol{\kappa}_0^T \boldsymbol{x})$, which has the wavenumber $\boldsymbol{\kappa}_0$.

Convolution

The *convolution* is one of the most important operations in signal processing. For continuous signals, it is defined by

$$(g * h)(\boldsymbol{x}) = \int_{-\infty}^{\infty} h(\boldsymbol{x}') g(\boldsymbol{x} - \boldsymbol{x}') \mathrm{d}^D x'. \tag{B.39}$$

In signal processing, the function $h(\boldsymbol{x})$ is usually zero except in a small range around the zero point. Such a function is then referred to as *convolution mask*. Convolution with $h(x)$ results in a new function $g'(x)$ whose values are a kind of weighted averaging of $g(x)$ in a small area around x. The signal is changed by the convolution with $h(x)$ in a defined way. For example, it can become smoother. This is why it is also referred to as *filtering*.

One- and multi-dimensional discrete convolutions are defined analogously to the continuous convolution in Eq. (6.39) by

$$g'_n = \sum_{n'=0}^{N-1} h_{n'} g_{n-n'}, \quad g'_{\boldsymbol{n}} = \sum_{\boldsymbol{n}'=0}^{N-1} h_{\boldsymbol{n}'} g_{\boldsymbol{n}-\boldsymbol{n}'}. \tag{B.40}$$

In the second formula, the vectorial indexing introduced in Section 10.3.1 was used as an elegant notation for complex operations with multidimensional signals.

The *convolution theorem* for the FT and DFT states:

Theorem B.4 (Convolution) *Let $\tilde{g}(\boldsymbol{\kappa})$ ($\tilde{\boldsymbol{G}}$) be the Fourier transform of $g(\boldsymbol{x})$ ($\boldsymbol{G}$) and $\tilde{h}(\boldsymbol{\kappa})$ ($\tilde{\boldsymbol{H}}$) the Fourier transform of $h(\boldsymbol{x})$, ($\boldsymbol{H}$). Then there are the following Fourier transform pairs:*

$$\begin{array}{llcl} \text{FT:} & h(\boldsymbol{x}) * g(\boldsymbol{x}) & \circ\!\!-\!\!\bullet & \tilde{h}(\boldsymbol{\kappa})\tilde{g}(\boldsymbol{\kappa}), \\ \text{DFT:} & \boldsymbol{H} * \boldsymbol{G} & \circ\!\!-\!\!\bullet & \prod_{d=1}^{D} N_d \tilde{\boldsymbol{H}}\tilde{\boldsymbol{G}}. \end{array} \tag{B.41}$$

A convolution of two functions in local space is equivalent to a complex multiplication in Fourier space. Conversely, a convolution in Fourier space is also equivalent to a multiplication in local space. Mathematically, the simplicity of convolution in Fourier space results from the fact that the basis functions of Fourier space, the complex exponential functions $\exp\left(2\pi \mathrm{i} \boldsymbol{\kappa}^T \boldsymbol{x}\right)$, the common *eigenfunctions all* convolution operations. A convolution with an arbitrary function leaves the exponential function unchanged except for one factor, the *eigenvalue*.

The following properties result directly from the convolution theorem. The convolution is

$$\begin{array}{ll} \text{commutative} & h * g = g * h, \\ \text{associative} & h_1 * (h_2 * g) = (h_1 * h_2) * g, \\ \text{distributive with respect to addition} & (h_1 + h_2) * g = h_1 * g + h_2 * g. \end{array} \tag{B.42}$$

The importance of convolution is further emphasized by the fact that two classes of operations that at first glance do not look like convolution operations, namely the shift and all derivative operations, can also be regarded as convolution operations. In both cases, the Fourier transform is only multiplied by a complex factor. For the shift, this results directly from the shift theorem (Theorem B.3). The convolution mask for a shift operator $\mathcal{S}$ is a shifted *δ-distribution*:

$$\mathcal{S}(\boldsymbol{s})g(\boldsymbol{x}) = \delta(\boldsymbol{x} - \boldsymbol{s}) * g(\boldsymbol{x}) \tag{B.43}$$

The δ-distribution can only be defined as the limit value of a sequence of functions, for example by

$$\delta(x) = \lim_{a \to 0} \left(\frac{\exp(-\pi x^2/a^2)}{a} \right) \tag{B.44}$$

with the important integral properties

$$\int_{-\infty}^{\infty} \delta(x)\mathrm{d}x = 1 \quad \text{and} \quad g(x_0) = \int_{-\infty}^{\infty} \delta(x - x_0)g(x)\mathrm{d}x. \tag{B.45}$$

Derivation
For the partial derivative of a function in space, the *derivative theorem* states:

Theorem B.5 (Derivation) *Let $g(\boldsymbol{x})$ be differentiable for all $\boldsymbol{x}$ and $\tilde{g}(\boldsymbol{\kappa})$ be the Fourier transform of $g(\boldsymbol{x})$. Then $2\pi \mathrm{i} k_d \tilde{g}(\boldsymbol{\kappa})$ is the Fourier transform of $\partial g(\boldsymbol{x})/\partial x_d$:*

$$\boxed{\frac{\partial g(\boldsymbol{x})}{\partial x_d} \quad \circ\!\!-\!\!\bullet \quad 2\pi \mathrm{i} k_d \tilde{g}(\boldsymbol{\kappa}).} \tag{B.46}$$

The derivative theorem follows directly from the definition of the inverse Fourier transform in Eq. (B.21) by swapping the integration with the partial derivative.

The Fourier transform of the first partial derivative in x_1 direction is $2\pi i k_1$. The inverse Fourier transform of $2\pi i k_1$, i. e. the associated convolution mask, is not an ordinary function, since $2\pi i k_1$ is not square integrable, but the derivative of the δ distribution:

$$2\pi i k \quad \bullet\!\!-\!\!\circ \quad \delta'(x) = \frac{\mathrm{d}\delta(x)}{\mathrm{d}x} = \lim_{a \to 0} \frac{2\pi x}{a^3} \exp(-\pi x^2/a^2) \tag{B.47}$$

The derivative of the δ-distribution, like all distributions, exists only as the limit of a sequence of functions, as shown in the equation above.

With this knowledge and the properties summarized in Eq. (B.42), we can draw some further important conclusions. Since any convolution operation can be interchanged with the shift operation, any convolution is a shift invariant operation. Furthermore, we get the same result if we first differentiate and then convolve a signal or perform these operations in reverse order. These properties are helpful for a deeper understanding of image processing operations and for a faster calculation of convolution operations.

Central Limit Theorem
The *central limit theorem* is known for its importance in probability theory and statistics [236]. However, it also plays an important role in signal processing, as it states under very general conditions that cascaded convolution operations lead to a convolution with the *Gaussian bell curve* ($\propto \exp(-ax^2)$). Since the Fourier transform of the Gaussian function is also a Gaussian function (≻ R14), this means that with continued convolution both the convolution mask and the transfer function take the form of a Gaussian bell curve.

Therefore, the central limit theorem is an expression of the central role of the Gaussian function for image processing. The necessary conditions for the validity of the central limit theorem can be formulated in different ways. Here we use the conditions from [236] and state the theorem in terms of convolution.

Theorem B.6 (Central limit theorem) *Given N functions $h_n(x)$ with vanishing mean value $\int_{-\infty}^{\infty} x h_n(x)\mathrm{d}x = 0$, the variance $\sigma_n^2 = \int_{-\infty}^{\infty} x^2 h_n(x)\mathrm{d}x$ and with $z = x/\sigma$, $\sigma^2 = \sum_{n=1}^{N} \sigma_n^2$. Then applies*

$$h = \lim_{N\to\infty} h_1 * h_2 * \ldots * h_N \propto \exp(-z^2/2) \tag{B.48}$$

on condition that

$$\lim_{N\to\infty} \sum_{n=1}^{N} \sigma_n^2 \to \infty \tag{B.49}$$

and that there is a number $\alpha > 2$ and a finite constant c such that

$$\int_{-\infty}^{\infty} x^{\alpha} h_n(x)\mathrm{d}x < c < \infty \quad \forall n. \tag{B.50}$$

The theorem is of great practical importance because — especially if the functions h_n are smooth — the Gaussian curve is approximated with sufficient accuracy even for very small values of N (e. g. 5).

Smoothness and Compactness
The smoother a function is, the more compact its Fourier transform is. This general rule can be quantified by expressing the smoothness by the number of derivatives that are continuous and the compactness by the asymptotic behavior for large values of the wavenumber κ: If a function $g(x)$ and its first $n-1$ derivatives are continuous, then its Fourier transform decays at least with $|\kappa|^{-(n+1)}$, i. e. $\lim_{|\kappa|\to\infty} |\kappa|^n g(\kappa) = 0$, for large wavenumbers κ.

As simple examples, we consider the rectangular and triangular functions (≻ R13). The rectangular function is discontinuous ($n = 0$). Therefore, its Fourier transform, the sinc function, decreases asymptotically with $|\kappa|^{-1}$. The triangular function is continuous, but its first derivative is not continuous. Therefore, its Fourier transform, the sinc^2 function, decreases more steeply with $|\kappa|^{-2}$.

We can also include impulsive functions such as the δ-distribution Eq. (B.44) in these considerations, since the derivative of a discontinuous function becomes impulsive at the points of discontinuity. Therefore, if the nth derivative of a function becomes impulsive, then the Fourier transform of the function decreases asymptotically with $|\kappa|^{-n}$.

We can understand the relationship between smoothness and compactness as an extension of the reciprocity between local and Fourier space. What is strongly localized in one space has a large extension in the other space and vice versa.

Uncertainty Principle

The reciprocity between spatial and Fourier space can also be found in another theorem, the classical *uncertainty principle*. The theorem relates the mean square width of a function and its Fourier transform to each other. The mean square width $(\Delta x)^2$ is defined as

$$(\Delta x)^2 = \frac{\int\limits_{-\infty}^{\infty} x^2 \left|g(x)\right|^2 \mathrm{d}x}{\int\limits_{-\infty}^{\infty} \left|g(x)\right|^2 \mathrm{d}x} - \left(\frac{\int\limits_{-\infty}^{\infty} x \left|g(x)\right|^2 \mathrm{d}x}{\int\limits_{-\infty}^{\infty} \left|g(x)\right|^2 \mathrm{d}x} \right)^2 . \tag{B.51}$$

This quantity is therefore the variance of $\left|g(x)\right|^2$ and thus a measure of the width of the distribution of the energy of the signal. The uncertainty principle states:

Theorem B.7 (Uncertainty principle) *For the product of the variance* $(\Delta x)^2$ *of* $\left|g(x)\right|^2$ *and the variance* $(\Delta\kappa)^2$ *of* $\left|\tilde{g}(\kappa)\right|^2$*, the following inequality applies*

$$\Delta x \Delta \kappa \geq \frac{1}{4\pi}. \tag{B.52}$$

Some criteria for the construction of linear (convolution) filter operators can be derived from the relationships between compactness and smoothness and from the uncertainty principle.

B.1.5 Alternative Definitions

Unfortunately, different definitions of the Fourier transform are used in the literature, which can easily lead to confusion and errors. One reason for this is the different definitions of the wavenumber. The definition of the wavenumber as a reciprocal wavelength ($\kappa = 1/\lambda$) is the most useful for signal processing, as the wavenumber directly indicates the number of wavelengths per unit length. In physics and electrical engineering, a definition of the wavenumber that includes the factor 2π is more common:

$k = 2\pi/\lambda$. With this notation for the wavenumber, the Fourier transform can be defined in two further ways, in the asymmetric form

$$\tilde{g}(k) = \left\langle \exp(\mathrm{i}kx) \,\middle|\, g(x) \right\rangle, \; g(x) = \frac{1}{2\pi} \left\langle \exp(-\mathrm{i}kx) \,\middle|\, \tilde{g}(k) \right\rangle \tag{B.53}$$

and in the symmetrical form

$$\tilde{g}(k) = \frac{1}{\sqrt{2\pi}} \left\langle \exp(\mathrm{i}kx) \,\middle|\, g(x) \right\rangle, \; g(x) = \frac{1}{\sqrt{2\pi}} \left\langle \exp(-\mathrm{i}kx) \,\middle|\, \tilde{g}(k) \right\rangle. \tag{B.54}$$

Since all three definitions of the Fourier transform are commonly used, it is easy for theorems and Fourier transform pairs to contain incorrect factors. The relationships between the transformation pairs for the three definitions can be derived directly from the definitions as follows:

$$\begin{array}{llll} \kappa = 1/\lambda & \text{(Eq. B.10)} & g(x) \;\circ\!\!-\!\!\bullet & \tilde{g}(\kappa) \\ k = 2\pi/\lambda & \text{(Eq. B.53)} & g(x) \;\circ\!\!-\!\!\bullet & \tilde{g}(k/2\pi) \\ k = 2\pi/\lambda & \text{(Eq. B.54)} & g(x) \;\circ\!\!-\!\!\bullet & \tilde{g}(k/2\pi)/\sqrt{2\pi}. \end{array} \tag{B.55}$$

B.2 Discrete Unitary Transformations

B.2.1 General Properties

In the sections 6.3.1 and 6.3.2 we learned that the discrete Fourier transform can be regarded as a linear transformation in a vector space. Thus, it is only one example of a large class of transformations, the *unitary transformations*. In this chapter, we discuss some general properties of unitary transformations. They will help us to gain a deeper insight into the basics of image processing. We will also discuss further examples of unitary transformations that have gained importance in image processing.

Unitary transformations are defined for vector spaces over the body of complex numbers for which an *inner product* or *scalar product* is defined. Both the continuous Eq. (B.10) and discrete Fourier transformations Eq. (B.17) use the scalar product to calculate the transformation.

The basic theorem for unitary transformations is

Theorem B.8 (Unitary transformation) *Let V be a finite-dimensional vector space with inner product. Furthermore, let $\boldsymbol{U}$ be a linear bijective transformation of V onto itself. Then the following statements are equivalent:*

1. *$\boldsymbol{U}$ is unitary.*
2. *$\boldsymbol{U}$ preserves the inner product, i. e. $\langle \boldsymbol{g} \,|\, \boldsymbol{h} \rangle = \langle \boldsymbol{U}\boldsymbol{g} \,|\, \boldsymbol{U}\boldsymbol{h} \rangle$, $\forall \boldsymbol{g}, \boldsymbol{h} \in V$.*
3. *The inverse transformation to $\boldsymbol{U}$, $\boldsymbol{U}^{-1}$, is the adjoint (transposed and conjugate-complex matrix) $\boldsymbol{U}^{\mathrm{T}}$ of $\boldsymbol{U}$: $\boldsymbol{U}\boldsymbol{U}^{\mathrm{T}} = \boldsymbol{I}$.*
4. *The row (and column) vectors of $\boldsymbol{U}$ form an orthonormal basis of the vector space V.*

This definition already includes one of the most important properties of unitary transformations, namely the preservation of the inner product. This implies that another important property, the *norm*, is also preserved:

$$\|g\|_2 = \langle g \,|\, g \rangle^{1/2} = \langle \boldsymbol{U}g \,|\, \boldsymbol{U}g \rangle^{1/2}. \tag{B.56}$$

We can consider the norm as *length* or *size* of the vector. The rotation in $\mathbb{R}^2$ or $\mathbb{R}^3$ is an example of a transformation where the conservation of the vector length is obvious (see also the discussion of homogeneous coordinates in Section 3.3.3).

The composition of two unitary transformations $\boldsymbol{U}_1\boldsymbol{U}_2$ is itself unitary. Since the identity operator $\boldsymbol{I}$ and the inverse of a unitary operator are also unitary, the set of all unitary transformations forms a *group* with respect to the composition. Practically, this means that we can compose complex unitary transformations from simpler or elementary transformations.

We will illustrate some of the properties of unitary transforms discussed in the context of discrete Fourier transforms. First, we consider the one-dimensional DFT in symmetric definition:

$$\tilde{g}_{\tilde{n}} = \frac{1}{\sqrt{N}} \sum_{n=0}^{N-1} g_n \mathrm{w}_N^{-n\tilde{n}}.$$

This equation can be expressed as a multiplication of the $N \times N$ matrix $\boldsymbol{W}_N$

$$(\boldsymbol{W}_N)_{n\tilde{n}} = \mathrm{w}_N^{-n\tilde{n}}$$

can be considered with the vector $\boldsymbol{g}$:

$$\tilde{\boldsymbol{g}} = \frac{1}{\sqrt{N}} \boldsymbol{W}_N \, \boldsymbol{g}. \tag{B.57}$$

Explicitly, the DFT of an eight-dimensional vector is given by

$$\begin{bmatrix} \tilde{g}_0 \\ \tilde{g}_1 \\ \tilde{g}_2 \\ \tilde{g}_3 \\ \tilde{g}_4 \\ \tilde{g}_5 \\ \tilde{g}_6 \\ \tilde{g}_7 \end{bmatrix} = \frac{1}{\sqrt{8}} \begin{bmatrix}
\mathrm{w}_8^0 & \mathrm{w}_8^0 & \mathrm{w}_8^0 & \mathrm{w}_8^0 & \mathrm{w}_8^0 & \mathrm{w}_8^0 & \mathrm{w}_8^0 & \mathrm{w}_8^0 \\
\mathrm{w}_8^0 & \mathrm{w}_8^7 & \mathrm{w}_8^6 & \mathrm{w}_8^5 & \mathrm{w}_8^4 & \mathrm{w}_8^3 & \mathrm{w}_8^2 & \mathrm{w}_8^1 \\
\mathrm{w}_8^0 & \mathrm{w}_8^6 & \mathrm{w}_8^4 & \mathrm{w}_8^2 & \mathrm{w}_8^0 & \mathrm{w}_8^6 & \mathrm{w}_8^4 & \mathrm{w}_8^2 \\
\mathrm{w}_8^0 & \mathrm{w}_8^5 & \mathrm{w}_8^2 & \mathrm{w}_8^7 & \mathrm{w}_8^4 & \mathrm{w}_8^1 & \mathrm{w}_8^6 & \mathrm{w}_8^3 \\
\mathrm{w}_8^0 & \mathrm{w}_8^4 & \mathrm{w}_8^0 & \mathrm{w}_8^4 & \mathrm{w}_8^0 & \mathrm{w}_8^4 & \mathrm{w}_8^0 & \mathrm{w}_8^4 \\
\mathrm{w}_8^0 & \mathrm{w}_8^3 & \mathrm{w}_8^6 & \mathrm{w}_8^1 & \mathrm{w}_8^4 & \mathrm{w}_8^7 & \mathrm{w}_8^2 & \mathrm{w}_8^5 \\
\mathrm{w}_8^0 & \mathrm{w}_8^2 & \mathrm{w}_8^4 & \mathrm{w}_8^6 & \mathrm{w}_8^0 & \mathrm{w}_8^2 & \mathrm{w}_8^4 & \mathrm{w}_8^6 \\
\mathrm{w}_8^0 & \mathrm{w}_8^1 & \mathrm{w}_8^2 & \mathrm{w}_8^3 & \mathrm{w}_8^4 & \mathrm{w}_8^5 & \mathrm{w}_8^6 & \mathrm{w}_8^7
\end{bmatrix} \begin{bmatrix} g_0 \\ g_1 \\ g_2 \\ g_3 \\ g_4 \\ g_5 \\ g_6 \\ g_7 \end{bmatrix}$$

When entering the elements in the matrix, the periodicity of the kernel of the DFT Eq. (B.29) was used so that the exponents of W are limited to the range between 0 and 7. The transformation matrix for the DFT is symmetrical ($\boldsymbol{W} = \boldsymbol{W}^T$), $\boldsymbol{W}^{*T}$ results in the reverse transformation.

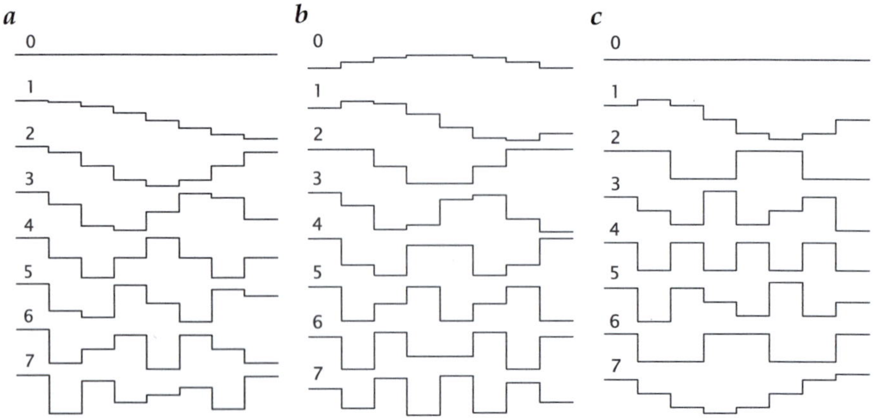

Figure B.1: *The basis functions of one-dimensional unitary transformations for N = 8:* ***a*** *cosine transform,* ***b*** *sine transform and* ***c*** *Hartley transform.*

The two-dimensional DFT can be described with similar equations if we map the $M \times N$ matrix to an MN-dimensional vector. However, there is also an easier way if we make use of the separability of the kernels of the DFT described in Eq. (B.25). Using the $M \times M$ matrix $\mathbf{W}_M$ and the $N \times N$ matrix $\mathbf{W}_N$, we can write Eq. (B.57) as follows, analogous to the one-dimensional DFT:

$$\tilde{g}_{uv} = \frac{1}{\sqrt{MN}} \sum_{m=0}^{M-1} \sum_{n=0}^{N-1} g_{mn} (\mathbf{W}_M)_{mu} (\mathbf{W}_N)_{nv} \tag{B.58}$$

or in matrix notation:

$$\underbrace{\tilde{\mathbf{G}}}_{M \times N} = \frac{1}{\sqrt{MN}} \underbrace{\mathbf{W}_M{}^T}_{M \times M} \underbrace{\mathbf{G}}_{M \times N} \underbrace{\mathbf{W}_N}_{N \times N} = \frac{1}{\sqrt{MN}} \mathbf{W}_M \, \mathbf{G} \, \mathbf{W}_N. \tag{B.59}$$

The physicist will remember the theoretical foundations of *quantum mechanics*, which are formulated in the infinite-dimensional vector space with inner product, the *Hilbert space*. In digital image processing, we can avoid the problems associated with infinite-dimensional vector spaces.

The discussion of the general properties is now followed by some examples of unitary transformations used in signal and image processing. However, their significance is not comparable to that of the Fourier transform.

B.2.2 Cosine, Sine and Hartley Transforms

It may be disturbing about the discrete Fourier transform that real images are mapped in a complex way by the transform. We can arrive at a real transformation if we consider the decomposition of the complex DFT into a real even and an imaginary odd part:

$$(\boldsymbol{W}_N)_{nv} = \cos\left(-\frac{2\pi nv}{N}\right) + \mathrm{i}\sin\left(-\frac{2\pi nv}{N}\right). \tag{B.60}$$

Neither the cosine nor the sine part can be used as a transformation kernel, as these functions do not form a complete basis of the vector space. Cosine and sine functions only span the subspaces of the even and odd functions of the entire vector space.

This problem can be eliminated by restricting the *cosine* and *sine transformation* to the positive half of the spatial and Fourier domains, because then the symmetry properties do not play a role:

$$\begin{aligned}
{}^c\tilde{g}(k) &= \int_0^\infty g(x)\,\sqrt{2}\cos(2\pi kx)\mathrm{d}x \quad \bullet\!\!-\!\!\circ \quad g(x) = \int_0^\infty {}^c\tilde{g}(k)\,\sqrt{2}\cos(2\pi kx)\mathrm{d}k,\\
{}^s\tilde{g}(k) &= \int_0^\infty g(x)\,\sqrt{2}\sin(2\pi kx)\mathrm{d}x \quad \bullet\!\!-\!\!\circ \quad g(x) = \int_0^\infty {}^s\tilde{g}(k)\,\sqrt{2}\sin(2\pi kx)\mathrm{d}k.
\end{aligned} \tag{B.61}$$

For the corresponding discrete transformations, basis vectors with the missing symmetry must be generated by adding trigonometric functions with half the wavelength in each case. This procedure is equivalent to doubling the base wavelength. Accordingly, the basis functions of the *cosine* and *sine transformation* are for an N-dimensional vector:

$$c_{n\tilde{n}} = \sqrt{\frac{2}{N}}\cos\left(\frac{\pi n\tilde{n}}{N}\right), \quad s_{n\tilde{n}} = \sqrt{\frac{2}{N+1}}\sin\left(\frac{\pi(n+1)(\tilde{n}+1)}{N+1}\right). \tag{B.62}$$

Fig. B.1a, b show the basis functions of the 1D cosine and 1D sine transformation. It is easy to see from the function graphs that the basis functions are orthogonal to each other. Due to the doubling of the periods, there are even and odd functions for both transformations. The basis functions with half the wavelength each complement the functions with the missing symmetry.

The cosine transformation has become important for *image data compression* [146]. It is part of the standard compression algorithm proposed by the *Joint Photographic Experts Group (JPEG)*.

The *Hartley transform (HT)* is a much more elegant solution than the cosine or sine transform to avoid a transformation with complex numbers. By adding the cosine and sine functions, a kernel is obtained that has no symmetry,

$$\operatorname{cas}(2\pi\kappa x) = \cos(2\pi\kappa x) + \sin(2\pi\kappa x) = \sqrt{2}\cos(2\pi(\kappa x - 1/8)), \tag{B.63}$$

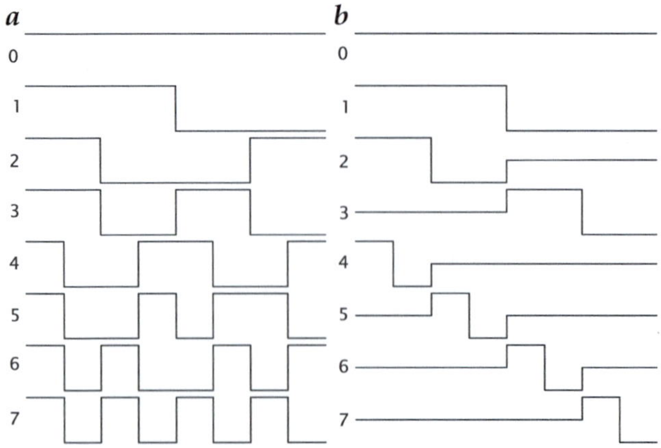

Figure B.2: The first 8 basis functions of one-dimensional unitary transformations for $N = 16$*:* ***a*** *Hadamard transform and* **b** *Haar transform.*

and which is therefore suitable for transforming the entire room:

$$ {}^h\tilde{g}(\kappa) = \int_{-\infty}^{\infty} g(x)\,\mathrm{cas}(2\pi\kappa x)\mathrm{d}x \quad \bullet\!\!-\!\!\circ \quad g(x) = \int_{-\infty}^{\infty} {}^h\tilde{g}(\kappa)\,\mathrm{cas}(2\pi\kappa x)\mathrm{d}\kappa. \tag{B.64} $$

The corresponding *discrete Hartley transform* (*DHT*) is:

$$ {}^h\tilde{g}_{\tilde{n}} = \frac{1}{\sqrt{N}}\sum_{n=0}^{N-1} g_n\,\mathrm{cas}(2\pi n\tilde{n}/N) \quad \bullet\!\!-\!\!\circ \quad g_n = \frac{1}{\sqrt{N}}\sum_{n=0}^{N-1} {}^h\tilde{g}_{\tilde{n}}\,\mathrm{cas}(2\pi n\tilde{n}/N). \tag{B.65} $$

The basis functions for $N = 8$ are shown in Fig. B.1c. Despite all the elegance of the Hartley transform for real image data, it has some disadvantages compared to the Fourier transform. In particular, the simple shift theorem B.3 of the Fourier transform no longer applies. Instead, shifts result in the mixing of the basis functions with positive and negative wavenumbers:

$$ \begin{array}{rcl} g(x - x_0) & \circ\!\!-\!\!\bullet & {}^h\tilde{g}(k)\cos(2\pi\kappa x_0) + {}^h\tilde{g}(-\kappa)\sin(2\pi\kappa x_0), \\ g_{n-n'} & \circ\!\!-\!\!\bullet & {}^h\tilde{g}_{\tilde{n}}\cos(2\pi n'\kappa/N) + {}^h\tilde{g}_{N-\tilde{n}}\sin(2\pi n'\tilde{n}/N). \end{array} \tag{B.66} $$

Similar complications arise with the convolution theorem for the Hartley transformation (≻ R16).

B.2.3 Hadamard Transform

The basis functions of the *Hadamard transform* are orthogonal binary patterns (Fig. B.2a). Some of these patterns are regular rectangular waves, others are not. The Hadamard transformation is characterized by a low computational effort, since its core consists only of the numbers 1 and –1. This means that only additions and subtractions are necessary for the calculation.

B.2.4 Haar Transform

The basis vectors of all previous transformations were *global* in nature, because the basis functions always extend over the entire vector or the entire image area. All local information is lost. For example, if we have two independent objects in an image, they are simultaneously decomposed into the global patterns. They are then no longer recognizable as individual objects in the transformation.

The *Haar transformation* is an example of a unitary transformation that partially preserves local information because its basis functions are pairs of impulses that are only non-zero at the position of the impulse (Fig. B.2b). The Haar transformation is advantageous for resolving the position of small structures.

Like the Hadamard transform, the Haar transform is also efficient in terms of computational effort, as its kernel only contains the numbers –1, 0 and 1.

C Statistics, Regression and Optimization

Contents

B. Jähne, *Digital Image Processing and Image Acquisition*,
https://doi.org/10.1007/978-3-032-04806-6

C.1 Introduction

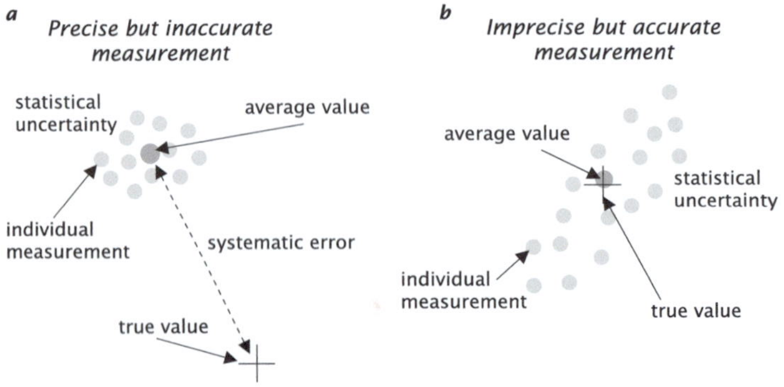

Figure C.1: *Illustration of **a** systematic error and **b** statistical error, which indicate the accuracy and precision of a measurement.The statistical error (precision) is determined by the distribution of the individual measurement points, while the systematic error (accuracy) refers to the difference between the actual value and the mean value of the measured values.*

In other words, our task is to determine the error propagation through the image processing operations. These questions are the subject of Section C.3.

In Section C.4 we deal with a number of probability distributions that are important for signal processing. In the last step, we turn to temporal and spatial fields of random variables, the *stochastic processes*, in Section C.5. This allows us to investigate stochastic processes in Fourier space.

Many modeling problems in signal processing, such as a linear regression in the simplest case, are overdetermined problems and therefore have no exact solution. They can only be solved in terms of minimizing an error measure. Therefore, there is a close connection between statistics and discrete modeling and the last two sections of this appendix deal with discrete modeling and inverse problems (Section C.6) and network models (Section 17.4). Digital image processing can be regarded as a sub-area of *digital signal processing*. Therefore, all methods for recording and analyzing measurements and their errors can also be applied to image processing. In particular, all measurements that we perform on an image, such as determining the size or position of an object or its mean gray value, are only meaningful if we can also estimate the measurement inaccuracy. This basic fact, familiar to every scientist and engineer, was often neglected in the early days of image processing. The use of empirical rather than scientifically based techniques made realistic error estimates impossible. Fortunately, knowledge in image processing has increased considerably. Today, many well-founded techniques are available that also enable error estimation.

In this context, it is necessary to distinguish between two important classes of errors. The *statistical error* describes the dispersion of the measured values when one and the same measurement is repeated (Fig. C.1). The width of the distribution of the measured values is a suitable measure of the statistical error and the center of the distribution is the mean value.

The deviation of the mean value from the actual value can occur independently of the statistical distribution and go beyond the statistical error limits. Such a deviation is called *systematic error*. Closely related to the difference between systematic and statistical error are the terms *precision* and *accuracy*. A measurement is precise but inaccurate if the statistical error is low but the systematic error is high (Fig. C.1a). Conversely, if the statistical error is high and the systematic error is low, the individual measurements are widely scattered, but the mean value is close to the actual value (Fig. C.1b).

It is easy — at least in principle — to obtain an estimate of the statistical error by repeating the same measurement many times. However, it is much more difficult to find systematic errors. They are often due to a lack of understanding of the measurement setup and the measurement procedure. Unknown or uncontrolled parameters that influence the measurement procedure can easily lead to systematic errors. Examples of systematic errors are *calibration error* and shifts due to temperature-dependent parameters in an experiment without temperature control.

In this chapter, we learn to treat image data as statistical quantities or random variables. We start with the statistical properties of the measured gray value at a single *sensor element* or *pixel* in Section C.2. Then we can use the statistical methods used in all scientific fields for measurements of a single quantity. This type of statistics is called *first-order statistics* because only the statistics at a single measurement point are considered.

Image processing operations calculate new quantities from measured gray values. In the simplest case, *point operations* (Section 7.1) only uses the gray value at a single pixel. In more complicated cases, new pixels are calculated from the gray values at many pixels. In each case, however, we need to know how the statistical properties, in particular the statistical error of the newly calculated values, depend on those of the initial values.

C.2 Random Variables

C.2.1 Probability Density and Histogram

Let's assume an experimental setup with which we record an object. The quantity measured at a specific point (pixel) in the image plane is the irradiance. Due to the statistical nature of the observed process, each measurement gives a different value.

This means that the observed radiation is not characterized by a single value, but by a *probability density function* (*PDF*) $f(g)$. In simple terms, this function indicates the probability with which a certain value g is observed. A measurand determined by random processes, such as radiation g, is referred to as *random variable* or *RV* for short.

In the following, we discuss continuous and discrete RVs and probabilities in parallel. Discrete probabilities are required because only discrete numbers can be processed with digital computers. Discrete values result from a process called quantization (see Section 5.6). Many formulas in this section contain continuous expressions on the left-hand side and their discrete counterparts on the right-hand side of the corresponding equations. In the continuous case, a certain gray value is measured in the interval g to $g + \mathrm{d}g$ with probability $f(g)\mathrm{d}g$. In the discrete case, we can only measure a finite number Q of gray values g_q ($q = 1, 2, \ldots, Q$) with probability f_q. Normally, the gray value of a pixel is stored in one byte. In this case, we can measure $Q = 256$ different gray values. Since the total probability of observing any gray value is one by definition, the probability density function must satisfy the following condition:

$$\int_{-\infty}^{\infty} f(g)\mathrm{d}g = 1, \qquad \sum_{q=1}^{Q} f_q = 1. \tag{C.1}$$

The definite integral of the PDF

$$F(g) = \int_{-\infty}^{g} f(g')\mathrm{d}g', \qquad F_q = \sum_{q'=1}^{q} f_{q'} \tag{C.2}$$

is called *distribution function*. As the PDF is a non-negative function, the distribution function increases monotonically from 0 to 1.

In general, the probability distribution is not known from the outset. Rather, it is estimated from measurements. If the observed process is *homogeneous*, i. e. independent of the position of the pixels in the image, there is a simple way to estimate the probability density, the so-called *histogram*.

The histogram of an image is a list (vector) that has one element for each quantization level. Each element contains the number of pixels whose gray value matches the index of the element. Histograms can be easily calculated for data of any dimension. First we set the whole histogram vector to zero. Then we sample all pixels of the image, take the gray value as the index of the list and increment the corresponding list element by one. The actual sampling algorithm depends on how the image is stored.

It is also possible to estimate the probability density for image data with a higher resolution, for example 16-bit images or floating point images. To do this, we divide the range of possible values into Q intervals of equal width. The value assigned to each interval is in the middle of the interval. However, when deciding which interval a value falls into, we must use the interval limits. If we do not make this distinction, values such as the mean, which we calculate from the histogram, will be distorted.

C.2.2 Mean Value, Variance and Moments

The two basic parameters that describe an RV g are the mean value μ (also known as *expected value* $E(g)$ or Eg) and the variance. The *mean* is defined as

$$\boxed{\mu = \int_{-\infty}^{\infty} g f(g) \mathrm{d}g, \quad \mu = \sum_{q=1}^{Q} g_q f_q, \quad \text{mean.}} \tag{C.3}$$

The mean value can also be estimated by P measurements and averaging without having to know the PDF explicitly:

$$\mu = \mathrm{E}\, g \approx \frac{1}{P} \sum_{p=1}^{P} g_p. \tag{C.4}$$

Since it is impossible to perform an infinite number of measurements, the determination of the mean value via Eq. (C.4) remains an estimate with a certain uncertainty, which depends on the form of the PDF, i. e. the type of random process and the number of measured values.

Using the estimated mean according to Eq. (C.4), the variance can also be estimated in a similar way:

$$\sigma^2 = \operatorname{var} g \approx \frac{1}{P-1} \sum_{p=1}^{P} (g_p - \mu)^2. \tag{C.5}$$

If there is only one measurement, the variance is unknown. It is therefore understandable that it is not divided by the number P of measured values, but by $P - 1$. Then Eq. (C.4) is undetermined due to the expression 0/0 if there is only one measurement.

The *variance* $\sigma^2 = \operatorname{var} g = \mathrm{E}\left((g - \mu)^2\right)$ is a measure of how far the measured values deviate from the mean value:

$$\boxed{\sigma^2 = \int_{-\infty}^{\infty} (g - \mu)^2 f(g) \mathrm{d}g, \quad \sigma^2 = \sum_{q=1}^{Q} (g_q - \mu)^2 f_q, \quad \text{variance.}} \tag{C.6}$$

The probability density function can be characterized more precisely by similar quantities as the variance, the *moments* of n-th order $\mu_n = E\left((g - \mu)^n\right)$:

$$\mu_n = \int_{-\infty}^{\infty} (g - \mu)^n f(g) \mathrm{d}g, \quad \mu_n = \sum_{q=1}^{Q} (g_q - \mu)^n f_q. \tag{C.7}$$

The first, central moment is — by definition — zero. The second, μ_2, corresponds to the variance and the third moment, the *skewness*, μ_3, is a measure of the asymmetry of the PDF around the mean value. If it is a function with even symmetry, $f(-(g - \mu)) = f(g - \mu)$, then the third and all other moments of odd order disappear.

C.2.3 Functions of Random Variables

Each image processing operation changes the values g at the individual pixels. In the simplest case, a function p: $h = p(g)$ transforms the value g into the value h. Such an operation is referred to in image processing as *point operator* (Section 7.2). Since g is an RV, so is h, and we need to know its PDF so that we can determine the statistical properties of the image after applying the point operation.

It is obvious that the PDF f_h of h has the same form as the PDF f_g of g if p is a linear function $h = a_0 + a_1 g$:

$$f_h(h) = \frac{f_g(g)}{|a_1|} = \frac{f_g((h - a_0)/a_1)}{|a_1|}. \tag{C.8}$$

We used the inverse function $g = p^{-1}(h) : g = (h - a_0)/a_1$ to write g as a function of h.

From Eq. (C.8) it is intuitively clear that in the general case of a nonlinear function $p(g)$ the slope a_1 is replaced by the first derivative $\dot{p}(g)$ of $p(g)$. Further difficulties arise when the inverse function is not unique. An important example is the quadratic function $h = g^2$, whose inverse function has the two branches $g_{1,2} = \pm\sqrt{h}$. In such a case, all branches of the inverse function must be taken into account to determine the PDF of h

Theorem C.1 (PDF of the function of a random variable) *Let f_g be the PDF of the RV g and p a differentiable function $h = p(g)$. Then the RV h has the PDF*

$$\boxed{f_h(h) = \sum_{s=1}^{S} \frac{f_g(g_s)}{|\dot{p}(g_s)|},} \tag{C.9}$$

where g_s represents the S branches of the multivalued inverse function $h = p(g)$.

A monotone function p has a single inverse function $p^{-1}(h)$. This simplifies Eq. (C.9) to

$$f_h(h) = \frac{f_g(p^{-1}(h))}{|\dot{p}(p^{-1}(h))|}. \tag{C.10}$$

In image processing, the following inverse problem often arises in relation to PDFs. The signal g with a PDF is to be transformed into a signal h in such a way that h has a specific PDF. This inverse problem has a surprisingly simple solution. The transformation

$$h = F_h^{-1}(F_g(g)) \tag{C.11}$$

converts the $f_g(g)$-distributed RV g into the $f_h(h)$-distributed RV h. The solution is particularly simple for a transformation into a signal with a *uniform distribution*, since then F^{-1} is a constant function and $h = F_g(g)$.

Now we can look at the mean and variance of functions of an RV. According to the definition of the mean value in Eq. (C.3), the mean value of h is given by

$$\mathrm{E}h = \mu_h = \int_{-\infty}^{\infty} h f_h(h) \mathrm{d}h. \tag{C.12}$$

We can also express it directly using the transformation function $p(g)$ and the PDF $f_g(g)$:

$$\mathrm{E}h = \mathrm{E}\left(p(g)\right) = \int_{-\infty}^{\infty} p(g) f_g(g) \mathrm{d}g. \tag{C.13}$$

Intuitively, we are tempted to assume that the mean value of h can also be calculated from the mean value of g with $Eh = p(Eg)$. However, this is only correct if p is a linear function. We assume that $p(g)$ can be approximated by a second-order polynomial:

$$p(g) = p(\mu_g) + \dot{p}(\mu_g)(g - \mu_g) + \ddot{p}(\mu_g)(g - \mu_g)^2/2 + \ldots. \tag{C.14}$$

Then follows

$$\mu_h \approx p(\mu_g) + \ddot{p}(\mu_g)\sigma_g^2/2. \tag{C.15}$$

This equation shows that $\mu_h \approx p(\mu_g)$ is only a good approximation if both the curvature of $p(g)$ and the variance of g are small, i. e. the function p in the interval $[\mu - 3\sigma, \mu + 3\sigma]$ can be well approximated by a linear function.

The variance of h results in a first-order approximation to

$$\boxed{\sigma_h^2 \approx \left|\dot{p}(\mu_g)\right|^2 \sigma_g^2.} \tag{C.16}$$

This expression is only exact for linear functions p.

Example C.1: Standard deviation of the power of an RV

The standard deviation can be easily calculated for the power of an RV:

$$h = g^\gamma \quad \rightsquigarrow \quad \frac{\mathrm{d}h}{\mathrm{d}g} = \gamma g^{\gamma-1} \quad \rightsquigarrow \quad \frac{\sigma_h}{h} = \gamma \frac{\sigma_g}{g}$$

The following simple and useful relationships for the mean and variance result directly from the previous considerations in this section (a is a constant):

$$E(ag) = aEg, \quad \mathrm{var}(ag) = a^2 \,\mathrm{var}\, g, \quad \mathrm{var}\, g = E(g^2) - (Eg)^2. \tag{C.17}$$

C.3 Multiple Random Variables

In image processing, we are not dealing with just one, but with (very) many random variables. As a rule, a new RV is calculated from RVs at many pixels. It is therefore important to examine the statistics of multiple RVs. In this section, we take the first step and examine the statistical properties of multiple RVs and of functions of multiple RVs.

C.3.1 Joint Probability Density Functions

First, we consider how we can describe the common statistical properties of several RVs. In general, for two RVs g_1 and g_2 these will not be described by their individual PDFs $f(g_1)$ and $f(g_2)$. Rather, it is necessary to define a *common probability density function* $f(g_1, g_2)$. Only if two RVs are *independent*, i. e. if the probability that g_1 assumes a certain value does not depend on the value of g_2, can we describe the joint PDF as the product of the individual PDFs, the *marginal PDFs*:

$$f(g_1, g_2) = f_{g_1}(g_1) f_{g_2}(g_2) \quad \Leftrightarrow \quad g_1, g_2 \text{ independent.} \tag{C.18}$$

Accordingly, for P RVs g_p, the random vector $\boldsymbol{g}$, the joint PDF is given by $f(g_1, g_2, \ldots, g_P) = f(\boldsymbol{g})$. P RVs are independent if and only if the joint PDF can be written as a product of the marginal PDFs:

$$f(\boldsymbol{g}) = \prod_{p=1}^{P} f_{g_p}(g_p) \quad \Leftrightarrow \quad g_p \text{ independent, } p = 1, \ldots, P. \tag{C.19}$$

C.3.2 Covariance and Correlation

The covariance indicates the extent to which the fluctuations of two RVs g_p and g_q are related to each other. In extension of the definition of *variance* in Eq. (C.6), the *covariance* is defined as

$$\sigma_{pq} = E\Big((g_p - \mu_p)(g_q - \mu_q)\Big) = E(g_p g_q) - E(g_p)E(g_q). \tag{C.20}$$

For P random variables, the covariances form a symmetric $P \times P$ matrix, the *covariance matrix* $\boldsymbol{\Sigma} = \operatorname{cov} \boldsymbol{g}$. The diagonal of this matrix contains the variances of the P RVs.

The *correlation coefficient* relates the covariance of two RVs to the corresponding variances:

$$c_{pq} = \frac{\sigma_{pq}}{\sigma_p \sigma_q} \quad \text{with} \quad |c| \leq 1. \tag{C.21}$$

Two RVs g_p and g_q are called *uncorrelated* if the covariance σ_{pq} is zero. According to Eqs. (C.20) and (C.21), the following relationships then apply for uncorrelated RVs:

$$\sigma_{pq} = 0 \Leftrightarrow c_{pq} = 0 \Leftrightarrow E(g_p g_q) = E(g_p)E(g_q) \Leftrightarrow g_p, g_q \text{ uncorrelated.} \tag{C.22}$$

From the last condition and Eq. (C.18) it follows that independent RVs are uncorrelated.

At first glance, it seems that only the statistical properties of independent RVs are easy to handle. In this case, we only need to consider the marginal PDFs of the individual variables and their means and variances. In the general case, on the other hand, we have to consider the dependencies of the individual RVs on each other, as given by the covariance matrix $\boldsymbol{\Sigma}$. However, since the covariance matrix is symmetrical, we can always find a coordinate system, i. e. a linear combination of the RVs, in which the covariance matrix is diagonal and the transformed RVs are therefore uncorrelated.

C.3.3 Linear Functions

Extending the discussion of functions of a single RV in Section C.2.3, we can express the mean value of a function of multiple RVs $h = p(g_1, g_2, \ldots, g_P)$ directly by the joint PDF:

$$Eh = \int\limits_{-\infty}^{\infty} p(g_1, g_2, \ldots, g_P) f(g_1, g_2, \ldots, g_P) \mathrm{d}g_1 \mathrm{d}g_2 \ldots \mathrm{d}g_P. \tag{C.23}$$

From this general relationship it follows that the mean value of any linear function

$$h = \sum_{p=1}^{P} a_p g_p \tag{C.24}$$

is given by a linear combination of the mean values of the RVs g_p:

$$E\left(\sum_{p=1}^{P} a_p g_p\right) = \sum_{p=1}^{P} a_p E\left(g_p\right). \tag{C.25}$$

This is a generally valid result that neither assumes the independence of the RVs nor depends on the PDF. As a special case, it contains the simple relationships

$$E(g_1 + g_2) = Eg_1 + Eg_2, \quad E(g_1 + a) = Eg_1 + a. \tag{C.26}$$

Unfortunately, the variance of functions of multiple RVs cannot be calculated so easily even in the linear case. Let $\boldsymbol{g}$ be a vector with P RVs, $\boldsymbol{h}$ a vector with Q RVs, which is a linear combination of the P RVs $\boldsymbol{g}$, $\boldsymbol{M}$ a $Q \times P$ matrix of coefficients and $\boldsymbol{a}$ a column vector with Q coefficients. Then the following applies

$$\boldsymbol{h} = \boldsymbol{M}\boldsymbol{g} + \boldsymbol{a} \quad \text{with} \quad E(\boldsymbol{h}) = \boldsymbol{M}E(\boldsymbol{g}) + \boldsymbol{a} \tag{C.27}$$

in generalization of Eq. (C.25). If $P = Q$, Eq. (C.27) can be understood as a coordinate transformation in a P-dimensional vector space. Therefore, it is not surprising that the symmetric covariance matrix transforms like a second level tensor [236]:

$$\boxed{\operatorname{cov}(\boldsymbol{h}) = \boldsymbol{M} \operatorname{cov}(\boldsymbol{g}) \boldsymbol{M}^T.} \tag{C.28}$$

We illustrate this important general relationship using three examples.

Variance of the Mean Value of RVs

First, we discuss the calculation of the variance of the *mean* $\overline{g}$ of P RVs with the same mean and the same variance σ^2. We also assume that the RVs are uncorrelated. Then the matrix $\boldsymbol{M}$ and the covariance matrix $\operatorname{cov} \boldsymbol{g}$ are given by

$$\boldsymbol{M} = \frac{1}{P}[1, 1, 1, \ldots, 1] \quad \text{and} \quad \operatorname{cov}(\boldsymbol{g}) = \begin{bmatrix} \sigma^2 & 0 & \ldots & 0 \\ 0 & \sigma^2 & \ldots & 0 \\ \vdots & \vdots & \ddots & \vdots \\ 0 & 0 & \ldots & \sigma^2 \end{bmatrix} = \sigma^2 \boldsymbol{I}.$$

When used in Eq. (C.28), the result is

$$\sigma^2_{\overline{g}} = \frac{1}{P}\sigma^2. \tag{C.29}$$

This means that the variance of the mean value $\sigma^2_{\overline{g}}$ is smaller than the variance of the individual values by a factor of P, and the *standard deviation* $\sigma_{\overline{g}}$ is reduced by a factor of $P^{-1/2}$. This means that we have to take four times more measurements to halve the standard deviation of the mean value.

However, this only applies to uncorrelated RVs. If the RVs are maximally correlated ($c_{pq} = 1, \sigma_{pq} = \sigma^2$), then it follows from Eq. (C.28) that the variance of the mean is equal to the variance of the individual values. In this case, it is not possible to reduce the variance by averaging.

Variance of the Sum and Difference of Uncorrelated RVs with Unequal Variances

We consider P uncorrelated RVs with *unequal* variances σ_p^2 and calculate the variance of the sum of the RVs. From Eq. (C.26) we already know that the mean of the sum is equal to the sum of the means (also for correlated RVs). From Eq. (C.28) it follows without any assumption about the PDFs that for uncorrelated RVs the variance of the sum is also equal to the sum of the variances:

$$\operatorname{var} \sum_{p=1}^{P} g_p = \sum_{p=1}^{P} \operatorname{var} g_p. \tag{C.30}$$

Similarly, Eq. (C.28) shows that the variances of the difference between two uncorrelated RVs add up:

$$\operatorname{var}(g_1 - g_2) = \operatorname{var} g_1 + \operatorname{var} g_2. \tag{C.31}$$

Linear Combination of Uncorrelated RVs

As a third example, we take Q RVs h_q, which are a linear combination of P uncorrelated RVs g_p with equal variances σ^2:

$$h_q = \boldsymbol{a}_q^T \boldsymbol{g}. \tag{C.32}$$

The vectors $\boldsymbol{a}_q^T$ form the rows of the $Q \times P$ matrix $\boldsymbol{M}$ in Eq. (C.27), and the covariance matrix of $\boldsymbol{h}$ results from Eq. (C.28) to

$$\operatorname{cov}(\boldsymbol{h}) = \sigma^2 \boldsymbol{M}\boldsymbol{M}^T = \sigma^2 \begin{bmatrix} \boldsymbol{a}_1^T\boldsymbol{a}_1 & \boldsymbol{a}_1^T\boldsymbol{a}_2 & \dots & \boldsymbol{a}_1^T\boldsymbol{a}_Q \\ \boldsymbol{a}_1^T\boldsymbol{a}_2 & \boldsymbol{a}_2^T\boldsymbol{a}_2 & \dots & \boldsymbol{a}_2^T\boldsymbol{a}_Q \\ \vdots & \vdots & \ddots & \vdots \\ \boldsymbol{a}_1^T\boldsymbol{a}_Q & \boldsymbol{a}_2^T\boldsymbol{a}_Q & \dots & \boldsymbol{a}_Q^T\boldsymbol{a}_Q \end{bmatrix}. \tag{C.33}$$

We can learn two things from this equation. Firstly, the variance of the RV h_q is given by $\boldsymbol{a}_q^T\boldsymbol{a}_q$, i. e. the sum of the squares of the coefficients:

$$\sigma^2(h_q) = \sigma^2 \boldsymbol{a}_q^T\boldsymbol{a}_q. \tag{C.34}$$

Secondly, the RVs h_p and h_q are correlated — although the RVs g_p are uncorrelated — if the scalar product of the coefficient vectors $\boldsymbol{a}_p^T\boldsymbol{a}_q$ does not vanish, i. e.if they are not orthogonal. From this we can draw the important conclusion that only with orthogonal transformation matrices $\boldsymbol{M}$ in Eq. (C.27) uncorrelated RVs remain uncorrelated. In the case of orthonormal transformations, the variances are also retained due to $\boldsymbol{a}_q^T\boldsymbol{a}_q = 1$. All transformations that we learned about in Appendix B.1 and B.2, including the Fourier transform, fulfill this condition.

For correlated RVs, we can draw the important conclusion that a suitable transformation $\boldsymbol{M}$ can always be used to find a linear combination of RVs that is uncorrelated. This follows directly from the elementary theorem of linear algebra that every symmetric matrix can be diagonalized by a suitable transformation, the so-called *principle axis transformation* [316, 351]. The axes of the principal coordinate system are called *eigenvectors* and the following applies

$$\operatorname{cov}(\boldsymbol{h})\boldsymbol{e}_p = \sigma_p^2 \boldsymbol{e}_p. \tag{C.35}$$

This means that the matrix-vector multiplication is reduced to a vector-scalar multiplication. The factor is called the *eigenvalue* to the eigenvector $\boldsymbol{e}_p$. For the covariance matrix, the pth eigenvalue is the variance σ_p^2 in the direction of the eigenvector $\boldsymbol{e}_p$.

C.3.4 Nonlinear Functions

We can generalize the analysis of the variance of functions of several RVs to nonlinear functions if they are sufficiently linear around the mean values. As in Section C.2.3, we develop the non-linear function $p_q(\boldsymbol{g})$ into a Taylor series around the mean:

$$h_q = p_q(\boldsymbol{g}) \approx p_q(\boldsymbol{\mu}) + \sum_{p=1}^{P} \frac{\partial p_q}{\partial g_p}(g_p - \mu_p). \tag{C.36}$$

The comparison with Eq. (C.27) shows that the matrix $\boldsymbol{M}$ must be replaced by the matrix

$$\boldsymbol{J} = \begin{bmatrix} \frac{\partial p_1}{\partial g_1} & \frac{\partial p_1}{\partial g_2} & \cdots & \frac{\partial p_1}{\partial g_P} \\ \frac{\partial p_2}{\partial g_1} & \frac{\partial p_2}{\partial g_2} & \cdots & \frac{\partial p_2}{\partial g_P} \\ \vdots & \vdots & \ddots & \vdots \\ \frac{\partial p_Q}{\partial g_1} & \frac{\partial p_Q}{\partial g_2} & \cdots & \frac{\partial p_Q}{\partial g_P} \end{bmatrix}. \tag{C.37}$$

which is known as *Jacobi matrix* of the transformation $\boldsymbol{h} = \boldsymbol{p}(\boldsymbol{g})$. Therefore, to a first approximation, the covariance of $\boldsymbol{h}$ is given by

$$\boxed{\mathrm{cov}(\boldsymbol{h}) \approx \boldsymbol{J}\,\mathrm{cov}(\boldsymbol{g})\boldsymbol{J}^T.} \tag{C.38}$$

C.4 Probability distributions

In the previous sections, we have derived a number of general properties of random variables that did not require knowledge of probability distributions. In this section, we discuss a number of probability distributions that are relevant to image processing.

As an introduction, we consider the PDFs of multiple RVs. We restrict ourselves to two simple cases. First, we consider the addition of two RVs. If two RVs g_1 and g_2 are independent, then the resulting PDF for additive superposition $g = g_1 + g_2$ is given by the *convolution integral*

$$p_g(g) = \int_{-\infty}^{\infty} p_{g_1}(h) p_{g_2}(g - h) \mathrm{d}h. \tag{C.39}$$

This property results from the multiplicative nature of probabilities. The probability $p_g(g)$ of measuring the value g is the product of the probabilities of measuring the values $g_1 = h$ and $g_2 = g - h$. The integral in Eq. (C.39) is necessary so that we take into account all combinations whose sum gives the value g.

The same procedure can be used for the product of two RVs if we convert the multiplication into an addition by forming the logarithm: $\ln g = \ln g_1 + \ln g_2$. The PDF of the logarithm of an RV can be calculated using Eq. (C.10).

C.4.1 Poisson Distribution

First we look at the image capture. An element of a semiconductor image sensor that receives a certain irradiance collects an average of N electrons

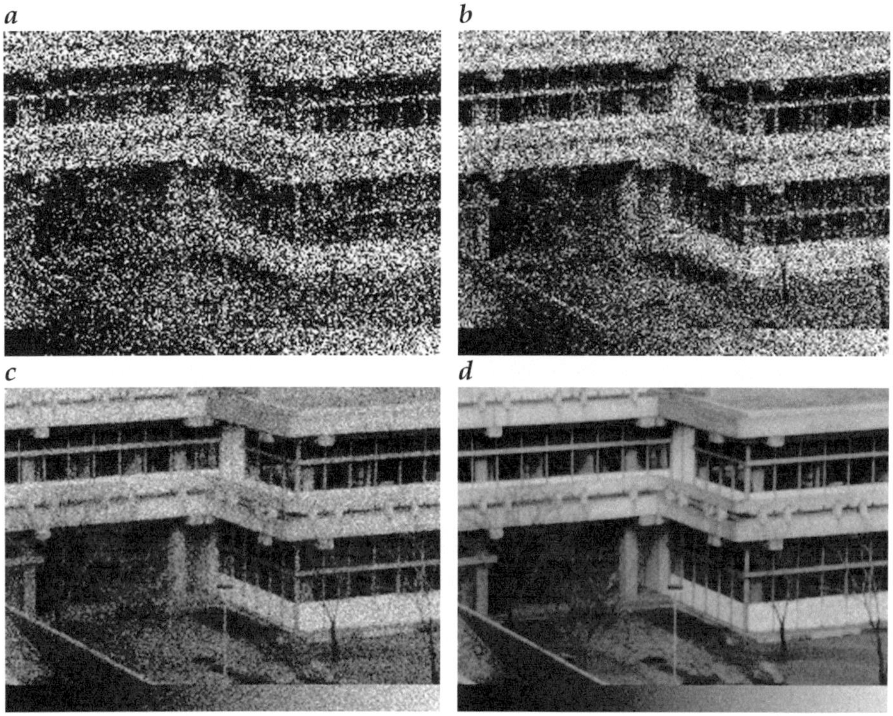

Figure C.2: *Simulation of images with Poisson noise at low irradiance, in which a maximum of **a** 3, **b** 10, **c** 100 and **d** 1000 charge carriers were collected during the exposure time. A linear gray wedge is shown at the bottom of the image.*

within a given *exposure time* Δt by photon absorption. Therefore, the average rate λ of the events is given by

$$\lambda = \frac{N}{\Delta t}. \tag{C.40}$$

Due to the random nature of the photon flux, a slightly different number of electrons will be collected with each exposure. A random process in which events occur on average $\lambda\Delta t$ times is called *Poisson process* $P(\lambda\Delta t)$. The discrete probability density function (PDF) f_n that n events take place is

$$\boxed{P(\lambda\Delta t): \quad f_n = \exp(-\lambda\Delta t)\frac{(\lambda\Delta t)^n}{n!}, \quad n \geq 0} \tag{C.41}$$

with the mean value and the variance

$$\boxed{\mu = \lambda\Delta t \quad \text{and} \quad \sigma^2 = \lambda\Delta t.} \tag{C.42}$$

Simulated images with Poisson noise at low irradiance shows Fig. C.2. For low mean values, the Poisson distribution is asymmetric with a longer

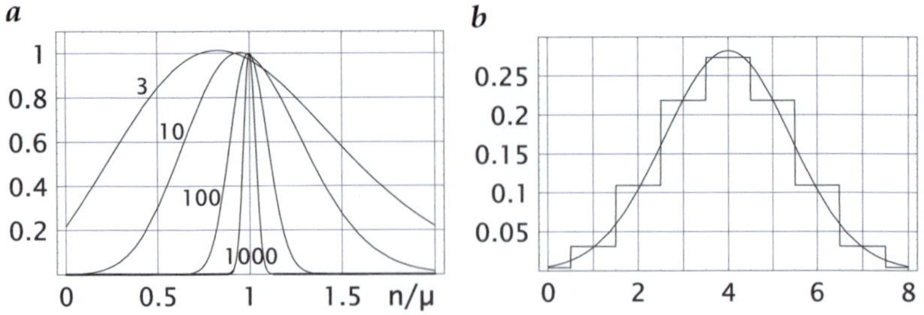

Figure C.3: ***a*** *Poisson probability density function* $P(\mu)$ *for the mean values* $\mu = 3, 10, 100$ *and 1000. The x-axis is normalized to a mean value of 1;* $P(\lambda\Delta t)$ *was multiplied by* $\sigma\sqrt{2\pi}$ *for normalization.* ***b*** *Discrete binomial distribution* $B(8, 1/2)$ *with a mean value of 4 and a variance of 2 and the corresponding normal distribution* $N(4, 2)$.

extension of the distribution towards higher values (Fig. C.3a). But even for moderate mean values (100), the distribution is surprisingly well symmetrical.

A typical CCD sensor element (Section D, > R8) collects around 10,000 or more electrons by absorbing photons. Therefore, the standard deviation is about 100 or 1% due to the photon noise. This figure clearly shows that even a perfect image sensor, where no additional electronic noise occurs, has a considerable noise level.

A Poisson process has the following important properties:

1. The standard deviation σ is not constant. Rather, it grows with the square root of the number of events. Therefore, the noise is signal-dependent.
2. Non-overlapping exposures are statistically independent events [236, Section. 3.4]. This means that images taken with the same sensor but at different times are independent RVs.
3. The Poisson process is additive. The sum of two independent Poisson distributed RVs with the mean values μ_1 and μ_2 is also Poisson distributed with the mean value and the variance $\mu_1 + \mu_2$.

C.4.2 Normal and Binomial Distributions

Many processes with continuous RVs can be well described by a *normal distribution* or *Gaussian distribution* $N(\mu, \sigma)$ with the mean value μ and the variance σ^2:

$$N(\mu, \sigma): \quad f(g) = \frac{1}{\sqrt{2\pi}\sigma} \exp\left(-\frac{(g-\mu)^2}{2\sigma^2}\right). \tag{C.43}$$

From Eq. (C.43) we see that the normal distribution is completely described by the mean and the variance.

The discrete analog to the normal distribution is the *binomial distribution*

$$B(Q,p): \quad f_q = \frac{Q!}{q!\,(Q-q)!} p^q (1-p)^{Q-q}, \quad 0 \le q < Q. \tag{C.44}$$

The natural number Q is the number of possible results and, together with the parameter $p \in]0,1[$, determines the mean value and the variance:

$$\mu = Qp \quad \text{and} \quad \sigma^2 = Qp(1-p). \tag{C.45}$$

In the often required symmetrical case with $p = 1/2$, the equations are simplified considerably:

$$\boxed{B(Q,1/2): \quad f_q = \frac{1}{2^Q} \frac{Q!}{q!\,(Q-q)!} \quad 0 \le q < Q, \quad \mu = \frac{Q}{2} \quad \text{and} \quad \sigma^2 = \frac{Q}{4}.} \tag{C.46}$$

Even for moderate values of Q, the symmetric binomial distribution comes very close to the normal distribution (Fig. C.3b).

As a generalization of Eq. (C.43), the joint PDF $N(\boldsymbol{\mu}, \boldsymbol{\Sigma})$ for multiple RVs, i. e. for the random vector $\boldsymbol{g}$ with the mean $\boldsymbol{\mu}$ and the covariance matrix $\boldsymbol{\Sigma}$, is given by

$$N(\boldsymbol{\mu}, \boldsymbol{\Sigma}): \ f(\boldsymbol{g}) = \frac{1}{(2\pi)^{P/2}\sqrt{\det \boldsymbol{\Sigma}}} \exp\left(-\frac{(\boldsymbol{g}-\boldsymbol{\mu})^T \boldsymbol{\Sigma}^{-1} (\boldsymbol{g}-\boldsymbol{\mu})}{2}\right). \tag{C.47}$$

At first glance, this expression looks very complex. But it is not. We only have to consider that the symmetric covariance matrix becomes a diagonal matrix when we rotate the coordinate system into its principal axis system. Then the joint PDF becomes a separable function

$$f(\boldsymbol{g}') = \prod_{p=1}^{P} \frac{1}{(2\pi\sigma_p^2)^{1/2}} \exp\left(-\frac{(g_p' - \mu_p)^2}{2\sigma_p^2}\right) \tag{C.48}$$

with variances σ_p^2 along the principal axes, and the components g_p' of the random vector become independent RVs.

For uncorrelated RVs with the same variance σ^2, the $N(\boldsymbol{\mu}, \boldsymbol{\Sigma})$ distribution is reduced to an isotropic normal distribution $N(\boldsymbol{\mu}, \sigma)$:

$$N(\boldsymbol{\mu}, \sigma): \quad f(\boldsymbol{g}) = \frac{1}{(2\pi\sigma^2)^{P/2}} \exp\left(-\frac{\left|(\boldsymbol{g}-\boldsymbol{\mu})\right|^2}{2\sigma^2}\right). \tag{C.49}$$

The great importance of the normal distribution results from the *central limit theorem* (Theorem B.6, p. 632), which we have already considered in relation to cascaded convolutions in Section 6.3.4 and Appendix B.1.4. The importance of random variables in image processing should be briefly emphasized here. The central limit theorem states that the PDF of a sum of

RVs tends to a normal distribution under very general conditions, which are almost always fulfilled in practical applications in image processing. As already discussed in Section C.3, weighted sums of many RVs are often formed in image processing. This is why these variables usually have a normal distribution.

C.4.3 Uniform Distribution

In a *uniform distribution* or *uniform distribution*, the probability density function is constant in an interval. This distribution is important for describing quantization (Section 5.6) if continuous measured values fall with equal probability into a quantization interval from $g_n - \Delta g/2$ to $g_n + \Delta g/2$. Then the density function is

$$U(g_n, \Delta g): \quad f(g) = \begin{cases} 1/\Delta g & g_n - \Delta g \leq g \leq g_n + \Delta g \\ 0 & \text{otherwise.} \end{cases} \tag{C.50}$$

Mean value and variance result in:

$$\mu = g_n, \quad \sigma^2 = \frac{1}{12}\Delta g^2. \tag{C.51}$$

C.4.4 Other Distributions

Despite the importance of the *normal distribution*, other distributions are also important in image processing. They mainly appear when RVs are combined by non-linear functions.

As a first example, we consider the conversion from *cartesian* to *polar coordinates*. To do this, we take the random vector $\boldsymbol{g} = [g_1, g_2]^T$ with independent $N(0, \sigma)$-distributed components. Then it can be shown [236, Sect. 6.3] that the magnitude of the vector $r = (g_1^2, g_2^2)^{1/2}$ and the angle $\phi = \arctan(g_2/g_1)$ are independent RVs. The absolute value has a *Rayleigh distribution*

$$R(\sigma): \quad f(r) = \frac{r}{\sigma^2}\exp\left(-\frac{r^2}{2\sigma^2}\right) \quad \text{für} \quad r > 0 \tag{C.52}$$

with the mean value and the variance

$$\mu_R = \sigma\sqrt{\pi/2} \quad \text{and} \quad \sigma_R^2 = \sigma^2\frac{4-\pi}{2}. \tag{C.53}$$

The angle ϕ has an *uniform distribution*

$$f(\phi) = \frac{1}{2\pi}. \tag{C.54}$$

We come to a generalization of the Rayleigh distribution by now considering the magnitude of a P-dimensional vector with P independent RVs with

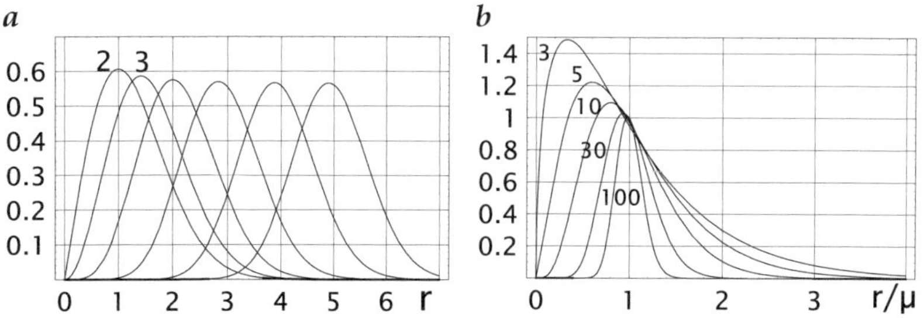

Figure C.4: ***a*** *Chi distribution with 2 (Rayleigh distribution), 3 (Maxwell distribution) and more degrees of freedom, as indicated;* **b** *chi square distribution in a normalized representation with a mean value of 1 and degrees of freedom, as indicated.*

an $N(0,\sigma)$-normal distribution, i. e.

$$r = \left(\sum_{p=1}^{P} g_p^2\right)^{1/2}. \tag{C.55}$$

This has a *chi distribution* with P degrees of freedom

$$\chi(P,\sigma): \quad f(r) = \frac{2r^{P-1}}{2^{P/2}\Gamma(P/2)\sigma^P}\exp\left(-\frac{r^2}{2\sigma^2}\right) \quad \text{for} \quad r > 0 \tag{C.56}$$

with the mean value

$$\mu_\chi = \sigma\frac{\sqrt{2}\,\Gamma(P/2+1/2)}{\Gamma(P/2)} \approx \sigma\sqrt{P-1/2} \quad \text{for} \quad P \gg 1 \tag{C.57}$$

and the variance

$$\sigma_\chi^2 = \sigma^2 P - \mu_\chi^2 \approx \sigma^2/2 \quad \text{for} \quad P \gg 1. \tag{C.58}$$

The mean value of the chi distribution increases with the square root of P, while the variance remains almost constant. For a high number of degrees of freedom, the chi distribution quickly converges to the normal distribution $N(\sigma\sqrt{P-1/2}, \sigma/\sqrt{2})$ (Fig. C.4a).

The PDF of the magnitude squared of a vector is a different function, since the squaring is a non-linear transformation function (Section C.2.3). Using Theorem C.1, the *chi square distribution* with P degrees of freedom can be calculated as

$$\chi^2(P,\sigma): \quad f(r) = \frac{r^{P/2-1}}{2^{P/2}\Gamma(P/2)\sigma^P}\exp\left(-\frac{r}{2\sigma^2}\right) \quad \text{for} \quad r > 0 \tag{C.59}$$

with the mean value and the variance

$$\mu_{\chi^2} = \sigma^2 P \quad \text{and} \quad \sigma_{\chi^2}^2 = 2\sigma^4 P. \tag{C.60}$$

The sum of squares of RVs is of particular importance because it can be used to calculate the error in determining the estimate of *variance*:

$$s^2 = \frac{1}{P-1}\sum_1^P (g_p - \mathrm{E}(g))^2 \quad \text{with} \quad \mathrm{E}(g) = \frac{1}{P}\sum_1^P g_p. \tag{C.61}$$

The normalized variance

$$\frac{(P-1)s^2}{\sigma^2} = \sum_1^P \left(\frac{g_p - \overline{g}}{\sigma}\right)^2 \tag{C.62}$$

has a *chi square distribution* with $P-1$ degrees of freedom [236, p. 200]. Therefore, the mean according to (Eq. C.60) is the estimate of the variance σ^2 (estimate without deviation) and the variance is $2\sigma^4/(P-1)$. For a few degrees of freedom, the chi-squared distribution shows significant deviations from the normal distribution (Fig. C.4b). For more than 100 degrees of freedom, however, it is normally distributed to a good approximation according to Fig. C.4b.

A reliable measurement of the variance requires many measurements. Of interest is not the variance of the estimate but the standard deviation relative to the estimated variance s^2. This results in

$$\frac{\sqrt{\mathrm{var}(s^2)}}{s^2} = \sqrt{\frac{2}{P-1}}. \tag{C.63}$$

Even with $P = 100$ degrees of freedom, the standard deviation of the variance is still around 14% (the relative standard deviation of the standard deviation is only half as large at 7%).

C.5 Stochastic Processes and Fields

Our previous considerations on statistics have in no way taken into account the temporal and spatial relationships between the individual points of a multidimensional signal. If we want to do this, we must consider the signal (image, image sequence) as a whole as a statistical quantity. This leads to the notions of a *stochastic field* for spatial data and a *stochastic process* for time series.

In the case of a $M \times N$ image, a stochastic field consists of a $M \times N$ matrix of random variables. This means that the joint PDF has MN variables. As with a single stochastic variable, see Eq. (C.3), the mean value of the stochastic field is the sum of all possible states q of the system:

$$\overline{G_{m,n}} = \sum_{q=1}^{Q^{MN}} f_q(\boldsymbol{G})\boldsymbol{G}_q. \tag{C.64}$$

In the equation, we have assumed Q quantization levels. This means that each individual pixel Q can assume different states and the stochastic $M \times N$ field in combination Q^{MN} can assume states $\boldsymbol{G}_q$. Thus, this general concept is completely unusable due to the combinatorial explosion of the number of possible states. Therefore, we are forced to find simpler concepts that allow us to adequately describe multidimensional signals as stochastic quantities. In this section we will present a practicable approach.

We start by determining the mean and variance of a stochastic field. In the same way as for a single random variable (see Eq. C.61), we can determine the mean $\boldsymbol{G}_p$ by P measurements under the same conditions by averaging as

$$\overline{\boldsymbol{G}} = \frac{1}{P}\sum_{p=1}^{P} \boldsymbol{G}_p. \tag{C.65}$$

Such a mean value is called *ensemble mean value*. Analogously, an estimated value of the *variance*, the *ensemble variance*, results as

$$S_G^2 = \frac{1}{P-1}\sum_{p=1}^{P}\left(\boldsymbol{G}_p - \overline{\boldsymbol{G}}\right)^2. \tag{C.66}$$

At this point, we already know the mean value and the variance at each pixel of an image. We can draw a number of interesting conclusions from these values. We can investigate the uniformity of the two quantities under given conditions, such as constant, homogeneous illumination.

C.5.1 Correlation and Covariance

In a second step, we can now correlate the gray values of two different positions. A measure for the correlation of values at different positions in the image is the *autocorrelation function*

$$R_{gg}(m,n;m',n') = \overline{G_{mn}G_{m'n'}}. \tag{C.67}$$

As in Eqs. (C.65) and (C.66), an ensemble mean is taken. The autocorrelation is usually of little use, since an image or signal usually consists of a superposition of a deterministic part with *mean-free noise*:

$$\boldsymbol{G}' = \boldsymbol{G} + \boldsymbol{N}, \quad \text{with} \quad \overline{\boldsymbol{G}'} = \boldsymbol{G} \quad \text{and} \quad \overline{\boldsymbol{N}'} = \boldsymbol{0}. \tag{C.68}$$

Then it is more useful to subtract the deterministic mean value to adequately characterize the noise:

$$C_{gg}(m,n;m',n') = \overline{(G_{mn} - \overline{G_{mn}})(G_{m'n'} - \overline{G_{m'n'}})}. \tag{C.69}$$

This function is called *autocovariance function*. For $m = m'$ and $n = n'$ it indicates the variance at the image point $[m,n]^T$, for all other values the *covariance* between the image points $[m,n]^T$ and $[m',n']^T$, which was introduced in Section C.3.2 Eq. (C.20). What is new here is that the autocovariance function now reflects the spatial relationships between different pixels. If the autocovariance is zero, the corresponding pixels are statistically uncorrelated.

The autocovariance function as defined in Eq. (C.69) is still unwieldy because it is four-dimensional. It can only be used sensibly if only a small number of pixels are correlated with each other, typically pixels that are close to each other.

The correlations are much simpler if the statistics do not explicitly depend on the position of the pixels. We then speak of a *homogeneous stochastic field*. Then the autocovariance function becomes *shift invariant*:

$$\begin{aligned} & C_{gg}(m+k, n+l; m'+k, n'+l) \\ &= C_{gg}(m,n;m',n') \\ &= C_{gg}(m-m', n-n'; 0,0) \\ &= C_{gg}(0,0; m'-m, n'-n). \end{aligned} \tag{C.70}$$

We obtain the last two relationships with $(k,l) = (-m', -n')$ and $(k,l) = (-m, -n)$. This also means that the variance $C_{gg}(m,n;m,n)$ no longer depends on the position in the image, but is the same for all pixels.

Since the autocovariance function now only depends on the distance between points, it is reduced from a four-dimensional to a two-dimensional function. Fortunately, many stochastic processes are homogeneous. Because of the shift invariance, the autocovariance function for a homogeneous stochastic field can also be obtained by spatial averaging:

$$C_{gg}(m,n) = \frac{1}{MN}\sum_{m'=0}^{M-1}\sum_{n'=0}^{N-1}(G_{m'n'} - \overline{G_{m'n'}})(G_{m'+m,n'+n} - \overline{G_{m'+m,n'+n}}). \tag{C.71}$$

In general, we cannot be sure that a spatial averaging leads to the same result as an averaging over an ensemble. A stochastic field that fulfills this condition is called *ergodic*.

A further difficulty arises with indexing. As soon as $(m,n) \neq (0,0)$, the indices in Eq. (C.71) are larger than the range of the matrix. However, as shown in Section 6.3.4, we can think of the matrix as periodically continued and then speak of a *cyclic correlation*.

Finally, we illustrate the meaning of the autocovariance function with a simple example. We consider an image with a deterministic part to which *mean-free homogeneous noise* is superimposed, see Eq. (C.68). We further assume that all pixels are statistically uncorrelated. The mean value then gives the deterministic part of the image and the autocovariance function disappears for all values except 0:

$$\boldsymbol{C}_{gg} = \sigma^2\, {}^{oo}\boldsymbol{P} \quad \text{or} \quad C_{gg}(m,n) = \sigma^2 \delta_m \delta_n. \tag{C.72}$$

At the zero point, the autocovariance function is equal to the variance of the noise, otherwise it is zero. The autocovariance function can therefore be used to determine whether the noise at pixels is correlated with each other. This is important because the degree of correlation between pixels affects the statistical properties of image processing operations (Section C.3.3).

In a similar way that we correlate an image with itself, we can also correlate two different images $\boldsymbol{G}$ and $\boldsymbol{H}$ with each other. These can be images from different scenes as well as images from a sequence taken at different times.

In analogy to Eq. (C.71), the *cross-correlation function* and the *cross-covariance function* are defined as

$$R_{gh}(m,n) = \frac{1}{MN}\sum_{m'=0}^{M-1}\sum_{n'=0}^{N-1} G_{m'n'} H_{m'+m,n'+n} \tag{C.73}$$

$$C_{gh}(m,n) = \frac{1}{MN}\sum_{m'=0}^{M-1}\sum_{n'=0}^{N-1} (G_{m'n'} - \overline{G_{m'n'}})(H_{m+m',n+n'} - \overline{H_{m+m',n+n'}}). \tag{C.74}$$

The cross-correlation is similar to *convolution* (Section 6.3.4, > R15). The only difference is the sign of the indices (m', n') in the second term.

C.5.2 Stochastic Fields in Fourier Space

In the last section we discussed stochastic fields in space. Because of the importance of the *Fourier transform* for signal and image processing (Section 6.3), we now turn to stochastic fields in Fourier space. For the sake of simplicity, we will restrict ourselves to the one-dimensional case. However, all statements discussed here can also be extended to higher dimensions.

The Fourier transformation requires complex numbers. However, this does not cause any problems for the statistical treatment, as we can consider the real and imaginary parts as independent stochastic variables. The definition of the mean value does not change, but we have to make a small change in the definition of the covariance compared to Eq. (C.20):

$$C_{pq} = E\left((g_p - \mu_p)^*(g_q - \mu_q)\right), \tag{C.75}$$

where the symbol $*$ indicates the formation of the conjugate complex. This definition ensures that the variance remains a real number:

$$\sigma_p^2 = E\left((g_p - \mu_p)^*(g_p - \mu_p)\right). \tag{C.76}$$

The 1D DFT maps the vector $\boldsymbol{g} \in \mathbb{C}^N$ to a vector $\tilde{\boldsymbol{g}} \in \mathbb{C}^N$. The components of $\tilde{\boldsymbol{g}}$ result as *scalar products* with the orthonormal basis vectors of the vector space $\mathbb{C}^N$ (compare Eqs. (6.17) and (6.18)):

$$\tilde{g}_v = \boldsymbol{b}_{\tilde{n}}^{*T}\boldsymbol{g} \quad \text{with} \quad \boldsymbol{b}_{\tilde{n}}^{*T}\boldsymbol{b}_{\tilde{n}'} = \delta_{\tilde{n}-\tilde{n}'}. \tag{C.77}$$

Therefore, the complex RVs in Fourier space are nothing more than linear combinations of the RVs in local space. If we assume that the RVs in the space are uncorrelated and have the same variance, i. e. represent a homogeneous stochastic field, then we can draw a far-reaching conclusion. According to Eq. (C.77), the vectors $\boldsymbol{b}_v$ are perpendicular to each other and have a magnitude of one. Therefore, from the considerations about functions of multiple RVs in Section C.3.3, especially Eq. (C.34), we can conclude that the RVs remain uncorrelated in Fourier space and have the same variance as in spatial space.

C.5.3 Spectra and Coherence

In Section C.5.1 we learned that homogeneous stochastic fields are described by the auto- and cross-correlation function. Now we look at the corresponding quantities in Fourier space.

Correlation in spatial space corresponds to a multiplication with the conjugate complex function in Fourier space (> R12):

$$\boldsymbol{G} \star \boldsymbol{G} \circ\!\!-\!\!\bullet\, P_{gg}(\boldsymbol{\kappa}) = \tilde{g}(\boldsymbol{\kappa})^* \tilde{g}(\boldsymbol{\kappa}) \tag{C.78}$$

and

$$\boldsymbol{G} \star \boldsymbol{H} \circ\!\!-\!\!\bullet\, P_{gh}(\boldsymbol{\kappa}) = \tilde{g}(\boldsymbol{\kappa})^* \tilde{h}(\boldsymbol{\kappa}). \tag{C.79}$$

In these equations, the correlation has been abbreviated with the $\star$ symbol, similar to how we abbreviate the convolution with the $*$ symbol. Because of the simpler notation, the functions in Fourier space are written as continuous functions. This corresponds to the transition to infinitely extended stochastic fields (Section 6.3.2, Table B.1).

The Fourier transform of the autocorrelation function is the *power spectrum* P_{gg}. It is a real-valued function. The name comes from the fact that for physical signals in Fourier space, it indicates the distribution of energy over the wavenumbers or frequencies if the square of the signal amplitude is proportional to the energy. If the power spectrum is averaged over several images, it represents a sum of the squares of independent RVs. If the RVs are uncorrelated in the spatial domain and have a *normal distribution*, then the averaged power spectrum has a *chi square distribution* according to the considerations in Section C.4.4.

The autocorrelation function of a field of uncorrelated RVs Eq. (C.72) is a δ distribution Eq. (B.44), since it vanishes everywhere except at the zero point. Therefore, the power spectrum is constant (> R15). This type of noise is called *white noise*.

The Fourier transform of the cross-correlation function is called *cross-correlation spectrum* P_{gh}. In contrast to the power spectrum, this function is complex-valued. The real and imaginary parts are called *co-* and *quad spectrum* respectively. To better understand the meaning of these functions, it is useful to define another function, the *coherence function* Φ:

$$\Phi^2(\boldsymbol{\kappa}) = \frac{|P_{gh}(\boldsymbol{\kappa})|^2}{P_{gg}(\boldsymbol{\kappa})P_{hh}(\boldsymbol{\kappa})}. \tag{C.80}$$

The coherence function indicates how similar two images are. To show this, we assume that the image $\boldsymbol{H}$ is a shifted copy of the image $\boldsymbol{G}$: $\tilde{h}(\boldsymbol{\kappa}) = \tilde{g}(\boldsymbol{\kappa})\exp(-\mathrm{i}\boldsymbol{\kappa}\boldsymbol{x}_s)$. In this case, the coherence function is one and the cross-correlation spectrum P_{gh} simplifies to

$$P_{gh}(\boldsymbol{\kappa}) = P_{gg}(\boldsymbol{\kappa})\exp(-\mathrm{i}\boldsymbol{\kappa}\boldsymbol{x}_s). \tag{C.81}$$

Since P_{gg} is a real-valued function, we can calculate the shift $\boldsymbol{x}_s$ between the two images from the phase factor $\exp(-\mathrm{i}\boldsymbol{\kappa}\boldsymbol{x}_s)$.

If there is no fixed phase relationship for a wavenumber vector between the two images, then the coherence decreases. If the phase shift from image to image in a sequence is randomly distributed, then the vectors of the cross-correlation function in the complex plane point in random directions and add up to zero. According to Eq. (C.80), the coherence then also becomes zero.

C.6 Discrete Modeling: Inverse Problems

In the second part of this appendix, we turn to discrete modeling. This can of course be derived as a direct discretization of variational approaches or numerical solutions of partial differential equations. Without saying it explicitly, we have already done this in Sections 15.2.4 and 17.3 through the iterative discrete schemes for diffusion.

However, it makes more sense to develop the discrete modeling independently, because then we get further insights into the modeling. So now we take another point of view and consider the modeling as a *linear discrete inverse problem*. As an introduction, we start with the well-known linear regression problem and develop the theory of inverse modeling based on this example.

C.6.1 A Simple Example: Linear Regression

The fit of a straight line through a set of experimental data x, y is a simple example of a discrete inverse problem. As shown in Fig. C.5, the variable y is measured as a function of the variable x. In this case, we only have two model parameters, the intercept a_0 and the slope a_1 of the line $y = a_0 + a_1 x$. With a set of Q data points $\left[x_q, y_q\right]$ we obtain the linear system of equations

$$\begin{bmatrix} 1 & x_1 \\ 1 & x_2 \\ \vdots & \vdots \\ 1 & x_Q \end{bmatrix} \begin{bmatrix} a_0 \\ a_1 \end{bmatrix} = \begin{bmatrix} y_1 \\ y_2 \\ \vdots \\ y_Q \end{bmatrix}, \tag{C.82}$$

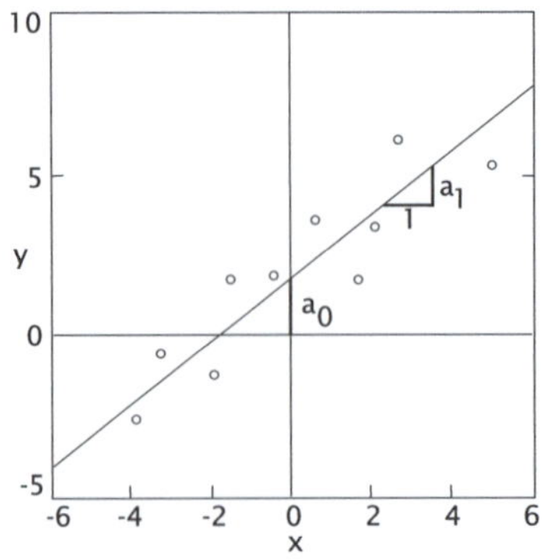

***Figure C.5:** Illustration of linear regression using the least squares method.*

that we can shorten with:

$$\boldsymbol{M}\boldsymbol{p} = \boldsymbol{d}. \tag{C.83}$$

The $Q \times 2$ matrix $\boldsymbol{M}$ is referred to as *design-* or *model matrix*. The matrix determines both the type of model (here a linear regression with two parameters) and the selected independent measurement points x_q. The *model-* or *parameter vector* $\boldsymbol{p}$ contains the parameters to be determined and the *data vector* $\boldsymbol{d}$ contains the measured data y_q.

If we only have two different points ($x_1 \neq x_2$), we obtain an exact solution to the linear system of equations. If there are more than two points, this results in more equations than unknowns. We refer to such a system of equations as *overdetermined inverse problem*. In such a case, it is no longer possible to obtain an exact solution. We can only calculate an estimate of the model parameters $\boldsymbol{p}_{\text{est}}$ where the deviation of the actual data $\boldsymbol{d}$ from the data predicted by the model $\boldsymbol{d}_{\text{pre}} = \boldsymbol{M}\boldsymbol{p}_{\text{est}}$ is minimal. This deviation can be expressed by an *error vector* $\boldsymbol{e}$:

$$\boldsymbol{e} = \boldsymbol{d} - \boldsymbol{d}_{\text{pre}} = \boldsymbol{d} - \boldsymbol{M}\boldsymbol{p}_{\text{est}}. \tag{C.84}$$

C.6.2 Error Norms

A suitable measure is required to minimize the error vector. For this purpose, we can use norms that we have already used for the vector spaces with inner product in Section 6.3.1. In general, the L_n-norm of the Q-dimensional vector $\boldsymbol{e}$ is defined as follows:

$$\|\boldsymbol{e}\|_n = \left(\sum_{q=1}^{Q} |e_q|^n\right)^{1/n}. \tag{C.85}$$

The L_∞ standard is a special case:

$$\|\boldsymbol{e}\|_\infty = \max_n |e_q|. \tag{C.86}$$

The L_2 norm is better known; in relation to an error vector, it represents the sum of squared deviations (method of *least squares*):

$$\|\boldsymbol{e}\|_2 = \left(\sum_{q=1}^{Q} (d_q - d_{\text{pre},q})^2 \right)^{1/2} . \tag{C.87}$$

Higher norms evaluate larger deviations with a stronger weighting. The statistics of the data determines which norm is the correct one. It can be shown that this is the L_2 norm for the normal distribution (Gaussian bell curve) [224].

C.6.3 Solution Using the Least Squares Method

The overdetermined linear inverse problem is solved in terms of a minimum L_2-norm of the error vector as follows:

$$\boxed{\boldsymbol{p}_{\text{est}} = \left(\boldsymbol{M}^T\boldsymbol{M}\right)^{-1}\boldsymbol{M}^T\boldsymbol{d} \quad \text{with} \quad \|\boldsymbol{e}\|_2^2 = \left\|\boldsymbol{d} - \boldsymbol{M}\boldsymbol{p}_{\text{est}}\right\| \rightarrow \text{minimum.}} \tag{C.88}$$

This solution can be made plausible as follows:

$$\begin{aligned} \boldsymbol{M}\boldsymbol{p}_{\text{est}} &= \boldsymbol{d} && \Big|\boldsymbol{M}^T \\ \boldsymbol{M}^T\boldsymbol{M}\boldsymbol{p}_{\text{est}} &= \boldsymbol{M}^T\boldsymbol{d} && \Big|\left(\boldsymbol{M}^T\boldsymbol{M}\right)^{-1} \\ \boldsymbol{p}_{\text{est}} &= \left(\boldsymbol{M}^T\boldsymbol{M}\right)^{-1}\boldsymbol{M}^T\boldsymbol{d} \end{aligned} \tag{C.89}$$

A solution requires that the inverse of $\boldsymbol{M}^T\boldsymbol{M}$ exists.

In the rest of this section, we derive the solution of the overdetermined discrete linear inverse problem Eq. (C.83), which minimizes the error vector $\boldsymbol{e}$ in Eq. (C.84) with the L_2-norm:

$$\|\boldsymbol{e}\|_2^2 = \sum_{q'=1}^{Q} \left(d_{q'} - \sum_{p'=1}^{P} m_{q'p'} p_{p'} \right) \left(d_{q'} - \sum_{p''=1}^{P} m_{q'p''} p_{p''} \right).$$

By multiplying out and swapping the sums we obtain:

$$\begin{aligned} \|\boldsymbol{e}\|_2^2 &= \underbrace{\sum_{p'=1}^{P}\sum_{p''=1}^{P} p_{p'} p_{p''} \sum_{q=1}^{Q} m_{qp'} m_{qp''}}_{A} \\ &\quad - \underbrace{2\sum_{p'=1}^{P} p_{p'} \sum_{q=1}^{Q} m_{qp'} d_q}_{B} + \sum_{q=1}^{Q} d_q d_q \end{aligned} \tag{C.90}$$

We determine a minimum for this expression by calculating the partial derivatives according to the parameters p_k to be optimized. Only the expressions A and B in Eq. (C.90) depend on p_k:

$$\begin{aligned}
\frac{\partial A}{\partial p_k} &= \sum_{p'=1}^{P}\sum_{p''=1}^{P}\left(\delta_{k-p''}p_{p'} + \delta_{k-p'}p_{p''}\right)\sum_{q'=1}^{Q} m_{q'p'}m_{q'p''} \\
&= \sum_{p'=1}^{P} p_{p'}\sum_{q'=1}^{Q} m_{q'p'}m_{q'k} + \sum_{p''=1}^{P} p_{p''}\sum_{q'=1}^{Q} m_{q'k}m_{q'p''} \\
&= 2\sum_{p'=1}^{P} p_{p'}\sum_{q'=1}^{Q} m_{q'p'}m_{q'k}, \\
\frac{\partial B}{\partial p_k} &= 2\sum_{q'=1}^{Q} m_{q'k}d_{q'}.
\end{aligned}$$

From this we obtain the total derivative and set it equal to zero:

$$\frac{\partial \|e\|_2^2}{\partial p_k} = 2\sum_{p'=1}^{P} p_{p'}\sum_{q'=1}^{Q} m_{q'k}m_{q'p'} - 2\sum_{q'=1}^{Q} m_{q'k}d_{q'} = 0.$$

To express the sums as matrix-matrix and matrix-vector multiplications, we replace the matrix $\boldsymbol{M}$ in two places by its transpose $\boldsymbol{M}^T$, so that

$$\sum_{p'=1}^{P} p_{p'}\sum_{q'=1}^{Q} m_{kq'}^{T} m_{q'p'} - \sum_{q'=1}^{Q} m_{kq'}^{T} d_{q'} = 0$$

and finally obtain the matrix equation

$$\underbrace{\underbrace{\underbrace{\boldsymbol{M}^T}_{P\times Q}\;\underbrace{\boldsymbol{M}}_{Q\times P}}_{P\times P}\;\underbrace{\boldsymbol{p}_{\text{est}}}_{P}}_{P} = \underbrace{\underbrace{\boldsymbol{M}^T}_{P\times Q}\;\underbrace{\boldsymbol{d}}_{Q}}_{P}\,. \tag{C.91}$$

This equation can be solved if the square and symmetrical $P \times P$ matrix $\boldsymbol{M}^T\boldsymbol{M}$ is invertible. Then the following applies:

$$\boldsymbol{p}_{\text{est}} = \left(\boldsymbol{M}^T\boldsymbol{M}\right)^{-1}\boldsymbol{M}^T\boldsymbol{d}. \tag{C.92}$$

The matrix $(\boldsymbol{M}^T\boldsymbol{M})^{-1}\boldsymbol{M}^T$ is called the *generalized inverse* $\boldsymbol{M}^{-g}$ of $\boldsymbol{M}$.

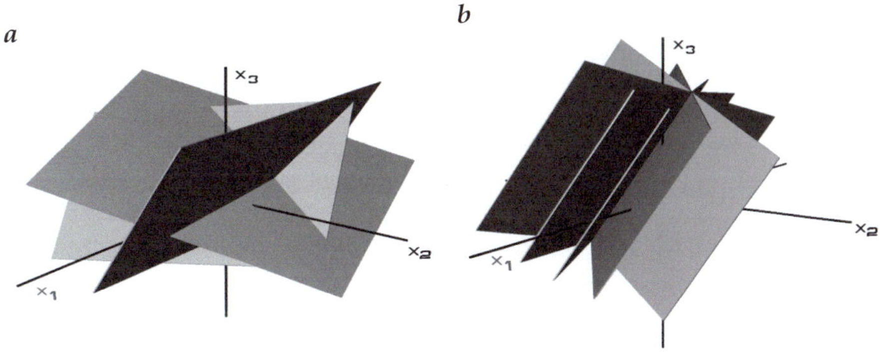

Figure C.6: *Geometry of the solution of a system of linear equations with three unknowns.* ***a*** *System of equations with exact solution;* ***b*** *overdetermined system of equations with non-unique solution.*

C.6.4 Geometric Consideration of Linear Systems of Equations

It is helpful to visualize linear systems of equations geometrically. The P model parameters $\boldsymbol{p}$ span a P-dimensional vector space. This can be regarded as the space of all possible solutions to an inverse problem with P model parameters. Now we ask ourselves what it means if we have a point d_q. According to Eq. (C.83), a point results in a linear equation that contains all model parameters $\boldsymbol{p}$:

$$\sum_{k=p'}^{P} m_{qp'} p_{p'} = d_q \quad \text{or} \quad \boldsymbol{m}_q \boldsymbol{p} = d_q. \tag{C.93}$$

This equation can be regarded as the scalar product of a row q of the model matrix $\boldsymbol{m}_q$ with the model vector $\boldsymbol{p}$. In the model space, this equation represents a $(P-1)$-dimensional *hyperplane* of all vectors $\boldsymbol{p}$ with a normal vector $\boldsymbol{m}_q$ and a distance d_q from the origin of the model space. The linear equation thus establishes a one-to-one relationship between a point in the data space and a $(P-1)$-dimensional hyperplane in the model space. This mapping of points into the model space is called *Hough transform* (Section 17.6.2).

Fig. C.6a illustrates the solution of a linear system of equations with three unknowns. With three equations, there are three planes in the model space that intersect at one point if the corresponding 3×3 model matrix is invertible. Even with an overdetermined system of equations, the solution does not necessarily have to be unique. Fig. C.6b shows the case of five planes that all intersect in a straight line. In this case, the solution is not unique, but only limited to the intersection line. If this straight line is oriented along an axis, the corresponding model parameter can take on any value. The other two parameters, however, are fixed. In the case of an arbitrarily oriented intersection line, things are more complicated. In this case, the parameter combinations perpendicular to the line are fixed, but not the parameter combinations described by the line. If we use the method of *singular value decomposition* to solve the system of linear equations, we can also solve the system of equations under such conditions and separate the determinable from the non-determinable parameter combinations [109, 248].

An overdetermined system of linear equations that does not have a unique solution is by no means a mathematical curiosity. In image processing, it is rather a frequently occurring problem. We have already encountered it, for example in the *aperture problem* of motion determination (Section 14.4.2).

C.6.5 Uncertainty of the Model Parameters

An overdetermined linear system of equations solved by minimizing the L_2-norm allows an error analysis. We can not only study the deviations between the modeled and measured data, but also estimate the errors of the calculated model parameter vector $\boldsymbol{p}_{\text{est}}$.

The mean deviation between the measured values and the values predicted by the model can be directly related to the norm of the error vector. The estimated value of the *variance* results in

$$\sigma^2 = \frac{1}{Q-P}\|\boldsymbol{e}\|^2 = \frac{1}{Q-P}\|\boldsymbol{d} - \boldsymbol{M}\boldsymbol{p}_{\text{est}}\|_2^2. \tag{C.94}$$

To avoid distorting the estimated value of the variance, the norm must be divided by the number of *degrees of freedom* $Q - P$ and not by Q.

According to Eq. (C.92), the calculated parameter vector $\boldsymbol{p}_{\text{est}}$ is a linear combination of the data vector $\boldsymbol{d}$. Therefore, we can apply the *law of error propagation* Eq. (C.28) from Section C.3.3. The *covariance matrix* (definition in Eq. C.20) of the parameter vector $\boldsymbol{p}_{\text{est}}$ results with $(\boldsymbol{A}\boldsymbol{B})^T = \boldsymbol{B}^T\boldsymbol{A}^T$ to

$$\operatorname{cov}(\boldsymbol{p}_{\text{est}}) = \left(\boldsymbol{M}^T\boldsymbol{M}\right)^{-1}\boldsymbol{M}^T \operatorname{cov}(\boldsymbol{d})\,\boldsymbol{M}\left(\boldsymbol{M}^T\boldsymbol{M}\right)^{-1}. \tag{C.95}$$

If the individual elements of the data vector $\boldsymbol{d}$ are uncorrelated with each other and have the same variance σ^2, i. e. $\operatorname{cov}(\boldsymbol{d}) = \sigma^2\boldsymbol{I}$, Eq. (C.95) simplifies to

$$\operatorname{cov}(\boldsymbol{p}_{\text{est}}) = \left(\boldsymbol{M}^T\boldsymbol{M}\right)^{-1}\sigma^2. \tag{C.96}$$

In this case, the inverse matrix $(\boldsymbol{M}^T\boldsymbol{M})^{-1}$ directly represents the covariance matrix of the parameter vector except for the factor σ^2. This means that the diagonal elements contain the variances of the individual parameters.

C.6.6 Regularization

In the previous considerations, the error functional Eq. (C.87) only contains a similarity term, but no regularization or smoothness term. For many discrete problems — such as, e. g. the linear regression discussed in Section C.6.1 — such a term makes no sense. However, if the parameters to be determined are the elements of a time series or points of an image, then such terms become important. A suitable smoothness parameter could then be the norm of the time series or the image convolved with a derivative filter:

$$\|r\|_2 = \left\|h * p\right\|_2^2 . \tag{C.97}$$

In the language of matrix algebra, the convolution can be expressed by a vector-matrix multiplication:

$$\|r\|_2 = \left\|Hp\right\|_2^2 . \tag{C.98}$$

Due to the convolution operation, the matrix H has a special shape. Only the coefficients around the diagonal are not equal to zero and all values in the diagonal direction are equal.

As an example, we consider the same smoothness term that we used in the variational approach in Section 17.2.4, the first derivative. This can be approximated by a forward difference, for example, which results in the matrix

$$H = \begin{bmatrix} -1 & 1 & 0 & 0 & \ldots & 0 \\ 0 & -1 & 1 & 0 & \ldots & 0 \\ 0 & 0 & -1 & 1 & \ldots & 0 \\ \vdots & \vdots & \ddots & \ddots & \ddots & \vdots \end{bmatrix} . \tag{C.99}$$

The minimization of the combined error functional with the L_2-norm

$$\boxed{\|e\|_2^2 = \underbrace{\left\|d - Mp\right\|_2^2}_{\text{similarity}} + \alpha^2 \underbrace{\left\|Hp\right\|_2^2}_{\text{smoothness}}} \tag{C.100}$$

leads to the following solution [224]:

$$\boxed{p_{\text{est}} = \left(M^T M + \alpha^2 H^T H\right)^{-1} M^T d.} \tag{C.101}$$

The structure of the solution is similar to that without the smoothness term in Eq. (C.88). This only causes the additional term $\alpha^2 H^T H$ in the matrix to be inverted.

Since images and other higher-dimensional signals can always be mapped to a vector, it is also possible to apply all the methods discussed here to this data. However, you then have to deal with an extremely large number of unknowns and direct solutions are no longer possible. Exemplary solutions to this problem are shown for tomography (Section 9.3.5) and restoration by inverse filtering (Section 17.5).

References for Further Reading

For an introduction to statistics, please refer to the textbooks, e. g. Rice [261]. More sophisticated and detailed descriptions can be found in Papoulis [236]. The textbook by Rosenfeld and Kak [273] provides a good introduction to stochastic processes with reference to image processing. Spectral analysis is dealt with in more detail by Marple Jr. [214]. The textbook by Strang [316] and the monograph by Golub and van Loan [109] give an excellent overview of matrix algebra. An easy-to-understand presentation of the discrete inverse theory can be found in Menke [224]. Other textbooks address curves and surface regression [192], total least squares methods [340] and parameter estimation and inverse problems [8].

D Reference Material

Elementary properties of electromagnetic waves (Section 2.3.1) R1

1. *frequency* ν (cycles per time) and *wavelength* λ (length of a period) are linked by the *phase velocity* c:

$$\lambda \nu = c$$

   In a vacuum the *velocity of light* c is 299,792,458 m/s.

2. Classification of electromagnetic waves from UV to radio waves (Fig. 2.16)

Name	Range	Remarks
EUV (Vacuum-UV)	10–121 nm	Strong absorption by air
VUV (Vacuum-UV)	100–180 nm	Strong absorption by air
UV-C	200–280 nm	CIE standard definition
UV-B	280–315 nm	CIE standard definition
UV-A	315–400 nm	CIE standard definition
Light	400–700 nm	Visible to the human eye
VNIR (very near IR)	$0.7 - 1.0\,\mu m$	IR wavelength range for which silicon sensors are sensitive
NIR (near IR)	$0.7 - 3.0\,\mu m$	
SWIR (short wave IR)	$0.9 - 2.5\,\mu m$	short-wave IR radiation
MWIR (mid wave IR)	$3 - 5.5\,\mu m$	medium-wave thermal radiation
LWIR (long wave IR)	$8 - 14\,\mu m$	long-wave thermal radiation
terahertz	30 µm – 3 mm	0.1 – 10 THz
microwaves	1 mm – 1 m	0.3 – 300 GHz
ultra-short waves (UKW)	1 – 10 m	30 – 300 MHz
short waves (KW)	10 – 100 m	3 – 30 MHz
medium waves (MW)	100 – 1000 m	0.3 – 3 MHz
long waves (LW)	1 – 10 km	30 – 300 kHz

3. Energy and momentum of particulate radiation such as β radiation (electrons), α radiation (helium nuclei), neutrons and photons (electromag-

B. Jähne, *Digital Image Processing and Image Acquisition*,
https://doi.org/10.1007/978-3-032-04806-6

netic radiation):

$$\begin{aligned} \nu &= E/h \quad \text{Bohr frequency condition,} \\ \lambda &= h/p \quad \text{de Broglie wavelength.} \end{aligned}$$

R2 Radiometric and photometric quantities (Section 2.4)

dA_0 is an area element of the surface, θ the angle of incidence, Ω the solid angle. The indices e, p and v are often used for quantities related to energy, photons and photometry.

Quantity	Energy-related	Photon rel.	Photometry
energy	radiant energy Q [Ws]	photon number [1]	luminous energy [lm s]
energy flux (power)	radiant flux $\Phi = \frac{dQ}{dt}$ [W]	photon flux [s^{-1}]	luminous flux [lumen (lm)]
incident energy flux density	irradiance $E = \frac{d\Phi}{dA_0}$ [W m^{-2}]	photon irradiance [m^{-2}s^{-1}]	illuminance [lm/m^2 = lux [(lx)]
excitant energy flux density	radiant excitance $M = \frac{d\Phi}{dA_0}$ [W m^{-2}]	photon excitance [m^{-2}s^{-1}]	luminous excitance [lm/m^2]
energy flux per solid angle	radiant intensity $I = \frac{d\Phi}{d\Omega}$ [Wsr^{-1}]	photon intensity [s^{-1}sr^{-1}]	luminous intensity [lm/sr = candela (cd)]
energy flux density per solid angle	radiance $L = \frac{d^2\Phi}{d\Omega dA_0 \cos\theta}$ [W m^{-2} sr^{-1}]	photon radiance [m^{-2}s^{-1}sr^{-1}]	lumincance [cd m^{-2}]
energy/area	irradiation [W s m^2]	photon irradiation [m^{-2}]	exposure [lm s m^{-2} = lx s]

R3 Spectral brightness sensitivity (Section 2.4)

Calculation of a photometric quantity from the corresponding radiometric quantity by the *spectral luminous intensity* $V(\lambda)$ in daytime vision (photopic vision conditions):

$$Q_v = 683 \frac{\text{lm}}{\text{W}} \int_{380\,\text{nm}}^{780\,\text{nm}} Q(\lambda) V(\lambda)\, \mathrm{d}\lambda$$

Table with the values standardized by the CIE in 1980 for the spectral luminous intensity $V(\lambda)$ for day vision

λ [nm]	$V(\lambda)$	λ [nm]	$V(\lambda)$	λ [nm]	$V(\lambda)$
380	0.00004	520	0.710	660	0.061
390	0.00012	530	0.862	670	0.032
400	0.0004	540	0.954	680	0.017
410	0.0012	550	0.995	690	0.0082
420	0.0040	560	0.995	700	0.0041
430	0.0116	570	0.952	710	0.0021
440	0.023	580	0.870	720	0.00105
450	0.038	590	0.757	730	0.00052
460	0.060	600	0,631	740	0.00025
470	0.091	610	0.503	750	0.00012
480	0.139	620	0.381	760	0.00006
490	0.208	630	0.265	770	0.00003
500	0.323	640	0.175	780	0.000015
510	0.503	650	0.107		

Color systems (Section 2.5.3) R4

1. Human color vision is based on three types of photoreceptor cells with maximum sensitivities at wavelengths of 445 nm, 535 nm and 575 nm (Fig. 2.18b).
2. RGB color system: additive color system with the three primary colors red, green and blue. These can either be monochromatic colors with the wavelengths 700 nm, 646,1 nm and 435,8 nm or red, green and blue phosphors as used in RGB screens (for example according to the European EBU standard). Not all colors can be created with the *RGB* color system (see Fig. 2.20a).
3. Color table: Reduction of the 3D color space to a 2D color plane by normalization with the intensity:
$$r = \frac{R}{R+G+B}, \quad g = \frac{G}{R+G+B}, \quad b = \frac{B}{R+G+B}.$$
It is sufficient to use the two components r and g, since $b = 1 - r - g$.
4. *XYZ* color system (Fig. 2.20c): additive color system with three virtual primary colors X, Y and Z, which can be used to create all possible colors. The *XYZ* values result from a linear transformation of the EBU-*RGB* color system.
$$\begin{bmatrix} X \\ Y \\ Z \end{bmatrix} = \begin{bmatrix} 0.490 & 0.310 & 0.200 \\ 0.177 & 0.812 & 0.011 \\ 0.000 & 0.010 & 0.990 \end{bmatrix} \begin{bmatrix} R \\ G \\ B \end{bmatrix}.$$

5. Color difference or *YUV* system: Color system with the origin at the white or achromatic point (Fig. 2.20b).
6. Hue and color saturation system (HSI): Color system with polar coordinates in a color difference system. The saturation is given by the radius and the color value by the angle.

R5 Thermal radiation (Section 2.6.1)

1. Spectral distribution of radiant intensity (Planck's radiation law)

$$M_e(\lambda, T) = \frac{2\pi hc^2}{\lambda^5} \frac{1}{\exp\left(\frac{hc}{k_B T\lambda}\right) - 1}$$

with

$h = 6.6262 \times 10^{-34}\,\mathrm{J\,s}$	Planck's constant,
$k_B = 1.3806 \times 10^{-23}\,\mathrm{J\,K^{-1}}$	Boltzmann constant and
$c = 2.9979 \times 10^{8}\,\mathrm{m\,s^{-1}}$	Speed of light in a vacuum.

2. Total radiant intensity (Stefan-Boltzmann law)

$$M_e = \frac{2}{15}\frac{k_B^4 \pi^5}{c^2 h^3} T^4 = \sigma T^4 \quad \text{with} \quad \sigma \approx 5.67 \cdot 10^{-8} \mathrm{W\,m^{-2}K^{-4}}$$

3. Wavelength of maximum radiant intensity (Wien's displacement law)

$$\lambda_m \approx \frac{2898\,\mathrm{K}\,\mu\mathrm{m}}{T}$$

R6 Interaction of radiation with matter (Section 2.6)

1. *Snell's law* of *refraction* of light at an interface of two optical media with refractive indices n_1 and n_2

$$\frac{\sin\theta_1}{\sin\theta_2} = \frac{n_2}{n_1}$$

θ_1 and θ_2 are the angle of incidence and angle of reflection respectively.
2. *Reflectivity* ρ: Ratio of reflected to incident radiance. *Fresnel equations* for reflectivity of
   (a) horizontally polarized light

$$\rho_{\|} = \frac{\tan^2(\theta_1 - \theta_2)}{\tan^2(\theta_1 + \theta_2)},$$

(b) vertically polarized light

$$\rho_\perp = \frac{\sin^2(\theta_1 - \theta_2)}{\sin^2(\theta_1 + \theta_2)},$$

(c) unpolarized light

$$\rho = \frac{\rho_\parallel + \rho_\perp}{2}.$$

3. Reflectivity at perpendicular incidence ($\theta_1 = 0$) for all polarization states

$$\rho = \frac{(n_1 - n_2)^2}{(n_1 + n_2)^2} = \frac{(n-1)^2}{(n+1)^2} \quad \text{with} \quad n = n_1/n_2$$

4. *Total reflection*: During the transition to an optically thinner medium with a lower refractive index, all radiation above a critical angle of incidence θ_c is reflected and none reaches the optically thinner medium.

$$\theta_c = \arcsin \frac{n_1}{n_2} \quad \text{with} \quad n_1 < n_2$$

Optical imaging (Section 3.3)

R7

1. Perspective projection with the *pinhole camera* model

$$x' = \frac{z'x}{z}, \quad y' = \frac{z'y}{z}$$

The central point is located at the origin of the world coordinate system $[x, y, z]^T$, z' is the distance of the image plane from the projection center and the z-axis is perpendicular to the image plane.

2. Newton's and Gauss's *imaging equation*

$$dd' = f^2 \quad \text{or} \quad \frac{1}{d' + f} + \frac{1}{d + f} = \frac{1}{f}$$

d and d' are the distances of the object or image from the object-side or image-side focal point of the optical system (see Fig. 3.7).

3. *Lateral image scale*

$$m_l = \frac{x'}{x} = \frac{f}{d} = \frac{d'}{f}$$

4. *Axial image scale*

$$m_a \approx \frac{d'}{d} = \frac{f^2}{d^2} = \frac{d'^2}{f^2} = m_l^2$$

5. The *f-number* k of an optical system is the ratio of the focal length to the diameter of the entrance pupil a

$$k = \frac{f}{a}$$

6. *Depth of focus* in geometric approximation

$$\Delta z' = 2k\left(1 + \frac{d'}{f}\right)u' = 2k(1 + m_l)u'$$

7. *Depth of field* in geometric approximation

General	Δz	$\approx$	$2k \cdot \dfrac{1 + m_l}{m_l^2} \cdot u'$
d_{min} for range to infinity	d_{min}	$\approx$	$\dfrac{f^2}{2ku'}$
Microscopy ($m_l \gg 1$)	Δz	$\approx$	$\dfrac{2u'}{NA}$

u' Diameter of the blur circle

8. Resolution of a diffraction-limited optical system

Minimum resolvable structure wavelength	λ_{min}	$=$	$k\lambda_e$
Edge widening	$\Delta x'_{\text{edge}}$	$\approx$	$1{,}05k\lambda_e$
Angular resolution	$\Delta\theta_0$	$=$	$1{,}22\dfrac{\lambda_e}{a}$
Lateral resolution in the image plane	$\Delta x'$	$=$	$0{,}61\dfrac{\lambda_e}{NA'}$
Lateral resolution in the object plane	Δx	$=$	$0{,}61\dfrac{\lambda_e}{NA}$

The resolution results from the *Rayleigh criterion* (Section 3.5.3); NA and NA' are the object-side and image-side numerical aperture of the light cone entering the optical system:

$$NA = n \sin\theta_0 \approx \frac{2n}{k};$$

n is the refractive index, k the f-number, and λ_e the wavelength of the radiation

9. Relationship between the irradiance (illuminance) in the image plane E' and the radiance (luminance) L of the object (Fig. 3.17)

$$E' = t\frac{\pi}{4}\frac{\cos^4\theta}{k^2(1 + m_l)^2}L \approx t\frac{\pi}{4}\frac{\cos^4\theta}{k^2}L \quad \text{for} \quad d \gg f$$

Selection of CCD image sensors (Section 4.4)

C: Fullwell capacity in electrons, eNIR: increased NIR sensitivity, FR: frame rate in s^{-1}, ID: image diagonal in mm, QE: peak quantum efficiency, Sensors from Sony (ICX. . .) and Kodak (KAI. . .), compiled from the manufacturers' data sheets

Chip	Format H × V	FR	ID	Pixel size H × V, μm	Remarks
Interlaced EIA video					
ICX278AL 1/4"	768 × 494	30	4.56	4.75 × 5.55	eNIR
ICX258AL 1/3"	768 × 494	30	6.09	6.35 × 7.4	eNIR
ICX248AL 1/2"	768 × 494	30	8.07	8.4 × 9.8	eNIR
ICX082AL 2/3"	768 × 494	30	11.1	11.6 × 13.5	
Interlaced CCIR video					
ICX279AL 1/4"	752 × 582	25	4.54	4.85 × 4.65	eNIR
ICX259AL 1/3"	752 × 582	25	6.09	6.5 × 6.25	eNIR
ICX249AL 1/2"	752 × 582	25	8.07	8.6 × 8.3	eNIR
ICX083AL 2/3"	752 × 582	25	10.9	11.6 × 11.2	
Progressive scanning interline					
ICX098AL 1/4"	659 × 494	30	4.61	5.6 × 5.6	
ICX424AL 1/3"	659 × 494	30	6.09	7.4 × 7.4	
ICX074AL 1/2"	659 × 494	40	8.15	9.9 × 9.9	C 32k, QE 0,43 @ 340 nm
ICX414AL 1/2"	659 × 494	50	8.15	9.9 × 9.9	C 30k, QE 0,40 @ 500 nm
ICX075AL 1/2"	782 × 582	30	8.09	8.3 × 8.3	
ICX204AL 1/3"	1024 × 768	15	5.95	4.65 × 4.65	
ICX205AL 1/2"	1360 × 1024	9.5	7.92	4.65 × 4.65	C 13k
ICX285AL 2/3"	1360 × 1024	10	11.0	6.45 × 6.45	eNIR, C 18k, QE 0.65 @ 500 nm
ICX085AL 2/3"	1300 × 1030	12.5	11.1	6.7 × 6.7	C 20k, QE 0.54 @ 380 nm
ICX274AL 1/1,8"	1628 × 1236	12	8.99	4.4 × 4.4	
KAI-0340DM 1/3"	640 × 480	200	5.92	7.4 × 7.4	C 20k, QE 0.55 @ 500 nm
KAI-1010M	1008 × 1018	30	12.9	9.0 × 9.0	QE 0,37 @ 500 nm
KAI-1020M	1000 × 1000	49	10.5	7.4 × 7.4	C 42k, QE 0.45 @ 490 nm
KAI-2001M	1600 × 1200	30	14.8	7.4 × 7.4	C 40k, QE 0.55 @ 480 nm
KAI-4020M	2048 × 2048	15	21.4	7.4 × 7.4	C 40k, QE 0.55 @ 480 nm
KAI-10000M	4008 × 2672	3	43.3	9.0 × 9.0	C 60k, QE 0.50 @ 500 nm

R9 Selection of early CMOS image sensors up to 2012 (Section 4.4)

C: Full well capacity in electrons, FR: Frame rate in s^{-1}, PC: Pixel clock rate in MHz, QE: peak quantum efficiency; compiled from manufacturers' data sheets

Chip	Format H × V	FR	PC	Pixel size H × V, μm	Remarks
Linear characteristic curve					
Micron MT9V403	656 × 491	200	66	9.9 × 9.9	QE 0.32 @ 520 nm
Fillfactory IBIS54-1300	1280 × 1024	30	40	6.7 × 6.7	QE 0.30–0.35 @ 600 nm, C 60k
Fillfactory IBIS4-4000	2496 × 1692	4,5		11.4 × 11.4	C 150k
CMOSIS CMV4000	2048 × 2048	180		5.5 × 5.5	C 8.5k, QE 0.57
Aptina MT9P	2592 × 1944	14		2.2 × 2.2	C 19k, QE 0.60
pco.edge sCMOS	2560 × 2160	100		6.5 × 6.5	C 27k, QE 0.57
High frame rate					
Fillfactory LUPA1300	1280 × 1024	450	40	12.0 × 12.0	16 parallel outputs
Micron MV40	2352 × 1728	240	80	7.0 × 7.0	16 × 10-bit outputs
Micron MT9M413	1280 × 1024	600	80	12.0 × 12.0	QE 0.27 @ 520 nm, C 63k, 10 × 10-bit outputs
Micron MV02	512 × 512	4000	80	16.0 × 16.0	16 × 10-bit outputs
pco.dimax	2016 × 2016	1279		11.0 × 11.0	12 bit, C 36k
Logarithmic characteristic curve					
IMS HDRC VGA	640 × 480	25	8	12 × 12	
PhotonFocus A1312	1312 × 1082	110	40	8.0 × 8.0	QE 0.50 @ 600 nm, C 140k, Linear characteristic curve with adjustable transition to logarithmic characteristic curve

Selection of current CMOS image sensors, available 2024 (Section 4.4) R10

C: Full well capacity in electrons, FR: Frame rate in s^{-1} (this is partly not limited by the sensor, but by the camera interface and depends on the bit depth), QE: peak quantum efficiency, data compiled from manufacturers' data sheets and own measurements

Chip	Format H × V	FR	Pixel size H × V, μm	Remarks
Global shutter				
OnSemi Python 500	832 × 632	286	4.8 × 4.8	C 6.4k, QE 0.53 @ 466 nm
Sony IMX174 Pregius Gen1	1920 × 1200	76	5.86 × 5.86	C 34k, QE 0.70 @ 466 nm
Sony IMX252 Pregius Gen2	2048 × 1536	57	3.45 × 3.45	C 10.8k, QE 0.66 @ 531 nm
Sony IMX273 Pregius Gen2	1456 × 1088	108	3.45 × 3.45	C 10.7k, QE 0.65 @ 531 nm
CMOSIS CMV12000	4088 × 3068	62	5.5 × 5.5	C 10.0k, QE 0.42 @ 466 nm
Rolling shutter				
OnSemi MT9P031	2592 × 1944	14	2.2 × 2.2	C 6.4k, QE 0.53 @ 536 nm
Sony IMX178 StarVis	3088 × 2028	27	2.4 × 2.4	C 14.6k, QE 0.79 @ 466 nm
Sony IMX226 StarVis	4024 × 3036	15	1.85 × 1.85	C 10.9k, QE 0.80 @ 466 nm
Sony IMX290 StarVis	1936 × 1096	63	2.90 × 2.90	C 11.2k, QE 0.81 @ 466 nm
Sony IMX411	14200 × 10600		3.76 × 3.76	C 48.4k, QE 0.81 @ 524 nm
Sony IMX415 StarVis	3840 × 2160	25	1.45 × 1.45	QE 0.82 @ 466 nm
Sony IMX662 StarVis	1920 × 1080	93	2.90 × 2.90	QE 0.84 @ 466 nm
High-speed cameras				
Sony IMX535 Pregius Gen4	4128 × 3000	192	2.74 × 2.74	C 9.2k, QE 0.69 @ 481 nm
Pharsighted E9.100S	640 × 480	326k	52 × 52	

R11 Image sensors for infrared radiation (IR, Section 4.4)

C: full well capacity in millions of electrons [Me], IT: integration time, NETD: noise equivalent temperature difference, QE: peak quantum efficiency, compiled from the manufacturer's data sheets

Chip	Format H × V	FR	PC	Pixel size H × V, μm	Remarks
Short-wave infrared (SWIR)					
Indigo InGaAs	320 × 256	345		30 × 30	0.9–1.68 μm, C 3.5 Me
Sony IMX990	1280 × 1024	71		5.0 × 5.0	0.35–1.65 μm, C 0.1 Me
Sony IMX992	2560 × 2048	70		3.45 × 3.45	0.35–1.65 μm
Medium-wave infrared (MWIR)					
AIM PtSi	640 × 486	50	12	24 × 24	3.0–5.0 μm, NETD < 75 mK @ 33 ms IT
Indigo InSb	320 × 256	345		30 × 30	2.0–5.0 μm, C 18 Me
Indigo InSb	640 × 512	100		25 × 25	2.0–5.0 μm, C 11 Me
AIM HgCdTe	384 × 288	120	20	24 × 24	3.0–5.0 μm, NETD < 20 mK @ 2 ms IT
IaF FhG bispectral InAs/GaSb, Type II Übergitter	384 × 288			24 × 24	3.0–4.0 & 4.0–5.0 μm, NETD 14.6 mK (3-4 μm), 9.0 mK (4-5 μm), QE 0.7
AIM/IaF FhG³ QWIP	640 × 512	30	18	24 × 24	3.0–5.0 μm, NETD < 15 mK @ 20 ms IT
IRCAM Velox 327kM HgCdTe	640 × 512	200			3.0–5.0 μm, NETD 20 mK
FLIR X8400sc InSb	1280 × 1024	100		15 × 15	1.5–5.1 μm, NETD 25 mK
ImageIR 9400hp InSb	1280 × 1024	180		10 × 10	2.0–5.5 μm, NETD 20 – 30 mK
ImageIR 10300 InSb	1920 × 1536	113		10 × 10	3.6–4.9 μm, NETD 22 – 35 mK
Long-wave infrared (LWIR)					
AIM HgCdTe	256 × 256	200	16	40 × 40	8–10 μm, NETD < 20 mK @ 0.35 ms IT
Indigo QWIP	320 × 256	345		30 × 30	8.0–9.2 μm, C 18 Me, NETD < 30 mK
AIM/IaF FhG³ QWIP	256 × 256	200	16	40 × 40	8.0–9.2 μm, NETD < 8 mK @ 20 ms IT
AIM/IaF FhG³ QWIP	640 × 512	30	18	24 × 24	8.0–9.2 μm, NETD < 10 mK @ 30 ms IT
Uncooled sensors					
FLIR A655sc Micro bolometer	640 × 480	50		17 × 17	7.5–13 μm, NETD 50 mK
PICO 640	640 × 480	60		17 × 17	7.5–13 μm, NETD 30 – 50 mK
ATTO 1280	1280 × 1024	60		12 × 12	7.5–13 μm, NETD 50 – 60 mK

Properties of the Fourier transform (Section 6.3.4)

R12

$g(\boldsymbol{x}) \circ\!\!-\!\!\bullet\; \tilde{g}(\boldsymbol{\kappa})$ and $h(\boldsymbol{x}) \circ\!\!-\!\!\bullet\; \tilde{h}(\boldsymbol{\kappa})$ are Fourier transform pairs:

$$\tilde{g}(\boldsymbol{\kappa}) = \int_{-\infty}^{\infty} g(\boldsymbol{x}) \exp\left(-2\pi \mathrm{i} \boldsymbol{\kappa}^T \boldsymbol{x}\right) \mathrm{d}^D x = \left\langle \exp\left(2\pi \mathrm{i} \boldsymbol{\kappa}^T \boldsymbol{x}\right) \middle| g(\boldsymbol{x}) \right\rangle;$$

$s \neq 0$ is a real number, a and b are complex numbers; $\boldsymbol{A}$ is a D×D matrix with $\det A \neq 0$, $\boldsymbol{R}$ is an orthogonal rotation matrix ($\boldsymbol{R}^{-1} = \boldsymbol{R}^T$, $\det R = 1$)

Property	Spatial domain	Fourier space
linearity	$a g(\boldsymbol{x}) + b h(\boldsymbol{x})$	$a \tilde{g}(\boldsymbol{\kappa}) + b \tilde{h}(\boldsymbol{\kappa})$
similarity	$g(s\boldsymbol{x})$	$\tilde{g}(\boldsymbol{\kappa}/s)/\lvert s\rvert^D$
generalized similarity	$g(\boldsymbol{A}\boldsymbol{x})$	$\tilde{g}\left((\boldsymbol{A}^{-1})^T \boldsymbol{\kappa}\right) / \lvert\det \boldsymbol{A}\rvert$
rotation	$g(\boldsymbol{R}\boldsymbol{x})$	$\tilde{g}\,(\boldsymbol{R}\boldsymbol{\kappa})$
separability	$\prod_{d=1}^{D} g_d(x_d)$	$\prod_{d=1}^{D} \tilde{g}_d(k_d)$
shift in x space	$g(\boldsymbol{x} - \boldsymbol{x}_0)$	$\exp(-2\pi \mathrm{i} \boldsymbol{\kappa}^T \boldsymbol{x}_0) \tilde{g}(\boldsymbol{\kappa})$
finite differences	$g(\boldsymbol{x} + \boldsymbol{x}_0/2) - g(\boldsymbol{x} - \boldsymbol{x}_0/2)$	$2\mathrm{i} \sin(\pi \boldsymbol{x}_0^T \boldsymbol{\kappa}) \tilde{g}(\boldsymbol{\kappa})$
shift in κ space	$\exp(2\pi \mathrm{i} \boldsymbol{\kappa}_0^T \boldsymbol{x}) g(\boldsymbol{x})$	$\tilde{g}(\boldsymbol{\kappa} - \boldsymbol{\kappa}_0)$
modulation	$\cos(2\pi \boldsymbol{\kappa}_0^T \boldsymbol{x}) g(\boldsymbol{x})$	$(\tilde{g}(\boldsymbol{\kappa} - \boldsymbol{\kappa}_0) + \tilde{g}(\boldsymbol{\kappa} + \boldsymbol{\kappa}_0))/2$
derivation in x space	$\dfrac{\partial g(\boldsymbol{x})}{\partial x_p}$	$2\pi \mathrm{i} k_p \tilde{g}(\boldsymbol{\kappa})$
derivation in κ space	$-2\pi \mathrm{i} x_p g(\boldsymbol{x})$	$\dfrac{\partial \tilde{g}(\boldsymbol{\kappa})}{\partial k_p}$
definite integral, mean value	$\int_{-\infty}^{\infty} g(\boldsymbol{x}')\mathrm{d}^D x'$	$\tilde{g}(\boldsymbol{0})$
moments	$\int_{-\infty}^{\infty} x_p^m x_q^n g(\boldsymbol{x})\mathrm{d}^D x$	$\left(\dfrac{\mathrm{i}}{2\pi}\right)^{m+n} \left(\dfrac{\partial^{m+n} \tilde{g}(\boldsymbol{\kappa})}{\partial k_p^m \partial k_q^n}\right)\Bigg\rvert_0$
convolution	$\int_{-\infty}^{\infty} h(\boldsymbol{x}') g(\boldsymbol{x} - \boldsymbol{x}')\mathrm{d}^D x'$	$\tilde{h}(\boldsymbol{\kappa}) \tilde{g}(\boldsymbol{\kappa})$
correlation	$\int_{-\infty}^{\infty} h(\boldsymbol{x}') g(\boldsymbol{x}' + \boldsymbol{x})\mathrm{d}^D x'$	$\tilde{g}^*(\boldsymbol{\kappa})\, \tilde{h}(\boldsymbol{\kappa})$
multiplication	$h(\boldsymbol{x}) g(\boldsymbol{x})$	$\int_{-\infty}^{\infty} \tilde{h}(\boldsymbol{\kappa}') \tilde{g}(\boldsymbol{\kappa} - \boldsymbol{\kappa}')\mathrm{d}^D k'$
inner product	$\int_{-\infty}^{\infty} g^*(\boldsymbol{x})\, h(\boldsymbol{x})\mathrm{d}^D x$	$\int_{-\infty}^{\infty} \tilde{g}^*(\boldsymbol{\kappa}) \tilde{h}(\boldsymbol{\kappa})\mathrm{d}^D k$

R13 Elementary transformation pairs of the continuous Fourier transform

2D and 3D functions are marked with † and ‡ respectively.

Spatial domain	Fourier space
delta, $\delta(x)$	constant, 1
constant, 1	delta, $\delta(k)$
$\cos(2\pi k_0 x)$	$\frac{1}{2}\left(\delta(k-k_0)+\delta(k+k_0)\right)$
$\sin(2\pi k_0 x)$	$\frac{\mathrm{i}}{2}\left(\delta(k-k_0)-\delta(k+k_0)\right)$
$\mathrm{sgn}(x) = \begin{cases} 1 & x \geq 0 \\ -1 & x < 0 \end{cases}$	$\frac{-\mathrm{i}}{\pi k}$
box, $\Pi(x) = \begin{cases} 1 & \lvert x\rvert < 1/2 \\ 0 & \lvert x\rvert \geq 1/2 \end{cases}$	$\mathrm{sinc}(k) = \frac{\sin(\pi k)}{\pi k}$
disk, † $\frac{1}{\pi r^2}\Pi\left(\frac{\lvert\boldsymbol{x}\rvert}{2r}\right)$	Bessel, $\frac{J_1(2\pi r\lvert\boldsymbol{\kappa}\rvert)}{\pi r\lvert\boldsymbol{\kappa}\rvert}$
sphere, ‡ $\Pi\left(\frac{\lvert\boldsymbol{x}\rvert}{2}\right)$	$\frac{\sin(\lvert\boldsymbol{\kappa}\rvert)-\lvert\boldsymbol{\kappa}\rvert\cos(\lvert\boldsymbol{\kappa}\rvert)}{\lvert\boldsymbol{\kappa}\rvert^3/(4\pi)}$
Bessel, $\frac{J_1(2\pi x)}{x}$	$2(1-k)^{1/2}\Pi\left(\frac{k}{2}\right)$
$\exp(-\lvert x\rvert)$, $\exp(-\lvert\boldsymbol{x}\rvert)^{\dagger}$	$\frac{2}{1+(2\pi k)^2}$, $\frac{2\pi}{(1+(2\pi\lvert\boldsymbol{\kappa}\rvert)^2)^{3/2}}^{\dagger}$

R14 Functions invariant under the Fourier transform

Spatial domain	Fourier space
Gaussian, $\exp\left(-\pi\boldsymbol{x}^T\boldsymbol{x}\right)$	Gaussian, $\exp\left(-\pi\boldsymbol{\kappa}^T\boldsymbol{\kappa}\right)$
$x_p\exp\left(-\pi\boldsymbol{x}^T\boldsymbol{x}\right)$	$-\mathrm{i}k_p\exp\left(-\pi\boldsymbol{\kappa}^T\boldsymbol{\kappa}\right)$
$\mathrm{sech}(\pi x) = \frac{1}{\exp(\pi x)+\exp(-\pi x)}$	$\mathrm{sech}(\pi k) = \frac{1}{\exp(\pi k)+\exp(-\pi k)}$
hyperbolic, $\lvert\boldsymbol{x}\rvert^{-D/2}$	$\lvert\boldsymbol{\kappa}\rvert^{-D/2}$
1D δ comb, $Ш(x) = \sum_{n=-\infty}^{\infty}\delta(x-n)$	$Ш(k) = \sum_{\tilde{n}=-\infty}^{\infty}\delta(k-\tilde{n})$

Properties of the 2D DFT (Section 6.3.4)

R15

$\boldsymbol{G}$ and $\boldsymbol{H}$ are complex M×N matrices, $\tilde{\boldsymbol{G}}$ and $\tilde{\boldsymbol{H}}$ are their Fourier transforms,

$$\begin{aligned}\tilde{g}_{\tilde{m}\tilde{n}} &= \frac{1}{MN}\sum_{m=0}^{M-1}\sum_{n=0}^{N-1} g_{m,n} w_M^{-m\tilde{m}} w_N^{-n\tilde{n}}, \quad w_N = \exp(2\pi i/N)\\ g_{mn} &= \sum_{\tilde{m}=0}^{M-1}\sum_{\tilde{n}=0}^{N-1} \tilde{g}_{\tilde{m}\tilde{n}} w_M^{m\tilde{m}} w_N^{n\tilde{n}},\end{aligned}$$

and a and b complex constants. Upsampling and replication by the factors $K, L \in \mathbb{N}$ result in KM×LN matrices. Proofs can be found in Poularikas [245].

Property	Spatial domain	Fourier space
mean value	$\frac{1}{MN}\sum_{m=0}^{M-1}\sum_{n=0}^{N-1} G_{mn}$	$\tilde{g}_{0,0}$
linearity	$a\boldsymbol{G} + b\boldsymbol{H}$	$a\tilde{\boldsymbol{G}} + b\tilde{\boldsymbol{H}}$
upward sampling	$g_{Km,Ln}$	$\tilde{g}_{\tilde{m}\tilde{n}}/(KL)$ $(\tilde{g}_{kM+\tilde{m},lN+\tilde{n}} = \tilde{g}_{\tilde{m}\tilde{n}})$
replication (frequency) up sampling)	g_{mn} $(g_{kM+m,lN+n} = g_{mn})$	$\tilde{g}_{K\tilde{m},L\tilde{n}}$
shift	$g_{m-m',n-n'}$	$w_M^{-m'\tilde{m}} w_N^{-n'\tilde{n}} \tilde{g}_{\tilde{m}\tilde{n}}$
modulation	$w_M^{\tilde{m}'m} w_N^{\tilde{n}'n} g_{mn}$	$\tilde{g}_{\tilde{m}-\tilde{m}',\tilde{n}-\tilde{n}'}$
finite differences	$(g_{m+1,n} - g_{m-1,n})/2$ $(g_{m,n+1} - g_{m,n-1})/2$	$i\sin(2\pi u/M)\tilde{g}_{\tilde{m}\tilde{n}}$ $i\sin(2\pi v/N)\tilde{g}_{\tilde{m}\tilde{n}}$
convolution	$\sum_{m'=0}^{M-1}\sum_{n'=0}^{N-1} h_{m'n'} g_{m-m',n-n'}$	$MN\tilde{h}_{\tilde{m}\tilde{n}}\tilde{g}_{\tilde{m}\tilde{n}}$
cross-correlation	$\sum_{m'=0}^{M-1}\sum_{n'=0}^{N-1} h^*_{m'n'} g_{m+m',n+n'}$	$MN\tilde{h}^*_{\tilde{m}\tilde{n}}\tilde{g}_{\tilde{m}\tilde{n}}$
product	$g_{mn}h_{mn}$	$\sum_{\tilde{m}'=0}^{M-1}\sum_{\tilde{n}'=0}^{N-1} h_{\tilde{m}'\tilde{n}'} g_{\tilde{m}-\tilde{m}',\tilde{n}-\tilde{n}'}$
inner product	$\sum_{m=0}^{M-1}\sum_{n=0}^{N-1} g^*_{mn}h_{mn}$	$MN\sum_{\tilde{m}=0}^{M-1}\sum_{\tilde{n}=0}^{N-1} \tilde{g}^*_{\tilde{m}\tilde{n}}\tilde{h}_{\tilde{m}\tilde{n}}$
norm	$\sum_{m=0}^{M-1}\sum_{n=0}^{N-1} \lvert g_{mn}\rvert^2$	$MN\sum_{\tilde{m}=0}^{M-1}\sum_{\tilde{n}=0}^{N-1} \lvert\tilde{g}_{\tilde{m}\tilde{n}}\rvert^2$

R16 Properties of the 1D Hartley transformation (Section B.2.2)

$g(x) \circ\!\!-\!\!\bullet\ \tilde{g}(k)$ and $h(x) \circ\!\!-\!\!\bullet\ \tilde{h}(k)$ are Hartley transformation pairs:

$$ {}^h\tilde{g}(k) = \int_{-\infty}^{\infty} g(x)\,\mathrm{cas}(2\pi kx)\mathrm{d}x \quad \circ\!\!-\!\!\bullet \quad g(x) = \int_{-\infty}^{\infty} {}^h\tilde{g}(k)\,\mathrm{cas}(2\pi kx)\mathrm{d}k $$

with

$$ \mathrm{cas}\,2\pi kx = \cos(2\pi kx) + \sin(2\pi kx). $$

s is a real number not equal to zero, a and b are real constants.

Property	Spatial domain	Fourier space
linearity	$ag(x) + bh(x)$	$a\tilde{g}(k) + b\tilde{h}(k)$
similarity	$g(sx)$	$\tilde{g}(k/s)/\lvert s\rvert$
shift in the spatial domain	$g(x - x_0)$	$\cos(2\pi kx_0)\tilde{g}(k) - \sin(2\pi kx_0)\tilde{g}(-k)$
modulation	$\cos(2\pi k_0 x)g(x)$	$(\tilde{g}(k - k_0) + \tilde{g}(k + k_0))/2$
spatial derivation	$\dfrac{\partial g(\boldsymbol{x})}{\partial x_p}$	$-2\pi k_p \tilde{g}(-k)$
defined integral, mean value	$\int_{-\infty}^{\infty} g(x')\mathrm{d}x'$	$\tilde{g}(0)$
convolution	$\int_{-\infty}^{\infty} h(x')g(x - x')\mathrm{d}x'$	$[\tilde{g}(k)\tilde{h}(k) + \tilde{g}(k)\tilde{h}(-k)$ $+\tilde{g}(-k)\tilde{h}(k) - \tilde{g}(-k)\tilde{h}(-k)]/2$
multiplication	$h(x)g(x)$	$[\tilde{g}(k) * \tilde{h}(k) + \tilde{g}(k) * \tilde{h}(-k)$ $+\tilde{g}(-k) * \tilde{h}(k) - \tilde{g}(-k) * \tilde{h}(-k)]/2$
autocorrelation	$\int_{-\infty}^{\infty} g(x')g(x' + x)\mathrm{d}x'$	$[\tilde{g}^2(k) + \tilde{g}^2(-k)]/2$

1. Computation the Fourier transform from the Hartley transform

$$ \tilde{g}(k) = \frac{1}{2}\left({}^h\tilde{g}(k) + {}^h\tilde{g}(-k)\right) - \frac{\mathrm{i}}{2}\left({}^h\tilde{g}(k) - {}^h\tilde{g}(-k)\right) $$

2. Computation of the Hartley transform from the Fourier transform

$$ {}^h\tilde{g}(k) = \mathrm{Re}[\tilde{g}(k)] - \mathrm{Im}[\tilde{g}(k)] = \frac{1}{2}\left(\tilde{g}(k) + \tilde{g}^*(k)\right) + \frac{\mathrm{i}}{2}\left(\tilde{g}(k) - \tilde{g}^*(k)\right) $$

Probability density functions (PDF, Section C.4) R17

Definition, mean values (MV) and variances

Name	Definition	MV	Variance
Discrete probability density functions f_n (PDF)			
Poisson $P(\mu)$	$\exp(-\mu)\frac{\mu^n}{n!},\ n \geq 0$	μ	μ
binomial $B(Q,p)$	$\frac{Q!}{n!\,(Q-n)!}p^n(1-p)^{Q-n},\ 0 \leq n < Q$	Qp	$Qp(1-p)$
Continuous probability density functions $f(x)$ (PDF)			
uniform $U(a,b)$	$\frac{1}{b-a}$	$\frac{a+b}{2}$	$\frac{(b-a)^2}{12}$
normal $N(\mu,\sigma)$	$\frac{1}{\sqrt{2\pi}\sigma}\exp\left(-\frac{(x-\mu)^2}{2\sigma^2}\right)$	μ	σ^2
Rayleigh $R(\sigma)$	$\frac{x}{\sigma^2}\exp\left(-\frac{x^2}{2\sigma^2}\right),\ x > 0$	$\sigma\sqrt{\pi/2}$	$\sigma^2(4-\pi)/2$
chi square $\chi^2(Q,\sigma)$	$\frac{x^{Q/2-1}}{2^{Q/2}\Gamma(Q/2)\sigma^Q}\exp\left(-\frac{x}{2\sigma^2}\right),\ x > 0$	$Q\sigma^2$	$2Q\sigma^4$

Addition theorems for independent random variables g_1 and g_2

PDF	g_1	g_2	$g_1 + g_2$
binomial	$B(Q_1,p)$	$B(Q_2,p)$	$B(Q_1+Q_2,p)$
Poisson	$P(\mu_1)$	$P(\mu_2)$	$P(\mu_1+\mu_2)$
normal	$N(\mu_1,\sigma_1)$	$N(\mu_2,\sigma_2)$	$N(\mu_1+\mu_2,(\sigma_1^2+\sigma_2^2)^{1/2})$
chi square	$\chi^2(Q_1,\sigma)$	$\chi^2(Q_2,\sigma)$	$\chi^2(Q_1+Q_2,\sigma)$

PDFs of functions of independent random variables g_n

PDF	Function	PDF of the function
g_n: $N(0,\sigma)$	$(g_1^2+g_2^2)^{1/2}$	$R(\sigma)$
g_n: $N(0,\sigma)$	$\arctan(g_2^2/g_1^2)$	$U(0,2\pi)$
g_n: $N(0,\sigma)$	$\sum_{n=1}^{Q} g_n^2$	$\chi^2(Q,\sigma)$

R18 Error propagation (Sections C.2.3, C.3.3 and 10.3.7)

f_g is the PDF of a random variable (RV) g, a and b are constants, $g' = p(g)$ is a differentiable monotone function with the derivative $\mathrm{d}p/\mathrm{d}g$ and the inverse function $g = p^{-1}(g')$.
$\boldsymbol{g}$ is a vector with P RVs with the covariance matrix $\mathrm{cov}(\boldsymbol{g})$, $\boldsymbol{g}'$ is a vector with Q RVs and the covariance matrix $\mathrm{cov}(\boldsymbol{g}')$, $\boldsymbol{M}$ a $Q \times P$ matrix and $\boldsymbol{a}$ a column vector with Q elements.

1. PDF, mean and variance of the linear function $g' = ag + b$

$$f_{g'}(g') = \frac{f_g((g'-a)/b)}{|a|}, \quad \mu_{g'} = a\mu_g + b, \quad \sigma_{g'}^2 = a^2\sigma_g^2$$

2. PDF of a monotone, differentiable function $g' = p(g)$

$$f_{g'}(g') = \frac{f_g(p^{-1}(g'))}{\left|\mathrm{d}p(p^{-1}(g'))/\mathrm{d}g\right|},$$

3. Mean value and variance of a differentiable non-linear function $g' = p(g)$

$$\mu_{g'} \approx p(\mu_g) + \frac{\sigma_g^2}{2}\frac{\mathrm{d}^2p(\mu_g)}{\mathrm{d}g^2}, \quad \sigma_{g'}^2 \approx \left|\frac{\mathrm{d}p(\mu_g)}{\mathrm{d}g}\right|^2 \sigma_g^2$$

4. Covariance matrix of a linear combination of RVs, $\boldsymbol{g}' = \boldsymbol{M}\boldsymbol{g} + \boldsymbol{a}$

$$\mathrm{cov}(\boldsymbol{g}') = \boldsymbol{M}\,\mathrm{cov}(\boldsymbol{g})\boldsymbol{M}^T$$

5. Covariance matrix of a nonlinear combination of RVs, $\boldsymbol{g}' = \boldsymbol{p}(\boldsymbol{g})$

$$\mathrm{cov}(\boldsymbol{g}') \approx \boldsymbol{J}\,\mathrm{cov}(\boldsymbol{g})\boldsymbol{J}^T \quad \text{with the Jacobi matrix } \boldsymbol{J}, \quad j_{q,p} = \frac{\partial p_q}{\partial g_p}.$$

6. Homogeneous stochastic field: Convolution of a random vector with a filter $\boldsymbol{h}$: $\boldsymbol{g}' = \boldsymbol{h} * \boldsymbol{g}$ (Section 10.3.7)
   (a) $\boldsymbol{g}$ has the autocovariance function $\boldsymbol{c}$

$$\boldsymbol{c}' = \boldsymbol{c} \star (\boldsymbol{h} \star \boldsymbol{h}) \quad \circ\!\!-\!\!\bullet \quad \tilde{c}'(k) = \tilde{c}(k)\left|\tilde{h}(k)\right|^2.$$

   (b) $\boldsymbol{g}$ has the autocovariance function $\boldsymbol{c} = \sigma^2\delta_n$ (uncorrelated elements)

$$\boldsymbol{c}' = \sigma^2(\boldsymbol{h} \star \boldsymbol{h}) \quad \circ\!\!-\!\!\bullet \quad \tilde{c}'(k) = \sigma^2\left|\tilde{h}(k)\right|^2.$$

Homogeneous point operations (Section 7.2)

R19

Point operation independent of the pixel position

$$G'_{mn} = P(G_{mn})$$

1. Negative

$$P_N(q) = Q - 1 - q$$

2. Detection of underflow and overflow using a $[r, g, b]$ pseudo color representation

$$P_{uo}(q) = \begin{cases} [0, 0, Q-1] & \text{(blue)} \quad q = 0 \\ [q, q, q] & \text{(gray)} \quad q \in [1, Q-2] \\ [Q-1, 0, 0] & \text{(red)} \quad q = Q-1 \end{cases}$$

3. Contrast enhancement of the gray value range $[q_1, q_2]$

$$P_{cs}(q) = \begin{cases} 0 & q < q_1 \\ \dfrac{(q - q_1)(Q-1)}{q_2 - q_1} & q \in [q_1, q_2] \\ Q - 1 & q > q_2 \end{cases}$$

Calibration

R20

1. Noise equivalization (Section 7.2.3)
   If the variance of the noise is dependent on the gray value, it can be made constant by the non-linear gray value transformation with two free parameters σ_h and C:

$$h(g) = \sigma_h \int_0^g \frac{\mathrm{d}g'}{\sqrt{\sigma^2(g')}} + C$$

   A linear variance function (Section 4.5.5)

$$\sigma_g^2(g) = \sigma_0^2 + \alpha g$$

   results for $g \in [0, g_{\max}] \mapsto h \in [0, \gamma g_{\max}]$ in the transformation function

$$h(g) = \gamma g_{\max} \frac{\sqrt{\sigma_0^2 + Kg} - \sigma_0}{\sqrt{\sigma_0^2 + Kg_{\max}} - \sigma_0}, \quad \sigma_h = \frac{\gamma K g_{\max}/2}{\sqrt{\sigma_0^2 + Kg_{\max}} - \sigma_0}.$$

2. Linear radiometric two-point calibration (Section 7.3.3)
   Two calibration images are taken, a dark image $\boldsymbol{B}$ without any illumination and a reference image $\boldsymbol{R}$ with constant irradiance. A calibrated

image that is corrected to the zero point ("fixed pattern noise") and the inhomogeneity in the sensitivity of the image sensor elements is obtained by

$$G' = c\frac{G - B}{R - B}.$$

R21 Interpolation (Section 5.5)

1. Interpolation of a continuous function from function values at grid points at a distance $\Delta \boldsymbol{x} = [x_1, \ldots, x_D]^T$ is a convolution operation

$$g_r(\boldsymbol{x}) = \sum_n g(\boldsymbol{x}_n) h(\boldsymbol{x} - \boldsymbol{x}_n).$$

The requirement to reproduce the function values at the interpolation points leads to the *interpolation condition*

$$h(\boldsymbol{x}_n) = \begin{cases} 1 & \boldsymbol{n} = \boldsymbol{0} \\ 0 & \text{otherwise} \end{cases}.$$

2. Ideal interpolation function

$$h(\boldsymbol{x}) = \prod_{d=1}^{D} \operatorname{sinc}(x_d/\Delta x_d) \quad \circ\!\!-\!\!\bullet \quad \tilde{h}(\boldsymbol{\kappa}) = \prod_{d=1}^{D} \Pi(\tilde{k}_d/2)$$

3. Discrete 1D filters for interpolating values between the existing grid points (resolution doubling)

Type	Mask	Transfer function
linear	$\begin{bmatrix} 1 & 1 \end{bmatrix}/2$	$\cos(\pi\tilde{k}/2)$
cubic	$\begin{bmatrix} -1 & 9 & 9 & -1 \end{bmatrix}/16$	$\dfrac{9\cos(\pi\tilde{k}/2) - \cos(3\pi\tilde{k}/2)}{8}$
cubic B-spline	$\begin{bmatrix} 1 & 23 & 23 & 1 \end{bmatrix}/48$ $\begin{bmatrix} 3 - \sqrt{3}, & \sqrt{3} - 2 \end{bmatrix}^{\dagger}$	$\dfrac{23\cos(\pi\tilde{k}/2) + \cos(3\pi\tilde{k}/2)}{16 + 8\cos(\pi\tilde{k})}$

†Recursive filter that is used successively in the forward and reverse direction, see Section 5.5.5

1D convolution filter (Section 10.3.5, Section 11.2 and Section 12.3) **R22**

1. Transfer function of a 1D filter with an odd number of coefficients ($2R+1$, $[h_{-R},\ldots,h_{-1},h_0,h_1,\ldots,h_R]$)
   (a) General
$$\tilde{h}(\underset{\sim}{k}) = \sum_{\tilde{n}=-R}^{R} h_{\tilde{n}} \exp(-\pi \mathrm{i} \tilde{n} \underset{\sim}{k})$$
   (b) Even symmetry ($h_{-\tilde{n}} = h_{\tilde{n}}$)
$$\tilde{h}(\underset{\sim}{k}) = h_0 + 2\sum_{\tilde{n}=1}^{R} h_{\tilde{n}} \cos(\pi \tilde{n} \underset{\sim}{k})$$
   (c) Odd symmetry ($h_{-\tilde{n}} = -h_{\tilde{n}}$)
$$\tilde{h}(\underset{\sim}{k}) = -2\mathrm{i}\sum_{\tilde{n}=1}^{R} h_{\tilde{n}} \sin(\pi \tilde{n} \underset{\sim}{k})$$
2. Transfer functions of a 1D filter with an even number of coefficients ($2R$, $[h_{-R},\ldots,h_{-1},h_1,\ldots,h_R]$, Convolution result lies on the intermediate grid)
   (a) Even symmetry ($h_{-\tilde{n}} = h_{\tilde{n}}$):
$$\tilde{h}(\underset{\sim}{k}) = 2\sum_{\tilde{n}=1}^{R} h_{\tilde{n}} \cos(\pi(\tilde{n}-1/2)\underset{\sim}{k})$$
   (b) Odd symmetry ($h_{-\tilde{n}} = -h_{\tilde{n}}$):
$$\tilde{h}(\underset{\sim}{k}) = -2\mathrm{i}\sum_{\tilde{n}=1}^{R} h_{\tilde{n}} \sin(\pi(\tilde{n}-1/2)\underset{\sim}{k})$$
3. Transfer functions of the two elementary filters
   (a) Averaging of two neighboring points
$$\boldsymbol{B} = [1\ \ 1]/2 \quad \circ\!\!-\!\!\bullet \quad \tilde{b}(\underset{\sim}{k}) = \cos(\pi \underset{\sim}{k}/2)$$
   (b) Difference between two neighboring points
$$\boldsymbol{D} = [1\ \ -1] \quad \circ\!\!-\!\!\bullet \quad \tilde{d}_1(\underset{\sim}{k}) = 2\mathrm{i}\sin(\pi \underset{\sim}{k}/2)$$

R23 Recursive 1D filters (Section 10.4)

1. General filter equation

$$g'_n = -\sum_{n''=1}^{S} a_{n''} g'_{n-n''} + \sum_{n'=-R}^{R} h_{n'} g_{n-n'}$$

2. General transfer function

$$\tilde{h}(\underset{\sim}{k}) = \frac{\sum\limits_{n'=-R}^{R} h_{n'} \exp(-\pi i n' \underset{\sim}{k})}{\sum\limits_{n''=0}^{S} a_{n''} \exp(-\pi i n'' \underset{\sim}{k})}$$

3. Factorization of the transfer function using the z-transform and the fundamental theorem of algebra

$$\tilde{h}(z) = h_{-R} z^R \frac{\prod\limits_{n'=1}^{2R} (1 - c_{n'} z^{-1})}{\prod\limits_{n''=1}^{S} (1 - d_{n''} z^{-1})}$$

4. Relaxation filter
   (a) Filter equation ($|\alpha| < 1$)

$$g'_n = \alpha g'_{n\mp1} + (1 - \alpha) g_n$$

   (b) Point spread function (PSF)

$$^{\pm}r_{\pm n} = \begin{cases} (1 - \alpha)\alpha^n & n \geq 0 \\ 0 & \text{else} \end{cases}$$

   (c) Transfer function of the symmetrical filter (cascaded application in forward and reverse direction)

$$\tilde{r}(\underset{\sim}{k}) = \frac{1}{1 + \beta - \beta \cos \pi \underset{\sim}{k}}, \quad \left(\tilde{r}(0) = 1, \tilde{r}(1) = \frac{1}{1 + 2\beta}\right)$$

   with

$$\beta = \frac{2\alpha}{(1 - \alpha)^2}, \quad \alpha = \frac{1 + \beta - \sqrt{1 + 2\beta}}{\beta}, \quad \beta \in] - 1/2, \infty[$$

5. Resonance filter with the transfer function one at the resonant wavenumber $\tilde{k}_0$ in the limiting case of low damping $1 - r \ll 1$
   (a) Filter equation (attenuation coefficient $r \in [0, 1[$, resonance wavenumber $\tilde{k}_0 \in [0, 1]$)

$$g'_n = (1 - r^2)\sin(\pi\tilde{k}_0)g_n + 2r\cos(\pi\tilde{k}_0)g'_{n\mp1} - r^2 g'_{n\mp2}$$

   (b) Point spread function (PSF)

$$h_{\pm n} = \begin{cases} (1 - r^2)r^n \sin[(n+1)\pi\tilde{k}_0] & n \geq 0 \\ 0 & n < 0 \end{cases}$$

   (c) Transfer function of the symmetrical filter (cascaded application in forward and reverse direction)

$$\tilde{s}(\tilde{k}) = \frac{\sin^2(\pi\tilde{k}_0)(1 - r^2)^2}{(1 - 2r\cos[\pi(\tilde{k} - \tilde{k}_0)] + r^2)(1 - 2r\cos[\pi(\tilde{k} + \tilde{k}_0)] + r^2)}$$

   (d) Approximation of the transfer function with low damping

$$\tilde{s}(\tilde{k}) \approx \frac{1}{1 + (\tilde{k} - \tilde{k}_0)^2 \Big/ \frac{(1-r^2)^2}{4r^2\pi^2}} \quad \text{for} \quad 1 - r \ll 1$$

   (e) Half-width Δk, defined by $\tilde{s}(\tilde{k}_0 \pm \Delta k) = 1/2$

$$\Delta k \approx (1 - r)/\pi$$

R24 Smoothing filters (Chapter 11)

1. General conditions for 1D filters with $2R + 1$ coefficients

Property	Spatial domain	Fourier space
preservation of mean value	$\sum_{n=-R}^{R} h_n = 1$	$\tilde{h}(0) = 1$
zero shift, even symmetry	$h_{-n} = h_n$	$\mathrm{Im}\,\tilde{h}(\kappa) = 0$
monotonously decreasing from one to zero	$\sum_{n=-R}^{R} (-1)^n h_n = 0$	$\tilde{h}(1) = 0$ and $\tilde{h}(\tilde{\kappa}_2) \leq \tilde{h}(\tilde{\kappa}_1)$ if $\tilde{\kappa}_2 > \tilde{\kappa}_1$, $\tilde{h}(\boldsymbol{\kappa}) \in [0, 1]$
isotropy	$h(\boldsymbol{x}) = h(\lvert\boldsymbol{x}\rvert)$	$\tilde{h}(\boldsymbol{\kappa}) = \tilde{h}(\lvert\boldsymbol{\kappa}\rvert)$

2. 1D box (rectangular) filter

Mask	Transfer function	Noise suppression†
${}^3\boldsymbol{R} = [1\ 1\ 1]/3$	$\frac{1}{3} + \frac{2}{3}\cos(\pi\tilde{\kappa})$	$\frac{1}{\sqrt{3}} \approx 0.577$
${}^4\boldsymbol{R} = [1\ 1\ 1\ 1]/4$	$\cos(\pi\tilde{\kappa})\cos(\pi\tilde{\kappa}/2)$	$1/2 = 0.5$
${}^R\boldsymbol{R} = \underbrace{[1\ \dots\ 1]}_{R \text{ times}}/R$	$\frac{\sin(\pi R\tilde{\kappa}/2)}{R\sin(\pi\tilde{\kappa}/2)}$	$\frac{1}{\sqrt{R}}$

†For white noise

3. 1D binomial filter

Mask	Transfer function	Noise suppression†
$\boldsymbol{B}^1 = [1\ 1]/2$	$\cos(\pi\tilde{\kappa}/2)$	$\sqrt{\frac{1}{2}} \approx 0.707$
$\boldsymbol{B}^2 = [1\ 2\ 1]/4$	$\cos^2(\pi\tilde{\kappa}/2)$	$\sqrt{\frac{3}{8}} \approx 0.612$
$\boldsymbol{B}^4 = [1\ 4\ 6\ 4\ 1]/16$	$\cos^4(\pi\tilde{\kappa}/2)$	$\sqrt{\frac{35}{128}} \approx 0.523$
$\boldsymbol{B}^R$	$\cos^R(\pi\tilde{\kappa}/2)$	$\frac{\sqrt{2R!}}{R!\,2^R}$

†For white noise

First-order derivative filter (Chapter 12)

1. General properties for a derivative filter in the direction x_d of a D-dimensional signal; d' denotes one of the possible directions and $\boldsymbol{n}$ vector indexing (Section 10.3.1)

Property	Spatial domain	Fourier space
zero mean value	$\sum_n h_{\boldsymbol{n}} = 0$	$\tilde{h}(\tilde{\boldsymbol{\kappa}})\big\|_{\tilde{\boldsymbol{\kappa}}=0} = 0$
zero shift, odd symmetry	$h_{n_1,\ldots,-n_d,\ldots,n_D} = -h_{n_1,\ldots,n_d,\ldots,n_D}$	$\mathrm{Re}\,\tilde{h}(\boldsymbol{\kappa}) = 0$
first order derivative	$\sum_n n_{d'} h_{\boldsymbol{n}} = \delta_{d'-d}$	$\left.\dfrac{\partial \tilde{h}(\tilde{\boldsymbol{\kappa}})}{\partial \tilde{\kappa}_d}\right\|_{\tilde{\boldsymbol{\kappa}}=0} = \pi \mathrm{i} \delta_{d'-d}$
isotropy		$\tilde{h}(\tilde{\boldsymbol{\kappa}}) = \pi \mathrm{i} \tilde{\kappa}_d \tilde{b}(\|\tilde{\boldsymbol{\kappa}}\|)$ with $\tilde{b}(0) = 1$, $\nabla_k \tilde{b}(\|\tilde{\boldsymbol{\kappa}}\|) = \mathbf{0}$

2. First-order derivative filter

Name	Mask	Transfer function
$\mathcal{D}_x$	$\begin{bmatrix} 1 & -1 \end{bmatrix}$	$2\mathrm{i}\sin(\pi\tilde{\kappa}_x/2)$
symmetric difference, $\mathcal{D}_{2x}$	$\begin{bmatrix} 1 & 0 & -1 \end{bmatrix}/2$	$\mathrm{i}\sin(\pi\tilde{\kappa}_x)$
cubic B-Spline $\mathcal{D}_{2x}\, {}^{\pm}\mathcal{U}$	$\begin{bmatrix} 1 & 0 & -1 \end{bmatrix}/2$, $\begin{bmatrix} 3-\sqrt{3}, & \sqrt{3}-2 \end{bmatrix}^{\dagger}$	$\mathrm{i}\dfrac{\sin(\pi\tilde{\kappa}_x)}{2/3 + 1/3\cos(\pi\tilde{\kappa}_x)}$

†Recursive filter ${}^{\pm}\mathcal{U}$, which is applied successively in the forward and reverse direction, see Section 5.5.5

3. Regularized first-order derivative filters

Name	Mask	Transfer function
2×2, $\mathcal{D}_x\mathcal{B}_y$	$\frac{1}{2}\begin{bmatrix} 1 & -1 \\ 1 & -1 \end{bmatrix}$	$2\mathrm{i}\sin(\pi\tilde{k}_x/2)\cos(\pi\tilde{k}_y/2)$
Sobel, $\mathcal{D}_{2x}\mathcal{B}_y^2$	$\frac{1}{8}\begin{bmatrix} 1 & 0 & -1 \\ 2 & 0 & -2 \\ 1 & 0 & -1 \end{bmatrix}$	$\mathrm{i}\sin(\pi\tilde{k}_x)\cos^2(\pi\tilde{k}_y/2)$
Scharr, $\mathcal{D}_{2x}(3\mathcal{B}_y^2 + \mathcal{I})/4$	$\frac{1}{32}\begin{bmatrix} 3 & 0 & -3 \\ 10 & 0 & -10 \\ 3 & 0 & -3 \end{bmatrix}$	$\mathrm{i}\sin(\pi\tilde{k}_x)(3\cos^2(\pi\tilde{k}_y/2) + 1)/4$

4. Performance indicators for edge detectors: Angular error, magnitude error and white noise rejection factor. The three values in the two columns with the errors refer to the wavenumber ranges of 0–0.25, 0.25–0.5, and 0.5–0.75.

Name	Angular error [°]	Magnitude error	Noise factor
$\mathcal{D}_x$	—	—	$\sqrt{2} \approx 1.414$
$\mathcal{D}_{2x}$	1.36 4.90 12.7	0.026 0.151 0.398	$1/\sqrt{2} \approx 0.707$
$\mathcal{D}_{2x}\,{}^{\pm}\mathcal{U}$	0.02 0.33 2.26	0.001 0.023 0.220	$\sqrt{3\ln 3/\pi} \approx 1.024$
$\mathcal{D}_x\mathcal{B}_y$	0.67 2.27 5.10	0.013 0.079 0.221	1
$\mathcal{D}_{2x}\mathcal{B}_y^2$	0.67 2.27 5.10	0.012 0.053 0.070	$\sqrt{3}/4 \approx 0.433$
$\mathcal{D}_{2x}(3\mathcal{B}_y^2 + \mathcal{I})/4$	0.15 0.32 0.72	0.003 0.005 0.047	$\sqrt{59}/16 \approx 0.480$

Second-order derivative filter (Chapter 12)

1. Summary of the general properties for a second-order derivative filter in the direction x_d of a D-dimensional signal; d' denotes one of the possible directions and $\boldsymbol{n}$ vector indexing (Section 10.3.1)

Property	Spatial domain	Fourier space
zero mean value	$\sum_n h_n = 0$	$\tilde{h}(\underset{\sim}{\boldsymbol{\kappa}})\Big\vert_{\underset{\sim}{\boldsymbol{\kappa}}=0} = 0$
zero slope	$\sum_n n_{d'} h_n = 0$	$\left.\dfrac{\partial \tilde{h}(\underset{\sim}{\boldsymbol{\kappa}})}{\partial \underset{\sim}{\kappa}_{d'}}\right\vert_{\underset{\sim}{\boldsymbol{\kappa}}=0} = 0$
zero shift, even symmetry	$h_{-n} = h_n$	$\mathrm{Im}\left(\tilde{h}(\boldsymbol{\kappa})\right) = 0$
2nd derivation	$\sum_n n_{d'}^2 h_n = 2\delta_{d'-d}$	$\left.\dfrac{\partial^2 \tilde{h}(\underset{\sim}{\boldsymbol{\kappa}})}{\partial \underset{\sim}{\kappa}_d^2}\right\vert_{\underset{\sim}{\boldsymbol{\kappa}}=0} = -2\pi^2\delta_{d'-d}$
isotropy		$\tilde{h}(\underset{\sim}{\boldsymbol{\kappa}}) = -(\pi \underset{\sim}{\kappa}_d)^2 \tilde{b}(\lvert\underset{\sim}{\boldsymbol{\kappa}}\rvert)$ with $\tilde{b}(0) = 1$, $\nabla_k \tilde{b}(\lvert\underset{\sim}{\boldsymbol{\kappa}}\rvert) = \mathbf{0}$

2. Second-order derivative filter

Name	Mask	Transfer function
1D Laplace, $\mathcal{D}_x^2$	$\begin{bmatrix} 1 & -2 & 1 \end{bmatrix}$	$-4\sin^2(\pi \underset{\sim}{k}_x/2)$
2D Laplace, $\mathcal{L}$	$\begin{bmatrix} 0 & 1 & 0 \\ 1 & -4 & 1 \\ 0 & 1 & 0 \end{bmatrix}$	$-4\sin^2(\pi \underset{\sim}{k}_x/2) - 4\sin^2(\pi \underset{\sim}{k}_y/2)$
2D Laplace, $\mathcal{L}'$	$\dfrac{1}{4}\begin{bmatrix} 1 & 2 & 1 \\ 2 & -12 & 2 \\ 1 & 2 & 1 \end{bmatrix}$	$4\cos^2(\pi \underset{\sim}{k}_x/2)\cos^2(\pi \underset{\sim}{k}_y/2) - 4$

R27 Gaussian and Laplacian pyramid (Section 15.3)

1. Construction of the *Gaussian pyramid* $\boldsymbol{G}^{(0)}, \boldsymbol{G}^{(1)}, \ldots, \boldsymbol{G}^{(P)}$ with $P+1$ planes by iterative smoothing and subsampling by a factor of two in all directions

$$\boldsymbol{G}^{(0)} = \boldsymbol{G}, \quad \boldsymbol{G}^{(p+1)} = \mathcal{B}_{\downarrow 2}\boldsymbol{G}^{(p)}$$

2. Condition for the smoothing filter to avoid overlapping effects (aliasing)

$$\tilde{b}(\underset{\sim}{\boldsymbol{\kappa}}) = 0 \;\; \forall \underset{\sim}{\kappa}_p \geq \frac{1}{2}$$

3. Construction of the *Laplacian pyramid* $\boldsymbol{L}^{(0)}, \boldsymbol{L}^{(1)}, \ldots, \boldsymbol{L}^{(P)}$ with $P+1$ planes from the Gaussian pyramid

$$\boldsymbol{L}^{(p)} = \boldsymbol{G}^{(p)} - \uparrow_2 \boldsymbol{G}^{(p+1)}, \quad \boldsymbol{L}^{(P)} = \boldsymbol{G}^{(P)}$$

   The last level of the Laplacian pyramid is the last level of the Gaussian pyramid.
4. Interpolation filter for upsampling $\uparrow_2$ (≻ R21)
5. Iterative reconstruction of the original image from the Laplacian pyramid. Calculate

$$\boldsymbol{G}^{(p-1)} = \boldsymbol{L}^{(p-1)} + \uparrow_2 \boldsymbol{G}^{(p)}$$

   starting with the highest level ($p = P$). If the same method is used for upward probing when constructing the Laplacian pyramid and reconstructing the original image, it is error-free except for rounding errors.
6. Directio-pyramidal decomposition into two directional components

$$\begin{aligned}
\boldsymbol{G}^{(p+1)} &= \downarrow_2 \mathcal{B}_x\mathcal{B}_y\boldsymbol{G}^{(p)} \\
\boldsymbol{L}^{(p)} &= \boldsymbol{G}^{(p)} - \uparrow_2 \boldsymbol{G}^{(p+1)} \\
\boldsymbol{L}_x^{(p)} &= 1/2(\boldsymbol{L}^{(p)} - (\mathcal{B}_x - \mathcal{B}_y)\boldsymbol{G}^{(p)}) \\
\boldsymbol{L}_y^{(p)} &= 1/2(\boldsymbol{L}^{(p)} + (\mathcal{B}_x - \mathcal{B}_y)\boldsymbol{G}^{(p)})
\end{aligned}$$

Bibliography

[1] Acharya, T. and Tsai, P.-S.: JPEG2000 Standard for Image Compression, Wiley, New York, 2005.

[2] Adelson, E. H. and Bergen, J. R.: Spatiotemporal energy models for the perception of motion, J. Opt. Soc. Am. A, 2, 284–299, doi: 10.1364/JOSAA.2.000284, 1985.

[3] Adelson, E. H. and Bergen, J. R.: The extraction of spatio-temporal energy in human and machine vision, in: Proceedings Workshop on Motion: Representation and Analysis, May 1986, Charleston, South Carolina, edited by Adelson, E. H. and Bergen, J. R., pp. 151–155, Washington, 1986.

[4] Adrian, R. J. and Westerweel, J.: Particle Image Velocimetry, Cambridge, doi: 10.1007/978-1-4419-0953-4, 2011.

[5] Aho, A. V., Hopcroft, J. E., and Ullman, J. D.: The Design and Analysis of Computer Algorithms, Addison Wesley, Reading, MA, 1974.

[6] Albrecht, H.-E., Borys, M., Damaschke, N., and Tropea, C.: Laser Doppler and Phase Doppler Measurement Techniques, Springer Verlag, Berlin, doi: 10.1007/978-3-662-05165-8, 2003.

[7] Arce, G. R.: Nonlinear Signal Processing, a Statistical Approach, John Wiley & Sons, Hoboken, NJ, 2005.

[8] Aster, R. C., Borchers, B., and Thurber, C. H.: Parameter Estimation and Inverse Problems, vol. 90 of *International Geophysics Series*, Academic Press, Boston, 2005.

[9] Ballard, D. H. and Brown, C. M.: Computer Vision, Prentice-Hall, Englewood Cliffs, NJ, 1982.

[10] Barrett, H. H. and Myers, K. J.: Foundations of Image Science, Wiley, 2004.

[11] Bass, M., ed.: Handbook of Optics, Vol. 1–5, McGraw Hill, 3 edn., 2010.

[12] Bauer, P. S.: Development of an imaging polarimeter for water wave slope measurements, Master's thesis, Institut für Umweltphysik, Heidelberg University, Germany, doi: 10.11588/heidok.00015899, 2013.

[13] Beauchemin, S. S. and Barron, J. L.: The computation of optical flow, ACM Computing Surveys, 27, 433–467, doi: 10.1145/212094.212141, 1995.

[14] Belbachir, A. N., ed.: Smart Cameras, Springer, New York, doi: 10.1007/978-1-4419-0953-4, 2010.

[15] Benedetto, J. J. and Ferreira, P. J. S. G., eds.: Modern Sampling Theory - Mathematics and Applications, Birkhäuser, doi: 10.1007/978-1-4612-0143-4, 2001.

B. Jähne, *Digital Image Processing and Image Acquisition*,
https://doi.org/10.1007/978-3-032-04806-6

[16] Beyerer, J., León, F. P., and Frese, C.: Machine Vision Automated Visual Inspection: Theory, Practice and Applications, Springer, doi: 10.1007/978-3-662-47794-6, 2016.

[17] Bähr, H.-P. and Vögtle, T., eds.: Digitale Bildverarbeitung. Anwendungen in Photogrammetrie, Fernerkundung und GIS, Wichmann, Heidelberg, 4 edn., URL http://www.vde-verlag.de/buecher/537409/digitale-bildverarbeitung.html, 2005.

[18] Biberman, L. M., ed.: Electro Optical Imaging: System Performance and Modeling, SPIE, Bellingham, WA, 2001.

[19] Bigün, J. and Granlund, G. H.: Optimal orientation detection of linear symmetry, in: ICCV, pp. 433–438, London, UK, URL http://urn.kb.se/resolve?urn=urn:nbn:se:liu:diva-103805, 1987.

[20] Bishop, C. M.: Pattern Recognition and Machine Learning, Springer, 2006.

[21] Blahut, R. E.: Fast Algorithms for Digital Signal Processing, Cambridge Univ. Press, Reading, MA, doi: 10.1017/CBO9780511760921, 2010.

[22] Born, M. and Wolf, E.: Principles of Optics — 60th anniversary edition, Cambridge, 7 edn., doi: 10.1017/9781108769914, 2019.

[23] Bracewell, R.: The Fourier Transform and its Applications, McGraw-Hill, New York, 2 edn., 1986.

[24] Brady, D. J.: Optical Imaging and Spectroscopy, Wiley, doi: 10.1002/9780470443736, 2009.

[25] Bredies, K. and Lorenz, D.: Mathematische Bildverarbeitung, Vieweg Teubner, doi: 10.1007/978-3-8348-9814-2, 2011.

[26] Breitenstein, O., Warta, W., and Langenkamp, M.: Lock-in Thermography, Springer Berlin Heidelberg, 2 edn., doi: 10.1007/978-3-642-02417-7, 2010.

[27] Brigham, E. O.: Fast Fourier Transform and its Applications, Prentice Hall, 1988.

[28] Broit, C.: Optimal registrations of deformed images, Diss., Univ. of Pennsylvania, USA, 1981.

[29] Bronstein, I. N.: Taschenbuch der Mathematik, Harri Deutsch, 7 edn., 2008.

[30] Bubba, T. A., Gaburro, R., Gazzola, S., Papafitsoros, K., Pereyra, M., and Schönlieb, C.-B., eds.: Scale Space and Variational Methods in Computer Vision: 10th International Conference, SSVM 2025, Dartington, UK, May 18–22, 2025, Proceedings, Part I, no. 15667 in Lecture Notes in Computer Science, Springer Nature, doi: 10.1007/978-3-031-92366-1, 2025.

[31] Bubba, T. A., Gaburro, R., Gazzola, S., Papafitsoros, K., Pereyra, M., and Schönlieb, C.-B., eds.: Scale Space and Variational Methods in Computer Vision: 10th International Conference, SSVM 2025, Dartington, UK, May 18–22, 2025, Proceedings, Part II, no. 15668 in Lecture Notes in Computer Science, Springer Nature, doi: 10.1007/978-3-031-92369-2, 2025.

[32] Burger, W. and Burge, M. J.: Digital Image Processing: An Algorithmic Introduction, Springer International Publishing, doi: 10.1007/978-3-031-05744-1, 2022.

[33] Burke, M. W.: Image Acquisition, Chapman & Hall, doi: 10.1007/978-94-009-0069-1, 1996.

[34] Burkhardt, H., ed.: Workshop on Texture Analysis, Albert-Ludwigs-Universität, Freiburg, Institut für Informatik, 1998.

[35] Burkhardt, H. and Siggelkow, S.: Invariant features in pattern recognition - fundamentals and applications, in: Nonlinear Model-Based Image/Video Processing and Analysis, edited by Kotropoulos, C. and Pitas, I., pp. 269–307, John Wiley & Sons, 2001.

[36] Burt, P. J.: Fast algorithms for estimating local image properties, Computer Vision, Graphics, and Image Processing, 21, 368–382, doi: 10.1016/S0734-189X(83)80049-8, 1983.

[37] Burt, P. J.: The pyramid as a structure for efficient computation, in: Multiresolution image processing and analysis, edited by Rosenfeld, A., vol. 12 of *Springer Series in Information Sciences*, pp. 6–35, Springer, New York, doi: 10.1007/978-3-642-51590-3_2, 1984.

[38] Busse, G., Wu, D., and Karpen, W.: Thermal wave imaging with phase sensitive modulated thermography, Journal of Applied Physics, 71, 3962–3965, doi: 10.1063/1.351366, 1992.

[39] Canny, J.: A computational approach to edge detection, IEEE Trans. Pattern Analysis Machine Intelligence, 8, 679–698, doi: 10.1109/TPAMI.1986.4767851, 1986.

[40] Castellano, F. N. and Lakowicz, J. R.: A water-soluble luminescence oxygen sensor, Photochemistry and Photobiology, 67, 179–183, doi: 10.1111/j.1751-1097.1998.tb05184.x, 1998.

[41] Chelappa, R., ed.: Digital Image Processing, IEEE Computer Society Press, Los Alamitos, CA, 1992.

[42] Chui, C. K., ed.: Wavelets: A Tutorial in Theory and Applications, Academic Press, Boston, MA, 1992.

[43] Close, C. M. and Frederick, D. K.: Modelling and Analysis of Dynamic Systems, Houghton Mifflin, Boston, 1978.

[44] Cooley, J. W. and Tukey, J. W.: An algorithm for the machine calculation of complex Fourier series, Mathematics of Computation, 19, 297–301, doi: 10.2307/2003354, 1965.

[45] Cormen, T. H., Leiserson, C. E., Rivest, R. L., and Stein, C.: Introduction to Algorithms, MIT Press, Cambridge, MA, 3 edn., 2009.

[46] Courant, R. and Hilbert, D.: Methoden der mathematischen Physik, Springer, Berlin, 4 edn., doi: 10.1007/978-3-642-58039-0, 1993.

[47] Crank, J.: The Mathematics of Diffusion, Clarendon Press, Oxford, 2nd edn., 1975.

[48] Czichos, H., Saito, T., and Smith, L., eds.: Springer Handbook of Materials Measurement Methods, Springer, doi: 10.1007/978-3-540-30300-8, 2006.

[49] Czycholl, G.: Solid State Theory, Volume 1: Basics: Phonons and Electrons in Crystals, Springer, doi: 10.1007/978-3-662-66135-2, 2023.

[50] Danielsson, P.-E., Lin, Q., and Ye, Q.-Z.: Efficient detection of second degree variations in 2D and 3D images, Tech. Rep. LiTH-ISY-R-2155, Department of Electrical Engineering, Linköping University, S-58183 Linköping, Sweden, 1999.

[51] Darmont, A.: High Dynamic Range Imaging: Sensors and Architectures, SPIE, 2 edn., doi: 10.1117/3.2512264, 2019.

[52] Davis, P. J.: Interpolation and Approximation, Dover, New York, 1975.

[53] DeCusaris, C., ed.: Handbook of Applied Photometry, Springer, New York, 1998.

[54] Demant, C., Streicher-Abel, B., and Garnica, C.: Industrial Image Processing. Visual Quality Control in Manufacturing, Springer, Berlin, 2 edn., doi: 10.1007/978-3-642-33905-9, includes CD-ROM, 2013.

[55] DeMarco, P., Pokorny, J., and Smith, V. C.: Full-spectrum cone sensitivity functions for X-chromosome-linked anomalous trichromats, J. Opt. Soc. Am., A9, 1465–1476, 1992.

[56] Dengler, J.: Methoden und Algorithmen zur Analyse bewegter Realweltszenen im Hinblick auf ein Blindenhilfesystem, Ph.D. thesis, Univiversity of Heidelberg, Heidelberg, Germany, 1985.

[57] Deriche, R.: Fast algorithms for low-level vision, IEEE Trans. Pattern Analysis Machine Intelligence, 12, 78–87, doi: 10.1109/34.41386, 1990.

[58] Diaspro, A., ed.: Confocal and Two-Photon Microscopy: Foundations, Applications, and Advances, Wiley, 2002.

[59] Diaspro, A., ed.: Nanoscopy and Multidimensional Optical Fluorescence Microscopy, CRC Press, 2010.

[60] Diaspro, A.: Expedition into the Nanoworld: An Exciting Voyage from Optical Microscopy to Nanoscopy, Springer, doi: 10.1007/978-3-030-94472-8, 2022.

[61] Diehl, N. and Burkhardt, H.: Planar motion estimation with a fast converging algorithm, in: Proc. 8th Int. Conf. Pattern Recognition, ICPR'86, October 27–31, 1986, Paris, pp. 1099–1102, IEEE Computer Society, Los Alamitos, 1986.

[62] Dittrich, P.-G., Bichra, M., Stiehler, D., Pfützenreuter, C., Radtke, L., Rosenberger, M., and Notni, G.: Extended characterization of multispectral resolving filter-on-chip snapshot-mosaic CMOS cameras, in: SPIE Proceeding, Algorithms, Technologies, and Applications for Multispectral and Hyperspectral Imagery XXV, edited by Velez-Reyes, M. and Messinger, D. W., vol. 10986, pp. 140–150, International Society for Optics and Photonics, doi: 10.1117/12.2518842, 2019.

[63] Dorf, R. C. and Bishop, R. H.: Modern Control Systems, Addison-Wesley, Menlo Park, CA, 8 edn., 1998.

[64] Drury, S. A.: Image Interpretation in Geology, Chapman & Hall, London, 2 edn., 1993.

[65] Dössel, O.: Bildgebende Verfahren in der Medizin, Springer Berlin Heidelberg, doi: 10.1007/978-3-642-54407-1, 2016.

[66] Dutton, N. A. W., Gyongy, I., Parmesan, L., Gnecchi, S., Calder, N., Rae, B. R., Pellegrini, S., Grant, L. A., and Henderson, R. K.: A SPAD-based QVGA image sensor for single-photon counting and quanta imaging, IEEE Transactions on Electron Devices, 63, 189–196, doi: 10.1109/ted.2015.2464682, 2016.

[67] Easton Jr., R. L.: Fourier Methods in Imaging, John Wiley & Sons, doi: 10.1002/9780470660102, 2010.

[68] Eldar, Y. C.: Sampling Theory - Beyond Bandlimited Systems, Cambridge University Press, doi: 10.1017/CBO9780511762321, 2015.

[69] Eldar, Y. C. and Kutyniok, G., eds.: Compressed Sensing, Cambridge University Press, doi: 10.1017/cbo9780511794308, 2012.

[70] Elmore, W. C. and Heald, M. A.: Physics of Waves, Dover Publications, New York, 1985.

[71] EMVA 1288 Working Group: EMVA Standard 1288 - Standard for Characterization of Image Sensors and Cameras, Release 4.0 Linear, open standard, European Machine Vision Association, URL https://www.emva.org/standards-technology/emva-1288/emva-standard-1288-downloads-2/, 2021.

[72] EMVA 1288 Working Group: EMVA Standard 1288 - Standard for Characterization of Image Sensors and Cameras, Release 4.0 General, open standard, European Machine Vision Association, URL https://www.emva.org/standards-technology/emva-1288/emva-standard-1288-downloads-2/, 2021.

[73] Erhardt, A., Zinser, G., Komitowski, D., and Bille, J.: Reconstructing 3D light microscopic images by digital image processing, Appl. Optics, 24, 194–200, doi: 10.1364/AO.24.000194, 1985.

[74] Everett, M. E.: Near-surface applied geophysics, Cambridge University Press, 2013.

[75] F. S. Crawford, J.: Waves, vol. 3 of *Berkely Physics Course*, McGraw-Hill, New York, 1965.

[76] Faugeras, O.: Three-Dimensional Computer Vision: A Geometric Viewpoint, MIT Press, 1993.

[77] Faugeras, O., Luong, Q.-T., and Papadopoulo, T.: The Geometry of Multiple Images, MIT Press, Cambridge, MA, 2001.

[78] Felsberg, M. and Sommer, G.: Structure Multivector for Local Analysis of Images, Tech. Rep. 2001, Christian-Albrechts-University of Kiel, 2000.

[79] Felsberg, M. and Sommer, G.: A New Extension of Linear Signal Processing for Estimating Local Properties and Detecting Features, in: Proceedings of the 22th DAGM Symposium on Pattern Recognition, 2000.

[80] Felsberg, M. and Sommer, G.: The monogenic signal, IEEE Trans. Signal Processing, 49, 3136–3144, doi: 10.1109/78.969520, 2001.

[81] Feynman, R.: Lectures on Physics, vol. 2, Addison-Wesley, Reading, Mass., 1964.

[82] Fischler, M. A. and Firschein, O., eds.: Readings in Computer Vision: Issues, Problems, Principles, and Paradigms, Morgan Kaufmann, Los Altos, CA, 1987.

[83] Fleet, D. J.: Measurement of Image Velocity, Kluwer Academic Publishers, Dordrecht, The Netherlands, doi: 10.1007/978-1-4615-3648-2, 1992.

[84] Fleet, D. J. and Jepson, A. D.: Hierarchical construction of orientation and velocity selective filters, IEEE Trans. Pattern Analysis Machine Intelligence, 11, 315–324, doi: 10.1109/34.21800, 1989.

[85] Fleet, D. J. and Jepson, A. D.: Computation of component image velocity from local phase information, Int. J. Computer Vision, 5, 77–104, doi: 10.1007/BF00056772, 1990.

[86] Fliege, N.: Multiraten-Signalverarbeitung, Teubner, Stuttgart, 1993.

[87] Foley, J. D., van Dam, A., Feiner, S. K., and Hughes, J. F.: Computer Graphics, Principles and Practice, Addison Wesley, Reading, MA, 2 edn., 1995.

[88] Forsyth, D. A. and Ponce, J.: Computer Vision: A Modern Approach, Pearson, 2 edn., 2011.

[89] Fossum, E., Ma, J., Masoodian, S., Anzagira, L., and Zizza, R.: The quanta image sensor: every photon counts, Sensors, 16, 1260, doi: 10.3390/s16081260, 2016.

[90] Fossum, E. R.: The quanta image sensor (QIS): concepts and challenges, in: Computational Optical Sensing and Imaging, Toronto, 10-14 July, 2011, Imaging and Applied Optics, OSA Technical Digest (CD), p. JTuE1, Optica Publishing Group, doi: 10.1364/cosi.2011.jtue1, 2011.

[91] Fossum, E. R.: Modeling the performance of single-bit and multi-bit quanta image sensors, IEEE Journal of the Electron Devices Society, 1, 166–174, doi: 10.1109/jeds.2013.2284054, 2013.

[92] Fowler, C. M. R.: The Solid Earth, Cambridge University Press, 2 edn., 2004.

[93] Freeman, W. T. and Adelson, E. H.: The design and use of steerable filters, IEEE Pattern Anal.Machine Intell., 13, 891–906, doi: 10.1109/34.93808, 1991.

[94] Friedl, F.: Investigating the Transfer of Oxygen at the Wavy Air-Water Interface under Wind-Induced Turbulence, Dissertation, Institut für Umweltphysik, Fakultät für Physik und Astronomie, Heidelberg University, doi: 10.11588/heidok.00014582, 2013.

[95] Frigo, M. and Johnson, S. G.: The design and implementation of FFTW3, Proc. IEEE, 93, 216–231, doi: 10.1109/JPROC.2004.840301, 2005.

[96] Friman, S. I. and Jähne, B.: Investigating SO_2 transfer across the air–water interface via LIF, Exp. Fluids, 60, 65, doi: 10.1007/s00348-019-2713-6, 2019.

[97] Förstner, W.: Image preprocessing for feature extraction in digital intensity, color and range images, in: Geomatic Method for the Analysis of Data in the Earth Sciences, edited by Dermanis, A., Grün, A., and Sansò, F., vol. 95 of *Lecture Notes on Earth Sciences*, pp. 165–189, Springer, doi: 10.1007/3-540-45597-3_4, 2000.

[98] Förstner, W. and Gülch, E.: A fast operator for detection and precise location of distinct points, corners and centres of circular features, in: Int. Soc. Photogr. Rem. Sens. Intercommission Workshop Interlaken on Fast Processing of Photogrammetric Data, pp. 281–305, 1987.

[99] Förstner, W. and Wrobel, B. P.: Photogrammetric Computer Vision, vol. 11 of *Geometry and Computing*, Springer, doi: 10.1007/978-3-319-11550-4, 2016.

[100] Fujiwara, H.: Spectroscopic Ellipsometry, Wiley, 2007.

[101] Garbe, C. S., Spies, H., and Jähne, B.: Estimation of surface flow and net heat flux from infrared image sequences, Journal of Mathematical Imaging and Vision, 19, 159–174, doi: 10.1023/A:1026233919766, 2003.

[102] Gaussorgues, G.: Infrared Thermography, Springer Netherlands, London, doi: 10.1007/978-94-011-0711-2, 1994.

[103] Geißler, P. and Jähne, B.: Measurements of bubble size distributions with an optical technique based on depth from focus, in: Air-Water Gas Transfer, Selected Papers, 3rd Intern. Symp. on Air-Water Gas Transfer, edited by Jähne, B. and Monahan, E., pp. 351–362, AEON, Hanau, doi: 10.5281/zenodo.10400,

1995.

[104] Geiler, T.: Polarisationsbildgebung in der industriellen Qualitätskontrolle, Dissertation, IWR, Fakultät für Physik und Astronomie, Heidelberg University, doi: `10.11588/heidok.00008533`, 2008.

[105] Gelles, J., Schnapp, B. J., and Sheetz, M. P.: Tracking kinesin driven movements with nanometre-scale precision, Nature, 331, 450–453, 1988.

[106] Girod, B., Rabenstein, R., and Stenger, A.: Einführung in die Systemtheorie: Signale und Systeme in der Elektrotechnik und Informationstechnik, B. G. Teubner, Wiesbaden, 2 edn., doi: `10.1007/978-3-322-99346-5`, 2003.

[107] Girosi, F., Verri, A., and Torre, V.: Constraints for the computation of optical flow, in: Proceedings Workshop on Visual Motion, March 1989, Irvine, CA, edited by Girosi, F., Verri, A., and Torre, V., pp. 116–124, Washington, doi: `10.1109/WVM.1989.47101`, 1989.

[108] Goldstein, H.: Classical Mechanics, Addison-Wesley Series in Physics, Addison-Wesley, Reading,MA, 2nd edn., 1980.

[109] Golub, G. H. and van Loan, C. F.: Matrix Computations, The Johns Hopkins University Press, Baltimore and London, 3 edn., 1996.

[110] Gonzalez, R. C. and Woods, R. E.: Digital image processing, Prentice Hall, Upper Saddle River, NJ, 2 edn., 2002.

[111] Gonzalez, R. C. and Woods, R. E.: Digital Image Processing, Pearson, 4 edn., 2018.

[112] Goodfellow, I., Bengio, Y., and Courville, A.: Deep Learning, MIT Press, 2017.

[113] Goodman, J. W.: Fourier Optics, Roberts & Company, Greenwood Village, CO, 3 edn., 2005.

[114] Granlund, G. H.: In search of a general picture processing operator, Computer Graphics and Image Processing, 8, 155–173, doi: `10.1016/0146-664X(78)90047-3`, 1978.

[115] Granlund, G. H. and Knutsson, H.: Signal Processing for Computer Vision, Kluwer, Dordrecht, The Netherlands, doi: `10.1007/978-1-4757-2377-9`, 1995.

[116] Griffin, L. D. and Lillhom, M., eds.: Scale Space Methods in Computer Vision, vol. 2695 of *Lecture Notes in Computer Science*, 4th Int. Conf. Scale-Space'03, Springer, Berlin, 2003.

[117] Groß, M.: Visual Computing, Springer, Berlin, doi: `10.1007/978-3-642-85023-3`, 1994.

[118] Haberäcker: Digitale Bildverarbeitung, Hanser, München, 1985.

[119] Hale, G. M. and Querry, M. R.: Optical constants of water in the 200 nm to 200 μm wavelength region, Appl. Optics, 12, 555–563, doi: `10.1364/AO.12.000555`, 1973.

[120] Harris, D. C. and Charles, L. A.: Quantitative Chemical Analysis, W. H. Freeman, 11 edn., 2025.

[121] Harris, G. and Stevens, M.: A combined corner and edge detector, in: Proceedings of the 4th Alvey Vision Conference, pp. 147–151, doi: `10.5244/c.2.23`, 1988.

[122] Harris, J. G.: The coupled depth/slope approach to surface reconstruction, Ph.D. thesis, Dept. Elec. Eng. Comput. Sci., MIT, Cambridge, Mass., URL

http://hdl.handle.net/1721.1/6859, 1986.

[123] Harris, J. G.: A new approach to surface reconstruction: the coupled depth/slope model, in: 1st Int. Conf. Comp. Vis. (ICCV), London, pp. 277–283, IEEE Computer Society, Washington, 1987.

[124] Hartley, R. and Zisserman, A.: Multiple View Geometry in Computer Vision, Cambridge Univ Press, 2 edn., 2003.

[125] Haußecker, H.: Messung und Simulation von kleinskaligen Austauschvorgängen an der Ozeanoberfläche mittels Thermographie, Dissertation, IWR, Fakultät für Physik und Astronomie, Heidelberg University, doi: 10.5281/zenodo.14789, 1996.

[126] Haußecker, H. and Fleet, D. J.: Computing optical flow with physical models of brightness variation, IEEE Trans. Pattern Analysis Machine Intelligence, 23, 661–673, doi: 10.1109/34.927465, 2001.

[127] Haußecker, H. and Spies, H.: Motion, in: Computer Vision and Applications—A Guide for Students and Practitioners, edited by Jähne, B. and Haußecker, H., chap. 10, pp. 347–395, Academic Press, doi: 10.1016/B978-012379777-3/50011-X, 2000.

[128] Hecht, E.: Optics, Pearson Education, Harlow, 5 edn., 2016.

[129] Heeger, D. J.: Optical flow using spatiotemporal filters, Int. J. Computer Vision, 1, 279–302, doi: 10.1007/BF00133568, 1988.

[130] Hertz, H.: Ueber sehr schnelle electrische Schwingungen, Annalen der Physik, 267, 421–448, doi: 10.1002/andp.18872670707, 1887.

[131] Hertz, H.: Ueber die Einwirkung einer geradlinigen electrischen Schwingung auf eine benachbarte Strombahn, Annalen der Physik, 270, 155–170, doi: 10.1002/andp.18882700510, 1888.

[132] Hertz, H.: Ueber die Ausbreitungsgeschwindigkeit der electrodynamischen Wirkungen, Annalen der Physik, 270, 551–569, doi: 10.1002/andp.18882700708, 1888.

[133] Hühnerfuss, H.: Basic physicochemical principles of monomolecular sea slicks and crude oil spills, in: Marine Surface Films, edited by Gade, M., Hühnerfuss, H., and Korenowski, G. M., pp. 21–35, Springer, doi: 10.1007/3-540-33271-5_4, 2006.

[134] Hildreth, E. C.: Computations underlying the measurement of visual motion, Artificial Intelligence, 23, 309–354, doi: 10.1016/0004-3702(84)90018-3, 1984.

[135] Hénon, S. and Meunier, J.: Microscope at the Brewster angle: Direct observation of first-order phase transitions in monolayers, Rev. Sci. Instrum., 62, 936–939, doi: 10.1063/1.1142032, 1991.

[136] Hoefflinger, B., ed.: High-Dynamic-Range (HDR) Vision — Microelectronics, Image Processing, Computer Graphics, vol. 26 of *Springer Series in Advanced Microelectronics*, Springer, Berlin, doi: 10.1007/978-3-540-44433-6, 2007.

[137] Hofmann, D.: Investigating Air-Water Gas Transfer via Fluorescence Imaging, Dissertation, Institut für Umweltphysik, Fakultät für Physik und Astronomie, Heidelberg University, doi: 10.11588/HEIDOK.00036916, 2025.

[138] Holst, G. C.: Electro-Optical Imaging System Performance, JCD Publishing and SPIE Press, 5 edn., 2008.

[139] Holst, G. C.: Testing and Evaluation of Infrared Imaging Systems, SPIE, Bellingham, WA, 3 edn., 2008.

[140] Holst, G. C. and Lomheim, T. S.: CMOS/CCD Sensors and Camera Systems, SPIE Press, Bellingham, Washington, USA, 2 edn., 2011.

[141] Horn, B. K. P.: Robot Vision, MIT Press, Cambridge, MA, 1986.

[142] Hornberg, A., ed.: Handbook of Machine Vision, Wiley-VCH, Weinheim, 2 edn., doi: 10.1002/9783527413409, 2017.

[143] Howell, S.: Handbook of CCD Astronomy, Cambridge University Press, Cambridge, 2000.

[144] Iizuka, K.: Engineering Optics, vol. 35 of *Springer Series in Optical Sciences*, Springer, Berlin, 2 edn., 1987.

[145] International Union of Pure and Applied Chemistry: Compendium of Chemical Terminology, Gold Book, Version 2.3.1, Tech. rep., IUPAC, URL http://goldbook.iupac.org/PDF/goldbook.pdf, 2012.

[146] Jain, A. K.: Fundamentals of digital image processing, Prentice Hall, New Jersey, 1989.

[147] Jain, R., Kasturi, R., and Schunck, B. G.: Machine Vision, McGraw-Hill, New York, 1995.

[148] Janesick, J., Klaasen, K., and Elliott, T.: CCD charge collection efficiency and the photon transfer technique, in: Solid State Imaging Arrays, edited by Dereniak, E. L. and Prettyjohns, K. N., vol. 570, pp. 7–19, Proc. SPIE, doi: 10.1117/12.950297, 1985.

[149] Janesick, J. R.: Scientific Charge-Coupled Devices, SPIE Press, Bellingham, Washington, USA, doi: 10.1117/3.374903, 2001.

[150] Janesick, J. R.: Photon Transfer, SPIE Press, Bellingham, Washington, USA, doi: 10.1117/3.725073, 2007.

[151] Jähne, B.: Image sequence analysis of complex physical objects: nonlinear small scale water surface waves, in: Proc. of 1st International Conference on Computer Vision, pp. 191–200, IEEE, London, UK, doi: 10.5281/zenodo.13126, 1987.

[152] Jähne, B.: Energy balance in small-scale waves: an experimental approach using optical slope measuring technique and image processing, in: Radar Scattering from Modulated Wind Waves, edited by Komen, G. J. and Oost, W. A., pp. 105–120, Kluwer Academic Publishers, Dordrecht, The Netherlands, doi: 10.1007/978-94-009-2309-6_10, 1989.

[153] Jähne, B.: Motion determination in space-time images, in: SPIE Proc.1135, Image Processing III, , international congress on optical science and engineering, Paris, 24-28 April 1989, pp. 147–152, doi: 10.1117/12.961657, 1989.

[154] Jähne, B.: From mean fluxes to a detailed experimental investigation of the gas transfer process, in: Air-Water Mass Transfer, selected papers from the 2nd International Symposium on Gas Transfer at Water Surfaces, September 11–14, 1990, Minneapolis, Minnesota, edited by Wilhelms, S. C. and Gulliver, J. S., pp. 244–256, ASCE, doi: 10.5281/zenodo.12204, 1991.

[155] Jähne, B.: Digital Image Processing - Concepts, Algorithms, and Scientific Applications, Springer, Berlin, doi: 10.1007/978-3-662-11565-7, 1991.

[156] Jähne, B.: Spatio-Temporal Image Processing: Theory and Scientific Applications, vol. 751 of *Lecture Notes in Computer Science*, Springer-Verlag, Berlin, doi: 10.1007/3-540-57418-2, 1993.

[157] Jähne, B.: Practical Handbook on Image Processing for Scientific and Technical Applications, CRC Press, Boca Raton, FL, 2 edn., doi: 10.1201/9780849390302, 2004.

[158] Jähne, B.: Digital Image Processing, Springer, Berlin, 6 edn., doi: 10.1007/3-540-27563-0, 2005.

[159] Jähne, B.: Almost lossless compression of noisy images, tm – Technisches Messen, 90, doi: 10.1515/teme-2023-0028, 2023.

[160] Jähne, B.: Digitale Bildverarbeitung und Bildgewinnung, Springer Vieweg, 8 edn., doi: 10.1007/978-3-662-59510-7, 2024.

[161] Jähne, B. and Haußecker, H., eds.: Computer Vision and Applications — A Guide for Students and Practitioners, Academic Press, URL http://www.sciencedirect.com/science/book/9780123797773, 2000.

[162] Jähne, B. and Schwarzbauer, M.: Noise equalisation and quasi loss-less image data compression – or how many bits needs an image sensor?, tm – Technisches Messen, 83, 16–24, doi: 10.1515/teme-2015-0093, published online 17.12.2015, 2016.

[163] Jähne, B., Klinke, J., and Waas, S.: Imaging of short ocean wind waves: a critical theoretical review, J. Opt. Soc. Am. A, 11, 2197–2209, doi: 10.1364/JOSAA.11.002197, 1994.

[164] Jähne, B., Massen, R., Nickolay, B., and Scharfenberg, H.: Technische Bildverarbeitung - Maschinelles Sehen, Springer, Berlin, doi: 10.1007/978-3-642-61403-3, 1995.

[165] Jähne, B., Geißler, P., and Haußecker, H., eds.: Handbook of Computer Vision and Applications, Academic Press, San Diego, 1999.

[166] Jähne, B., Scharr, H., and Körkel, S.: Principles of filter design, in: Handbook of Computer Vision and Applications, edited by Jähne, B., Haußecker, H., and Geißler, P., vol. 2, pp. 125–151, Academic Press, 1999.

[167] Jähne, B., Klar, M., and Jehle, M.: Data analysis, in: Handbook of Experimental Fluid Mechanics, edited by Tropea, C., Foss, J., and Yarin, A., chap. 25, pp. 1437–1491, Springer, Berlin, doi: 10.1007/978-3-540-30299-5_25, 2007.

[168] Jähne, B., Mester, R., Barth, E., and Scharr, H., eds.: Complex Motion, Proceedings of the 1st Workshop, Günzburg, October 2004, vol. 3417 of *Lecture Notes in Computer Science*, Springer, Berlin, doi: 10.1007/978-3-540-69866-1, 2007.

[169] Jiang, X. and Bunke, H.: Dreidimensionales Computersehen, Springer, Berlin, doi: 10.1007/978-3-642-61447-7, 1997.

[170] Jost, W.: Diffusion in Solids, Liquids, Gases, Third Printing with Addendum, Acadamic Press, 1960.

[171] Kajiya, J. T.: The rendering equation, Computer Graphics, 20, 143–150, 1986.

[172] Kass, M. and Witkin, A.: Analysing oriented patterns, Comp. Vis. Graph. Im. Process., 37, 362–385, doi: 10.1016/0734-189X(87)90043-0, 1987.

[173] Kass, M., Witkin, A., and Terzopoulos, D.: Snakes: Active Contour Models, Int. J. Computer Vision, 1, 321–331, 1987.

[174] Kasturi, B. Y. and Jain, R. C., eds.: Computer Vision: Principles, IEEE Computer Society, Los Alamitos, 1991.

[175] Kasturi, B. Y. and Jain, R. C., eds.: Computer Vision: Advances and Applications, IEEE Computer Society, Los Alamitos, 1991.

[176] Kazmierczak, H., ed.: Erfassung und maschinelle Verarbeitung von Bilddaten, Springer, URL `https://link.springer.com/book/9783709122563`, 1980.

[177] Kearney, J. K., Thompson, W. B., and Boley, D. L.: Optical flow estimation: an error analysis of gradient-based methods with local optimization, IEEE Trans. Pattern Analysis Machine Intelligence, 9, 229–244, doi: `10.1109/TPAMI.1987.4767897`, 1987.

[178] Kenyon, I. R.: The Light Fantastic: A Modern Introduction to Classical and Quantum Optics, Oxford Univ Press, 2008.

[179] Kerckhove, M., ed.: Scale-Space and Morphology in Computer Vision, vol. 2106 of *Lecture Notes in Computer Science*, 3rd Int. Conf. Scale-Space'01, Vancouver, Canada, Springer, Berlin, 2001.

[180] Kimmel, R., Sochen, N., and Weickert, J., eds.: Scale-Space and PDE Methods in Computer Vision, vol. 3459 of *Lecture Notes in Computer Science*, 5th Int. Conf. Scale-Space'05, Springer, Berlin, doi: `10.1007/b107185`, 2005.

[181] Kirchhoff, G.: Ueber das Verhältniss zwischen dem Emissionsvermögen und dem Absorptionsvermögen der Körper für Wärme und Licht, Annalen der Physik und Chemie, 185, 275–301, doi: `10.1002/andp.18601850205`, 1860.

[182] Kittel, C.: Introduction to Solid State Physics, John Wiley & Sons, 8 edn., 2005.

[183] Klette, R.: Concise Computer Vision, Springer, doi: `10.1007/978-1-4471-6320-6`, 2014.

[184] Klette, R., Koschan, A., and Schlüns, K.: Computer Vision, Räumliche Information aus digitalen Bildern, Vieweg, Brauschweig, 1996.

[185] Knutsson, H.: Filtering and reconstruction in image processing, Diss., Linköping Univ., 1982.

[186] Knutsson, H.: Representing local structure using tensor, in: The 6th Scandinavian Conference on Image Analysis, pp. 244–251, 1989.

[187] Knutsson, H. E., Wilson, R., and Granlund, G. H.: Anisotropic nonstationary image estimation and its applications: part I – restoration of noisy images, IEEE Trans. COMM., 31(3), 388–397, doi: `10.1109/tcom.1983.1095832`, 1983.

[188] Koenderink, J. J. and van Doorn, A. J.: Generic neighborhood operators, IEEE Trans. Pattern Analysis Machine Intelligence, 14, 597–605, doi: `10.1109/34.141551`, 1992.

[189] Korn, A.: Bildverarbeitung durch das visuelle System, Springer Berlin Heidelberg, doi: `10.1007/978-3-642-81902-5`, 1982.

[190] Kubitscheck, U.: Fluorescence Microscopy: From Principles to Biological Applications, Wiley-VHC, 2 edn., 2017.

[191] Kunz, J. and Jähne, B.: Investigating small scale air-sea exchange processes via thermography, Front. Mech. Eng., 4, 4, doi: `10.3389/fmech.2018.00004`, 2018.

[192] Lancaster, P. and Salkauskas, K.: Curve and Surface Fitting. An Introduction, Academic Press, London, 1986.

[193] Lanser, S. and Eckstein, W.: Eine Modifikation des Deriche-Verfahrens zur Kantendetektion, in: Proc. 13. DAGM-Symposium zur Mustererkennung

1991, München, edited by Radig, B., vol. 290 of *Informatik Fachberichte*, pp. 151–158, 13. DAGM Symposium, München, Springer, Berlin, doi: 10.1007/978-3-662-08896-8_18, 1991.

[194] Lauterborn, W. and Kurz, T.: Coherent Optics, Springer, Berlin, 2 edn., 2003.

[195] Lehmann, T., Oberschelp, W., Pelikan, E., and Repges, R.: Bildverarbeitung für die Medizin. Grundlagen, Modelle, Methoden, Anwendungen, Springer, Berlin, doi: 10.1007/978-3-642-60487-4, 1997.

[196] Lenz, R. K.: Linsenfehlerkorrigierte Eichung von Halbleiterkameras mit Standardobjektiven für hochgenaue 3D-Messungen in Echtzeit, in: Proc. 9. DAGM Symposium Mustererkennung, vol. 149 of *Informatik Fachberichte*, pp. 212–216, Springer, doi: 10.1007/978-3-662-22205-8_55, 1987.

[197] Lenz, R. K. and Tsai, R. Y.: Techniques for calibration of the scale factor and image center for high accuracy 3D machine vision metrology, IEEE Trans. Pattern Analysis Machine Intelligence, 10, 713–720, doi: 10.1109/34.6781, 1988.

[198] Levine, M.: Vision in Man and Machine, McGraw-Hill, New York, 1985.

[199] Li, F.-C., Kawaguchi, Y., Segawa, T., and Suga, K.: Simultaneous measurement of turbulent velocity field and surface wave amplitude in the initial stage of an open-channel flow by PIV, Exp. Fluids, 39, 945–953, doi: 10.1007/s00348-005-0995-3, 2005.

[200] Lim, J. S.: Two-dimensional Signal and Image Processing, Prentice-Hall, New Jersey, 1990.

[201] Lindeberg, T.: Scale-Space Theory in Computer Vision, Kluwer Academic Publishers, Dordrecht, The Netherlands, doi: 10.1007/978-1-4757-6465-9, 1994.

[202] Linscott, R. B., ed.: The Photonics Design and Applications Handbook, Laurin, Pittsfield, MA, 40 edn., URL http://www.photonics.com/edu/Handbook.aspx, 1994.

[203] Litfin, G., ed.: Technische Optik in der Praxis, Springer, Berlin, 3 edn., 2005.

[204] Liu, T., Sullivan, J. P., Asai, K., Klein, C., and Egami, Y.: Pressure and Temperature Sensitive Paints, Springer International Publishing, doi: 10.1007/978-3-030-68056-5, 2021.

[205] Luhmann, T., ed.: Photogrammetrie — Laserscanning: Anwendung für As-Built-Dokumentation und Facility Management, Wichmann, 2002.

[206] Luhmann, T.: Nahbereichsphotogrammetrie: Grundlagen, Methoden und Anwendungen, Wichmann, Heidelberg, 5 edn., 2023.

[207] Luhmann, T. and Schuhmacher, C., eds.: Photogrammetrie — Laserscanning — Optische 3D-Messtechnik: Beiträge der Oldenburger 3D-Tage 2020, VDE Verlag, 2020.

[208] Luhmann, T. and Schuhmacher, C., eds.: Photogrammetrie — Laserscanning — Optische 3D-Messtechnik: Beiträge der Oldenburger 3D-Tage 2022, VDE Verlag, 2022.

[209] Luhmann, T. and Sieberth, T., eds.: Photogrammetrie — Laserscanning — Optische 3D-Messtechnik: Beiträge der Oldenburger 3D-Tage 2022, VDE Verlag, 2025.

[210] Luhmann, T., Robson, S., Kyle, S., and Boehm, J.: Close-Range Photogrammetry and 3D Imaging, De Gruyter, 4 edn., doi: 10.1515/9783111029672, 2023.

[211] Ma, S., Gupta, S., Ulku, A. C., Bruschini, C., Charbon, E., and Gupta, M.: Quanta burst photography, ACM Transactions on Graphics, 39, doi: 10.1145/3386569.3392470, 2020.

[212] Mallot, H. A.: Computational Vision: Information Processing in Perception and Visual Behavior, The MIT Press, Cambridge, MA, 2000.

[213] Markandey, V. and Flinchbaugh, B. E.: Multispectral constraints for optical flow computation, in: Proc. of ICCV, pp. 38–41, IEEE Computer Society, Osaka, Japan, 1990.

[214] Marple Jr., S. L.: Digital Spectral Analysis with Applications, Prentice-Hall, Englewood Cliffs, NJ, 1987.

[215] Marr, D.: Vision, W. H. Freeman and Company, New York, 1982.

[216] Marr, D. and Hildreth, E. C.: Theory of edge detection, Proc. R. Soc. Lond. B, 207, 187–217, doi: 10.1098/rspb.1980.0020, 1980.

[217] Maruyama, Y., Terada, T., Yamazaki, T., Uesaka, Y., Nakamura, M., Matoba, Y., Komori, K., Ohba, Y., Arakawa, S., Hirasawa, Y., Kondo, Y., Murayama, J., Akiyama, K., Oike, Y., Sato, S., and Ezaki, T.: 3.2-MP back-illuminated polarization image sensor with four-directional air-gap wire grid and 2.5 μm pixels, IEEE Transactions on Electron Devices, 65, 2544–2551, doi: 10.1109/ted.2018.2829190, 2018.

[218] Maxwell, E. A.: General Homogeneous Coordinates in Space of Three Dimensions, Cambdridge Univ. Press, 1951.

[219] Mayinger, F. and Feldmann, O., eds.: Optical Measurements, Springer, 2 edn., doi: 10.1007/978-3-642-56443-7, 2001.

[220] McCann, J. J. and Rizzi, A.: The Art and Science of HDR Imaging, Wiley, 2012.

[221] McCluney, W. R.: Introduction to Radiometry and Photometry, Artech House, Boston, 1994.

[222] McCreery, R. L.: Raman Spectroscopy for Chemical Analysis, vol. 157 of *Chemical Analysis*, Wiley-Interscience, 2000.

[223] Mead, C.: Analog VLSI and Neural Systems, Addison-Wesley, Reading, MA, 1989.

[224] Menke, W.: Geophysical Data Analysis and Inverse Theory with MATLAB and Python, Academic Press, 5 edn., doi: 10.1016/c2022-0-02745-8, 2024.

[225] Nagel, H.: Displacement vectors derived from second-order intensity variations in image sequences, Computer Vision, Graphics, and Image Processing (GVGIP), 21, 85–117, doi: 10.1016/S0734-189X(83)80030-9, 1983.

[226] Nagel, L., Krall, K. E., and Jähne, B.: Comparative heat and gas exchange measurements in the Heidelberg Aeolotron, a large annular wind-wave tank, Ocean Sci., 11, 111–120, doi: 10.5194/os-11-111-2015, 2015.

[227] Nagel, L., Krall, K. E., and Jähne, B.: Measurement of air-sea gas transfer velocities in the Baltic Sea, Ocean Sci., 15, 235–247, doi: 10.5194/os-15-235-2019, 2019.

[228] Nakamura, J., ed.: Image Sensors and Signal Processing for Digital Still Cameras, CRC Press, 2006.

[229] Nakayama, Y. and Tanida, Y., eds.: Atlas of Visualization III, CRC, Boca Raton, FL, 1997.

[230] Nalwa, V. S.: A Guided Tour of Computer Vision, Addison-Wesley, Reading, MA, 1993.

[231] Nielsen, M., Maas, R., Niessen, W., Florack, L., and Romeny, B. M.: Binocular Stereo from Grey-Scale Images, Journal of Mathematical Imaging and Vision, 10, 103–122, 1999.

[232] Nishihara, H. K.: Practical real-time stereo matcher, Optical Eng., 23, 536–545, 1984.

[233] Nitsche, W. and Dobriloff, C., eds.: Imaging Measurement Methods for Flow Analysis, Results of the DFG Priority Programme 1147 Imaging Measurement Methods for Flow Analysis 2003-2009, vol. 106 of *Notes on Numerical Fluid Mechanics and Multidisciplinary Design*, Springer, doi: 10.1007/978-3-642-01106-1, 2009.

[234] Oppenheim, A. V. and Schafer, R. W.: Discrete-Time Signal Processing, Pearson, Upper Saddle River, NJ, 3 edn., 2010.

[235] Oppenheim, A. V., Schafer, R. W., and Buck, J. R.: Zeitdiskrete Signalverarbeitung, Pearson Studium, 2 edn., 2004.

[236] Papoulis, A.: Probability, Random Variables and Stochastic Processes, McGraw-Hill, 3 edn., 1991.

[237] Papst, M.: Development of a method for quantitative imaging of air-water gas exchange, Master's thesis, Institut für Umweltphysik, Universität Heidelberg, Germany, doi: 10.11588/heidok.00027271, 2019.

[238] Paragios, N., Chen, Y., and Faugeras, O., eds.: Handbook of Mathematical Models in Computer Vision, Springer US, doi: 10.1007/0-387-28831-7, 2006.

[239] Perona, P. and Malik, J.: Scale-space and edge detection using anisotropic diffusion, IEEE Trans. Pattern Analysis Machine Intelligence, 12, 629–639, doi: 10.1109/34.56205, 1990.

[240] Petrou, M. and Petrou, C.: Image Processing, The Fundamentals, John Wiley & Sons, 2 edn., 2010.

[241] Pharr, M. and Humphreys, G.: Physically Based Rendering From Theory to Implementation, Morgan Kaufmann, 2 edn., 2010.

[242] Pitas, I. and Venetsanopoulos, A. N.: Nonlinear Digital Filters. Principles and Applications, Kluwer Academic Publishers, Norwell, MA, 1990.

[243] Planck, M.: Zur Theorie des Gesetzes der Energieverteilung im Normalspectrum, Verhandl. Deutsche Physikalische Gesellschaft, pp. 237–245, doi: 10.1002/phbl.19480040404, 1900.

[244] Platt, U. and Stutz, J.: Differential Optical Absorption Spectroscopy, Principles and Applications, Physics of Earth and Space Environments, Springer, doi: 10.1007/978-3-540-75776-4, 2008.

[245] Poularikas, A., ed.: The Transforms and Applications Handbook, CRC Press, 3 edn., doi: 10.1201/9781315218915, 2010.

[246] Pratt, W. K.: Digital image processing, PIKS Inside, Wiley, New York, 3 edn., 2001.

[247] Pratt, W. K.: Introduction to Digital image Processing, CRC Press, New York, 2014.

[248] Press, W. H., Teukolsky, S. A., Vetterling, W. T., and Flannery, B. P.: Numerical Recipes in C: The Art of Scientific Computing, Cambridge Univ. Press, 3 edn., URL http://www.nr.com/, 2007.

[249] Prince, S. J. D.: Computer vision: models, learning and inference, Cambdridge Univ. Press, 2012.

[250] Proakis, J. G. and Manolakis, D. G.: Digital Signal Processing. Principles, Algorithms, and Applications, Prentice Hall, New York, 4 edn., 2006.

[251] Puente León, F.: Messtechnik: Grundlagen, Methoden und Anwendungen, Springer, 11 edn., doi: 10.1007/978-3-662-59767-5, 2019.

[252] Quam, L. H.: Hierarchical warp stereo, in: Proc. DARPA Image Understanding Workshop, October 1984, New Orleans, LA, pp. 149–155, 1984.

[253] Råde, L. and Westergren, B.: Springers Mathematische Formeln, Springer, Berlin, 3 edn., doi: 10.1007/978-3-642-57239-5, 2000.

[254] Raffel, M., Willert, C. E., Wereley, S. T., and Kompenhans, J.: Particle Image Velocimetry: A Practical Guide, Springer-Verlag, Heidelberg, Berlin, 2 edn., doi: 10.1007/978-3-540-72308-0, 2007.

[255] Raith, M. M. and Raase, P.: Leitfaden zur Dünnschliffmikroskopie, Universität Bonn, URL https://www.steinmann.uni-bonn.de/pdf/leitfaden-zur-duennschliffmikroskopie, 2009.

[256] Rao, A. R.: A Taxonomy for Texture Description and Identification, Springer, New York, doi: 10.1007/978-1-4613-9777-9, 1990.

[257] Rao, A. R. and Schunck, B. G.: Computing oriented texture fields, in: Proceedings CVPR'89, San Diego, CA, edited by Rao, A. R. and Schunck, B. G., pp. 61–68, Washington, DC, doi: 10.1109/CVPR.1989.37829, 1989.

[258] Rehm, R.: Dual-Color InAs/GaSb Superlattice Detectors for Missile Approach Warning Systems / Bispektrale InAs/GaSb-Übergitterdetektoren für Raketenwarnsensorik, Tech. rep., Jahresbericht 2011-2012, Fraunhofer-Instituts für Angewandte Festkörperphysik IAF, Freiburg, S. 90–93, 2012.

[259] Rehm, R., Walther, M., Rutz, F., Schmitz, J., Wörl, A., Masur, J.-M., Scheibner, R., Wendler, J., and Ziegler, J.: Dual-color InAs/GaSb superlattice focal-plane array technology, Journal of Electronic Materials, 40, 1738–1743, doi: 10.1007/s11664-011-1674-1, 2011.

[260] Reiss, T. H.: Recognizing Planar Objects Using Invariant Image Features, no. 676 in Lecture notes in computer science, Springer, 1993.

[261] Rice, J. A.: Mathematical Statistics and Data Analysis, Duxbury Press, 2 edn., 1995.

[262] Richards, A.: Alien Vision: Exploring the Electromagnetic Spectrum with Imaging Technology, SPIE, Bellingham, WA, 2 edn., URL www.alienvision.org, 2011.

[263] Richards, J. A. and Jia, X.: Remote Sensing Digital Image Analysis, Springer, Berlin, doi: 10.1007/978-3-642-30062-2, 1999.

[264] Riedl, M. J.: Optical Design Fundamentals for Infrared Systems, SPIE, Bellingham, 2 edn., 2001.

[265] Rieke, G.: Detection of Light: From the Ultraviolett to the Submillimeter, Cambridge University Press, 2 edn., doi: 10.1017/CBO9780511606496, 2003.

[266] Riemer, K.: Analyse von Wasseroberflächenwellen im Orts-Wellenzahl-Raum, Dissertation, Institut für Umweltphysik, Fakultät für Physik und Astronomie, Heidelberg University, 1991.

[267] Riemer, K., Scholz, T., and Jähne, B.: Bildfolgenanalyse im Orts-Wellenzahl-Raum, in: Proc. 13. DAGM-Symposium zur Mustererkennung 1991, München, Springer, doi: 10.1007/978-3-662-08896-8_28, dAGM main award, 1991.

[268] Robinson, I. S.: Measuring the Oceans from Space — The Principals and Methods of Satellite Oceangraphy, Springer, Berlin, 2004.

[269] Rocholz, R., Wanner, S., Schimpf, U., and Jähne, B.: Combined visualization of wind waves and water surface temperature, in: Gas Transfer at Water Surfaces 2010, edited by Komori, S., McGillis, W., and Kurose, R., pp. 496–506, doi: 10.5281/zenodo.14957, 2011.

[270] Rosenfeld, A.: Picture Processing by Computer, Academic Press, New York, 1969.

[271] Rosenfeld, A., ed.: Multiresolution Image Processing and Analysis, Springer, doi: 10.1007/978-3-642-51590-3, 1984.

[272] Rosenfeld, A. and Kak, A. C.: Digital Picture Processing, Academic Press, San Diego, 1976.

[273] Rosenfeld, A. and Kak, A. C.: Digital Picture Processing, vol. I and II, Academic Press, San Diego, 2 edn., 1982.

[274] Rumble, J. R.: CRC Handbook of Chemistry and Physics, CRC Press, 98 edn., 2017.

[275] Russ, J.: The Image Processing Handbook, CRC Press, 6 edn., 2011.

[276] Ryan, O.: Linear Algebra, Signal Processing, and Wavelets — A Unified Approach: Python Version, Springer Nature Switzerland, doi: 10.1007/978-3-030-02940-1, 2019.

[277] Sackewitz, M., ed.: Handbuch zur industriellen Bildverarbeitung, Fraunhofer Verlag, 3 edn., 2017.

[278] Saleh, B.: Introduction to Subsurface Imaging, Cambridge, doi: 10.1017/CBO9780511732577, 2011.

[279] Saleh, B. and Teich, M.: Fundamentals of Photonics, John Wiley & Sons, Inc., 2 edn., doi: 10.1002/0471213748, 2007.

[280] Saleh, B. E. A. and Teich, M. C.: Fundamentals of Photonics, 2 Volume Set, Part I: Optics, Part II: Photonics, John Wiley & Sons, Inc., 2019.

[281] Saleh, B. E. A. and Teich, M. C.: Optik und Photonik, Wiley-VCH, 3 edn., 2019.

[282] Samet, H.: The Design and Analysis of Spatial Data Structures, Addison-Wesley, Reading, MA, 1990.

[283] Samet, H.: Applications of Spatial Data Structures: Computer Graphics, Image processing, and GIS, Addison-Wesley, Reading, MA, 1990.

[284] Scharr, H.: Optimale Operatoren in der Digitalen Bildverarbeitung, Dissertation, IWR, Fakultät für Physik und Astronomie, Heidelberg University, Heidelberg, Germany, doi: 10.11588/heidok.00000962, 2000.

[285] Scharr, H. and Uttenweiler, D.: 3D anisotropic diffusion filtering for enhancing noisy actin filaments, in: Pattern Recognition, 23rd DAGM Stmposium, Munich, edited by Radig, B. and Florczyk, S., vol. 2191 of *Lecture Notes in Computer Science*, pp. 69–75, Springer, Berlin, doi: 10.1007/3-540-45404-7_10, 2001.

[286] Scharr, H. and Weickert, J.: An anisotropic diffusion algorithm with optimized rotation invariance, in: Proceedings of the 22th DAGM Symposium on Pattern Recognition, pp. 460–467, Kiel, Germany, doi: 10.1007/978-3-642-59802-9_58, 2000.

[287] Scharr, H., Körkel, S., and Jähne, B.: Numerische Isotropieoptimierung von FIR-Filtern mittels Querglättung, in: Proceedings of the 19th DAGM Symposium on Pattern Recognition, Braunschweig, edited by und F. M. Wahl, E. P., pp. 199–208, doi: 10.1007/978-3-642-60893-3_39, 1997.

[288] Scherzer, O., Grasmair, M., Grossauer, H., Haltmeier, M., and Lenzen, F.: Variational Methods in Imaging, Springer New York, doi: 10.1007/978-0-387-69277-7, 2009.

[289] Scheuermann, T., Pfundt, G., Eyerer, P., and Jähne, B.: Oberflächenkonturvermessung mikroskopischer Objekte durch Projektion statistischer Rauschmuster, in: Proc. 17. DAGM-Symposium Mustererkennung, Bielefeld, 13.-15. September 1995, pp. 319–326, 1995.

[290] Schilling, H.: Light Field Camera Arrays Calibration and Depth Estimation, Dissertation, IWR, Fakultät für Physik und Astronomie, Heidelberg University, doi: 10.11588/HEIDOK.00032553, 2022.

[291] Schölkopf, B. and Smola, A. J.: Learning with Kernels — Support Vector machines, Regularization, Optimization, and Beyond, The MIT Press, Cambridge, MA, 2002.

[292] Schmundt, D., Stitt, M., Jähne, B., and Schurr, U.: Quantitative analysis of the local rates of growth of dicot leaves at a high temporal and spatial resolution, using image sequence analysis, Plant Journal, 16, 505–514, doi: 10.1046/j.1365-313x.1998.00314.x, 1998.

[293] Schnörr, C. and Weickert, J.: Variational Image Motion Computation: Theoretical Framework, Problems and Perspectives, in: Proceedings of the 22th DAGM Symposium on Pattern Recognition, pp. 476–487, Kiel. Germany, 2000.

[294] Schott, J. R.: Remote Sensing. The Image Chain Approach, Oxford University Press, New York, 1997.

[295] Schürmann, J.: Pattern Classification — A Unified View of Statistical and Neural Approaches, John Wiley & Sons, New York, 1996.

[296] Schwarte, R., Xu, Z., Heinol, H.-G., Olk, J., and Buxbaum, B.: New optical four-quadrant phase-detector integrated into a photogate array for small and precise 3D-cameras, SPIE Proceedings - Three-Dimensional Image Capture, 3023, 119–128, doi: 10.1117/12.269749, 1997.

[297] Schwarte, R., Häusler, G., and Malz, R. W.: Three-dimensional imaging techniques, in: Computer Vision and Applications — A Guide for Students and Practitioners, edited by Jähne, B. and Haußecker, H., chap. 7, pp. 177–208, Academic Press, San Diego, doi: 10.1016/B978-012379777-3/50008-X, 2000.

[298] Sedgewick, R.: Algorithms in C, Part 1–4, Addison-Wesley, Reading, MA, 3 edn., 1997.

[299] Seitz, P.: Solid-state image sensing, in: Computer Vision and Applications — A Guide for Students and Practitioners, edited by Jähne, B. and Haußecker, H., chap. 5, pp. 111–151, Academic Press, San Diego, doi: 10.1016/B978-012379777-3/50006-6, 2000.

[300] Serra, J.: Image analysis and mathematical morphology, Academic P, London, 1982.

[301] Serra, J. and Soille, P., eds.: Mathematical Morphology and its Applications to Image Processing, vol. 2 of *Computational Imaging and Vision*, Kluwer, Dordrecht, 1994.

[302] Simoncelli, E. P., Freeman, W. T., Adelson, E. H., and Heeger, D. J.: Shiftable Multiscale Transforms, IEEE Transactions on Information Theory, 38, doi: 10.1109/18.119725, 1992.

[303] Simonds, R. M.: Reduction of large convolutional kernels into multipass applications of small generating kernels, J. Opt. Soc. Am. A, 5, 1023–1029, doi: 10.1364/JOSAA.5.001023, 1988.

[304] Singh, A.: Optic Flow Computation: A Unified Perspective, IEEE Computer Society Press, Los Alamitos, California, USA, 1991.

[305] Skudrzyk, E.: Die Grundlagen der Akustik, Springer, doi: 10.1007/978-3-7091-5830-2, 1954.

[306] Smith, A. T. and Snowden, R. J., eds.: Visual Detection of Motion, Academic Press, London, 1994.

[307] Smith, S. M. and Brady, J. M.: SUSAN – A New Approach to Low Level Image Processing, Int. J. Computer Vision, 23, 45–78, 1997.

[308] Smith, W. J.: Modern Optical Engineering, SPIE, New York, 4 edn., 2007.

[309] Soille, P.: Morphological Image Analysis, Springer Verlag, Berlin, 2 edn., 2002.

[310] Sommer, G., ed.: Geometric Computing with Clifford Algebras, Springer, Berlin, 2001.

[311] Sonka, M., Hlavac, V., and Boyle, R.: Image Processing, Analysis, and Machine Vision, CenGage Learning, 4 edn., 2015.

[312] Starck, J.-L., Murtagh, F., and Fadili, J. M.: Sparse Image and Signal processing - Wavelets, Curvelets. Morphological Diversity, Cambridge University Press, doi: 10.1017/CBO9780511730344, 2010.

[313] Steger, C., Ulrich, M., and Wiedemann, C.: Machine Vision Algorithms and Applications, Wiley-VCH, Weinheim, 2008.

[314] Steurer, J., Giebel, H., and Altner, W.: Ein lichtmikroskopisches Verfahren zur zweieinhalbdimensionalen Auswertung von Oberflächen, in: Proc. 8. DAGM-Symp. Mustererkennung 1986, Informatik-Fachberichte 125, edited by Hartmann, G., pp. 66–70, DAGM, Springer, Berlin, doi: 10.1007/978-3-642-71387-3_12, 1986.

[315] Stewart, R. H.: Methods of Satellite Oceanography, University of California Press, Berkeley, 1985.

[316] Strang, G.: Introduction to Linear Algebra, Cambridge University Press, 6 edn., 2023.

[317] Strat, T. M.: Recovering the camera parameters from a transformation matrix, in: Proc. DARPA Image Understanding Workshop, pp. 264–271, 1984.

[318] Strzodka, R. and Garbe, C. S.: Real-time motion estimation and visualization on graphics cards, in: Proceedings IEEE Visualization 2004, pp. 545–552, doi: 10.1109/VISUAL.2004.88, 2004.

[319] Sze, S.: Semiconductor Devices, John Wiley and Sons, New York, 1985.

[320] Szeliski, R.: Computer Vision Algorithms and Applications, Springer, London, doi: 10.1007/978-1-84882-935-0, 2011.

[321] Telford, W. M., Geldart, L. P., and Sheriff, R. E.: Applied Geophysics, Cambridge University Press, 2 edn., 1990.

[322] ter Haar Romeny, B., Florack, L., Koenderink, J., and Viergever, M., eds.: Scale-Space Theory in Computer Vision, vol. 1252 of *Lecture Notes in Computer Science*, 1st Int. Conf., Scale-Space'97, Utrecht, The Netherlands, Springer, Berlin, 1997.

[323] Terzopoulos, D.: Regularization of inverse visual problems involving discontinuities, IEEE Trans. Pattern Analysis Machine Intelligence, 8, 413–424, doi: 10.1109/TPAMI.1986.4767807, 1986.

[324] Terzopoulos, D.: The Computation of Visible-Surface Representations, IEEE Trans. Pattern Analysis Machine Intelligence, 10, 417–438, 1988.

[325] Terzopoulos, D., Witkin, A., and Kass, M.: Symmetry-seeking models for 3D object reconstruction, in: Proc. 1st Int. Conf. Comp. Vis. (ICCV),, pp. 269–276, London, UK, 1987.

[326] Teubner, U. and Brückner, H. J.: Optical Imaging and Photography: Imaging Optics, Sensors and Systems, De Gruyter, 2 edn., doi: 10.1515/9783110789966, 2023.

[327] Theuwissen, A. J.: Solid-State Imaging with Charge-Coupled Devices, Kluwer, Dordrecht, doi: 10.1007/0-306-47119-1, 1995.

[328] Thévenaz, P., Blu, T., and Unser, M.: Interpolation revisited, IEEE Transactions on Medical Imaging, 19, 739–758, doi: 10.1109/42.875199, 2000.

[329] Tinnefeld, P., Eggeling, C., and Hell, S. W.: Far-Field Optical Nanoscopy, Springer, doi: 10.1007/978-3-662-45547-0, 2015.

[330] Towne, D. H.: Wave Phenomena, Dover, New York, 1988.

[331] Trottenberg, U., Oosterlee, C. W., and Schuller, A.: Multigrid, Academic Press, 2000.

[332] Ulaby, F. T. and G.Long, D.: Microwave Radar and Radiometric Remote Sensing, The University of Michigan Press, Ann Arbor, URL http://mrs.eecs.umich.edu, 2014.

[333] Ullman, S.: High-level Vision. Object Recognition and Visual Cognition, The MIT Press, Cambridge, MA, 1996.

[334] Umbaugh, S. E.: Computer Vision and Image Processing: A Practical Approach Using CVIPTools, Prentice Hall PTR, Upper Saddle River, NJ, 1998.

[335] Umbaugh, S. E.: Digital Image Processing and Analysis with MATLAB and CVIPtools, CRC Press, 3 edn., doi: 10.1201/9781351228374, 2017.

[336] Unbehauen, R.: Systemtheorie 2. Mehrdimensionale, adaptive und nichtlineare Systeme, Oldenbourg, München, 7 edn., 1998.

[337] Unser, M.: Sampling — 50 years after Shannon, Proceedings of the IEEE, 88, 569–587, doi: 10.1109/5.843002, 2000.

[338] Unser, M., Aldroubi, A., and Eden, M.: Fast B-spline transforms for continuous image representation and interpolation, IEEE Trans. Pattern Analysis Machine Intelligence, 13, 277–285, doi: 10.1109/34.75515, 1991.

[339] van der Heijden, F.: Image Based Measurement Systems. Object Recognition and Parameter Estimation, Wiley, Chichester, England, 1994.

[340] Van Huffel, S. and Vandewalle, J.: The Total Least Squares Problem: Computational Aspects and Analysis, Society for Industrial and Applied Mathematics, Philadelphia, doi: 10.1137/1.9781611971002, http://www.netlib.org/vanhuffel/, 1991.

[341] Verri, A. and Poggio, T.: Against quantitative optical flow, in: Proc. of 1st International Conference on Computer Vision, pp. 171–180, London, 1987.

[342] Verri, A. and Poggio, T.: Motion field and optical flow: qualitative properties, IEEE Trans. Pattern Analysis Machine Intelligence, 11, 490–498, doi: 10.1109/34.24781, 1989.

[343] Vollmer, M. and Möllmann, K.-P.: Infrared Thermal Imaging: Fundamentals, Research and Applications, Wiley-VCH, Berlin, 2 edn., doi: 10.1002/9783527693306, 2017.

[344] von Helmholtz, H.: Handbuch der Physiologischen Optik, Leopold Voss, doi: 10.1037/13536-000, 1867.

[345] Voss, K.: Theoretische Grundlagen der digitalen Bildverarbeitung, Akademie der Wissenschaften der DDR, Zentralinstitut für Kybernetik und Informationsprozesse, Berlin, 1988.

[346] Wahl, F. M., ed.: Digitale Bildsignalverarbeitung, Springer Berlin Heidelberg, doi: 10.1007/978-3-642-82314-5, 1984.

[347] Wandell, B. A.: Foundations of Vision, Sinauer Ass., Sunderland, MA, 1995.

[348] Watt, A.: 3D Computer Graphics, Addison-Wesley, Workingham, England, 3 edn., 1999.

[349] Weickert, J.: Anisotropic Diffusion in Image Processing, Ph.D. thesis, Faculty of Mathematics, University of Kaiserslautern, 1996.

[350] Weickert, J.: Anisotropic Diffusion in Image Processing, Teubner-Verlag, Stuttgart, Germany, URL https://www.mia.uni-saarland.de/weickert/book.html, 1998.

[351] Weisstein, E. W.: CRC Concise Encyclopedia of Mathematics, CRC, Boca Raton, FL, 2 edn., 2002.

[352] Wells III, W. M.: Efficient synthesis of Gaussian filters by cascaded uniform filters, IEEE Trans. Pattern Analysis Machine Intelligence, 8, 234–239, doi: 10.1109/TPAMI.1986.4767776, 1986.

[353] Wendland, B.: Fernsehtechnik I: Grundlagen, Hüthig, Heidelberg, 1988.

[354] Widrow, B. and Kollar, I.: Quantization Noise — Roundoff Error in Digital Computation, Signal Processing, Control, and Communications, Cambridge Univ Press, URL www.cambridge.org/9780521886710, 2008.

[355] Wolberg, G.: Digital Image Warping, IEEE Computer Society, Los Alamitos, CA, 1990.

[356] Woodham, R. J.: Multiple light source optical flow, in: Proc. of ICCV, pp. 42–46, Osaka, Japan, 1990.

[357] Xu, Z., Schwarte, R., Heinol, H.-G., Buxbaum, B., and Ringbeck, T.: Smart pixel : photonic mixer device (PMD); new system concept of a 3D-imaging camera-on-a-chip, in: Int. Conf. Mechatronics and Machine Vision in Practice (M2VIP1998), Nanjing, 10–12. Sept. 1998, 1998.

[358] Yaroslavsky, L. P.: Digital Picture Processing, Springer, doi: 10.1007/978-3-642-81929-2, 1985.

[359] Yin, X., Wee, A. T. S., and Tang, C. S.: Introduction to Spectroscopic Ellipsometry of Thin Film Materials Instrumentation, Data Analysis and Applications, Wiley, 2022.

[360] Zamperoni, P.: Methoden der digitalen Bildsignalverarbeitung, Vieweg, 1989.

[361] Zappa, C., Banner, M., Schultz, H., Corrada-Emmanuel, A., Wolff, L., and Yalcin, J.: Retrieval of short ocean wave slope using polarimetric imaging, Meas. Sci. Technol., 19, 055 503 (13pp), doi: 10.1088/0957-0233/19/5/055503, 2008.

[362] Zhang, B., Zerubia, J., and Olivo-Marin, J.-C.: Gaussian approximations of fluorescence microscope point-spread function models, Applied Optics, 46, 1819–1829, doi: 10.1364/AO.46.001819, 2007.

[363] Zickler, T., Belhumeur, P., and Kriegman, D.: Helmholtz stereopsis: exploiting reciprocity for surface reconstruction, Int. J. Comp. Vis., 49, 215–227, doi: 10.1023/A:1020149707513, 2002.

Index

B. Jähne, *Digital Image Processing and Image Acquisition*,
https://doi.org/10.1007/978-3-032-04806-6

MIX
Papier aus verantwortungsvollen Quellen
Paper from responsible sources
FSC® C105338

If you have any concerns about our products,
you can contact us on
ProductSafety@springernature.com

In case Publisher is established outside the EU,
the EU authorized representative is:
Springer Nature Customer Service Center GmbH
Europaplatz 3, 69115 Heidelberg, Germany

Printed by Libri Plureos GmbH
in Hamburg, Germany